Handbook of Experimental Pharmacology

Volume 89

Antiarrhythmic Drugs

Contributors

J. Brugada, A. J. Camm, R. W. F. Campbell, T. J. Campbell, P. A. Chiale, S. M. Cobbe, P. Coumel, C. Cowan, D. W. Davies, D. W. G. Harron, S. Herzig, L. Hondeghem, L. N. Horrowitz, D. G. Julian, W. Kobinger, H. Lüllmann, J. C. McGrath, O. D. Mjøs, E. N. Moore, J. Morganroth, A. Munoz, S. B. Olsson, J. R. Parratt, S. G. Priori, P. Puech, M. Rosen, M. B. Rosenbaum, P. Rossi, A. Sassine, P. J. Schwartz, R. G. Shanks, B. N. Singh, P. Sleight, J. Smallwood, J. C. Somberg, J. F. Spear, W. Spinelli, M. I. Steinberg, P. Taggart, E. M. Vaughan Williams

Editor

E. M. Vaughan Williams

Assistant Editor

T. J. Campbell

Springer-Verlag
Berlin Heidelberg New York
London Paris Tokyo

E. M. Vaughan Williams, Professor Dr.

Hertford College, Oxford University
153 Woodstock Road
Oxford OX2 7NA, Great Britain

T. J. Campbell

School of Physiology and Pharmacology
The University of New South Wales
P. O. Box 1
Kensington, New South Wales
Australia 2033

With 113 Figures

ISBN-13:978-3-642-73668-1 e-ISBN-13:978-3-642-73666-7
DOI: 10.1007/978-3-642-73666-7

Library of Congress Cataloging-in-Publication Data
Antiarrhythmic drugs / contributors, J. Brugada ... [et al.] ...; editor, E. M. Vaughan Williams, assistant editor, T. J. Campbell. p.cm. – (Handbook of experimental pharmacology; v. 89) Includes bibliographies. ISBN-13:978-3-642-73668-1
(U.S.) 1. Arrhythmia – Chemotherapy. 2. Myocardial depressants. I. Brugada, J. II. Vaughan Williams, E. M. III. Campbell, T. J. (Terry J.), 1949–. IV. Series. QP905.H3 vol. 89 [RC685.A65] 615'.1 s – dc19 [616.1'28061]

Softcover reprint of the hardcover 1st edition 1989

2122/3020 – 543210 – Printed on acid-free paper

List of Contributors

J. Brugada, Laboratoire de Physiologie 1, Boulevard Henri IV, F-34060 Montpellier Cedex, France

A. J. Camm, The Department of Cardiological Sciences, St. George's Hospital Medical School, Cranmer Terrace, Tooting, London SW17 ORE, Great Britain

R. W. F. Campbell, Academic Department of Cardiology, Freeman Hospital, Newcastle-upon-Tyne NE7 7DN, Great Britain

T. J. Campbell, School of Physiology, The University of New South Wales, P. O. Box 1, Kensington, New South Wales, Australia 2033

P. A. Chiale, Service of Cardiology, Pabellon de Cardiologia Louis H. Inchaspe, J. M. Ramos Mejia Hospital, General Urquiza 609, 1221 Buenos Aires, Argentina

S. M. Cobbe, Department of Medical Cardiology, Royal Infirmary, University of Glasgow, 10, Alexandra Parade, Glasgow G31 2ER, Great Britain

P. Coumel, Hôpital Lariboisière, Salle Charles-Lambry, 2, rue Ambroise-Paré, F-75010 Paris, France

C. Cowan, Department of Cardiovascular Studies, The University, Leeds, LS2 9JT, Great Britain

D. W. Davies, Department of Cardiology, St. Bartholomew's Hospital, West Smithfield, London EC1A 7BE, Great Britain

D. W. G. Harron, Department of Therapeutics and Pharmacology, The Queen's University of Belfast, The Whitla Medical Building, 97 Lisburn Road, Belfast BT9 7BL, N. Ireland

S. Herzig, Department of Pharmacology, University of Kiel, Hospitalstr. 4–6, D-2300 Kiel 1, W. Germany

L. Hondeghem, CC 2209 MCN Vanderbilt University, Stahlman Cardiovascular Research Program, School of Medicine, Nashville, TN 37232, USA

L. N. Horrowitz, The Philadelphia Heart Institute of Presbyterian-University of Pennsylvania, Medical Center, Philadelphia, PA 19102, USA

D. G. Julian, Department of Cardiology, Medical School, University of Newcastle-upon-Tyne, Framlington Place, Newcastle-upon-Tyne NE2 4HH, Great Britain

W. Kobinger, Ernst Boehringer Institut für Arzneimittelforschung, Bender & Co., Ges.mbH Wien, Dr. Boehringer-Gasse 5–11, A-1121 Wien, Austria

H. Lüllmann, Department of Pharmacology, University of Kiel, Hospitalstr. 4–6, D-2300 Kiel 1, W. Germany

J. C. McGrath, Autonomic Physiology Unit, Institute of Physiology, University of Glasgow, Glasgow G12 8QQ, Great Britain

O. D. Mjøs, Institute of Medical Biology, University of Tromsø, P. O. Box 977, N-9001 Tromsø, Norway

E. N. Moore, Department of Animal Biology, University of Pennsylvania, School of Veterinary Medicine, 3800 Spruce Street, Philadelphia, PA 19104, USA

J. Morganroth, Cardiac Research and Development Department of Medicine and Division of Cardiology, The Graduate Hospital, One Graduate Plaza, Philadelphia, PA 19146, USA

A. Munoz, Service de Cardiologie "B", Hôpital Saint-Eloi, F-3400 Montpellier Cedex, France

S. B. Olsson, Division of Cardiology, Department of Medicine I, University of Göteborg, Sahlgrenska Hospital, S-413 45 Göteburg, Sweden

J. R. Parratt, Department of Physiology and Pharmacology, Royal College, University of Strathclyde, Glasgow G1 1XW, Great Britain

S. G. Priori, Ospiedale Maggiore, Centro di Fisiologia Clinica e Ipertensione, Clinica Medica IV dell'Università di Milano, Via F. Sforza 35, I-20122 Milano, Italy

P. Puech, Service de Cardiologie "B", Hôpital Saint-Eloi, F-34000 Montpellier Cedex, France

M. Rosen, Departments of Pharmacology and Pediatrics, Columbia University, College of Physicians and Surgeons, 630 W. 168th Street, New York, NY 10032, USA

M. B. Rosenbaum, Servicio de Cardiologia, Pabellon de Cardiologia Louis H. Inchaspe, J. M. Ramos Mejia Hospital, General Urquiza 609, 1221 Buenos Aires, Argentina

P. Rossi, Cardiac Department, John Radcliffe Hospital, Headington, Oxford OX3 9DU, Great Britain

A. Sassine, Laboratoire de Physiologie 1, Boulevard Henri IV, F-34060 Montpellier Cedex, France

P. J. Schwartz, Istituto di Clinica Medica IV Generale, Pad. Sacco, Via F. Sforza 35, I-20122 Milano, Italy

R. G. Shanks, Department of Therapeutics and Pharmacology, The Queen's University of Belfast, Whitla Medical Building, 97 Lisburn Road, Belfast BT9 7BL, N. Ireland

B. N. Singh, Department of Cardiology 691/111 E., Wadsworth Veterans Administration Hospital, Wilshire and Sawtelle Blvds., Los Angeles, CA 90073, USA

P. Sleight, Cardiac Department, University of Oxford, John Radcliffe Hospital, Headington, Oxford OX3 9DU, Great Britain

J. Smallwood, Lilly Research Laboratories, Eli Lilly and Company, Lilly Corporate Center, Indianapolis, IN 46285, USA

J. C. Somberg, Chicago Medical School, 33333 Green Bay Road, North Chicago, IL 60064/3095, USA

J. F. Spear, Department of Animal Biology, School of Veterinary Medicine, University of Pennsylvania, 3800 Spruce Street, Philadelphia, PA 10104, USA

W. Spinelli, Division of Developmental Pharmacology, Departments of Pharmacology and Pediatrics, Columbia University, College of Physicians and Surgeons, 630 W. 168th Street, New York, NY 10032, USA

M. I. Steinberg, Lilly Research Laboratories, Eli Lilli and Company, Lilly Corporate Center, Indianapolis, IN 46285, USA

P. Taggart, Department of Cardiology, The Middlesex Hospital, Mortimer Street, London W1N 8AA, Great Britain

E. M. Vaughan Williams, Hertford College, Oxford University, 153 Woodstock Road, Oxford OX2 7NA, Great Britain

Preface

The development of a new antiarrhythmic drug involves many people with disparate skills. The organic chemist who makes it is guided not only by the structure-action relations of previous compounds, but by anticipation of a requirement for a particular type of action. In fact several of the best-known antiarrhythmics, including lidocaine, mexiletine, amiodarone and verapamil, were originally synthesized for other purposes. Physicians have to determine whether the new drug works, and pharmacologists how it works. For some years I have believed that there was room for a work which could be understood by all these groups and which could enlighten each about the point of view of the others. Thus when I was invited by Springer-Verlag to prepare a volume in their series *Handbook of Experimental Pharmacology*, I already had a firm conception of what its form should be.

In any multi-author work there are two objectives which cannot always readily be reconciled. The first is to select topics which would relate to each other in a coherent manner, to give a logical and orderly shape to the volume as a whole. The second is to offer authors the greatest possible freedom to express themselves as they wish. When the general design was complete, prospective contributors were invited to write specific chapters, being provided with a complete list of their coauthors and chosen topics, so that they could avoid overlap. The page of contents glitters with the names of a truly international team, from Argentina, Australia, Austria, Belgium, England, France, Germany, Ireland, Italy, Norway, Scotland, Sweden and the United States. Not only have the contributions expertly covered the selected fields, but also fit elegantly into the pattern of the book as a whole.

The classification of antiarrhythmic actions has often been critized as of little use to practising physicians. Nothing is useful unless it is used, and rational use requires that the scientific basis be understood. Pharmacologists endeavour to interpret the mode of action of drugs in terms of the modulation of some physiological process, but this implies that the process is already elucidated. Unfortunately conflicting opinions about cardiac electrophysiology have been, and still are, held. For example, the transient repolarization providing a notch between the upstroke and plateau of some Purkinje cell potentials was attributed to a chloride current, then to a potassium current. Even more confusing was the history of "the pacemaker current", Ik2, said to be a pure potassium current switched on during systole and decaying during diastole. Changes in the reversal potential of this current were used to suggest that at therapeutic concentrations cardiac glycosides might *increase* the activity of the Na/K pump (NOBLE 1980) and papers were written interpreting the mode of action of drugs in terms of their effects on Ik2. Later evidence suggested that this outward potassium current activated by depolarization did not exist. To avoid controversy, therefore, it has been thought advisable simply to present evidence, together with an interpretation where there is wide consensus, but where there is not, to let investigators speak for themselves, by quoting their actual words.

Introductory chapters on electrophysiology and classification are followed by an account of experimental arrhythmias in animals and actual arrhythmias in man, the latter by Coumel, who not only has used original techniques to identify arrhythmias but has pioneered the rational use of antiarrhythmic drugs. Two chapters then estimate realistically the value of antiarrhythmic drugs in balancing benefit against risk. Terry Campbell subdivides class I drugs using beating preparations; at the other extreme Hondeghem studies effects on sodium channels isolated as far as possible from other currents and influences. The clinical pharmacology of the class Ia, b, and c compounds is reviewed in the next three chapters. Turning to class II agents, after an assessment of arrhythmias induced by stress in normal subjects, the acute and longterm effects of antiadrenergic agents are evaluated, especially in relation to the long QT syndrome. The next four chapters concern class III action, assessing its validity as an antiarrhythmic principle and the practical value of drugs which delay repolarization. Then follow chapters on drugs acting primarily on nodal tissue, the class IV calcium antagonists, adenosine, and the newly discovered (class V) specific bradycardic agents. The remaining chapters are concerned mainly with possible mechanisms of arrhythmogenesis, which involve discussing the mode of action of cardiac glycosides, adrenergic mechanisms, prostaglandins, lipids, free radicles and abnormal electrophysiology engendered by Chagas' disease, ischaemia, etc. Finally Camm critically examines antiarrhythmic drug therapy in comparison with other methods of treatment, including implantation of pacemakers and surgical ablation.

Even the most enthusiastic researcher has to recognize that drug therapy is palliative. Patients with arrhythmias have a disability, from congenital abnormality or disease (often progressive), which may be alleviated but not cured. Nevertheless there can be no doubt that the concerted efforts of chemists, pharmacologists and physicians have enabled some patients to survive longer with an improved quality of life, to achieve which is, after all, the ultimate aim of all medical science.

Oxford, Summer 1988 E. M. Vaughan Williams

Reference

Noble D (1980) Mechanisms of action of therapeutic levels of cardiac glycosides. Cardiovasc Res 14: 495–514

Contents

Introduction

CHAPTER 2

Classification of Antiarrhythmic Actions

CHAPTER 3

Acute and Chronic Animal Models of Cardiac Arrhythmias

CHAPTER 4

Classification of Human Arrhythmias

CHAPTER 5

Successes and Limitations of Antiarrhythmic Drug Therapy

Antiarrhythmic Therapy
Class I Agents

CHAPTER 8

Interaction of Class I Drugs with the Cardiac Sodium Channel

CHAPTER 9

Clinical Use of Class Ia Antiarrhythmic Drugs

CHAPTER 10

Clinical Use of Class Ib Antiarrhythmic Drugs

Class II Agents

Class III Agents

CHAPTER 16

Sotalol

Class V Agents

CHAPTER 19

Specific Bradycardic Agents

Other Therapies

CHAPTER 20

Use of Adenosine as an Antiarrhythmic Agent

Factors Involved in Arrhythmogenesis

CHAPTER 23

Adrenergic Arrhythmogenesis and the Long Q-T Syndrome

CHAPTER 24a

Effects of Cardiac Glycosides at the Cellular Level

CHAPTER 24b

Clinical Efficacy of Cardiac Glycosides for Arrhythmias

CHAPTER 25

Eicosanoids and Arrhythmogenesis

Introduction

CHAPTER 1

Cardiac Electrophysiology

E. M. Vaughan Williams

The heart may be considered simply as an anatomical network of electrically interconnected cells, through which waves of depolarizing and repolarizing current travel from the sinus node in an orderly and appropriate manner to ensure that all parts contract at the correct time. Clinical electrophysiologists can indentify disorders of rhythm from ECG records and invasive studies with intracardiac recording and stimulating electrodes (Chap. 4). Such investigations, however, tell us nothing of the nature of the ionic currents flowing in and out of individual cells. The heart is regarded merely as a wiring diagram. Alternatively an attempt may be made to identify in vitro the sources of current in terms of concentration differences of ions inside and outside the cells, and to examine how individual currents are switched on and off. Such studies ignore the anatomy of the heart and the vitally important role of the intercellular connections, the "gap" junctions, the number and distribution of which determine the pathway of conduction. In this chapter both approaches to cardiac electrophysiology will be discussed in the hope that a coherent pattern may emerge.

A. The Heart as an Electrical Network

I. Conduction

Messages were first sent by submarine cable across the Atlantic in 1858.

The cable, 2500 miles long, was taken in equal portions on board HMS Agammemnon (91 guns) and the US frigate Niagara, spliced in mid-ocean and finally landed. The results were not encouraging. The current obtained through the wire was so weak that a congratulatory message from the Queen to the President, consisting of 90 words, took 67 minutes to transmit. After a few more messages the cable became useless (Chambers 1888).

Ledermüller had concluded in 1761 that nerves were tubes (designed to contain the flow of "animal spirits") (Lidell 1960), and the similarity between nerves and cables—a central conducting core, the axoplasm, an insulating sheath, the myelin—attracted physiologists to study the "cable-like" properties of excitable tissues. In an ideal cable with perfect insulation no current would leak through the wall, and the voltage along the core would fall linearly with distance x, ($dV/dx = k$), so that a voltage of 10 units applied at one end would have fallen to 5 units at the other, assuming for simplicity that the resistance of the return pathway through recording apparatus and earth was equal to that of the cable core (Fig. 1 a). If the distal half of the cable became leaky, the voltage would fall rapidly to zero, declining exponentially if a constant fraction of current were lost through the insulation (or "membrane" resistance R_m) per unit distance (Fig. 1 b). The current in the segment of the core b (dV/dx,b) would be less than that in segment a (dV/dx,a) by exactly the amount of current I_m leaking through R_m (inset). Thus the leakage current is proportional to the second derivative of voltage against distance ($I_m = d^2V/dx^2$). The distance along a leaky cable from a point at which the voltage V has fallen to V/e is its space constant, by convention assigned the symbol lambda (Fig. 1 b). The higher the insulation resistance the longer lambda will be ($\lambda \propto R_m$). The resistance beyond the leak is the sum of the core resistance, R_i, and that of the return pathway, R_e, and the higher this sum relative to R_m the greater the current shunted through the leak and the shorter lambda will be ($\lambda \propto 1/(R_i + R_e)$. The product of both sides yields $\lambda^2 \propto R_m/(R_i + R_e)$ or $\lambda \propto SQR\ (R_m/(R_i + R_e))$.

Since a submarine cable consists of a conductor, the core, separated by insulation from another conductor, the ocean, it behaves as a capacitor (Fig. 2 a).

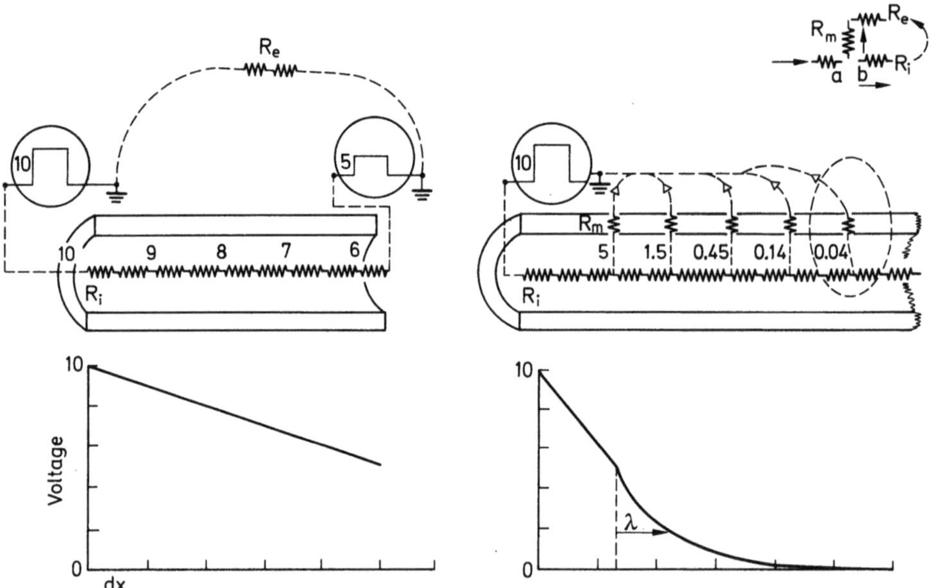

a b

Fig. 1 a, b. Cable conduction. **a** Assuming perfect insulation, and resistance in the recording system and return pathway (R_e) equal to that of the internal core conductor (R_i), a signal of 10 units at the input would be 5 units at the far end, voltage dropping linearly with distance (x), as shown in the *lower panel*. If the insulation became leaky, current would be lost through the wall, and voltage (V) would decline exponentially from the start of the leak, falling to V/e at a distance lambda, which defines the space constant of the cable. *Inset* The current through the wall at a point is equal to the difference between the core current before (dV/dx a) and after (dV/dx b)

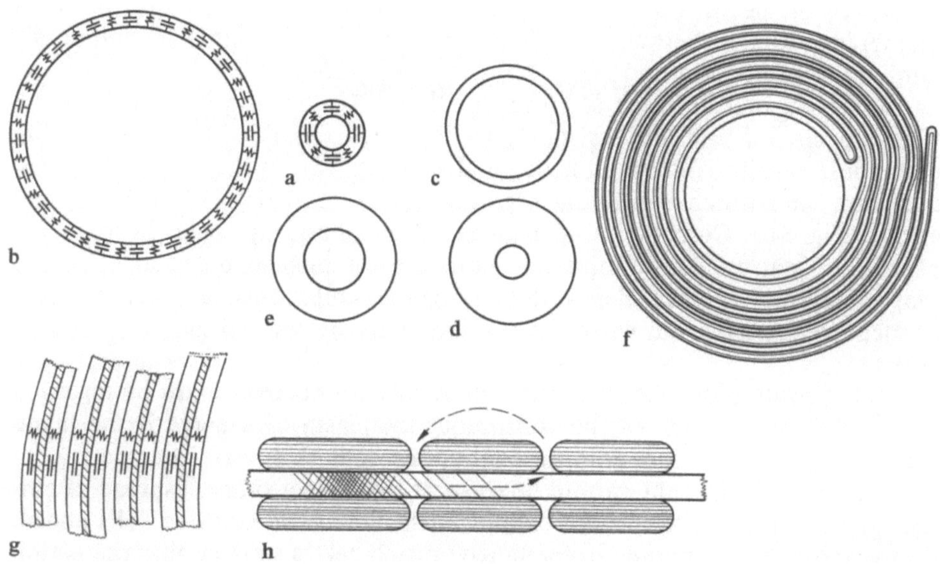

Fig. 2a–h. Myelinated nerve. The thickness of the wall of axons **a** and **b** is the same, but the cross-sectional area of **b** is 13 times that of **a**, reducing its internal resistance R_i. In myelinated nerves Schwann cells wind themselves round an axon (**f**), increasing the resistance of the wall, but reducing its capacity (**g**). The optimal ratio of axon diameter to wall thickness is as in **e**, too small in **d**, too large in **c**. In myelinated nerves current from resting to depolarized axon is saltatory, passing out and in at the nodes **h**

Applied current has first to fill this capacity before a signal reaches its full voltage, hence the delay in transmitting the Queen's message. The insulation of a cable, therefore, must be designed to maximize its space constant (R_m large, R_i small) and minimize its capacity (C_m). In myelinated nerves the Schwann cells achieve these objectives by wrapping themselves around the core conductor, the axon (Fig. 2f), providing layer after layer of resistance, the total being the sum of the resistance in series ($R = r + r + r...$). The total capacity is reduced with each layer in series ($1/C = 1/c + 1/c + 1/c...$) (Fig. 2g). For a nerve of given diameter *D*, there must be an optimum ratio between the thickness of the myelin and the diameter of the axon, *d*. If *d* is large (Fig. 2c) R_i will be small, but R_m will be low and C_m high. Conversely if the myelin is thick (Fig. 2d) although R_m will be large and C_m small, R_i will be high. RUSHTON (1951) calculated the optimum ratio of *d/D* to be between 0.47 (Fig. 2e) and 0.74, peaking at 0.6, which is the ratio actually observed in most myelinated nerves.

The space constant of even the largest myelinated nerves is only a few millimetres, yet the signal (action potential) at the far end is as large as at the beginning. In the 1930s (HODGKIN (1937a,b) demonstrated that although "cable-like" (electrotonic) flow of current to an active region of nerve from the inactive region ahead of it was responsible for initiating activity in the latter, the activity itself represented the triggering of a signal of constant value, hence the "all-or-none" phenomenon. In myelinated nerves, due to the high resistance of the myelin, virtually all this current passes through the axon membrane situated between the Schwann cells, i. e. the signal is "saltatory", jumping from node to node (Fig. 2h) (STÄMPFLI 1954).

In cardiac muscle there is no myelin at all, so that the membrane has a high capacitance and low resistance to leakage, as in Fig. 2b. In developing embryonic cardiac cells small hexagonal spots appear on the surface (Fig. 3a) and form lines which eventually aggregate (GROS et al. 1982) into hexagonal arrays (Fig. 3b). Corresponding regions on neighbouring cells become apposed, and narrow communicating paths are established between cells, the "gap junctions" (Fig. 3c). Before this, individual embryonic cells beat at independent frequencies, but when the gap junctions are formed they beat in synchrony.

A long column of cells in series, electrically connected, might be regarded as a cable, the function of the continuous axoplasm of a nerve being taken over by the gap junctions joining contiguous cells (Fig. 4a). The space constant of such a leaky cable would be so short that "electrotonic" spread of current to an active cell would not extend further than two or three cells ahead. Furthermore the membrane capacitance of each cell is so large that the action potential would be triggered in each successive cell after an appreciable delay (Fig. 4b). In a large myelinated nerve conduction velocity can be as high as 120 m/s, whereas in cardiac muscle cell-to-cell conduction is transmitted at a velocity of 0.3-0.6 m/s.

Myocardial cells are not only connected in series, but gap junctions exist between neighbours on each side, ensuring lateral synchronization, so that propagation proceeds as a smooth advancing wavefront. The space constant is

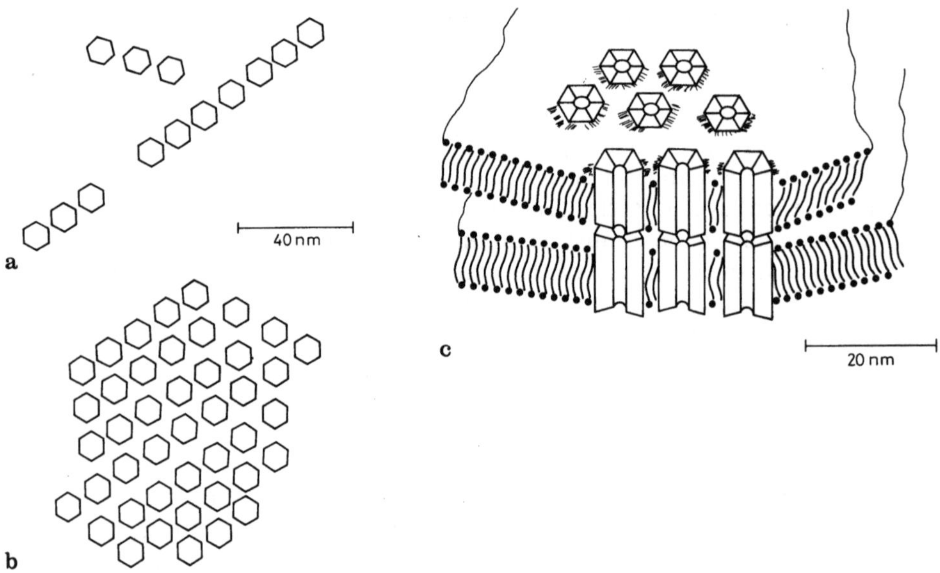

Fig. 3a–c. Development of gap junctions. Hexagonal spots appear on the surface of a myocyte and line up in rows (**a**), then areas (**b**). When cells come into contact with neighbours the areas are aligned to make intercellular electrical connections (**c**)

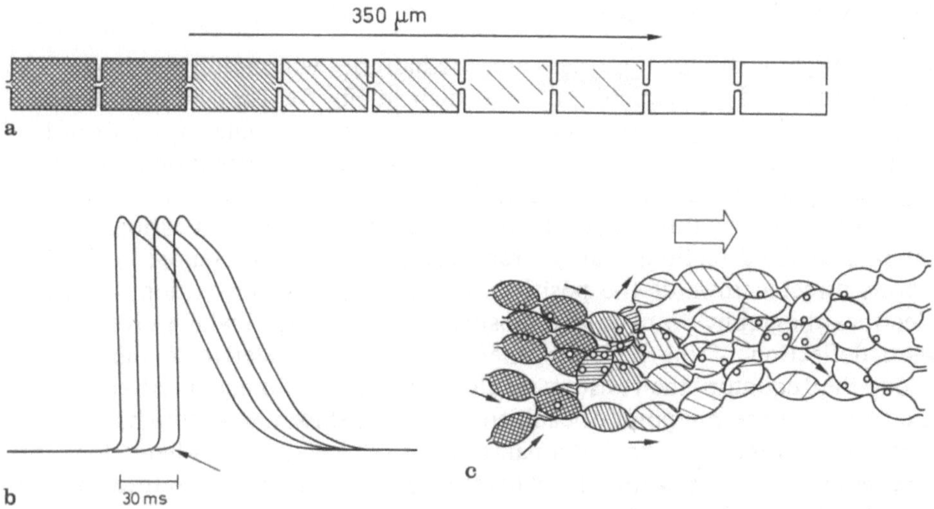

Fig. 4a–c. Cell-to-cell conduction. A row of electrically connected cardiac cells could be likened to a very leaky cable **a**, so that a signal would spread electrotonically over the length of only a few cells. The action potential is regenerated in each cell, after the delay imposed by the need for current to flow through the gap junction and charge the next cell's capacity. Even so, the action potential duration (APD) is so long that even when the potential of every tenth cell only is monitored, **b**, the depolarizations overlap, so that large areas remain depolarized simultaneously. **c** Since cells are connected side-to-side as well as end-to-end depolarization advances as a smooth wave front

so small that there is a very sharp demarcation between excited and as yet unexcited regions, electrotonic spread of current extending over only a few cells (Fig. 4c). Within the normal myocardium, both atrial and ventricular, there are no "preferential pathways" of conduction. On the contrary such pathways, by permitting the development of salients, would be arrhythmogenic, because neighbouring bundles would conduct independently without lateral synchronization. The increased probability of the development of reentry when lateral synchronization is lost and disjunctures occur between conduction velocities in neighbouring muscle tracts has been elegantly demonstrated in theoretical models by KRINSKY (1981). Unfortunately desynchronized pathways do occur, either as congenital malformations, as in the Wolf-Parkinson-White abnormality, or as a consequence of deletion of conducting regions by ischaemic or toxic damage, and these constitute an arrhythmogenic factor by providing an anatomical substrate for re-entry. Specialized pathways between the sinoatrial and atrioventricular nodes and left and right atria have been detected histologically (JAMES 1963), and some cells in the crista terminalis have larger and more rapidly depolarizing action potentials. Such cells provide "preferential" conduction only in the sense that they may be more resistant to anoxia, and continue to beat at lower temperatures when exposed to cold experimentally, but they do not give rise to salients of faster conduction in vivo (JANSE and ANDERSON 1974).

II. The Electrocardiogram

If a flashlight is dropped into seawater the batteries will rapidly discharge. The conducting fluid, as depicted in a flat dish in Fig. 5a, may be regarded as a family of resistances between the positive and negative poles of a pair of batteries connected in series inside an insulating sheath. Current density will be highest in the fluid closest to the flashlight, and weaker at a distance, but current will be flowing throughout the dish, carried by anions and cations migrating in opposite directions, but "current flow" has been represented in Fig. 5 by arrows directed towards the positive pole, because by convention vectors in electrocardiography point to the "positive" lead (the lead which gives an upward deflection of the recorder when it is positive relative to the others). Along the "equator" the resistors would be bisected into equal halves, so that the voltage at the midpoint of each conducting pathway would be half (1.5 V) that of the source (3 V), so that a pair of electrodes placed anywhere along this line would record nothing. This illustrates one axiom of electrocardiography: *1. No signal is recorded from electrodes situated at right angles to the axis of the source.*

If a pair of bipolar electrodes were positioned in the fluid to the east of the source, as illustrated in Fig. 5b, the more northerly of the pair would be nearer to the negative pole of the battery and so would be negative relative to the other. If, however, the pair were shifted, as indicated, to a position directly north of the negative pole, then, because current flows in all directions outward from the pole of the battery towards the north, the more southerly of the pair would be nearer to the negative pole, and so would be negative relative to the other, the reverse of what it was before. Thus little information about the nature of the source could be deduced by dipping a bipolar electrode at random into the fluid.

In Fig. 5c the flashlight has been rotated through −45°, and the north-south notation has been replaced by a clock. If a pair of electrodes were placed successively at 12 and 6 o'clock, 1 and 7, 2 and 8 and so on round the clock, the voltage differences could be plotted, as imagined in Fig. 5f. It is apparent from the graph that the maximum voltage would have been recorded with electrodes at 10.30 and 4.30, and the minimum at 1.30 and 7.30. According to the axiom above, it could be deduced that the axis of the source must be along a line drawn between 4.30 and 10.30. By interpolation the maximum voltage was about 22 units on the arbitrary numerical scale at the left. A parameter with magnitude and direction can be represented as a vector (22 at

---

▶

Fig. 5a–f. Batteries discharging in a homogeneous conducting medium. **a** Current flow through the fluid represented as through a family of concentric resistances. **b** An isopotential line (*dashed*) drawn through the midpoint of each resistance. The voltage difference between a pair of electrodes dipped into the fluid depends upon which of them is nearer, in electrical terms, to the negative or positive pole. **c** A clockface substituted for the NESW notation. **d, e** No current flows in the fluid unless the batteries are in series. **f** Imaginary plot of voltage differences "around the clock", between electrodes at *12* and *6*, *1* and *7* etc.

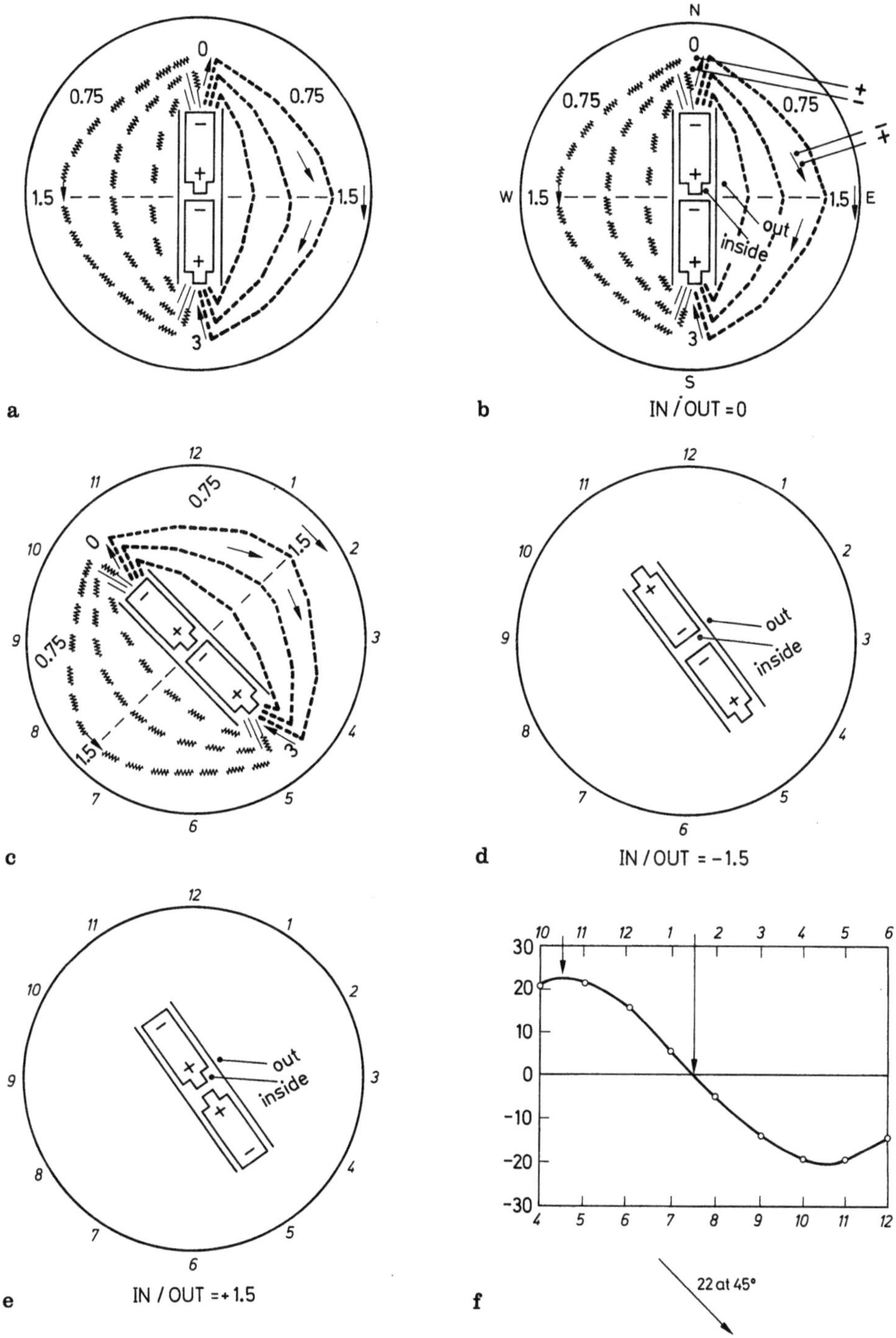
0
0.75
0.75
1.5
1.5
3
a
N
0
+
-
0.75
0.75
-
+
W
1.5
1.5
E
out
inside
3
S
b
IN / OUT = 0
12
11
1
0.75
10
0
1.5
2
9
0.75
3
8
4
1.5
3
7
5
6
c
12
11
1
10
2
out
9
inside
3
8
4
7
5
6
d
IN / OUT = −1.5
12
11
1
10
2
out
9
inside
3
8
4
7
5
6
e
IN / OUT = + 1.5
10
11
12
1
2
3
4
5
6
30
20
10
0
−10
−20
−30
4
5
6
7
8
9
10
11
12
22 at 45°
f

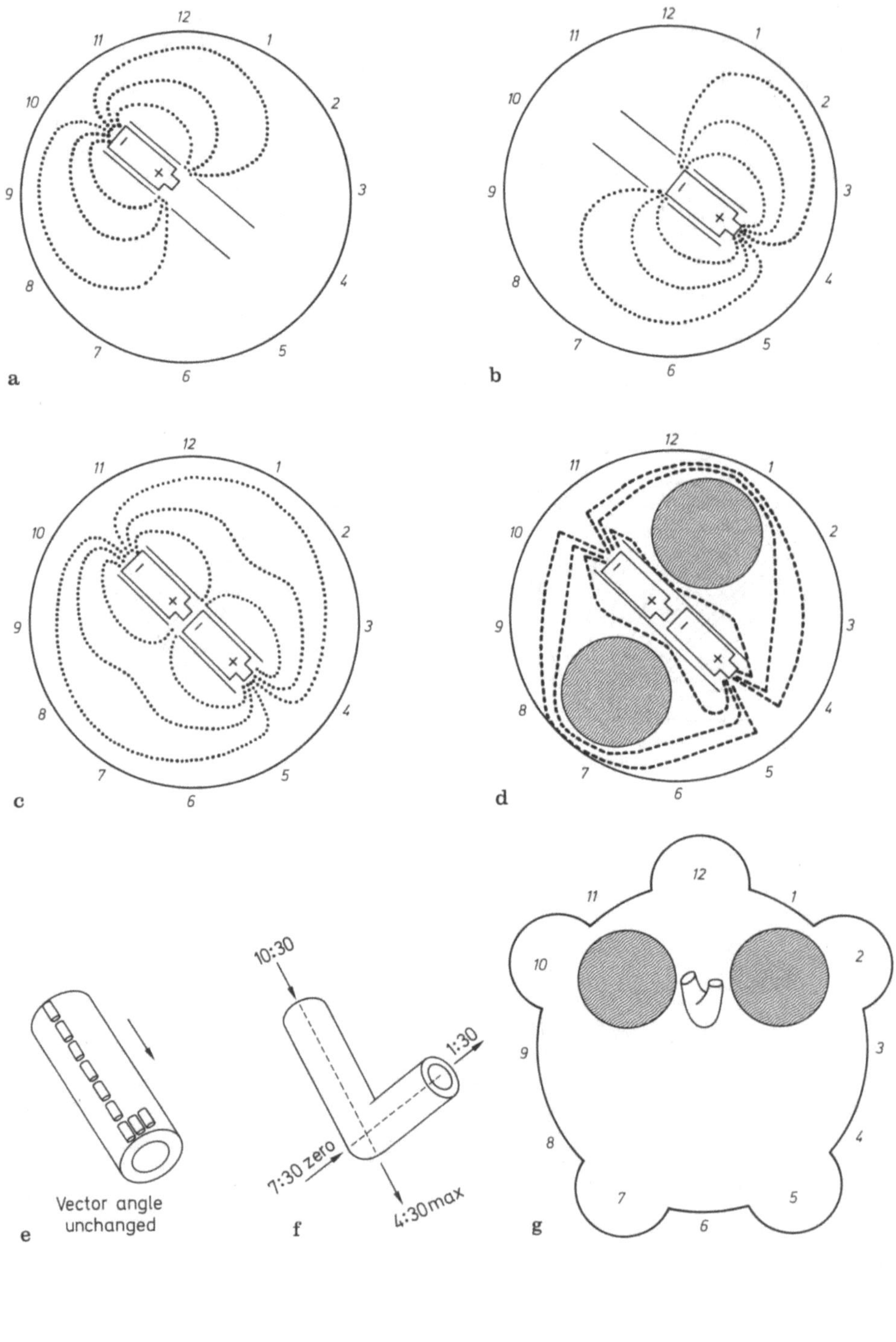
a
b
c
d
Vector angle unchanged
e
10:30
1:30
7:30 zero
4:30max
f
g

45). Incidentally, half of the information in Fig. 5f is redundant, because the positive and negative records are mirror images.

If electrodes were placed, as in Fig. 5b, inside and outside the insulating sheath of the flashlight at its midpoint, no voltage difference would be recorded, both electrodes being at 1.5 V. If the upper battery were reversed, as in Fig. 5d, a voltage would be recorded of 1.5 V, negative inside, but no current would flow in the dish, because both ends of the battery would be at the same (positive) voltage. This would represent the situation existing in a resting heart in which all intracellular elements were at a negative potential. If, however, we reversed the lower battery instead of the upper (Fig. 5e), then the inside would be at 1.5 V positive to the outside, but still no current would flow in the dish, because both poles would be at the same negative potential. This would represent the situation in a heart in which all cells were simultaneously in their active depolarized state. This illustrates another axiom of electrocardiography: *2. No surface potentials can be detected unless current is flowing from one part of the heart to another.*

If the insulation were pierced at the midpoint of the sheath and the lower battery were removed, as in Fig. 6a, current would flow mainly in the upper part of the dish, but the axis derived from recording round the clock would be unchanged. Similarly, flow would be mainly in the lower half of the dish if the upper battery were removed (Fig. 6b). If both batteries were present, some current would flow through the leaking insulation, but recordings at the periphery would be very little changed, the electrodes being too distant to be affected by the local current flow in the immediate vicinity of the leak. This illustrates another axiom (not quite valid, as will be seen): *3. Records at the surface can be attributed to a single dipole as source at any given instant.*

Thus a changing signal can be split up in time, and resolved into a series of instantaneous vectors succeeding each other in sequence (e.g. the situation of Fig. 6a could be followed by that in Fig. 6b).

It has been assumed that the dish contained a homogeneous conducting fluid. Introduction of a couple of bubbles of air or oil into the dish, as depicted in Fig. 6d, would oblige current to flow around them. Thus the presence of an insulating medium (fat, cartilage, bone) would have the effect of increasing the current density in the better conducting pathways. This would distort the pattern of records from the surface, from which it may be concluded: *4. The electrical axis may not coincide with the geometric axis of the source in a heterogeneous medium.*

◀

Fig. 6a, b. Current flow from a single battery in the tube would show a change of distribution but not of orientation. **c** Puncturing the insulation would affect current flow close to the source, but little change would be observed at the periphery. **d** Introduction of poorly conducting obstructions would alter both distribution and apparent orientation of the source deduced from surface recording. **e, f** No records would be obtained from electrodes at right angles to the tube in **e**, but would appear when conduction "rounded the corner" in **f**. **g** Dish represented as homuncule

If multiple arrays of small batteries were placed end to end in the wall of a tube, as in Fig. 6e, and switched on in turn, a series of vectors would be recorded, all with the axis along a line drawn between 10.30 and 4.30, but no signal would be obtained between 7.30 and 1.30, at right angles to this axis. If we now put a bend in the tube, as in Fig. 6f, the axis would start off between 10.30 and 4.30 as before, but would finish between 7.30 and 1.30.

The heart starts its development as a straight tube, which is ultimately bent and divided into chambers. The cells in its walls may be regarded as little batteries, switched on in series as conduction proceeds. In the ventricles the septum is activated first, then the apex, conduction finally "turning the corner" to the left ventricular free wall. The heart lies in a heterogeneous medium, containing lungs, bone, cartilage, etc., so that records at the surface will differ (to a surprisingly minor extent) from what would be expected if the body were a perfectly homogeneous conductor. Nevertheless, as can be seen from the final illustration (Fig. 6g) records from the surface "around the clock" should make it possible to resolve a changing signal into a series of individual vectors as sources.

III. Surface Leads

Eindhoven recorded the surface electrocardiogram with leads on the right arm and the left arm (I) or left leg (II) and on the left arm and leg (III). Wilson argued that a lead common to all three Eindhoven leads (i.e. with contacts on LA, RA and LL attached via series resistances to a single point) would give a "zero" or indifferent lead, and he recorded between this common (V) lead and the RA, LA and LL as "monopolar" leads (Fig. 7a). There is no "zero", however, in a differential amplifier, which simply records a difference of potential between two leads, and the observer can choose to regard either of them as zero, the other being positive or negative to it. Moreover, it is clear from Fig. 7a that the record from the left arm is partly short-circuited by another attachment to the left arm going to the common lead. Severing the shunt connection must, therefore, increase the signal, the arrangement being as in Fig. 7b, the "augmented" limb lead aVL. The recording axis is thus between the left arm (2 o'clock) and the resultant of records from the right arm (10 o'clock) and left foot (6 o'clock), i.e. between 2 and 8. Similarly, placing the positive lead on the right arm or left foot, and using the other limbs as common, provides the augmented leads aVR and aVF respectively. As illustrated in Fig. 8, the six (three Eindhoven, three augmented) limb leads record along axes pointing to a (positive) pole at 2 (aVL), 3 (I), 5 (II), 6 (aVF), 7 (III) and 10 o'clock (aVR), giving a good sample of round the clock records in a vertical (coronal) plane. If the polarity of recording leads for aVR is reversed, then the axis would point to 4 instead of 10 o'clock, which is the convention actually used. It is also customary to refer to axes in degrees of rotation clockwise from a leftward pointing horizontal (0), rather than at hours around the clock, but the principle is the same, as illustrated in Fig. 8a–f. According to this notation, recording from aVL, in the axis between 2 and 8

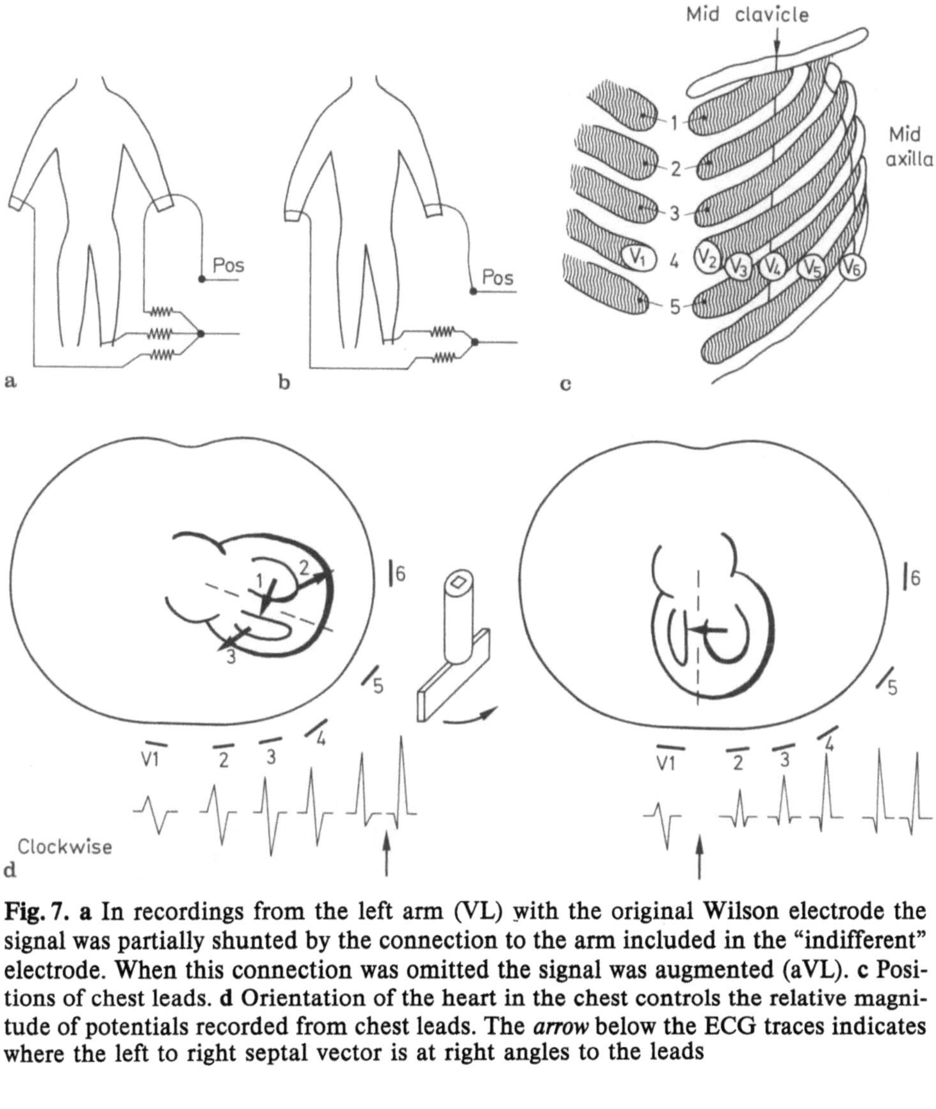

Fig. 7. a In recordings from the left arm (VL) with the original Wilson electrode the signal was partially shunted by the connection to the arm included in the "indifferent" electrode. When this connection was omitted the signal was augmented (aVL). **c** Positions of chest leads. **d** Orientation of the heart in the chest controls the relative magnitude of potentials recorded from chest leads. The *arrow* below the ECG traces indicates where the left to right septal vector is at right angles to the leads

o'clock, as described above, would be at −30°. Thus the other five limb leads provide recording axes at 30° intervals, the next being at 0° (I), the resultant of the left arm, −30°, and right arm, +30° (because the negative is at −150° on the right arm). Likewise, it can be seen from Fig. 8 how the other leads provide recording axes at 30° (aVR), 60° (II), 90° (aVF) and 120° (III). Typical ECG records laid out in this sequence are shown in Fig. 9a.

The limb leads record in the vertical plane. Additional information is obtained by using the Wilson lead as one electrode and going round the clock in a horizontal plane. By convention the horizontal monopolar (chest or V) leads progress from V1, just to the right of the sternum at the fourth intercostal space, to V6 in the left midaxillary line, as illustrated in Fig. 7c. Normally cur-

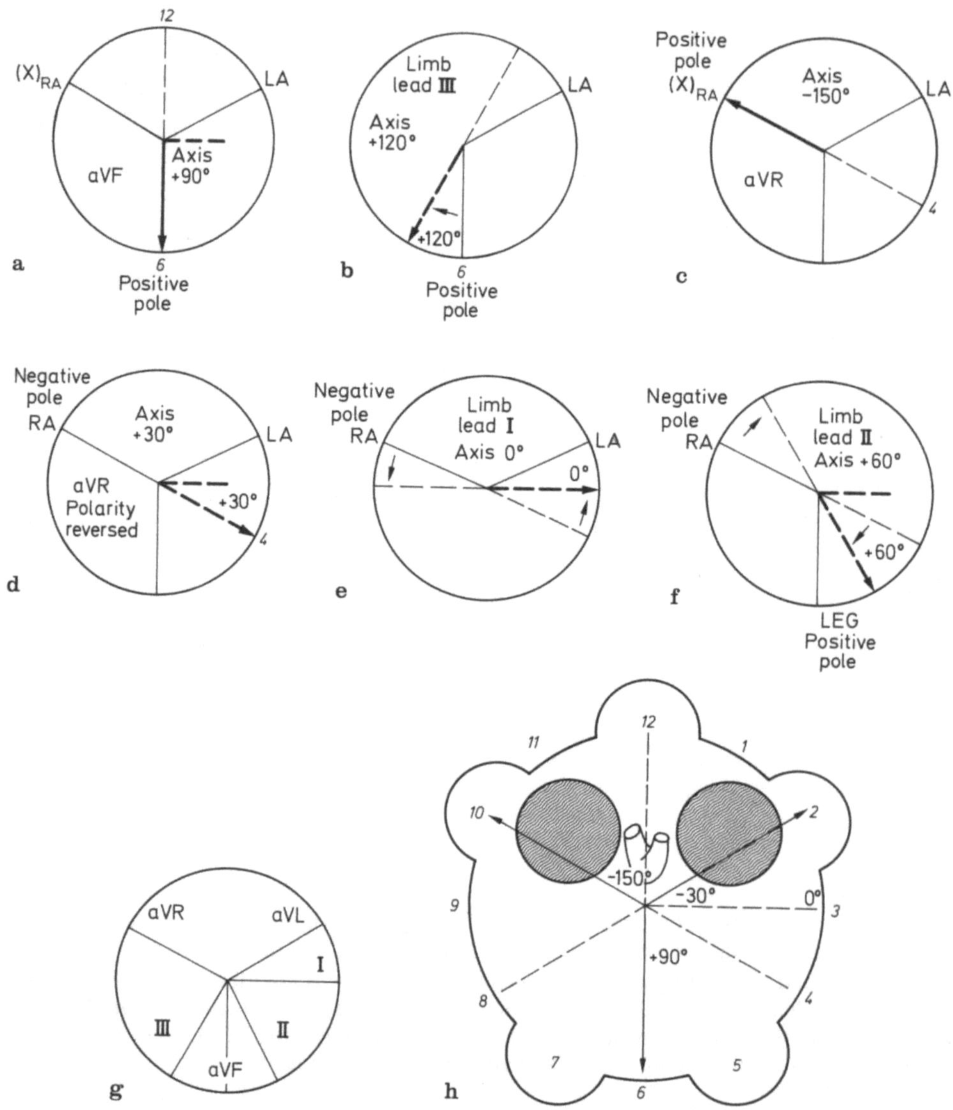

Fig. 8. a–f indicate the axes of vectors from which signals will be maximally recorded by the six limb leads, summarized in **g**, and in the homuncule in **h**

rent flow from one part of the heart to another is greatest when half of the mass of the left ventricle is already depolarized, the other half (the free left ventricular wall) not. At this time the axis of the vector is the arrow labelled 2 in Fig. 7d. The first part of the ventricle to depolarize, however, is the interventricular septum, the wave of activation spreading from the left bundle branch through the septum from left to right (the axis of the vector being labelled 1 in Fig. 7d). If the heart were orientated as shown in the right panel of

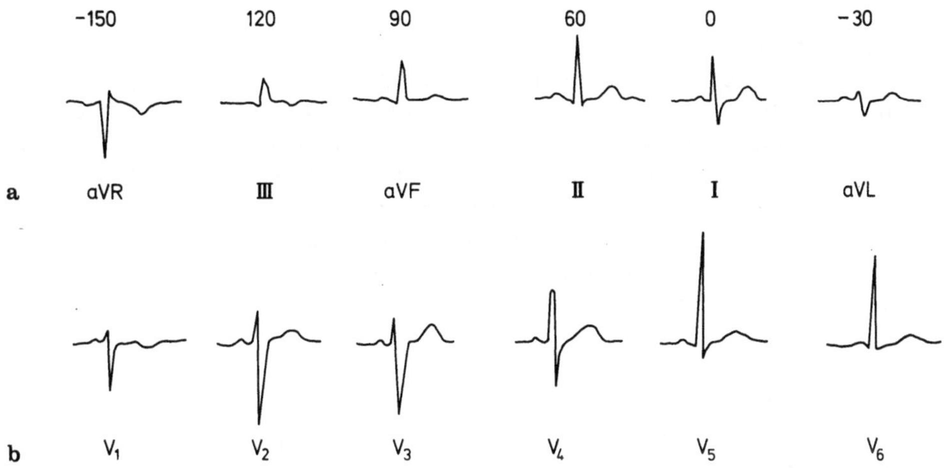

Fig. 9. a ECG traces from the limb leads presented "round the clock". **b** ECG traces from chest leads

Fig. 7d, this axis would be at right angles to the sternum, so that initial wave would be recorded positive by leads to the right of the sternum (V1) and negative in leads to its left (V2-6). Conversely, if the heart had been rotated clockwise by an imaginary hand thrust upwards through the diaphragm to the position depicted in the left panel of Fig. 7d, the first (trans-septal) vector would be pointing towards V3, but away from V6, in which the initial wave would be negative. The second vector, representing depolarization spreading into the left ventricular free wall, would be maximally recorded by this lead in the mid-axillary line, V6. By such simple analysis of the relative magnitudes of the different phases of the electrocardiogram in the V leads recorded around the clock the horizontal orientation of the heart in the chest can be deduced. Figure 9b illustrates a set of records.

IV. Vectorcardiography

In a typical lead II ECG the first (septal) vector is the Q wave, the second (LV) the R wave. The base of the right ventricle depolarizes last, the axis of the vector being labelled 3 in Fig. 7d, synchronous with the S wave. Thus the whole QRS complex can be plotted as a series of vectors following each other in time, until the records end with the arrival of the isoelectric interval between the QRS and T waves, this interval representing the period when the entire ventricle is depolarized. If the tips of the arrows of each of the successive vectors are joined by a line, the plot from start to finish describes a loop, as illustrated in Fig. 10a. Application of suitable electrodes at other points of the body surface can provide records in the anteroposterior (sagittal) and coronal planes also (Fig. 10b), and all three can then be combined to give a resultant loop orientated inside a three-dimensional representation of the chest as in Fig. 10c.

For more detailed analysis (TACCARDI and DE AMBROGGI 1975), multiple recordings from arrays of electrodes on the inside of a waistcoat (Fig. 11 a) are fed to a computer which constructs isopotential maps for appropriate moments throughout the cardiac cycle, as illustrated in front and back views in Fig. 11 b,c. Heterogeneity of the conducting pathways between heart and surface causes the isopotential lines to be more widely separated on the back than at the front. Secondly, two centres of negativity appear on the anterior chest wall, i.e. at the time the record was taken the surface potential cannot be considered to originate from a single dipole (contradicting axiom 3 above, derived from Craib's work, KRIKLER 1987).

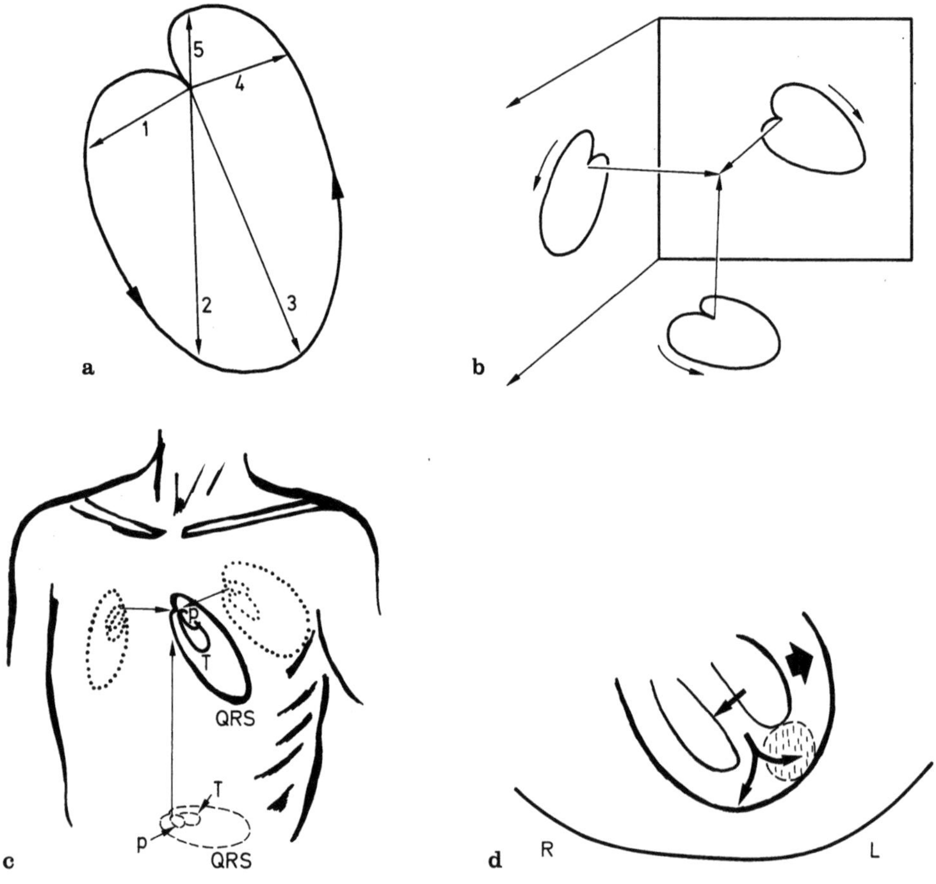

Fig. 10. a Joining the tips of successive vectors generated by the QRS complex forms a loop. **b** Suitably placed pairs of electrodes in three axes at right angles to each other permit construction of three vector loops, which can be resolved **c** to a single loop in its appropriate orientation in the chest. **d** The ventricular repolarization vector is in the same direction as the depolarization vector, because the epicardium repolarizes before the endocardium (see Fig. 13). The cells which depolarize last, repolarize first. As repolarization retreats from the apices of left and right ventricles, the vectors could cancel each other, so that the end of the Twave need not signal the end of electrical activity

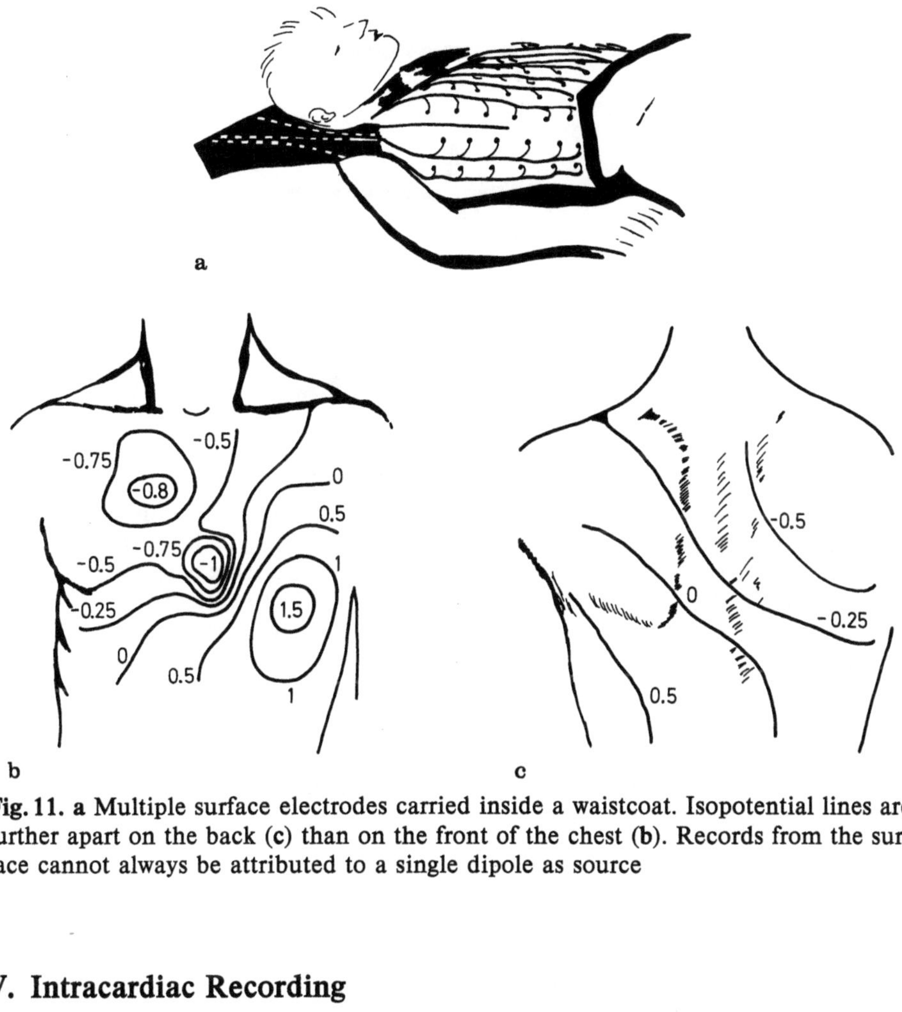

Fig. 11. a Multiple surface electrodes carried inside a waistcoat. Isopotential lines are further apart on the back (**c**) than on the front of the chest (**b**). Records from the surface cannot always be attributed to a single dipole as source

V. Intracardiac Recording

Skilled cardiologists can, by experienced pattern recognition, extract information necessary for diagnosis from inspection of traces from a few leads of the conventional ECG (Chap. 4) or vectorcardiogram. Much of the information from multiple leads is redundant and limited by the "inverse problem"; i.e. if the pathway of intracardiac conduction were known, surface potentials could be predicted by a model, but calculating back from the surface to the pathway is more difficult because, by mutual cancellation of vectors, a similar surface picture could be produced by different intracardiac conduction pathways (Nelson and Geselowitz 1976). Some information about the latter can be obtained from recording and stimulating catheter electrodes introduced into the heart itself. Monophasic potentials can be observed with catheters pressed or sucked onto the endocardial surface (Fig. 12a; & Chap. 14). A catheter with multiple electrodes apposed to the septum from the right ventricle (Fig. 12c) provides records from which the conduction time from atrium to His bundle (A-H) and His bundle to ventricle (H-V) can be measured (Fig. 12b). Other

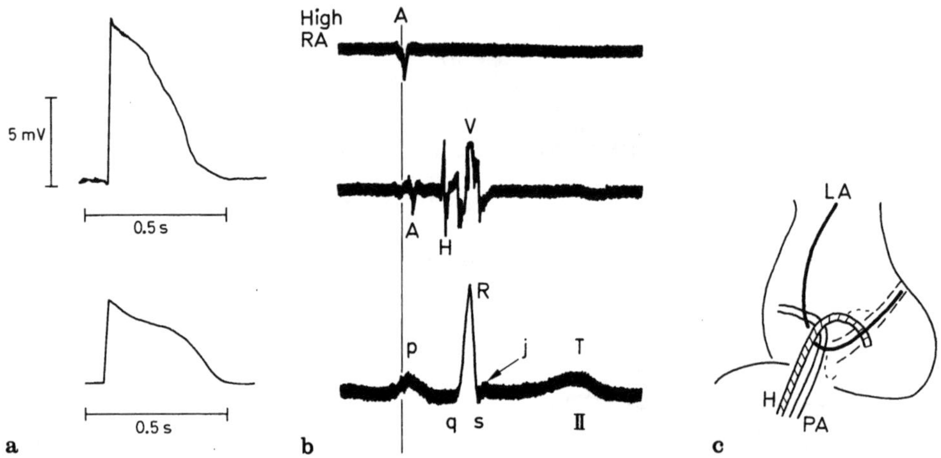

Fig. 12. a Tracings of monophasic action potentials (MAPs) recorded with suction electrodes from human atrium at two sweep speeds. **b** Intracavitary records from a high right atrial catheter electrode and from a His bundle recording catheter, with a simultaneously recorded lead II ECG. A, atrial potential; H, His bundle potential; V, ventricular potential; J, the junction between the QRS and isoelectric interval, the J-T interval being the time between this point and the end of the Twave. **c** Positioning of intracavitary electrodes for atrial pacing (PA), His bundle recording (H) or measurement of left atrial recordings from the coronary sinus (LA)

intracardiac electrodes are introduced to pace the atrium or right ventricle, or to record from within the coronary sinus (Fig. 11c).

VI. The T Wave

Multiple recordings with plunge electrodes in an isolated human heart (DURRER et al. 1970) demonstrated that ventricular depolarization spreads from the endocardium via a series of concentric leaflets outwards to the epicardium. When the endocardium is depolarized current will flow towards it from the as-yet resting epicardium. In the left centre panel of Fig. 13 the solid line represents an endocardial intracellular action potential and the dotted line an epicardial potential, on the assumption that the latter is identical to the former, but starts later. The upper panel indicates that initially current flows from inactive epicardium to active endocardium, but when the whole wall is depolarized (both action potentials being simultaneously in their plateau phase) current flow ceases (isoelectric interval). Finally, if the endocardium repolarizes first, current will flow from it towards the still depolarized epicardium. In the centre panel the two action potentials have been joined by arrows to represent the direction and magnitude of the potential difference between them throughout the cycle, and these arrows have been replotted below on a horizontal axis. The sets of arrows depict an R wave and discordant T wave, idealized, of course, because they are derived from single "sample" action potentials. In reality, in the normal heart, the epicardial action potential is considerably shorter than that of the endocardium, as illustrated in the right-hand panels, so that the epicardium both depolarizes last and repolarizes first. The arrows show that the repolarization vector is in the same direction as the depolarization vector, providing a concordant T wave.

Ventricular action potentials are shortened by a number of factors, including hypoxia or ischaemia, hyperthyroidism, and halogen anaesthesia. Endocardial ischaemia would have two effects. The height of the endocardial action potential would be reduced, so that during the plateau the epicardium and endocardium would no longer be isopotential, and some current would continue to flow during the phase of full ventricular activation, shifting the ST segment. Secondly, because the endocardial action potential would be shorter, the concordant T wave would be reduced in magnitude, or in an extreme case

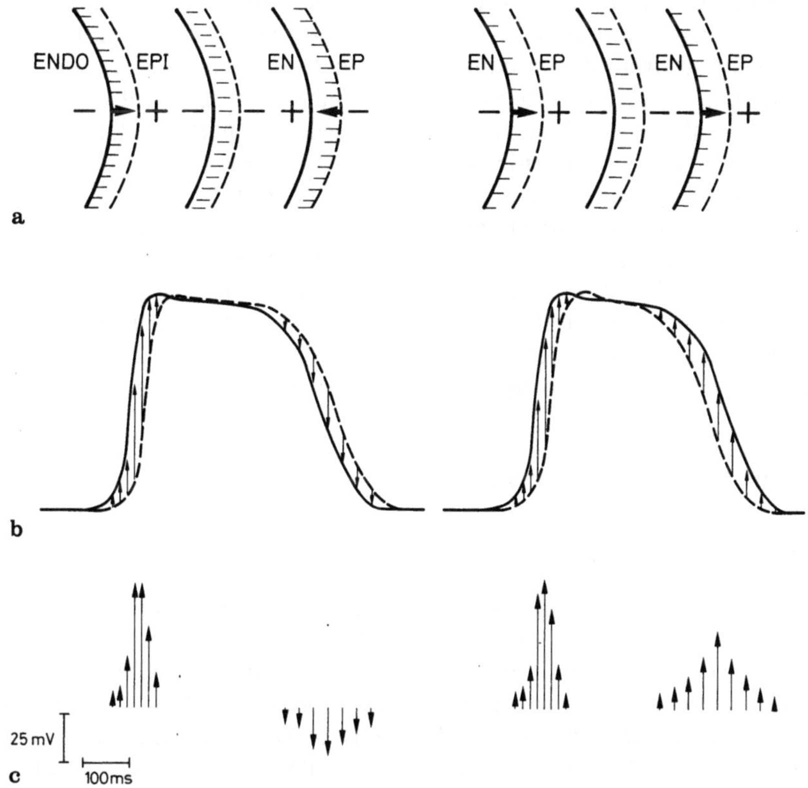

Fig. 13. a Diagram of progress of the invasion of the left ventricular free wall, as current flows first from the epicardium to endocardium, then ceases when both are depolarized, and finally flows from endocardium to epicardium if the endocardium repolarizes first. **b** Representative intracellular potentials from a cell in the endocardium (*solid line*) and epicardium (*dotted line*) on the assumption that the potentials were identical. The magnitude and direction of current flow is indicated by the *arrows joining the potentials*. These *arrows* are plotted below on a horizontal axis of time. *On the right*, the epicardial APD is shown as shorter than the endocardial APD, which makes the depolarization and repolarization vectors concordant in sign

might even become discordant, if the endocardial action potential duration (APD) became as short as the epicardial APD. If, however, it was the epicardium which became ischaemic, the epicardial APD would become even shorter than before, so that the height of the concordant T wave would be increased.

If invasion of the ventricle is slowed the QRS complex widens and the Q-T interval will accordingly be prolonged. There may be no change at all in APD of ventricular cells, which repolarize later only because they are depolarized later. Cibenzoline, for example, a class 1C antiarrhythmic drug, prolongs the Q-T interval by exactly the amount by which the QRS is widened. Thus the J-T interval (J being the junction between the end of QRS and the beginning of the isoelectric interval, Fig. 12b) is unaltered. Conversely amiodarone lengthens APD uniformly in myocardial cells, but in normoxic muscle has no effect on the width of the QRS complex, so that the J-T interval is prolonged.

In a heart with multiple lesions, whether from previous myocardial infarcts or toxic or congential abnormalities, conduction is slowed because a pathway has to be found around non-conducting blocks, or through ischaemic or damaged tissue, causing late depolarization and prolongation of Q-T. Since ischaemia also shortens APD, there will be a wide dispersion of repolarization times, a highly arrhythmogenic factor. Thus if prolongation of Q-T interval is caused by heterogeneity of conduction and repolarization, it is an indication of myocardial damage, and prognosis is poor.

Adrenergic stimulation (by mechanisms described in Chap. 12) can also cause heterogeneity of conduction and repolarization. The situation is particularly acute if the distribution of sympathetic nerves is itself heterogeneous, either through congenital defect (long Q-T syndrome), or because sympathetic nerves have been destroyed by infarction or disease.

B. Cellular Electrophysiology

I. Electromotive Force

Nobody knows how life began but an alternative to the Garden of Eden is that complex molecules began to reproduce themselves, profiting from external energy. To keep the hereditary system together, and to exclude undesirable components of the environment, they became surrounded by a membrane, the frontier between self and non-self. Not only cells but intracellular organelles are walled off by a bimolecular leaflet, the hydrophilic surfaces of which enclose between them a semifluid lipophilic layer (Fig. 3c; GOMPERTS 1977). Cell membranes are not thin smooth containers assumed in some electrical models, but are traversed by numerous structures, embedded in them and continually removed and replaced, the function of which is to react appropriately to contacts with external entities which collide with them and which are recognized.

At the beginning of the century it was noted that cells contained a high concentration of potassium ions, and it was suggested that this could be the source of electromotive force (EMF) in excitable cells. A pure potassium chloride solution isotonic with mammalian tissue would contain 155 mmol/litre K^+. If the membrane were freely permeable to K, to retain this intracellular K in equilibrium with an external concentration of 4.5 mmol/litre it would be necessary to apply a negative voltage inside the cell (calculated from the Nernst equation) of $E = RT/F \times \log(n)[K]o/[K]i$, or, converting to log (10) at 37 °C, $E = 60.5 \times \log(10)\ 4.5/155$, $E = -95$ mV. Early attempts to measure the actual intracellular voltage involved placing an electrode on an injured part of a nerve or muscle cell to make contact with the interior, and measuring the "injury" current between this and an electrode on a distant intact part. Even with high-impedance instruments not more than −30 to −40 mV was observed: this could reach −70 mV if the current was "backed off" potentiometrically. The action potential was thought to be due to a temporary shortcircuit at an "active" segment of membrane. It was then found that the action potential was larger than the maximum resting potential observed, theoretically impossible, but confirmed by direct measurement across the membrane of a squid axon by internal and external electrodes (Hodgkin and Huxley 1939, 1945). It was later discovered that the intracellular negativity is maintained by the Na/K pump, situated in most cell membranes, which utilizes ATP to extrude three Na ions and take up two K ions, thus transferring one negative charge to the cell interior per cycle. There were, therefore, two sources of EMF, a sodium battery and a potassium battery. In cardiac muscle there are in fact four major sources of EMF (Table 1).

Table 1. Calculated equilibrium potentials

Ion	Inside concentration (approx)	Outside concentration	Ratio outside/inside	Equilibrium potential
1. Ions distributed so that increased flux would depolarize the cell				
Sodium	15 m*M*	125 m*M*	8.3	$E_{Na} = +55.7$ mV
Calcium	1 − 0.1 μ*M*	2.2 m*M*	2200–22000	$E_{Ca} = +101 + 131$ mV
2. Ions distributed so that increased flux would repolarize the cell				
Potassium	158 m*M*	4.5 m*M*	0.028	$E_K = -93.5$ mV
Chloride	30 m*M*	135 m*M*	4.5	$E_{Cl} = -39.5$ mV

NB: Since the calcium ion has two charges the voltage required to maintain a ten-fold concentration difference ist 30.25 not 60.5 mV

II. Voltage Clamp

The squid axon is almost 1 mm in diameter, and it was found possible to introduce two wires, wound spirally around a fine glass rod, for a considerable distance. The axon was threaded through holes in four plates, sealed with vaseline, so the exterior solution of artificial seawater was divided into three compartments insulated from one another (Fig. 14). In the central compart-

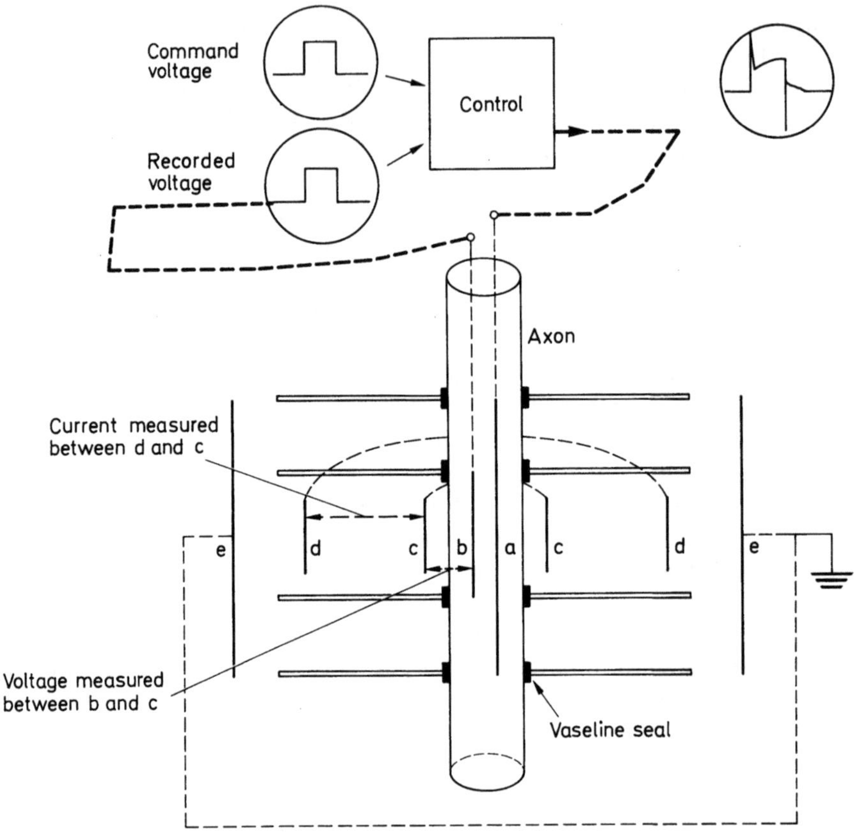

Fig. 14. Voltage clamp. Diagram of a squid axon threaded through holes in four plates and sealed with Vaseline to divide the extracellular artificial sea water into three compartments. Voltage across the membrane in the central compartment was measured between an intracellular wire, b, insulated except for the section opposite this compartment, and an external concentric electrode, c, close to the surface. Current was passed between another internal wire and an earthing electrode (e). The current flowing in the central compartment only was measured, by monitoring the voltage drop between c and an outer concentric electrode d. A control device, on receipt of a command signal, passed current until the voltage recorded across the membrane equalled the command voltage. The current required (*inset right*) after an initial surge as the membrane capacity was charged might then continue to increase if the membrane resistance fell with time

ment were two concentric electrodes, one (c) close to the axon, the other more distant (d). One of the internal wires (b) was insulated except for the portion opposite the central compartment, and the voltage across the membrane was measured between it and the external electrode c. Current could be injected through the wire a, but only that part flowing through the membrane in the central compartment was measured, by recording the voltage difference between the external electrodes c and d. It was thus ensured that voltage was measured across the same segment of membrane as that through which current passed.

The initial part of any applied current would charge the capacity of the membrane, causing voltage to lag behind current. Current was therefore injected by a device which drove current very rapidly until a selected desired voltage across the membrane was achieved, and continued to pass current as needed to maintain that voltage constant for as long as required, i.e. the membrane voltage was clamped. The response of the device to a command was very fast, so that the membrane capacity was rapidly "swamped" by an early current surge. If the current required to maintain a constant voltage changed thereafter (Fig. 14, top right), the membrane resistance must have changed in response to the voltage change.

Families of voltage steps were imposed, in solutions with sodium or in which sodium was replaced with choline, and it was concluded that the membrane contained ion-selective pathways with characteristics described by a set of now celebrated equations (HODGKIN and HUXLEY 1952). The essential features of a such a pathway are:

1. It is relatively selective for a particular ion.
2. There is a voltage range in which it passes from a non-conducting to a conducting state. On one side of this range the pathway is closed, on the other the probability of its being open is high.
3. The sodium pathway closes again rapidly even when the voltage is maintained beyond the opening level, changing to an "inactivated" state. The voltage must return negative before it can re-open.
4. Transitions do not follow voltage shifts instantaneously, i.e. they are both *time* and *voltage* dependent.

It was concluded that ions passed through the pathways by diffusion down their concentration and voltage gradients. The "driving force" was thus the difference between the applied voltage E and the equilibrium potential for the ion, which could be calculated from the Nernst equation if the activities of the ion outside and inside were known. Outside concentrations were fixed as desired, and the internal concentration of potassium was readily measured because the concentration was high. The intracellular concentration of sodium was obtained by activation analysis, the axoplasm being squeezed out and exposed to radiation in a nuclear reactor. The conductance g (reciprocal of resistance $= 1/R$) of the potassium pathway, in sodium-free solution, could then be calculated from $I(\mathrm{K}) = g\mathrm{K}(E - E(\mathrm{K}))$. The sodium current was obtained by subtracting the potassium current from total current observed after returning sodium to the solution, and the conductance calculated from $I(\mathrm{Na}) = g\mathrm{Na}(E - E(\mathrm{Na}))$.

It was concluded that a conducted action potential was initiated by passive current flow from an inactive towards an approaching active region of nerve, the consequent depolarization shifting its voltage into the range for transition of sodium pathways into their conducting state. Sodium entering the fibre depolarized the membrane further, so that "positive feedback" rapidly opened all channels. The voltage inside became positive, leading to inactivation of the pathways and sodium current ceased. Since the membrane was permeable to potassium, which was retained by the negative resting potential, as soon as this restraint was removed potassium flowed outward along its concentration gradient recharging the membrane to a negative voltage. In the squid axon, however, this process was accelerated by an additional mechanism. Potassium selective pathways, non-conducting at the negative resting potential, became conducting when the inside was positive, i.e. they acted as a rectifier. They did not respond immediately, but after a brief delay; this mechanism of "delayed rectification" ensured that the whole process of depolarization and repolarization was completed in a few milliseconds, permitting the nerve to transmit action potentials at a high frequency.

It was naturally hoped that it might be possible to extend the technique of voltage clamping to cardiac muscle, but from a comparison of the relative diameters of squid axons and of cardiac cells (Fig. 15 a) it was obvious that very different methods would have to be used. Two micropipettes were introduced into neighbouring cells, current being injected through one and voltage measured with the other. It was clear, however, that true voltage/current relations were not being studied because current could be affected by factors other than the voltage maintained across the membrane of a distant cell containing the voltage electrode (JOHNSON and LIEBERMAN 1971). The situation was improved by using a single electrode to inject current and measure voltage, but the small dimensions of cardiac cells (3.5 × 6 µm for the cell containing an electrode in Fig. 15 b), and the fact that in such multicellular preparations the extracellular environment consisted of narrow clefts, meant that the extracellu-

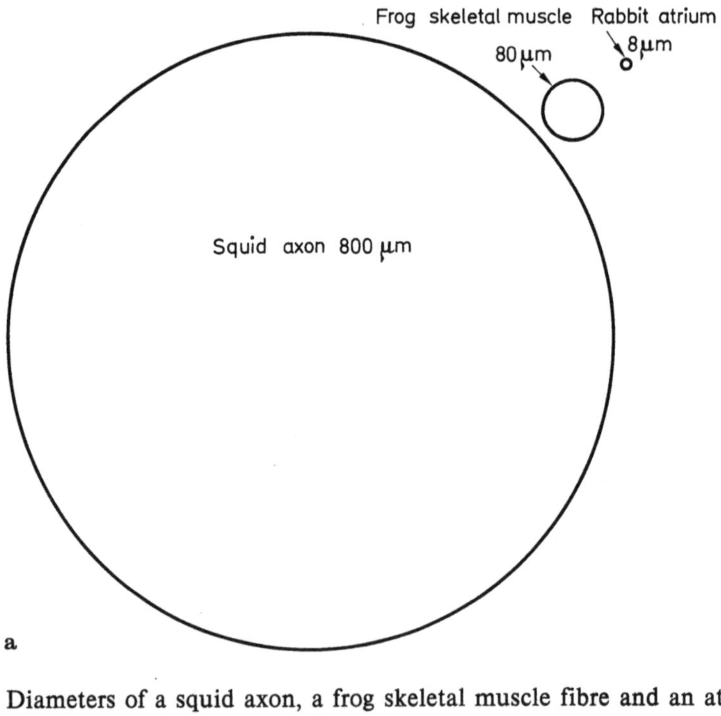

a

Fig. 15 a. Diameters of a squid axon, a frog skeletal muscle fibre and an atrial myocardial cell illustrated on the same scale

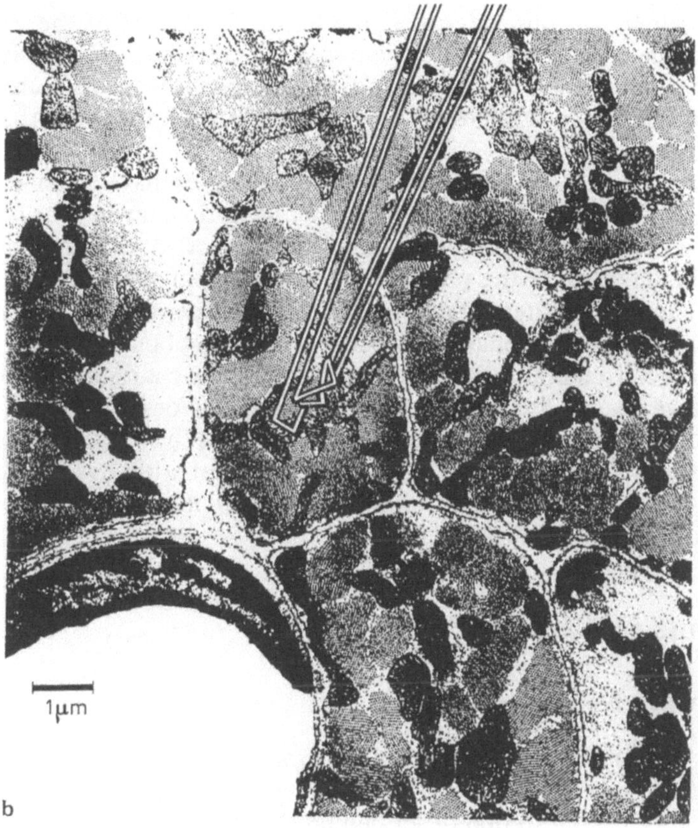

Fig. 15 b. Electron micrograph of rabbit ventricular cells, with a diagram of a glass capillary electrode with tip diameter 0.5 µm inserted into one cell. The extracellular spaces are connected with a capillary (*bottom left*) by extremely narrow clefts

lar concentrations of ions could not be assumed to be the same as those of the bath fluid, and could moreover be changed by current injection employed to clamp the membrane potential. If the extracellular K was altered, E(K) would also be altered, and gK could not, therefore, be calculated.

The problem may be illustrated quantitatively. A cylindrical sinus node cell 5 µm in diameter and 10 µm in length would have a surface of 354, or allowing for folding say 500 µm^3. At 1 µF/cm^2, the capacity would be 5 pF, requiring a charge of 0.1 pC to shift its voltage by 20 mV. With a volume of 200 µm^3 and at a concentration of 0.15 mol/litre K, the charge which could be carried by the total K in the cell would be 3 nC (3000 times the charge required to shift its membrane voltage by 20 mV). In voltage-clamp studies of sinus node cells Brown et al. (1979) injected a current of 30 nA to shift the measured voltages 20 mV negative, which had risen to 50 nA after 0.4 s, from which it was concluded that the pulse had activated a voltage-controlled time dependent current. A current of 50 nA for 0.4 s represents transfer of charge of 20 nC. Assuming that half was carried by K ions entering the KCl-filled electrode, and half by Cl ions leaving it, the entire K content of the cell would have been replaced 3 times, and Cl content 20 times, during the pulse. These huge currents must have been passing into neighbouring cells and exiting through their membranes into the clefts, altering equilibrium potentials over a substantial area.

More recently cellular electrophysiological studies have been performed on single cells isolated from neighbours by proteolytic enzymes, or on a patch of membrane sucked into the mouth of a micropipette. Although such techniques eliminate some of the problems encountered in multicellular preparations, they encounter others. After proteolytic treatment, what has happened to the gap junctions previously connecting with neighbouring cells? Has the treatment left ion channels entirely unchanged? Do channels in membrane patches behave exactly as they did in the intact cell? For all these reasons cellular electrophysiology has been much more difficult to study in cardiac muscle than in squid axons. Nevertheless a mass of information has been accumulated by a variety of techniques which has greatly illuminated the mode of action of antiarrhythmic drugs.

C. Currents in Cardiac Cells

I. Fast Inward Sodium Current

WEIDMANN (1955) showed that depolarization in cardiac Purkinje cells involved a sodium-selective pathway activated by a voltage-and-time-dependent mechanism very similar to that described by Hodgkin and Huxley in squid axons. The characteristics of the cardiac sodium channel are discussed in Chap. 5, 8 and 9.

II. Residual or Plateau Sodium Current

In Purkinje cells, and perhaps also in nodal cells, not all sodium channels remain inactivated at potentials positive to −50 mV (as they do in squid axon). CORABOEUF and DEROUBAIX (1978) observed that the selective sodium-channel poison tetrodotoxin (TTX) shortened action potential duration, and concluded "in P fibres the TTX-sensitive steady state current flows in a larger potential range than in the myocardium". This residual sodium current contributes to maintaining the plateau, and its abolition by the class 1b antiarrhythmic drugs in high concentration (see below) may explain why they shorten APD.

III. Inward Rectification

The high intracellular K concentration is established by the Na/K pump. This will not function unless both K sites and all three Na sites are occupied, and there is an adequate supply of ATP. The pump can be stopped experimentally by reducing $[K]_o$ below 1 mM or replacing $[Na]_i$ (e.g. by *Li*), but neither of

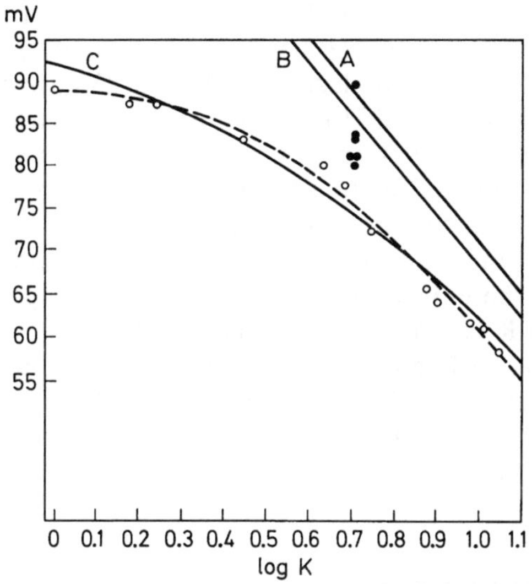

Fig. 16. Inward rectification. Relation between the resting potential (*RP*) of isolated rabbit atria at 32 °C and [K]*o* on a logarithmic scale. The lines A, B plot *E*(K) when [K]*i* is given respectively the maximum and minimum values actually measured (155 and 138 mmol/litre). The line C is drawn to $E = 59.5 \log(10)\ (P[\mathrm{K}]o/(p[\mathrm{K}]i+[\mathrm{Na}]o))$ where *P* is 49.5, the ratio of K to Na permeability chosen for best fit to the points (*open circles*). The *dotted line* plots a similar equation, giving a significantly better fit to the points, when *P* was not a constant, but K permeability was reduced as *RP* deviated progressively from *E*(K) (inward rectification). The *filled circles* show values of *RP* observed in the presence of acetylcholine, which pushed the potential towards *E*(K)

these circumstances would be compatible with life. In the squid axon delayed rectification rapidly restores the resting potential, but in cardiac muscle depolarization is maintained in the plateau phase, presumably to ensure a long refractory period. K would flow out of the cell during the plateau, and have to be pumped back in again, so that it would clearly be advantageous if K loss could be reduced by ensuring that K permeability fell when the membrane was depolarized (i.e. the opposite to delayed rectification). During diastole, K permeability is high, Na permeability low, and if the ratio of permeabilities, *P*, remained constant the relation between resting potential and changes in extracellular K (which governs *E*(K)) should be given by $RP = 60.5 \times \log(10)\ (P \times [\mathrm{K}]o + [\mathrm{Na}]i/P \times [\mathrm{K}]i + [\mathrm{Na}]o]$, Since *P* is about 50 and [Na]*i* is small the latter term can be neglected. In Fig. 16 the circles represent the means of more than 2000 measurement of resting potential at 32 °C in isolated rabbit atria, plotted against log (10) [K]o over a range from 1 m*M* (to avoid Na/K pump inhibition at lower concentrations) and 11 m*M* (to avoid sinus node arrest at higher concentrations). The solid line *C* is plotted to $E = 59.5 \times \log(10)\ (P \times [\mathrm{K}]o/P \times [\mathrm{K}]i + [\mathrm{Na}]\mathrm{o})$, *P* being chosen for best fit (49.5). The solid lines *A* and *B* give *E*(K) plotted against log [K]*o* on the basis of the highest and lowest values respectively of the measured [K]*i*. The main point of interest was that a significantly better fit to the results was obtained (dotted line) if *P* was not constant, but if it was arranged that potassium permeability was reduced when the *RP* became increasingly positive to *E*(K), the condition in which *K* would be lost from the cell (VAUGHAN WILLIAMS 1959). In the same year HODGKIN and HOROWITZ noted a similar phenomenon in frog skeletal muscle; since a reduction of K permeability on depolarization was the opposite of the delayed rectification found in squid axon, it was called "anomalous" rectification, but the presently preferred term is inward rectification.

The existence of inward rectification has three important consequences:

1. In conditions where K would be lost [when *E* is positive to *E*(K)] the permeability to K is reduced and K conserved.
2. Reduced outflow of K contributes to maintainance of the plateau
3. During repolarization as *E* again approaches *E*(K) the inward rectification is switched off, K permeability rises and repolarization is rapidly restored by positive feedback.

IV. Second Inward Current

Quinidine, the first antiarrhythmic drug, reduces fast inward sodium current. Even very high concentrations, however, several times the maximum usable clinically, did not reduce the resting potential nor abolish conduction, and it was suggested that in cardiac muscle there was a second inward current. The actual description of the evidence (Fig. 17a) was: "the action potential developed a 'step'. A fast phase of depolarization was cut off, yet depolarization proceeded at a slower rate to an overshoot. The early phase of the step represents the operation of the sodiumcarrier, and the second slower phase is pro-

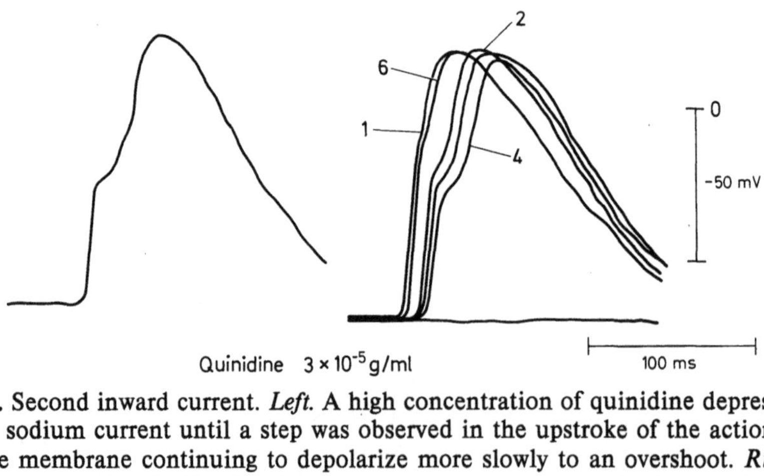

Fig. 17. Second inward current. *Left.* A high concentration of quinidine depressed fast inward sodium current until a step was observed in the upstroke of the action potential, the membrane continuing to depolarize more slowly to an overshoot. *Right.* Responses to six stimuli. The fast inward current was progressively depressed, and the step became more obvious, until the fifth stimulus, which elicited no response. In the double interval between the fourth and sixth stimuli, sodium channels had time to recover from inactivation, so that the response to the sixth stimulus was as fast as to the first

duced by current carried by other ions. The sodium carrier might, therefore, be the fastest but is not necessarily the only mechanism for transporting depolarizing current" (VAUGHAN WILLIAMS 1958). Responses to six stimuli are shown in Fig. 17. The fast inward current was progressively depressed, and there was no response, to the fifth stimulus. In the double interval between the fourth and sixth stimuli the sodium channels had time to recover from inactivation, and the response to the sixth stimulus was as fast to the first.
The existence of two inward currents in cardiac muscle was not widely accepted. NOBLE (1962) demonstrated that Purkinje cell action potentials could be simulated by a computer model based on modifications of Hodgkin-Huxley (H-H) equations for voltage and time-dependent conductances of Na and K only. Although inward rectification was included in the model, there was no second inward current. Nevertheless the simulation was very close to observed potentials, but this could not be taken to imply that second inward current did not exist or was physiologically unimportant. REUTER (1967) presented evidence for the existence in Purkinje cells of a second inward current carried by calcium ions. It was activated at more positive potentials, and apparently much more slowly, than the fast inward sodium current, and was called "slow" inward current. More recently, however, it has been suggested that the second inward current carried by calcium has three components.

1. BEAN (1985) using patch pipettes on canine atrial cells concluded: "Two components of Ca channels could be distinguished. One (*I* fast) was present only if cells were held at negative potentials, was most prominent for relatively small depolarizations, and inactivated within tens of milliseconds.".
2. "The other (*I* slow), corresponding to the Ca current previously reported in single cardiac cells, persisted even at relatively positive holding potentials, required stronger depolarizations for maximum current, and inactivated more slowly". Both currents were unaffected by tetrodotoxin and both were reduced by cobalt.
3. LEE et al. (1984) observed: "A small and very slow calcium current has been identified in isolated single ventricular cells using TTX and Cd^{2+} to block the sodium and fast calcium currents. Activation requires about 300 ms at the threshold potential of −60 mV, decreasing to 80 ms at the peak current voltage of −30 mV. Inactivation is five to ten times longer". They named it "Isi3".

LEE et al. (1984) proposed that Isi3 "contributes significant current to help maintain a major portion of the long ventricular action potential". It is doubtful, however, whether the kinetics of a current studied at room temperature in a cell isolated by proteolytic enzymes and exposed to TTX and cadmium can be extrapolated to those of a beating heart. The action potential of the isolated cell had a duration of 1 s, several times longer than the natural APD.

REUTER (1967) had concluded that the calcium channels he studied were inactivated by a voltage-and-time-dependent process described by H-H Equations. Evidence from other cells (e.g. snail neurones) suggested that calcium channels were inactivated, not by positive voltage, but by a rise in intracellular Ca concentration (direct negative feedback). LEE et al. (1985) concluded that both voltage and [Ca]*i* were involved in the inactivation of inward calcium current in mammalian heart cells.

The selectivity of the channels carrying second inward current has been questioned, and it has been suggested that sodium ions also can pass through. HESS et al. (1986) provided evidence that monovalent ions could traverse the channel only when calcium was absent. They suggested that the channel did not contain a selectivity "filter" of the kind believed to exist in Na- and K-selective channels, but that occupancy of a site within the ion pathway with high affinity for Ca "by only one Ca ion is sufficient to block the pore's high conductance for monovalent ions like Na. Rapid permeation by Ca ions depends upon double occupancy which only becomes significant at millimolar [Ca]*o*. Ions pass through the pore in single file, interacting with multiple binding sites along the way".

V. Excitation-Contraction Coupling

The majority of studies of E-C coupling have been concerned with skeletal muscle, but cardiac musle is basically similar, in that calcium is released from a store, binds to troponin (two sites skeletal, one cardiac) leading to movement of tropomyosin to uncover an actin site for attachment of an apposed myosin head. ATP is split, tension is developed and movement of myosin relative to actin occurs, the process being repeated so long as calcium is attached to troponin and the supply of ATP is maintained. Contraction is maximal when [Ca]*i* is 2 μmol/litre, and ceases when it is less than 0.1 μmol/litre. In skeletal muscle the store of calcium is in the cisternae adjacent to the transverse tubules, the signal for its release having been attributed directly to depolarization of the T tubule linked in some way to the cisterna, or to ion flow across the tubular wall. Changes in [Ca]*o* have little effect on contraction, so that Ca is released, recaptured and restored entirely within the cell. In cardiac muscle contraction is exquisitively sensitive to [Ca]*o*, so that extrusion and re-entry of Ca must be involved in some way in the overall control. There are cisternae not only apposed to the T tubules but also between the sarcoplasmic reticulum (SR) and surface sarcolemma. The ratio of sarcolemmal to tubular membrane surface area is 50 times higher in cardiac than skeletal muscle, suggesting that SR may be an additional site for Ca release in the heart (PHILLIPSON et al. 1981). It has been suggested that Ca entering as second inward current may induce further Ca release from the SR, but although "Ca-induced Ca release" has been demonstrated in skinned fibres (FABIATO and FABIATO 1979), "a major problem is whether the methods used for skinning cardiac cells could induce in the SR properties it does not have in intact cardiac muscle".

VI. Destinations of Intracellular Calcium

All the calcium entering a cell in systole must be extruded before the end of diastole, otherwise Ca would be accumulated indefinitely. There may be wide variations in calcium flux per beat, but a new steady state must soon be established, the amount extruded equalling that entering. There are many intermediate intracellular destinations. In addition to troponin, myosin ATP-ase and light meromyosin chains and several other proteins can bind calcium. Parvalbumin is a calcium buffer; calmodulin binds calcium in regulating adenylate cyclase, phosphodiesterase, phosphorylase b kinase and other systems. Mitochondria both take up and extrude calcium, and it has been suggested that rises of intramitochondrial Ca may be important regulators of oxidative metabolism keeping ATP production in step with the demands of contraction. An ATP-driven pump avidly takes up Ca into the SR, where it is bound by calsequestrin, putting it beyond the reach of probes (aequorin, dyes) requiring ionized Ca. All these sites are intracellular, however, and excess calcium accumulated must eventually find a pathway to the exterior.

The main route is via a sodium/calcium exchange, three or four sodium ions entering and one calcium leaving, so that each cycle depolarizes the membrane by one or two charges (Mullins 1981). It has been suggested that net depolarizing Na/Ca current may contribute to the plateau, but since the sodium ions are expelled by the Na/K pump, which is electrogenic in the opposite sense, there is ultimately no net depolarization. If, however, the [Na]*o*/[Na]*i* ratio falls for any reason (lowering [Na]*o*, accumulating [Na]*i* by pacing at high frequency or inhibiting the Na/K pump), the driving force for Ca extrusion will be diminished, with two major consequences. First, competition for Ca will favour SR uptake instead of Ca extrusion, so that more Ca will be available for release by a subsequent action potential, which accounts for the positive inotropic actions of reducing [Na]*o*, of digitalis and of double-pacing (postextrasystolic potentiation). Secondly, the depolarizing Na/Ca exchange current will be delayed, causing a prolongation of (or even depolarization during) the plateau ("early after-depolarization"). If repolarization proceeds to the point at which return from inward rectification reinforces rapid repolarization, the depolarizing Na/Ca exchange may outlast the decay of "activated" repolarizing current (see below) and so persist into diastole (late after-depolarization), as may be observed after digitalis intoxication of after rapid trains of action potentials in depressed myocardium (triggered activity), as discussed in Chap. 20 and 26.

In some tissues an ATP-driven Ca extrusion pump has been described but its presence in the sarcolemma of cardiac cells has been disputed, because of the possible contamination of "sarcolemmal" preparations with fragments of SR (Fabiato and Fabiato 1979).

VII. Repolarizing Current

The appropriate duration of a cardiac action potential will depend on an animal's size and natural cardiac frequency, being shorter in mice than in men.

Delayed rectification of a kind which assures rapid repolarization in squid axon does not therefore occur in the heart of larger mammals; on the contrary, a long plateau is needed to activate contraction and prevent rapid re-excitation. Indeed, activation of an outward current is not necessary for repolarization, because when inward current ceases, passive outflow of K from its high intracellular concentration would cause repolarization, accelerated by disappearance of inward rectification. Even in some myelinated nerves (rabbit) delayed rectification is absent, and from voltage clamp experiments on sheep ventricle GIEBISCH and WEIDMANN (1971) concluded "the results suggest that normal repolarization depends on a time-dependent decrease of inward current (Na, Ca) rather than on a time dependent increase of outward current (K). In cat ventricle, however, MCDONALD and TRAUTWEIN (1978 a,b) concluded that an outward current carried by K, which they called Ik, was activated in the plateau region with a time constant of 370 ms. They appreciated, however, that long-lasting clamp currents causing alterations of [K]*o* made estimation of reversal potential insecure. In Purkinje cells, NOBLE and TSIEN (1969 a, b) identified two outward currents, I_{X1} and I_{X2}, described by H-H-type equations, activated in the range -50 to $+20$ mV, and with time constants at plateau voltages of 0.5 and 4 s respectively.

D. Currents of Uncertain Physiological Significance

I. Transient Outward Current (Dynamic Current, Iqr)

In some Purkinje cells a brief repolarization, or notch, is observed between the fast phase of depolarization and the plateau. CARMELIET (1961) concluded "that chloride ions contribute little to the total membrane conductance at the resting potential, but become important as electrical charge carriers when the membrane is depolarized". DUDEL et al. (1967) desribed an outward current, which they called dynamic current, which was reduced by removal of extracellular chloride. FOZZARD and HIRAOKA (1973) also concluded that a "positive dynamic current" governed by H-H-type variables was carried by Cl ions, and MCALLISTER et al. (1975) accepted it as a chloride current in their reconstruction of a cardiac action potential, calling it "transient outward current", *Iqr*.

KENYON and GIBBONS (1979 a,b) reinvestigated what they now called "early outward current" in Purkinje cells and concluded that it was carried by K ions, because it was blocked by substances blocking K channels, and was not reduced in solutions containing only 8.6% of normal Cl. CARMELIET and VERDONCK (1977) found that substitution of Cl by other anions reduced K permeability by 38%, which might explain why the transient current was reduced in low Cl even if carried by K ions.

II. Calcium-Activated Potassium Current

ISENBERG (1977) injected current into Purkinje cells through a micropipette containing 1 mol/litre $CaCl_2$ and noted that for a few seconds subsequently action potentials were shorter. From this and other experiments it was concluded that repolarization was controlled by raised [Ca]*i* activating an outward K current. From a variety of experiments designed to vary [Ca]*i* (rapid stimulation, exposure to cyanide, reduction of [Na]*o*) in sheep and calf ventricular strips, BASSINGTHWAITE et al. (1976) also concluded that APD may be controlled by "[Ca]*i* at the inner side of the membrane" which "sets the level of the background current". In invertebrate cells and red blood corpuscles there is evidence that K permeability is increased by raised [Ca]*i*, but the existence in cardiac muscle of Ca-activated delayed rectification has been disputed (KASS 1982).

III. ATP-Regulated Potassium Current

Hypoxia induces shortening af APD in cardiac cells, in association with increased K efflux (VLEUGELS and CARMELIET 1976). NOMA and TOHRU (1985) found that in guineapig ventricular cells depleted of ATP by 1 mmol/litre cyanide K conductance was increased and "concluded that the ATP-regulated channels are responsible for the increase in outward current and the shortening of action potential duration in various anoxic conditions". The increase in outward current was not observed unless the ATP was reduced to below 2 mmol/litre. Even in total hypoxia, however, provided that glucose is present there is very little change in ATP in cardiac cells, yet substantial shortening of APD occurs in the presence of glucose and 20% oxygen, so that ATP-regulated K channels could not be responsible (VAUGHAN WILLIAMS 1987).

IV. Calcium-Activated Inward Current

COLQUHOUN et al. (1981) described an inward current activated by raised [Ca]*i* in cultured cardiac cells. It is not known whether such currents have physiological significance in adult mammalian cardiac muscle, but if so the effect of raised [Ca]*i* on cardiac cellular electrophysiology would be hard to predict, since both repolarizing and depolarizing Ca-activated currents have been claimed.

E. Initiation of the Heartbeat

I. Normal Beats. The Sinoatrial Node.

The histology of the SA node has been extensively studied, notably by LEGATO (1973) and JAMES (1966, 1977). The most characteristic cells were pale (P cells), polyhedral, 5–10 μm in diameter, containing few mitochondria or other organelles, myofibrils being sparse and randomly orientated. The P cells were grouped in clusters, and between these and the myocardium were transitional cells, of which "the principal function was probably distribution of the sinus impulse. There are numerous infoldings which form intercellular clefts, so that the clusters have a sponge-like arrangement relative to the extracellular space". Although P cells were held together by desmosomes, "no fusion of apposing plasma membranes in the form of a nexus (= gapjunction) could be identified. The paucity of specialized intercellular contact points is an explanation for retarded intranodal conduction". "Conduction between P cells is probably too slow to permit sufficiently rapid "entrainment" of all of them. Rate of firing of different clusters may be random".

Although other authors did find a few gapjunctions between P cells and considered them sufficient to allow synchronization within a group, all are agreed the coupling between adjacent P cells is very much weaker (< 1/10) than between P cells and the surrounding transitional cells. In a motoneurone the cell body is bombarded by depolarizations and hyperpolarizations spreading into it electrotonically from dendrites. The action potential is not initiated in the soma, however, but at the axon hillock which "fires" first, sending an action potential not only peripherally down the axon, but also back into the soma which reaches full depolarization later. The sinus node may operate in an analogous fashion. Clusters of P cells, poorly coupled to each other but

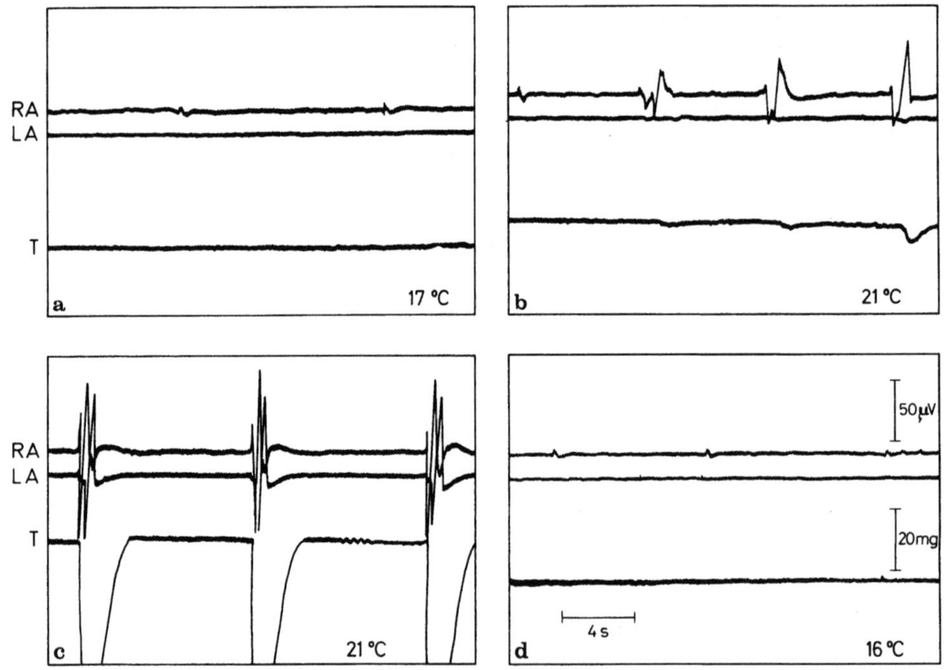

Fig. 18. a–d Effect of cooling on the initiation of the heartbeat. A bipolar electrode recorded from the right atrium (RA) close to the sinus node; another recorded from the left atrium (LA). Contractions were recorded with a high gain transducer (T). **a** At 17 °C heartbeats were no longer initiated, but the sinus node continued to produce potentials at irregular intervals. **b** On warming to 21 °C the sinus node potentials began to spread to the surrounding atrium, and heartbeats recommenced. **c** Eventually the potentials spread to the left atrium also, with contractions now off-scale. **d** On recooling, the heartbeats again ceased, but sinus node potentials continued

closely coupled to surrounding transitional cells, slowly depolarize, acting as a sink. As soon as one of the transitional cells fires, a wave of depolarization encircles the island of P cells, spreading inwards to complete its depolarization. The wave also spreads outwards to engulf the other islands of P cells and to travel concentrically to the rest of the atrium. Thus the synchronization of the node as a whole is achieved, not by the poor intercellular coupling between P cells themselves, but from outside via the much closer coupling of transitional cells both to each other and to their P cell neighbours.

The *initiating* event is, therefore, an *interaction* between a pair of cells: a P cell depolarizing spontaneously and a transitional neighbour. Many such cell pairs can exist within the node, so that the "dominant pacemaker site" can jump from one zone to another as a result of modification in either member of the pair, e.g. by hyperpolarization of the transitional cell by acetylcholine, or by altered coupling between cells induced by changes in extracellular calcium, which controls the conductance of gap junctions (DAHL and ISENBERG 1980).

Coupling between the sinus node and its surroundings can be broken by lowering the temperature. The heartbeat is no longer initiated, but the sinus node continues to function. The potentials produced, however, no longer occur at regular intervals, as though first one cluster, then another, took priority. Recording electrodes (Fig. 18) were placed on the sinus node, and on the left

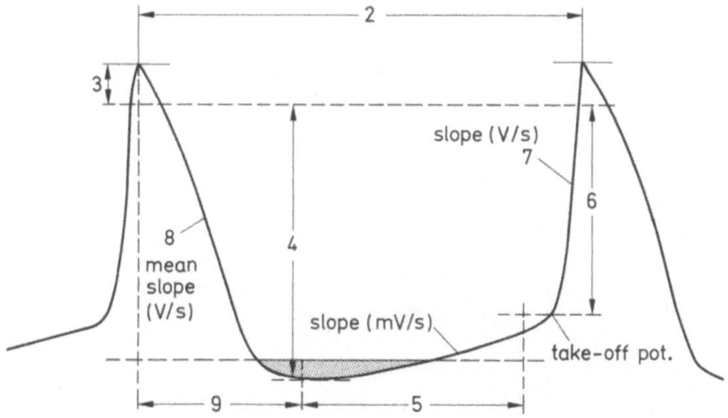

Fig. 19. Sinus node potentials. The parameters measured were labelled as follows: 2, cycle length; 9, repolarization time from the peak potential to the maximum diastolic potential; 3, overshoot potential, from zero to peak; 4, maximum diastolic potential (MDP), from zero to lowest point; 1, action potential amplitude, the sum of 3 and 4; 7, the maximum rate of depolarization (MRD) obtained by differentiation of the upstroke; 8, mean rate of repolarization, obtained by a tangent to the most linear portion of the repolarizing limb; 5, slope of the slow diastolic potential, measured as the mean slope of the middle third of the almost linear portion of the segment marked 5 between the MDP and the start of the upstroke; 6, take-off potential, obtained by extrapolating forwards along 5, and backwards along 7. All measurements were made by a computer programme on digitized recordings of potentials. The *shaded area* indicates the length of time the potential is negative to -50 mV, the activation threshold for If

atrium of isolated rabbit atria, and contractions were recorded with a high-gain transducer. At 17 °C no heartbeats were initiated, but potentials were recorded at varying intervals from the node. On warming, coupling was re-established at 21 °C and action potentials, simultaneously with contractions, appeared in the right atrium, and spread rapidly to the left atrium.

A typical sinus node potential is illustrated in Fig. 19, with the various parameters usually measured marked by numbers. Attempts to identify the point of origin of the initiation of the heartbeat involved construction of maps with "isochrone" contours. What fixes the zero? Masson-Pevet et al. (1979) decided arbitrarily: "As the activation time of a cell we chose the moment at

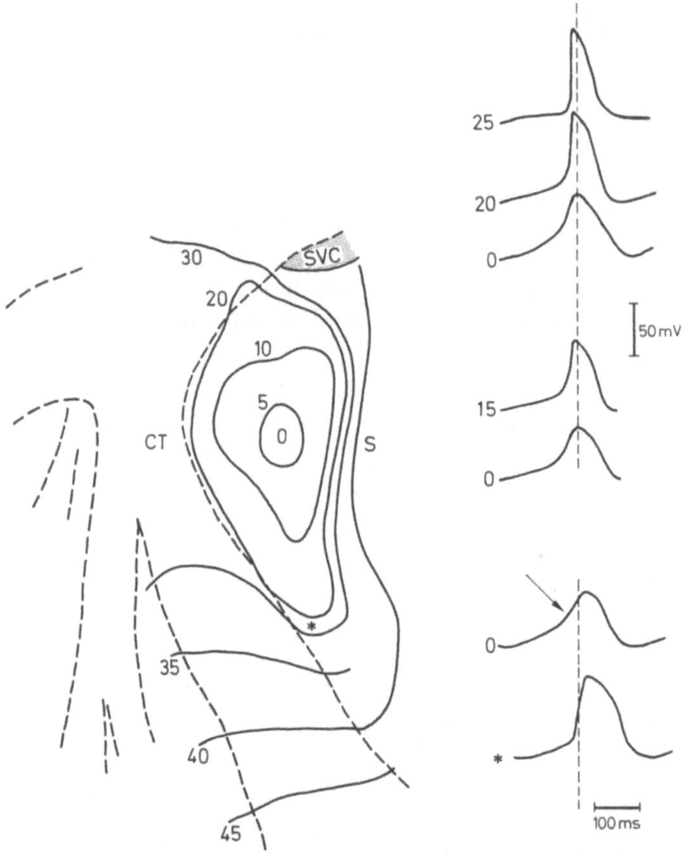

Fig. 20 Map of rabbit sinus node with isochrone contours in milliseconds from the earliest midpoint between MDP and peak of potentials recorded with an exploring microelectrode. SVC, superior vena cava; CT, crista terminalis. *On the right,* action potentials from the cells in the isochrone regions indicated. In the upper two sets the *dotted line* is drawn through the peak of the action potential in the zero zone, indicating that the peaks in the zones marked 15, 20 and 25 actually occurred earlier than in the "zero" zone. In the lower pair, the *dotted line* is drawn through the midpoint of the potential from the cell marked with an *asterisk* in the 30-ms zone to show that the peak in the zero zone occurred later, although the mid point (arrow) between MDP and peak occurred earlier than in the 30 ms zone

which the transmembrane potential reached half the amplitude of the action potential. The activation moment of the impaled cells was timed with respect to an atrial electrogram". A representative map constructed on this principle is illustrated in Fig. 20. Three sets of records from the isochrone regions indicated are presented on the right. In the upper two sets the vertical dotted line has been drawn to synchronize with the *peaks* of the action potentials recorderd from the "zero" area. It is apparent that the peaks of the potentials from the 20- and 25-ms areas (above) and from a 15-ms region (centre) occurred *earlier* than in the zero region, indicating that full depolarization of the central region was assisted to completion by current flow from the surrounding transitional cells. The slow diastolic potential of cells in the central region merges into the upstroke of the action potential, making it impossible to measure accurately the time of the "take-off" potential. For the lowest pair of potentials the dotted line has been drawn to coincide with the halfway point (as defined by Masson-Pevet, above) of the action potential from a site marked with an asterisk in the 30-ms isochrone. The halfway point of the potential from the zero zone is marked with an arrow, and occurs earlier, although its peak occurs later, than in the 30-ms zone. If the action potential peak had been accepted as the criterion, the heartbeat would be shown as being initiated in a circle corresponding approximately to the 15-ms isochrone of the map.

Similar maps (Bleeker et al. 1980; Masson-Pevet et al. 1982) show the wave of depolarization proceeding rapidly down the crista terminalis, but towards the atrial septum conduction is so slow as to represent a virtual block. Bleeker et al could find "no morphological basis, such as, for example, a zone of connective tissue" to explain the block, but it must be remembered that in utero the septum contained the foramen ovale, around which conduction had to proceed above and below, and after closure the impulse would be likely to follow the same pathways. Action potentials from this region often have double peaks, implying the development of some electrical coupling between originally disparate bundles.

II. Ionic Currents in Nodal Cells

Resting potentials in the central zone are −40 to −55 mV, and the maximum rate of depolarization (MRD) of the upstroke is less than 2 V/s, and the peak is rounded. If fast sodium channels exist, they would be inactivated, and the upstroke current is probably carried exclusively by Ca ions (Kohlhardt et al. 1976). Further out resting potentials increase, so that fast Na channels begin to operate, being the predominant charge carriers in contracting atrial fibres.

III. Ik

Repolarization by electrogenic Na/K pumping would only occur in cells in which Na entered, and it is widely believed that repolarization is due mainly to activation (between −50 and +10 mV) of a time-dependent current Ik (Brown 1982).

IV. Ik2

The nature of the current responsible for the slow diastolic depolarization is still controversial. NOBLE (1975/1979) suggested that the slow diastolic potential in Purkinje cells was associated with the decay of an outward current Ik2, carried exclusively by K ions, which had been activated during systole. Many papers were written describing the effects of drugs, especially cardiac glycosides, and catecholamines in relation to modification of this current. The evidence for its existence, however, involved use of long-lasting voltage clamp pulses in multicellular preparations which modified ionic concentrations in the extracellular space. "It is concluded that Ik2 is not, as had previously been thought, a pure K current, but is rather an inward current activated during hyperpolarizations negative to about −50 mV" (DI FRANCESCO 1981).

V. Ih (If)

An *inward current* activated by hyperpolarization (Ih) in small isolated strips of rabbit SA node (0.2 × 1 mm) was described by NOMA et al. (1977) and YANAGIHARA and IRISAWA (1980). Although the ions carrying the current were not identified, it was unaffected by exposure to barium, which blocks K channels, and it was suggested that the chargecarriers might be chloride ions, because the reversal potential was close to the equilibrium potential for chloride, E(Cl), and the magnitude of the current was reduced when Cl ions were replaced by acetate. The activation range was −50 to −100 mV, but the activation illustrated at −50 mV was insignificant. Even at −70 mV activation was extremely slow, with a timeconstant of 2-4 s. It was concluded that since the slow diastolic potential in sinus node cells ranges from −70 mV maximum diastolic to −40 mV "in this voltage range Ih has too long a time constant to show dynamic changes during each cardiac cycle. It is unlikely that Ih has a significant effect on the normal action potential pattern." BROWN et al. (1979) described a current, which they called If, with characteristics so similar to those of Ih that it is probable that both groups were investigating the same phenomen. BROWN et al. (1979) took the view, however, that the current was functionally important in the generation of the slow diastolic potentials. Nevertheless, as already noted, activation required long-lasting and large currents at potentials near −50 mV and below. In an actual sinus node potential the time during which the potential remains negative to −50 mV (Fig. 19, shaded area) is so short that it seems unlikely that the current could be important in controlling sinus rhythm.

The maximum diastolic potential of P cells is several millivolts positive to the MDP of neighbouring transitional cells, so that K will pass through gap junctions from P to transitional cells, and Cl ions in the opposite direction, both tending to hyperpolarize the P cell. Yet the P cell continues to depolarize during diastole, so that substantial depolarizing current must be flowing into it from the extracellular space. Although fast sodium channels are inactivated at voltages more positive than −50 mV, some Na could enter through the "residual" or plateau Na current channels. Such current cannot be quantitatively important, however, because TTX has a very small effect on sinus rhythm, prolonging cycle length by only a few milliseconds, and the slope of the slow diastolic depolarization is not detectably altered (YAMAGISHI and SANO 1966).

DE MELLO (1963) concluded that Cl ions contributed substantially to depolarizing current in the SA node. "When the membrane is repolarized, and the membrane potential exceeds (i.e. is negative to) E(Cl) (-37 mV), there is an outward movement of Cl ions which contributes to diastolic depolarization". More recently SEYAMA (1977, 1979), studied small preparations of rabbit SA node in various anionic solutions, and concluded that there was "a relatively large contribution of anion conductance to the total membrane conductance in the sinoatrial cell".

To summarize, the P cell reaches maximum diastolic potential when K ions flow out (Ik) through channels activated during depolarization. As Ik declines, the P cell depolarizes slowly, in spite of repolarizing current from more negative neighbours, because (according to the evidence cited above) Cl conductance is high. Eventually the membrane becomes sufficiently positive (-40 mV) for the second inward (calcium) current to be activated. Meanwhile, although weakly coupled to other P cells, close coupling to transitional cells causes these to depolarize rapidly, partly by fast inward current, but mainly by second inward current also. These reach peak depolarization before the P cells, so that the heartbeat is initiated from the ring of transitional cells around one cluster of P cells, a wave of depolarization engulfing the other clusters and spreading out from the sinus node as a whole.

VI. Sinus Node Recovery Time

If the peripheral atrium is paced by an intracardiac electrode, close coupling with the sinus node ensures that the latter is depolarized by incoming action potentials. At the end of the pacing, the interval before the first sinus beat, the sinus node recovery time (SNRT), is longer than a normal cycle, even in the presence of atropine (so vagal stimulation is not involved). It has been suggested that during pacing the spare capacity of the Na/K pump is mobilized by increased Na entry. In the P cells there is no Na entry, however, but accumulation of extracellular K could activate the pump in accordance with the evidence of NOMA and IRISAWA (1974). If P cell clusters are destroyed by disease, the area of depolarizing "sink" into which current can flow from surrounding transitional cells will be reduced and SNRT prolonged. The large number of independent clusters of P cells gives the node a high safety factor. In theory a single P cell/transitional couple could initiate a heartbeat. In practice some patients with sinus node disease, known to have been in sinus rhythm before death, have been found at postmortem to have lost 80% of the node. Thus prolongation of SNRT has limited prognostic value, but at least warns that caution should be exercised in exhibiting drugs which depress SAN function.

VII. Abnormal Heartbeats

Excised Purkinje cells may beat spontaneously in vitro. Trabeculae containing Purkinje cells have been much used for electrophysiological study, because they are easy to dissect, resist anoxia, and survive well in slaughterhouse material. They are, however, the cells least appropriate to study, expecially for investigations of contraction, because they are specialized for conduction, contribute nothing to cardiac output, and have few mitochondria. Catabolism in contracting myocardium is principally oxidative, mitochondra occupying up to one-third of intracellular volume.

Mammalian left atria, papillary muscles and strips of ventricular muscle, if carefully dissected, are quiescent in vitro. In dogs, after AV block, escape rhythms invariably originated in the proximal part of the specialized ventricular conducting system (SVCS), 60% arising in the left bundle branch (LBB), 28% in the RBB and 9% in the bundle of His (HOPE et al. 1976). They never originated in distal Purkinje cells. If APD is measured in the latter in situ, it is prolonged by about 100 ms if the Purkinje cells are cut free from neighbouring tissue. After experimental myocardial infarction abnormal impulses arise, not from Purkinje cells, but from ventricular myocardium in the neighbourhood of the infarct. It seems doubtful, therefore, whether the pacemaking of excised Purkinje cells has any relevance to human arrhythmias. The AV node is embryologically the "sinus node" of the left heart. It is possible, therefore, that in the proximal SVCS there are residues of P cell tissue which provide slowly depolarizing sinks for generation of ventricular escape pacemakers after AV block, a view supported by pharmacological evidence to be described later.

VIII. Transient Inward Current (Ti)

It has already been noted that in circumstances in which [Na]i is increased, a late afterpotential may occur early in diastole, attributed to delayed electrogenic Na/Ca exchange. Under extreme conditions designed to reduce Ca extrusion further, potentials during diastole (i.e. after repolarization) may oscillate, in association with fluctuations (about 70–80 ms later in phase) of contraction and in the emission of light from cells into which aequorin had been introduced. It was believed that these oscillations could be caused by intermittent release of Ca from the SR, which in turn caused fluctuations of membrane potential through electrogenic Na/Ca excange (ALLEN et al. 1985; EISNER and VALDEOLMILLOS 1986). CANNELL and LEDERER (1986), also using sheep Purkinje cells, exposed them to 0 m*M* K and Na-free isotonic $CaCl_2$, and found that about 300 ms after repolarization from a depolarizing pulse, oscillations of membrane potential (followed by contractions) still occurred, which could not be due to Na/Ca exchange because [Na]*o* was zero. They concluded that the current fluctuations were due to the "activation of a non-selective cation channel by intracellular calcium". The charge carriers could be K, Na or Ca, (permeability Na = K > Ca). They also suggested (from intensity of aequorin light measurements) that [Ca]*i* could rise during a normal action potential to the level required to activate Ti.

The relevance of experiments on sheep Purkinje cells in zero K and Na, and isotonic $CaCl_2$, to human arrhythmias is hard to assess. It has already been noted that Ca-selective channels can become permeable to Na when $[Ca]o$ is reduced, and it seems possible that in the extreme conditions required to demonstrate oscillatory fluctuations of $[Ca]i$ and contraction, normally selective membrane channels become non-selective. It is, perhaps, premature to name Ti *the* arrhythmogenic current. Many arrhythmias occur in circumstances in which there is no reason to believe that the intra- or extracellular ionic concentrations are in any way abnormal.

IX. Conclusion

In the clinical situation one may question whether intracellular biochemical changes and their electrophysiological correlates are as important in the development of serious arrhythmias as alterations in the pathway of conduction. The electrical and biochemical events associated with adrenergic stimulation (tachycardia, high oxygen demand, raised $[Ca]i$ and cAMP) are part of normal physiological control, and may lead to severely abnormal electrocardiograms being recorded in healthy subjects under stress without any progression to life-threatening arrhythmia (Chap. 12). Even in unstressed normal people occasional extrasystoles are common, so why are such events innocuous when they may precipitate ventricular fibrillation in a patient? It is a reasonable assumption that in subjects at risk, as a result of congenital abnormality or previous pathology, preferential pathways exist for re-entry to develop. The reproducibility, sometimes over years, of arrhythmias inducible during electrophysiological study implies the existence of an underlying anatomical substrate increasing the probability of re-entry. The high incidence of sudden death in persons not known to have any cardiac abnormality suggests that subclinical lesions may exist which make such subjects prone to respond with a life-threatening arrhythmia to stresses which in others cause no symptoms or disability. If human arrhythmias are multifactorial in origin the search for "the arrhythmogenic current" may be fruitless.

Many books and thousands of papers have been, and will continue to be, written about cardiac electrophysiology. A single chapter can do no more than expose the bare bones of the topic. It is hoped that this skeleton will suffice to support the flesh of the volume, the study of the mode of action and clinical use of antiarrhythmic drugs.

References

Allen DG, Eisner DA, Pirolo JS, Smith GL (1985) The relationship between intracellular calcium and contraction in calcium-overloaded ferret papillary muscles. J Physiol (Lond) 364:169-182

Bassingthwaite JB, Fry CH, McGuigan JAS (1976) Relationship between internal calcium and outward current in mammalian ventricular muscle: a mechanism for the control of action potential duration? J Physiol (Lond) 262:15-37

Bean BP (1985) Two kinds of calcium channels in canine atrial cells. Differences in kinetics, selectivity, and pharmacology. J Gen Physiol 86:1-30
Bleeker WK, Mackay JC, Masson-Pévet M, Bouman LN, Becker AE (1980) Functional and morphological organization of the rabbit sinus node. Circ Res 46:11-22
Brown HF (1982) Electrophysiology of the sinus node. Physiol Rev 62:505-530
Brown HF, DiFrancesco D, Noble SJ (1979) How does adrenaline accelerate the heart? Nature (Lond) 280:235-236
Cannell MB, Lederer WJ (1986) The arrhythmogenic current Iti in the absence of electrogenic sodium-calcium exchange in sheep cardiac Purkinje fibres. J Physiol (Lond) 374:201-220
Carmeliet E (1961) Chloride ions and the membrane potential of Purkinje fibres. J Physiol (Lond) 156:375-388
Carmeliet E, Verdonck F (1977) Reduction of potassium permeability by chloride substitution in cardiac cells. J Physiol (London) 265:193-206
Chambers Encyclopedia (1888) Atlantic telegraph. Lippincott, Philadelphia, vol 1; p 545
Colquhoun D, Neher E, Reuter H, Stevens CF (1981) Inward current channels activated by intracellular Ca in cultured cardiac cells. Nature 294:752-754
Coraboeuf E, Deroubaix E (1978) Shortening effect of tetrodotoxin on action potentials of the conducting system in the dog heart. J Physiol (Lond) 280:24P
Dahl G, Isenberg G (1980) Decoupling of heart muscle cells: correlations with increased cytoplasmic calcium activity and with changes of nexus ultrastructure. J Membr Biol 53:63-75
De Mello WC (1963) Role of chloride ions in cardiac action and pacemaker potentials. Am J Physiol 205:567-575
DiFrancesco D (1981) A new interpretation of the pacemaker current in Purkinje fibres. J Physiol (Lond) 314:359-376
Dudel J, Peper K, Rudel R, Trautwein W (1967) The dynamic chloride component of membrane current in Purkinje fibres. Pflugers Arch 295:197-212
Durrer D, van Dam RT, Freud GE, Janse MJ, Meijler FL, Arzbaecher RC (1970) Total excitation of the isolated human heart. Circulation 41:899-912
Eisner DA, Valdeolmillos M (1986) Measurement of intracellular calcium during the development and relaxation of tonic tension in sheep Purkinje fibres. J Physiol (Lond) 375:269-281
Fabiato A, Fabiato F (1979) Calcium and cardiac excitation-contraction coupling. Annu Rev Physiol 41:473-484
Fozzard HA, Hiraoka M (1973) The positive dynamic current and its inactivation properties in cardiac Purkinje fibres. J Physiol (Lond) 234:569-586
Giebisch G, Weidmann S (1971) Membrane currents in mammalian ventricular heart muscle fibers using a voltage clamp technique. J Gen Physiol 57:290-296
Gomperts B (1977) The plasma membrane. Academic, New York
Gros D, Lee I, Challice CE (1982) Formation and growth of myocardial gap junctions: in vivo and in vitro studies. In: Bouman LN, Jongsma HJ(eds) Cardiac rate and rhythm. Nijhoff, The Hague, pp 243-264
Hess P, Lansman JB, Tsien R (1986) Calcium channel selectivity for divalent and monovalent cations. Voltage- and time-dependence of single channel current in guinea-pig ventricular heart cells. J Gen Physiol 88:293-319
Hodgkin AL (1937a) Evidence for electrical transmission in nerve. Part 1. J Physiol (Lond) 90:183-210
Hodgkin AL (1937b) Evidence for electrical transmission in nerve. Part 2. J Physiol (Lond) 90:211-232

Hodgkin AL, Horowitz P (1959) The influence of potassium and chloride ions on the membrane potential of single muscle fibres. J Physiol (Lond) 148:127–160
Hodgkin AL, Huxley AF (1939) Action potentials recorded from inside a nerve fibre. Nature 144:710
Hodgkin AL, Huxley AF (1945) Resting and action potentials in single nerve fibres. J Physiol (Lond) 104:195
Hodgkin Al, Huxley AF (1952) A quantitative description of membrane current and its application to conduction and excitation in nerve. J Physiol (Lond) 117:500-544
Hope RR, Scherlag BJ, El-Sharif N, Lazzard R (1976) Hierarchy of ventricular pacemakers. Circ Res 39:883-888
Isenberg G (1977) Cardiac Purkinje fibres: [Ca]i controls steady state potassium conductance. Pflugers Arch 371:71-76
James TN (1963) The connecting pathways between the sinus node and A-V node and between the right and left atrium in the human heart. Am Heart J 66:498-508
James TN (1977) The sinus node. Am J Cardiol 40:965-986
James TN, Sherf L, Fine G, Morales AR (1966) Comparative ultrastructure of the sinus node in man and dog. Circulation 34:139-163
Janse MJ, Anderson RA (1974) Specialized internodal atrial pathways: fact or fiction? Eur J Cardiol 2:117–136
Johnson EA, Lieberman M (1971) Heart: excitation and contraction. Annu Rev Physiol 33:479-532
Kass RS (1982) Delayed rectification is not a calcium activated current in Purkinje fibers (Abstr). Biophys J 37:342
Kenyon JL, Gibbons WR (1979a) Influence of chloride, potassium and tetraethylammonium on the early outward current of sheep cardiac Purkinje fibers. J Gen Physiol 73:117-138
Kenyon JL, Gibbons WR (1979b) 4-Aminopyridine and the early outward current of sheep cardiac Purkinje fibers. J Gen Physiol 73:139-157
Kohlhardt M, Figulla H-R, Tripathi O (1976) The slow membrane channel as the predominant mediator of the excitation process of the sinoatrial pacemaker cell. Basic Res Cardiol 71:17-26
Kreitner D (1978) Effects of polarization and of inhibitors of ionic conductances on the action potentials of nodal and perinodal fibers in rabbit sinoatrial node. In: Bonke FIM(ed) The sinus node. Nijhoff, The Hague, pp 270-278
Krikler DM (1987) Jubilee editorial: electrocardiography then and now: what next? Br Heart J 57:113-117
Krinsky VI (1981) Mathematical models of cardiac arrhythmias (spiral waves). In: Szekeres L(ed) Pharmacology of antiarrhythmic agents. Pergamon, Oxford, pp 105-124
Lee KS, Noble D, Lee E, Spindler AJ (1984) A new calcium current underlying the plateau of the cardiac action potential. Proc Soc Lond [Biol] 223:5-48
Lee KS, Marban E, Tsien RW (1985) Inactivation of calcium channels in mammalian heart cells: joint dependence on membrane potential and intracellular calcium. J Physiol (Lond) 364:395-411
Legato MJ (1973) Ultrastructure of the atrial, ventricular and Purkinje cell, with special reference to the genesis of arrhythmias. Circulation 47:178-189
Liddell EGT (1960) The discovery of reflexes. University Press, Oxford
Masson-Pèvet M, Bleeker WK, Gros D (1979) The plasma membrane of leading pacemaker cells in the rabbit sinus node. Circ Res 45:621-629
Masson-Pèvet M, Bleeker WK, Besselsen E, Mackay AJC, Jongsma HJ, Bouman LN

(1982) On the ultrastructural identification of pacemaker cell types within the sinus node. In: Bouman LN, Jongsma HJ(eds) Cardiac rate and rhythm. Nijhoff, The Hague, pp 19-34

McAllister RE, Noble D, Tsien RW (1975) Reconstruction of the electrical activity of cardiac Purkinje fibres. J Physiol (Lond) 251:1-60

McDonald TF, Trautwein W (1978a) Membrane currents in cat myocardium: separation of inward and outward components. J Physiol (Lond) 274:193-216

McDonald TF, Trautwein W (1978b) The potassium current underlying delayed rectification in cat ventricular muscle. J Physiol (Lond) 274:217-246

Mullins LJ (1981) Ion transport in the heart. Raven, New York

Nelson CV, Geselowitz DB (1976) The theoretical basis of electrocardiography. Clarendon, Oxford

Noble D (1962) A modification of the Hodgkin-Huxley equations applicable to Purkinje fibre action and pacemaker potentials. J Physiol (Lond) 160:317-352

Noble D (1975/1979) The initiation of the heartbeat. Oxford University Press, Oxford

Noble D, Tsien RW (1969a) Outward membrane currents activated in the plateau range of potentials in cardiac Purkinje fibres. J Physiol (Lond) 200:205-232

Noble D, Tsien RW (1969b) Reconstruction of the repolarization process in cardiac Purkinje fibres based on voltage clamp measurements of membrane current. J Physiol (Lond) 200:233-254

Noma A, Irisawa H (1974) Electrogenic pump in rabbit sinoatrial node cell. Pflugers Arch 357:177-182

Noma A, Tohru S (1985) Properties of adenosine-triphosphate-regulated potassium channels in guinea-pig ventricular cells. J Physiol (Lond) 363:463-480

Noma A, Yanagihara K, Irisawa H (1977) Inward membrane currents in the rabbit sinoatrial node cell. Pflugers Arch 372:43-51

Philipson KD, Bers DM, Langer GA (1981) The role of sarcolemmal Ca^{2+} in myocardial contactility. In: Grinnell AD, Brazier MAB(eds) The regulation of muscle contraction. Excitation-contraction coupling. Academic, New York, pp 215-226

Reuter H (1967) The dependence of slow inward current in Purkinje fibres on the extracellular calcium concentration. J Physiol (Lond) 192:479-492

Rushton WAH (1951) A theory of the effects of fibre size in medullated nerve. J Physiol (Lond) 115:101-122

Seyama I (1977) The effect of Na, K and Cl ions on the resting membrane potential of sinoatrial node cell of the rabbit. Jpn J Physiol 27:577-588

Seyama I (1979) Characteristics of the anion channel in the sinoatrial node of the rabbit. J Physiol (Lond) 294:447-460

Stämpfli R (1954) Saltatory conduction in nerve. Physiol Rev 34:101-112

Taccardi B, de Ambroggi L (1975) Le elektromappe cardiache In: Anguissola AB, Puddu V(eds) Cardiologia oggi. Edizioni Medico Scientifice, Torino, pp 361-381

Vaughan Williams EM (1958) The mode of action of quinidine on isolated rabbit atria interpreted from intracellular potential records. Br J Pharmacol 13:276-287

Vaughan Williams EM (1959) The effect of changes in extracellular potassium concentration on the intracellular potentials of isolated rabbit atria. J Physiol (Lond) 146:411-427

Vaughan Williams EM (1987) Is phosphodiesterase inhibition arrhythmogenic? Electrophysiological effects in pithed rats and in normoxic and hypoxic rabbit atria of enoximone, a new cardiotonic agent. J Clin Pharmacol, 27:91-100

Vleugels A, Carmeliet E (1976) Hypoxia increases potassium efflux from mammalian myocardium. Experientia 32:483-484

Weidmann S (1955) The effect of the cardiac membrane potential on the rapid availability of the sodium-carrying system. J Physiol (Lond) 127:213-224
Yamagishi S, Sano J (1966) Effect of tetrodotoxin on the pacemaker action potential of the sinus node. Proc Jpn Acad 42:1194-1196
Yanagihara K, Irisawa H (1980) Inward current activated during hyperpolarization in the rabbit sinoatrial node. Pflugers Arch 385:11-19

CHAPTER 2

Classification of Antiarrhythmic Actions

E. M. VAUGHAN WILLIAMS

A. Functions of Selective Ionic Currents

In cardiac cells there are four main sources of EMF, as noted in the preceding chapter, engendered by concentration differences across the cell membrane of Na, Ca, K and Cl ions, which would be in equilibrium at intracellular potentials of approximately +56, +120, −94 and −40 mV respectively. In order that current from these sources may be used for physiological functions, ion-selective pathways through the membrane can be opened and closed. The function of sodium current is to depolarize atrial and ventricular muscle cells and Purkinje cells, and of calcium current to depolarize nodal cells. Slow depolarization in the central SA node cells permits faster depolarization in the surrounding transitional cells, when they reach threshold, to overtake them and synchronize firing of the node as a whole, from which a ring of excitation spreads outwards in all directions. Slow depolarization of the AV node provides the delay of conduction required while blood is transferred from atrium to ventricle. Calcium current is also involved in excitation-contraction coupling, and can be increased by the opening of an additional set of pathways under the control of adrenergic stimulation. The main function of potassium current is repolarization, but the function of chloride current is uncertain. High Cl permeability in P cells (pale sinoatrial node (SAN) cells) may provide a sink for depolarizing neighbours in the pacemaking process. Chloride-bicarbonate exchange may be concerned in the control of intracellular pH (VAUGHAN-JONES 1979).

The existence of selective ion channels with different functions offers the possibility of pharmacological modulation, and groups of drugs have been developed capable of restricting with varying degrees of selectivity sodium, potassium and calcium currents in the heart. There is some evidence that restriction of chloride current may be involved in the action of the "specific bradycardic agents" (SBAs, Chap. 19), but this is unproven. Calcium channels dependent on adrenergic stimulation can be modulated by antisympathetic drugs.

During the past 15 years antiarrhythmic drugs have been presented at a frequency of one new compound every few months. Some were developed for quite different roles—mexiletine, for example, as an antiepileptic, melperone as a tranquillizer—and were later discovered to have antiarrhythmic properties in the heart. All of them, however, with a single exception, the SBAs, have been found to possess one or more of four actions upon which a classification of antiarrhythmic actions was based originally in 1970, updated at intervals (Vaughan Williams 1970, 1980, 1981, 1984). Several of the drugs had more than one of the four actions, so that it deserves emphasis that the classification was not so much a categorization of drugs in relation to chemical structure or physical properties, but rather a description of four ways in which abnormal cardiac rhythms can be corrected or prevented. Drugs in the same class from the point of view of their antiarrhythmic action could appear very different in the total clinical environment, due to differences in kinetics of action, distribution, especially into the CNS, metabolism, and sideeffects unrelated to actions on the heart.

In experimental studies of antiarrhythmic actions two main strategies have been adopted. Abnormalities of cardiac rhythm can be induced in animals by a variety of procedures (aconitine, digitalis, barium, programmed stimulation, coronary ligation, etc.), and the effects of drugs investigated empirically. Unfortunately such procedures do not induce consistent results (Chap. 3) and, even if they did, doubt would still remain concerning the relevance of such artefacts to human cardiac arrhythmias. In this field the only important target is man. An alternative approach is to study the detailed electrophysiological and pharmacological actions of known antiarrhythmic agents in the hope of finding some common properties to which their efficacy could be attributed, and which might throw some light on the probable origin of the arrhythmias themselves.

B. Class I Antiarrhythmic Action

The first class of action was that exerted by quinidine and a number of other remedies (Szekeres and Vaughan Williams 1962) which, at 10–100 times their antiarrhythmic concentrations, incidentally also behaved as local anaesthetics on nerves. These compounds, though differing in other respects, had in common the property of "interfering specifically with the process by which depolarizing charge is transferred across the membrane", i.e. by fast inward sodium current. The effect was revealed as a depression of the maximum rate of depolarization (MRD), unless the interstimulus interval was so long that it permitted full recovery between beats (Chap. 1, Fig. 17). Sodium channels are rapidly inactivated after depolarization and remain in the inactivated state until repolarization proceeds to voltages more negative than about −55 mV. It was suggested that the (class I) drugs interfered with the process by which Na channels were "reactivated in response to repolarization" and that in consequence they "extended the effective refractory period to a point long after the time at which repolarization was already complete". The concept that these drugs were combined with the sodium channels in their inactivated form (i.e. after depolarization) was supported by the very early finding that the compounds were more potent the shorter the diastolic interval, as already noted.

That this, historically the first, class of action was responsible for the antiarrhythmic effects was not universally accepted, and Bigger and Mandel (1970) concluded from their own work "the results caused us to reject the hypothesis that lidocaine exerts electrophysiological effects essentially like those of procainamide or quinidine". DAVIS and TEMTE (1969), discussing the effects of lidocaine and diphenylhydantoin (DPH), concluded that "a reduction in rising velocity is not a necessary feature for antiarrhythmic activity". On the contrary they concluded that "the most significant effect of lidocaine with regard to its antiarrhythmic action is the prevention of decremental conduction in Purkinje fibers". They considered that DPH and propranolol (!) acted in the same way as lidocaine. BASSETT and HOFFMAN (1971) also took the view that "DPH and lidocaine may abolish a re-entrant rhythm by *improving* conduction. DPH either *increases* or does not substantially alter membrane responsiveness and conduction velocity".

There are two reasons why it might be concluded that therapeutic concentrations of lidocaine would be insufficient to reduce MRD. The first is that if the extracellular potassium concentration is low (and the above authors employed a $[K]_o$ of 2.7 mmol/litre, little more than half the normal plasma level), the resting potential is hyperpolarized and the depressant effect on MRD counteracted (SINGH and VAUGHAN WILLIAMS 1971). The second is that lidocaine delays recovery from inactivation, but for a brief period only, so that if the interval between action potentials is long, no effect on MRD will be seen (VAUGHAN WILLIAMS 1980). Both these points are of clinical significance, because lidocaine may be ineffective in patients in whom, perhaps as a consequence of diuretic therapy, serum potassium has fallen (WATANABE et al. 1963). Secondly, even when no effect on HV conduction time or QRS width is apparent in sinus rhythm, the class I effect of lidocaine could nevertheless be responsible for depressing or eliminating conduction of *premature* ventricular beats or for slowing a ventricular tachycardia. EL-SHERIF et al. (1977a, b; EL-SHERIF and LAZZARA 1978, 1979) studied experimentally induced ventricular arrhythmias in the late myocardial infarction period in dogs, and found that at the time when lidocaine and DPH exerted an antiarrhythmic action the effect of both drugs was to *slow* conduction. They concluded that "there is currently no basis to substantiate the concept that both lidocaine and DPH can abolish re-entrant rhythms by improving conduction in the re-entrant pathway".

Another complicating factor is that some drugs, notably lidocaine and mexiletine, shorten APD in the ventricular conduction pathway, especially in the preterminal Purkinje fibres in which APD is normally much longer than in the His bundle or ventricular muscle.

ROSEN et al. (1973) observed that procainamide also acts on APD in the ventricular conducting system "in such a way that a greater similarity of contour between proximal, gate (preterminal) and distal sites developed than had existed under control." While heterogeneity of APD within a block of myocardium may be arrhythmogenic, the long APD in preterminal Purkinje cells may have a protective function against retrograde "backfiring". Thus abolition of the disparity of APD between preterminal and terminal cells is not necessarily beneficial. Nevertheless the finding of the differential shortening of APD in preterminal Purkinje cells by lidocaine led WITTIG et al. (1973) to suggest that the shortening of APD could constitute an antiarrhythmic action.

Premature action potentials, conducting more slowly because they took off from as yet not fully repolarized cells, would conduct more rapidly after the drug, because they would, as a result of the shortening of APD, now take off from fully repolarized tissue.

There are several reasons for not accepting this hypothesis. (1) Although it is simple enough experimentally to select an appropriate interstimulus (S1-S2) interval to ensure that the premature response to S2 takes off from partially repolarized membrane in the stimulated region, to select the S1-S2 interval when stimuli are applied to the bundle of His and ensure that the second action potential at a remote site (preterminal fibre) still takes off from the repolarizing limb of the first becomes difficult because the second action potential conducts more slowly and lags further and further behind the first. (2) There is no evidence that premature action potentials occurring in man, in contrast to those artificially initiated with paired stimuli, behave in this way; the occurrence of "R on T" in ECG records, although it indicates that repolarization starts in some cells before it finishes in others, cannot be taken to imply that depolarization starts in a single cell before that cell itself has fully repolarized, because the surface ECG is the resultant vector of currents generated in millions of cells. (3) There is abundant evidence in animals (Vaughan Williams 1980) and man (Olsson et al. 1971) that shortening of APD is an *arrhythmogenic* factor. (4) The class 1 action of lidocaine provides a sufficient explanation for its antiarrhythmic effects. Thus lidocaine is antiarrhythmic in spite of, rather than because of, the APD shortening.

What is the mechanism by which lidocaine shortens APD? Arnsdorf and Bigger (1972) concluded from voltage clamp studies of sheep Purkinje cells that lidocaine, at a concentration of 21.4 μmol/litre (taken as equivalent to therapeutic levels), "increased chord conductance for the potassium ion (gK)". In contrast, Carmeliet and Saikawa (1982), also using voltage-clamped sheep Purkinje cells, while noting that in the presence of 18.5-37 micromol/litre lidocaine "steady-state currents were shifted outwards", found that the "outward shift of current at the plateau level by lidocaine was not observed in the presence of tetrodotoxin (TTX): in Na-free medium the effect of lidocaine was totally suppressed". "Local anaesthetics do not increase the K inward-rectifier current at potentials negative to −60 mV." It would appear, therefore, that in Purkinje cells there is a population of sodium channels not inactivated at plateau potential. Coraboeuf et al. (1979) came to a similar conclusion from voltage clamp studies of canine Purkinje cells, finding that TTX shortened APD and stating that "the effect is probably due to the fact that in Purkinje fibers the TTX-sensitive steady-state sodium current flows in a larger potential range than in the myocardium". Colatsky (1982) reported similar results with rabbit cardiac Purkinje cells.

I. Subdivision of Class I Antiarrhythmic Agents

By definition class I drugs share the property of restricting fast inward sodium current, but may differ in other respects.

1. Some compounds, notably quinidine and disopyramide, are anticholiner-

gic. Apart from noncardiac complications of antimuscarinic actions, there are several cardiac consequences. In a patient with substantial vagal activity sinus tachycardia may follow administration, or, more seriously, an atrial fibrillation or supraventricular tachycardia which was previously innocuous because impulses were blocked at the AV node may develop into ventricular tachycardia when AV conduction is improved by blockade of vagal tone. Secondly, the fact that the drugs depress conduction through their class I action may be masked by the simultaneous improvement of conduction by the anticholinergic effect, with no net change in AV conduction time. In a patient with little vagal background tone, however, and with depressed AV conduction, the full class I depression by quinidine or disopyramide will be revealed, and serious AV block may occur (BIRKEAD and VAUGHAN WILLIAMS 1977).

2. Distribution into the CNS varies greatly between class I drugs, and more information is required concerning the extent to which this is influenced by fat solubility, p*Ka* and molecular size. Certainly dizziness and other CNS effects, even convulsions, limit the use of several class I agents, including lidocaine, mexiletine and lorcainide.
3. Metabolism and rates of excretion vary enormously, both factors influencing elimination. Rapid metabolism by the liver makes oral administration of lidocaine impractical, since its half-life may be only ½ h in some subjects. At the other extreme aprindine has an elimination half-life of 2 ½ days. The half-lives and volumes of distribution of nine class I drugs are given in Table 1 (RONFELD 1980).
4. Some compounds, especially lidocaine, mexiletine and tocainide, shorten APD, as already discussed.
5. There are wide variations in the rapidity with which the drugs become attached to and released from the sodium channels.

Table 1. Elimination half-lives and volumes of distribution

	Elimination half-life (h)	Volume of distribution (litres/kg)
Lidocaine	1.8	1.6
Mexiletine	13.0	6.6
Tocainide	13.0	2.8
Disopyramide	7.0	0,5
Quinidine	6.3	2.5
Procainamide	3.0	2.9
Flecainide	14.0	
Lorcainide	7.7	7.9
Aprindine	50.0	4.0

Table 2. Clinical subdivision of class I antiarrhythmic drugs

Effecton	Group B	A	C
	Lidociane Mexiletine Tocainide	Quinidine Procainamide Disopyramide	Lorcainide Encainide Flecainide
1. QRS	None in sinus rhythm	Widen at high concentration	Widen at low concentration
2. Conduction	None in sinus rhythm	Slowed at high concentration	Slowed at low concentration
3. ERP	Lengthened in relation to APD	Lengthened absolutely and relative to APD	Very little change
4. J-T Interval	Shortened	Lengthened at high concentration	Very little change

The clinical differences between class I drugs necessitated their subdivision. The older compounds, quinidine, procainamide and disopyramide, were classified as group la. Lidocaine, mexiletine and tocainide (group Ib) shorten APD in vitro and monophasic action potentials (MAPs) and J-T intervals in vivo. They do not lengthen H-V conduction time or widen QRS in sinus rhythm, but prolong effective refractory period (ERP) as measured by programmed stimulation (a premature stimulus S2, interpolated every *n*th pacing stimulus S1). The most recently introduced class I drugs, including flecainide, encainide, lorcainide and propafenone (group Ic), have little effect on APD or J-T interval, and do not prolong ERP on programmed stimulation, but increase H-V conduction time and widen QRS even in sinus rhythm. Quinidine, procainamide and disopyramide are intermediate, moderately prolonging ERP and J-T interval, and widening QRS at high concentration (indicating possible overdosage). The anticholinergic and APD-*lengthening* effects of quinidine are additional properties unrelated to its class I action. Quinidine delays repolarization by restriction of outward potassium current (COLATSKY 1982). The differences between the class Ia, Ib and Ic groups are summarized in Table 2.

The maximum rate of depolarization (MRD) upon stimulation at increasing intervals after repolarization can be taken as an approximate measure of the number of sodium channels which have recovered from inactivation, although simultaneous or overlapping outward current can complicate the picture to some extent (Chaps. 7, 8). The heights of the columns in the lower left panel of Fig. 1 represent the magnitude of MRD in response to stimuli at various times after the beginning of the atrial action potential depicted in the upper panel. The solid columns depict normal responses, the dashed columns the responses to be expected if 40 % of the channels had been permanently eliminated by a drug. The remaining 60 %, without drug attached, would be normal and so would recover at the same time as the controls. Electrical threshold would be raised, and conduction velocity reduced, because fewer channels would be available to carry inward sodium current. ERP would be little changed, because, provided sufficient drug-free channels remained to sustain a propagated impulse at all, the available channels would recover from inactivation at the normal time. In the lower right panel, the dotted line depicts the "envelope" of normal responses. The columns represent responses to be expected if no channels had been permanently eliminated, but if *all* channels had been put temporarily out of action at the beginning of diastole by drug attachment during systole, but recovered rapidly (solid columns) or more slowly (dashed columns) as the drug became detached from more and more channels

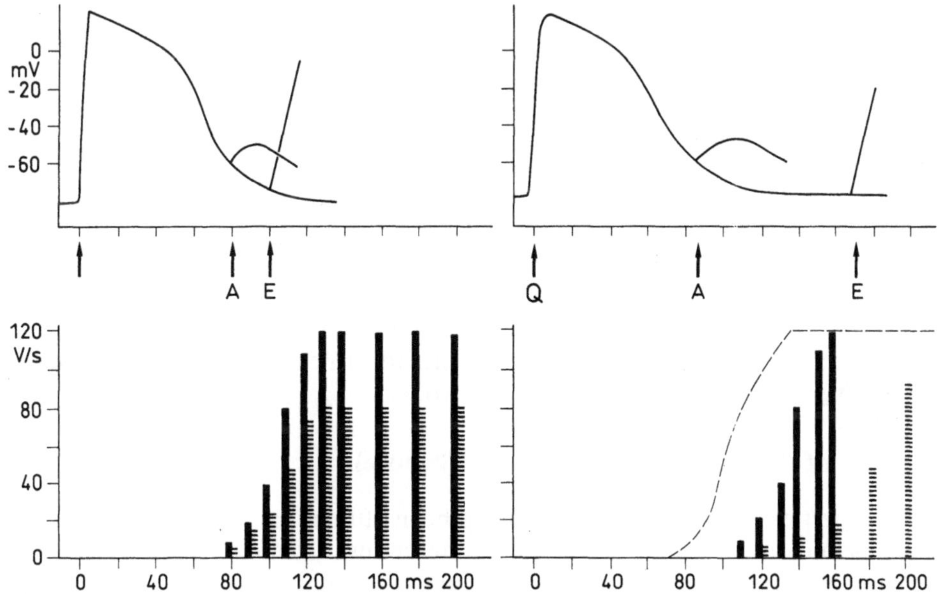

Fig. 1. Recovery of MRD in response to stimuli at intervals after an action potential. *Upper panels*, intracellular atrial potentials, and responses to stimuli at *A*, the absolute refractory period, when sufficient sodium channels had recovered from inactivation to support a local but not a propagated response; and at *E*, the effective refractory period, when enough channels had recovered to support a propagated response. *Left*, controls; *Right*, in the presence of quinidine. *Lower panels*, heights of columns indicate MRD in response to stimuli applied at intervals after the start of the potential *in the upper panel. Left, solid lines*, controls; *dashed lines*, on assumption that 40 % of channels had been permanently rendered non-conducting. *Right dotted line* depicts envelope of normal responses, as *in left panel*; columns depict MRD on assumption that no channels were permanently eliminated, but that at the start of diastole all channels were attached to a drug (which kept them inactivated), but that after repolarization the drug was rapidly (*solid columns*) or more slowly (*dashed columns*) dissociated from the channels

after repolarization, permitting them to recover from inactivation. At the end of diastole, however, *all* of the channels would have been freed from the rapidly dissociating drug, so that MRD (and conduction velocity) would be little changed at sinus rhythm. The more slowly dissociating drug would depress MRD in sinus rhythm to some extent.

According to this interpretation, the action of the drug depicted on the left of Fig. 1 (published in VAUGHAN WILLIAMS 1980) would correspond to that of a class Ic compound. The fast dissociating drug on the right would be a class Ib compound, and the drug with intermediate kinetics would be in group Ia. The rapidity of dissociation of class I drugs from sodium channels after repolarization can be tested by administering to cardiac muscle trains of stimuli of a frequency and duration sufficient to induce a steady-state depression of MRD. Stimulation is arrested, and MRD is then measured in response to a single stimulus applied at increasing intervals after the end of the train. As a result of many such studies (Chap. 7) the rate of detachment and attachment to sodium channels of many class I drugs was investigated. It was found that drugs categorized on clinical grounds as belonging to class Ib had very fast onset-offset kinetics, which explains why H-V time and QRS width are unaffected in sinus rhythm, a normal diastole being of sufficent duration to permit detachment of the drug from most sodium channels. ERP is prolonged because at the *beginning* of diastole nearly all the channels have drug attached and so are non-conducting. The class Ic drugs, in contrast, detached so slowly from sodium channels that the number non-conducting at the end of diastole was almost the same as at the beginning. Consequently H-V time was lengthened and QRS widened. The kinetics of the Ia group were intermediate between those of Ib and Ic. Thus an objective electrophysiological study of onset-offset kinetics of class I drugs in vitro was able to provide an explanation for the subdivision imposed on clinical grounds in vivo.

II. Site of Attachment to Sodium Channels

The properties of sodium channels have been studied in a variety of preparations, squid axons, frog nerves, snail neurones, isolated single cells, and patches of membrane from various tissues. Although it is reasonable to assume that all voltage-dependent channels have features in common, an ion-permeable core, a selectivity filter, a charged "voltage sensor", and structures controlling the opening and inactivation of the pathway, it is possible that cardiac channels differ in some respects from those found in other organs, and that the effect of a local anaesthetic on a nerve is not identical to that of a class I drug on a myocardial cell. For example, although TTX blocks sodium channels in rabbit Purkinje cells, the concentration required is 1000-10000 times that sufficient to inhibit conduction in nerve (COLATSKY 1975). Local anaesthetics have long been supposed to cross nerve membranes in their un-ionized state (the concentration being a function of their pKa and the extracellular pH), but to block sodium channels by entering the ion pathway from the inside and binding to it in their ionized state.

Many years ago a different mechanism was suggested for sodium channel

Table 3. Fat solubility[a] and dissociation constant of some class 1 drugs

Drug	Fat solubility	pK_a
Lignocaine	2.76	7.85
Mexiletine	1.3	9.3
Tocainide	0.8	7.8
Disopyramide	1.8	8.4
ORG 6001	3.93	8.0
Quinidine	3.6	8.8
Procainamide	0.8	9.2
Flecainide	1.15	9.3
Encainide	0.91	10.2
Lorcainide	4.16	9.44
CCI 22277	7.19	9.5

[a] Log of octanol/water partition coefficient

block by class I drugs in cardiac cells (Vaughan Williams 1972). "They have a variety of chemical structures but are mostly lipophilic" and "are taken up into the primary membrane, from where they press their attack on the sodium channels. Class I action by the local anaesthetic type of drug may involve restriction of the freedom of charged elements controlling the ion gates to move in response to changes in voltage across the membrane". Hille, while still retaining the view that local anaesthetics must bind to the interior of the ion pathway to block conduction, suggested that lipophilic compounds entered the membrane and passed from there through the wall of the channel to reach their attachment site in the pore (Hille 1977). The pKa and fat solubility of several class Ia, Ib and Ic compounds are given in Table 3. More recently Katz et al. (1986) demonstrated with ligand probes that various compounds were indeed located within the lipophilic interior of the membrane. Although the authors were unaware of my 1972 paper, their description is so similar that it is worth quoting their actual words.

The lipophilic properties of many cardiac drugs appear to play an important role in their interaction with specific receptor sites in the cardiac sarcolemmal membrane. Recent evidence suggests that some of these drugs approach their receptors by a two-step process in which the ligand first partitions into the bulk lipid bilayer of the sarcolemmal membrane, which has a relatively high lipid content. Such drugs can then become precisely oriented in the bilayer, within which they diffuse laterally through the bulk lipids to reach a specific protein receptor site also within the bilayer.

Haydon et al. (1980; Haydon and Kimura 1981; Haydon and Urban 1984) in a series of papers studied the effects of many compounds on squid axons. They assumed the validity of H-H equations, and determined by successive approximations how the various factors had to be altered to make the equations fit the data after exposure to the agents. They concluded that many lipophilic compounds lined up along the inner border of the membrane, increasing its thickness and reducing its capacity, thus altering the field influencing the sensing element in the channel.

The two mechanisms of action, a physical effect and binding to a specific site, are not mutually exclusive, and it is possible that both are involved in varying proportion in the action of different class I agents. The topic has generated a mass of experiment and speculation, and this brief introduction serves merely to put in historical perspective the five chapters (Chaps. 7-11) devoted to various aspects of the mode of action and clinical use of class I drugs.

C. Class II Antiarrhythmic Action

I. Acute Antisympathetic Effects

For many years it had been realized that cardiac arrhythmias could be initiated or exacerbated by stress or emotion (see Chap. 12), but until the advent of the beta-blockers no potent drugs were available to control adrenergically induced tachycardia. It was soon shown that the first clinically useful beta adrenoceptor blocker, pronethalol, was antiarrhythmic in animals (VAUGHAN WILLIAMS and SEKIYA 1963) and man (PAYNE and SENFIELD 1964). Pronethalol (and its successor propranolol) both turned out to be local anaesthetics on nerve, and to have a class I action in cardiac muscle (SEKIYA and VAUGHAN WILLIAMS 1963; GILL and VAUGHAN WILLIAMS 1964), and some authors questioned whether the antisympathetic action was responsible for the antiarrhytmic effects. The evidence was discussed in great detail elsewhere (VAUGHAN WILLIAMS 1981) and may briefly be summarized as follows. (1) The *l*-isomers of beta-blockers, which have the same local anaesthetic potency but much greater beta-blocking potency than the *d*-isomers, are more effective antiarrhythmics. (2) Beta-blockers lacking class I action are nevertheless antiarrhythmic. (3) Drugs which block release of noradrenaline from sympathetic nerves are antiarrhythmic. It was concluded that acute antisympathetic action, whether by receptor blockade or presynaptic effect, constituted a second class of antiarrhythmic action in its own right.

II. Long-term Antisympathetic Effects

Beta-blockers have been in use for more than 20 years as a treatment for hypertension, yet the mechanism by which blood pressure is lowered has still not been fully explained. A question which aroused early interest was why the hypotensive action of beta-blockers was often delayed in onset and why weeks or months of therapy might sometimes be required to achieve an optimal result. The mode of action of a drug is customarily analysed in acute animal experiments, yet patients are treated with beta-blockers for years, often for the rest of their lives. It seemed logical to study the drugs' effects in animals which also had been treated for prolonged periods, and a programme was initiated to study the long-term effects of a number of beta-blockers, with and without the subsidiary properties of beta-1 selectivity, intrinsic sympathomimetic activity (ISA) or local anaesthetic action. First a dosage regime was established in rabbits to correspond in beta-blocking potency to that obtaining in clinical practice (VAUGHAN WILLIAMS 1984). With propranolol, for example, rabbits were injected twice daily for several weeks with 4 mg/kg s.c., which we had shown would provide clinical levels of blockade for at least two-thirds of each day. The treatment was stopped a sufficient time before the animals were killed (24 h for propranolol) to permit the drug to be eliminated from the body. The hearts were then removed and set up for study in vitro.

There was no difference in responses to isoprenaline in the hearts of these animals, in comparison with hearts from saline-treated litter mates, indicating the absence both of residual blockade and of any hypersensitivity induced by the treatment. The most striking difference was that the long-term treatment induced a large prolongation of APD in atria and ventricles paced at constant frequency. Since no drug was present this represented a secondary adaptation to the treatment. A uniform prolongation of ventricular APD should lengthen the Q-T interval and it was found possible to monitor the rate of onset of the adaptation non-invasively by measurement of ECG records (RAINE and VAUGHAN WILLIAMS 1981). The effect was not related to ISA, beta-1 selectivity or local anaesthetic action, nor was it due to the associated bradycardia (VAUGHAN WILLIAMS et al. 1986).

Several studies have established that a similar prolongation of Q-T interval, persisting for days after cessation of treatment, is induced by long-term beta-blockade in man (VAUGHAN WILLIAMS et al. 1980; EDVARDSSON and OLSSON 1981). These observations can explain both the prevalence of arrhythmias in the long Q-T syndrome (Chap. 23) and the efficacy of therapy with beta-blockers or by left-sided sympathetic cardiac denervation. It was noted in the previous chapter that the end of the Q-T interval does not necessarily signal the end of cardiac electrical activity, because oppositely orientated late repolarization vectors may cancel each other (Chap. 1, Fig. 13 d). Beta adrenergic stimulation shortens (but alpha adrenergic stimulation lengthens) APD and refractory period (GOVIER et al. 1966; GIOTTI et al. 1973; DUKES and VAUGHAN WILLIAMS 1984). In the heart beta-adrenoceptors predominate (LEFKOWITZ and HABER 1971). In the long Q-T syndrome a paucity of right stellate innervation, causing a reduced heart rate response to exercise, is combined with a compensatory excess of left stellate activity (SCHWARTZ et al. 1975). Thus high sympathetic activity in the left ventricle could shorten a late repolarization vector which normally balanced a similar late repolarization vector in the right ventricle, so that a T wave which had previously been "silent" could become apparent; hence the long Q-T. Furthermore, the permanent loss of right ventricular sympathetic tone would have caused an adaptive lengthening of APD in the RV. In exercise APD would shorten in the LV under sympathetic drive, but the denervated right side would not respond similarly, so that heterogeneity of repolarization times would be exacerbated, a highly arrhythmogenic factor. Long-term beta-blockade or left-sided cardiac sympathetic denervation would cause an adaptive lengthening of APD (and reduced response to sympathetic drive) on the left side also, thus reducing the imbalance. The mutual cancellation of vectors might be restored, so that the *lengthening* of left-sided APD could paradoxically *shorten* the surface Q-T interval.

III. Beta-Blockade After Myocardial Infarction

One of the exciting developments of the past few years has been the demonstration that long-term beta-blockade protects a significant proportion of postinfarction patients from reinfarction and sudden death (WILHEMSSON et al. 1974; MULTICENTRE INTERNATIONAL STUDY 1977; NORWEGIAN MULTICENTER STUDY GROUP 1981; BETA-BLOCKER HEART TRIAL RESEARCH GROUP 1982). The reason for the protection is unknown, but since a uniform prolongation of APD is antiarrthythmic (class III), this could contribute to protection against sudden death. Furthermore, myocardial infarction involves loss not only of muscle but of sympathetic nerves, so that healthy myocardium distal to the infarct, (depicted as a dotted circle in Chap. 1, Fig. 13 d), becomes sympathetically denervated, inducing a sort of "mini long Q-T syndrome". This would create an arrhythmogenic heterogeneity of repolarization times, which would be reduced by long-term beta-blockade. Finally, there is evidence that in rabbits prolonged beta-blockade induces increased capillarity of the ventricle (TASGAL and VAUGHAN WILLIAMS 1981), an effect possibly due to the associated bradycardia (HUDLICKA and WRIGHT 1977). Blood flow in the ventricular microcirculation occurs during diastole, and a longer diastole would cause increased flow, which, according to HUDLICKA et al. (1983), is the primary stimulus to capillary proliferation. With the advent of selective bradycardic agents (SBAs), which reduce heart rate by a non-adrenergic mechanism, HUDLICKA's hypothesis could be tested. Increased capillarity, if it occurs in man also, could contribute to the resistance to reinfarction during long-term beta-blockade.

D. Class III Antiarrhythmic Action

The class I drugs delay the recovery of sodium channels from inactivation for varying lengths of time after repolarization. It was obvious that an alternative way of delaying recovery would be to prolong the duration of repolarization itself. Indeed, many years ago WEST and AMORY (1966) maintained that the antiarrhythmic action of quinidine *was* due to delayed repolarization. They employed very low stimulation frequencies, however, and although quinidine does slightly increase APD at high concentration, it is now clear that the primary antiarrhythmic action at normal or faster than normal cardiac frequencies is due to an effect on sodium channels. Nevertheless the concept that uniform delay of repolarization could be antiarrhythmic remained valid, and was supported by the finding that in the hypothyroid state, in which human cardiac arrhythmias are rare, a uniform prolongation of APD occurs in both atria and ventricles in rabbits (FREEDBERG et al. 1970), an effect confirmed by monophasic action potential recording in man (COTOI et al. 1972).

Amiodarone was produced a quarter of a century ago as an antianginal agent, and was but one of a series of vasodilator benzofurans, chosen because it was believed at the time that coronary artery vasodilatation was required for the treatment of angina pectoris. The beta-blockers were introduced in 1962, and were antianginal but not vasodilators. The Labaz group considered, however, that it would be preferable to induce only a partial antagonism to adrenergic stimulation, for fear of compromising a possibly vital sympathetic support, and amiodarone was selected because it reduced by about a third effects mediated both by alpha- and beta-adrenoceptors, partly by a non-competitive antagonism (CHARLIER 1970), partly by diminishing noradrenaline release (BACQ et al. 1976). Amiodarone delays A-V conduction, an effect attributable to its antisympathetic action, not to blockade of calcium current associated with excitation-contraction coupling because the drug is not negatively inotropic. On the other hand, it has been shown recently (DUKES and VAUGHAN WILLIAMS 1984) that nodal calcium current can be distinguished pharmacologically from calcium current inducing contraction. Amiodarone may, therefore, delay AV conduction by a selective nodal calcium antagonist action, in addition to its class II effect. CHARLIER et al. (1969) found that amiodarone had antiarrhythmic activity in a number of acute animal models of arrhythmia (aconitine, $CHCl_3$, coronary ligation, etc.), but it was not potent, with about one-sixth the activity of quinidine, and it was unclear whether the antiarrhythmic effect was simply a reflection of its antisympathetic action or involved some other property.

In my own laboratory the acute antiarrhythmic action of amiodarone was confirmed, attributable to its antisympathetic effect, in protecting guinea pigs against ouabain-induced arrhythmias [removal of sympathetic influences by reserpine (CIOFALO et al. 1967) or by adrenalectomy and sympathectomy (MENDEZ et al. 1961) was long ago shown to prevent digitalis-induced ventricular fibrillation]. There were two further observations, however, which made amiodarone *qualitatively* different from other antiarrhythmic drugs then available (SINGH and VAUGHAN WILLIAMS 1970). First, amiodarone induced a prolongation of action potential duration uniformly in atria and ventricles at doses which had no effect on MRD. Secondly, the effect was progressive, increasing, on continued administration of the drug, over a period of several weeks. Although this was similar to the effect of thyroidectomy, amiodarone was not directly antithyroid, and did not alter the weight of the thyroid gland. Thus we appeared to have found a drug with the "third class" of antiarrhythmic action which had previously been no more than a theoretical prediction. Since that time amiodarone has become established as probably the most potent antiarrhythmic drug in use today (Chap. 15; HARRIS and RONCUCCI 1986), and many other class III drugs are currently being evaluated (Chaps. 14, 16, 17).

E. Class IV Antiarrhythmic Action

Verapamil, like amiodarone, was originally introduced as a coronary dilator for the treatment of angina pectoris. Verapamil did not block the effects of isoprenaline on ventricular stroke volume (ROSS and JORGENSEN 1967) or on bronchial muscle (HILLS 1970), and it was not a competitive antagonist of catecholamines (NAYLER et al. 1968; KAUMANN and ARAMENDIA 1968). Soon after its introduction verapamil was discovered to have antiarrhythmic properties (MELVILLE et al. 1964; SCHMID and HANNA 1967) but at that time the evidence for the existence of a second inward current in cardiac muscle, presented in 1958 and described in the previous chapter, had not been widely accepted. Furthermore, although the charge carrier of the second inward current had been attributed to "ions other than sodium", their identity had not been established, although attention was drawn at the time to the evidence of FATT and KATZ (1953) and of FATT and GINSBORG (1958) that calcium ions could carry depolarizing charge in other tissues. Indeed, it was not until almost a decade later that REUTER (1967) showed the dependence of the second inward current on calcium. The concept of "calcium antagonism" was originally concerned with inhibition of Ca-activated ATP-ase. In their review in 1967 FLECKENSTEIN et al. stated "As to the inhibitors of utilization they seem either to block the movements of Ca *from the excited membrane to the myofibrils or to compete with Ca for the active sites in the contractile system where ATP is split*". The list of substances cited as "capable of producing contractile failure of the heart by acting as *Ca-antagonists in excitation-contraction coupling*" included "a number of beta-receptor blocking substances and *related compounds (Isoptin, Segontin)*. Some local anesthetics have similar effects". There was no mention of a second inward current, or of its blockade by calcium antagonists.

The idea of a fourth class of antiarrhythmic action due to block of the second inward current carried by calcium ions was based on electrophysiological studies in spontaneously beating rabbit atria and in paced papillary muscles (SINGH and VAUGHAN WILLIAMS 1972), an action subsequently confirmed by voltage clamp data (KOHLHARDT et al. 1972; KOHLHARDT and MNICH 1978). Verapamil and diltiazem are most efficacious, as would be expected, in controlling supraventricular arrhythmias involving re-entry in and around the SA and AV nodes, in which the depolarizing current is carried by calcium ions. The more recently introduced "calcium antagonists" such as nifedipine, although being about 5 times more potent than verapamil on cardiac muscle, are perhaps 20 times more potent on vascular smooth muscle. Thus they can be used to lower blood pressure at concentrations which have little depressant action on the SA or AV nodes. Indeed, reflex sympathetic responses to hypotension may induce tachycardia. In practical terms the dihydropyridines with high selectivity for vascular smooth muscle cannot be regarded as antiarrhythmic agents.

Although all cardiac cells, apart from nodal cells, are normally depolarized by fast inward sodium current, it was suggested that in ischaemia or pathological situations causing partial diastolic depolarization and persistent inactivation of sodium channels, second inward current could "take over" the depolarizing function, permitting "slow action potentials" to conduct at a low velocity, thus encouraging the development of re-entry. The antiarrhythmic action of calcium antagonists was then attributed to abolition of such slow action potentials.

There are several reasons for doubting this hypothesis. (1) There is no evidence that slow action potentials occur in Purkinje or ventricular cells in vivo. Even in vitro when sodium current is depressed by class I drugs, conduction fails altogether unless sufficient fast current remains to depolarize the cell to the critical level (−40 mV) for activation of calcium current (Chap. 1, Fig. 17). To induce "slow responses" special procedures are required; depolarization by exposure to 18 mmol/litre K, or to long-lasting depolarizing clamp currents, usually accompanied by high concentrations of adrenaline also, and application of strong stimuli with large electrodes. Such circumstances do not occur in life, and even if an area existed locally with high K and catecholamines, it is doubtful whether an incoming action potential from a normal area could provide an adequate stimulus to trigger a calcium potential in a cell deprived of its fast sodium current. (2) EL-SHERIF and LAZZARA (1979) showed that in experimentally induced arrhythmias in the late myocardial infarction period, at the time that verapamil and D-600 exerted an antiarrhythmic action, conduction velocity was *increased*, the contrary of what would have been expected if conduction was dependent on slow responses depressed by calcium-entry blockers. SINGH and VAUGHAN WILLIAMS (1972) also found that verapamil increased conduction velocity, in association with a shortening of the action potential "foot". Intracellular calcium may reduce conductance of gap junctions, and the increased conduction velocity could be explained by better cell-to-cell coupling in the presence of reduced $[Ca]i$ (VAUGHAN WILLIAMS 1984). (3) If ventricular arrhythmias involved re-entry maintained by slow responses, calcium antagonists should be especially effective in ventricular arrhythmias associated with ischaemia or myocardial damage. In fact they are of little value in such ventricular arrhythmias, and in those associated with cardiomyopathy may actually be deleterious (MCKENNA et al. 1981; DONALDSON and FOX 1983).

F. Class V Antiarrhythmic Action

Alinidine, the *N*-allyl derivative of clonidine (KOBINGER and LILLIE 1984; Chap. 19) has a selective bradycardic action on the SA node in vivo and in isolated atria. Its properties are summarized as follows:

1. Reduces slope of slow diastolic depolarization in the SA node in clinical dose range
2. Effect persists in presence of atropine
3. Prolongs APD in atrium, ventricle, and SAN transitional cells
4. Does not increase K conductance
5. Does not block sodium channels
6. Clinical doses have no negative inotropic action
7. Does not block positive inotropic effect of raised extracellular calcium

8. Does not block beta-adrenoceptors
9. Bradycardic action greater if Cl replaced by smaller anion Br
10. Bradycardic action diminished if Cl replaced by larger anion, methyl sulphate

All these experimental observations by MILLAR and VAUGHAN WILLIAMS (1981) have been confirmed by other investigators. Thus there has been no dispute concerning the effects of alinidine, but opinions have differed about their interpretation. Totally contrary conclusions may be drawn from the same experiments by different authors, or even by the same authors at different times, as is exemplified by the history of the current Ik2 (Chap. 1). The evidence for the high chloride conductance of SAN cells, already discussed, combined with the greater efficacy of alinidine in solutions in which chloride was substituted by a more permeant anion, and the reduced efficacy when Cl was substituted by a less permeant anion, led to the suggestion that alinidine might restrict the flow of current through anion-selective channels.

Another hypothesis is that alinidine blocks the time-dependent inward current activated by hyperpolarization (I_h or I_f), but the concentrations employed were very much greater than those occurring in man, and alinidine still caused bradycardia at clinical concentrations in the presence of caesium sufficient to block I_f (DENNIS and VAUGHAN WILLIAMS 1986). In any case it is the timing of depolarization and repolarization in neighbouring cells which is the most important factor in arrhythmogenesis and antiarrhythmic action rather than the species of ion which transfers charge. If alinidine is regarded as a potentially antiarrhythmic drug, one may consider, in the light of the effects of alinidine outlined above, irrespective of the nature of the ion-channels affected, whether it is possible to predict what its antiarrhythmic action, if any, would be.

The absence of effect on sodium channels would make it unlikely that the SBA would be able to control re-entry arrhythmias of ventricular origin. Since SBAs act selectively on nodal tissues, which have low resting potentials and are depolarized primarily by calcium current, it would be anticipated that their main therapeutic efficacy would be in the treatment of supraventricular arrhythmias involving re-entry in and around the SA und AV nodes. This type of arrhythmia is at present best controlled by verapamil or adenosine (Chap. 20), which differ both from each other and from alinidine in respect of the species of ion channel modulated, which illustrates that it is the target which is important, not the manner in which activity is suppressed. Indeed, the effect of adenosine on K channels is the opposite of that of alinidine, gK being increased by adenosine and decreased by alinidine.

Extrasystoles due to ectopic activity in specialized pacemaking cells, such as are scattered in the neighbourhood of the AV node, bundle of His and proximal bundle branches, would also be expected to be suppressed by SBAs. In the presence of AV block it is these specialized cells which provide ventricular pacemakers, not the Purkinje cells in the moderator band and terminal Purkinje fibres. Although the latter are much used for study and can pacemake when excised, they are usually electrically silent in situ. Thus, after an inferior myocardial infarction, often associated with bradycardia, the SBAs could be positivelvy deleterious.

To what extent does the actual evidence correlate with these predictions? On the whole, remarkably well. HARRON et al. (1982) showed that alinidine suppressed arrhythmias induced by halothane and adrenaline, i. e. involving specialized pacemaker cells, but *not* those occurring after coronary ligation in dogs, which are associated with re-entry in ventricular tissue. SENGES et al. (1983) found that the SBA AQ-A39, like nifedipine, prevented AV nodal re-entrant tachycardia in isolated rabbit preparations, and NICHOLLS et al. (1983) found that alinidine reduced human tachycardias induced by hydralazine, i.e. again involving reflex responses of nodal pacemakers. SIMOONS and HUGENHOLTZ (1982) showed that alinidine could control sinus tachycardia in patients, and AUBOECK et al. (1982) observed a similar effect on tachycardia induced by mental stress. Thus it is suggested that *if we consider only an antiarrhythmic action associated with the unique effect by which the drugs are defined as selective bradycardic agents*, then they would be likely to be effective only in arrhythmias involving specialized pacemaking cells, and would be unlikely to be useful in the treatment of ventricular arrhythmias involving re-entry in cells depolarized by sodium current.

In attempting to interpret the antiarrhythmic action of a drug in terms of its local electrophysiological effect it is important to employ concentrations within the range occurring clinically, and to study drug effects on all cardiac tissues, in order that such effects may be related to the part of the heart in which the arrhythmia occurs. HARRON and SHANKS (1985) showed that a significant lowering of heart rate could be achieved in man by plasma concentrations of alinidine between 43 and 160 ng/ml, and after 40 mg orally t. i. d. the steady plasma levels were betweeen 42 and 180 ng/ml (0.12–0.51 μmol/litre). If it is maintained that block of I_f is responsible for the bradycardic action of alinidine in man, it would be necessary to show a significant effect on I_f at similar concentrations experimentally.

In conclusion, even if the exact mechanism by which SBAs induce bradycardia still needs to be elucidated, they differ from the other known agents so profoundly that it seems justifiable to attribute to them a fifth class of antiarrhythmic action.

G. Digitalis

Many general practitioners and some cardiologists prescribe their preferred cardiac glycoside to elderly patients with minor irregularity of cardiac rhythm. The mean heart rate falls, and in this respect it is reasonable to consider digitalis among the antiarrhythmic agents. The predominant effect of cardiac glycosides when fully active is to shorten APD. Electrogenic Na/K pumping does, however, contribute to repolarization of the normal action potential, and the *initial* effect of digitalis is to *lengthen* APD in isolated cardiac muscle in vitro (DUDEL and TRAUTWEIN 1958) and in man in vivo, as confirmed by monophasic action potential recording with suction electrodes. Cardiac glycosides enter the CNS and increase the vagal output, directly slowing the SAN and increasing the AV conduction time. Vagal activity has an indirect effect also, vagal endings impinging on sympathetic nerves and inhibiting sympathetic activity. A mild positive inotropic action may also contribute to a virtuous circle of improved output reducing reflex sympathetic drive, though the positive inotropic action of cardiac glycosides in non-failing hearts has been questioned. Paradoxically, to the experimental pharmacologist, digitalis is a useful tool for *inducing* arrhythmias, and the prevalence of coupled extrasystoles and even ventricular tachycardia limits its therapeutic utility. The mode of action and clinical use of cardiac glycosides are fully discussed in Chap. 24.

H. Conclusion

New antiarrhythmic drugs continue to be presented at frequent intervals, and already in 1987 two more class Ic compounds have been announced (Donoghue et al. 1987; Sawada et al. 1987). It is remarkable that all the new drugs introduced since 1970 have turned out to possess one or more of the properties on which the original classification was based, with the addition of SBA as a fifth class. This implies that although the original classification was founded on empirical observation, it revealed actions of fundamental nature, substantiated by more recent electrophysiological evidence. There are four main sources of EMF, the concentration gradients of Na, Ca, K and Cl. The class I compounds restrict Na current, class II the adrenergically dependent Ca current. Class III action probably involves restriction of K current, and class IV restriction of non-adrenergically dependent Ca current. Although restriction of Cl current by class V compounds has not been proved, it has not been disproved either, and there is evidence which favours this interpretation. The classification of the existing antiarrhythmic agents has led to much more detailed electrophysiological studies of new drugs being undertaken before they are presented, as evidenced by the two new drugs mentioned above.

The clinical utility of the classification has been questioned, and at a recent meeting a cardiologist announced his own personal classification, presenting a slide with 40 antiarrhythmic drugs in alphabetical order. He was asked whether he found it useful and administered the drugs in alphabetical order to his patients. Whether or not the original classification is helpful in the study of the mode of action and clinical use of antiarrhythmic drugs will become apparent in the pages which follow.

References

Arnsdorf MF, Bigger JT (1972) Effect of lidocaine hydrochloride on membrane conductance in mammalian cardiac Purkinje fibers. J Clin Invest 51:2253–2263

Auboeck J, Konzett H, Olbrich E (1982) The effect of alinidine (ST 567) on emotionally induced tachycardia. Eur J Clin Pharmacol 21:467–471

Bacq ZM, Blakeley AGH, Summers RJ (1976) The effects of amiodarone, an alpha

and beta receptor antagonist, on adrenergic transmission in the cat spleen. Biochem Pharmacol 25:1195-1199
Bassett AL, Hoffman BF (1971) Anti-arrhythmic drugs: electrophysiological actions. Annu Rev Pharmacol 11:143-170
Beta-Blocker Heart Attack Trial Research Group (1982) A randomized trial of propanolol in patients with acute myocardial infarction. JAMA 247:1707-1714
Bigger JT, Mandel WJ (1970) Effect of lidocaine on the electrophysiological properties of ventricular muscle and Purkinje fibers. J Clin Invest 49:63-77
Birkhead JS, Vaughan Williams EM (1977) Dual effect of disopyramide on atrial and atrioventricular conduction and refractory periods. Br Heart J 39:657-660
Carmeliet E, Saikawa T (1982) Shortening of the action potential and reduction of pacemaker activity by lidocaine, quinidine and procainamide in sheep cardiac Purkinje fibers. Circ Res 50:257-272
Charlier R (1970) A new antagonist of adrenergic excitation not producing competitive receptor blockade. Br J Pharmacol 39:668-674
Charlier R, Delaunois G, Bauthier J, Deltour G (1969) Dans la série des benzofurannes. XL. Propiètès antiarrhythmiques de l'amiodarone. Cardiologia 54:83-90
Ciofalo F, Levitt B, Roberts J (1967) Some factors affecting ouabain induced ventricular arrhythmia in the reserpine treated cat. Br J Pharmacol 30:143-154
Colatsky TJ (1975) Sodium channels in rabbit Purkinje fibres. Nature 278:265-268
Colatsky TJ (1982) Mechanism of action of lidocaine and quinidine on action potential duration in rabbit cardiac Purkinje fibers. Circ Res 50:17-27
Coraboeuf E, Deroubaix E, Coulomb A (1979) Effect of tetrodotoxin on action potentials of the conducting system in dog heart. Am J Physiol 236:H561-H567
Cotoi S, Constantinescu L, Gavrilescu S (1972) The effect of thyroid state on monophysic action potentials in human heart. Experientia 28:797-798
Davis LD, Temte JV (1969) Electrophysiological actions of lidocaine on canine ventricular muscle and Purkinje fibers. Circ Res 24:639-655
Dennis PD, Vaughan Williams EM (1986) Further studies of alinidine induced bradycardia in the presence of caesium. Cardiovasc Res 20:375-378
Donaldson RM, Fox FM (1983) Calcium antagonists in cardiovascular disease. Prescribers J 23:16-20
Donoghue S, Payne PN, Allen G (1987) Effects of BW A256C, a novel class I antiarrhythmic agent, on the maximum rate of depolarization of cardiac action potentials in vitro: frequency, use and voltage dependence. J Cardiovasc Pharmacol 9:12-18
Dudel J, Trautwein W (1958) Elektrophysiologische Messungen zur Strophanthinwirkung am Herzmuskel. Arch Pathol Pharmakol 232:393-407
Dukes ID, Vaughan Williams EM (1984) Effects of selective alpha-1, alpha-2, beta-1 and beta-2 adrenoceptor stimulation on potentials and contractions in the rabbit heart. J Physiol (Lond) 355:523-546
Edvardsson N, Olsson SB (1981) Effects of acute and chronic beta-receptor blockade on ventricular repolarization in man. Br Heart J 45:628-636
El-Sherif N, Lazzara R (1978) Re-entrant ventricular arrhythmias in the late myocardial infarction period. 5. Mechanism of action of diphenylhydantoin. Circulation 57:465-472
El-Sherif N, Lazzara R (1979) Re-entrant ventricular arrhythmias in the late myocardial infarction period. 7. Effect of verapamil and D-600 and the role of the "slow channel". Circulation 60:605-615
El-Sherif N, Scherlag BJ, Lazzara R, Hope RR (1977a) Re-entrant ventricular arrhythmias in the late myocardial infarction period. 1. Conduction characteristics in the infarction zone. Circulation 55:687-718

El-Sherif N, Scherlag BJ, Lazzara R, Hope RR (1977b) Re-entrant ventricular arrhythmias in the late myocardial infarction period. 4. Mechanism of action of lidocaine. Circulation 56:395-405

Fatt P, Ginsborg BL (1958) The production of regenerative responses in crayfish muscle fibres by the action of calcium, strontium and barium. J. Physiol (Lond) 140:59P-60P

Fatt P, Katz B (1953) The electrical properties of crustacean muscle fibres. J Physiol (Lond) 120:171-204

Fleckstein A, Döring HJ, Kammermeier H (1967) Experimental heart failure due to inhibition of utilization of high-energy phosphate. In: Marchetti G, Taccardi B(eds) Coronary circulation and energetics of the myocardium. Karger, Basel, pp 220-236

Freedberg As, Papp JG, Vaughan Williams EM (1970) The effect of altered thyroid state on atrial intracellular potentials. J Physiol (Lond) 207:357-370

Gill EW, Vaughan Williams EM (1964) Local anaesthetic activity of the beta-receptor antagonist, pronethalol. Nature 201:199

Giotti A, Ledda F, Mannaioni PF (1973) Effects of noradrenaline and isoprenaline in combination with alpha- and beta-receptor blocking substances on the action potential of cardiac Purkinje fibres. J Physiol (Lond) 229:99-113

Govier WC, Mosal NC, Whittington P, Broom AH (1966) Myocardial alpha and beta adrenergic receptors as demonstrated by atrial functional refractory period changes. J Pharmacol Exp Ther 154:255-263

Harris L, Roncucci R (1986) Amiodarone. Médecine et Sciences Internatinales, Paris

Harron DWG, Shanks RG (1985) Pharmacology, clinical pharmacology and potential therapeutic uses of the specific bradycardic agent, alinidine. Eur Heart J 6:722-729

Harron DWG, Allen J, Wilson R, Shanks RG (1982) Effect of alinidine on experimental cardiac arrhythmias. J Cardiovasc Pharmacol 4:221-225

Haydon DA, Kimura JE (1981) Some effects of *n*-pentane on the sodium and potassium currents of the squid giant axon. J Physiol (Lond) 312:57-70

Haydon DA, Urban BW (1984) The actions of hydrocarbons and carbon tetrachloride on the sodium current of the squid giant axon. J Physiol (Lond) 338:435-450

Haydon DA, Requena J, Urban BW (1980) Some effects of aliphatic hydrocarbons on the electrical capacity and ionic current of the squid axon membrane. J Physiol (Lond) 309:229-245

Hille B (1977) Local anaesthetics: hydrophilic and hydrophobic pathways for the drug-receptor reaction. J Gen Physiol 69:497-515

Hills EA (1970) Iproveratril and bronchial asthma. Br J Clin Pract 24:116-117

Hudlicka O, Wright AJA (1977) Capillary growth and increased heart performance induced by chronic bradycardial pacing. J Physiol (Lond) 271:41P

Hudlicka O, Tyler KR, Wright AJA, Ziada AM (1983) The effect of long-term vasodilatation on capillary growth and performance in rabbit heart and skeletal muscle. J Physiol (Lond) 334:49P

Katz AM, Rhodes DG, Herbette LG (1986) Role of membrane bilayer in ligand-receptor interactions. J Mol Cell Cardiol [Supp 1] 18:A33

Kaumann AJ, Aramendia P (1968) Prevention of ventricular fibrillation induced by coronary ligation. J Pharmacol Exp Ther 164:326-332

Kobinger W, Lillie C (1984) Alinidine. N Drugs Annu 2:193-210

Kohlhardt M, Mnich Z (1978) Studies of the inhibitory effect of verapamil on the slow inward current in mammalian ventricular myocardium. J Mol Cell Cardiol 10:1037-1052

Kohlhardt M, Bauer P, Krause H, Fleckenstein A (1972) Differentiation of the Na and

Ca channels in mammalian cardiac fibres by the use of specific inhibitors. Pflugers Arch 335:309-322
Lefkowitz RJ, Haber E (1971) A fraction of the ventricular myocardium that has the specificity of the cardiac beta-adrenergic receptor. Proc Natl Acad Sci USA 68:1773-1777
McKenna WJ, Harris L, Perez G, Krikler DM, Oakley C, Goodwin JF (1981) Arrhythmia in hypertrophic cardiomyopathy. II. Comparison of amiodarone and verapamil in treatment. Br Heart J 46:173-178
Melville KI, Shister HE, Huq S (1964) Iproveratril: experimental data on coronary dilation and antiarrhythmic action. Can Med Assoc J 90:751-770
Mèndez C, Aceves J, Mèndez R (1961) Inhibition of adrenergic cardiac acceleration by cardiac glycosides. J Pharmacol Exp Ther 131:191-198
Millar JS, Vaughan Williams EM (1981) Pacemaker selectivity. Effect on rabbit atria of ionic environment and of alinidine, a possible anion channel antagonist. Cardiovasc Res 15:335-350
Multicentre International Study (1977) Reduction in mortality after myocardial infarction with long-term beta-adrenoceptor blockade. Br Med J 2:419-421
Nayler WG, McInnes I, Swann JB, Price JM, Carson V, Race D, Lowe TE (1968) Some effects of iproveratril (isoptin) on the cardiovascular system. J Pharmacol Exp Ther 161:247-261
Nicholls DF, Harron DWG, Shanks RG (1983) Cardiovascular effects of alinidine and propranolol alone and in combination with hydralazine in normal man. Br J Clin Pharmacol 17:21-29
Norwegian Multicenter Study Group (1981) Timolol-induced reduction in mortality and reinfarction in patients surviving acute myocardial infarction. N Engl J Med 304:801-807
Olsson SB, Cotoi S, Varnauskas E (1971) Monophasic action potential and sinus rhythm stability after conversion of atrial fibrillation. Acta Med Scand 190:381-388
Payne JP, Senfield RM (1964) Pronethalol in the treatment of ventricular arrhythmias during anaesthesia. Br Med J 1:603-604
Raine AEG, Vaughan Williams EM (1981) Adaptation to prolonged beta-blockade of rabbit atrial, Purkinje and ventricular potentials, and of papillary muscle contraction. Time-course of development of, and recovery from, adaptation. Circ Res 48:804-812
Reuter H (1967) The dependence of slow inward current in Purkinje fibres on the extracellular calcium concentration. J Physiol (Lond) 192:479-492
Ronfeld RA (1980) Comparative pharmacokinetics of new antiarrhythmic drugs. Am Heart J 100:978-983
Rosen MR, Merker C, Gelbland H, Hoffman BF (1973) Effects of procainamide on the electrophysiologic properties of the canine ventricular conducting system. J Pharmacol Exp Ther 185:438-446
Ross G, Jorgensen CR (1967) Cardiovascular actions of iproveratril. J Pharmacol Exp Ther 158:504-509
Sawada K, Shoji T, Igarashi T, Hiraoka M (1987) Voltage- and rate-dependent depression of *V-max* of action potentials by a new antiarrhythmic agent, E-0747, in swine cardiac Purkinje fibers. J Cardiovasc Pharmacol 9:51-56
Schmid JR, Hanna C (1967) A comparison of the antiarrhythmic actions of two new synthetic compounds, iproveratril and MJ 1999, with quinidine and pronethalol. J Pharmacol Exp Ther 156:331-338
Schwartz PJ, Periti M, Malliani A (1975) The long Q-T syndrome. Am Heart J 89:378-390

Sekiya A, Vaughan Williams Em (1963) A comparison of the antifibrillatory actions and effects on intracellular cardiac potentials of pronethalol, disopyramide and quinidine. Br J Pharmacol 21:473-481

Senges J, Rizos I, Brachmann J, Anders G, Javernig R, Hamman HD, Kübler W (1983) Effect of nifedipine and AQ-A39 on the sinoatrial and atrioventricular nodes of the rabbit and their antiarrhythmic action on atrioventricular nodal tachycardia. Cardiovasc Res 17:132-144

Simoons ML, Hugenholtz PG (1982) Alinidine, a new drug for the treatment of sinus tachycardia in myocardial infarction or cardiogenic shock. Am J Cardiol 49:980

Singh BN, Vaughan Williams EM (1970) The effect of amiodarone, a new anti-anginal drug, on cardiac muscle. Br J Pharmacol 39:657-668

Singh BN, Vaughan Williams EM (1971) Effect of altering potassium concentration on the action of lidocaine and diphenylhydantoin on rabbit atrial and ventricular muscle. Circ Res 29:286-295

Singh BN, Vaughan Williams EM (1972) A fourth class of antiarrhythmic action? Effect of verapamil on ouabain toxicity, on atrial and ventricular intracellular potentials, and on other features of cardiac function. Cardiovasc Res 6:109-119

Szekeres L, Vaughan Williams EM (1962) Antifibrillatory action. J Physiol (Lond) 160:470-482

Tasgal J, Vaughan Williams EM (1981) The effect of prolonged propranolol administration on myocardial transmural capillary density in young rabbits. J Physiol (Lond) 315:353-367

Vaughan-Jones RD (1979) Regulation of chloride in quiescent sheep-heart Purkinje fibres studied using intracellular chloride and pH-sensitive micro-electrodes. J Physiol (Lond) 295:111-137

Vaughan Williams EM (1970) Classification of anti-arrhythmic drugs. In: Sandoe E, Flensted-Jensen E, Olesen KH(eds) Symposium on cardiac arrhythmias. Astra, Södertälje, pp 449-472

Vaughan Williams EM (1972) Biophysical background to beta-blockade. In: Burley DM(ed) New perspectives in beta-blockade. CIBA, Horsham, pp 11-39

Vaughan Williams EM (1975) Classification of antiarrhythmic drugs. Pharmacology & Therapeuties B. Pergamon. 1:115-138

Vaughan Williams EM (1980) Antiarrhythmic Action. Academic Press, London

Vaughan Williams EM (1981) Classification of antiarrhythmic drugs. In: Szekeres L(ed) Pharmacology of antiarrhythmic drugs. Pergamon, Oxford, pp 125-150

Vaughan Williams EM (1984) A classification of antiarrhythmic actions reassessed after a decade of new drugs. J Clin Pharmacol 24:129-147

Vaughan Williams EM, Sekiya A (1963) Prevention of arrhythmias due to cardiac glycosides by block of beta sympathetic receptors. Lancet 2:420-421

Vaughan Williams EM, Hassan MO, Floras JS, Sleight P, Jones JV (1980) Adaptation of hypertensives to treatment with cardio-selective and non-selective beta blockers. Absence of correlation between bradycardia and blood-pressure control, and reduction in slope of the Q-T/R-R relation. Br Heart J 44:473-487

Vaughan Williams EM, Dennis PD, Garnham C (1986) Circadian rhythm of heart rate in the rabbit; prolongation of action potential duration by sustained beta adrenoceptor blockade is not due to associated bradycardia. Cardiovasc Res 20:528-535

Watanabe Y, Dreifus LS, Likoff W (1963) Electrophysiological antagonism and synergism of potassium and antiarrhythmic agents. Am J Cardiol 12:702-710

West TC, Amory DW (1960) Single fiber recording of the effects of quinidine at atrial and pacemaker sites in the isolated right atrium of the rabbit. J Pharmacol Exp Ther 130:183-193

Wilhelmsson C, Vedin JA, Whilhelmsen L, Tibblin G, Werkö L (1974) Reduction of sudden deaths after myocardial infarction by treatment with alprenolol: preliminary results. Lancet 2:1157-1160
Wittig JH, Harrison LA, Wallace AG (1973) Electrophysiological effects of lidocaine on distal Purkinje fibers of canine heart. Am Heart J 86:69-78

CHAPTER 3

Acute and Chronic Animal Models of Cardiac Arrhythmias

E. N. MOORE and J. F. SPEAR

A. Introduction

Animal models have played a major role in the development of new antiarrhythmic drugs as well as contributed to our understanding of the mechanisms of antiarrhythmic drug action. The number and diversity of experimental animal models used to screen and evaluate potential antiarrhythmic agents points out the inadequacy of any one model to reproduce the malignant arrhythmias which occur in man. An ideal animal model would both closely simulate a human counterpart and permit control of most variables. "All animal models are wrong, but some are useful" because by selecting the appropriate animal model valuable information can be obtained. Most animal models are nonatherogenic, nonprimate, and arrhythmias occur in ischemic or infarcted hearts with otherwise normal coronary anatomy. In animals drug dosages and metabolites differ from those in humans and concomitant pulmonary, renal and hepatic disease is not present. In spite of their limitations, animal models still have to be used to develop and evaluate pharmacological, surgical and electrical pacing techniques for controlling and preventing malignant arrhythmias.

The choice of animal model depends on whether one is screening for a potential antiarrhythmic compound or is testing and evaluating an already proven antiarrhythmic agent. In screening, one may have hundreds of compounds to analyze, and therefore an inexpensive, simple and quick animal model is required. However, when one is testing an agent that has already been demonstrated to have antiarrhythmic efficacy then a more sophisticated animal model is required. By employing in vivo animal models, it is possible to evaluate the favorable and unfavorable effects of an agent on blood pressure, myocardial contractility, central and autonomic nervous system, and gastrointestinal system as well as to define potential undesirable side effects.

Nonmammalian animal models will not be discussed in this chapter and primary emphasis will be upon canine models. We will include: (a) cellular electrophysiological models, (b) AV conduction models, (c) ventricular fibrillation threshold models, (d) neural models, (d) drug models, (e) acute coronary ligation reperfusion models, (f) subacute infarct models, and (g) chronic infarct models.

B. Cellular Electrophysiological Models

In vitro models have a number of advantages including the ability to control many factors that cannot be controlled in the intact animal such as the influence of the autonomic nervous system, hormonal influences, and drug metabolite effects. Until recently, it was thought that most arrhythmias were due to alteration of normal cardiac automaticity or alterations of conduction leading to reentry. Records obtained with intracellular microelectrodes led to the view that slow channel currents could be responsible for slow channel action potentials, triggered automaticity, and early and late afterdepolarizations (Zipes et al. 1980). Such experimentally induced sources of abnormal automaticity, coupled with the recent demonstration that even reentry can develop without unidirectional block via reflection occuring within a single Purkinje strand, have increased the number of possible mechanisms of cardiac arrhythmias. The fact that more than one mechanism can cause arrhythmias in the same heart complicates antiarrhythmic therapy and makes it likely that no single antiarrhythmic agent alone will be effective in preventing arrhythmias in all patients.

Microelectrode studies are often performed using normal ventricular and/or free-running Purkinje fiber tissue removed from normal canine ventricles. Standard parameters measured include resting membrane potential, action potential amplitude, maximum rate of depolarization, action potential duration, and rate of diastolic depolarization (automaticity). Membrane responsiveness is determined by evaluating the maximum rate of depolarization as a function of the membrane potential from which the action potential is evoked. Evaluation of the effects of antiarrhythmic agents on these electrophysiological parameters has been used to predict whether a given agent is likely to be an antiarrhythmic agent as well as to determine its probable mode of action.

Reentry depends upon critical interrelationships between refractoriness, conduction, and excitability in order for unidirectional block to occur and for the reentrant circuit to be maintained. Conduction velocity can be determined by surface recording, and the rate of depolarization (dv/dt) and maximum action potential amplitude can be measured with microelectrodes. Besides conduction velocity, refractoriness also influences whether a circus movement arrhythmia is maintained or terminated; action potential duration helps predict refractoriness of the tissue. Refractoriness and conduction are associated with the inactivation-reactivation of the channels responsible for the genesis of the action potential. Membrane "excitability" can be evaluated by measuring the threshold current applied through a single microelectrode. The effects of a potential antiarrhythmic agent on depolarizing currents and on recovery current kinetics measured by microelectrodes have been used as a screening method for predicting new antiarrhythmic agents, since there is a large data base of information on cellular effects of known antiarrhythmic agents. Figure 1 presents results from a microelectrode experiment that predicted the antiarrhythmic efficacy of tocainide (Moore et al. 1978). The effects of 40 and 60 µg/mol tocainide on the duration of the transmembrane ac-

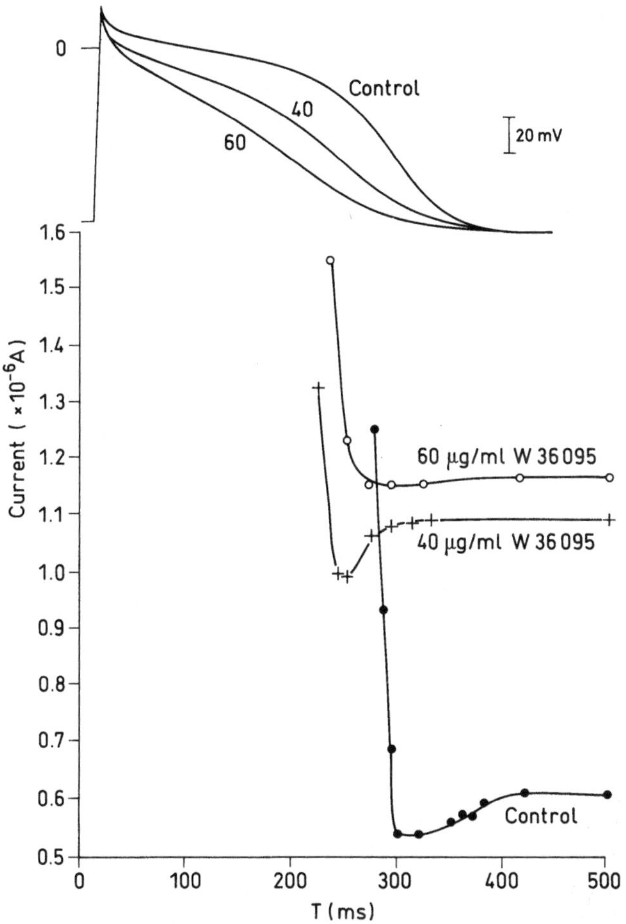

Fig. 1. The effect of tocainide on dog Purkinje fiber excitability. The action potentials were recorded during a control period and following superfusion with tocainide at 40 and 60 μg/ml (*top*). The graph (*bottom*) indicates the threshold current required to elicit an all-or-none action potential using current injected through the recording microelectrode (Reproduced by permission from MOORE et al. 1978)

tion potential is shown above and the graph below indicates the minimum current required to elicit an action potential during various times within the repolarization phase. The abbreviation of the action potential duration by tocainide is associated with shortening of the refractory period. In the excitability curve during the control period, there was a reduced current requirement to excite the fiber during the late period of repolarization. This is the "supernormal" period of excitability. Following administration of 40 and 60 μg/ml tocainide, the current required to excite the fiber during diastole was increased. Furthermore, associated with the increase in current required to excite the fiber, there also was a reduction in the supernormal period of excitability. Although not shown in Fig. 1, it was found that the rate of diastolic depolarization (phase 0) was depressed at higher levels of tocainide. Thus this study suggested that tocainide was an antiarrhythmic agent.

To evaluate the possibility that a drug has active metabolites, interacts with the autonomic system, or requires a long period before antiarrhythmic activity develops (cf. amiodarone), acute experiments on isolated tissues are insufficient. It is, therefore, necessary to utilize more than one model for evaluating and identifying new antiarrhythmic agents.

Until recently most studies were performed on normal ventricular muscle and Purkinje fibers. However, the realization that in the infarcted heart one is dealing with diseased tissues exposed to abnormal extracellular fluids and anoxia, has encouraged investigations on tissues exposed to abnormal media or obtained from previously infarcted hearts (SPEAR and MOORE 1982). Microelectrode studies in such preparations have shown that both atrial and ventricular myocardial fibers can exhibit automaticity under these abnormal circumstances, although normal myocardial fibers within the atrium and ventricle do not beat spontaneously. The existence of experimentally induced triggered automaticity and early and delayed afterdepolarizations has led to speculation that such mechanisms might be involved in the genesis of cardiac arrhythmia (cf. Chap. 28).

C. Monophasic Action Potential Technique

The modification of the monophasic action potential (MAP) technique in which a pressure-directed catheter was used instead of a suction MAP technique to record a MAP has provided the first direct demonstration that early afterdepolarizations and triggered activity can be induced experimentally (by cesium) in situ in the intact beating canine heart (LEVINE et al. 1985). Studies have demonstrated that the time course of repolarization of MAPs correlate directly with those of simultaneously recorded transmembrane action potentials during a variety of conditions. Not only have early afterdepolarizations and triggered activity been demonstrated, but MAPs reliably reproduced delayed afterdepolarizations as well as diastolic depolarization in normal automatic fibers.

Difficulties interpreting the MAP recordings can occur due to instability in positioning of the MAP electrode leading to mechanical artifacts. Also, it can be difficult to position the MAP electrode at the exact location where abnormal electrophysiological activity is present. For example, although the MAP technique can record delayed afterdepolarizations and triggered activity in isolated Purkinje fibers made toxic with digitalis, it nevertheless has been impossible to duplicate these experiments in the intact in situ beating heart.

Figure 2 shows canine ventricular ectopy which developed within minutes of the administration of cesium chloride. A MAP was recorded with a contact catheter electrode in the upper trace and a lead II electrocardiogram in the lower trace. In the lead II electrocardiogram a ventricular extrasystole occurs at the solid arrow (lower right hand corner). In the MAP tracing, early after-

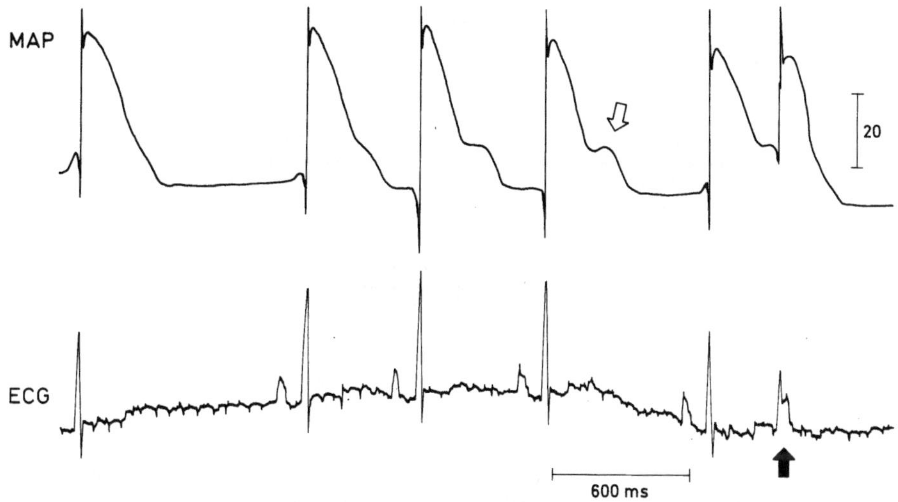

Fig. 2. Ventricular ectopy developed within minutes of the administration of cesium chloride. In this example, afterdepolarizations developed after cesium was administered (*open arrow*) and progressively increased in size before the development of the ectopic beat (*large arrow*)

depolarizations begin and become progressively more prominent (at the open arrow) until finally an all-or-none response occurs in the MAP recording resulting in the premature ventricular complex in the ECG tracing (arrow, right-hand side of figure). When cesium toxicity progressed still further, the electrocardiogram resembled "Torsade de pointes". All regions explored by the MAP contact electrode on both the epicardium and endocardium exhibited early afterdepolarization following cesium toxicity. It is not known whether such triggered activity and early and late afterdepolarizations induced by cesium are relevant to the genesis of clinical arrhythmias in man, but the similarity of the records to those occurring in the long QT syndrome suggests that poisoning with cesium chloride could provide an experimental model to evaluate the efficacy of antiarrhythmic agents against torsade de pointes.

D. Atrioventricular Conduction

In testing an antiarrhythmic agent, it is necessary to study its effects on normal pacemaker activity and upon atrial, atrioventricular, and ventricular conduction. Although the surface electrocardiogram provides direct evidence of sinus node pacemaker activity, atrial activation, and ventricular activation/recovery, it does not show whether AV conduction delays occur above, within, or below the AV node. Bundle of His electrograms recorded simultaneously with atrial and ventricular electrograms can do so. Figure 3 illustrates how recording of atrial, His bundle, and ventricular electrograms together with a lead II electrocardiogram can indicate how a drug affects conduction.

Autonomic nerve activity modifies both impulse formation and conduction, and may exacerbate or modulate arrhythmias associated with drug appli-

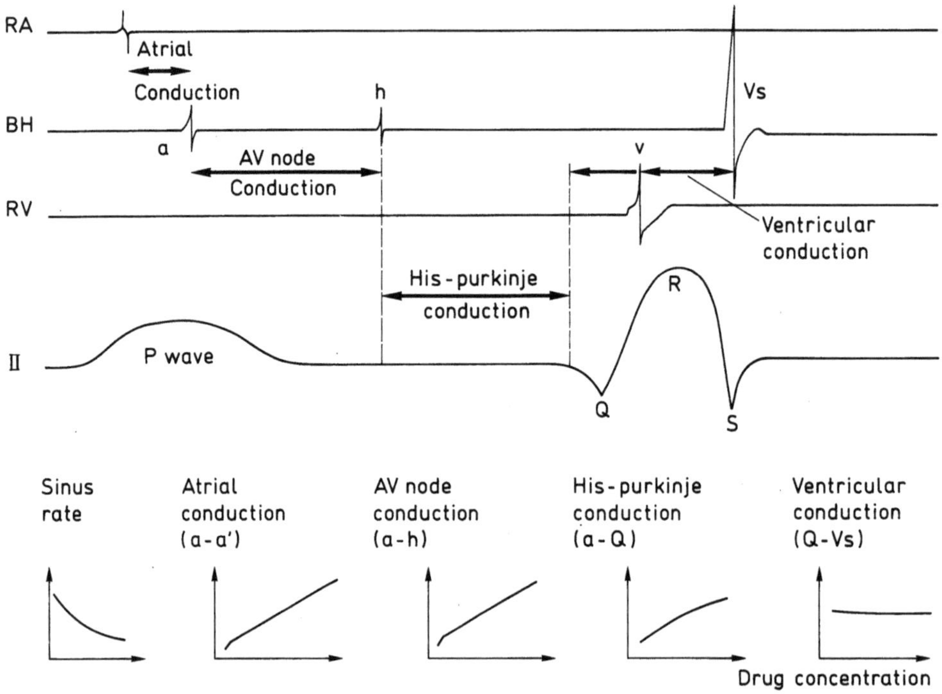

Fig. 3. A right atrial (RA), bundle of His (BH), and right ventricular (RV) electrogram are schematically drawn together with a lead II electrocardiogram. The method of measuring atrial conduction (a–a′), AV node conduction (a–h), His-Purkinje conduction (h–v), and ventricular conduction (V–Vs) are noted. Immediately below is noted how different intervals, i. e., *a–a′* for atrial conduction, can describe conduction of increasing drug concentrations

cation as well as during coronary occlusion (SPEAR and MOORE 1973). The parasympathetic system can influence sinus rhythm, ectopic atrial rhythm, junctional rhythm, as well as ectopic ventricular rhythms. Parasympathetic induced slowing of sinus rhythm, ectopic atrial rhythm, and AV junctional rhythm occurs within a few hundred milliseconds of the start of an increase in vagal activity and ceases within about 4 s after its end. In contrast, an ectopic ventricular rhythm is slowed only after 1 s of vagal stimulation, and the slowing persists for 10 s afterwards. Vagal effects on the ventricle may, therefore, be indirect, by inhibition of the sympathetic system, because acceleration by the sympathetic system of sinus rhythms, ectopic atrial rhythms, junctional rhythms, and ectopic ventricular arrhythmias is associated with a slow onset. The effects of enhanced sympathetic drive are not observed until more than 1 s following a burst of sympathetic activity and last for 12 or more seconds.

Figure 4 shows how brief vagal activity can modify conduction and cause a rhythm disturbance. Right atrial, bundle of His, and right ventricular electrograms were recorded simultaneously with a lead II electrocardiogram. In the lead II electrocardiogram a constant PR interval can be observed with a sudden dropped beat after the third P wave (Mobitz type II block). When Mobitz type II block is associated with infranodal block within the His-Pur-

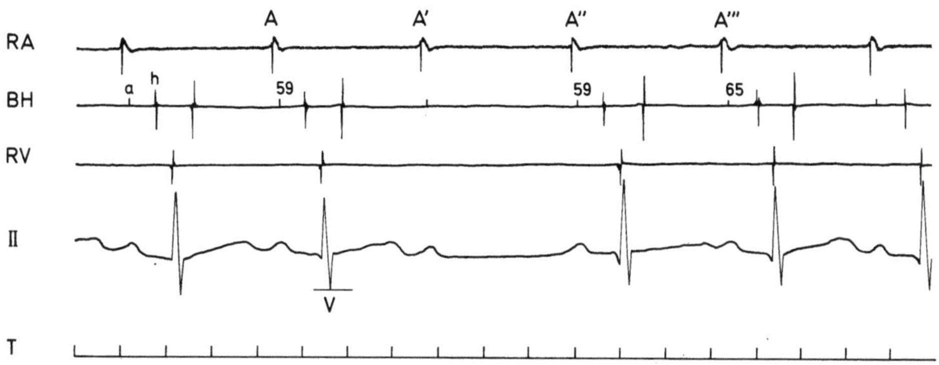

Fig. 4. The influence of vagally delayed AV conduction on the conduction time of subsequent beats. Bipolar electrograms were recorded from the right atrium (RA), the bundle of His (BH), and the right ventricle (RV) simultaneously with a lead II electrocardiogram (II). In the right atrium electrogram, A indicates the beats initiated before vagal stimulation, and A, A′, A″, and A‴ indicate successive beats after vagal stimulation. V indicates the timing of a 100-ms train of stimuli delivered to the left of vagus. The timing signal (T) denotes 100-ms intervals

kinje conduction system it frequently progresses to complete AV block requiring implantation of a cardiac pacemaker. However, the His bundle electrogram in Fig. 4 shows that AV block of the third atrial response (A') occurred within the AV node, since no His bundle deflection followed the A' atrial depolarization. This type of block within the AV node would not be prognostic of subsequent complete AV block, and Fig. 4 illustrates how AV block by vagal activity may be distinguished in this preparation from delayed conduction below the bundle of His.

E. Ventricular Fibrillation Threshold Technique

Ventricular fibrillation (VF) is the primary cause of sudden death. It has often been stated that it may be more important for cardiac drugs to be "antifibrillatory" than simply to have properties capable of suppressing premature ventricular beats. Ventricular fibrillation may be preceded by a rapid ventricular tachyarrhythmia (VT) or by a premature ventricular beat (PVB) falling within the vulnerable phase of the preceding beat (R on T phenomenon). The possibility of predicting the probability of VF from prior ECG recordings is discussed in Chap. 6. Wiggers in the 1940s found that an appropriately timed electrical impulse of sufficient energy delivered to the ventricles in late systole could induce ventricular fibrillation (Wiggers and Wegria 1940), and developed a semiquantitative method to measure ventricular "vulnerability". The ventricular fibrillation threshold (VFT) is defined as the minimum current required to initiate VF. Although the mechanism by which VF is initiated is not precisely known, passing current in the high milliampere range during the vulnerable period probably induces heterogeneity of recovery during the vulnerable period between adjacent ventricular fibers. Our lab has re-

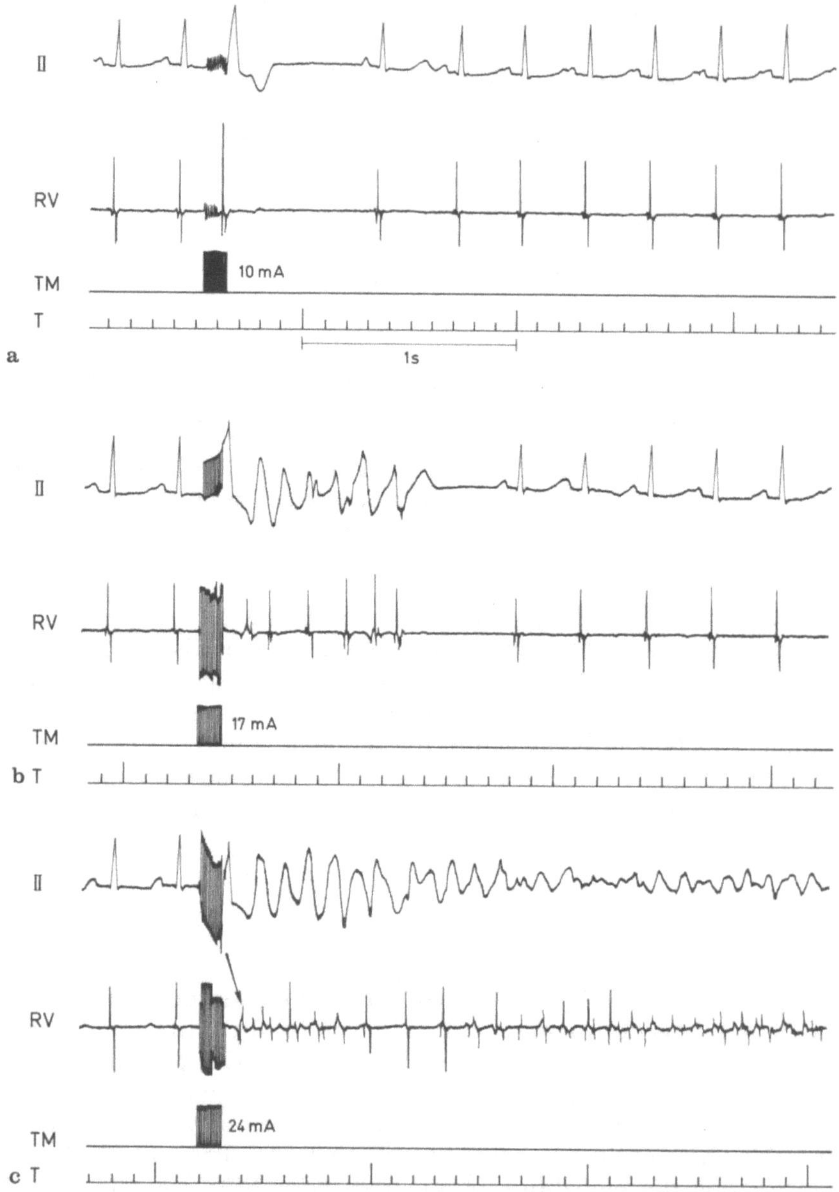

Fig. 5 a–c. Ventricular fibrillation threshold technique. A lead II electrocardiogram is recorded simultaneously with a right ventricular (RV) electrogram. TM indicates when the VFT train was deliverd. **a** A 10-mA, 120-ms-duration, 100-Hz train of current evoked a single extrasystole. **b** the current was increased to 17 mA and a nonsustained ventricular tachycardia occurred. **c** When the VFT constant current train was increased to 24 mA, ventricular fibrillation was initiated cently demonstrated that fractionated continuous activity occurs in the region close to the electrode through which the current is applied. The fact that continous fractionated electrical activity occurs around the VFT electrode, and is also sometimes observed in the presence of myocardial infarction and ischemia, suggests that the VFT current may induce micro- and macroreentry in the region of applied current.

Many modifications of the original single-pulse VFT technique have been used to access the vulnerability of the ventricles to fibrillation. All techniques have in common that a relatively high current of 15–40 mA is delivered during the vulnerable period of a single cardiac cycle, or if less intense currents are used, then the current train must span the vulnerable periods of *2* or more sequential cardiac cycles. The latter technique of using the duration of time that a constant current train must be administered to cause fibrillation is fraught with many uncertainties. While the fibrillating train is being delivered, multiple premature ventricular complexes (PVC) result which in turn invariably alter the electrical stability of the myocardium. It is well known that PVCs are associated with increased electrical instability and that with each sequential PVC there is a further decrease in ventricular stability. A VFT technique that is widely used today is demonstrated in Fig. 5 and was suggested by HAN (1969). The Han VFT method uses a brief train (120 ms) of constant current pulses (4 ms pulses at 100 Hz) delivered across the vulnerable period. This technique reduces the time involved in making a single measurement of the ventricular fibrillation threshold since it is only necessary to increase the intensity of the train of pulses in a stepwise fashion until fibrillation ensues, rather than having to scan the vulnerable period with each increase in current intensity. Figure 5 is a demonstration of the VFT train technique. In A, B, and C the lead II electrocardiogram (II) was recorded simultaneously with a right ventricular electrogram located within less than 1 cm of the VFT electrode where the 10-mA, 120-Hz train of current stimuli was applied. The 10-mA current (A) resulted in only a single premature ventricular beat (PVB). In B, the intensity of the current was increased to 17 mA and a nonsustained ventricular tachyarrhythmia resulted. The minimum current that resulted in fibrillation is presented in C where 24 mA induced ventricular fibrillation. The VFT is, therefore, 24 mA.

The VFT technique has been employed by many investigators to evaluate new antiarrhythmic agents. The method can assess the influence of single factors on ventricular vulnerability in a system free of some variables associated with fribrillation induced by coronary ligation in the otherwise intact animal. Nevertheless, many variables must be controlled for reproducible VFT results to be obtained, including heart rate, pH, pCO_2, pO_2, body temperature, and autonomic tone. It is also necessary to adjust for QT interval changes and to ascertain that the train of stimuli does not evoke a PVB, in order to avoid the possibility of timing the train for the PVB rather than for a beat of supraventricular origin. Many drugs found in the laboratory to increase VFT have proved to be effective against human ventricular arrhythmias. A variation of the VFT method, with programmed stimuli, has been used in man and results correlate well with results obtained in the canine lab (HOROWITZ et al. 1977).

F. Drug Models

There is a number of cardioactive drugs which in addition to preventing cardiac arrhythmias may also cause them. The statement "All cardiac drugs are poisons with desirable side effects" attests to their deleterious as well as their beneficial effects. Three drugs which reliably produce arrhythmias and which have been used to provide animal models to evaluate antiarrhythmic agents are digitalis, aconitine, and beta-adrenoceptor agonists. The most commonly used drug to establish an arrhythmia model for testing antiarrhythmic agents is digitalis. In one procedure used in dogs or guinea pigs a loading dose of acetylstrophanthidin is followed by incremental doses given every 10 min until 85% of the ventricular beats are ectopic. Potential antiarrhythmic agents can then be administered to determine whether ectopy can be suppressed or reduced. In guinea pigs a quantitative and reliable assay of the amount of ouabain required to induce various stages of arrhythmia depends on an intermittent automated infusion (VAUGHAN WILLIAMS and SEKIYA 1983). Toxic concentrations of cardiac glycosides induce automaticity and delayed afterdepolarizations in laboratory experiments, but the importance of these mechanisms in man is unknown. Steady-state levels of digitalis toxicity are difficult to maintain, so that at the time that antiarrhythmic agents are administered, digitalis intoxication may be increasing or decreasing, meaning that drug effects may be indirectly modulated by spontaneous arrhythmia variability.

Another laboratory model of arrhythmia involves intoxication by barium chloride, which induces automaticity by blocking potassium (repolarizing) current. In the barium model the abnormal automaticity is stable for hours.

Aconitine when applied via small soaked cotton pledgets to either the atrium or ventricle can lead to flutter and/or fibrillation. The mechanism for aconitine's action appears to be failure of inactivation of sodium current. As with digitalis, it is difficult to establish a stable level of ectopic activity with aconitine and the mechanism of arrhythmogenesis differs from those responsible for inducing arrhythmia in man.

Catecholamines have been used to initiate arrhythmias in cyclopropane or chloroform-sensitized animals. Even in the absence of these sensitizing agents, catecholamines can still result in arrhythmias but only at very high doses where they predominantly exaggerate normal automaticity. In diseased cardiac tissues, catecholamines may improve or exacerbate arrhythmias. For all these reasons the drug-induced models of arrhythmia are unsatisfactory.

G. Neural Models

Evidence that the central nervous system and psychological stress can influence the devolopment of lethal arrhythmias has been presented by Skinner et al. in the pig (SKINNER et al. 1975) and in the dog (CORBALAN et al. 1974). Using the psychological stress of an unfamiliar environment, Skinner was able to demonstrate an increase in incidence of ventricular fibrillation in conscious pigs upon coronary arterial occlusion in unfamiliar versus familiar environments. Corbalan also used psychological stress to demonstrate an increase incidence of ventricular fibrillation in dogs following myocardial ischemia. Fibrillation was three times more frequent in animals placed in a Pavlovian sling where the animals had undergone shock conditioning than in a nonstressful enviroment where they were relaxed. The mechanism by which psychological stress facilitates the development of ventricular fibrillation is not understood. Such psychological stress models are of unknown relevance to man because of the wide variation in individual responses to stress.

The effects of the autonomic nervous system on the heart are considered in detail elsewhere in this volume, but it may be recalled that left stellate stimulation has been found to render the ventricles more vulnerable to fibrillation as measured by the VFT technique, whereas ablation of the left stellate decreases the incidence of lethal arrhythmias after coronary occlusion (SCHWARTZ et al. 1978).

The parasympathetic system also influences the incidence of lethal arrhythmias not only by reducing heart rate, but also by other means. VERRIER and LOWN (1974) found that even when heart rate was maintained constant vagal stimulation still elevated ventricular fibrillation threshold. It may also be recalled that vagal activity inhibits sympathetic nerves, and whereas acetylcholine had little effect on normal Purkinje fibers, it depressed automaticity in diseased Purkinje fibers. Thus the balance between the sympathetic and parasympathetic nervous system could influence the incidence or consequences, of arrhythmias, and neural interactions are the topic of much current research.

H. Acute Coronary Ligation-Reperfusion Models

When the left descending coronary artery is acutely ligated for a period of 20-30 min, a high percentage of animals exhibit ventricular fibrillation during the occlusion period and if not, then upon reperfusion. The incidence of ventricular fibrillation is variable but 80 % of the animals will fibrillate either during occlusion or upon reperfusion. One cannot predict the time of onset of arrhythmias, however, nor even whether arrhythmias will or will not develop during the occlusion or reperfusion period, so that it is necessary to use a large number of animals to obtain statistical significance. Commonly, arrhythmias occur 2-12 min and/or 13-30 min after occlusion (KAPLINSKY et al. 1979). It has been suggested that the earlier arrhythmias are due to reentry and the delayed arrhythmias are due to ectopic automaticity. Variability and the efficacy of antiarrhythmic agents in this model depends on whether the drug is administered prior to or at the time of occlusion. Other influential factors include proximity of occlusion, manipulation prior to occlusion, neural/humoral state, anesthesia, blood electrolytes, blood gases, duration of ligation, and anatomy of the collatoral circulation which is species dependent. A disadvantage of the model is that higher blood levels of antiarrhythmic agents are required to depress ventricular arrhythmias, and their duration of action is shorter than found subsequently in man.

The increasing use of streptokinase, PTA, and angioplasty to reopen occluded coronary arteries, has concentrated interest on the arrhythmias that occur upon reperfusion of a previously occluded coronary artery. Reperfusion arrhythmias in the dog have been shown to be highly malignant and resistant to most antiarrhythmic drugs.

Evidence that alpha-adrenergic blockade has antiarrhythmic efficacy in the cat against reperfusion arrhythmias is discussed elsewhere in this volume (Chap. 22). Preliminary work suggesting that oxygen-free-radical scavengers can decrease reperfusion arrhythmias is of obvious relevance in this context (see Chap. 26).

J. Subacute Coronary Artery Ligation Models

Nearly 30 years ago Sidney Harris found that most dogs survived a partial occlusion of the left coronary artery (HARRIS and ROJAS 1943). After complete occlusion at a second intervention arrhythmias increased in frequency after about 6 h, reached a peak 24–36 h later, and subsided after 72 h. These spontaneous arrhythmias are rarely malignant, do not lead to death, and respond to drugs. They are usually associated with enhanced automaticity within the Purkinje system at the interface between normal and infarcted tissue (SPIELMAN et al. 1978). Triggered activity has also been induced in isolated subendocardial Purkinje fibers removed 24 h after two-stage coronary ligation. However, since the arrhythmias can usually be overdriven by electrical pacing and can be unmasked by slowing of the heart rate by increasing vagal tone, it is thought that the predominant rhythm is due to enhanced automaticity in the Purkinje system. Arrhythmias that occur 24-48 h after two-stage coronary ligation resemble accelerated idioventricular rhythms observed in man following a recent myocardial infarct.

The two-stage Harris coronary ligation model is often used to evaluate new antiarrhythmic agents, but drug levels necessary to suppress the ectopic rhythms are often higher than those required for suppression of ventricular ectopic activity in man and the duration of action is shorter than in man. This may relate not only to the differences in mechanisms underlying the arrhythmia but also to species variations.

Variability in the effectiveness of antiarrhythmic agents using this model can result from not taking into account the overall heart rate and percentage of beats that are ectopic. For example, a model that has a higher incidence of malignant PVBs may be a more severe test than one having fewer PVBs. Advantages of the subacute Harris coronary ligation model are that it is possible to do studies in the awake animal, oral administration of drug can be tested and potential side effects on blood pressure, myocardial contractility, CNS, GI and other effects can be monitored. Also, it has been found recently that 5-7 days following two-stage coronary ligation, when spontaneous ventricular tachyarrhythmias have disappeared, it is often possible to initiate ventricular tachyarrhythmias by programmed electrical stimulation (EL-SHERIF et al. 1977). These tachyarrhythmias are associated with delayed fractionated activity similar to that recorded in man during reentrant rhythms. Thus it appears that for a short period following the cessation of spontaneous PVBs there is a period when reentrant arrhythmias can be induced.

K. Chronic Myocardial Canine Infarct Models

Chronic animal models having electrically inducible ventricular tachyarrhythmias similar to the malignant arrhythmias that contribute to sudden death in patients with ischemic heart disease have proven very difficult to develop. However, between 1977 and 1981 several laboratories found that the use of occlusion-reperfusion (MICHELSON et al. 1980; KARAQUEZIAN et al. 1979; MYERBURG et al. 1977) and/or multiple ligation of distal coronary arteries can result in the canine and feline hearts having ventricular tachyarrhythmias inducible by programmed electrical stimulation. These chronic infarct models have proven useful for investigating the mechanism of action of known antiarrhythmic agents and for predicting efficacy of new antiarrhythmic agents. Although spontaneous ventricular tachyarrhythmias seldom occur, ventricular tachyarrhythmias can be reproducibly initiated by appropriate programmed electrical stimulation. Interestingly, spontaneous ventricular arrhythmias were reported to occur in cats with chronic myocardial infarcts (MYERBURG et al. 1977).

In the occlusion-reperfusion canine chronic infarct model developed in our lab we occlude the anterior descending coronary artery just proximal or distal to the first large diagonal branch using a two-stage Harris procedure, followed 2 h later by the complete release of the occlusion (MICHELSON et al. 1980). In animals with extensive collaterals or anastomosing vessels we also ligate permanently at least two or three of the more prominent anastomosing epicardial vessels just at the cardiac apex. This procedure results in dogs with heterogeneous mottled infarctions with interspersing of normal and abnormal tissue. While these dogs do not develop spontaneous ventricular tachyarrhythmias it is possible to initiate by programmed electrical stimulation ventricular tachycardia and/or fibrillation in about 60 %–70 % of the animals. An example of programmed stimulation initiating and terminating a ventricular tachycardia is presented in Fig. 6. The ability of programmed stimulation to induce a sustained ventricular tachycardia is still present 5 years after the initial occlusion-reperfusion surgical procedure, and low-level late potentials associated with fractioned activation can also be observed (unpublished observation). Chronic canine infarct animals can be investigated acutely in order to obtain detailed electrophysiological data at multiple sites including strength-interval curves, conduction times, refractoriness, site of origin of the arrhythmia, and biophysical membrane properties of infarcted cells. It is also possible to study these animals with chronically implanted electrodes in the unanesthetized state and to use each animal as its own control over a prolonged period of investigation.

The models still have limitations. The metabolism of a drug and its duration of action may be different in dogs and man. Metabolites may have antiarrhythmic activity in the dog but not in man. Nevertheless many antiarrhythmic agents effective in man are also effective in the chronic canine infarct model.

The operated dogs tolerate chronically implanted electrodes, and could, therefore, be used to study drugs with a slow onset of action, i.e. amiodarone. Since in some animals ventricular fibrillation can reliably be initiated by pro-

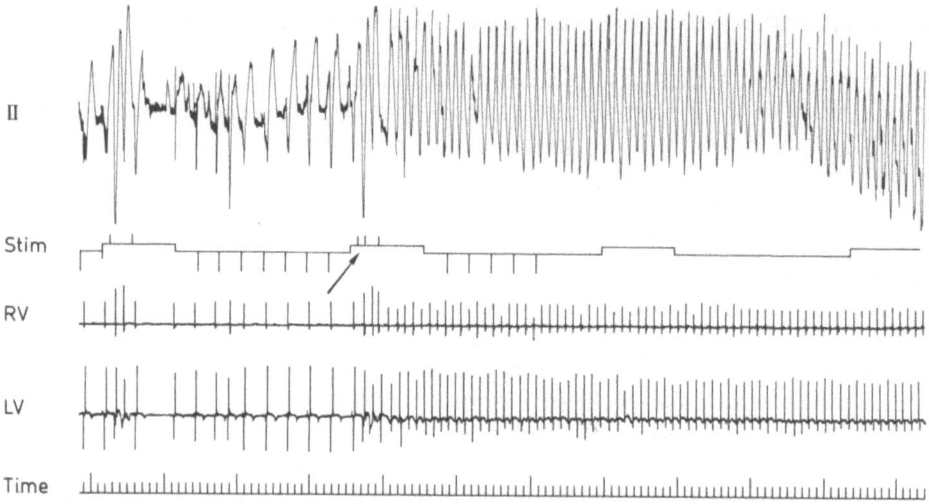

Fig. 6. Demonstration of the initiation of sustained ventricular tachycardia in a chronic infarct dog. The lead II electrocardiogramm (II) was recorded together with stimulus artifacts (*stim*) and right and left (RV and LV) ventricular electrograms. The *arrow* indicates the time of three aearly premature stimuli (upward-going stimulus artifacts) which initiated a sustained ventricular tachycardia

grammed electrical stimulation, it might be possible to study whether antiarrhythmic agents which do not eliminate all ectopic activity might still protect against VF.

In addition, since in man complicating pulmonary, renal or hepatic diseases are often present besides myocardial infarction, these chronic canine models should prove valuable in evaluating drug interactions, hypo- and hyperkalemia, left ventricular dysfunction, and other factors associated with chronic disease states in man.

Recent attempts at further modification of chronic myocardial infarction in canines has included closed chest coronary occlusion using coronary catheters, the addition of methyl prednisolone and low-protein diets to produce ventricular aneurysms, and the addition of acute infarction to a prior chronic infarct. Inducing a second acute infarct in a dog with a chronic myocardial infarct might be regarded as simulating some features of multivessel disease in man.

L. Supraventricular Models

Although supraventricular arrhythmias are not frequently life threatening, they nevertheless represent arrhythmias that may require antiarrhythmic therapy. Three supraventricular tachycardia (SVT) models that have been suggested to evaluate new antiarrhythmic agents include the aconitine model, the Frame-Hoffman modification of the Rosenbluth anatomical reentrant atrial model, and the recently reported Waldo model of atrial flutter. The aconitine model has the disadvantages of inconstancy of the arrhythmia and noncorrespondence of the mechanism of arrhythmogenesis with that of human SVT. In the Frame model an anatomical reentrant circuit is established surgically, which allows atrial flutter to be initiated and sustained in a stable state for a prolonged period (Frame et al. 1986). The reproducibility of this anatomical model with its underlying reentrant mechanism makes this a promising model for antiarrhythmic drug testing against supraventricular tachycardia. The Waldo model of atrial flutter is produced by introducing talcum powder into the pericardial sac (Page et al. 1986). This insult results in changes in the atrium, enabling programmed electrical stimulation to initiate and terminate atrial flutter.

M. Noncanine Models

There are a number of noncanine animal models that have been used for screening as well as testing antiarrhythmic agents (Moore et al. 1981). One screening model that has often been used is the mouse chloroform arrhythmia model where chloroform sensitizes the myocardium to catecholamines. The model has been used to test for both antiarrhythmic as well as beta-blocking activity. In this model, skeletal muscle tremors may cause ECG artifacts which must be distinguished from true ventricular fibrillation.

The rat has been employed for many toxicology studies and also for coronary occlusion and reperfusion investigations. There are several problems in using the rat, including that the rat electrocardiogram is different from that of most other mammals, i.e., no ST segment is present, rendering the rat unsuitable for electrophysiological studies of antiarrhythmic agents.

Pigs have also been used for electrophysiological and electropharmacological studies, but the pig heart has a terminal end artery coronary system without anastomoses, and the Purkinje system is more extensive than in man or the dog, and has different characteristics. The electrocardiogram in all ungulates including the pig differs from that of the dog, cat, and man. It has been demonstrated, however, that pigs with coronary arteries damaged by irradiation, and fed a high-cholesterol diet, do develop sudden cardiac death, preceded by lethal arrhythmias.

N. Summary

Animal arrhythmia models have been useful not only to probe and demonstrate new mechanisms of arrhythmogenesis but also to analyze the effectiveness and mechanism of antiarrhythmic efficacy. Unfortunately no single model can predict the effectiveness of a cardioactive agent in man. However, by using a series of different animal models including cellular electrophysiological models, acute and chronic myocardial infarct animal models, ventricular fibrillation threshold models, as well as neural and drug arrhythmia models, it is possible to identify new antiarrhythmic agents.

Acknowledgments. The authors thank Ralph Iannuzzi and Bejay Moore for preparation of the manuscript and illustrations. Supported in part by grants from the National Heart Lung and Blood Institute and W. W. Smith Charitable Trust.

References

Corbalan R, Verrier RL, Lown B (1974) Psychologic stress and ventricular arrhythmia during myocardial infarction in the conscious dog. Am J Cardiol 34:692

El-Sherif N, Hope RR, Scherlag BJ, Lazzara R (1977) Reentrant arrhythmias in the late myocardial infarction period. II. Patterns of initiation and termination of reentry. Circulation 55:702

Frame LH, Page RL, Hoffman BF (1986) Atrial reentry around an anatomical barrier with a partial refractory gap: a canine model of atrial flutter. Circ Res 58:495–511

Han J (1969) Ventricular vulnerability during acute coronary occlusion. Am J Cardiol 24:857–864

Harris AS, Rojas AG (1943) Initiation of ventricular fibrillation due to coronary occlusion. Exp Med Surg 1:105

Horowitz LN, Spear JF, Josephson ME, Kastor JA, MacVaugh H, Moore EN (1977) Ventricular fibrillation threshold in man. Am J Physiol 39:274

Kaplinsky E, Ogawa S, Balke CW, Dreifus LS (1979) Two periods of early ventricular arrhythmia in the canine acute myocardial infarction model. Circulation 60:397–403

Karaqueuzian HS, Fenoglio JJ, Weiss MB, Wit AL (1979) Protracted ventricular tachycardia induced by premature stimulation of the canine heart after coronary artery occlusion and reperfusion. Circ Res 44:833–846

Levine JH, Spear JF, Guarnieri T, Weisfeldt ML, DeLangen CDJ, Becker LC, Moore EN (1985) Cesium chloride-induced long QT syndrome: demonstration of afterdepolarizations and triggered activity in vivo. Circulations 72 (5):1092–1103

Michelson El, Spear JF, Moore EN (1980) Electrophysiologic and anatomic correlates of sustainend ventricular tachyarrhythmias in a model of chronic myocardial infarction. Am J Cardiol 45:583–590

Moore EN, Spear JF, Feldman HS, Moller R (1978) Electrophysiological properties of a new antiarrhythmic drug—Tocainide. Am J Physiol 41:703

Moore EN, Spear JF, Michelson EL (1981) Non-canine animal models for evaluating antiarrhythmic efficacy. In: Morganroth J, Moore EN, Dreifus LS, Michelson EL (eds) The evaluation of new antiarrhythmic drugs. Nijhoff, The Hague

Myerburg RJ, Gelband H, Nilsson K, Sung RJ, Thurer RJ, Morales AR, Bassett AL (1977) Long-term electrophysiological abnormalities resulting from experimental myocardial infarction in cats. Circ Res 41:73–84

Page PL, Plumb VJ, Okumura K, Waldo AL (1986) A new animal model of atrial flutter. J Am Coll Cardiol 8(4):872–879

Schwartz PJ, Brown AM, Malliani A, Zanchetti A (1978) Neural mechanisms in cardiac arrhythmias. Raven, New York

Skinner JE, Lie JT, Entman ME (1975) Modification of ventricular fibrillation latency

following coronary artery occlusion in the conscious pig: the effect of psychological stress and beta adrenergic blockade. Circulation 51:656
Spear JF, Moore EN (1973) The influence of brief vagal and stellate nerve stimulation on pacemaker activity and conduction within the atrioventricular conduction system of the dog. Circ Res 32:27–41
Spear JF, Moore EN (1982) Mechanisms of cardiac arrhythmias. Annu Rev Physiol 44:485
Spielman SR, Michelson EL, Horowitz LN, Spear JF, Moore EN (1978) The limiations of epicardial mapping as a guide to the surgical therapy of ventricular tachycardia. Circulation 57:666–670
Vaughan Williams EM, Sekiya A (1963) Prevention of arrhythmias due to cardiac glycosides by block of β-sympathetic receptors. Lancet 1:420–421
Verrier RL, Lown B (1978) Sympathetic-parasympathetic interactions and ventricular electrical stability. In: Schwartz PJ, Brown AM, Malliani A, Zanchetti A (eds) Neural mechanisms in cardiac arrhythmias. Raven, New York
Wiggers CJ, Wegria R (1940) Ventricular fibrillation due to single localized induction and condenser shock applied during the vulnerable phase of ventricular systole. Am J Physiol 128:500–505
Zipes DP, Bailey JC, Elhararrar B (eds) (1980) The slow inward current and cardiac arrhythmias. Nijhoff, The Hague

CHAPTER 4

Classification of Human Arrhythmias

P. COUMEL

There are many ways to classify human arrhythmias, each of them having its particular interest in terms of theoretical comprehension, clinical management or therapeutic consequences. Attempts have been made to obtain some consensus for defining and classifying cardiac arrhythmias and conduction disturbances (WHO/ISFC TASK FORCE 1978, 1979). With so many possibilities, any classification may seem to be the best from a particular point of view, without being in fact the correct one. Arrhythmias can be classified according to their electrophysiological mechanism, their point of origin in the myocardium, the presence of a causal disease or its apparent absence, the presence or absence of impaired myocardial function, the very simple clinical aspect of tachycardia or bradycardia, the electrocardiographic pattern, the relationships with the autonomic nervous system and the sensitivity to the various categories of drugs. Our purpose in the present chapter will be to list human arrhythmias according to the classical ECG criteria referring mainly to arrhythmia location, and then to examine the possible advantages of other approaches.

A. Classification of Arrhythmias According to Their Origin

I. Atrial Arrhythmias

1. Extrasystoles and Tachyarrhythmias

Figure 1 summarizes the various arrhythmias which can be observed at the atrial level. Atrial extrasystole (Fig. 1a) is the most trivial arrhythmia, usually asymptomatic. Although atrial extrasystoles in themselves can be considered as benign, they often precede more serious tachyarrhythmias, so that it is important to monitor them.

Typical atrial flutter (Fig. 1b, c) is characterized by the absence of an isoelectric line between the deflections of the atrial activity at a fixed rate of 300/min. Atypical forms (Fig. 2) do not display the sawtooth appearance in leads II and III, but since they may be observed in the same patients with the same rate as the typical forms, they probably have a common re-entrant mechanism, as is suggested by their possible interruption by stimulation. Distinguishing between the atypical forms of atrial flutter and atrial tachycardias with a rapid rate is not always possible. However, in those atrial tachycardias

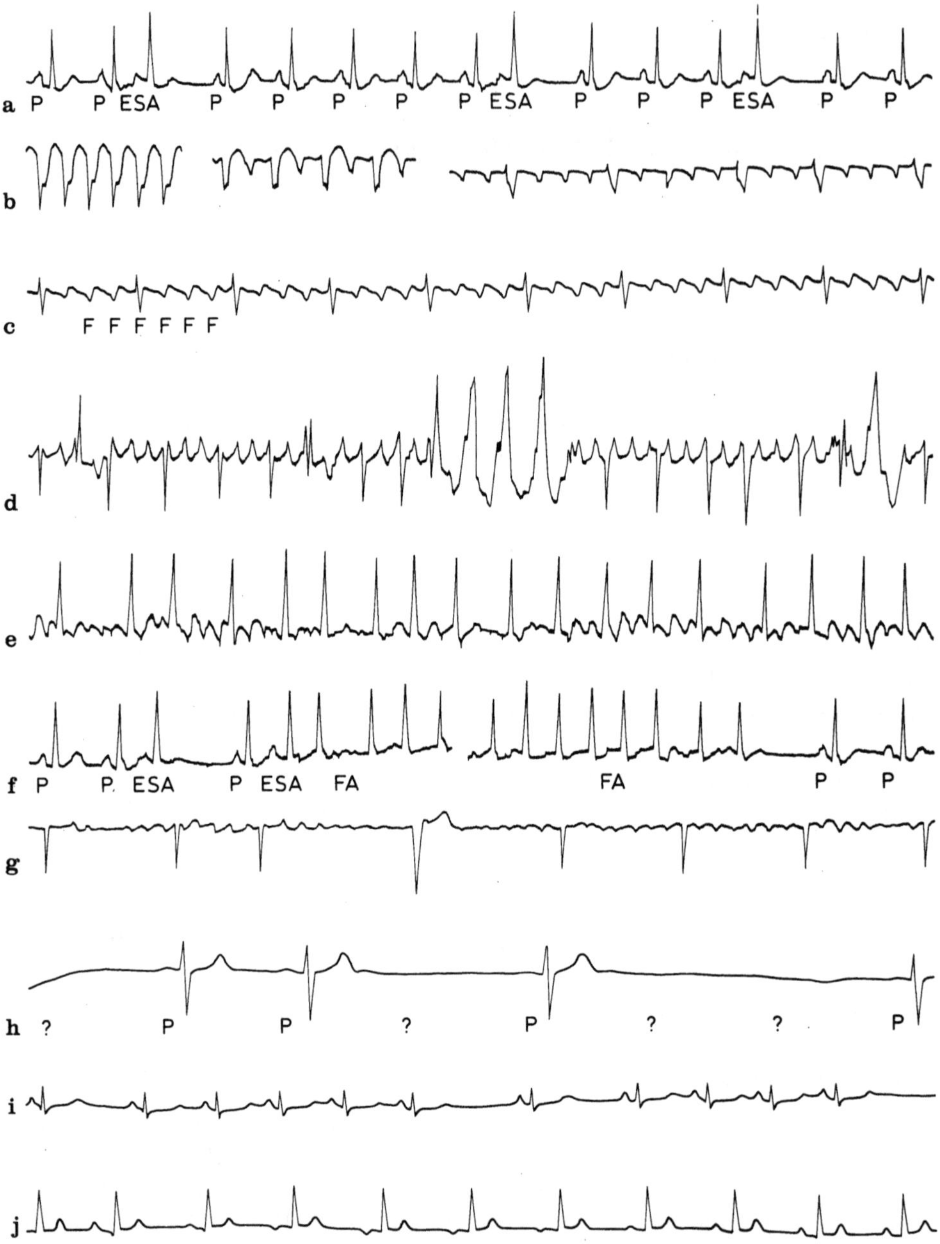

Fig. 1. Atrial arrhythmias. The various forms of arrhythmias occuring at the atrial level are displayed: atrial extrasystoles (**a**), typical flutter in lead 2 with various atrioventricular conduction ratio (**b, c**), atrial tachycardia with rapid positive P waves in V_1 (**d**), atrial fibrillation (**e**) with its initiation and termination (**f**) and an advanced degree of atrioventricular block (**g**), second-degree sinoatrial block (**h**) with Wenckebach periods (**i**), atrial wandering pacemaker (**j**). (ESA, atrial extrasystoles; F, flutter waves; FA, atrial fibrillation; P, sinus P waves; ?, dropped P waves

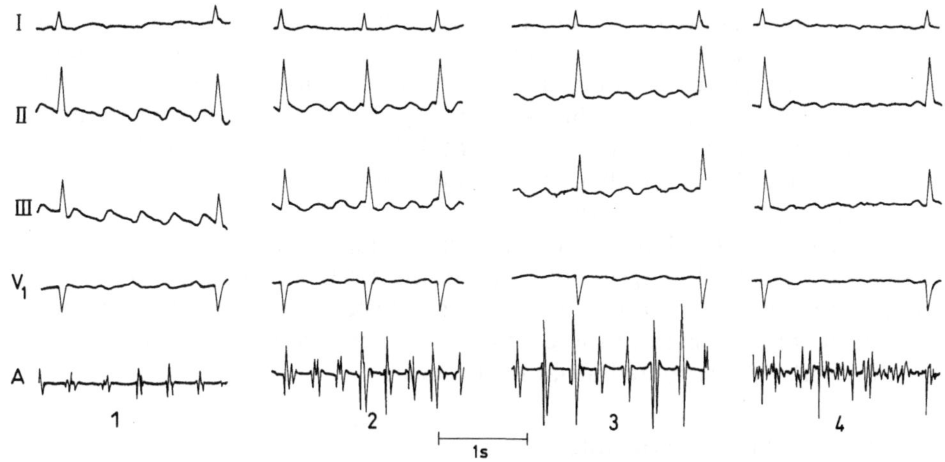

Fig. 2. Various aspects of typical and atypical atrial flutter and fibrillation in the same patient, studied with peripheral standard leads, precordial lead V_1, and intra-atrial bipolar recordings (A)

which can be attributed to a mechanism of automatic activity rather than re-entry, the amplitude of the electrical activity is usually more marked in lead V_1 (Fig. 1d) while in the peripheral leads the isoelectrical line can be defined between the atrial waves. In such cases the diagnostic problem may be further complicated by the mode of the ventricular response: often 2:1, rarely 1:1, and the best way to evidence the atrial electrical activity is to alter the conditions of the atrioventricular transmission by drugs or vagal manoeuvres. More recently, atrial re-entry has been individualized: it has the morphological characteristics of atrial tachycardias, but with a much slower rate of the order of 100-200/min. Frequently re-entry seems to occur in the area of the sinoatrial node, so that the P waves of the tachycardia are almost similar to sinus P waves and the term "sinoatrial re-entry" is then used.

In atrial fibrillation (Fig. 1e-g) the P waves are absent and the baseline consists of irregular waveforms which continuously change in shape, duration, amplitude and duration. If this applies to peripheral leads, in fact endocavitary recordings show that this may not be true when one considers local bipolar electrograms in which the activity can well be regular at a rate usually exceeding 300/min: the existence of intra-atrial conduction disturbances explains such discrepancies between the surface and endocavitary ECG pattern.

2. Atrial Bradycardias

The absence of P waves may be due to the sinus node standstill or to a complete sinoatrial block, and the distinction between these two possibilities can now be made on the basis of precise endocavitary recordings in the sinus node region. Frequently, however, a second-degree sinoatrial block can be diagnosed in the surface ECG on the basis of regularly dropped P waves (Fig. 1h).

An interesting variety of sinoatrial block with the existence of Wenckebach periods is reflected by the progressive shortening of the atrial cycle alternating with pauses which are not a multiple of the atrial cycle but define a 6/5, 5/4 or 4/3 sinoatrial second-degree block as in Fig. 1i. The combination of brady- and tachyarrhythmias in the same patient defines the "bradytachy syndrome", a term which is less used than "sick sinus syndrome" though it gives a more complete image of the arrhythmia.

The aspect of wandering atrial pacemaker (Fig. 1j) classically attributed to a progressive shift of the point of origin of the atrial activation, is in fact most often if not always due to the competition between the sinus activation and a subsidiary focus located in the coronary sinus region, thus resulting in more or less regularly distributed atrial fusion beats.

II. Junctional Arrhythmias

Invasive electrophysiology has flourished in the past 2 decades in the exploration of the AV junction, with the extensive use of programmed stimulation for tachycardias, and precise endocavitary recordings for bradycardias. It would be presumptuous and fruitless to try to summarize in a few lines the numerous studies dealing with junctional arrhythmias. A few general statements will be sufficient for the non-specialist in this field, and the specialist can easily skip them.

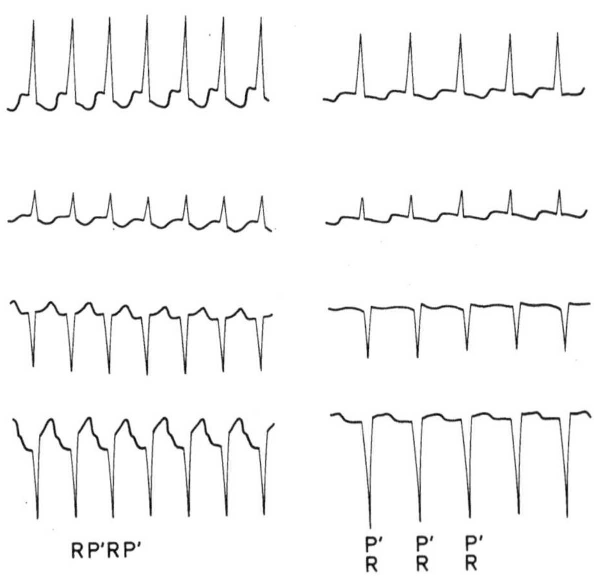

Fig. 3. Paroxysmal supraventricular tachycardias. Peripheral standard leads and the oesophageal recording (*bottom trace*) are shown in two cases of paroxysmal reciprocating junctional tachycardia. *Left panel*, retrograde P′ waves follow the R waves indicating the probable presence of a pre-excitation pathway as the retrograde limb of the circuit. *Right panel*, P′ and R waves are superimposed indicating that the circuit is intranodal

1. Junctional Tachycardias

Paroxysmal junctional tachycardias are all due to re-entry. Electrophysiological investigation permits precise location of the re-entry circuit, which may be restricted to the atrioventricular node or, more frequently, can involve both the normal AV nodal-His bundle conducting tract and accessory pathways: most often, a direct atrioventricular connection (Kent bundle) is inculpated but in other cases atrionodal and nodoventricular (Mahaim) fibres have been demonstrated. In practice the clinician should bear in mind that, although precise and sophisticated endocavitary studies may be required if surgical division is contemplated of a tiny, but uncomfortable or indeed life-threatening bypass tract, quite often (15 %) multiple, in most cases a correct analysis of surface tracings, sometimes combined with an oesophageal recording is sufficient to provide an accurate diagnosis.

By definition, a 1:1 supraventricular tachycardia is both junctional and re-entrant if no second-degree atrioventricular block can be obtained without also stopping the tachycardia. In this case the classical term of paroxysmal atrial tachycardia is inappropriate, because depressing the atrio-ventricular nodal conduction by any manoeuvre would not influence the atrial activity if the tachycardia is not re-entrant but located in the atrium. Figure 3 gives the two main features which, in the setting of a junctional tachycardia, make it possible to differentiate intranodal circuits from circuits involving extranodal pathways: P and R waves are superimposed in the former, whereas P waves follow the T waves in the latter. The ECG pattern displayed in Fig. 4, including P' waves that precede the QRS and are negative not only in leads II, III and VF but also in the left precordial leads, is highly characteristic of the special form of permanent reciprocating tachycardia which includes a slow-conducting ret-

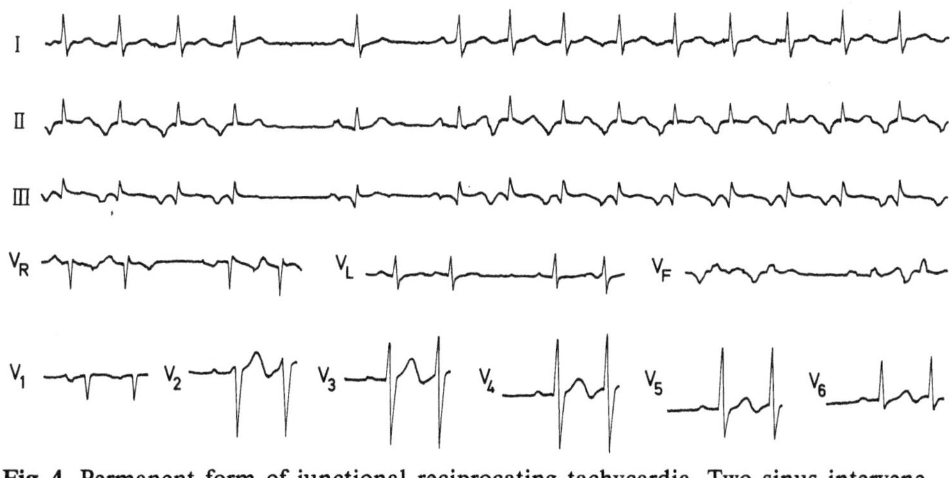

Fig. 4. Permanent form of junctional reciprocating tachycardia. Two sinus intervene between the spontaneous termination and reinitiation of the reciprocating tachycardia. Typical retrograde P′ waves are visible in the inferior peripheral leads and in the left precordial leads, with an R-P′ interval which is longer than the P′-R interval, thus indicating the presence of a slowly conducting retrograde pathway

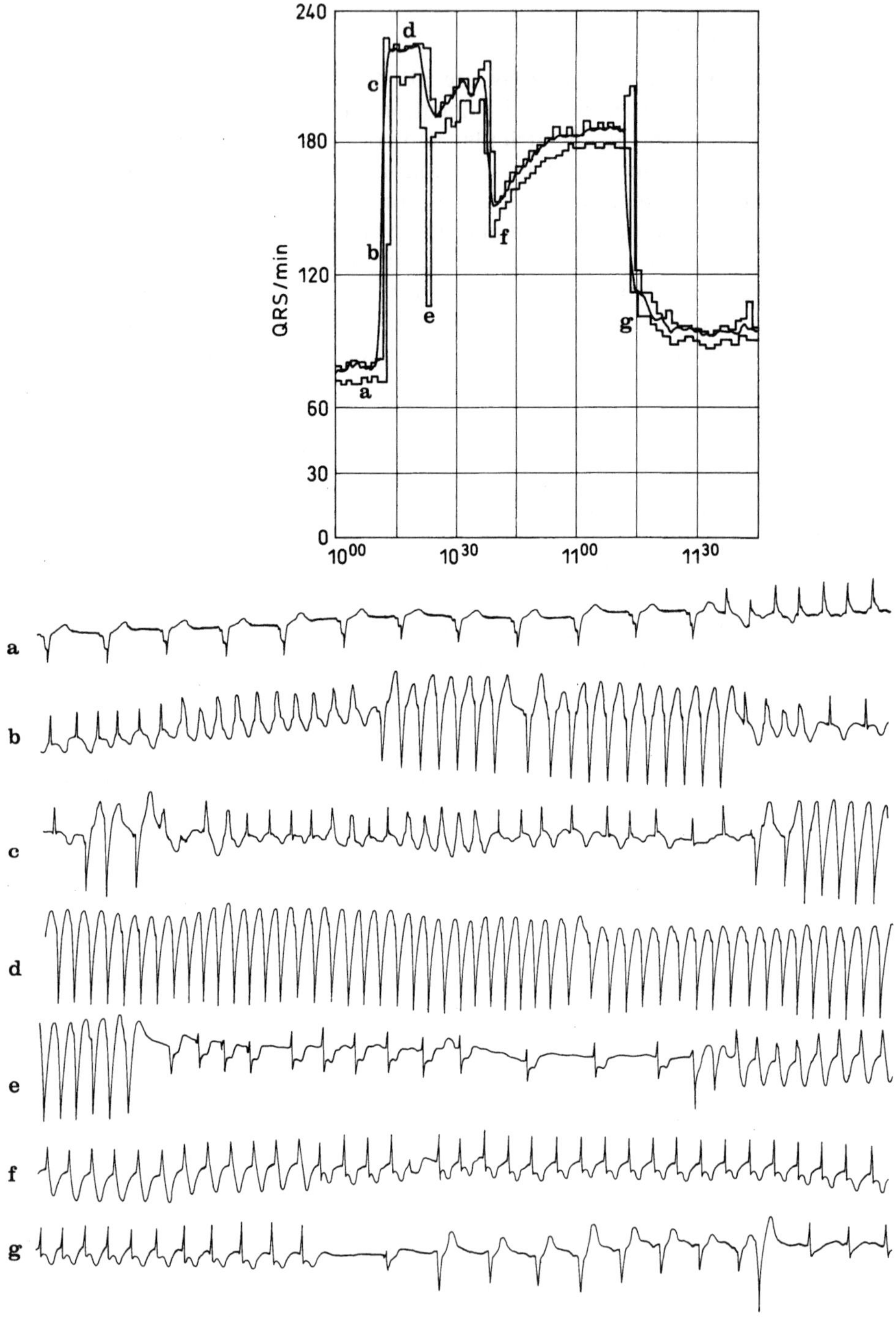

Fig. 5. Multiple patterns of supraventricular tachycardias. During a 2-h Holter recording, multiple aspects of slower or faster tachycardia are recorded in this patient in whom various arrhythmias are alternating in the setting of a pre-excitation syndrome. All possible forms of atrial tachycardias and junctional reciprocating tachycardia with or without aberrant conduction were observed. Only endocavitary recordings would allow a precise diagnosis in such complex cases

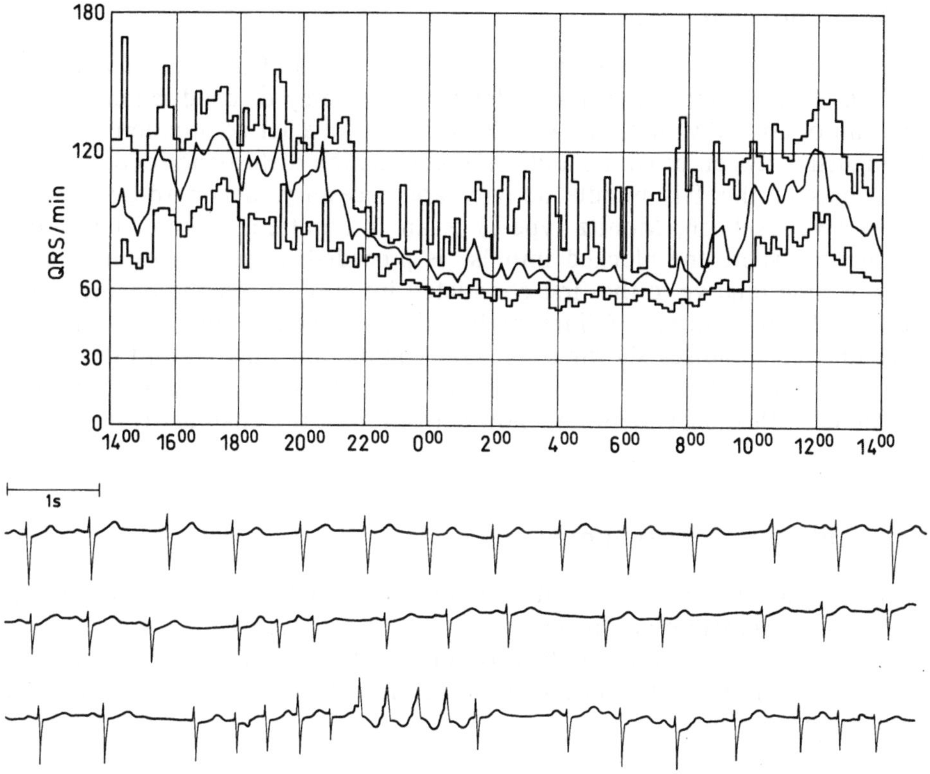

Fig. 6. His bundle tachycardia. The sinus rhythm is often replaced by a regular, rather slow junctional rhythm which may suddenly accelerate; then the QRS complexes may widen with a right bundle branch block pattern related to an aberrant intraventricular conduction. The *upper diagram* shows the trends of maximal, mean and minimal heart rate during the 24-h recording in this 12-year-old patient

rograde accessory accessory pathway. All these different forms are amenable to surgery, and some of them can now even be treated by endocavitary fulguration. These rather simple interpretations of classical surface electrocardiography contrast with the complexity of some other manifestations requiring more sophisticated analysis. For example, Fig. 5 displays in a single case the combination of a patent Wolff-Parkinson-White syndrome in sinus rhythm, and various aspects of automatic atrial or re-entrant junctional tachycardia with maximal pre-excitation pattern, or transmission through the normal AV conducting pathway with or without right or left functional bundle branch block.

Cases of true atrial automatic rhythms are, in fact rare when compared with the frequency of re-entrant arrhythmias in the atrioventricular junction. However, they do exist, most often as semiactive rhythms favoured by some degree of sinus bradycardia. Still, tachyarrhythmias related to the presence of active junctional foci may be observed in children, as in the case of Fig. 6. Though rare, they are sometimes poorly tolerated and difficult to control with drugs.

2. Atrioventricular Block

Many studies have categorized various disturbances of atrioventricular conduction. First-degree AV block is defined as a delay of conduction beyond the limit of normal values, but with conservation of a 1:1 conduction ratio. To distinguish between proximal and distal conduction disturbance endocavitary recordings of the His bundle activity are necessary. Second-degree block type I (Wenckebach, Mobitz type I) is an intermittent failure of impulse conduction where the blocked impulse is preceded by prolongation of conduction time relative to the first conducted impulse. In second-degree block type II, the intermittent failure of impulse propagation follows a number of conducted impulses showing constant conduction times. Third-degree or complete block involves complete failure of impulse propagation, usually associated with independent and slow activation of the chamber distal to the site of block.

III. Ventricular Arrhythmias

Ventricular arrhythmias are usually classified, as atrial arrhythmias, according to their three possible components of extrasystole, tachycardia and fibrillation. However, these three simple attributes cannot provide a description adequate either for an understanding of the nature of the arrhythmia or as a basis for its management.

1. Ventricular Extrasystoles

Ventricular extrasystoles are defined according to their prematurity with respect to the basic cycle length, and to their width, which reveals their ventricular origin. The significance of these two fundamentally different types of information is often neglected. The morphology of right or left bundle branch block pattern can indicate a left or right ventricular origin (ROSENBAUM 1969) related to the causal disease and has both prognostic and therapeutic implications. Interpretation is more difficult if the causal disease is a myocardial infarction (JOSEPHSON et al. 1981, COUMEL et al. 1984) although it is still important to localize the arrhythmia origin because of therapeutic implications. Fixed or variable coupling intervals are probably indicative of different mechanisms, but the classical opposition between re-entry and parasystole appears oversimplified now that it is demonstrated that parasystole may be electrotonically modulated in such a way that it mimics the re-entry criteria (JALIFE and MOE 1979). Similarly, in the past it has been customary to attribute a potentially dangerous prognostic implication to short coupling intervals only. Although this is sometimes a reality, it is the variability of the coupling interval as opposed to its relative fixity which probably has the more general prognostic significance.

2. Ventricular Tachyarrhythmias

Just considering Fig. 7 gives an idea of the wide variety of phenomena covered by the term "ventricular tachyarrhythmia", and at the same time explains why descriptions can often be used in a qualitative sense only. When uniform, the morphology is important as it can indicate the location of the arrhythmogenic area. Two or more monomorphic ventricular tachycardias may be present in the same patient and the term "pleiomorphic" is sometimes used in this situation: it is by no means equivalent to the term polymorphic, which supposes the interplay between arrhythmias of different origins or at least a changing propagation of the ventricular activity. It may become difficult in some cases to differentiate between polymorphic ventricular tachycardias and the now widely accepted entity of torsade de pointes in which the QRS axis progressively changes so that in some leads the ventricular complex appears to twist around the isoelectric line. Initially, the description applied to specific etiological factors like bradycardia, various drug intoxications and the congenital long QT syndrome, all of them having in common the presence of a long coupling interval of the initial beat of the tachycardia. It is now clear, however, that arrhythmias with features typical of torsade de pointes are also observed in other situations with a strikingly *short* coupling interval of less than

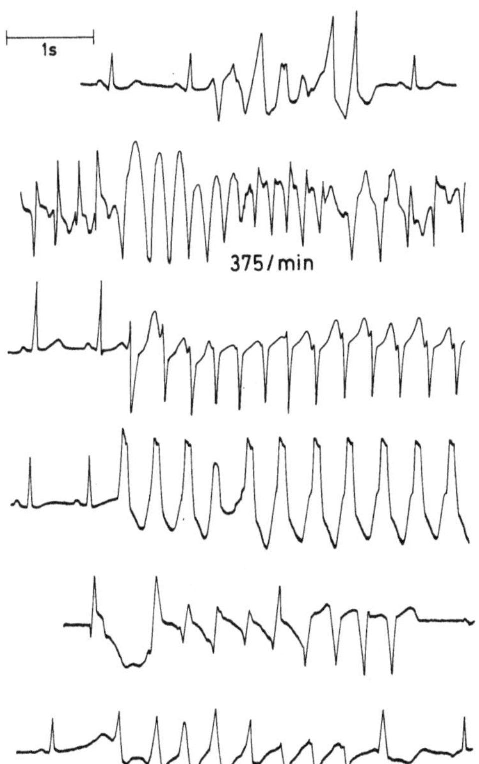

Fig. 7. Ventricular tachyarrhythmias. The six tracings are recorded in different patients; they may all be labelled as ventricular tachycardia though occurring in completely different contexts. *From top to bottom*: short-coupled torsade de pointes in a 40-year-old patient with no apparent cardiac pathology; adrenergic ventricular in a child; slightly polymorphic and irregular ventricular tachycardia; monomorphic and slower ventricular tachycardia; and two cases of torsade de pointes with a long coupling interval in the setting of an atrioventricular block and a congenital long QT syndrome

300 ms (COUMEL et al. 1985 a). In any case, it remains important to differentiate this ECG pattern clearly from that of ventricular fibrillation, as the latter implies a totally desynchronized ventricular electrical activity which cannot terminate spontaneously. Finally, the term of ventricular flutter should be limited to the relatively rare condition in which the tracing is still organized, with a rapid and regular activity of the order of 250/min or more, but in which components of QRS and T cannot be identified or separated.

Ventricular tachyarrhythmias cannot be characterized by consideration of their morphology alone. It is essential to specify not only the coupling interval of the initial beat but also the regularity of the tachycardia and its duration: the terms "sustained" and "non-sustained" ventricular tachycardia provide a convenient distinction. However, it should be kept in mind that these terms separate only different manifestations of basically identical arrhythmias; either may be observed in the same patients, in the same diseases.

B. Other Current Classifications of Arrhythmias

As a consequence of the devolopment of various techniques of investigation, much new information about human cardiac arrhythmias has been obtained concerning their possible mechanism, prognosis, relationships with the autonomic nervous system or reaction to drugs.

I. Invasive Electrophysiology and the Mechanisms of Arrhythmias

In the 1960s, two major electrical tools once reserved for animal experiments became available to clinicians: cardiac defibrillation by external countershock, and stimulation. Though very different in essence, they were complementary in that the former, by eliminating the ever-present risk of lethal electrically induced ventricular fibrillation, made possible the routine use of programmed stimulation in humans. This new discipline of clinical electrophysiology allowed impressive progress to be made in the comprehension of arrhythmias as it became possible to reproduce at will a variety of paroxysmal tachycardias. The supraventricular arrhythmias were first explored, and it happened that the type most amenable to the study were junctional tachycardias: their re-entrant mechanism was thus proved and characterized. Progressively the technique was extended to the exploration of arrhythmias at atrial and ventricular levels, with the implication that whenever an arrhythmia could be initiated and terminated by artificial stimulation it was related to the existence of a pathway for reentry, whereas non-inducible arrhythmias were probably induced by other mechanisms, such as ectopic pacemakers. We now realize, however, that such conclusions represented an oversimplification.

At the atrial level, the majority of arrhythmias cannot be induced and terminated at will; albeit one of these criteria can be occasionally found in many of them. Atrial fibrillation can be induced by programmed stimulation, but this has not been proved reliable in detecting patients at risk of this arrhythmia occurring spontaneously: the induction of "non-clinical" atrial arrhythmias or the non- inducibility of clinical arrhythmias are frequent. In other words, the specificity and the sensitivity of programmed stimulation is poor at this level. Typical atrial flutter can be converted, which tends to confirm its long-supposed re-entrant mechanism, but usually it cannot be re-induced at will. Atrial tachycardias can rarely be manipulated by programmed stimulation, and if some of them were labelled "atrial re-entry" it was precisely because of their inducibility rather than the undoubted demonstration of the presence of an anatomical circuit. On the other hand, the induction of "functional" anatomical re-entry in the absence of a specific reentrant circuit has been demonstrated experimentally. Furthermore, other possible electrophysiological mechanisms like triggered activity can perfectly mimic re-entry in terms of inducibility, so that inducibility alone cannot be used as a criterion to classify reliably arrhythmias in humans.

At the ventricular level the debate remains largely open, to the point that to avoid perpetual controversy about mechanisms, it becomes preferable to speak in terms of inducible and non-inducible, rather than re-entrant and non-reentrant, tachycardias. Inducibility or non-inducibility may then have different significance in relation to different disease states responsible for the arrhythmia. Almost all postmyocardial ventricular tachycardias can be induced and terminated, and precise studies have now determined the stimulation protocols that are most reliable (Brugada and Wellens 1984) in terms of clinical significance of the provoked arrhythmias: regular, monomorphic ventricular tachycardias (excluding ventricular flutter) induced by one or two premature stimulations may be considered to have a consistent clinical significance. However, interpretation of invasive electrophysiological study is more difficult in other diseases like cardiomyopathies, mitral valve prolapse and arrhythmogenic right ventricular dysplasia, and even more so in the absence of any apparent cardiac pathology. In this situation any attempt to classify human arrhythmias according to a hypothetic mechanism related to their inducibility fails altogether. Triggered activity (Hoffman and Rosen 1981) and reentry, though entirely different as arrhythmogenic mechanisms, meet almost the same criteria in respect of inducibility. Conversely, in the setting of the same disease, with an identical pattern of ventricular tachycardia, sometimes the arrhythmia is inducible sometimes not, although the arrhythmogenic mechanism is the same. Thus inducibility cannot be used as an indicator of the mechanism of arrhythmogenesis. In such circumstances the more important question is to determine what is the basis for the differences in responses to stimulations.

II. Classification of Arrhythmias and Holter Monitoring Technique

Dynamic electrocardiography is utilized with a philosophy which completely differs from that of invasive electrophysiological studies. By definition, only spontaneous arrhythmias are considered, and they are classified in terms of isolated premature beats, and repetitive activity ranging from couplets to sustained tachycardias. Managing a considerable amount of data makes it tempting to rely on their *quantitative* evaluation to assess the prognostic value of the events or the significance of the effects of drugs. However, such an approach totally ignores the real meaning of the arrhythmic events. In contrast the Lown's classes are mainly *qualitative*, based on the criteria of repetitive activity, morphology and coupling interval of arrhythmias: though they reflect the daily clinical experience, they are somewhat arbitrary and not quantitative enough for satisfactory evaluation. For that reason many authors tried to create a balance between the two tendencies, thus distinguishing "simple" and "complex" arrhythmias according to mixed qualitative and quantitative parameters, and improving the correspondence with the findings of invasive methods. Recently for instance it was proposed (Gradman et al. 1985) to consider the ratio between premature beats that were isolated or were followed by repetitive activity. This seems very relevant as there are indeed reasons to think that different mechanisms operate in the two cases (COUMEL et al. 1985b). We do know that at least two components are necessarily implicated in any arrhythmia: an initiating factor (schematically the isolated premature beat) and a substrate (responsible for the repetitive activity, whether or not it is inducible).

C. Arrhythmia Environment: Other Possible Classifications

A common characteristic of the preceding purely electrocardiographic or electrophysiological classifications is to consider only the arrhythmia itself, without taking into account the conditions in which it arises or its behaviour in relation to the basic heart rate. The rate dependence of arrhythmias was first recognized by LANGENDORF et al. (1955) when they described the "rule of bigeminy", and the influence of the autonomic nervous system on arrhythmias is clear not only from clinical experience, but is also involved in practically all experimental models of arrhythmias. The difficulty is that, in the clinical conditions, since heart rate is itself controlled by the autonomic system, it is difficult to separate the influence of autonomic activity per se from effects which are secondary to changes in heart rate. Nevertheless the two influences can be differentiated to some extent because of differences in timing of response, and a few examples of such a differentiation can be given.

I. Arrhythmias and the Autonomic Nervous System

Considering the heart rate during the minutes or even tens of minutes preceding paroxysmal tachyarrhythmias provides relevant indications concerning their mechanism. For instance, paroxysmal atrial fibrillation may systematically occur after a progressive heart rate increase in some patients, whereas a consistent deceleration precedes the arrhythmia in others (COUMEL 1984): this strongly suggests that sympathetic as well as vagally dependent atrial fibrillations are clinical realities, thus coinciding with the two classical methods of obtaining them experimentally. This example shows that a well-defined arrhythmia like atrial fibrillation does not constitute a single entity and that classifying atrial fibrillations indiscriminately may well be misleading, parti-

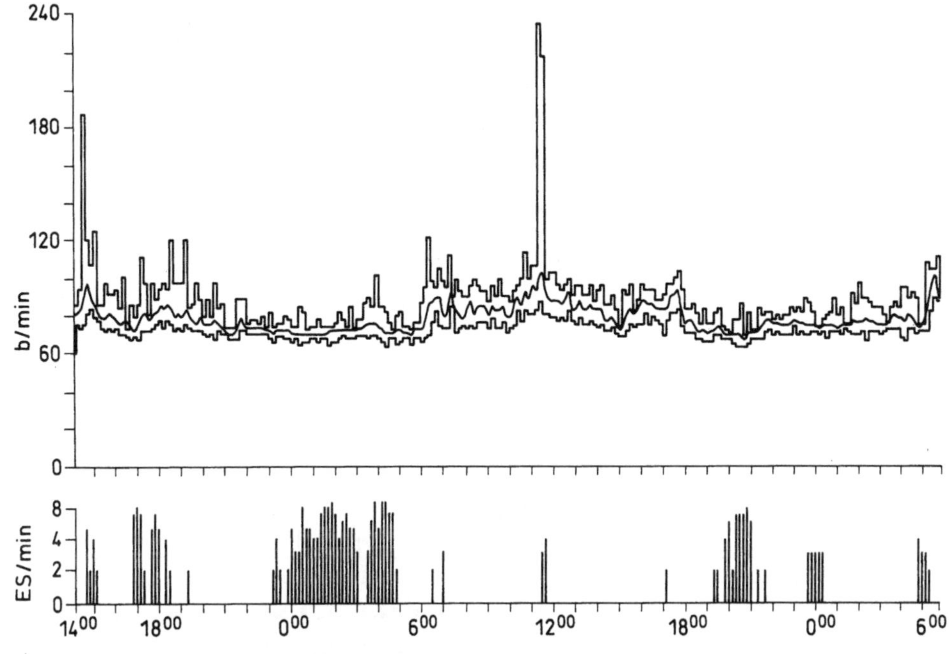

Fig. 8. Ventricular arrhythmias and heart frequency. In this 40-h tracing, two attacks of ventricular tachycardia were recorded at about 15.00 on day 1, and at 11.30 on day 2; both episodes coincide with the progressive heart rate increase during the hours and minutes which precede them, as indicated by careful examination of the mean heart rate trend (*middle curve*). In other words, the ventricular tachycardia occurs only in the context of an increased sympathetic drive. In contrast, the ventricular extrasystoles (*bottom diagram*) occur preferably in the setting of a relatively slow heart rate

cularly if the therapeutic consequences are considered: the effect of beta-blocking therapy is beneficial or deleterious according to the mechanism of the arrhythmia. Many analogous examples of this problem can be found in other arrhythmias and they are by no means surprising: actually, it is difficult to imagine that an arrhythmia has no relationship whatsoever with the autonomic nervous system. The example in Fig. 8 relates to a postmyocardial infarction ventricular tachycardia, which occurred in this patient only in the context of an elevated mean heart rate, whereas the isolated premature beats became frequent only at rest. In the example in Fig. 9, opposite behaviour of the ventricular premature beat was observed: the correlation between the number of premature beats per minute and the basic heart rate was positive. Thus in these two patients extrasystoles can only be assessed in the context of heart rate, which is relevant both to prognostic significance and treatment.

II. Rate Dependence of Arrhythmias

Figure 10 deals with a trivial case of bigeminal ventricular extrasystoles. It illustrates the reality of their rate dependence and the possibilities of a very pre-

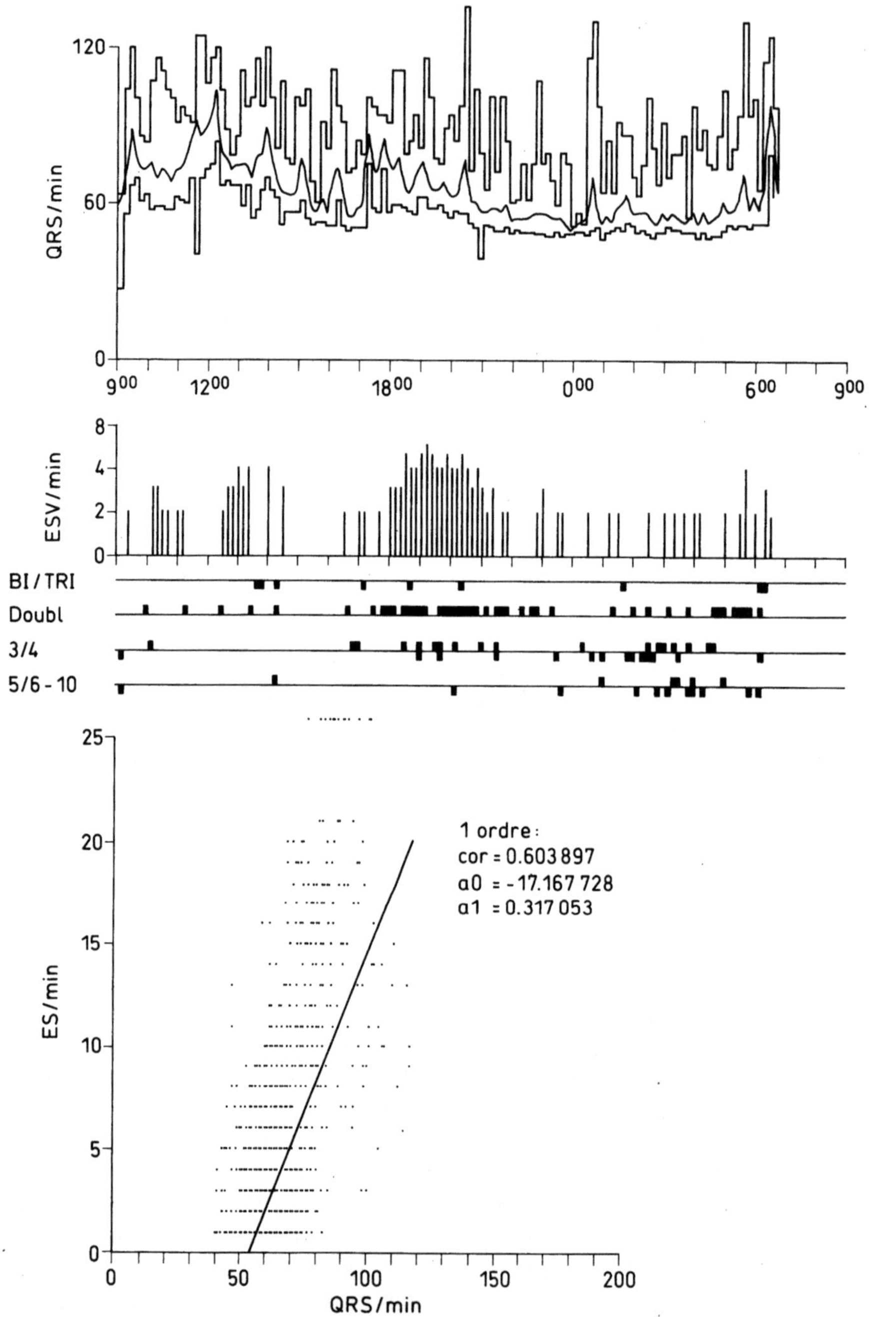

Fig. 9. Extrasystolic rate and cardiac frequency. A 24-h recording including numerous ventricular extrasystoles is analyzed in terms of (*from top to bottom*): maximal, mean and minimal heart rate (QRS/min), extrasystolic rate (ESV/min), presence of bi and trigeminy (BI/TRI), pairs (*Doubl*), salvos three of to ten extrasystoles (3/4, 5/6–10). The *lower diagram* correlates the number of premature beats/min with the heart rate/min: the first-order regression line clearly indicates that the frequency of the premature beats increases with the sympathetic drive

cise electrophysiological analysis by Holter monitoring, a technique which is not usually utilized for such an approach. The patient's arrhythmia, during the 3-h recording extending from 2.00 a. m. to 5.00 a. m., was exclusively formed of unifocal ventricular premature beats, always isolated and often grouped into bigeminy. In Fig. 10, panels A to C, the extrasystoles (arrows in the diagrams, cycle labelled "0" in the bottom line) are collected, including the nine cycles that precede (−9 to −1) and follow (+1 to +9) them. The analysis was performed by the ATREC II system (Coumel et al. 1985b) in three different ways. In A, extrasystoles not preceded by any other event during at least nine cycles were collected: they formed a total of 262. In B all extrasystoles were collected without any restriction, forming a total of 954. In C only extrasystoles not followed by any other event during the nine next cycles were collected: unsurprisingly their number was indentical to that in panel A. The mean value ± SD is given for each population of cycles (in milliseconds), and in the diagrams the mean value ± SEM of the cycles is shown.

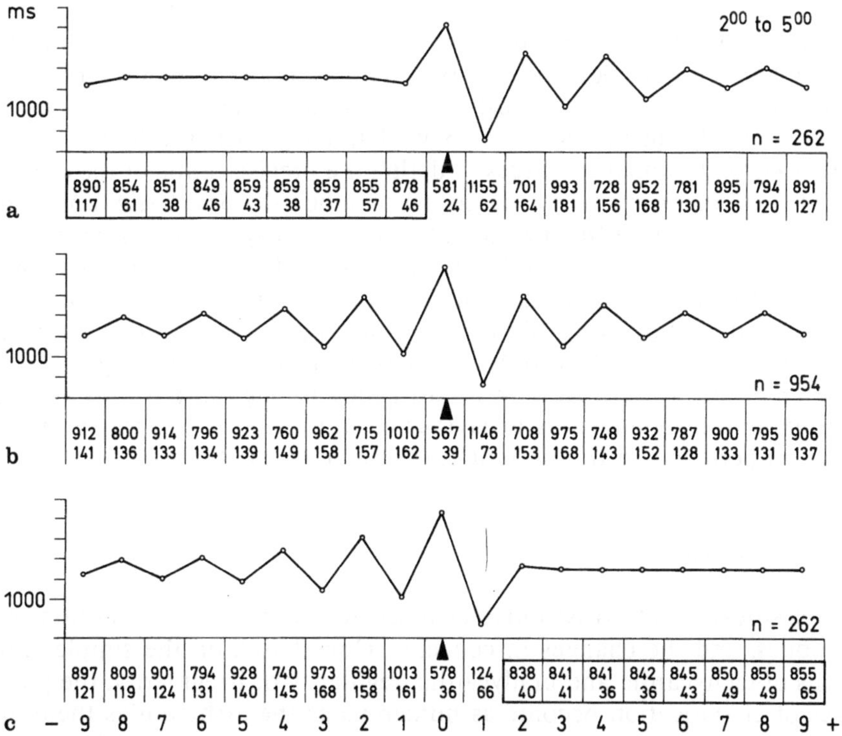

Fig. 10. Study of the initial and terminal extrasystoles. In a 3-h ambulatory recording, the nine cardiac cycles surrounding the extrasystoles are collected in different ways according to the absence of events before **(a)** or after **(c)**, or without taking into account this factor **(b)**. The main findings (see text) are that the sinus rate is slower before than after the extrasystoles, and the coupling interval (cycle No. 0) is longer for the first **(a)** and the last beat **(c)** of a series of bigeminy than for the global population. Cycle durations are indicated in milliseconds (mean ± SD in the tables, mean ± SEM in the diagrams), and the number of events (*n*) is given

The 262 extrasystoles collected in A are either isolated or they are the first of a sequence of bigeminy continuing for two to six events. Beats −9 to −1 are all of sinus origin. The average duration is quite stable for cycles −8 to −2 (about 850 ms). It is longer for cycle −9 which includes some postextrasystolic pauses (presence of an extrasystole forming cycle −10). Sinus cycle −1 is longer than the preceding ones ($P < 0.01$, analysis of variance) and apparently this phenomenon leads to the extrasystole occurrence. The value of this cycle (878 ± 46 ms) is important to consider: it means that 95 % of the extrasystoles are immediately preceded by a sinus cycle length of between 786 and 972 ms (mean ± 2 SD) and 99 % by a sinus cycle ranging from 740 to 1018 ms (mean ± 3 SD). In contrast, the sinus cycles which follow the last extrasystole of a series of bigeminy (panel C) are significantly shorter: about 840 ± 40 ms.

These observations demonstrate that, as the sinus rate decreases, it "enters" a frequency zone that favours extrasystoles. The average upper limit of this zone has a mean value situated somewhere between the mean cycle length of the −2 and −1 sinus beats in panel A, i.e. 855 and 878 ms respectively. In contrast with panel A, panel C deals with the termination of the extrasystolic phenomenon, bigeminal or not and verifies the rule of bigeminy that implies the disappearance of extrasystoles as the rate accelerates. As a matter of fact, after the +1 postextrasystolic sinus pause, the +2 cycle of 838 ms is clearly shorter than the above-defined "upper threshold".

It is beyond human resources to define the precise limits of these electrophysiological mechanisms by measuring on a beat-by-beat basis thousands of RR intervals of surface tracings at a 25-mm/s paper speed. A computer can do so, however, with a suitable statistical programme, to an accuracy of a few milliseconds from recording tapes running at 1 or 2 mm/s. Even invasive electrophysiological studies using 100-mm/s, paper speed cannot attain such precision and cannot be used to study spontaneous behaviour. An appropriate use of dynamic electrocardiography is worth developing, as the only means of scrutinizing the spontaneous behaviour of various types of arrhythmias over prolonged periods and in different environmental conditions.

D. Conclusion

The classification of arrhythmias changes according to the evolution of the techniques, and the changes in concepts. Once based on the simple clinical notions of tachycardia and bradycardia, regularity or irregularity, the possible modes of classification become as numerous as the arrhythmias themselves. This means that our vocabulary is progressively becoming too limited to avoid speaking of quite different things using a single word. We must be conscious of the practical importance of the problem which is not restricted to theoretical concepts. An investigator cannot reasonably state that this or that arrhythmia has a good or poor prognosis, or a good or poor sensitivity to a particular category of drugs, if the emphasis has not been put on defining precisely the particular arrhythmia he was dealing with in a particular study. This probably explains the frequent discrepancies between studies which are apparently devoted to similar problems and provide different conclusions. "Classification is the beginning of science": much further progress is required in the field of human cardiac arrhythmias.

References

Brugada P, Wellens HJJ (1984) Standard diagnostic programmed electrical stimulation protocols in patients with paroxysmal recurrent tachycardias. Pace 7:1121–1128

Coumel P (1984) Atrial fibrillation. In: Surawicz B, Reddy CP, Prystowsky EN (eds) Tachycardias Nijoff, The Hague, pp 231–244

Coumel P, Leclercq JF, Attuel P, Maisonblanche P (1984a) The QRS morphology in post-myocardial infarction ventricular tachycardia. A study of 100 tracings compared with 70 cases of idiopathic ventricular tachycardia. Eur Heart J 5:792–805

Coumel P, Leclercq JF, Maisonblanche P, Attuel P, Cauchemez B (1985b) Computerized analysis of dynamic electrocardiograms: a tool for comprehensive electrophysiology. A description of the ATREC II system. Clin Prog Electrophysiol Pacing 3:181–201

Coumel P, Leclerq JF, Lucet V (1985a) Possible mechanisms of the arrhythmias in the long QT syndrome. Eur Heart J 6D:115–129

Coumel P, Leclerq JF, Slama R (1985c) Repetitive monomorphic idiopathic ventricular tachycardia. In: Zipes DP, Jalife J (eds) Cardiac electrophysiology and arrhythmias. Grune and Stratton, New Jork, pp 457–468

Gradman AH, Batsford WP, Rieur EC, Leon L, van Zetta AM (1985) Ambulatory electrocardiographic correlates of ventricular inducibility during programmed electrical stimulation. J Am Coll Cardiol 5:1087–1093

Hoffman BF, Rosen MR (1981) Cellular mechanisms for cardiac arrhythmias. Circ Res 49:1–15

Jalife J, Moe GK (1979) A biological model of parasystole. Am J Cardiol 43:761–772

Josephson ME, Horowitz LN, Waxman HL, Cain ME, Spielman SR, Greenspan AM, Marchlinski FE, Ezri MD (1981) Sustained ventricular tachycardia: role of the 12-lead electrocardiogram in localizing site of origin. Circulation 64:257–272

Langendorf R, Pick A, Winternitz M (1955) I. Appearance of ectopic beats dependent upon length of the ventricular cycle, the "rule of bigeminy" Circulation 11:422–430

Rosenbaum MB (1969) Classification of ventricular extrasystoles according to form. J Electrocardiol 2:289–298

WHO/ISFC Task Force (1978) Definition of terms related to cardiac rhythm. Eur J Cardiol 8:127–144

WHO/ISFC Task Force (1979) Classification of cardiac arrhythmias and conduction disturbances. Am Heart J 98:263–267 most important problems and provide different classifications. [illegible]

References

[illegible]

CHAPTER 5

Successes and Limitations of Antiarrhythmic Drug Therapy

D. G. JULIAN and J. C. COWAN

A. Introduction

This book testifies to the growing fund of knowledge, both at the basic scientific and at the clinical level, concerning the electrophysiological effects of antiarrhythmic drugs. A large number of antiarrhythmic agents are available for clinical use. All are of proven efficacy. Yet the question remains: is a particular agent going to benefit a particular patient?

In this chapter indications for antiarrhythmic therapy are discussed together with the possible disadvantages and limitations of treatment.

B. Haemodynamic Effect of Arrhythmias

Disturbances of rhythm and conduction affect cardiac function in a number of different ways.

I. High Rate

Healthy hearts can tolerate heart rates of 150–200 beats/min for long periods without adverse consequences. However, tachycardia imposes added burdens on the heart. It is associated with increased contractility and a substantial increase in myocardial oxygen requirement. Because diastole is shortened, there is less time available for ventricular filling and for coronary blood flow. In the compromised heart, this can lead rapidly to cardiac failure and to myocardial ischaemia. Soon after the onset of tachycardia, the immediate fall in flow and pressure is reversed by reflex responses. This reflex adaptation is dependent on an intact sympathetic nervous system (NAKANO and MCGLOY 1970) and is therefore attenuated or absent in patients on beta-blockers.

II. Loss of Atrial Transport Function

In the normal heart, some 20 % of ventricular filling is achieved by atrial contraction, the exact figure depending on the PR interval (LEINBACH et al. 1969). If atrial contraction is out of phase (as in atrioventricular dissociation) or ineffective (as in atrial fibrillation) there is a fall in cardiac output, especially on exercise.

III. Abnormal Sequence of Ventricular Activation

It has been found that stroke volume may be decreased by 10% as a consequence of an abnormal pattern of ventricular contraction, as may be seen in ventricular pacing (SOYEUR 1986).

C. Symptoms and Complications of Arrhythmias

It is not surprising that haemodynamically important arrhythmias lead to dizziness, dyspnoea, chest pain and syncope, but many patients with arrhythmias never experience symptoms or do so only occasionally. Why this should be is by no means clear. It is remarkable that so few patients complain of symptoms which can be attributed to arrhythmias during the course of acute myocardial infarction, when such arrhythmias are almost invariable. Likewise, Holter monitoring has revealed that patients may have paroxysmal tachycardia at rates of up to 200/min without being aware of it. Some patients, on the other hand, may be uncomfortably aware of their heart when there is only a slight sinus tachycardia.

I. Palpitation

An uncomfortable awareness of the heart's action is a common complaint and may take many different forms. With ectopic beats, the patient may be conscious of an occasional thump in the chest which is associated with the strong beat which follows a postectopic pause. The latter phemonenon may give rise to a feeling that "the heart has stopped". Paroxysmal tachycardias give rise to a sensation of the heart racing, usually with a very abrupt onset, particularly if the episode is supraventricular in origin. Patients with atrial fibrillation may be aware of the fast and irregular action of the heart, especially when they first experience the arrhythmia, but they often cease to note this when the disorder becomes established.

II. Dyspnoea

Dyspnoea is likely to occur when a haemodynamically significant arrhythmia complicates heart disease, especially when the patient is close to or in heart failure already. Even in the absence of heart disease, prolonged episodes of paroxysmal tachycardia may give rise to breathlessness.

III. Dizziness and Syncope

If the arrhythmia causes a fall in cardiac output and blood pressure, dizziness and syncope may occur. This may happen virtually at the onset of the arrhythmia, particularly if the heart rate is very fast, but adaptations may take place which correct the blood pressure and restore cerebral blood flow. Syncope is especially prone to occur with ventricular tachycardia at high rates and in self-terminating episodes of torsade de pointes and ventricular fibrillation.

IV. Chest Pain

Angina is most likely to complicate arrhythmias when they develop in conditions in which the coronary blood supply is already inadequate, such as coronary atherosclerosis and aortic stenosis. However, even individuals with otherwise normal hearts may experience chest tightness during paroxysmal tachycardia, if this is prolonged.

V. Pulmonary and Systemic Embolism

Embolism as a complication of arrhythmias is virtually confined to atrial fibrillation. Thrombosis in the atria develops as a result of stasis, and is particularly common in mitral stenosis. It does, however, occur in other contexts such as ischaemic heart disease and the sick sinus syndrome.

D. Arrhythmias in the Normal Population

The widespread availability of ambulatory electrocardiographic monitoring has led to the realization that arrhythmias are a common occurrence, even in normal, asymptomatic individuals. Rhythm disturbances have been shown to occur in 93% of healthy, asymptomatic middle-aged men (HINKLE et al. 1969). The mere occurrence of an arrhythmia on a 24-h tape cannot therefore be said to be abnormal. Rather, both qualitative and quantitative judgement needs to be exercised concerning the type and frequency of the arrhythmia. Limits of normality for long-term ECG recording have been suggested (BJERREGAARD 1983).

The decision that a particular individual's 24-h tape is abnormal is not necessarily an indication for treatment. If a patient is asymptomatic and has no other manifestations of heart disease, treatment is seldom, if ever, indicated.

E. Indications for Antiarrhythmic Therapy

Which arrhythmias, then, should be treated? The advantages of treatment need to be weighed against the disadvantages. First, we will consider the various clinical situations that may necessitate treatment:

I. Symptoms as an Indication for Treatment

The most common reason for treating the relatively benign arrhythmias, such as supraventricular tachycardia and ventricular ectopic beats, is the complaint of palpitation. Very often with reassurance and, if necessary, anxiolytic drugs, antiarrhythmic drugs are not required. Before giving them, very serious consideration must be given to their potential side effects and risks.

II. Termination of Tachycardias

There may be an urgent need to terminate a tachycardia, if the rate is fast and, more particularly, if it is occurring in a patient with serious organic heart disease. Intravenous drug therapy (e. g. lignocaine in ventricular tachycardia) may be the easiest to administer quickly, although there is an important role for DC shock therapy in this context. When the situation is less urgent, it may be reasonable, on occasion, to delay therapy in the expectation that the arrhythmia will be self-terminating, but in general it is necessary to give drugs orally or, occasionally, to utilize pacing to terminate the episode.

III. Prevention of Tachycardias

Much antiarrhythmic therapy is directed at the prevention of arrhythmias, either because of their previous occurrence or because the clinical situation suggests they are likely to develop. An example of the former is the prevention of supraventricular tachycardia in the patient with recurrent episodes of that condition. An example of the latter is the use of antiarrhythmic drugs on a routine basis in an attempt to prevent ventricular fibrillation in the context of acute myocardial infarction.

F. Side Effects of Antiarrhythmic Therapy

Antiarrhythmic agents are not without problems and the possible adverse effects of therapy should be considered.

I. Non-cardiac Side Effects

Antiarrhythmic drugs, as a group, carry a particularly high incidence of side effects. Specific side effects are easily recognized. Examples include drug-induced lupus with procainamide, thrombocytopenia with quinidine and anticholinergic side effects with disopyramide. Amiodarone deserves particular mention, as it is more prone to side effects than any other antiarrhythmic agent.

Less specific side effects such as nausea and dizziness are commonplace. Other problems such as tiredness or mental slowing are less apparent and may well be underrated by physician and patient alike—patients stopping antiarrhythmic therapy frequently remark on the return of a sense of general well being, even when the therapy has been considered trouble-free.

It seems probable that such side effects are unrelated to the mechanism of antiarrhythmic action, and it is to be hoped that newer agents may be developed which cause fewer problems. For example, the most obvious need in the antiarrhythmic armamentarium is for a new class III agent, sharing amiodarone's antiarrhythmic action, but with less troublesome side effects.

II. Adverse Haemodynamic Effects

Most antiarrhythmic agents have adverse haemodynamic effects (BOURKE et al. 1987). If left ventricular function is normal, these may be inconsequential, but in patients with impaired left ventricular function adverse haemodynamic effects may be of considerable significance.

With some classes of antiarrhythmic agent, negative inotropism may be an inevitable accompaniment of the mechanism of drug action. This seems probable for class II and class IV agents. In the case of class I drugs, the situation is less clear. All current class I agents have, to a greater or lesser extent, a negative inotropic action (BOURKE et al. 1987). Indeed, there is some evidence that negative inotropism may be an inevitable accompaniment of inward sodium current inhibition (SHEU and LEDERER 1985). However, the considerable variation in the degree of negative inotropism among different class I agents suggests that negative inotropism is not solely a function of sodium current inhibition.

III. Arrhythmogenic Effects

It has been appreciated for many years that quinidine occasionally has an arrhythmogenic action (ROSKETH and STORSTEIN 1963). The early recognition of this arrhythmogenic effect owes much to the unusual nature of the resultant arrhythmia, namely torsade de pointes. Other instances of arrhythmogenesis generally result in a less distinctive arrhythmia and until recently the arrhythmogenic action of quinidine was regarded as unique amongst antiarrhythmic agents. In the past decade, there has been a growing realization that this is far from the case. We now appreciate that many antiarrhythmic agents (other than the beta-blockers) can, potentially, have an arrhythmogenic effect. If a patient on antiarrhythmic therapy for a known history of ventricular arrhythmias dies suddenly, it was formerly natural to assume that this was despite rather than because of therapy. It has now been clearly demonstrated that some such cases are attributable to therapy (RUSKIN et al. 1983).

Our understanding of arrhythmogenic actions of drugs depends upon the degree of our comprehension of the mechanism of the arrhythmia itself and how it is altered by antiarrhythmic agents. When arrhythmia mechanisms are understood, it is possible to comprehend how arrhythmogenic effects may arise. For example, in Wolff-Parkinson-White syndrome, a drug preferentially blocking anterograde conduction over the accessory pathway, without blocking retrograde conduction, may facilitate the development of reciprocating tachycardia. Such an action would not represent a drug side effect, but rather a manifestation of the drug's antiarrhythmic action, which happens, in that particular patient, to facilitate the arrhythmia.

In the case of most ventricular arrhythmias, the mechanism of arrhythmogenesis is unknown. Arrhythmogenic effects of antiarrhythmic agents often occur with plasma drug concentrations within the therapeutic range and therefore cannot be said to be a manifestation of drug toxicity (VELEBIT et al. 1982). It seems probable that the same electrophysiological properties which prove advantageous in arrhythmia control in one patient may result in arrhythmia facilitation in another. This is an important point, as it suggests that an arrhythmogenic capability is an inevitable accompaniment of some classes of antiarrhythmic action and that all future antiarrhythmic agents of that class will share the problem of arrhythmogenesis.

G. Risk-Benefit Ratio of Antiarrhythmic Therapy

Before commencing any patient on antiarrhythmic therapy, the relative risks and benefits should be considered. The benefits of controlling symptomatic arrhythmias and prolonging life expectancy need to be set against the disadvantages of possible side effects, diminished quality of life, impairment of left ventricular function and possible arrhythmogenic effects.

Unfortunately, this is not an easy equation to solve, as benefits and risks are interdependent. The arrhythmogenic capability of antiarrhythmic agents varies considerably between patient groups. For example the incidence of arrhythmogenic effects of the class I agent, encainide, has been found to vary between 0 % in normal individuals, 0 % in patients treated for supraventricular arrhythmias, 2 % in patients treated for ventricular ectopic beats and 11 % in patients with ventricular tachycardia or fibrillation (WINKLE et al. 1981). Similarly, in the case of flecainide, the incidence of arrhythmogenic effects varied between 4 % in the treatment of ventricular ectopic beats and 12 % in the treatment of ventricular tachycardia or fibrillation (MORGANROTH and HOROWITZ 1984). The risk of arrhythmogenesis therefore increases in parallel with the severity of the original rhythm disturbance. The risk of arrhythmia facilitation has similarly been shown to be greater in patients with left ventricular failure (MORGANROTH and HOROWITZ 1984).

Thus, a patient with severe impairment of left ventricular function following myocardial infarction may be at high risk of sudden death and hence stand most to benefit from antiarrhythmic therapy; yet, it is precisely this patient who is most likely to be disadvantaged by antiarrhythmic treatment as he will be particularly susceptible to arrhythmogenic and negative inotropic effects. It is therefore important to be able to evaluate the success or failure of antiarrhythmic therapy in each individual patient.

H. Evaluation of the Success of Antiarrhythmic Therapy

If antiarrhythmic therapy has been commenced for a simple indication such as palpitation, then the symptomatic response to therapy may be the most important means of assessing therapeutic success. Similarly, in the control of ventricular rate response in atrial fibrillation, response to therapy is easily assessed. However, if the indication for therapy is the prevention of a life-threatening arrhythmia, it is obviously important to have means of assessment other than arrhythmia occurrence.

Both ambulatory electrocardiography and invasive programmed stimulation studies are widely employed in evaluating antiarrhythmic therapy. The use of either strategy involves the basic premise that these tests identify patients at subsequent high risk of arrhythmic death. This issue is discussed in detail in Chap. 6.

Evaluation by ambulatory electrocardiography is based largely on the control of ventricular ectopic beats. This is dependent on the assumption that control of these ectopics will indicate successful control over more serious arrhythmias. If ventricular ectopic beats provide the trigger for more serious arrhythmias, it would certainly be anticipated that their control would provide a valuable monitor of the success of therapy. It is possible, however, that ventricular ectopic beats and more serious arrhythmias are independent manifestations of an underlying arrhythmogenic propensity. Prevention of ectopic beats would not then necessarily indicate prevention of more serious arrhythmias.

Invasive programmed stimulation is widely used as an alternative. The electrophysiological stimulation protocol provides the trigger, enabling pharmacological effects on the underlying arrhythmogenic propensity to be assessed. There is a wide consensus of agreement that, when an antiarrhythmic agent renders a previously inducible arrhythmia non-inducible, this is indicative of a good prognosis and that the risk of spontaneous recurrence on that treatment will be low (Friedman and Yusuf 1986).

The converse, however (that is, failure to prevent arrhythmia induction during invasive electrophysiological testing), does not necessarily indicate that the agent concerned will be of no value in preventing the recurrence of spontaneous arrhythmia. This problem has been most clearly documented in the case of amiodarone. Successful long-term therapy has been demonstrated, despite failure to prevent arrhythmia induction on acute drug testing (Nademanee et al. 1982). The same consideration applies, although to a lesser extent, with other antiarrhythmic agents. There are several possible reasons for such a discrepancy. Firstly, the acute effects of a drug following intravenous dosing may not accurately represent its chronic effects. Secondly, programmed stimulation tests only the underlying arrythmogenic propensity. Suppression of *spontaneous* trigger beats may be a valuable antiarrhythmic mechanism, which would remain undetected by programmed stimulation.

Both ambulatory electrocardiography and programmed stimulation involve another assumption, namely that modification of test outcome by the antiarrhythmic agent under investigation is causally related to any modification of subsequent prognosis. This remains unproven. For example, if an antiarrhythmic agent renders a previously inducible arrhythmia non-inducible to programmed electrical stimulation, this is, as discussed above, indicative of a good prognosis; but it remains uncertain whether therapy has in fact altered prognosis or whether success on testing merely selects patients with a good prognosis. Similar considerations apply to therapy assessment by ambulatory electrocardiography.

Unfortunately, both ambulatory electrocardiography and programmed stimulation are so widely accepted in the evaluation of antiarrhythmic drugs that their value is unlikely ever to be fully assessed. To do so would require a prospective study comparing the value of placebo, empirical antiarrhythmic therapy and the antiarrhythmic drug indicated by therapeutic testing (FRIEDMAN and YUSUF 1986). The widespread assumption that testing can select an appropriate drug would make such a study ethically difficult.

When assessing antiarrhythmic therapy, evaluation of possible arrhythmogenic effects is as important as assessing antiarrhythmic benefit. Once again evaluation is based on ambulatory electrocardiography and programmed stimulation (COWAN et al. 1987). Just as the value of these methods in assessing therapeutic benefit is uncertain, the relevance of "technical" arrhythmogenesis, during drug testing, to the occurrence of spontaneous, life-threatening arrhythmias is unclear.

J. Alternatives to Antiarrhythmic Drug Therapy

Antiarrhythmic drug therapy, when successful, controls and ameliorates an underlying arrhythmogenic propensity, but does not cure it. It is, at best, an imperfect management strategy. Treatment aimed at abolishing the underlying cause of arrhythmias would certainly be more desirable. This may be extremely simple, for example the correction of any underlying biochemical disturbance, such as hypokalaemia.

In recent years, a range of curative remedies have become available, in particular a surgical approach to arrhythmia management. The most successful example is surgical division of the accessory pathway in patients with Wolff-Parkinson-White syndrome. This offers the attraction of a definitive cure for the arrhythmia, obviating the need for long-term antiarrhythmic therapy. A high success rate and low mortality have been reported for this procedure (GALLAGHER et al. 1986). Surgical methods are being increasingly employed in the management of serious ventricular arrhythmias and again offer the opportunity of a definitive cure (GALLAGHER et al. 1986).

Less invasive definitive procedures are also available. Catheter ablation of the atrioventricular node is being used increasingly in the management of intractable supraventricular arrhythmias (SCHEINMAN and DAVIS 1986). This technique is similarly proving successful in the ablation of some accessory pathways.

The relative merits of these various methods await evaluation. In each case, the success rate, mortality and morbidity of the method need to be compared with that of antiarrhythmic drug therapy. Hitherto, these definitive methods have in general only been used when conventional treatment, that is drug therapy, has failed. This is the correct approach as we do not at present know the possible long-term complications of these procedures. However, it seems possible that, in the future, some of these definitive procedures may be considered as a first-line alternative to antiarrhythmic drug therapy.

Yet another approach is to manage arrhythmias as they arise, with an antitachycardia pacemaker or defibrillator. The use of such devices implies an admission that other techniques have failed or are likely to fail. The role of these physical devices and the role of surgical management in general are discussed in detail in Chap. 21.

K. Role of Antiarrhythmic Drugs in the Treatment of Specific Arrhythmias

I. Supraventricular Ectopic Beats

These, in general, require no treatment. If they are giving rise to intolerable palpitation, it may be appropriate to give an anxiolytic drug or a beta-blocker.

II. Acute Atrial Fibrillation

At the onset of atrial fibrillation, the ventricular response is often rapid, and it is common for the patient with serious underlying heart disease to develop cardiac failure and, very often, pulmonary oedema. Usually the most urgent requirement is to slow the ventricular rate by drugs, but consideration must be given to the use of DC shock if drugs are not immediately effective. This has the advantage of restoring the atrial contribution to ventricular filling. It must be borne in mind that an episode of atrial fibrillation occuring during the course of an acute myocardial infarction or a pulmonary infection is usually self-terminating within hours or, at most, days and if the arrhythmia is well-tolerated haemodynamically it may be sufficient to control the rate until sinus rhythm is spontaneously restored. The risk of emboli is considerable and it is usually wise to administer anticoagulant drugs.

III. Chronic or Relapsing Atrial Fibrillation

If sinus rhythm cannot be restored or maintained, the main aim is to slow conduction through the AV node so that the ventricular rate is slowed and prevented from rising excessively on exertion. Anticoagulants are indicated in most cases, particularly in the presence of mitral valve disease. If drug therapy fails to control ventricular response rate, atrioventricular node ablation and subsequent permanent pacing may be considered as an alternative.

IV. Bradycardia-Tachycardia Syndrome

In this condition, the rhythm varies between sinus bradycardia and supraventricular arrhythmias of various types, including atrial fibrillation, atrial flutter and atrial tachycardias. Minor forms of the disorder may require no treatment, but it is often necessary to combine drugs aimed to control the atrial arrhythmias with pacing to control the bradycardia, which may be aggravated by the antiarrhythmic therapy.

V. Paroxysmal AV Nodal Re-entrant Tachycardias

Individual episodes may be terminated by non-pharmacological means, such as the Valsalva manoeuvre or carotid sinus massage. If these fail, intravenous drug therapy is indicated if symptoms are severe. If the patient is liable to symptomatic recurrences, long-term drug therapy may be necessary, but this should not be undertaken if the episodes are infrequent or asymptomatic. In severe cases which fail to respond to treatment, antitachycardia pacemaker devices should be considered.

VI. Supraventricular Tachycardias in the Wolff-Parkinson-White Syndrome

Treatment depends upon the particular arrhythmia. In all but the mildest cases, it is probably wise to determine the choice of therapy, depending upon electrophysiological studies. Particular care is necessary in the presence of atrial fibrillation, because high ventricular rates are often encountered which may be aggravated by antiarrhythmic drugs, especially digitalis glycosides. In cases where the arrhythmia is life-threatening or the patient fails to respond to drug therapy, surgical division of the accessory pathway should be considered.

VII. Ventricular Ectopic Beats

These are considered in detail by Campbell in Chap. 6. In general, it may be said that they do not require treatment, unless they are giving rise to symptoms. Even then, it is usually best to avoid drugs, if possible, and if they are used to try beta-blockers first.

VIII. Ventricular Tachycardia

Brief non-sustained episodes usually do not require treatment. Sustained episodes require treatment, often urgently if they occur in the presence of severe organic heart disease. Thus, in myocardial infarction, intravenous drug therapy (usually with lignocaine or another class I agent) is necessary, although termination may necessitate the use of DC shock or pacing.

The prevention of recurrences is very important in acute and chronic ischaemic heart disease and in the cardiomyopathies. Prolonged therapy with one of the class I or III drugs is usually indicated. Where these fail, surgical treatment may be successful.

IX. Ventricular Fibrillation

An individual episode requires treatment with DC shock. The routine administration of antiarrhythmic drugs (such as lignocaine) may be effective in acute myocardial infarction, but the evidence of overall benefit from this therapy is questionable. For repetitive ventricular fibrillation, long-term therapy with class I or III agents may be effective. When this is not so, the patient should be considered for surgery or an implantable defibrillator.

L. Sudden Death

Sudden death perhaps represents the greatest challenge to the arrhythmologist. Some 50% of those dying of ischaemic heart disease do so suddenly, and in about half of these the victim has not complained to his doctor of symptoms suggestive of coronary disease. It is true that of those for whom sudden death is the first recognized manifestation of coronary disease, a substantial proportion have exhibited evidence of a propensity to this disorder in that they have had such conditions as hypertension and diabetes, but these risk factors in themselves have a low predictive value and would not, per se, justify the use of antiarrhythmic drugs in the hope of preventing ventricular fibrillation.

Patients in the acute and late phase of myocardial infarction identify themselves as being at risk of sudden arrhythmic death. Naturally, much of the work on prevention of sudden arrhythmic death has centred on these two situations.

M. Antiarrhythmic Therapy in Acute Myocardial Infarction

Ventricular fibrillation is the most important arrhythmia complicating acute myocardial infarction and accounts for the majority of early deaths. The role of prophylactic lignocaine in its prevention remains controversial (KERTES and HUNT 1984), with differing practices in different centres.

Within the coronary care unit, the role of prophylactic lignocaine can be seriously questioned. It is unclear whether the occurrence of ventricular fibrillation actually matters if adequate resuscitation facilities are at hand—episodes of ventricular fibrillation can be treated as they arise by DC cardioversion. The prognosis of resuscitated patients is controversial. While early studies suggested that this patient group has a comparable prognosis to patients without ventricular fibrillation (LAWRIE et al. 1968), more recent studies have suggested that this patient group is at increased risk (SCHWARTZ et al. 1985; DUBOIS et al. 1986). However, it remains uncertain whether ventricular fibrillation is an independent prognostic indicator rather than a reflection of other risk factors, such as increased infarct size (DEWHURST et al. 1984). If ventricular fibrillation is demonstrated to be an independent prognostic indicator, then there may be a value in prophylaxis. However, without this evidence, there is at present no clear justification for lignocaine prophylaxis within the coronary care unit.

Different considerations apply to out-of-hospital prophylaxis. Resuscitation facilities are not likely to be readily at hand and every episode of ventricular fibrillation prevented may represent a life saved. A recent large study suggested that lignocaine was indeed effective in preventing out-of-hospital ventricular fibrillation (KOSTER and DUNNING 1985), although the benefit observed was relatively minor. Of 6024 patients with suspected acute myocardial infarction randomized to receive intramuscular lignocaine or no treatment, 2 treated and 12 control patients developed primary ventricular fibrillation during the period between 15 and 60 min following randomization, when plasma lignocaine levels were thought to be adequate. An increased incidence of asystolic episodes was observed in lignocaine-treated patients, although none of these proved fatal. It seems probable, therefore, that intramuscular lignocaine is of benefit in the prophylaxis of out-of-hospital ventricular fibrillation, although the benefit is fairly marginal.

N. Antiarrhythmic Therapy in the Late Phase of Myocardial Infarction

Particular interest has focussed for two reasons on patients who have survived a myocardial infarction; first, because they form a group who are known to have a high risk of sudden death, particularly during the 1st year after the event; secondly, because they form a "captive" group, who are able and eager to participate in studies.

Several predictors of death in this population have been demonstrated, of which the most powerful appear to be those related to the extent of myocardial damage, and the presence of complex or frequent ventricular arrhythmias (Moss et al. 1979). The importance of the latter is considered in detail by Campbell in Chap. 6. Amongst indicators of ventricular damage may be included a history of pulmonary oedema or shock during the acute phase, cardiomegaly, a low ejection fraction, and a poor haemodynamic response to an exercise test.

There have been a number of studies that have sought to determine whether class I antiarrhythmic drugs can reduce the risk of death in post infarction patients. All the studies, the results of which are itemized in Table 1, were small and it is not surprising that none of them had a statistically significant result. Furthermore, most of them were not concerned with treating a high-risk group, although the study by Chamberlain et al. was confined to patients with clinical evidence of poor ventricular function, and the Gent/Rotterdam study recruited only patients with ventricular arrhythmias. As will be seen, only in the latter study was there a strong trend towards benefit in those treated with the antiarrhythmic agent; indeed, several of the studies exhibited a slightly but non-significantly higher mortality in the treated group.

In all five of the trials in which Holter monitoring was undertaken, the antiarrhythmic agent reduced the incidence of arrhythmias. An interesting and, perhaps, important observation in the Gent/Rotterdam study was that there was no mortality in those patients in whom aprindine succeeded in suppressing the arrhythmias, all the deaths in the treated group being in those who continued to exhibit arrhythmias. While this might be taken as evidence that the suppression of arrhythmias can lead to the prevention of death, it might

Table 1. Mortality in postinfarction antiarrhythmic studies

Trial name	Control group		Intervention group		
	Number of patients	% died	Drug	Number of patients	% died
Collaborative group	285	8.1	Phenytoin	283	9.2
PETER et al. (1978)	76	18.4	Phenytoin	74	24.3
RYDEN et al. (1980)	56	8.9	Tocainide	56	8.9
BASTIAN et al. (1980)	74	4.1	Tocainide	72	5.6
CHAMBERLAIN et al. (1980)	163	11.7	Mexiletine	181	13.3
IMPACT RESEARCH GROUP (1984)	313	4.8	Mexiletine	317	7.6
HUSENHOLTZ et al. (1978)	152	12.5	Aprindine	153	7.8

also be noted that it is easier to suppress arrhythmias in those with good left ventricular function, a group whose prognosis is anyway relatively good.

There is an urgent need to undertake trials which answer the question whether the suppression of ventricular arrhythmias by antiarrhythmic drugs will prevent death; it may well be that they do, but that in the trials so far undertaken a beneficial effect has been obscured by an adverse arrhythmogenic effect in those whose arrhythmias were not controlled. If this is the case, the great need for the future is to be able to differentiate these two groups of patients.

It must be emphasized that the trials described above did not include patients who had had severe or symptomatic ventricular arrhythmias, nor those who had been resuscitated from "sudden death" outside hospital, two groups who have a high risk of ventricular fibrillation. As might be expected, no controlled trials of antiarrhythmic drug therapy have been undertaken in such patients, but there is strong circumstantial evidence, considered in detail elsewhere in this book, that patients in whom a drug has been shown to suppress sustained ventricular arrhythmias, provoked electrophysiologically, will, on continued therapy with that drug, be at low risk of arrhythmia recurrence. It is less clear whether drugs can prevent a subsequent "sudden death".

In contrast with the apparent failure of prophylactic class I antiarrhythmic drug therapy, it is now fairly clear that drugs which block cardiac beta-adrenoceptors (class II) do reduce the risk of death after myocardial infarction, especially sudden death. It should be noted, also, that there is no evidence that class II drugs are arrhythmogenic. Two major trials of these drugs showed a highly significant reduction in mortality in a period up to 3 years after myocardial infarction (NORWEGIAN MULTICENTRE RESEARCH GROUP 1981; BETA-BLOCKER HEART ATTACK TRIAL RESEARCH GROUP 1982) and when all the large secondary prevention trials with beta-blockers are pooled, there is seen to be a 20%–25% reduction in mortality (YUSUF et al. 1985). It should be pointed out that these trials excluded patients with severe ventricular dysfunction, as manifested by features of left ventricular failure, and thereby excluded those at highest risk. However, of those who were included, the most substantial benefit was observed in those who had had electrical or mechanical complications during the preceding acute phase of myocardial infarction. It is unproven whether beta-blockade achieves any reduction in mortality in the asymptomatic post myocardial infarction patient who has had an uncomplicated course of recovery.

O. Conclusions

The need for effective antiarrhythmic therapy is well established. However, the clarity of this need should not obscure the fact that many currently available antiarrhythmic drugs are imperfect and may have side effects which are potentially serious and even life-threatening. It is therefore important with any arrhythmia in any patient to consider carefully the rationale for therapy and assess the relative risks and benefits of treatment. In many instances we unfortunately remain uncertain of how best to assess benefit and risk. This problem has limited the success of antiarrhythmic drug therapy in the past, particularly antiarrhythmic prophylaxis after infarction. Distinguishing advantageous from disadvantageous effects of antiarrhythmic agents represents the greatest challenge for the future.

References

Bastian BC, McFarlane PW, McLauchlan JH, Ballantyne D, Clark R, Hillis WS, Rae AP, Hutton I (1980) A prospective randomised trial of tocainide in patients following myocardial infarction. Am Heart J 100:1017-1022

Beta-Blocker Heart Attack Trial Research Group (BHAT) (1982) A randomised trial of propranolol in patients with acute myocardial infarction. JAMA 247:1707-1714

Bjerregaard P (1983) Premature beats in healthy subjects 40-79 years of age. Eur Heart J 3:493-503

Bourke JP, Cowan JC, Tansuphaswadikul S, Campbell RWF (1987) Antiarrhythmic drug effects on left ventricular performance. Eur Heart J [Suppl A], 8:105-111

Chamberlain DA, Jewitt DE, Julian DG, Campbell RWF, Boyle DMcC, Shanks RG (1980) Oral mexiletine in high-risk patients after myocardial infarction. Lancet 2:1324-1327

Collaborative Group (1971) Phenytoin after recovery from myocardial infarction: controlled trial in 568 patients. Lancet 2:1055-1057

Cowan JC, Bourke J, Campbell RWF (1987) Arrhythmogenic effects of antiarrhythmic drugs. Eur Heart J [Suppl A], 8:133-136

Dewhurst NG, Hannan WJ, Muir AL (1984) Ventricular performance and prognosis after primary ventricular fibrillation complicating acute myocardial infarction. Eur Heart J 5:275-281

Dubois C, Smeets JP, Demoulin C, Foidart G, Henrard L, Tulippe C, Preston L, Carlier J, Kulbertus HE (1986) Incidence, clinical significance and prognosis of ventricular fibrillation in the early phase of myocardial infarction. Eur Heart J 7:945-951

Friedman L, Yusuf S (1986) Does therapy directed by programmed electrical stimulation provide a satisfactory clinical response? Circulation [Suppl 2] 73:59-66

Gallagher JJ, Sealy WC, Selle JG, Svenson RH, Zimmern SH (1986) Role of ablation techniques and surgery in the treatment of cardiac arrhythmias. In: Kulbertus HE (ed) Medical management of cardiac arrhythmias. Livingstone, Edinburgh, pp 253-283

Hinkle LE, Carver ST, Stevens M (1969) The frequency of asymptomatic disturbances of cardiac rhythm and conduction in middle-aged men. Am J Cardiol 24:629-650

Hugenholtz PG, Hagemeijer F, Lubsen J, Glaser B, van Durme JP, Bogaert MG (1978) One year follow-up in patients with persistent ventricular dysrhythmias after myocardial infarction treated with aprindine and placebo. In: Sandoe E, Julian DG, Bell J (eds) Management of ventricular tachycardia—role of mexiletine. Excerpta Medica, Amsterdam, pp 572-578

IMPACT Research Group (1984) International mexiletine and placebo antiarrhythmic coronary trial. I. Report on arrhythmia and other findings. J Am Coll Cardiol 4:1148-1163

Kertes P, Hunt D (1984) Prophylaxis of primary ventricular fibrillation in acute myocardial infarction. The case against lignocaine. Br Heart J 52:241-247

Koster RW, Dunning AJ (1985) Intramuscular lidocaine for the prevention of lethal arrhythmias in the prehospitalization phase of acute myocardial infarction. N Engl J Med 313:1105-1110

Lawrie DM, Higgins MR, Godman MJ, Oliver MF, Julian DG, Donald KW (1968) Ventricular fibrillation complicating acute myocardial infarction. Lancet 2: 523-528

Leinbach RC, Chamberlain DA, Kastor JA, Harthrone JW, Sanders CA (1969) A comparison of the haemodynamic effects of ventricular and sequential A-V pacing in patients with heart block. Am Heart J 78:502-508

May GS, Eberlein KA, Furberg CD, Passamani ER, DeMets DL (1982) Secondary prevention after myocardial infarction: a review of long-term trials. Prog. Cardiovasc Dis 24:331-352

Morganroth J, Horowitz LN (1984) Flecainide: its proarrhythmic effect and expected changes on the surface electrocardiogram. Am J Cardiol 53:89B-94B

Moss AJ, Davis HT, DeCamilla J, Bawer LW (1979) Ventricular ectopic beats and their relation to sudden and non-sudden cardiac death after myocardial infarction. Circulation 60:998-1003

Nademanee K, Hendrickson J, Hannan R, Singh BN (1982) Antiarrhythmic efficacy and electrophysiologic actions of amiodarone in patients with life-threatening ventricular arrhythmias: potent suppression of spontaneous ventricular tachyarrhythmias vs inconsistent abolition of induced VT. Am Heart J 103:950-959

Nakano J, McCloy RB (1970) Effects of atrial and ventricular tachycardia on systemic and coronary circulations, and myocardial oxygen consumption in control dogs and in dogs with adrenergic blockade. Cardiovasc Res 4:180-187

Norwegian Multicentre Research Group (1981) Timolol-induced reduction in mortality and reinfarction in patients surviving acute myocardial infarction. N Engl J Med 304:801-807

Peter T, Ross D, Duffield A, Luxton M, Harper R, Hunt D, Sloman G (1978) Effect on survival after myocardial infarction of long-term treatment with phenytoin. Br Heart J 40:1356-1360

Rokseth R, Storstein O (1963) Quinidine therapy of chronic ventricular fibrillation. The occurrence and mechanisms of syncope. Arch Intern Med 111:102-107

Ruskin JN, McGovern B, Garan H, DiMarco JP, Kelly E (1983) Antiarrhythmic drugs: a possible cause of out-of-hospital cardiac arrest. N Engl J Med 309:1302-1306

Ryden L, Arnman K, Conradson T-B, Hofvendsahl S, Mortensen O, Smedsard P (1980) Prophylaxis of ventricular tachyarrhythmias with intravenous and oral tocainide in patients with and recovering from acute myocardial infarction. Am Heart J 100:1006-1012

Scheinman MM, Davis JC (1986) Catheter ablation for treatment of tachyarrhythmias: present role and potential promise. Circulation 73:10-13

Schwartz PJ, Zara A, Grazi S, Lombardo M, Lotto A, Sbressa C, Zappa P (1985) Effect of ventricular fibrillation complicating acute myocardial infarction on long-term prognosis: importance of the site of infarction. Am J Cardiol 56:384-389

Sheu SS, Lederer WJ (1985) Lidocaine's negative inotropic and antiarrhythmic actions. Circ Res 57:578-590

Soyeur D (1986) Haemodynamic consequences of cardiac arrhythmias. In: Kulbertus HE (ed) Medical management of cardiac arrhythmias. Livingstone, Edinburgh, pp 26-40

Velebit V, Podrid P, Lown B, Cohen BH, Graboys TB (1982) Aggravation and provocation of ventricular arrhythmias by antiarrhythmic drugs. Circulation 65:886-894

Winkle RA, Mason JW, Griffin JC, Ross D (1981) Malignant ventricular arrhythmias associated with the use of encainide. Am Heart J 102:857-864

Yusuf S, Peto R, Lewis J, Collins R, Sleight P (1985) Beta-blockade during and after myocardial infarction: an overview of the randomised trials. Prog Cardiovasc Dis 27:335-371

CHAPTER 6

Distinguishing Potentially Lethal from Benign Arrhythmias

R. W. F. CAMPBELL

A. Introduction

Benign cardiac arrhythmias are those which cause neither symptoms nor haemodynamic upset and which have no prognostic significance. Such arrhythmias include sinus arrhythmia, isolated atrial ectopic beats and infrequent ventricular ectopic beats. Arrhythmias associated with lethality are of two types: those which cause such severe haemodynamic upset as immediately to jeopardize life by their presence and those which betoken an adverse prognosis. The former are characterized by their electrocardiographic patterns and their threat is largely independent of the cardiovascular setting in which they occur. Prognostically important arrhythmias are more complex. The basic arrhythmia is often a transient disturbance of which the patient is unaware and it is the underlying cardiovascular pathology which is the major determinant of prognosis. The majority of prognostically important arrhythmias are merely markers of risk and do not in themselves jeopardize life.

B. Arrhythmias Which can be Directly Fatal

Any sustained tachyarrhythmia or bradyarrhythmia which markedly reduces cardiac output may kill.

I. Sustained Tachyarrhythmias

1. Ventricular Fibrillation

The electrocardiogram of ventricular fibrillation shows disorganized electrical activity. Experimental evidence suggests more coherent myocardial activation in the form of multiple interlacing re-entrant wavelets (MOE 1975). Nonetheless, ventricular fibrillation produces no useful mechanical activity and cardiac output falls precipitously. Whilst it has long been considered that ventricular fibrillation is always fatal unless reversed by countershock or exceptionally by drugs (SANNA and ARCIDICACONA 1973) or by a blow to the chest (LOWN and TAYLOR 1970), there are instances of spontaneous termination of ventricular fibrillation.

Ventricular fibrillation occurs predominantly in patients with ischaemic heart disease. Subtypes of ventricular fibrillation complicating myocardial infarction are recognized. Primary and secondary ventricular fibrillation are defined respectively by the absence or presence of shock or cardiac failure (OLIVER et al. 1967). The separation is relevant as the success rate for defibrillation of primary ventricular fibrillation is high with a good subsequent prognosis whilst the success rate for reversion of secondary ventricular fibrillation is no better than 30% with a 1-year mortality of 90% for those in whom sinus rhythm can be re-established (GOLDBERG et al. 1978; ADGEY et al. 1971). Improved understanding of the pathophysiology of myocardial infarction suggests that there may be other clinically useful subdivisions of ventricular fibrillation. Ventricular fibrillation which occurs with acute occlusion of the coronary artery, ventricular fibrillation which occurs postocclusion (probably the type most often seen in coronary care units) and ventricular fibrillation which occurs on coronary reperfusion (either natural or as a consequence of thrombolytic therapy) are recognizable entities which have differing immediate and late prognostic implications.

2. Ventricular Tachycardia

Ventricular tachycardia is not consistently defined. Three or more consecutive ventricular ectopic complexes at a rate equal to or greater than 120 beats/min is a common definition but, particularly with the advent of invasive electrophysiological testing, there has been a need to separate brief unsustained runs of ventricular ectopic beats from sustained ventricular tachycardia (usually defined as of at least 30-s duration). Salvoes of ventricular ectopic beats frequently complicate the acute phase of myocardial infarction but they rarely cause symptoms, appear not to have prognostic implications and are not lethal. By contrast, sustained ventricular tachycardia complicating any form of cardiovascular disease is associated with a substantial immediate and late mortality (DIMARCO et al. 1985; MARCHLINSKI et al. 1983). The risk posed by sustained ventricular tachycardia depends upon its rate and the functional state of the myocardium. Sustained ventricular tachycardia occurring in patients with cardiomyopathies or in patients recovering from the acute phase of myocardial infarction is not well tolerated. As with ventricular fibrillation, subtypes of ventricular tachycardia probably exist although, as yet, little work has been undertaken to identify them.

3. Torsade de Pointes

In 1966 DESSERTENNE described an unusual form of polymorphic ventricular tachycardia which he called torsade de pointes. Its electrocardiographic features are of a rapid ventricular tachyarrhythmia with continual and dramatic vector shifts (KRIKLER and CURRY 1976). It can resemble ventricular flutter or fibrillation and often has the same implications for cardiac output as these arrhythmias. Torsade de pointes characteristically spontaneously terminates but recurrences are common and the risk of death during an individual event is high. Torsade de pointes occurs in two important clinical contexts; the congenital long QT syndromes and as a feature of drug toxicity.

Torsade de pointes is the classical manifestation of arrhythmogenicity (the capacity of drugs to aggravate an existing arrhythmias or to create a new one). Torsade de pointes can be caused by many drugs including prenylamine (GRENADIER et al. 1980) and lidoflazine (KENNELLY 1977) and occurs with many antiarrhythmic drugs including quinidine (DENES et al. 1981), disopyramide (TZIVONI et al. 1981), amiodarone (McGOVERN et al. 1983), flecainide (NATHAN et al. 1984), sotalol (KONTOPOULOS et al. 1981) and encainide (WINKLE et al. 1981). This variant of ventricular tachycardia is important as its recognition should immediately alert attention to the possibility of drug toxicity. If its implications are not appreciated, increased or continued drug dosing may be fatal.

II. Severe Bradyarrhythmias

Ventricular tachyarrhythmias are not the only disturbances of cardiac rhythm which dramatically reduce cardiac output and threaten life. Ventricular asystole, profound bradycardia and complete heart block are dangerous and potentially fatal arrhythmias. Ambulatory ECG recordings obtained from patients who die suddenly reveal that bradyarrhythmias are a not uncommon cause of death (ROELANDT et al. 1984). Most cannot be predicted although some forms of bundle branch block have prognostic significance (FISCH et al. 1980). Many may be agonal rhythms complicating events such as massive pulmonary embolism, cardiogenic shock or cardiac rupture.

C. Arrhythmias Associated with an Adverse Prognosis

A variety of cardiac arrhythmias are associated with an adverse prognosis. The arrhythmias in themselves often have no immediate impact for the patient, causing neither symptoms nor appreciable haemodynamic upset. The clinical circumstances in which these rhythm disturbances occur is the critical element in defining prognosis.

I. Ventricular Ectopic Beats

Ventricular eçtopic beats complicate a wide variety of cardiac diseases and have long been feared, particularly during myocardial infarction (JULIAN et al. 1964; LOWN et al. 1967). In fact, in acute unstable ischaemic conditions, ventricular ectopic beats are relatively benign and have no short- or long-term prognostic implications (CAMPBELL et al. 1981; EL SHERIF et al. 1976; LIE et al. 1975), but in most other clinical contexts they are important. Twenty-four-hour ambulatory ECG recordings have identified the prognostic implications of ventricular ectopic beats complicating the post acute and chronic phases of myocardial infarction (Multicenter Postinfarction Research Group 1983; Moss 1980; BIGGER et al. 1977), cardiomyopathies (MARON et al. 1981; MCKENNA et al. 1981a), mitral valve prolapse (CAMPBELL et al. 1976), aortic stenosis (CHIZNER et al. 1980) and heart failure (CHAKKO and GHEORGHIADE 1985). Even in apparently normal healthy asymptomatic individuals, ventricular ectopic beats may not be innocuous (CULLEN et al. 1982). In all these situations, the frequency of the ventricular ectopic beats is of critical relevance. In the setting of overt cardiovascular pathology, a ventricular ectopic beat frequency of 10/h is the threshold level at which prognostic implications become apparent. In apparently normal individuals, the threshold frequency is probably ten times higher (CAMPBELL 1984).

Ventricular ectopic beats do not in themselves threaten life. The mechanism of their prognostic association operates through either their ability to trigger latent life-threatening arrhythmias or their acting as a marker of severe underlying disease.

II. Ventricular Fibrillation

Primary ventricular fibrillation (VF in the absence of shock and or cardiac failure) has a good immediate prognosis if treated swiftly and has been considered to have an excellent long-term prognosis with quality and quantity of life for afflicted patients being similar to that for patients whose infarction was not so complicated (GEDDES et al. 1969; KUSHNIR et al. 1975; GOLDBERG et al. 1978). Recent work suggests that the long-term prognosis is much poorer when ventricular fibrillation complicates anterior myocardial infarction as compared with inferior myocardial infarction (SCHWARTZ et al. 1985). If substantiated, there would be a case for energetic prevention of ventricular fibrillation in acute infarction although, even if accomplished, prognosis might not be altered.

III. Second-Degree AV Block

Second-degree heart block is not in itself life threating but it appears to presage more profound disturbances of AV conduction resulting in ventricular asystole. The adverse prognostic implications of Mobitz II and high-grade forms of second-degree AV block can be ameliorated by permanent pacemaking. Until recently, Wenckebach block was considered relatively benign but new evidence prompts reappraisal of this assumption (SHAW et al. 1985). The prognosis for unpaced patients with Wenckebach block is no better than that for the other forms of second-degree AV block. Permanent pacing is indicated and does improve prognosis (SHAW et al. 1985).

IV. Arrhythmias Associated with the Pre-excitation Syndromes

Atrioventricular nodal bypasses are an important basis for arrhythmias (GALLAGHER et al. 1978). Reciprocating tachycardia, the commonest pre-excitation arrhythmia, is macro re-entrant and involves the atrium, the AV node, the ventricle and bypass tissue. It commonly causes symptoms but rarely is it a threat to life. Its rate is usually no greater than 220 beats/min and the activation pattern and the performance of the ventricle are normal. In some patients, reciprocating tachycardia may so disorganize atrial activation as to cause atrial fibrillation (CAMPBELL et al. 1977). Atrial fibrillation may also occur because of haemodynamic upset or because of associated pathology (eg. *Ebstein's anomaly*). In the presence of a short anterograde refractory period AV nodal bypass, atrial fibrillation can be dangerous. Without the ventricular rate-limiting effects of the AV node, ventricular responses in atrial fibrillation may be in excess of 300 beats/min and ventricular fibrillation may ensue. Any patient with a history of *syncope* and with evidence of anterograde function in an AV nodal bypass should be considered at high risk of lethal arrhythmias. Acute intravenous drug testing with ajmaline (WELLENS et al. 1980) or procainamide (WELLENS et al. 1982) has been recommended to identify "high-risk" accessory pathways in patients with Wolff-Parkinson-White syndrome. These agents abolish anterograde pre-excitation in the majority of long (and therefore "safe") refractory period pathways. A more reliable assessment of risk may be obtained with invasive electrophysiological testing (MORADY et al. 1983); the characteristics of the bypass can be accurately determined and atrial fibrillation can be induced in secure circumstances with resuscitation equipment available. The shortest RR interval between consecutive pre-excited complexes in atrial fibrillation (either spontaneous or induced) correlates reasonably well with risk, intervals of less than 205 ms being dangerous (SELLERS et al. 1977). Sympathetic drive may markedly alter pre-excitation syndrome arrhythmias and on occasion "safe" pathways determined by either acute drug testing or by invasive electrophysiology may prove dangerous during exercise or emotional upset. Complete evaluation of risk should include electrophysiolgical testing during isoprenaline administration.

D. Arrhythmias Which are Benign

An alternative to defining potentially lethal arrhythmias is to establish those which are incontrovertibly benign. Low-frequency ventricular ectopic beats in apparently normal individuals and almost all types of ventricular arrhythmias other than sustained VT and ventricular fibrillation occurring in acute myocardial infarction should be considered innocuous (CAMPBELL et al. 1981). The simple occurence of R-on-T ventricular ectopic beats does not presage ventricular fibrillation but the close temporal relationship of these two arrhythmias in acute infarction prompts caution before ascribing early cycle ventricular ectopic beats as having no prognostic importance (Editorial 1986). Supraventricular ectopic beats complicating most forms of cardiac disease are benign although they may predict individuals likely to develop atrial fibrillation.

E. Situations of High Risk for Fatal Arrhythmias

High-risk clinical situations for developing fatal arrhythmias are identifiable.

I. Myocardial Ischaemia

1. Acute

Myocardial ischaemia is a potent arrhythmogenic substrate. Its acute form in the early minutes of myocardial infarction is responsible for the high incidence of ventricular fibrillation at this time (ADGEY et al. 1971). Similarly in patients with *Prinzmetal variant angina*, acute ischaemia is associated with serious arrhythmias (SCRUTINO et al. 1984). In chronic conditions of myocardial ischaemia, arrhythmias are much less frequent but, particularly with exercise, fluctuations in ischaemia may provoke rhythm disturbances. Exercise-induced, ischaemia-related ventricular tachycardia and ventricular fibrillation require critical perfusional conditions which rarely can be reproduced by electrophysiological testing unless fast drive rates are employed. These arrhythmias, which usually do not respond to antiarrhythmic therapy other than beta-adrenoreceptor blocking drugs, are probably best treated by improving myocardial perfusion with coronary artery surgery or angioplasty.

2. Chronic

The majority of serious ventricular tachyarrhythmias which occur in patients with chronic ischaemic heart disease are not actively dependent upon the ischaemia itself but are supported by the myocardial fibrosis substrate created by previous infarction. Left ventricular *aneurysms* are the best-recognized form of ischaemic myocardial damage which supports life-threatening arrhythmias (COHEN et al. 1983). Characterized by a history of syncope, palpitations and often breathlessness, left ventricular aneurysms usually are associated with electrocardiographic changes of an extensive Q wave infarction and persisting ST segment elevation. The chest X-ray may be normal or may show a discrete bulge of the cardiac border or cardiomegaly. Left ventricular angiography is the optimal method of diagnosis but echocardiography and radionucleide angiography may be helpful. Ventricular tachyarrhythmias complicating left ventricular aneurysms usually are not exercise related, are typically sustained mono- or multimorphic ventricular tachycardia and are inducible by programmed stimulation. Individually prescribed antiarrhythmic therapy is useful but in many patients the rhythm is refractory to drugs. It is for this group of patients that surgical treatment has become an important therapeutic option (MILLER and JOSEPHSON 1985). Activation or fragmentation map directed surgery (CAMPBELL 1986) has a high success rate for abolishing the arrhythmia. Surgical management is expanding to encompass all forms of ventricular tachyarrhythmias including polymorphic ventricular tachycardia and ventricular fibrillation and is also appropriate for arrhythmias unrelated to aneurysms.

II. Resuscitated Survivors of Out-of-Hospital Sudden Death

Resuscitated patients with potentially lethal arrhythmias are at high risk of recurrences (LIBERTHSON et al. 1974). This is particularly so if there has been no evidence of myocardial infarction (RUSKIN et al. 1980). In the majority of these patients there is severe underlying coronary artery disease and it is likely that ischaemia plays a major role in arrhythmogenesis. Only a small proportion of such patients have sustained inducible arrhythmias at electrophysiology testing.

III. Long QT Syndromes

The congenital long QT syndromes are associated with a high incidence of sudden arrhythmic death. The usual arrhythmia is a form of torsade de pointes which presumptively on fatal occasions does not self-terminate. The condition is inherited as either an autosomal dominant—the ROMANO WARD syndrome (ROMANO et al 1963; WARD 1964) or a recessive trait—the JERVELL-LANGE-NEILSON syndrome (JERVELL and LANGE-NEILSON 1957). Prognostically bad features are the degree of QT prolongation on the surface ECG, young age at the onset of syncopal attacks and a family history of sudden death at an early age. The condition appears to be caused by either a sympathetic imbalance on the heart or a differential myocardial sensitivity to sympathetic input. Beta-adrenoreceptor blocking drugs or sympathectomy seem to improve prognosis (SCHWARTZ 1983; Chap. 23). "Excessive" QT prolongation during the Valsalva manoeuvre may identify latent and high-risk cases (MITSUTAKI et al. 1981). Invasive electrophysiological testing is of little value in identifying individuals at high risk.

IV. Cardiac Failure

Up to 30% of patients with cardiac failure die suddenly (PACKER 1985). Patients in heart failure have a high incidence of ventricular ectopic beats and unsustained ventricular tachycardia and there is evidence of an association between arrhythmias and mortality. Arrhythmia suppression has not been shown to improve prognosis and, as with other conditions described, ventricular ectopic beats may merely be a marker of risk.

V. Antiarrhythmic Therapy

As previously discussed, antiarrhythmic drugs may paradoxically aggravate existing arrhythmias or produce new rhythm disturbances. Arrhythmogenicity is a serious problem and carries a substantial mortality. Any patient on antiarrhythmic drugs should be considered at risk. The relative importance of this adverse reaction is unknown but it is a recognized risk factor for out-of-hospital sudden death (BIGGER et al. 1985).

F. Identifying Latent High-Risk Arrhythmias

The pathophysiology of cardiovascular disease is at least as important for prognosis as arrhythmia expression, suggesting the existence of latent arrhythmogenic substrates which can be acted upon by triggers such as ventricular ectopic beats. The arrhythmia substrate is difficult to characterize in man but some features are clinically accessible.

I. Standard ECG Features

The conventional 12-lead surface electrocardiogramm has only a small role to play in identifying high-risk patients. The association of anterior myocardial infarction with right bundle branch block and axis shift has a bad prognosis. Death occurs not by heart block and asystole as initially thought but by late ventricular fibrillation (LIE et al. 1978; HAUER et al. 1982). The suggested management is a prolonged stay in hospital as the risk of ventricular fibrillation is almost exclusive to the first 6 weeks following the infarction.

The QT_c prolongs in most patients in the first few days after infarction thereafter gradually returning to normal. Persistent and excessive QT prolongation has been associated with a high incidence of sudden and presumptively arrhythmic death (AHNVE et al. 1984).

II. Exercise Stress Testing

Exercise stress testing has a useful but limited role in identifying those patients who generate ischaemia-related arrhythmogenic substrates (DENNISS et al. 1985).

III. Signal Averaging

The fixed stable arrhythmogenic substrate in patients with fibrotic infiltration and replacement of the myocardium can be detected by high-gain signal averaging of the surface ECG. Late low-voltage potentials—late potentials—at the end of the QRS complex which are not normally evident on the standard ECG can be resolved by averaging (SIMSON 1981) and their presence is both specific and sensitive for identifying postinfarction patients at high risk either of sudden death or of developing sustained ventricular tachycardia (ZIMMERMANN et al. 1985; BREITHARDT et al. 1985). Signal averaging does not help predict patients who will develop ventricular fibrillation during acute infarction (KERTES et al. 1984) and is of less consistent value when used in patients with non-ischaemic arrhythmias.

However, the very nature of ensemble averaging restricts the utility of this technique to identifying electrophysiological features which occur at regular times with respect to the QRS complex. Further progress in this field is likely with developments in beat-to-beat averaging techniques.

IV. Programmed Stimulation

The most important investigative tool to identify latent arrhythmogenic substrates is programmed stimulation. Multi-catheter electrophysiology testing by providing trigger phenomena tests the myocardial substrate and its ability to sustain arrhythmias. Invasive electrophysiology has proved disappointing in investigating the congenital long QT syndromes and other conditions in which either autonomic modulation is important or in which dynamic ischaemia contributes. The major roles for programmed stimulation are in identifying high-risk patients with AV nodal bypass tracts and in investigating and managing patients with sustained ventricular tachyarrhythmias.

Programmed stimulation can identify latent arrhythmogenic substates particularly in patients postinfarction but false-positive results may occur if the stimulation protocol is overly aggressive (DENNISS et al. 1985; WASPE et al. 1985). The significance of unsustained polymorphic ventricular tachycardia and ventricular fibrillation provoked at electrophysiological testing is uncertain but is considered by many to be a "non-specific" response. The induction of sustained monomorphic ventricular tachycardia is a reliable positive result with prognostic implications.

G. Management

The management strategies for potentially lethal arrhythmias depend upon whether the arrhythmia can be directly fatal or whether it operates as a marker of risk. Arrhythmias with immediate lethal effects require a direct antiarrhythmic intervention chosen empirically, by acute drug testing or by programmed stimulation. Arrhythmias which are associated with an adverse prognosis identify patients in whom disease modification is necessary.

I. Ventricular Ectopic Beats Postinfarction

Suppression of ventricular ectopic beats by antiarrhythmic drugs given to survivors of acute myocardial infarction has not improved prognosis (MAY et al. 1982). This result might argue that ventricular ectopic beats act as markers and that their removal would therefore do little to improve the patient's outlook. However, in these studies, ventricular ectopic beats were reduced but not completely suppressed, leaving the possibility that ventricular ectopic beats may still operate as triggers and that benefit would be obtained only by their total abolition. The role for antiarrhythmic therapy in survivors of myocardial infarction has not been adequately evaluated. All studies performed to date have been small and in only one did the study population comprise patients with manifest arrhythmias (STOEL and HAGEMEIJER 1980), the others including high-risk patients defined by a variety of non-arrhythmic clinical criteria. By contrast with the performance of antiarrhythmic drugs used in groups of patients postmyocardial infarction, important prognostic benefit has been demonstrated with prophylactic beta-adrenoreptor blocking drugs given to relatively unselected survivors of infarction (YUSUF et al. 1985). These agents have modest antiarrhythmic properties against ventricular ectopic beats but it is widely believed that the protection they offer is mediated either by prevention of ischaemia or, more likely, by prophylaxis of ventricular fibrillation. Analysis of subgroups of the BHAT study suggests that prognostic benefit occurred in patients who had suffered electrical rather than mechanical complications of their acute infarction (FURBERG et al. 1984). However, in another study of patients treated with metoprolol, it was only those whose ventricular ectopic beats were suppressed by the drug that benefited (OLSSON and REHNQVIST 1986).

The mechanisms of interaction between ventricular ectopic beats and prognosis are ill understood. Clinically, beta-adrenoreceptor blocking drugs are the only proven pharmacological intervention which improves the outlook for survivors of acute myocardial infarction.

II. Ventricular Ectopic Beats in Non-ischaemic Cardiovascular Diseases

Little is known of therapeutic strategies to improve prognosis for patients with other types of cardiovascular pathology in which ectopic beats have implications. Amiodarone may benefit patients with hypertrophic cardiomyopathy (MCKENNA et al. 1981b) but this strategy should be reserved only for selected high-risk patients. There is no evidence that ectopic beat suppression in patients with mitral valve prolapse, aortic stenosis, heart failure or in apparently normal individuals improves prognosis and, at present, general use of antiarrhythmic therapy in these conditions cannot be recommended.

III. Long QT Syndromes

Sudden death is well recognized in the congenital long QT syndromes and, despite energetic use of antiarrhythmic drugs, little impact on mortality has occurred. However, beta-adrenoreceptor blocking therapy or left cervical sympathectomy significantly improves prognosis for afflicted patients (SCHWARTZ 1983).

H. Conclusions

Lethal arrhythmias are not as simple as might first appear. Sustained tachy- and bradyarrhythmias have immediate importance but in most clinical circumstances their occurrence is unpredictable. Their management requires the ready availability of experienced personnel and often of sophisticated equipment.

Prognostically relevant arrhythmias are extremely important as they identify high-risk patients but much work must be done if we are to understand the complex relationship between these rhythm disturbances and subsequent mortality. Successful risk reduction for patients with these arrhythmias probably requires modification of the underlying arrhythmogenic substrate rather than a direct antiarrhythmic intervention.

References

Adgey AAJ, Allen JD, Geddes JS, James RGG, Webb SW, Zaida SA, Pantridge JF (1971) Acute phase of myocardial infarction. Lancet 2:501-504

Ahnve S, Gilpin E, Madsen EB, Froelicher V, Henning H, Ross J (1984) Prognostic importance of QTc interval at discharge after myocardial infarction: a multicenter study of 865 patients. Am Heart J 108:395-400

Bigger JT, Dresdale RJ, Heissenbuttel RH, Weld FM, Wit AL (1977) Ventricular arrhythmias in ischaemic heart disease: mechanism, prevalence, significance and management. Prog Cardiovasc Dis 19:255-300

Bigger JT, Fleiss JL, Rolnitzky LM, Merab JP, Ferrick KJ (1985) Effect of digitalis treatment on survival after myocardial infarction. Am J Cardiol 55:623-630

Breithardt G, Borgreffe M, Haerten K, Trampisch HJ (1985) Prognostic significance of programmed ventricular stimulation and non invasive detection of ventricular late potentials in the post infarction period. Z Kardiol 74:389-396

Campbell RWF (1984) Ventricular ectopic activity and its relevance to aircrew licencing. Eur Heart J 5 (Suppl A):95-98

Campbell RWF (1986) Mapping of ventricular arrhythmias. Cardiol Clin 4:497-505

Campbell RWF, Godman MJ, Fiddler GI, Marquis RMM, Julian DG (1976) Ventricular arrhythmias in syndrome of balloon deformity of mitral valve. Definition of possible high risk group. Br Heart J 38:1053-1057

Campbell RWF, Smith RA, Gallagher JJ, Pritchett ELC, Wallace AG (1977) Atrial fibrillation in the pre-excitation syndrome. Am J Cardiol 40:514-520

Campbell RWF, Murray A, Julian DG (1981) Ventricular arrhythmias in first 12 hours of acute myocardial infarction. Natural history study. Br Heart J 46:351-357

Chakko S, Gheorghiade M (1985) Ventricular arrhythmias in severe heart failure: incidence, significance and effectiveness of therapy. Am Heart J 109:497-504

Chizner MA, Pearle DL, de Leon AC (1980) The natural history of aortic stenosis in adults. Am Heart J 99:419-424

Cohen M, Packer M, Gorlin R (1983) Indication for left ventricular aneurysmectomy. Circulation 67:717-22

Cullen K, Stenhouse NS, Wearne KL, Cumpston GN (1982) Electrocardiograms and 13 year cardiovascular mortality in Bussleton study. Br Heart J 47:209-212

Denes P, Gabster A, Huang SK (1981) Clinical, electrocardiographic and follow-up observations in patients having ventricular fibrillation during Holter monitoring. Role of quinidine therapy. Am J Cardiol 48:9-16

Denniss AR, Baaijens H, Cody DV, Richards DA, Russell PA, Young AA, Ross DL, Uther JB (1985) Value of programmed stimulation and exercise testing in predicting one year mortality after acute myocardial infarction. Am J Cardiol 56:213-220

Dessertenne F (1966) La tachycardie ventriculaire a deux foyers opposes variables. Arch Mal Coeur 59:263

Di Marco JP, Lerman BB, Kron IL, Sellers TD (1985) Sustained ventricular tachyarrhythmias within two months of acute myocardial infarction; results of medical and surgical therapy in patients resuscitated from the initial episode. J Am Coll Cardiol 6:759-668

Editorial (1986) R-on-T ventricular ectopic beats. Lancet ii:902-903

El-Sherif N, Myerburg RJ, Scherlag BJ, Befeler B, Aranda JM, Castellanos A, Lazzara R (1976) Electrocardiographic antecedents of primary ventricular fibrillation. Br Heart J 38:415-422
Fisch GR, Zipes DP, Fisch C (1980) Bundle branch block and sudden death. Prog Cardiovasc Dis 23:187-224
Furberg CD, Hawkins CM, Lischstein E (1984) Effect of propranolol in postinfarction patients with mechanical or electrical complications. Circulation 69:761-765
Gallagher JJ, Pritchett ELC, Sealy WC, Kasell J, Wallace AG (1978) The pre-excitation syndromes. Prog Cardiovasc Dis 20:285-327
Geddes JS, Adgey AAJ, Pantridge JF (1969) Prognosis after recovery from ventricular fibrillation complicating ischaemic heart disease. Lancet 2:273-275
Goldberg R, Szklo M, Kennedy H, Tonascia J (1978) Short and long term prognosis of myocardial infarction complicated by ventricular fibrillation or cardiac arrest. Circulation 58 (Suppl 11):89
Grenadier E, Keidar S, Alpan G, Marmor A, Palant A (1980) Prenylamine induced ventricular tachycardia and syncope controlled by ventricular pacing. Br Heart J 44:330-334
Hauer RNW, Lie KI, Liem KL, Durrer D (1982) Long-term prognosis in patients with bundle branch block complicating acute anteroseptal infarction. Am J Cardiol 449:1581-1585
Jervell A, Lange-Neilson F (1957) Congenital deaf—mutism, functional heart disease with prolongation of the QT interval and sudden death. Am Heart J 54:59-60
Julian DG, Valentine PA, Miller GG (1964) Disturbances of rate, rhythm and conduction in acute myocardial infarction. Am J Med 37:915-927
Kennelly BM (1977) Comparison of lidoflazine and quinidine in prophylactic treatment of arrhythmias. Br Heart J 39:540-546
Kertes PJ, Glabus M, Murray A, Julian DG, Campbell RWF (1984) Delayed ventricular depolarisation—correlation with ventricular activation and relevance to ventricular fibrillation in acute myocardial infarction. Eur Heart J 5:974-983
Kontopoulos A, Filindris A, Manoudis F, Metaxas P (1981) Sotalol induced Torsade de pointes. Postgrad Med J 47:321-323
Krikler DM, Curry PVL (1976) Torsade de pointes, an atypical ventricular tachycardia. Br Heart J 38:117-120
Kushnir B, Fox KM, Tomlinson IW, Portal RW, Aber CP (1975) Primary ventricular fibrillation and resumption of work, sexual activity and driving after first myocardial infarction. Br Med J 4:609-613
Liberthson RR, Nagel EL, Hirschman JC, Nussenfeld SR (1974) Prehospital ventricular defibrillation. Prognosis and follow up course. N Engl J Med 291:317-321
Lie KI, Wellens HJJ, Downar E, Durrer D (1975) Observation on patients with primary ventricular fibrillation complicating acute myocardial infarction. Circulation 52:755-759
Lie KI, Liem KL, Schuilenberg RM, David GK, Durrer D (1978) Early identification of patients developing late in-hospital ventricular fibrillation after discharge from the coronary care unit. A 5.5 year retrospective study of 1897 patients. Am J Cardiol 41:674-677
Lown B, Fakhro AM, Hood WB, Thorn GW (1967) The coronary care unit. New perspectives and directions. J Am Med Assoc 19:188-198
Lown B, Taylor J (1970) Thump version. N Engl J Med 283:1223
Marchlinski FE, Waxman HL, Buxton AE, Josephson ME (1983) Sustained ventricular tachyarrhythmias during the early post infarction period: electrophysiological findings and prognosis for survival. J Am Coll Cardiol 2:240-50

Maron BJ, Savage DD, Wolfson JK, Epstein SE (1981) Prognostic significance of 24 hour ambulatory electrocardiographic monitoring in patients with hypertrophic cardiomyopathy: a prospective study. Am J Cardiol 48:252-257
May GS, Eberlein KA, Fubers CD, Passamani ER, DeMets DL (1982) Secondary prevention after myocardial infarction. A review of long term trials. Prog Cardiovasc Dis 24:331-352
McGovern B, Garan H, Kelly, Ruskin JN (1983) Adverse reactions during treatment with amiodarone hydrochloride. Br Med J 287:175-180
McKenna WJ, England D, Doi YL, Deanfield JE, Oakley CM, Goodwin JF (1981) Arrhythmia in hypertrophic cardiomyopathy: I. Influence on prognosis. Br Heart J 46:168-172
McKenna WJ, Harris L, Perez G, Krikler DM, Oakley C, Goodwin JF (1981b) Arrhythmias in hypertrophic cardiomyopathy. II. Comparison of amiodarone and verapamil in treatment. Br Heart J 46:173-178
Miller JM, Josephson ME (1985) Malignant ventricular arrhythmias early after myocardial infarction: brighter prospects. J Am Coll Cardiol 6:796-771
Mitsutaki A, Takeshita A, Kuroiwa A, Nakamura M (1981) Usefulness of the Valsalva manoeuvre in management of the long QT syndrome. Circulation 63:1029-1035
Moe GK (1975) Evidence for reentry as a mechanism of cardiac arrhythmias. Rev Physiol Biochem Pharmacol 72:55-81
Morady F, Sledge C, Shen E, Sung RJ, Gonzales R, Scheinman MM (1983) Electrophysiological testing in the management of patients with the Wolff-Parkinson-White syndrome and atrial fibrillation. Am J Cardiol 51:1623-1628
Moss AJ (1980) Clinical significance of ventricular arrhythmias in patients with and without coronary artery disease. Prog Cardiovasc Dis 23:33-52
Multicenter Postinfarction Research Group (1983) Risk stratification and survival after myocardial infarction. N Engl J Med 309:331-336
Nathan AW, Hellestrand KJ, Bexton RS, Banim SO, Spurrell RAJ, Camm AJ (1984) Proarrhythmic effect of the new antiarrhythmic agent flecainide acetate. Am Heart J 107:222-228
Oliver MF, Julian DG, Donald KW (1967) Problems in evaluating coronary care units. Their responsibility and their relation to the community. Am J Cardiol 20:465-474
Olsson G, Rehnqvist N (1986) Identification of long term survivors after myocardial infarction by evaluation of initial antiarrhythmic response to metoprolol (abstr). Eur Heart J 6:44
Packer M (1985) Sudden unexpected death in patients with congestive heart failure: a 2nd frontier. Circulation 72:681-685
Roelandt J, Klootwijk J, Lumsen J, Janse MJ (1984) Sudden death during long-term ambulatory monitoring. Eur Heart J 5:7-20
Romano C, Gemme G, Pongiglione R (1963) Aritmie cardiache rare dell'eta pediatrica. Clin Pediatr (Phila) 45:656-683
Ruskin JN, DiMarco JP, Garan H (1980) Out-of-hospital cardiac arrest. N Engl. J Med 30:607-613
Sanna G, Arcidicacono R (1973) Chemical ventricular defibrillation of the human heart with bretylium tosylate. Am J Cardiol 32:982-987
Schwartz PJ (1983) The idiopathic long QT syndrome. The need for a prospective registry. Eur Heart J 4:529-531
Schwartz PJ, Zaza A, Grazi S, Lombardo M, Lotto L, Sbressa C, Zappa P (1985) Effects of ventricular fibrillation complicating acute myocardial infarction on long term prognosis. Influence of site of infarction. Am J Cardiol 57:384-389
Scrutino D, de Toma L, Mangini SG, Lagoioa R, Accettua D, Ricca A, Rizzon P

(1984) Ischaemia related ventricular arrhythmias in patients with variant angina pectoris. Eur Heart J 5:1013-1022

Sellers TD, Campbell RWF, Bashore TM, Gallagher JJ (1977) Effects of procainamide and quinidine sulphate in the Wolff-Parkinson-White syndrome. Circulation 55:15-22

Shaw DB, Kekwick CA, Veake D, Gowers J, Wistance T (1985) Survival in second degree atrioventricular block. Br Heart J 53:587-593

Simson MB (1981) Use of signals in the terminal QRS complex to identify patients with ventricular tachycardias after myocardial infarction. Circulation 64:235-242

Stoel I, Hagemeijer F (1980) Aprindine: a review. Eur Heart J 1:147-156

Tzivoni D, Keren A, Stern S, Gottlieb S (1981) Disopyramide induced Torsade de pointes. Arch Intern Med 141:946-947

Ward OC (1964) New familial cardiac syndrome in children. J Ir Med Assoc 103-106

Waspe LE, Seinfeld D, Ferrick A, Kim SG, Matos JA, Fisher JD (1985) Predictors of sudden death and spontaneous ventricular tachycardia in survivors of complicated myocardial infarction: value of the response to programmed stimulation using a maximal of three ventricular extra stimuli. J Am Coll Cardiol 5:1292-1301

Wellens HJJ, Bar FW, Gorgels AP, Vanagt EJ (1980) Use of ajmaline in patients with the Wolff-Parkinson-White syndrome to disclose a short refractory period of the accessory pathway. Am J Cardiol 45:130-133

Wellens HJJ, Braat S, Brugada P, Gorgels APM, Bar FW (1982) Use of procainamide in patients with the Wolff-Parkinson-White syndrome to disclose a short refractory period of the accessory pathway. Am J Cardiol 50:1087-1089

Winkle RA, Mason JW, Griffin JC, Ross D (1981) Malignant ventricular tachyarrhythmias associated with the use of encainide. Am Heart J 102:854-864

Yusuf S, Peto R, Lewis J, Collins R, Sleight P (1985) Beta blockade during and after myocardial infarction: an overview of the randomised trial. Prog Cardiovasc Dis 27:335-371

Zimmermann M, Adamec R, Simonin P, Richez J (1985) Prognostic significance of ventricular late potentials in coronary artery disease. Am Heart J 109:725-732

Antiarrhythmic Therapy

Class I Agents

CHAPTER 7

Subclassification of Class I Antiarrhythmic Drugs

T. J. Campbell

A. Introduction

Class I comprises a very large number of compounds with widely disparate chemical structures and clinical electrophysiological properties (see Chaps. 2, 9-11). They are grouped together because they share the ability to block the fast inward sodium current in cardiac muscle. It is not surprising, therefore, that there have been a number of attempts to subclassify these agents into smaller, more homogeneous groups. Indeed, as was discussed in Chap. 2, there was at one time considerable disagreement regarding the inclusion of lidocaine and diphenylhydantoin as class I drugs at all.

Diphenylhydantoin is no longer widely used as an antiarrhythmic agent and lidocaine is now generally accepted as being a class I drug. Nevertheless, as early as 1974 Singh and Hauswirth proposed subdivision into class Ia (quinidine, procainamide) and Ib (lidocaine, diphenylhydantoin) based on the fact that the sodium-channel blocking effect of the Ia agents is apparent in normal tissues at therapeutic concentrations whereas levels of lidocaine or diphenylhydantoin in the therapeutic range only suppress the sodium current in abnormal (depolarized) myocardium.

The other obvious difference between these putative subgroups was that whereas the Ia agents produce lengthening of action potential duration (APD), the Ib agents tend to shorten it.

In the 1960s and 1970s, a number of new antiarrhythmic drugs with class I properties came into wide clinical use. Some of these fell neatly into subgroups Ia (e.g. disopyramide, 1963) or Ib (e.g. mexiletine, 1973), but others did not. In particular, there appeared a group of agents which markedly depressed the sodium curent in both normal and depolarized myocardium, but which had little effect on APD. These were labelled, naturally, class Ic by Harrison (Harrison et al. 1980, 1981; Harrison 1985; Estes et al. 1985; Table 1).

B. Outline of Differences Between Subgroups

In therapeutic concentrations, quinidine, disopyramide and procainamide depress conduction (measured clinically as His-Purkinje conduction time or H-V interval) of both sinus beats and premature beats. They also tend to prolong repolarization, measured in vitro as APD or clinically as the JT interval

Table 1. Subclassification of class I drugs

	Drugs	Effects on action potential	Effects on clinical electrophysiological parameters
Class Ia	Quinidine Procainamide Disopyramide	Depress rate of depolarization and increase duration of repolarization	Moderate depression of intracardiac conduction velocity and prolongation of refractory periods
Class Ib	Lidocaine Tocainide Mexiletine	Minimal effect on upstroke of normal action potentials, but depress rate of depolarization of action potentials in ischaemic tissues; accelerate repolarization	Relatively selective depression of conduction in ischaemic tissues; usually no change or shortening of refractory periods
Class Ic	Flecainide Encainide Lorcainide	Markedly depress rate of depolarization; minimal effects on repolarization	Marked depression of conduction; small to moderate increase in refractory periods

on the electrocardiogram. The Ib agents, lidocaine, tocainide and mexiletine, usually have little effect on H-V intervals of sinus beats but depress conduction of premature beats or conduction through ischaemic tissue. They shorten the duration of repolarization. The Ic drugs, flecainide, encainide and lorcainide, markedly depress conduction of normal and premature beats (leading to increased QRS duration on the electrocardiogram in therapeutic concentrations). Acute exposure to Ic agents generally produces no effect or a small increase in the duration of repolarization, though this increase may be slightly more marked during chronic dosage.

The three subgroups differ also in their effects on atrial and ventricular refractory periods as measured by the usual, single-extrastimulus technique (JOSEPHSON and SEIDES 1979). The Ia compounds prolong the effective refractory periods (ERPs), as would be expected from their prolongation of APD (Chap. 9). The Ic drugs produce less marked slowing of repolarization and less prolongation of ERP (HELLESTRAND et al. 1982; SAMI et al. 1979; BAR et al. 1981). Variable effects on ERP have been reported for Ib drugs (JOSEPHSON et al. 1972; ANDERSON et al. 1978; MCCOMISH et al. 1977). In clinical concentrations they often reduce refractory periods as might be expected from their reduction of repolarization time. However, in higher concentrations, both in vivo and in vitro they are capable of prolonging ERP. Whether or not the refractory period is lengthened in a given case, it has been known for many years (SINGH and HAUSWIRTH 1974) that these drugs can markedly prolong the ERP relative to the APD. This phenomenon cannot be explained simply on the basis of steady-state blockade of a fixed number of sodium channels (CAMPBELL 1983a). An understanding of this phenomenon requires a consideration of recent studies on the kinetics of interaction of class I drugs with the sodium channel (see below and Chap. 8).

Finally, although sinoatrial node cells have virtually no functioning fast inward sodium channels (BROWN 1982), class I antiarrhythmic drugs, especially in toxic concentrations, have been reported to produce depression of these cells (LIPPESTAD and FORFANG 1971; ROOS et al. 1976; BIRKHEAD and VAUGHAN WILLIAMS 1977; DHINGRA and ROSEN 1979; KIM and FRIEDMAN 1979; SINGH et al. 1984; DIBIANCO et al. 1982). There are some differences between the various subgroups in terms of their effects on sinus node action potentials and these will be discussed below.

C. Fundamental Bases for Subclassification

Given then, that the long list of class I antiarrhythmic drugs can be condensed into three reasonably well defined subgroups, what do we know of the fundamental reasons for these differences? This question will be answered by considering in turn the three main electrophysiological grounds for discrimination between these compounds, namely, their interaction with the sodium channel, their effects on APD in atrial, ventricular and Purkinje fibers and their actions on sinoatrial node potentials.

I. Interaction with Sodium Channel

1. Rate-Dependent Block

Most of the evidence available is consistent with the view that the cardiac sodium channel is similar to its much better studied counterpart in nerve (FOZZARD et al. 1985) although recent single-channel studies with the new technique of patch-clamping may lead to alterations in that view (ALDRICH et al. 1983). For our purposes at present, the cardiac sodium channel can be thought of as existing in one of three possible states: active or open, resting and inactivated. Only the first of these will conduct sodium ions (HODGKIN and HUXLEY 1952; see Chap. 8 for more detail). Resting channels are available for activation by a depolarizing stimulus which reaches threshold. Inactivated channels cannot be activated without first returning to the resting phase, a time- and voltage-dependent process.

A study of the kinetics of interaction of class I drugs with the cardiac sodium channel would ideally be based on a detailed voltage-clamp analysis in which sodium current would be measured directly. That no such definitive study has yet been published is due to the great technical difficulties involved in attempting accurately to voltage-clamp the sodium current in cardiac tissues (FOZZARD et al. 1985). In the absence of such data, many workers have relied on the maximum rate of depolarization ($\dot{V}_{max}$) of cardiac action potentials as a measure of the sodium current. There are theoretical difficulties with this approach (FOZZARD et al. 1985, GRANT et al. 1984; COHEN et al. 1984; COURTNEY 1985; HONDEGHEM 1985), but it has nonetheless provided a number of interesting insights into the mode of action of these drugs, which we will now review.

These studies flowed from the observation that the degree of depression of $\dot{V}_{max}$ (presumed to be a function of sodium current blockade), produced by class I drugs depended on the rate of stimulation of the cardiac tissue (rate or use dependence). This phenomenon was first reported for quinidine by JOHNSON and MCKINNON (1957). They found that the maximum rate of depolarization ($\dot{V}_{max}$) of guinea pig ventricular fibres in control solution was not altered by increases in driving rate over the range 0.1-2 Hz. Beyond that, and up to the maximum rate of 10 Hz, $\dot{V}_{max}$ fell progressively to between 50% and 90% of its initial value. They detected no fall in resting potential. In the presence of quinidine sulphate, 10 µg/ml, the effect on $\dot{V}_{max}$ was markedly exaggerated and began to appear at rates as low as 0.5 Hz. A sudden increase in rate, in the presence of quinidine, resulted in a gradual decline of $\dot{V}_{max}$ to a new plateau level over about 5 s.

Very similar findings for the action of quinidine on rabbit atrial tissue were reported by WEST and AMORY in 1960 and, more than a decade later, such behaviour was also documented for a number of drugs with class I activity, including lidocaine, procainamide, pronethanol and propranolol (HEISTRACHER 1971; TRITTHART et al. 1971; MANDEL and BIGGER 1971). Since then, rate dependence has been found in a large number of local anaesthetics and class I antiarrhythmic drugs and appears likely to be a property common to all such compounds (see, for example, COURTNEY 1979, 1980a, b, c, 1983; SADA and BAN 1980, 1981; CAMPBELL 1983b; CAMPBELL and VAUGHAN WILLIAMS 1983).

2. Kinetics of Onset and Offset of Rate-Dependent Block— A Basis for Subclassification

Figure 1 shows three examples of this phenomenon with three different class I drugs. It can be seen that in the absence of stimulation, the drugs produced little or no depression of $\dot{V}_{max}$ as estimated from the first action potential of a train. When present, this small degree of depression is designated resting block. With subsequent stimulation, at a cycle length of 300 ms, $\dot{V}_{max}$ fell progressively to a new plateau level in the presence of each drug. The degree of this fall is referred to as rate-dependent block (RDB) and may be expressed as a percentage of the initial value of $\dot{V}_{max}$. Recovery of $\dot{V}_{max}$ towards its initial (resting) value occurs at a predictable rate at the end of such a train of stimuli. This can be studied by adding single extra stimuli at varying intervals after a series of identical trains (CAMPBELL 1982a, b; Fig. 2).

For any class I drug studied in this way, the degree of rate-dependent block increases with increasing concentration as does the rate of onset of rate-dependent block. The rate of recovery from rate-dependent block, however, is independent of drug concentration. These findings have been confirmed for a large number of agents by several different workers (CAMPBELL 1983a, b, c; CAMPBELL and VAUGHAN WILLIAMS 1983; COURTNEY 1980b, c).

When the nine class I drugs in the putative subclasses Ia, b, c were studied

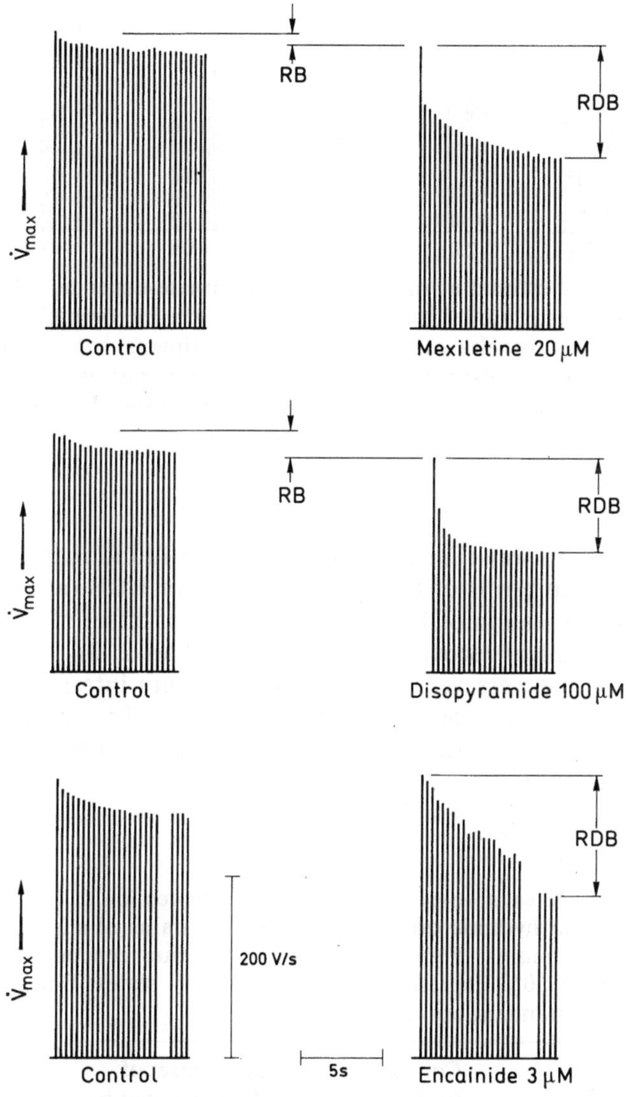

Fig. 1. The effect on V_{max} of a train of action potentials in previously quiescent tissue in control solution (*left-hand panels*) and in the presence of 20 μ*M* mexiletine (*top, right*), 100 μ*M* disopyramide (*middle, right*) and 3 μ*M* encainide (*bottom right*). The interstimulus interval in all cases is 300 ms. The spikes represent the V_{max} of successive action potentials. There is minor rate-dependent depression of V_{max} (RDB) in control solution in each case, and marked and approximately equal depression in the presence of each drug. RDB develops rapidly with mexiletine, slowly with encainide (so that only the first 20 and last 4-beats of 60-beat trains are shown) and at an intermediate rate with disopyramide. There is also minor resting depression of V_{max} (resting block, RB) in the presence of mexiletine and disopyramide, but not encainide. Vertical calibration, 200 *V*/s; horizontal calibration, 5 s (calibrations apply to all three panels). (CAMPBELL 1983b; reproduced with permission)

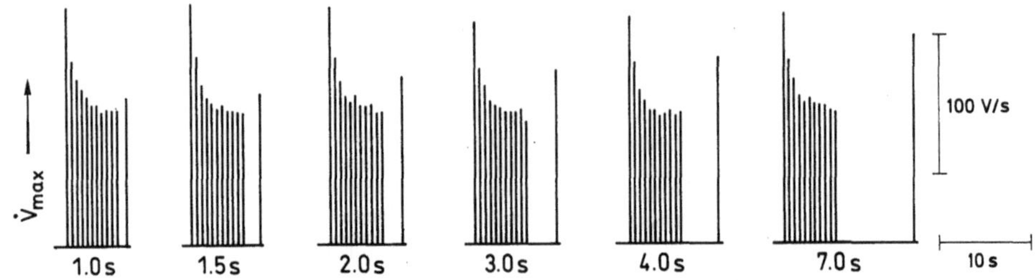

Fig. 2. The method of estimating rate of recovery from rate-dependent block. The drug in this case is an investigational steroid, Org 6001 and the interstimulus interval of the trains is 600 ms. *Numbers below each train* represent the time (s) between the last beat of the train and the single extrastimulus. At least 1 min rest was allowed between trains. Vertical calibration, 100 *V*/s, horizontal calibration, 10 s (CAMPBELL 1982; reproduced with permission)

systematically in this way (CAMPBELL 1983 a), it was found that they fell neatly into three groups in terms of their rates of onset of RDB (at concentrations producing 50% depression of $\dot{V}_{max}$) and their rates of offset of RDB estimated as the time constant of recovery of $\dot{V}_{max}$ (τ_{re}) (Tables 2, 3). Three drugs had "fast" kinetics, three were very "slow" and three were intermediate. Furthermore these three groups corresponded to subclasses Ib, Ic and Ia, respectively, of HARRISON's classification, which is based on quite different criteria (HARRISON 1985; Table 1).

Table 2. Rates of onset of rate-dependent depression of $\dot{V}_{max}$ for nine antiarrhythmic drugs (estimated from trains of stimuli at an interstimulus interval of 300 ms at a drug concentration producing 50% depression of $\dot{V}_{max}$ at steady state). Data are means ± standard error. (CAMPBELL 1983 a)

Drug	Concentration (μ*M*)	Rate of onset of Rate-dependent block (AP^{-1})
Lidocaine	200	> 0.6
Mexiletine	20	> 0.6
Tocainide	300	0.277 ± 0.029
Disopyramide	100	0.112 ± 0.007
Quinidine	20	0.068 ± 0.005
Procainamide	180	0.055 ± 0.003
Flecainide	5	0.029 ± 0.002
Encainide	3	0.025 ± 0.006
Lorcainide	2	0.022 ± 0.004

AP, action potential

Table 3. Time constant of recovery from rate-dependent depression of $\dot{V}_{max}$ (τ_{re}) for nine antiarrhythmic drugs

Drug	τ_{re}	Reference
Lidocaine	300 ms	Courtney 1980b
Mexiletine	471 ms	Campbell 1983b
Tocainide	200 ms	Courtney 1980c
Quinidine	4.7 s	Grant et al. 1982
Procainamide	2.3 s	Courtney 1980b
Disopyramide	2.2 s	Campbell 1983c
Lorcainide	13.2 s	Campbell 1983c
Flecainide	15.5 s	Campbell 1983c
Encainide	20.3 s	Campbell 1983c

3. Possible Mechanism of Rate-Dependent Block

As early as 1958 the observation was made that quinidine was capable of prolonging the effective refractory period (ERP) beyond the duration of the action potential (JOHNSON and ROBERTSON 1958; VAUGHAN WILLIAMS 1958). SZEKERES and VAUGHAN WILLIAMS (1962) showed similar effects with procainamide and concluded that "antifibrillatory drugs could act by interfering with the process by which the carrier (of Na^{+}) is reactivated in response to repolarization". These concepts of inactivation and reactivation (removal of inactivation) were borrowed from the terminology introduced by HODGKIN and HUXLEY (1952) in their classic description of the time- and voltage-dependent gating processes of the fast sodium channel of the squid axon. WEIDMANN (1955) had by this time produced evidence that procainamide did not appear to act by simply blocking a certain proportion of sodium channels, since the depression of $\dot{V}_{max}$ that they produced could be reversed by hyperpolarizing prepulses. This led him to propose that these drugs acted by "shifting to the right" the equilibrium of the three possible states of the sodium channel:

$$\text{resting} \longrightarrow \text{active} \longrightarrow \text{inactivated} \quad (\text{inactivated} \longrightarrow \text{resting})$$

Now this would not, of itself, explain the gradual decline in $\dot{V}_{max}$ seen in response to an increase of rate in the presence of these drugs. The mechanism proposed by SZEKERES and VAUGHAN WILLIAMS (1962) could produce such rate dependence, however, but only if the prolongation of the time course of the inactivated → resting state transition were such that this reactivation process was not completed by the time of the next action potential. This would lead to a progressive pooling of channels in the inactivated (i.e. non-conductive state), an effect which would be enhanced at faster rates and eventually disappear at sufficiently slow rates.

4. Experimental Evidence for Prolongation of Reactivation Time by Class I Drugs

Evidence that class I antiarrhythmic drugs do, in fact, prolong the recovery from inactivation of sodium channels has come from three types of experimental approach. Some workers have used voltage clamping in an attempt to measure directly the time constants of the inactivation and reactivation processes and have produced evidence that the latter process (but not the former) is markedly prolonged by quinidine (LUCKSTEAD and TARR 1972; DRIOT and GARNIER 1972; DUCOURET 1976), lidocaine (BROWN et al. 1980; WELD and BIGGER 1975; DUCOURET et al. 1981), procainamide and disopyramide (DUCOURET et al. 1981). A second approach, which avoids the technical and theoretical problems of voltage clamping, has been to use the time constant of recovery of $\dot{V}_{max}$ from rate-dependent block. The data in Table 2 were obtained in this way.

At normal resting potentials, in the absence of drug, the time constant of recovery (τ_{re}) from rate-dependent block for $\dot{V}_{max}$ is less than 100 ms (GETTES and REUTER 1974; BROWN et al. 1981). As can be seen from Table 2, class I drugs can prolong this enormously so that for all but the Ib drugs only partial diastolic recovery of $\dot{V}_{max}$ is possible at clinically relevant heart rates.

The third approach to assessing the effects of local anaesthetic drugs on recovery from inactivation is to observe them directly by studying the capacitative "gating" currents produced on activation of the sodium channels. This has not been technically feasible in cardiac muscle because it requires internal perfusion and voltage clamping and most of the work has been done in squid axons (CAHALAN et al. 1980; ARMSTRONG 1981; KHODOROV 1981). Briefly, these workers have confirmed prolongation of reactivation by these drugs, and have demonstrated that procedures which destroy the inactivation process but leave activation intact (such as internal perfusion with pronase) markedly reduce their effect on sodium conductance. CAHALAN et al. (1980) suggest that the action of these drugs is entirely due to this "inactivation enhancement", and imply that the inactivation "gate" may be the drug receptor.

By the mid-1970s, there was an obvious need for a model of the sodium channel and its interaction with antiarrhythmic drugs. Preferably, this model would not only explain the mechanism of action of these drugs on cardiac sodium channels and the differences between various drugs, but would also explain their local anaesthetic properties on nerve sodium channels. In fact, when such a model first appeared in 1977, it depended heavily on extensive peripheral nerve studies that had been going on concurrently, and it is therefore appropriate to pause at this point and briefly review this work.

5. The Sodium Channel in Nerve: Local Anaesthetic Action

It was known by this time that lidocaine, procaine and quinidine "blocked" the sodium channels of squid axons from within the axoplasm, and that the cationic forms were apparently the active species, but that the neutral molecules were essential for initial access to the axoplasm (NARAHASHI et al. 1970;

FRAZIER et al. 1970; NARAHASHI and FRAZIER 1975; YEH and NARAHASHI (1976). Using microiontopheretic and other techniques, this has now been demonstrated for myocardium too (GLIKLICH and HOFFMAN 1978).

STRICHARTZ (1973) studied two quaternary (i. e. permanently cationic) derivatives of lidocaine and found that, in the frog myelinated nerve, these cations only blocked the sodium current if they were applied internally. Furthermore, membrane depolarizations of sufficient size to activate the sodium channels greatly enhanced these otherwise small effects on fast sodium current. He termed this "voltage-dependent" block and proposed the following model to explain his findings: there is a single receptor for local anaesthetics which lies within the sodium channel. The drugs cannot gain access to it from outside because of a narrow ion-selectivity pore (HILLE 1971) near the outer end of the channel. Access via the lipid phase of the membrane is also denied to cationic drugs and thus their only route to the receptor is from the cytosol by passing through the activation gate, which, he proposed, lay at the inner end of the channel. Thus depolarization, which opens this gate, greatly facilitates access to the receptor.

COURTNEY (1975) took the work a step further. He referred to Strichartz's "voltage dependence" as "frequency" or "use dependence" since frequent action potentials (producing frequent activation of sodium channels) enhanced local anaesthetic action. He showed that the same effect could be seen with a tertiary lidocaine derivative of low lipophilicity (GEA 968) and found that slow recovery from use-dependent block induced by this drug occurred even if the membrane was held at normal resting potential (presumably due to drug escaping through the lipid phase of the membrane in its uncharged from). COURTNEY also found that when a molecule of GEA 968 bound to its receptor it not only blocked the sodium channel but shifted the voltage dependence of the inactivation curve in the hyperpolarizing direction, as had been suggested many years before for quinidine acting on the cardiac Purkinje fibre (WEIDMANN 1955).

6. Model for the Sodium Channel and Its Interaction with Antiarrhythmic Drugs

On the basis of this work, very similar models for the interaction of local anaesthetics with sodium channels in nerve (HILLE 1977a, b, 1978) and myocardium (HONDEGHEM and KATZUNG 1977a, b, 1984) were proposed by independent groups. The myocardial model (the only one which need concern us and which is discussed in detail in the next chapter) accepts the concept of the sodium channel outlined above. The drugs were considered capable of binding to (and blocking) the channel in any of the three states (active, inactivated, resting), but to have much greater affinity for the first two of these. Once bound to a channel the drug also shifted the steady-state inactivation curve for that channel in the hyperpolarizing direction. Since the rate constants for binding and unbinding depend on the state of the channel, this model is termed a "modulated receptor hypothesis". The greater binding to non-resting states and the shift of the inactivation curve imply greater binding at faster rates because the channel is spending proportionately more time in the active and inactivated states. (This model also provides an explanation for the fact that cardiac sodium channels, which have long action potentials and hence spend a relatively large amount of time inactivated, are far more sensitive to local anaesthetic drugs than are nerve channels, with very short action potentials.)

Some at least of the differences in kinetics of interaction with the sodium channel can also be explained by this model (HONDEGHEM and KATZUNG 1980, 1984; BEAN et al. 1983; MATSUBARA and HONDEGHEM 1983; GINTANT et al. 1980; MATSUKI et al. 1981). It is proposed that drugs with slow or intermediate onset of rate-dependent block have significant affinity for the open (active) channel state only. Since channels are open for only 1-3 ms per action potential, these drugs require a number of beats at a given cycle length to reach equilibrium. "Fast" drugs, on the other hand, are thought to bind with high affinity to the inactivated channel state which occupies most of the action potential duration. Equilibrium binding can thus be established within one or a few beats after a change in heart rate. Whether binding occurs preferentially to inactivated channels or first to open channels which subsequently become inactivated, it seems that the binding process tends markedly to slow the recovery of the drug-associated sodium channels from inactivation. The time constant of recovery of $\dot{V}_{max}$ from rate-dependent depression is thought to be a function of the unbinding rate constants of the various drugs. If these hypotheses are correct then one would expect that the onset kinetics of a particular drug would be concentration dependent but the offset kinetics (τ_{re}) would not. This is indeed the case (HEISTRACHER 1971; COURTNEY 1980b, c; SADA and BAN 1981; SADA et al. 1979; OSHITA et al. 1980; GRANT et al. 1982; CAMPBELL 1983c). The factors determining τ_{re} for a given drug are presently unknown but there is a considerable body of data as to the correlation between various physicochemical properties of the compounds and their rates of recovery (COURTNEY 1984). In particular, low molecular weight seems to correlate well with fast recovery kinetics (EHRING and HONDEGHEM 1981; SADA and BAN 1981; CAMPBELL 1983c).

A detailed consideration of these data is beyond the scope of this article. For now, let us consider briefly some of the differences between class I drugs in terms of their effects on parameters other than the sodium current (or $\dot{V}_{max}$).

II. Effects on Action Potential Duration

The three subgroups of class I drugs as have been outlined above differ in their effects on the duration of atrial and ventricular action potentials. The Ia compounds tend to prolong their duration, the Ib drugs to shorten it and the Ic drugs usually produce small prolongation or have no effect at least in acute use (HARRISON 1985; ESTES et al. 1985; CAMPBELL 1983a).

1. Ionic Bases for the Differences

An agent which acts solely to depress sodium influx would be expected to shorten APD, as indeed occurs with tetrodotoxin (CORABOEUF et al. 1979; CARMELIET and SAIKAWA 1982). This is because a small inward sodium conductance, either "leak" current or "window" current (residual fast sodium channel conductance still present at plateau potentials; see ATTWELL et al. 1979) or both, contributes to the duration of the plateau and hence to APD (CARMELIET and SAIKAWA 1982). This inward component of the plateau currents is depressed by tetrodotoxin and the plateau is consequently shortened. The same phenomenon is seen with even very low concentrations of lidocaine (CARMELIET and SAIKAWA 1982). This, and not an enhancement of outward potassium current, as once thought (BIGGER and MANDEL 1970; ARNSDORF and BIGGER 1972), is almost certainly the basis of the APD shortening seen with all Ib drugs.

On the other hand, there is good evidence from both nerve and cardiac studies that Ia drugs depress repolarizing potassium currents (NAWRATH 1981; NAWRATH and ECKEL 1979; NISHIMURA et al. 1982). This effect presumably overrides the tendency to shorten APD by reduction of plateau sodium currents and leads to the prolongation of repolarization that is a feature of this subgroup.

Finally, the limited evidence available suggests that Ic agents may share this ability to block potassium currents but it is only significant at concentrations well above the therapeutic range (COWAN and VAUGHAN WILLIAMS 1981; CARMELIET 1980).

III. Effects on Sinus Node

All the class I drugs are capable of increasing the spontaneous cycle length of the sinus node in sufficient concentration. This is not a common clinical problem except in patients with pre-existent sinus node disease. The effects of nine class I drugs on action potentials have recently been studied during continuous recording from guinea pig sinus node cells (CAMPBELL 1987).

The Ia drugs all produced significant slowing of spontaneous rate in therapeutic concentrations. The Ib agents did so only in concentrations well above therapeutic levels, and Ic drugs were of intermediate potency. All nine drugs markedly slowed the repolarization rate and this was the major mechanism of sinus slowing for the Ia and Ic compounds. The Ib drugs shared this effect but prolongation of phase 4 by reduction of the slope of diastolic depolarization was also a prominent feature of their action.

The ionic mechanisms of these effects were not studied but it seems likely that the slowing of repolarization was produced by depression of the outward potassium current i_K (BROWN 1982). Because of the complex and still controversial nature of the currents underlying pacemaking (BROWN 1982; D. NOBLE and S.J. NOBLE 1984), any attempt to explain the phase 4 effects would be pure speculation.

Whatever the ultimate explanation of these actions the differences observed between the various drugs provide further support for the concept of subclassification.

D. Clinical Implications of Subclassification

I. Relevance of Differing Onset and Offset Kinetics

The major significance of an understanding of the differences between class I drugs in terms of their speed of interaction with the sodium channel has not been so much in opening up new clinical therapeutic manoeuvres, as in explaining some of the known differences in clinical electrophysiological properties between the various subgroups.

1. Selectivity for Ischaemic Myocardium

Drugs which bind preferentially to inactivated sodium channels but unbind very quickly from resting channels (Ib drugs) would be expected to display selectivity for ischaemic tissue. Ischaemic myocardium is usually partly depolarized and hence has a higher proportion of its sodium channels in the inactivated state between action potentials than does normal myocardium. This encourages binding and discourages unbinding. It is not surprising, therefore, that such selectivity has been reported (HONDEGHEM et al. 1974). Thus it is possible to give lidocaine, for example, in concentrations which markedly depress ischaemic and potentially arrhythmogenic myocardium but have little effect on remaining normal heart muscle. Most of the drug binding that occurs during an action potential is reversed during the subsequent diastolic interval provided the resting potential is normal. Ia and Ic drugs unbind slowly even at normal resting potentials so that concentrations which depress ischaemic tissue are likely also to markedly affect normal cells.

2. Selectivity for Ventricular Cells

Ib antiarrhythmic agents are relatively less effective against atrial than ventricular arrhythmias (MATSUBARA and HONDEGHEM 1983), a property not generally seen with Ia or Ic drugs. This phenomenon can be explained by a similar argument to that given in the preceding section. Atrial action potentials are shorter than ventricular potentials so that at a given heart rate sodium channels in atrial cells spend a smaller proportion of their time in the inactivated state and correspondingly more time in the resting state than do their ventricular counterparts. Thus Ib drugs which bind quickly to inactivated channels but also unbind rapidly when those channels are repolarized will depress atrial cells less than ventricular cells at a given heart rate and drug concentration. This differential effect would not be expected to be so great for drugs exhibiting slower kinetics (Ia and Ic).

3. Differing Effects on Refractory Periods

Cardiac refractory periods are normally measured clinically by interpolating single premature extrastimuli every eight or ten beats in a basic drive cycle. The interval between normal drive beat and extrastimulus is reduced in steps until no response is produced by the extrastimulus (refractoriness). This procedure can be thought of as periodically presenting the heart with a sudden increase in rate which happens to last for only a single cycle. It seems reasonable to hypothesize that the ability of a class I drug to increase refractoriness measured in this way should be largely determined by the drug's ability to respond to a sudden increase in stimulation frequency by producing a rapid additional depression of the cardiac sodium channels (provided any drug-induced changes in action potential duration are also taken into account). This hypothesis has recently been tested for nine class I drugs in guinea pig ventricle and found to be largely true (CAMPBELL 1983a). The "fast" drugs (Ib) were able markedly to prolong the effective refractory period (ERP) relative to the duration of the action potential (APD), the "slow" (Ic) drugs had only minor effects on ERP-APD and the "intermediate" (Ia) compounds produced small to moderate increases in ERP relative to APD. In addition the Ia drugs significantly prolonged APD, which was shortened by the Ib group. Thus, by studying differences in onset kinetics and effects on APD, it was possible to explain the differing actions of the nine agents on ERP. Similar findings for lidocaine, mexiletine, procainamide and disopyramide have since been reported by others using dog Purkinje fibres (VARRO et al. 1985). These workers found that quinidine and flecainide did prolong ERP relative to APD in Purkinje fibres but not in ventricular myocardium.

4. Differential Depression of Premature Beats

Just as differing rates of onset of rate-dependent blockade of sodium channels can explain differences in effects on refractory periods, so it seems reasonable to suggest that variation in offset kinetics (unbinding from sodium channels) might explain differential drug effects on the conduction velocity of premature beats ("ectopics"). Thus Ib drugs with very fast recovery kinetics between action potentials could be given in concentrations which have little effect on $\dot{V}_{max}$ (and hence conduction) of normal heart beats but selectively depress conduction of premature beats and tachycardia. Ia and Ic drugs with very long recovery time constants relative to normal diastolic intervals would be expected to suppress $\dot{V}_{max}$ essentially equally at all clinically relevant rates. Hence concentrations which are antiarrhythmic must also produce depression of conduction of normal beats. A degree of selectivity for early diastolic beats has been documented for tocainide and lidocaine (MAN and DRESEL 1979) and although this hypothesis has not been fully tested for a large number of class I drugs it probably provides the explanation for the clinical observation that therapeutic concentrations of Ib drugs do not usually depress conduction of normal beats, whereas Ia drugs usually produce some increase in conduction times and Ic drugs do so very markedly.

II. Relevance of Effects on Action Potential Duration

This question has already been touched upon in reference to the effects of class I drugs on refractoriness. The overall effect of a given drug on refractory period is a function both of its ability to prolong refractoriness relative to APD and its direct effects on APD. Thus in the study of ERP and APD referred to above (CAMPBELL 1983a), tocainide did not significantly increase the effective refractory period (ERP) despite prolonging ERP relative to APD, be cause it simultaneously shortened APD. At concentrations within the therapeutic range, quinidine and disopyramide prolonged ERP more effectively than any of the Ib drugs despite the greater ability of the latter group to enhance ERP relative to APD. This was simply because a marked class III effect (increased APD) was seen with quinidine and disopyramide whereas the Ib drugs all shortened APD.

It should be noted at this point, however, that in vitro data such as those just discussed cannot always be translated exactly to the in vivo situation. For instance there is evidence that the class III effects of quinidine and disopyramide are diminished in vivo because of the anticholinergic properties of these agents (MIRRO et al. 1980a, b, 1981). Furthermore it seems from clinical studies that some Ic drugs in chronic use may produce significant APD lengthening and associated increases in refractory period (JACKMAN et al. 1982). Whether this is an effect of the parent compounds or of metabolites remains to be determined.

Two further subjects that should be discussed in this section are the role of APD prolongation in the proarrhythmic effects of antiarrhythmic drugs and the question of whether APD shortening (by Ib drugs) is proarrhythmic or antiarrhythmic.

1. Action Potential Prolongation as a Proarrhythmic Effect

This matter will be dealt with in terms of individual drugs in later chapters (especially Chap. 9). A small percentage of patients on chronic antiarrhythmic therapy develop new, severe ventricular arrhythmias, particularly, though not exclusively, a bizarre form of ventricular tachycardia called "torsade de pointes" (KRIKLER and CURRY 1976). These arrhythmias may be fatal. They are often associated with prolongation of the QT interval on the electrocardiogram which can be due to widening of QRS, increased APD in the cardiac cells or both, and they reverse on drug withdrawal. They are seen most frequently with drugs that markedly prolong APD in vitro especially the Ia agents (see Chap. 9) and the class III drugs sotalol and amiodarone, but have also been reported for Ic drugs (LUI et al. 1982; WINKLE et al. 1981). There is evidence that the arrhythmias may be on the basis of oscillatory electrical activity occurring during the prolonged plateau of the cardiac action potentials ("early after depolarization"). There is also very recent evidence that low serum potassium levels and concurrent digitalis therapy may be risk factors for this side effect, which commonly appears within the first few days of therapy if it occurs at all (MINARDO et al. 1986).

2. Action Potential Shortening—Proarrhythmic or Antiarrhythmic?

There has been considerable debate over whether Ib drugs are antiarrhythmic partly because they shorten APD, or despite this effect (WITTIG et al. 1973; VAUGHAN WILLIAMS 1978). WITTIG et al. (1973) showed in dog hearts that lidocaine shortened APD throughout the ventricle but that this effect was greatest where APD had previously been longest, namely at the distal Purkinje fibres. They suggested that this action contributed to the antiarrhythmic effect of lidocaine by reducing non-uniformity of recovery of excitability, shortening the long functional refractory period of the Purkinje fibre-ventricular muscle junction zone and abolishing decremental conduction of early premature beats. They suggested that such enhancement of conduction in particular would abolish re-entrant circuits. VAUGHAN WILLIAMS (1978) strongly disagreed with this hypothesis on theoretical grounds. WALD et al. (1980) attempted to test the hypothesis that some antiarrhythmic drugs (including lidocaine) might abolish re-entrant circuits by improving conduction through abnormal areas or by converting unidirectional block (conducive to re-entry) to bidirectional conduction. Reduction of APD (with its accompanying increase in diastolic recovery time at a constant heart rate) was one of the mechanisms by which it was thought this effect might be produced. In fact, they found that the usual response was for unidirectional block to be converted to bidirectional block, an effect almost certainly mediated by fast sodium channel blokkade rather than actions on APD.

Thus, there is as yet no convincing experimental evidence favouring an antiarrhythmic role for the APD reduction seen with Ib drugs. These drugs can certainly provoke or aggravate arrhythmias in some patients (PODRID 1985) and it is possible that the tendency to reduced refractory times associated with decreased APD may contribute to these iatrogenic arrhythmias.

E. Conclusions

We have reviewed the evidence for subdividing class I antiarrhythmic drugs into three groups according to their interaction with cardiac sodium channels, their effects on action potential duration and their effects on sinus node function.

In particular it seems possible to subclassify class I drugs, according to the kinetics of their binding to and unbinding from sodium channels, into fast, intermediate and slow groups. Further, the "fast" drugs share the property of shortening action potential duration, the "intermediate" drugs lengthen it and the "slow" drugs have little effect.

The clinical relevance of this subclassification lies in its ability to explain some of the differences in the electrophysiological properties of these agents already noted by clinicians. Indeed the subgroups defined on cellular criteria coincide exactly with those already suggested on clinical grounds.

With new class I agents appearing at frequent intervals it is likely that the borders between these subgroups will become blurred. Nevertheless the concepts involved in subclassifying them will still repay careful consideration and will no doubt continue to enhance our understanding of this very useful group of drugs.

References

Aldrich RW, Corey DP, Stevens CF (1983) A reinterpretation of mammalian sodium channel gating based on single channel recording. Nature 306:436-441

Anderson JL, Mason JW, Winkle RA, Meffin PJ, Fowles RE, Peters F, Harrison DC (1978) Clinical electrophysiological effects of tocainide. Circulation 57:685-691

Amstrong CM (1981) Sodium channels and gating currents. Physiol Rev 61:644-683

Arnsdorf MF, Bigger JT (1972) Effect of lidocaine hydrochloride on membrane conductance in mammalian cardiac Purkinje fibres. J Clin Invest 51:2252-2263

Attwell D, Cohen I, Eisner D, Ohba M, Ojeda C (1979) The steady state TTX-sensitive ("window") sodium current in cardiac Purkinje fibres. Pflugers Arch 379:137-142

Bär FW, Farre J, Ross D, Vanagt EJ, Gorgels AP, Wellens HJJ (1981) Electrophysiological effects of lorcainide, a new antiarrhythmic drug. Br Heart J 45:292-298

Bean BP, Cohen CJ, Tsien RW (1983) Lidocaine block of cardiac sodium channels. J Gen Physiol 81:613-642

Bigger JT, Mandel WJ (1970) Effect of lidocaine on conduction in canine Purkinje fibers and at the ventricular muscle-Purkinje fiber junction. J Pharmacol Exp Ther 172:239-254

Birkhead JS, Vaughan Williams EM (1977) Dual effect of disopyramide on atrial and atrioventricular conduction and refractory periods. Br Heart J 39:657-660

Brown AM, Giles W, Hume JR, Lee KS (1980) Voltage clamp analysis of lidocaine and quinidine induced depression of the sodium current in isolated rat ventricular cells. J Physiol (Lond) 307:62-63

Brown AM, Lee KS, Powell T (1981) Sodium current in single rat heart muscle cells. J Physiol (Lond) 318:479-500

Brown HF (1982) Electrophysiology of the sinoatrial node. Physiol Rev 62:505-530

Cahalan M, Shapiro BI, Almers W (1980) Relationship between inactivation of sodium channels and block by quaternary derivatives of local anesthetics and other compounds. In: Fink BR (ed) Molecular mechanisms of anesthesia. Raven, New York, pp 17-33 (Progress in anesthesiology, vol 2)

Campbell TJ (1982a) Voltage- and time-dependent depression of maximum rate of depolarisation of guinea-pig ventricular action potentials by two steroidal antiarrhythmic drugs, CCI22277 and Org6001. Br J Pharmacol 77:541-548

Campbell TJ (1982b) Studies on the mode of action of cardioactive drugs in animals and man. D. Phil thesis, Oxford University

Campbell TJ (1983a) Kinetics of onset of rate-dependent effects of Class I antiarrhythmic drugs are important in determining their effects on refractoriness in guinea-pig ventricle, and provide a theoretical basis for their subclassification. Cardiovasc Res 17:344-352

Campbell TJ (1983b) Resting and rate-dependent depression of maximum rate of depolarisation ($\dot{V}_{max}$) in guinea-pig ventricular action potentials by mexiletine, disopyramide and encainide. J Cardiovasc Pharmacol 5:291-296

Campbell TJ (1983c) Importance of physico-chemical properties in determining the kinetics of the effects of Class I antiarrhythmic drugs on maximum rate of depolarization in guinea-pig ventricle. Br J Pharmacol 80:33-40

Campbell TJ (1987) Differing electrophysiological effects of Class IA, IB and IC antiarrhythmic drugs on guinea-pig sinoatrial node Br J Pharmacol 91:395-401

Campbell TJ, Vaughan Williams EM (1983) Voltage and time-dependent depression of maximum rate of depolarisation of guinea-pig ventricular action potentials by two new antiarrhythmic drugs, flecainide and lorcainide. Cardiovasc Res 17:251-258

Carmeliet E (1980) Electrophysiological effects of encainide on isolated cardiac muscle and Purkinje fibers on the Langendorff-perfused guinea-pig heart. Eur J Pharmacol 61:247-262

Carmeliet E, Saikawa T (1982) Shortening of the action potential and reduction of pacemaker activity by lidocaine, quinidine, and procainamide in sheep cardiac Purkinje fibers: an effect on Na or K currents. Circ Res 50:257-272

Cohen CJ, Bean BP, Tsien RW (1984) Maximal upstroke velocity as an index of available sodium conductance. Circ Res 54:636-651

Coraboeuf E, Deroubaix E, Coulombe A (1979) Effect of tetrodotoxin on action potentials of the conducting system in the dog heart. Am J Physiol 236:H561-H567

Courtney KR (1975) Mechanism of frequency-dependent inhibition of Na currents in frog myelinated nerve by the lidocaine derivative GEA 968. J Pharmacol Exp Ther 195:255-36

Courtney KR (1979) Fast frequency-dependent block of action potential upstroke in rabbit atrium by small local anesthetics. Life Sci 24:1581-1588

Courtney KR (1980a) Structure-activity relations for frequency-dependent sodium channel block in nerve by local anesthetics. J Pharmacol Exp Ther 213:114-119

Courtney KR (1980b) Interval-dependent effects of small antiarrhythmics drugs on excitability of guinea-pig myocardium. J Mol Cell Cardiol 12:1273-86

Courtney KR (1980c) Antiarrhythmic drug design: frequency-dependent block in myocardium. In: Fink BR (ed) Molecular mechanisms of anesthesia. Raven, New York, pp 111-118 (Progress in anesthesiology, vol 2)

Courtney KR (1983) Quantifying antiarrhythmic drug blocking during action potentials in guinea-pig papillary muscle. J Mol Cell Cardiol 15:749-747

Courtney KR (1984) Size-dependent kinetics associated with drug block of sodium current. Biophys J 45:42-44

Courtney KR (1985) Letter. Circ Res 57:194-195

Cowan JC, Vaughan Williams EM (1981) Characterization of a new oral antiarrhythmic durg, flecainide (R818). Eur J Pharmacol 73:333-342

Dhingra RC, Rosen KM (1979) Procainamide and the sinus node. Chest 76:620-662

DiBianco R, Fletcher RD, Cohen AI, Gottdiener JS, Singh SN, Katz RJ, Bates HR, Sauerbrunn B (1982) Treatment of frequent ventricular arrhythmia with encainide: assessment using serial ambulatory electrocardiograms, intracardiac electrophysiologic studies, treadmill exercise tests, and radionucelide cineangiographic studies. Circulation 65:1134-1147

Driot P, Garnier D (1972) Analyse en courant et voltage imposé des propriétés antiarrhythmiques de la Quinidine appliquée au myocarde de la Grenouille. C R Acad Sci Paris (D) 274:3421-3424

Ducouret P (1976) The effect of quinidine on membrane electrical activity in frog auricular fibres studied by current and voltage clamp. Br J Pharmacol 57:163-84

Ducouret P, Gargouil YM, Poindessault JP (1981) Heart fast initial current reactivaton and antidysrhythmic agents. J Physiol (Lond) 320:29-30

Ehring GR, Hondeghem LM (1981) Antiarrhythmic structure activity relationships in a series of lidocaine and procainamide derivatives. Proc West Pharmacol Soc 24:221-224

Estes NAM, Garan H, McGovern B, Ruskin JN (1985) Class I antiarrhythmic agents: classification, electrophysiologic considerations, and clinical effects. In: Reiser HJ, Horowitz LN (eds), Mechanisms and treatment of cardiac arrhythmias; relevance of basic studies to clinical management. Urban and Schwarzenberg, Baltimore, pp 183-199

Fozzard HA, January CT, Makielski JC (1985) New studies of the excitatory sodium currents in heart muscle. Circ Res 56:475-485

Frazier DT, Narahashi T, Yamada M (1970) The site of action and active form of local anesthetics II: experiments with quaternary compounds. J Pharmacol Exp Ther 171:45-51

Gettes LS, Reuter H (1974) Slow recovery from inactivation of inward currents in mammalian myocardial fibres. J Physiol (Lond) 240:703-724

Gintant GA, Naylor RE, Hoffman BF (1980) Interaction of local anesthetic antiarrhythmic agents with sodium channels (abstr). Circulation 62 (Suppl II): III-137

Gliklich JI, Hoffman BF (1978) Sites of action and active forms of lidocaine and some derivatives on cardiac Purkinje fibers. Circ Res 43:638-51

Grant AO, Trantham JL, Brown KK, Strauss HC (1982) pH-dependent effects of quinidine on the kinetics of dV/dt_{max} in guinea-pig ventricular myocardium. Circ Res 50:210-217

Grant AO, Starmer CF, Strauss HC (1984) Antiarrhythmic drug action: blockade of the inward sodium current. Circ Res 55:428-439

Harrison DC (1985) Is there a rational basis for the modified classification of antiarrhythmic drugs? In: Morganroth J, Moore EN (eds), Cardiac arrhythmias: new therapeutic drugs and devices. Nijhoff, Boston, pp 36-48

Harrison DC, Winkle R, Sami M, Mason J (1980) Encainide: a new and potent antiarrhythmic agent. Am Heart J 100:1046-1054

Harrison DC, Winkle RA, Sami M, Mason JW (1981) Encainide: a new and potent antiarrhythmic agent. In: Harrison DC (ed) Cardiac arrhythmias: a decade of progress. Hall Medical, Boston, pp 315-330

Heistracher P (1971) Mechanism of action of antifibrillatory drugs. Naunyn-Schmiedebergs Arch Pharmacol 269:199-212

Hellestrand KJ, Bexton RS, Nathan AW, Camm AJ (1982) Acute electrophysiological effects of flecainide acetate on cardiac conduction and refractoriness in man. Br Heart J 48:140-148

Hille B (1971) The permeability of the sodium channel to organic cations in myelinated nerve. J Gen Physiol 58:599-619

Hille B (1977a) The pH-dependent rate of action of local anesthetics on the node of Ranvier. J Gen Physiol 69:475-496

Hille B (1977b) Local anesthetics: hydrophilic and hydrophobic pathways for the drug-receptor reaction. J Gen Physiol 69:497-515

Hille B (1978) Local anesthetic action on inactivation of the sodium channel in nerve and skeletal muscle: possible mechanisms for antiarrhythmic agents. In: Morad M (ed) biophysical aspects of cardiac muscle. Academic, New York, pp 55-74

Hondeghem LM (1985) Letter. Circ Res 57:192-193

Hondeghem LM, Katzung BG (1977a) Time and voltage-dependent interactions of antiarrhythmic drugs with cardiac sodium channels. Biochim Biophys Acta 472:373-398

Hondeghem LM, Katzung BG (1977b) A unifying molecular model for the interaction of antiarrhythmic drugs with cardiac sodium channels: application to quinidine and lidocaine. Proc West Pharmacol Soc 20:253-256

Hondeghem LM, Katzung BG (1980) Test of a model of antiarrhythmic drug action. Circulation 61:1217-1224

Hondeghem LM, Katzung BG (1984) Antiarrhythmic agents: the modulated receptor mechanisms of actions of sodium and calcium channel-blocking drugs. Annu Rev Pharmacol Toxicol 24:387-423
Hondeghem LM, Grant AO, Jensen RA (1974) Antiarrhythmic drug action: selective depression of hypoxic cardiac cells. Am Heart J 87:602-605
Hodgkin AL, Huxley AF (1952) A quantitative description of membrane current and its application to conduction and excitation in nerve. J Physiol (Lond) 117:500-544
Jackman WM, Zipes DP, Rinkenberger RL, Heger JJ, Prystowsky EN (1982) Electrophysiology of oral encainide. Am J Cardiol 49:1270-1278
Johnson EA, McKinnon MG (1957) The differential effect of quinidine and pyrilamine on the myocardial action potential at various rates of stimulation. J Pharmacol Exp Ther 120:460-468
Johnson EA, Robertson PA (1958) The stimulatory action of acetylcholine on isolated rabbit atria. Br J Pharmacol 13:304-7
Josephson ME, Seides SF (1979) Clinical cardiac electrophysiology: techniques and interpretations. Lea and Febiger, Philadelphia, p 46
Josephson ME, Caracta AR, Lau SH, Gallager JJ, Damato AN (1972) Effects of lidocaine on refractory periods in man. Am Heart J 84: 778-786
Kim HG, Friedman HS (1979) Procainamide-induced sinus node dysfunction in patients with chronic renal failure. Chest 76:699-700
Krikler DM, Curry PVL (1976) Torsade de pointes, an atypical ventricular tachycardia. Br Heart J 38:117-120
Lippestad CT, Forfang K (1971) Production of sinus arrest by lignocaine. Br Med J 1:537
Luckstead EF, Tarr M (1972) Comparson of quinidine and bretylium tosylate effects on cardiac ionic currents (abstr). Fed Proc 31:818
Lui HK, Lee G, Dietrich P, Low RI, Mason DT (1982) Flecainide-induced QT prolongation and ventricular tachycardia. Am Heart J 103:567-569
Man RYK, Dresel PE (1979) A specific effect of lidocaine and tocainide on ventricular conduction of mid-range extrasystoles. J Cardiovasc Pharmacol 1:329-342
Mandel WJ, Bigger JT (1971) Electrophysiologic effects of lidocaine on isolated canine and rabbit atrial tissue. J Pharmacol Exp Ther 178:81-93
Matsubara T, Hondeghem L (1983) Mechanism for preferential effectiveness of lidocaine against ventricular arrhythmias (abstr). Circulation 68 (Suppl III):295
Matsuki N, Quandt F, Yeh J, Ten Eick R (1981) Rate-dependent block of sodium channels by phenytoin and quinidine. Circulation 64 (Suppl IV):126
McComish M, Robinson C, Kitson D, Jewitt DE (1977) Clinical electrophysiological effects of mexiletine. Postgrad Med J 53 (Suppl 1):85-91
Minardo JD, Heger JJ, Zipes DP, Miles WM, Prystowsky EN (1986) Drug associated ventricular fibrillation: analysis of clinical features and QT_C prolongation (abstr). I Am Coll Cardiol 7:158a
Mirro MJ, Watanabe AM, Bailey JC (1980a) Electrophysiological effects of disopyramide and quinidine on guinea pig atria and canine cardiac Purkinje fibers. Circ Res 46:660-668
Mirro MJ, Manalan AS, Bailey JC, Watanabe AM (1980b) Anticholinergic effects of disopyramide and quinidine on guinea pig myocardium. Circ Res 47:855-865
Mirro MJ, Watanabe AM, Bailey JC (1981) Electrophysiological effects of the optical isomers of disopyramide and quinidine in the dog. Circ Res 48:867-874
Narahashi T, Frazier DT (1975) Site of action and active form of procaine in squid giant axons. J Pharmacol Exp Ther 194:506-513

Narahashi T, Frazier DT, Yamada M (1970) The site of action and active form of local anesthetics I. J Pharmacol Exp Ther 171:32-44
Nawrath H (1981) Action potential, membrane currents and force of contraction in mammalian heart muscle fibers treated with quinidine. J Pharmacol Exp Ther 216:176-181
Nawrath H, Eckel L (1979) Electrophysiological study of human ventricular heart muscle treated with quinidine: interaction with isoprenaline. J Cardiovasc Pharmacol 1:415-425
Nishimura M, Yamada S, Watanabe Y (1982) Electrophysiologic effects of disopyramide phosphate on the spontaneous action potential and membrane current systems of the rabbit atrioventricular node. Am J Cardiol (Abstr) 49:921
Noble D, Noble SJ (1984), A model of sino-atrial node electrical activity based on a modification of the DiFrancesco-Noble (1984) equations. Proc Roy Soc Lond [Biol] 222:295-304
Oshita S, Sada H, Kojima M, Ban T (1980) Effects of tocainide and lidocaine on the transmembrane action potentials as related to external potassium and calcium concentrations in guinea-pig papillary muscles, Naunyn Schmiedebergs Arch Pharmacol 314:67-82
Podrid PJ (1985) Aggravation of ventricular arrhythmia: a drug-induced complication. Drugs 29 (Suppl 4):33-44
Roos JC, Paalman ACA, Dunning AJ (1976) Electrophysiological effects of mexiletine in man. Br Heart J 38:1262-1271
Sada H, Ban T (1980) Effects of acebutolol and other structurally related beta adrenergic blockers on transmembrane action potential in guinea-pig papillary muscles. J Pharmacol Exp Ther 215:507-514
Sada H, Ban T (1981) Effects of various structurally related beta-adrenoceptor blocking agents on maximum upstroke velocity of action potential in guinea-pig papillary muscles. Naunyn Schmiedebergs Arch Pharmacol 317:245-251
Sada H, Kojima M, Ban T (1979) Effects of procainamide on transmembrane action potentials in guinea-pig papillary muscles as affected by external potassium concentration. Naunyn Schmiedebergs Arch Pharmacol 309:179-190
Sami M, Mason JW, Peters F, Harrison DC (1979) Clinical electrophysiological effects of encainide, a newly developed antiarrhythmic agent. Am J Cardiol 44:526-532
Singh BN, Hauswirth D (1974) Comparative mechanisms of action of antiarrhythmic drugs. Am Heart J 87:367-382
Singh SN, DiBianco R, Kostroff LI, Fletcher RD (1984) Lorcainide for high-frequency ventricular arrhythmia: preliminary results of a short-term double-blind and placebo-controlled crossover study and long-term follow-up. Am J Cardiol 54:22B-28B
Strichartz GR (1973) The inhibition of sodium currents in myelinated nerve by quartenary derivatives of lidocaine. J Gen Physiol 62:37-57
Szekeres L, Vaughan Williams EM (1962) Antifibrillatory action. J Physiol (Lond) 160: 470-482
Tritthart H, Fleckenstein B, Fleckenstein A (1971) Some fundamental actions of antiarrhythmic drugs on the excitability and contractility of single myocardial fibres. Naunyn Schniedebergs Arch Pharmacol 269:212-219
Varro A, Elharrar V, Surawicz B (1985) Frequency-dependent effects of several Class I antiarrhythmic drugs on $\dot{V}_{max}$ of action potential upstroke in canine cardiac Purkinje fibers. J Cardiovasc Pharmacol 7:482-492
Vaughan Williams EM (1958) The mode of action of quinidine on isolated rabbit atria interpreted from intracellular potential records. Br J Pharmacol 13:267-287

Vaughan Williams EM (1978) Some factors that influence the activity of antiarrhythmic drugs. Br Heart J 40 (Suppl):52-61
Wald RW, Waxman MB, Downar E (1980) The effect of antiarrhythmic drugs on depressed conduction and unidirectional block in sheep Purkinje fibers. Circ Res 46:612-619
Weidmann S (1955) Effects of calcium ions and local anaesthetics on electrical properties of Purkinje fibres. J Physiol (Lond) 129:568-582
Weld FM, Bigger JT (1975) Effect of lidocaine on the early inward transient current in sheep cardiac Purkinje fibers. Circ Res 27:630-639
West TC, Amory DW (1960) Single fiber recording of the effects of quinidine at atrial and pacemaker sites in the isolated right atrium of the rabbit. J Pharmacol Exp Ther 130:183-193
Winkle RA, Mason JW, Griffin JC, Ross D (1981) Malignant ventricular tachyarrhythmias associated with the use of encainide. Am Heart J 102:857-864
Wittig J, Harrison LA, Wallace AG (1973) Electrophysiological effects of lidocaine on distal Purkinje fibers of canine heart. Am Heart J 86:69-78
Yeh JZ, Narahashi T (1976) Mechanism of action of quinidine on squid axon membranes. J Pharmacol Exp Ther 196:62-70

CHAPTER 8

Interaction of Class I Drugs with the Cardiac Sodium Channel

L. M. HONDEGHEM

A. Introduction

In the preceding chapters it was shown that, based upon electrophysiological principles, arrhythmias can be classified according to mechanisms, and antiarrhythmic agents also can be grouped into several classes and subclasses (see Chaps. 4, 2). In the present chapter, I will describe the electrophysiological basis for subclassification of sodium channel blocking (class I) antiarrhythmic agents and review current models for their mechanism of action. Interactions of multiple agents with a common receptor and clinical implications will be emphasized. Although the chapter will be limited to the sodium channel receptor, it should be noted that these principles can be applied equally well to other classes of antiarrhythmic agents (HONDEGHEM and KATZUNG 1984; RODEN et al. 1986).

B. Historical Background

The various models of antiarrhythmic drug action are based mainly upon three key observations. JOHNSON and MCKINNON (1957) showed that in heart the action of sodium channel blockers is augmented by stimulation: when the preparation was rested for a long time, even in the presence of quinidine, there was little drug effect. However, upon starting stimulation the maximum upstroke velocity ($\dot{V}_{max}$) of the cardiac action potential declined on a beat-by-beat basis until a steady level of block was attained. This important observation became only fully appreciated when STRICHARTZ (1973) showed that in nerve, too, the action of local anesthetic agents is strongly use-dependent.

A second key observation was made by WEIDMANN (1955), who observed that in the presence of antiarrhythmic agents or sodium channel blockers the relationship between $\dot{V}_{max}$ and the membrane potential was shifted to more negative potentials. He proposed that these agents appear to shift sodium channel inactivation to more negative potentials.

Third, JENSEN and KATZUNG (1970) demonstrated that small increases in external potassium concentrations could enhance the effect of phenytoin. A similar observation was demonstrated by SINGH and VAUGHAN WILLIAMS (1971) for lidocaine. Additionally, antiarrhythmic agents selectively depress the sodium current in hypoxic (HONDEGHEM et al. 1974) and ischemic tissues (HOPE et al. 1974). It was, however, only in 1975 with the work of CHEN,

GETTES and KATZUNG that it became fully understood that antiarrhythmic selectivity was largely based upon a voltage-dependent action of these drugs: in depolarized tissues recovery from block occurs much more slowly and less completely than in well-polarized cells (CHEN et al. 1975). Therefore, conditions that are accompanied by depolarization such as hypoxia, ischemia or increased external potassium (HILL and GETTES 1980) increase the effect of the antiarrhythmic agents.

C. Classifications of Antiarrhythmic Agents

Although several classifications of antiarrhythmic drugs have been proposed, only two have been widely accepted. One was proposed by HOFFMAN and BIGGER (1971) and classifies antiarrhythmic agents into two groups: one group slows conduction and reduces $\dot{V}_{max}$ (prototype quinidine), while they believe that the second group has little effect on conduction and $\dot{V}_{max}$ or actually may increase them (e.g., lidocaine, phenytoin). This antidepressant concept was especially appealing as their laboratory had shown that lidocaine could restore blocked conduction across the Purkinje myocardial junction (BIGGER and MANDEL 1970) and that phenytoin could protect the tissue against the depressant actions of hypoxia (BASSETT et al. 1970). Unfortunately, at more normal potassium concentrations than used by BIGGER and MANDEL (1970) or BASSET et al. (1970), phenytoin (JENSEN and KATZUNG 1970) and lidocaine (SINGH and VAUGHAN WILLIAMS 1971) always reduce $\dot{V}_{max}$. Moreover, the antidepressive action of phenytoin in hypoxic tissue could not be reproduced, not even when the potassium concentration was kept at 3 m*M*. To the contrary, under properly controlled conditions phenytoin selectively depresses hypoxic tissue (HONDEGHEM et al. 1974).

The other classification of antiarrhythmic drugs is based upon the various mechanisms of action of antiarrhythmic agents (VAUGHAN WILLIAMS 1970). This classification is now widely accepted and appears to be a practical way of grouping clinically useful antiarrhythmic agents. More recently, the sodium channel blocking or class I agents have been further subdivided into subclasses (see Chaps. 2, 6): class Ia is typified by quinidine, class Ib by lidocaine, and class Ic by flecainide. Recently, a fourth subclass, Id (e.g., transcainide), has been suggested (BENNETT et al. 1987a). As relates to block of sodium channels, members of class Ib have fast onset of block and fast recovery from block. Therefore, under normal physiological conditions, a large fraction of the sodium channels becomes blocked during each action potential, but this block dissipates so quickly during diastole that little or no reduction of sodium channel availability remains at the time of the regular heart beat. Thus, conduction of the normal heart beat is not slowed and the QRS duration is not widened. In contrast, agents in class Ic are characterized by a slow onset of block and a slow recovery from block, so that even at normal heart rates there is a significant degree of block at end diastole. As a result, in effective concentrations these agents always reduce sodium channel availability, slow conduction, and widen the QRS complex. Class Ia agents have intermediary kinetics of block development and recovery, and hence have intermediary effects upon conduction and QRS duration. Finally, at clinically relevant heart rates class Id agents exert a fixed block (little or no use-dependent block) and thus there is also no recovery from block during diastole (BENNETT et al. 1987a).

Grouping of drugs into classes must not imply that all members of a group have identical actions. Indeed, drugs may belong to more than one class. For example, amiodarone although classified as an agent that lengthens action potential duration (class III; SINGH and VAUGHAN WILLIAMS 1970) also has sodium channel blocking or class I properties (MASON et al. 1984), antagonizes adrenergic class II effects (POLSTER and BROEKHUYSEN 1976), and reduces calcium currents (class IV) as well (NISHIMURA et al. 1986; NOKIN et al. 1986). It is important to note that the most relevant action will depend upon the circumstances and type of cardiac tissue. For example, whereas amiodarone has very marked effects upon action potential duration at slow heart rates or after long rest periods, at fast heart rates the lengthening of action potential duration is much less marked. Conversely, block of sodium channels is most marked in tachycardias and may be minimal at slow heart rates (MASON et al. 1984). Also, sodium channel block would be most important in myocardial and conducting tissues, but less relevant in nodal tissues where block of calcium channels may be more important.

Members of the same subclass may become substantially different under pathological conditions. For example, over the −90- to −65-mV range, quinidine is characterized by a fixed percentage block (CHEN et al. 1975) and only a small change in recovery time constant (5-7.5 s; WELD et al. 1982). In contrast, for procainamide (another class Ia agent) upon depolarization the amount of block is markedly enhanced and recovery from block is substantially slowed (SADA et al. 1979; EHRING et al. 1987).

Thus, drugs may belong to different classes and subclasses as the circumstances are changed. It is important to note that drugs are classified based primarily upon their actions in normal conditions rather than in abnormal conditions that are arrhythmogenic. In the latter clinically most relevant conditions, their actions may be not only quite different, they are frequently unknown. Nevertheless, just as it is useful to subdivide light into primary colors (even though it has a continuous spectrum), grouping of antiarrhythmic agents into classes and subclasses similarly has useful clinical implications (see below).

D. Models of Sodium Channel Block

I. Strichartz-Courtney Model

In 1973, STRICHARTZ demonstrated that quaternary local anesthetic agents interact with sodium channels primarily when they are open. He therefore proposed that binding and unbinding of the drugs to the sodium channel occur when the channels are in the open state, and that both drug-free and drug-associated channels cycle between rested, open and inactivated states in a scheme as described by HODGKIN and HUXLEY (1952). However, the drug-associated channels do not conduct sodium (i.e., are blocked) and behave as if their inactivation gates are sensing less negative potentials. This model was extended by COURTNEY (1975) to the interaction of tertiary compounds with the sodium channel. He also provided more direct experimental evidence for the altered voltage dependence of drug-associated channels.

II. Modulated Receptor Model

1. Experimental Basis

Although the Strichartz-Courtney model can nicely account for use-dependent blocking and unblocking, it can not explain the voltage-dependent recovery of sodium channels as described by CHEN et al. (1975). To account for this observation, HONDEGHEM and KATZUNG (1977) postulated that drug unbinding from inactivated channels could also occur: ID to I transition in Fig. 1. In addition, since block of inactivated sodium channels at depolarized potentials also could develop under steady state conditions (at very slow driving rates), the I to ID transition also was required. Furthermore, in order to account for the voltage-dependent action described by WEIDMANN (1955), it was also necessary to allow interactions between the R and RD states. Independently and for somewhat different reasons, HILLE (1977) came to similar conclusions for local anesthetic agents in nerve.

2. Formulation

As a result of the above considerations, HILLE (1977) and HONDEGHEM and KATZUNG (1977) independently formulated the following hypothesis: sodium channel blockers associate and dissociate from sodium channels in all three of their primary states: rested (R), open (O) and inactivated (I). Each of these states has a characteristic set of association (k) and dissociation (l) rate constants. Drug-associated channels do not conduct sodium and their voltage dependence of inactivation is shifted to more negative potentials. Because the affinity of the blocker with the receptor is modulated by channel state, this hypothesis is generally referred to as the modulated receptor hypothesis.

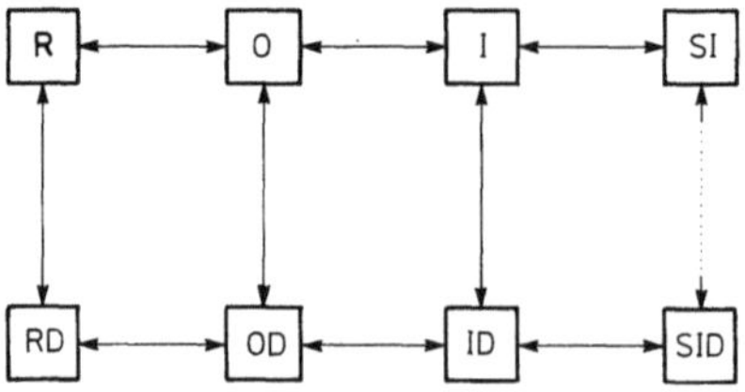

Fig. 1. Schematic of modified modulated receptor hypothesis. R, O and I represent the standard Hodgkin-Huxley rested, open, and inactivated sodium channels. SI represents the slow inactivated state. RD, OD, ID, and SID represent the respective drug-associated channels

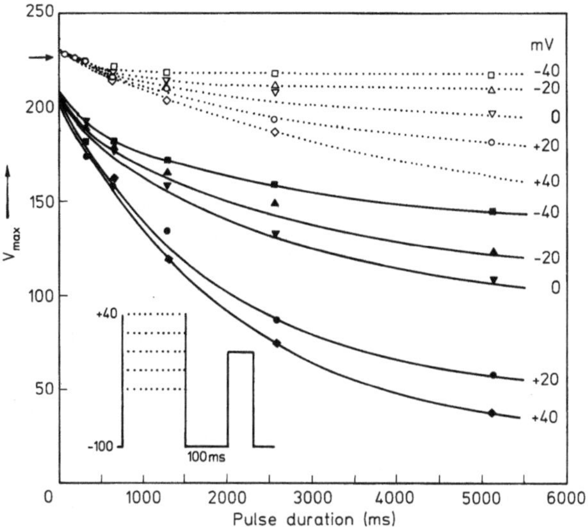

Fig. 2. Effect of quinidine upon slow inactivation at different voltages. *Open symbols, control; filled symbols,* quinidine (10 µ*M*); *squares,* +40 mV; *diamonds,* 0 mV; *circles,* −40 mV. These experiments were carried out in conventional solutions using the single electrode voltage clamp protocol (Hondeghem and Matsubara 1984). As the conditioning pulse duration or amplitude was increased (see *inset*), the amplitude of V_{max} of the response to the test pulse declined. However, extrapolation to zero time on the ordinate indicates that some block appears to develop "instantaneously" in synchrony with channel opening (open channel block). The slower phase of block development (during inactivation) parallels slow inactivation and therefore is thought to be a channel block while in the slow inactivated state. (Reprinted from Hondeghem and Matsubara (1987) with permission of the Br T Pharmacol

The modulated receptor hypothesis has been widely tested and utilized to explain experimental results in several hundred scientific papers (Hondeghem and Katzung 1984). The original proposal set the association and dissociation rate constants independent of voltage, because at that time there was no compelling experimental data to support voltage dependence. Recent experimental evidence strongly suggests that open channel interactions may be voltage dependent (Cahalan 1978; Yeh and Narahashi 1977; Clarkson et al. 1985). The inactivated channel interactions appear to be independent of voltage (Mason et al. 1984; Sanchez-Chapula et al. 1983; Bean et al. 1983). It is not known if interactions with rested states are voltage dependent.

In addition to the R, O, and I states, cardiac sodium channels also exhibit slow inactivation (SI) and ultra slow inactivation (Saikawa and Carmeliet 1982; Clarkson et al. 1984). In 1984, Hondeghem and Matsubara (1984) showed that for long depolarizing pulses reduction of sodium channel availability is augmented in the presence of quinidine. Moreover, voltage and time dependence of this reduction was similar to that of slow inactivation (see Fig. 2). Similar effects have been observed in Purkinje fibers (Weld et al. 1982). It therefore appears that quinidine binds to SI channels or promotes the SI state (Hondeghem and Matsubara 1987). Although alteration of slow inactivation may be very important for drug action, to date it has only been studied to a very limited extent. Similarly, no effects of antiarrhythmic agents upon ultra slow inactivation have yet been described.

3. Implementation

In the absence of drug when sodium channels are kept at membrane potentials more negative than −80 mV, the channels are in the rested state (R-state in Fig. 1). The rested state is a closed state that can be opened upon depolarization. When the channels are maintained at depolarized potentials (more positive than −50 mV), they are inactivated (I state in Fig. 1) or in a closed state from which channels cannot be opened, no matter how strong the depolarization. Before they can open, they first must be reactivated by repolarization, i.e., returned to the rested state. Only when channels are depolarized in the rested state will they transiently open (O state in Fig. 1) upon depolarization. This description for drug-free channels was formulated into differential equations by HODGKIN and HUXLEY (1952) and has formed the basis for numerous quantitative descriptions of the sodium channel. This description correctly approximates the fraction of channels in the rested, open and inactivated states, but the interpretation for the mechanism has changed: it has been pointed out by single channel work (ALDRICH et al. 1983; KUNZE et al. 1985) that activation is relatively slow and inactivation is relatively fast. This interpretation is just the opposite from the proposal by HODGKIN and HUXLEY (1952). However, as we pointed out in 1977, as long as the drug-free model accurately describes the time occupancy of the three states, the precise details of any particular model are not very important for the simulation of drug action.

Each blocker has a characteristic set of association and dissociation rate constants for each of the three primary channel states. We have published a set of differential equations incorporating these rate constants together with the Hodgkin-Huxley assumptions (HONDEGHEM and KATZUNG 1977). For useful clinical antiarrhythmic agents, the rested state usually has a low affinity for the drug (dissociation constant, m*M* range), whereas the open and inactivated state has a higher affinity (μ*M* range) for the drug (HONDEGHEM and KATZUNG 1984).

The drug-associated channels differ from drug-free channels in two main respects. They do not conduct sodium and behave as if their inactivation gate senses a less negative potential. Thus, whereas drug-free channels are mostly in the rested state at −80 mV, drug associated channels may be mostly in the inactivated state. As a consequence, each time the channels are opened or inactivated and drug block develops, upon repolarization the drug-associated channels remain trapped in the ID state, from which they slowly reprime the R state. If the diastolic time is too short for full repriming then block in the ID state accumulates. Thus the modulated receptor hypothesis naturally accounts for use-dependent block. Indeed, the more frequently the channels are used, the more time the channels spend in the open and inactivated states and therefore the more extensive the block and vice versa.

4. Application of Modulated Receptor Hypothesis

a. Class Ia

The interactions of class Ia (e.g., quinidine) with the sodium channel occur rather slowly so that it takes about 5- 20 action potentials before a steady level of block is achieved. The primary interaction state for most of these agents appears to be the open state, and recovery from the ID state appears to be rather slow ($1\ s < \tau < 10\ s$). Because block by these drugs accumulates especially at fast heart rates, class Ia agents are very effective against tachycardias (HONDEGHEM and KATZUNG 1984). In addition, quinidine interacts with the SI state (Fig. 2). The relative importance of the latter interaction deserves further study.

b. Class Ib

These agents interact so quickly with the sodium channel receptor that a steady level of block is reached in less than four action potentials. Actually, for lidocaine (the prototype of this class), the steady level of block is frequently reached during the first action potential of a train. Recovery from block also is fast ($\tau < 1$ s). As a result these agents are most effective early during diastole (HONDEGHEM and KATZUNG 1984) and in depolarized tissue (CHEN et al. 1975). The latter may result from the fact that class Ib agents

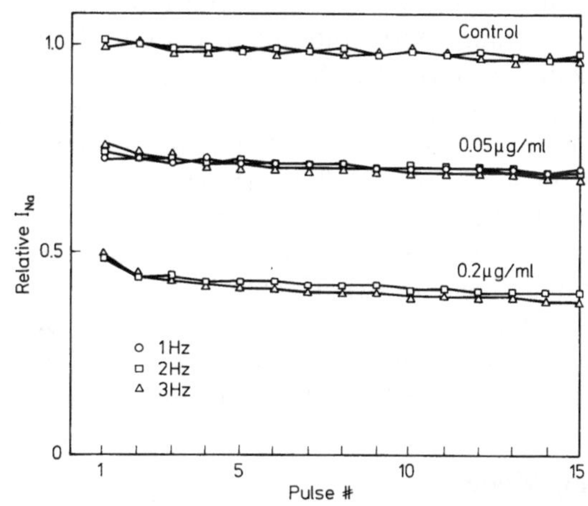

Fig. 3. Effects of transcainide upon the sodium current of pulse trains at various frequencies; 20 ms pulses to −10 mV were applied from a holding potential of −120 mV. The solution for cell isolation and storage was Joklik-modified minimum essential medium (Gibco Laboratories, Chagrin Falls, OH). Solutions were equilibrated with 95% O_2, 5% CO_2. The pH was adjusted to 7.40 at 37 °C with NaOH. Recordings were made in a bath solution containing in millimoles: 20 NaCl, 110 CsCl, 2 $MgCl_2$, 1 $CaCl_2$, 3 $CoCl_2$, 10 D-glucose, 10 HEPES, 10 TEA-Cl; pH was adjusted with CsOH to 7.35 at 22 °C. The pipette solution contained in millimoles: 110 CsF, 10 NaF, 30 CsCl, 2 $MgCl_2$, 10 HEPES, 2 EGTA; pH was adjusted to 7.35 with CsOH at 22 °C (Reprinted from BENNETT et al. (1987a), with permission of the Cardiovasc Pharmacol.)

block primarily inactivated sodium channels, although they also block open channels.

c. Class Ic

These drugs interact very slowly with the sodium channel receptor, requiring more than 20 action potentials before reaching a steady level of block. Block development per action potential is small and recovery from block has a time constant exceeding 10 s.

d. Class Id: Drugs That Lack Use-Dependent Block

The interaction of these agents with the sodium channel is not use-dependent, i.e., these drugs elicit a tonic block only. Even when rested for a long time these drugs exhibit little or no recovery from block, and there is also no development of use-dependent block no matter how fast the driving rate (Fig. 3). (To determine whether at the molecular level these drugs have identical affinities for all states of the channel, or whether they interact infinitely slow or fast with one or more of the channel states so that they would always appear to be at steady state block will require more detailed single-channel studies).

III. Simplified Versions of the Modulated Receptor Model

The full implementation of the differential equations upon a minicomputer (PDP 11-10) requires extraordinary amounts of computation time. For example, the original model as proposed by HONDEGHEM and KATZUNG (1977) implemented in BASIC language required about 20 min per action potential. Even after the program was rewritten in C language, extensively optimized and executed on faster hardware (PDP 11/45 with floating point processor), computation still required about 1 min per action potential. Such slow computations do not lend themselves very well for least square error fitting of the modulated receptor hypothesis to experimental data.

For this and other reasons, several laboratories have been looking for a simplification to the full modulated receptor model that will retain its essential properties but that would compute much faster.

1. Kappa-Repriming Model

The simplest of these models was described by COURTNEY (1983): a fraction ($\varkappa$) of block develops during each action potential and is followed by an exponential repriming between action potentials. This scheme basically provides a description of the modulated receptor hypothesis for a fixed action potential duration and diastolic membrane potential. Although it cannot simulate potential-dependent drug effects and does not provide estimates for the drug affinities, it is, nevertheless, very useful: (1) since it is only the sum of exponential terms, the coefficients of the equations are easily computed; and (2) obviously, these coefficients must have a fixed relation to drug rate constants, so that meaningful quantitative structure analyses can be done using these coefficients (COURTNEY 1983). This particular simplification is especially useful for comparing drugs at a fixed membrane potential. Unfortunately, since recovery from block kinetics is strongly voltage-dependent, this scheme cannot be used to predict drug actions at different potentials.

2. Guarded Receptor Model

STARMER et al. (1983, 1984) proposed that the sodium channel receptor has a constant affinity for the drug but that the accessibility of the receptor is guarded by the channel gates so that the rate constants for the open and inactivated state are different although the affinity is the same. In addition, drug-associated channels retain the same voltage dependence as drug-free channels. In their implementation, traffic between rested, open and inactivated states is not allowed except at potential changes, so that the differential equations reduce to three consecutive sets of first-order equations. Of course, the guarded receptor hypothesis could be implemented with free and continuous transitions between the states, but then it would compute as slowly as the full modulated receptor hypothesis. Needless to say, computation of three consecutive differential equations can be done very quickly. Unfortunately, channels do not obey these limitations: (1) even at fixed potentials they redistribute between states, especially in the −60 to −90 mV range. (2) Drug-associated channels do not have voltage-dependent kinetics that are similar to those of drug-free channels. Indeed, at any potential where their kinetics can be monitored (−70 to −160 mV; using voltage clamp techniques), the redistribution between the channel states (RD, OD and ID) occurs more slowly. Moreover, the steady-state voltage dependence of the distribution between the states is shifted to more negative potentials (WEIDMANN 1955) and in a way that is not possible to simulate with the guarded receptor proposal (SNYDERS and HONDEGHEM 1987). (3) The affinity for open and inactivated state is not equal or constant. Some drugs have a high affinity for the inactivated state but exhibit little or no binding to open channels, e.g., bupivacaine (CLARKSON and HONDEGHEM 1985a) and amiodarone (MASON et al. 1984). Other drugs preferentially bind to open channels, e.g., quinidine (WELD et al. 1982). It could be argued that they have similar affinities for the open and inactivated states, but that the rate constants for one of the states is very slow. However, lidocaine binds quickly to both open and inactivated channels. At low concentrations (μM), block of open channels is minimal at a time when block of inactivated channels reaches about 50% (BEAN et al. 1983). At higher concentrations, open channel block reaches a steady value very quickly (in about ten depolarizations), but the inactivation block that develops more slowly exceeds this level of block (MATSUBARA et al. 1987). (4) As discussed above, whereas block of inactivated channels is independent of membrane potential between −40 and +40 mV, block of open channels appears voltage dependent. Clearly, affinity of open and inactivated channels for lidocaine could only be equal at *one* potential, at best. (5) In the guarded receptor model, it is difficult to visualize how the rested (guarded) channels could be associated with the drug (STARMER and COURTNEY 1986) but compelling evidence for the existence of the RD state has been provided by several laboratories (STRICHARTZ 1973; GINTANT and HOFFMAN 1984; SNYDERS and HONDEGHEM 1987).

3. Analytical Solution of Modulated Receptor Model

Recent information obtained from single-channel recordings indicates that sodium channels open on average for approximately 1 ms at 20 °C (GRANT et al. 1983; KUNZE et al. 1985). At 37 °C the open time of a single channel is probably much shorter. Hence, drug binding to open sodium channels, if it occurs at all, must be very fast, i. e., a fraction of the channels will appear to block (or unblock) nearly instantaneously with each channel opening. Such fractional treatment is not new, as it was originally proposed by STRICHARTZ (1973) and COURTNEY (1975). However, if one eliminates open channel binding/unbinding from the modulated receptor at all times, except during fast depolarization when fractional binding/unbinding is considered, the model has an analytical solution (MOYER 1985) which computes very quickly (about 1 ms/action potential on an IBM PC/AT).

This analytical solution to the modulated receptor model behaves very much like the full model, at least for lidocaine and quinidine (MOYER 1985).[1]

This analytical implementation of the modulated receptor is useful as an educational tool as well as for making initial estimates of the modulated receptor affinities. However, any time that open channel interactions may play an important role outside of the action potential upstroke (i. e., when channels flicker or provide a window current), this model will mandatorily fail. At other potentials, the analytical model can accurately simulate the effect of a large series of compounds (EHRING et al. 1987) which renders it an important tool in quantitative structure analysis.

E. Multiple Drug Interactions with the Cardiac Sodium Channel Receptor

The modulated receptor hypothesis implies that there exists a specific receptor site associated with each sodium channel. If this is the case, then in the presence of more than one drug that binds to the receptor site, one expects that these drugs will compete with each other for the receptor. Indeed, there exist now numerous examples of such competitive interactions for the sodium channel receptor.

---

[1] To obtain a copy of the analytical version of the modulated receptor hypothesis for the IBM PC, mail a formatted diskette and a self-addressed and stamped floppy disk mailer to Dr. Hondeghem.

I. Charged and Neutral Drug Form

Many of the clinically useful antiarrhythmic agents are weak bases with pK_a between 7 and 10. Accordingly, the fraction of drug molecules that are in the charged and neutral forms will vary with pH. At body pH both charged and neutral species will be present to variable extents. The charged species of the antiarrhythmic agent is more hydrophilic than the neutral form, which is more lipophilic. As pointed out by HILLE (1977), there appear to be two access pathways to the sodium channel receptor: a hydrophilic and a hydrophobic route. It is expected that the cationic or charged form of the local anesthetic will preferentially access the sodium channel receptor through the hydrophilic pathway whereas the neutral species will preferentially interact through the lipophilic pathway. It should be noted, however, that the pK_a of the drug-receptor complex may be substantially different from that of the drug in solution (MOORMAN et al. 1986). Also, as illustrated by observations of STRICHARTZ (1973), the cationic drug form nearly exclusively interacts with the sodium channel receptor when the channel is open. In contrast, lipophilic agents such as amiodarone and bupivacaine hardly interact with open sodium channels at all and mainly interact with the inactivated closed channel presumably through the lipophilic pathway.

COURTNEY (1983) pointed out that molecular size is the most important determinant of recovery from block, but that lipid solubility is also important: the neutral form of the drug is more lipid soluble and recovery from block is faster than for the charged species. Hence, since pH modifies the ratio of charged to neutral form, one can expect that: (1) the cationic species of the drug form will become more prevalent and consequently there will be more hydrophilic or open channel interactions with the sodium channel receptor, (2) recovery from block will be slowed, and (3) distribution of the drug between intracellular phase, membrane phase, and extracellular phase will be altered (GRANT 1980).

II. Two Different Drugs Interacting with the Sodium Channel Receptor

In the presence of two drugs that have different electrophysiological properties, one can expect synergistic and antagonistic interactions. When the combination of two drugs results in more effect than is achieved by a single drug, I shall consider it a synergistic action; but when addition of a second drug reduces the level of block established by another drug, I shall describe it as an antagonistic interaction. "Synergistic" and "antagonistic" drug interactions can be accounted for in terms of the modulated receptor theory.

1. Synergistic Interactions

In 1980, HONDEGHEM and KATZUNG computed that it is possible to combine quinidine and lidocaine in such a way that the effect of the two agents could provide extra diastolic block that could not be attained with either agent alone. The principles involved in this synergistic action are illustrated in Fig. 4.

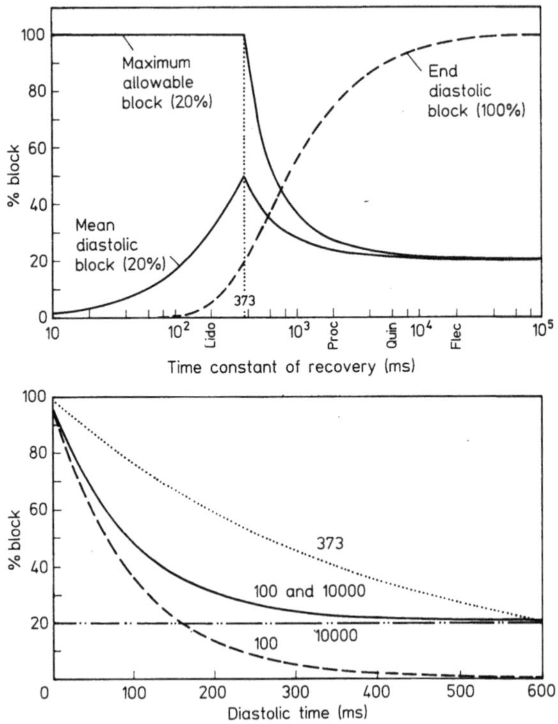

Fig. 4. Block of sodium channels as a function of the time constant of recovery (*top panel*) and recovery from maximum allowable block (*bottom panel*) for a time constant of recovery of 100, 373, and 10000; and for the combination of two drugs having time constants of recovery of 100 and 10000 ms (for explanation see text)

Assuming that 100 % of sodium channels were blocked at the end of the action potential, after a diastolic interval of 600 ms, the percentage of channels that remains blocked increases with the time constant of recovery (dashed line in the top panel of Fig. 4). If one arbitrarily assumes that for non-toxic concentrations only 20 % of channels can be blocked after a 600 ms diastolic interval, then the maximum percentage of channels that could be blocked at the end of the action potential would be 100 % for time constants shorter than 373 ms, but the allowable block would abruptly decline for longer time constants (see Fig. 4). Assuming that the maximum allowable block was achieved, one can compute the mean diastolic block resulting from this maximum as a function of the time constant of recovery. It is obvious that the most block on average can be achieved with a time constant of 373 ms. For shorter time constants, the average declines because the block does not persist long enough: (bottom panel Fig. 4, see $\tau = 100$ ms). For longer time constants, average diastolic block declines because not enough block can be achieved without precipitation of toxicity at the end of diastole (Fig. 4, see $\tau = 10\,000$ ms). Although for different durations of the diastolic interval a different curve would result, there always would exist a single time constant that would provide maximum diastolic block.

Obviously it would be difficult to guess this optimal time constant under clinical conditions. However, when combining a drug that has a time constant that is too long with one that is too short, one can obtain more diastolic block (for a given end diastolic block) than with either drug alone. Thus, the drug combination can approach the maximum achievable (dotted line in Fig. 4) better than either drug alone. Obviously, as the time constants of the two drugs combined approach the optimal time constant, a higher average block can be achieved.

Thus, the modulated receptor theory predicts that a combination of two drugs usually can reduce, without precipitation of toxicity, the available sodium current throughout diastole to a larger extent than either drug alone. To the extent that reduction of sodium current is a mechanism for arrhythmia suppression, one would predict that combinations could provide more diastolic protection against arrhythmias than one drug alone.

This prediction has been tested experimentally as well as clinically: in vitro it was found that the combination of quinidine plus lidocaine can suppress an early diastolic extrasystole more markedly than either quinidine or lidocaine by itself (HONDEGHEM and KATZUNG 1980). The modulated receptor predictions have also been confirmed in several clinical studies. For example, DUFF et al. (1983) showed that, when combining quinidine and mexiletine, arrhythmias could be suppressed much more effectively and with fewer side effects than could be done by either drug alone. Similar observations have been made by several other laboratories and for other drug combinations (BREITHARDT et al. 1981; ANDERSON 1985). It should be noted that such synergistic interactions will probably be most prevalent when drugs belonging to different subclasses are combined. Indeed, if the combined drugs are too similar in their electrophysiological properties, one would not expect much benefit (HONDEGHEM 1987). In the extreme, this combination of two drugs that have identical electrophysiological properties would be exactly the same as simply increasing the dose of one drug. And under those circumstances, no extra protection would be possible at all.

2. Antagonistic Interactions

When combining drugs they will compete for the receptors. As a result, at certain cycle lengths and for a combination of drugs with certain kinetics, addition of an antiarrhythmic drug to another drug can actually result in less block. The competitive displacement with reduced drug effect is predicted by modulated receptor hypothesis and was first experimentally verified by CLARKSON and HONDEGHEM (1985b). They showed that when toxic concentrations of the bupivacaine were combined with a high concentration of lidocaine there was a reversal of the sodium current block by the bupivacaine. In the range of physiological heart rates, bupivacaine markedly reduces $\dot{V}_{max}$, whereas lidocaine reduces $\dot{V}_{max}$ relatively little. The reason for this difference is that, whereas both drugs extensively block sodium channels during each action potential, the block by the bupivacaine takes several seconds to dissipate, while the block by lidocaine dissipates in just a few hundred milliseconds. As a result, when adding lidocaine to the bupivacaine, lidocaine will compete with bupivacaine for the receptor. To the extent that lidocaine succeeds in occupying receptors instead of bupivacaine, receptors occupied by lidocaine will relinquish the drug much more quickly during diastole than those occupied by bupivacaine. Therefore, recovery from total block will proceed faster than in the presence of bupivacaine alone and over physiological cycle lengths addition of lidocaine may attenuate the block observed by bupivacaine alone. Similar antagonistic interactions have been demonstrated for other drugs (SANCHEZ-CHAPULA 1985; KOHLHARDT and SEIFERT 1985).

3. Parent-Metabolite Interaction

It is now well established that small modifications to antiarrhythmic agents can, and often do, result in substantial modifications of the electrophysiological properties (COURTNEY 1983). It is also well known that most antiarrhythmic agents are metabolized. Therefore, one anticipates that in the presence of the parent compound and its metabolite synergistic and antagonistic interactions between two or more drugs will result.

In some cases, the metabolite is more potent than the parent compound, so that most of the antiarrhythmic action may be assigned to the metabolites, e.g., encainide (ELHARRAR and ZIPES 1982).

In other cases the parent compound is metabolized to an agent of lesser potency: lidocaine has two major metabolites, mono-ethyl-glycylxylidide (MEGX) and glycylxylidide (GX). The potency of MEGX is similar to that of lidocaine but GX is much less potent than lidocaine (BROUGHTON et al. 1984). We predicted (BENNETT et al. 1987b) that if GX were to accumulate to any extent in a patient (HANDEL et al. 1983), this agent of lesser potency might compete with lidocaine and consequently reduce its effectiveness. In voltage-clamp studies of the sodium current, we observed that over the −80 to −110 mV range, the recovery from block for GX is faster than for lidocaine (Fig. 5). Therefore, each time a GX molecule displaces a lidocaine molecule, the channel occupied by GX can recover from block more quickly than that occupied by lidocaine. Consequently, addition of GX to lidocaine can result in a reduction of sodium channel block (right panel, bottom of Fig. 5). If this phenomenon occurs *in vivo*, then accumulation of GX might also undo the antiarrhythmic action of lidocaine. Note that at more negative potentials, recovery from GX is slower than for lidocaine. Thus, in these conditions, whenever a GX molecule rather than a lidocaine molecule occupies a channel, block will persist longer. Thus, at more negative potentials addition of GX to lidocaine is expected to increase the block (left panel, bottom of Fig. 5).

F. Summary

Sodium channel blocking antiarrhythmic agents (class I) can be subdivided into subclasses that, in terms of the modulated receptor theory, exhibit different kinetics of interaction with sodium channels. For any particular situation

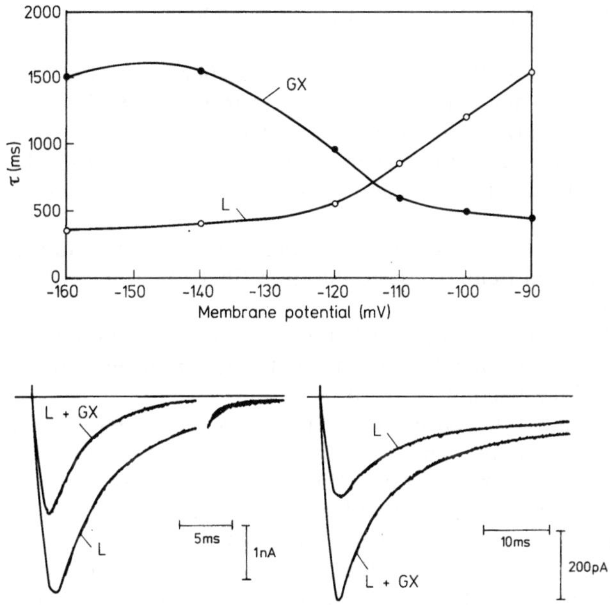

Fig. 5. Time constant of recovery from block for lidocaine (L) and glycylxylidide (GX) as a function of membrane potential (*top panel*). At −120 mV addition of GX to L reduces the sodium current (*bottom left*); at −100 mV addition of GX to L *increases* the sodium current. Solutions used as for Fig. 3. (Unpublished observations kindly provided by Dr. Bennett.)

there are optimum kinetics of drug action. Although it is difficult to estimate this optimum for any clinical condition, it can be approximated by a combination of drugs. Under certain conditions, drugs can, by competitive displacement, reduce another drug's effect. This can be used to reduce toxicity; but at times a metabolite might, through this mechanism, reduce the effectiveness of the parent compound. At times the metabolite may be more potent than the parent compound.

Acknowledgments. This work was supported by a grant from the National Institutes of Health (HL-36020). The author wishes to thank Drs. Paul B. Bennett and James A. Johns for their numerous helpful suggestions, and Mrs. Betty Hondeghem and Mrs. Mimi Klein for their editorial assistance.

References

Aldrich RW, Corey DP, Stevens CF (1983) A reinterpretation of mammalian sodium channel gating based on single channel recording. Nature 306:436-441

Anderson JL (1985) Rationale of combination antiarrhythmic drug therapy. Cardiovasc Clin 16(1):307-327

Bassett AL, Bigger JT, Hoffman BF (1970) Protective action of diphenylhydantoin on canine Purkinje fibers during hypoxia. J Pharmacol Exp Ther 173:336-343

Bean BP, Cohen CJ, Tsien RW (1983) Lidocaine block of cardiac sodium channels. J Gen Physiol 81:613-642

Bennett PB, Stroobandt R, Kesteloot H, Hondeghem LM (1987a) Sodium channel block by a potent, new antiarrhythmic agent, transcainide, in guinea pig ventricular myocytes. J Cardiovasc Pharmacol 9:661-667

Bennett PB, Woosley RL, Hondeghem LM (1987b) Competition between lidocaine and one of its metabolites, glycylxylidide, for cardiac sodium channels. Circulation in press

Bigger JT, Mandel WJ (1970) Effect of lidocaine on conduction in canine Purkinje fibers and at the ventricular muscle-Purkinje fiber junction. J Pharmacol Exp Ther 172:239-254

Breithardt G, Seipel L, Abendroth RR (1981) Comparison of antiarrhythmic efficacy of disopyramide and mexiletine against stimulus-induced ventricular tachycardia. J Cardiovasc Pharmacol 3:1026-1037

Broughton A, Grant AO, Starmer CF, Klinger JK, Stambler BS, Strauss HC (1984) Lipid solubility modulates pH potentiation of local anesthetic block of V_{max} reactivation in guinea pig myocardium. Circ Res 55:513-523

Cahalan MD (1978) Local anesthetic block of sodium channels in normal and pronase-treated squid giant axons. Biophys J 23:285-311

Chen C, Gettes LS, Katzung BG (1975) Effects of lidocaine and quinidine on steady-state characteristics and recovery kinetics of $(dV/dt)_{max}$ in guinea pig ventricular myocardium. Circ Res 37:20-29

Clarkson CW, Matsubara T, Hondeghem LM (1984) Slow inactivation of V_{max} in guinea pig ventricular myocardium. Am J Physiol 27:645-654

Clarkson CW, Hondeghem LM (1985a) Mechanism for bupivacaine depression of cardiac conduction: fast block of sodium channels during the action potential with slow recovery from block during diastole. Anesthesiology 62:396-405

Clarkson CW, Hondeghem LM (1985b) Evidence for a specific receptor site for lidocaine, quinidine, and bupivacaine associated with cardiac sodium channels in guinea pig ventricular myocardium. Circ Res 56:496-506

Clarkson CW, Follmer CH, Yeh JZ, Ten Eick RE, Hondeghem LM (1985) Evidence for two components of sodium channel block by lidocaine in single isolated cardiac myocytes. Circulation 72 (Suppl III):38

Courtney KR (1975) Mechanism of frequency-dependent inhibition of sodium currents in frog myelinated nerve by the lidocaine derivative GEA 968. J Pharmacol Exp Ther 195:225-236

Courtney KR (1983) Quantifying antiarrhythmic drug blocking during action potentials in guinea-pig papillary muscle. J Mol Cell Cardiol 15:749-757

Duff HJ, Roden D, Primm RK, Oates JA, Woosley RL (1983) Mexiletine in the treatment of resistant ventricular arrhythmias: enhancement of efficacy and reduction of dose-related side effects by combination with quinidine. Circulation 67:1124-1128

Ehring GR, Moyer JW, Hondeghem LM (1987) Quantitative structure activity studies of antiarrhythmic properties in a series of lidocaine and procainamide derivatives. J Pharmacol Exp Ther (to be published)

Elharrar V, Zipes DP (1982) Effects of encainide and metabolites (MJ14030 and MJ9444) on canine cardiac Purkinje and ventricular fibers. J Pharmacol Exp Ther 220:460-447

Gintant GA, Hoffman BF (1984) Use-dependent block of cardiac sodium channels by quaternary derivatives of lidocaine. Pflugers Arch 400:121-129

Grant AO, Strauss LJ, Wallace AG, Strauss HC (1980) The influence of pH on the electrophysiological effects of lidocaine in guinea pig ventricular myocardium. Circ Res 47:542-550

Grant AO, Starmer CF, Strauss HC (1983) Unitary sodium channels in isolated cardiac myocytes of rabbit. Circ Res 53:823-829
Handel F, Luzzi FA, Wenger TL, Barchowsky A, Shand DG, Strauss H (1983) Lidocaine and its metabolites in canine plasma and myocardium. J Cardiovasc Pharmacol 5:44-50
Hill JL, Gettes LS (1980) Effect of acute coronary occlusion on local myocardial extracellular K^+ activity in swine. Circulation 61:768-778
Hille B (1977) Local anesthetics: hydrophilic and hydrophobic pathways for the drug-receptor reaction. J Gen Physiol 69:497-515
Hodgkin AL, Huxley AF (1952) A quantiative description of membrane current and its application to conduction and excitation in nerve. J Physiol (Lond) 117:500-544
Hoffman BF, Bigger JT (1971) Antiarrhythmic drugs. In: DiPalma J (ed) Drill's pharmacology in medicine, 4th edn. McGraw-Hill, New York p 824
Hondeghem LM (1987) Antiarrhythmic agents: modulated receptor applications. Circulation 75:514-520
Hondeghem LM, Katzung BG (1977) Time- and voltage-dependent interactions of antiarrhythmic drugs with cardiac sodium channels. Biochim Biophys Acta 472:373-398
Hondeghem LM, Katzung BG (1980) Test of a model antiarrhythmic drug action: effects of quinidine and lidocaine on myocardial conduction. Circulation 61:1217-1224
Hondeghem LM, Katzung BG (1984) Antiarrhythmic agents: the modulated receptor mechanism of action of sodium and calcium channel blocking drugs. Annu Rev Pharmacol Toxicol 24:386-423
Hondeghem LM, Matsubara T (1984) Quinidine and lidocaine: activation and inactivation block. Proc West Pharmacol Soc 27:19-21
Hondeghem LM, Grant AO, Jensen RA (1974) Antiarrhythmic drug action: selective depression of hypoxic cardiac cells. Am Heart J 87:602-605
Hondeghem LM, Matsubara T (1988) Quinidine blocks cardiac sodium channels during opening and slow inactivation in guinea-pig papillary muscle. Br J Pharmacol 93:311-318
Hope RR, Williams DO, El-Sherif N, Lazzara R, Scherlag BJ (1974) The efficacy of antiarrhythmic agents during acute myocardial ischemia and the role of heart rate. Circulation 50:507-514
Jensen R, Katzung B (1970) Electrophysiological actions of diphenylhydantoin on rabbit atria. Circ Res 26:17-27
Johnson EA, McKinnon MG (1957) The differential effect of quinidine and pyrilamine on the myocardial action potential at various rates of stimulation. J Pharmacol Exp Ther 120:460-468
Kohlhardt M, Seifert C (1985) Properties of V_{max} block of I_{na}-mediated action potentials during combined application of antiarrhythmic drugs in cardiac muscle. Naunyn Schmiedeberg's Arch Pharmacol 330:235-244
Kunze DL, Lacerda AE, Wilson DL, Brown AM (1985) Cardiac Na currents and the inactivating, reopening, and waiting properties of single cardiac Na channels. J Gen Physiol 86:691-719
Mason JW, Hondeghem LM, Katzung BG (1984) Block of inactivated sodium channels and of depolarization-induced automaticity in guinea pig papillary muscle by amiodarone. Circ Res 55:277-285
Matsubara T, Clarkson C, Hondeghem L (1987) Lidocaine blocks open and inactivated cardiac sodium channels. Naunyn Schmiedeberg's Arch Pharmacol 336:224-231

Moorman JR, Yee R, Bjornsson T, Starmer CF, Grant AO, Strauss HC (1986) pK_a does not predict pH potentiation of sodium channel blockade by lidocaine and W6211 in guinea pig ventricular myocardium. J Pharmacol Exp Ther 238:159-166
Moyer JW (1985) The interaction of a series of aprindine derivatives with cardiac sodium channels. Dissertation, University of California, San Francisco
Nishimura M, Follmer CH, Cigan AL, Yah JZ (1986) Amiodarone blocks Ca^{++} current in guinea pig ventricular myocytes. Circulation 74 (Suppl II):169
Nokin P, Clinet M, Swillens S, Delisee C, Meysmans L, Chatelain P (1986) Allosteric modulation of [^{3}H] nitrendipine binding to cardiac and cerebral cortex membranes by amiodarone. J Cardiovasc Pharmacol 8:1051-1057
Polster P, Broekhuysen J (1976) The adrenergic antagonism of amiodarone. Biochem Pharmacol 25:131-134
Roden DM, Bennett PB, Hondeghem LM (1986) Quinidine blocks cardiac potassium channels in a time- and voltage-dependent fashion. Biophys J 49:352a
Sada H, Kojima M, Ban T (1979) Effect of procainamide on transmembrane action potentials in guinea-pig papillary muscles as affected by external potassium concentration. Arch Pharmacol 309:179-190
Saikawa T, Carmeliet E (1982) Slow recovery of the maximal rate of rise (V_{max}) of the action potential in sheep cardiac Purkinje fibers. Pflugers Arch 394:90-93
Sanchez-Chapula J (1985) Electrophysiological interactions between quinidine-lidocaine and quinidine-phenytoin in guinea-pig papillary muscle. Naunyn Schmiedeberg's Arch Pharmacol 331:369-375
Sanchez-Chapula J, Tsuda Y, Josephson IR (1983) Voltage- and use-dependent effects of lidocaine on sodium current in rat single ventricular cells. Circ Res 52:557-565
Singh BN, Vaughan Williams EM (1970) The effect of amiodarone, a new antianginal drug, on cardiac muscle. Br J Pharmacol 39:657-667
Singh BN, Vaughan Williams EM (1971) Effect of altering potassium concentration on the action of lidocaine and diphenylhydantoin on rabbit atrial and ventricular muscle. Circ Res 24:286-295
Snyders DJ, Hondeghem LM (1987) Drug-associated channels inactivate and reactivate at more negative potentials than drug-free channels. Proc West Pharmacol Soc 30 (in Press):149-151
Starmer CF, Courtney KR (1986) Modelling ion channel blockade at guarded binding sites: application to tertiary drugs. Am J Physiol 251:H848-856
Starmer CF, Grant AO, Strauss HC (1983) A model of the interaction of local anesthetics with Na channels. Biophys J 41:145a
Starmer CF, Grant AO, Strauss HC (1984) Mechanisms of use-dependent block of sodium channels in excitable membranes by local anesthetics. Biophys J 46:15-27
Strichartz GR (1973) The inhibition of sodium currents in myelinated nerve by quaternary derivatives of lidocaine. J Gen Physiol 62:37-57
Vaughan Williams EM (1970) Classification of anti-arrhythmic drugs. In: Sandoe E, Flensted-Jensen E, Olesen KH (eds) Symposium on cardiac Arrhythmias. Astra, Elsinore, Denmark, pp 449-501
Weidmann S (1955) Effects of calcium ions and local anesthetics on electrical properties of Purkinje fibers. J Physiol (Lond) 129:568-582
Weld FM, Coromilas J, Rottman JN, Bigger JT (1982) Mechanisms of quinidine-induced depression of maximum upstroke velocity in ovine cardiac Purkinje fibers. Circ Res 50:369-376
Yeh JZ, Narahashi T (1977) Kinetic analysis of pancuronium interaction with sodium channels in squid axon membranes. J Gen Physiol 69:2293-2323

CHAPTER 9

Clinical Use of Class Ia Antiarrhythmic Drugs

T. J. CAMPBELL

A. Historical

Quinidine is by far the oldest agent developed solely as an antiarrhythmic drug. It is an isomer of quinine, derived from cinchona bark, which was reported as being useful for palpitations (presumably atrial fibrillation) in 1749 (WILLIUS and KEYS 1942). The regular prescription of quinidine for atrial fibrillation dates largely from the work of Wenckebach, Frey, Lewis and Love, between 1914 and 1926 (VAUGHAN WILLIAMS 1980). Quinidine was also the first drug recognized effective against ventricular tachycardia (SELZER 1982), and despite the large number of alternative agents now available, it is still one of the most popular antiarrhythmic drugs in many countries, including the United States. MAUTZ (1936) first demonstrated antiarrhythmic properties for procaine. This compound is, however, too rapidly hydrolyzed to be of clinical use. A study of related substances led to the introduction of procainamide (MARK et al. 1951). It remains an extremely widely prescribed antiarrhythmic drug particulary in the management of ventricular arrhythmias. Its efficacy in supraventricular arrhythmias has also been amply demonstrated (WOSKE et al. 1953; FENSTER et al. 1983). Disopyramide is an example of a deliberate search for an antiarrhythmic agent as an alternative to quinidine and procainamide. It was found to have significant efficacy against arrhythmias in animals (MOKLER and VAN ARMAN 1962) and man (KATZ et al. 1963).

B. Clinical Pharmacology

I. Quinidine

Quinidine is usually administered orally as the sulphate or gluconate or in various long-acting forms. The usual dose of quinidine sulphate is 300-600 mg four times a day and doses about 30% higher are used for the gluconate. Bioavailability is about 70% for both forms but peak plasma levels are reached earlier for the sulphate (60-90 min) than for the gluconate. Protein binding is 70%-80% and the elimination half-life is 5-8 h, with most of the drug (80%) being metabolized in the liver and the remainder excreted unchanged in the urine. Therapeutic plasma levels are generally quoted as 3-8 μg/ml (SOKOLOW and BALL 1956). These figures are based on assays which do not distinguish quinidine from its major metabolites mono- and dihydroxyquinidine, and it has been suggested that the therapeutic range for modern assays specific for quinidine should be revised to 0.72-5.92 μg/ml (CARLINER et al. 1980). The metabolites have generally been regarded as essentially inactive but this view has recently been questioned (DRAYER et al. 1978; HOLFORD et al. 1981).

Sustained release formulations of both quinidine sulphate and gluconate are available and have been shown to produce adequate plasma concentrations for at least 8 h (TAGGART and HOLYOAK 1983).

Quinidine can also be given intramuscularly or intravenously. The intramuscular route is painful and may produce a sterile abscess. The intravenous route has been unpopular for many years because of reports of severe hypotension. This is largely due to the fact that the drug has both alpha-adrenergic blocking and negative inotropic effects. Intravenous quinidine has recently returned to favour, however, at least in the setting of clinical electrophysiological studies (see below) and it has proved possible to give the drug safely in this way provided certain precautions are taken (SWERDLOW et al. 1983).

Finally, mention should be made of the fact that quinidine interacts with digoxin to produce a clinically significant increase (50%-100%) in the plasma digoxin level. This seems to be the result of a combination of displacement of digoxin from tissues and decreases in both renal and non-renal clearance of digoxin (DOERING 1979; WOODCOCK and RIETBROCK 1982).

II. Procainamide

Procainamide also is usually given orally to a total dose of 3-6 g/day. Bioavailability is high and peak plasma levels are achieved 1-2 h after tablet ingestion. Protein binding is only 10%-20% and the elimination half-life is quite short (3-5 h). Therapeutic plasma levels are 5-15 μg/ml (18-55 μ*M*; ROSEN et al. 1972; MYERBURG et al. 1981a).

Procainamide is eliminated by renal excretion and hepatic metabolism (DREYFUSS et al. 1972; GIARDINA et al. 1976, 1977). The major metabolite is *N*-acetylprocainamide (NAPA), which has potent class III and some class I antiarrhythmic activity (ELISON et al. 1975; RODEN et al. 1980; JAILLON et al. 1981). In fast acetylators (REIDENBERG et al. 1975) or in renal failure, as much as 40% of a dose of procainamide may be excreted as NAPA and blood concentrations of NAPA can exceed those of the parent drug.

Sustained release preparations are available and have been proved effective in ventricular arrhythmias (VLASSES et al. 1983; GIARDINA et al. 1980). They result in an effective doubling of the dosing interval required.

Procainamide may also be given intravenously. As with quinidine, the risk of significant hypotension is minimized by slow administration and regular monitoring of blood pressure.

III. Disopyramide

Disopyramide may be given orally or intravenously. Oral biovailability is about 80% and peak plasma levels occur at 1-2 h (HINDERLING and GARRETT

1976). The usual oral dose is 300-600 mg/day in three or four divided doses. Unlike most other antiarrhythmic drugs, protein binding of disopyramide shows non-linear, saturable characteristics (DAVID et al. 1980; MEFFIN et al. 1979). This is of clinical importance since apparently small increases in total plasma level within the therapeutic range (2.8-7.5 μg/ml or 8-22 μM; NIARCHOS 1976) may mask larger rises in free (active) drug concentration (MEFFIN et al. 1979). Furthermore, disopyramide binds significantly to alpha$_1$-acid glycoprotein, levels of which rise during many acute illnesses, including myocardial infarction (ROUTLEDGE et al. 1980). The elimination half-life in seven healthy volunteers was found to be 4.5 h (HINDERLING and GARRETT 1976). This increases markedly in renal failure, as 50%-80% of the drug is normally excreted unchanged in the urine. There are several metabolites. These are not thought to contribute to the antiarrhythmic effects but one of them may contribute to the side effects because it has 24 times the anticholinergic potency of the parent (BAINES et al. 1976). Long-acting forms of disopyramide are available and have been proved effective (FECHTER et al. 1983; ZEMA 1984).

When administered intravenously, disopyramide is usually given as a bolus (1.5-2 mg/kg) followed by an infusion (0.4 mg/kg/h). It less frequently produces hypotension than do quinidine or procainamide. This may be because its negative inotropic effects are masked by its tendency to produce peripheral vasoconstriction (HEEL et al. 1978).

C. Cardiac Electrophysiological Effects

The cellular electrophysiological effects and basic mechanisms of action of the Ia drugs have been covered in detail in Chaps. 2, 7, 8. The emphasis in this section will be on changes seen in surface electrocardiograms and intracardiac recordings in the presence of these agents. These are summarized in Table 1.

All three drugs tend to increase spontaneous sinus rate in vivo (JOSEPHSON et al. 1973, 1974a, b). This is the opposite of their direct in vitro actions (BIRKHEAD and VAUGHAN WILLIAMS 1977; DHINGRA and ROSEN 1979; WEST and AMORY 1960; CAMPBELL 1987) and is due to a combination of the anticholinergic properties of the drugs (MIRRO et al. 1980a, b, 1981) and reflex increases in sympathetic tone due to their tendency to decrease blood pressure. As might be expected, significant slowing of sinus rate has been reported in patients with pre-existent sinus node disease (BIRKHEAD and VAUGHAN WILLIAMS 1977; LABARRE et al. 1979; KIM and FRIEDMAN 1979), in those with transplanted (and hence denervated) hearts (MASON et al. 1977; BEXTON et al. 1982), or in the presence of toxic blood levels of one of these agents. Most studies have shown no significant change in PR interval and only minor increases in QRS duration at therapeutic levels but the QT intervals and rate-corrected QT (QT_c) are generally increased as expected from the ability of these drugs to prolong action potential duration (JOSEPHSON et al. 1973, 1974a, b; FLECAINIDE-QUINIDINE RESEARCH GROUP 1983; MORGANROTH et al. 1986; LERMAN et al. 1983; CAMPBELL and MORGAN 1980; GAVAGHAN et al. 1985; HULTING and JANSSON 1977). These effects also are more pronounced in

Table 1. Clinical electrophysiological effects of therapeutic concentrations of class Ia antiarrhythmic drugs

Sinus rate	Small increase
PR interval	No change
QRS duration	No change or minor increase
QT interval	Moderate increase
AH[a] interval	No change
HV[b] interval	Moderate increase

[a] Conduction time from low right atrium to His bundle
[b] Conduction time from His bundle to start of QRS

the transplanted heart as are the actions on intracardiac conduction time and refractory periods discussed below (MASON et al. 1977; BEXTON et al. 1982).

Atrial and ventricular refractoriness is prolonged, conduction across the atrioventricular node (AH interval) is not changed (prolonged in transplants) and His-Purkinje conduction (HV time) is moderately slowed (JOSEPHSON et al. 1973, 1974a, b; Ross et al. 1978). The risk of aggravating pre-existent atrioventricular node disease or bundle branch block with these agents appears to be slight (WILKINSON et al. 1982; DESAI et al. 1979). No systematic study has been done to compare quantitatively the electrophysiological effects of the Ia drugs.

D. Efficacy

Class Ia drugs, particularly quinidine and procainamide, were widely accepted as being of value for supraventricular and ventricular arrhythmias before randomized clinical trials became common practice in assessing putative antiarrhythmic drugs. Consequently much of our present information regarding their clinical efficacy comes from recent trials in which one of these agents was used as the standard against which an investigational drug was being assessed.

In this section we will consider the efficacy of Ia drugs in each of three situations, namely supraventricular arrhythmias, ventricular ectopy (including non-sustained ventricular tachycardia) and sustained ventricular tachycardia and/or fibrillation. The bulk of the discussion will be devoted to the ventricular arrhythmias since these are generally of greater clinical concern and often more difficult to manage.

The role, if any, of Ia drugs in the postmyocardial infarct patient and the questions of interchangability of Ia drugs and their usefulness in combination with each other or with other antiarrhythmics will also be covered. The increasingly recognized problem of the potential proarrhythmic actions of these drugs will be covered in detail in a subsequent section.

I. Supraventricular Arrhythmias

1. Quinidine

As noted earlier, quinidine was first used for the reversion of atrial fibrillation (WENCKEBACH 1914). It was much later studied in comparison with other antiarrhythmics for re-entrant supraventricular tachycardia in patients with Wolff-Parkinson-White syndrome (SELLERS et al. 1977). This study documented its efficacy though no placebo control group was used.

Despite this lack of "hard" data, quinidine is in frequent, routine use in many countries today for reversion and prophylaxis of all types of supraventricular arrhythmia. It is particularly useful in atrial fibrillation/flutter both for acute reversion and for maintaining sinus rhythm (SODERMARK et al. 1975; GRANDE et al. 1986). Its popularity for treating re-entrant supraventricular tachycardias has been considerably eroded by the advent of verapamil.

When using quinidine (or disopyramide) to treat atrial fibrillation or flutter it is common practice to begin therapy with digitalis first to reduce the risk of producing a clinically significant degree of enhanced atrioventricular conduction due to the anticholinergic properties of these drugs (BIGGER and HOFFMAN 1980).

2. Procainamide

As noted earlier, procainamide is infrequently used to treat supraventricular arrhythmias although its efficacy for this indication has been known for many years (WOSKE et al. 1953). Recently FENSTER and colleagues (1983) obtained reversion to sinus rhythm during a 1 h infusion of procainamide in 15/26 patients awaiting electrical cardioversion for atrial fibrillation.

3. Disopyramide

A number of uncontrolled studies suggested disopyramide to be a useful drug for reversion of atrial fibrillation or flutter (DEANO et al. 1977; CAMM et al. 1980; VANDENBOSCH et al. 1975) or for maintenance of sinus rhythm following electrical cardioversion (HARTEL et al. 1974; MILLAR CRAIG and RAFTERY 1977). In a series of controlled trials, intravenous disopyramide (in combination with digoxin) has been shown to be more effective than digoxin alone and equally effective as intravenous sotalol, in reversion of atrial fibrillation or flutter following open heart surgery (CAMPBELL and MORGAN 1980; GAVAGHAN et al. 1985; CAMPBELL et al. 1985). It has also been shown to be more effective than placebo in maintaining patients in sinus rhythm after reversion from atrial fibrillation (KARLSON et al. 1986; SCHENK-GUSTAFSSON et al. 1986). Disopyramide has also been successful in either reverting or rendering non-inducible more than 50% of patients with paroxysmal atrioventricular nodal tachycardia (SWIRYN et al. 1981; BRUGADA and WELLENS 1984) and in the treatment of supraventricular tachycardia associated with Wolff-Parkinson-White syndrome (SPURRELL et al. 1975; BENNETT 1978). Although there is a conspicuous lack of controlled studies comparing Ia drugs with calcium antagonists (class IV) or beta-blockers (class II), the class IV drugs and class II drugs are generally regarded as first and second choice agents for both these types of supraventricular tachycardia (HOROWITZ 1985a). Both groups are contraindicated, however, in patients who have atrial fibrillation with atrioventricular conduction via a bypass tract. Class Ia drugs are the agents of choice in this situation (WELLENS et al. 1984), and the Ic drug flecainide is also of potential benefit (HELLESTRAND et al. 1984).

II. Ventricular Ectopy and/or Non-sustained Ventricular Tachycardia

There are two unresolved questions which need to be noted in any discussion of antiarrhythmic therapy for ventricular ectopy. These are which patients with ectopics should be treated; and, having decided to treat, how can we be sure the chosen drug is being effective?

Premature ventricular complexes are seen during prolonged monitoring in 15%-76% of active, asymptomatic and apparently "normal" individuals (RAFTERY and CASHMAN 1976; Chap. 12a). These arrhythmias may include bigeminy, multimorphological ectopics, very premature beats ("R-on-T" phenomenon) and, less frequently, non-sustained ventricular tachycardia (VT).

In the absence of cardiovascular abnormalities, especially coronary artery disease, ventricular ectopics of all grades, including non-sustained VT, seem to carry a largely benign prognosis (RODSTEIN et al. 1971; MCHENRY et al. 1976; CALIFF et al. 1982; KENNEDY et al. 1980). They probably do not require treatment. Indeed, even recurrent, sustained ventricular tachycardia in patients with otherwise "normal" hearts carries a good prognosis for long-term survival (WELLENS et al. 1982; DEAL et al. 1986; LEMERY et al. 1986), although such patients are frequently treated with a view to improvement ot their symptoms.

The story is quite different, however, for patients with cardiac disease, particularly coronary artery disease (HINKLE et al. 1969; WELLENS et al. 1982; PODRID et al. 1985; Chaps. 5, 6). VISMARA and colleagues (1975) reported complex ventricular ectopic activity to be twice as frequent in those who died following acute myocardial infarction as in survivors. Moss et al. (1979) made similar findings in a group of 978 subjects with definite or suspected myocardial infarction (MI). A 5-year follow-up of 1739 men with at least one MI revealed a 5-year mortality of 5% in those without ventricular ectopy compared with 25% in those with Lown grade 4A or above ectopic activity. Similar findings were reported for 430 patients who survived the coronary care unit phase of acute MI (BIGGER et al. 1981). Mortality during the subsequent year was 38% in those who had runs of three or more ventricular ectopics on a 24-h electrocardiographic recording compared with 12% in those who did not (54% vs, 19% at 3 years). Nor was the increased risk confined to those with high-grade arrhythmias. Patients with either paired ventricular ectopics or simple ventricular ectopics at an average frequency of >30/h also had significantly increased mortality at 1 year. Two later studies of 866 and 533 post-myocardial infarction patients extended this and showed that the risk of death in the first 3 years after an infarct increases significantly for ectopic rates above 3/h (BIGGER et al. 1984; MUKHARJI et al. 1984). Ectopic frequency, repetitive ectopic activity and left ventricular dysfunction (ejection fraction <30%) were each shown to be independent predictors of increased mortality over this period and especially in the first 6 months.

In the face of this evidence linking ectopic activity with mortality in patients with coronary disease it has become common practice to attempt to suppress this activity with drugs. It should be pointed out, however, that there is, as yet, no general agreement to support the hypothesis that a drug effective against ectopic beats will necessarily be effective in the prevention of life-threatening arrhythmias. Total supression of ectopic activity is not a practical goal in most patients and may not be necessary (MYERBURG et al. 1981a, b; BIGGER and ROLNITZKY 1985). Therapy with various beta-adrenergic blocking agents has been shown to reduce the incidence of sudden death following acute MI despite the frequent lack of marked efficacy of these agents in abolishing ectopic activity (YUSUF et al. 1985).

There is evidence emerging, however, that patients at risk from malignant arrhythmias have a greatly improved prognosis if placed on an antiarrhythmic drug regimen which abolishes at least non-sustained VT and very premature (R-on-T) ectopic beats (GRABOYS et al. 1982; HOFFMAN et al. 1984). It is to be hoped that subsequent studies will confirm this finding. Indeed a pilot study (CAPS; Cardiac Arrhythmia Pilot Study) is currently in progress and will be followed by a larger-scale study (CATS; Cardiac Arrhythmia Therapy Study) aimed at testing the hypothesis that reduction of total arrhythmia frequency by >70% and elimination of couplets and runs of VT in post-MI patients will improve prognosis (ANDERSON 1986; BIGGER 1985; CARDIAC ARRHYTHMIA PILOT STUDY INVESTIGATORS 1986). These trials have already run into some difficulties with high apparent "efficacy" rates for placebo (PRATT et al. 1986) and will have to allow for high predicted withdrawal rates (23%–70% over 2 years; SALERNO et al. 1986). Another full-scale study, TEST (timolol, encainide, sotalol trial), has also been running along similar lines since 1984 (BIGGER 1985).

Let us now briefly consider the ways available for assessing drug efficacy in patients on chronic antiarrhythmic therapy for suppression of ventricular ectopic beats. This subject has recently been reviewed extensively (MORGANROTH 1984; BIGGER and ROLNITZKY 1985). Because of wide spontaneous variations in ectopic frequency (depending on the type of analysis used), a given drug must show 70%–85% reduction in ectopic beats on a 24-h ECG recording (compared with a similar control recording) in order to be judged effective. For patients with an average of less than 20–30 ectopics/h in the control record, the reduction required to prove efficacy is more like 90%. Estimates for required reduction of frequency of runs of ventricular tachycardia range from about 65% to more than 90% (BIGGER and ROLNITZKY 1984).

Bearing these criteria in mind (and also remembering the lack of evidence linking anti-ectopic efficacy with enhanced life expectancy), how effective are Ia drugs in suppressing ventricular ectopic beats and non-sustained ventricular tachycardia?

Table 2. Efficacy of Ia drugs in suppressing ventricular ectopic activity and non-sustained ventricular tachycardia

Drug	No. of patients	Efficacy[a] (%)	Reference
Quinidine	139	57	Flecainide-Quinidine Research Group 1983
	187	67	Morganroth et al. 1986
	20	44	Sami et al 1981
	90	52	Searle Laboratories 1977
Procainamide	10	70	Winkle et al. 1978
	33	76	Giardina et al. 1980
	33	39	Lidell et al. 1985
Disopyramide	10	30[b]	Caron et al. 1985
	27	22[b]	Pratt et al. 1984
	25	32[b]	Kjekshus et al. 1984
	90	54	Searle Laboratories 1977

[a] Efficacy defined as > 75 % reduction of ectopic frequency
[b] In these studies efficacy was regarded as > 80 % reduction of ectopic frequency

1. Quinidine

As can be seen from Table 2, quinidine has been reported effective (>75 % reduction of VEBs) in 44 %–67 % of a total of nearly 450 patients. In three of these studies, quinidine was found to be less effective than the class Ic agent (flecainide or encainide) with which it was being compared and in another trial it was reported equally effective as disopyramide (52 % vs, 54 % of 90 patients showed >75 % reduction in ectopics; SEARLE LABORATORIES 1977). In addition complete suppression of non-sustained ventricular tachycardia was seen in 55 % of 67 patients taking quinidine in the quinidine-flecainide study, 9 out of 10 patients in the study by SALERNO and colleagues (1983) and 5 out of 7 patients in another trial (SINGH et al. 1984). MORGANROTH and associates (1986) reported at least 75 % reduction in ventricular tachycardia beats in 78 % of 154 patients and at least 75 % reduction in total ectopic frequency in 67 % of patients.

2. Procainamide

In three controlled studies with a total of 76 patients (Table 2), procainamide therapy was associated with at least 75 % reduction in ventricular ectopic activity in 39 %, 70 % and 76 % of cases. In one of these studies a comparison was made with quinidine. Quinidine reduced ectopics by at least 80 % in 10 of 13 patients and procainamide in 7 of 10 (WINKLE et al. 1978).

3. Disopyramide

In three recent, controlled studies, disopyramide reduced ventricular ectopic activity by >80 % in only 30 %, 22 % and 32 % of patients (Table 2; CARON et al. 1985; PRATT et al. 1984; KJEKSHUS et al. 1984) An earlier trial comparing quinidine and disopyramide has already been mentioned in which > 75% reduction of ectopics was achieved in 52% and 54% of cases respectively (SEARLE LABORATORIES 1977).

III. Sustained Ventricular Tachyarrhythmias

Patients who have had documented episodes of sustained ventricular tachycardia (VT) or of ventricular fibrillation (VF) outside the setting of acute MI are at very high risk of recurrence and of death. A subgroup of these patients have already been resuscitated from at least one episode of "sudden death" and have at least a 20%-40% chance of recurrence in the 1st year (LOWN 1979). Sudden cardiac death is the most frequent cause of death in the Western world (approximately 350000 victims/year in the United States alone).

The majority of patients at high risk of VT/VF have coronary heart disease. Less frequent pathologies include congestive and hypertrophic cardiomyopathies and long QT syndrome (congenital or acquired). The bulk of available research data have been obtained from the coronary disease group and the subsequent discussion refers largely to these patients.

There are at least three ways of approaching patients at high risk of VT/VF. The first is some sort of community-based bystander resuscitation programme. In the best of hands, this is associated with long-term survival figures of 40% or better (EDITORIAL 1979; EISENBERG et al. 1982). A second approach is being developed in several centres at present. This is the implantable defibrillator (MIROWSKI et al. 1980, 1983) and is beyond the scope of this article. The third approach, which we will now consider, is attempted prophylaxis using various pharmacological regimens.

Apart from the well-documented special case of beta-blockade post-MI (YUSUF et al. 1985), there is as yet no general agreement regarding the value of antiarrhythmic therapy for patients with recurrent VT or VF. Certainly treating these patients with a drug selected more or less at random does not work (DENNISS et al. 1986). In the data reported from the Seattle Heart Watch Program over 75% of patients who had a second episode of out-of-hospital sudden death were on antiarrhythmic drugs at the time (SCHAFFER and COBB 1975).

The picture is further clouded by the fact that there is not even general agreement as how best to select appropriate antiarrhythmic drugs for a given patient and monitor the effects of the chosen regimen (PLATIA and REID 1984; KIM et al. 1985; GRADMAN et al. 1985). One approach is to use serial ambulatory ECG recordings as discussed above. This has the disadvantage of relying on the assumption that suppression of ectopic activity implies a decreased risk of sustained ventricular arrhythmias. It also relies on the presence of relatively frequent ectopic activity in patients with spontaneous sustained VT. This is not always present (COOPER et al. 1985). On the other hand, it is non-invasive and does not require hospitalization, unlike the main alternative approach which is the use of repeated intracardiac electrophysiological studies. In this latter situation, various drugs are administered in an attempt to render the patient's ventricular arrhythmia non-inducible by programmed electrical stimulation (HOROWITZ 1985b; MASON and WINKLE 1978; HOROWITZ et al. 1978; GREENE et al. 1981; MASON et al. 1981). There is disagreement as to the acceptability of end points other than non-inducibility. Some workers feel strongly that non-inducibility is the only reliable index of drug efficacy (HOROWITZ 1985b) whereas others feel that suppression of repetitive ventricular responses (non-sustained VT) can be a useful guide (PODRID et al. 1983; SCHAEFFER et al. 1978). The possible significance of increased difficulty in induction of VT is also under investigation (GREENE et al. 1981; LOMBARDI et al. 1986). A multicentre trial (ESVM; electrophysiological study versus electrocardiographic monitoring) is currently underway to test these two approaches for guiding therapy of VT/VF (ANDERSON 1986).

A third approach which may prove to be of value is the use of the signal-averaged surface electrocardiogram. This techniue can identify with considerable accuracy patients at risk of sustained VT after MI (KANOVSKY et al. 1984; DENNISS et al. 1985; KUCHAR et al. 1986). It remains to be seen whether it can be used to predict drug efficacy.

There is now emerging good evidence that "directed" drug therapy, whether selected on the basis of invasive or non-invasive studies, is associated with a marked improvement in prognosis in patients with life-threatening ventricular arrhythmias (PODRID 1985; HOROWITZ et al. 1978; MASON and WINKLE 1978; DI MARCO et al. 1985; BENDITT et al. 1983; PLATIA and REID 1984; RUSKIN et al. 1980; SWERDLOW et al. 1983; MITCHELL et al. 1986). PODRID (1985) reported on 123 patients with VT/VF studied non-invasively (Holter monitoring) and 52 similar patients studied invasively. Among 98 patients in the first group deemed "controlled" (suppression of salvos of VT on Holter recording) only 6 died suddenly during a follow-up of 31.5 months (2.3% annual mortality). Of the 25 considered not-controlled, 17 died suddenly in the same period (43.6% annual mortality). Of the 45 patients in the second group who had an inducible arrhythmia, 36 (80%) were "controlled" (abolition of repetitive ventricular response) by antiarrhythmic therapy. Only one patient (2.8%) has had a recurrence after 22 months of follow-up. On the other hand, five of the nine (56%) patients not controlled by drugs have had recurrence. Similar findings have emerged from many of the other studies referred to above, although not all agree that Holter monitoring is a useful measure of prognostic benefit (SKALE et al. 1986) and a small randomized study has suggested invasive studies to be a better prognostic indicator than ambulatory monitoring (MITCHELL et al. 1986). It must be stated, however, that studies of this kind do not prove that drug therapy in "responders" leads to decreased mortality. It may be that drug response somehow selects "survivors" versus "non-survivors". Testing of the former hypothesis would require randomizing a group of drug responders into treatment (with the successful regimen) and non-treatment subgroups and following progress. There are obvious ethical objections to such a trial. Recent support for the usefulness of pharmacological testing comes from a study which compared the outcome of patients whose ventricular arrhythmia was converted from inducible to non-inducible by an antiarrhythmic drug (PLATIA 1986). Those who continued to take this drug for 1–3 years did far bet-

Table 3. Efficacy of I a drugs in suppressing sustained ventricular tachycardia/fibrillation

Drug	No. of patients	How assessed	Efficacy %	Reference
Quinidine	89	Invasive[a]	34	Di Marco et al. 1983
	24	Invasive	25	Ferrick et al. 1982
	23	Invasive	26	Senges et al. 1982
Procainamide	126	Invasive	33	Waxman et al. 1983
Disopyramide	31	Non-invasive[b]	55	Puech 1970
	7	Non-invasive	86	Vismara et al. 1977
	50	Invasive	34	Lerman et al. 1983

[a] Invasive assessment: efficacy defined as ability to render ventricular arrhythmia non-inducible by programmed ventricular extrastimuli
[b] Non-invasive: efficacy defined as abolition of clinical/ambulatory ECG episodes of sustained ventricular arrhythmia

ter (1 sudden death and 1 arrhythmia recurrence out of 30) than those who stopped or altered their antiarrhythmic regimen (11 sudden deaths and 14 recurrences out of 33; $P < 0.001$).

What do we know of the relative efficacies of individual antiarrhythmic drugs? Figures vary but in general a drug chosen at random seems to have only about a 15%–35% chance of being effective in a given patient with recurrent sustained ventricular arrhythmias. Amiodarone may be an exception (>50% success rate; Haffajee 1985). There is also a significant chance of proarrhythmic effects (discussed later). Let us briefly consider the individual Ia antiarrhythmic drugs in this context (Table 3).

1. Quinidine

Di Marco and coworkers (1983) studied 89 patients with sustained VT or VF using invasive electrophysiology. Oral quinidine alone (under blood level control) rendered 30 (34%) of these patients non-inducible (all were inducible initially). A further eight patients became non-inducible on a combination of quinidine with mexiletine (seven) or propranolol (one). On continued therapy, 32 of these 38 remained clinically asymptomatic during 24 ±3 months of follow-up (three relapsed and three withdrew because of side effects). In two smaller studies with similar groups of patients, acute therapy with intravenous quinidine produced non-inducibility in 6/24 (25%) and 6/23 (26%) cases respectively (Ferrick et al. 1982; Senges et al. 1982).

2. Procainamide

In a large study of 126 patients with inducible and spontaneous sustained VT, Waxman and colleagues found that procainamide led to non-inducibility in 42 cases (33%) and, further, that 60 of the 69 "failures" also failed all other conventional antiarrhythmic regimes tested. They suggested that "response to procainamide accurately predicted response to other conventional antiarrhythmic drugs during electrophysiologic study".

This concept has been tested recently (RAE et al. 1985). Of 59 patients with spontaneous sustained VT who failed to respond to procainamide at electrophysiological study, only 11 (19%) were rendered non-inducible by any other antiarrhythmic drug or drug combination tried by the authors of the study.

An earlier electrophysiological study (GREENSPAN et al. 1980) of 35 patients with recurrent sustained VT found procainamide the most effective drug in 16 cases. A follow-up study of these 16 showed that all 11 patients whose plasma procainamide levels remained at or above the acute effective level remained asymptomatic for up to 4 years. All five with low blood levels had early recurrences.

3. Disopyramide

Early reports of the efficacy of disopyramide include a study which showed clinical improvement in 17/31 patients with recurrent VT (PUECH 1970) and a study of 7 patients with VT refractory to conventional drugs in which disopyramide was clinically successful in acute and chronic treatment in 6 (VISMARA et al. 1977). A recent, larger trial (LERMAN et al. 1983) involved 50 patients with inducible sustained VT or VF. Following 3 days of oral therapy with disopyramide, 17 (34%) became non-inducible. Eleven of these were followed on disopyramide for 19 ±9 months and nine remained free of VT (although two of these developed clinical heart failure).

It appears from the studies presented above that each of the Ia drugs can be expected to be effective in about 25%–35% of patients with recurrent sustained VT or VF. In addition there is some circumstantial evidence that response to procainamide may predict response to other antiarrhythmic drugs (WAXMAN et al. 1983).

IV. Postinfarction Prophylaxis

We have seen that there is evidence for prognostic benefit of "directed" antiarrhythmic therapy in those "at high risk" of VT/VF (as determined by Holter or electrophysiological findings). Considerable interest has also been shown by a number of workers in the possibility of using antiarrhythmic drugs as a more generalized prophylactic therapy following myocardial infarction, in an attempt to reduce the incidence of sudden death in these patients (MAY et al. 1982; FURBERG 1983).

While a suggestion of mortality reduction has been reported in small-scale studies for disopyramide and procainamide (ZAINAL et al. 1977; KOSOWSKY et al. 1973), none of the seven major long-term controlled trials using Ia agents has shown any significant reduction in postinfarct mortality (MAY et al. 1982; FURBERG 1983). As the authors of these reviews note, however, the study designs have been inadequate to detect modest (but clinically relevant) effects on death rates. The only drugs thus far demonstrated to reduce mortality following MI are the beta-blockers (YUSUF et al. 1985).

E. Proarrhythmic Effects of Antiarrhythmic Drugs

An association has long been recognized between quinidine and malignant ventricular arrhythmias (DENES et al. 1981; PRATT et al. 1983). In one recent review of 3307 consecutive Holter monitor records, ventricular fibrillation was recorded in 5 patients, all 5 of whom had recently commenced taking quinidine (DENES et al. 1981). Procainamide and disopyramide have also been implicated in causing VT/VF (STRASBERG et al. 1981; NICHOLSON et al. 1979; WALD et al. 1981). It was VELEBIT and his colleagues (1982), however, who first identified the scale of the problem. In a series of papers, they have found definite evidence of aggravation of arrhythmias by antiarrhythmic agents in 11% of over 1000 drug trials (245 patients) using non-invasive monitoring, and 13% of 216 invasive electrophysiological drug tests (63 patients; POSER et al. 1985; PODRID 1984, 1985; SLATER et al. 1986). A large range of drugs, including all three of the Ia agents under discussion, showed this side effect in 5%–16% of cases. Definite proarrhythmic effects were defined on Holter recordings as the occurrence of a fourfold increase in ventricular ectopic frequency, a tenfold increase in repetitive forms, or the first emergence of sustained ventricular tachycardia coincident with the time course of action of the drug under study. At electrophysiological study, definite aggravation was assumed if a non-sustained VT was converted to sustained VT or if a given end point which required three extrastimuli to provoke before drug, could be produced by only one extrastimulus in the presence of drug (VELEBIT et al. 1982; POSER et al. 1985).

Of the Ia drugs, quinidine gave the highest proarrhythmia frequency (15.4%) and procainamide (9.1%) and disopyramide (5.9%) had lower, but still significant, incidences. The arrhythmias generally occurred when plasma drug levels were in the "therapeutic range". No clinical predictors of patients likely to exhibit this side effect have been identified except that it seems more frequent in those with the most severe presenting arrhythmias (sustained VT or VF; SLATER et al. 1986). Proarrhythmic effects at electrophysiological study appear to be a good predictor of future arrhythmia aggravation (COOPER et al. 1986.

I. Torsade de Pointes

One of the most frequent and most serious of the proarrhythmic effects reported is the occurrence of a bizarre form of VT which may progress to VF and which has been labelled "torsade de pointes". Torsade de pointes and/or drug-induced VF appear to be more likely to occur in patients with low serum potassium (<3.5 mmol/litre) or patients also taking digitalis (MINARDO et al. 1986; BAUMAN et al. 1984; WOOSLEY et al. 1984; RODEN et al. 1986). If torsade de pointes or VF occur they usually do so within the 1st week of treatment (median = 3 days). The QT interval is usually prolonged by the drug but this is also true of patients who do not develop drug-induced arrhythmias and, unless extreme, is probably not a reliable predictor of risk of VT/VF (MINARDO et al. 1986; WOOSLEY et al. 1984). There is also some crossover risk, at least with the three Ia agents (MINARDO et al. 1986).

There is considerable evidence for the view that drug-induced torsade de pointes and VF (at least in the case of quinidine) are due to early afterdepolarizations interrupting abnormally long ventricular action potentials associated with long QT intervals or following long pauses (heart block or postectopic pauses) especially in the presence of low extracellular potassium levels (RODEN and HOFFMAN 1985; WOOSLEY et al. 1984; HENNING and WIT 1982). Management of this side effect should include withdrawal of the suspected drug or drugs and correction of any hypokalemia and or bradycardia. The use of manoeuvres which shorten action potential duration, such as Ib antiarrhythmic drugs, isoprenaline infusion, magnesium infusion and overdrive pacing have all been reported of value (KHAN et al. 1981; STEINBRECHER and FITCHETT 1980; STRASBERG et al. 1980; TZIVONI et al. 1984; ANDERSON and MASON 1978).

F. Concordance Among Class Ia Drugs

The classification of antiarrhythmic drugs into groups and subgroups (see Chap. 2, 7) is ultimately of little value to the clinician unless a degree of concordance can be demonstrated within the subgroups. If it could be shown, for instance, that all arrhythmias which fail to respond to quinidine will also fail to respond to other Ia agents, then the subclassification has been of clinical value, but the distinction is complicated by actions of these agents, e.g. anticholinergic, unrelated to effects on sodium channels. If each patient has to be given a trial of each Ia drug in turn, then the subclassification is no real help to the cardiologist in selecting a drug (although, as we have seen, it may be useful in predicting a crossover of side effects within the subclass; MINARDO et al. 1986).

It must be said that there are few data suggesting concordance of drug efficacy within the Ia subgroup. There is evidence (discussed above) that response to procainamide in patients with VT/VF tends to predict response to other antiarrhythmic drugs and failure of response similarly predicts likely failure to find any successful agent (WAXMAN et al. 1983; RAE et al. 1985). This concordance was not, however, exclusive to class Ia or even to class I drugs. A small study in patients with atrioventricular re-entrant tachycardia found that 8 of 9 patients who responded to procainamide, versus 2 of 17 who failed to respond, were also responsive to quinidine ($P < 0.01$). There was no significant concordance between procainamide and disopyramide, nor between quinidine and disopyramide (BAUERNFEIND et al. 1983).

G. Combinations of Ia Drugs with Other Class I Agents

Because of their relatively slow offset kinetics (see Chap. 7), Ia drugs in therapeutic concentrations tend to depress ectopic activity throughout diastole whereas Ib drugs with faster kinetics can be given in concentrations which selectively depress early diastolic activity (CAMPBELL 1983). It has therefore been suggested that drugs from these two subclasses might usefully be combined (HONDEGHEM and KATZUNG 1980).

This possibility is yet to be explored systematically but at least two studies have shown therapeutic benefit from combinations of mexiletine (Ib) with disopyramide (BREITHARDT et al. 1981) and quinidine (DUFF et al. 1983).

H. Conclusions

The class Ia drugs we have discussed comprise a group of well-established antiarrhythmic agents which still have an important place in the therapeutic armamentarium, despite the plethora of newer compounds available.

Quinidine, disopyramide and procainamide have similar effects on clinical electrophysiological parameters. In therapeutic doses they tend to increase the sinus rate slightly and to produce some increase in the QT and HV intervals.

All three agents but particularly the first two are useful for reversion of supraventricular arrhythmias. The Ia drugs are also quite effective against ventricular ectopic activity and many cases of ventricular tachycardia. In common with virtually all present antiarrhythmic drugs (except perhaps amiodarone), the Ia agents are less often effective in preventing the more severe forms of ventricular tachycardia, and ventricular fibrillation. Furthermore, there is a definite incidence of proarrhythmic effects when using these drugs in such patients.

It seems likely, however, that Ia agents will continue to play a role in the management of cardiac arrhythmias for the foreseeable future.

References

Anderson JL (1986) Sudden cardiac death, ventricular arrhythmias and antiarrhythmia therapy: current trials and tribulations. Aust NZ J Med 16:409-412

Anderson JL, Mason JW (1978) Successful treatment by overdrive pacing of recurrent quinidine syncope due to ventricular tachycardia. Am J Med 64:715-718

Anderson JL, Mason JW, Winkle RA, Meffin PJ, Fowles RE, Peters F, Harrison DC (1978) Clinical electrophysiological effects of tocainide. Circulation 57:685-691

Baines MW, Davies JE, Keilett DN, Munt PL (1976) Some pharmacological effects of disopyramide and a metabolite. J Int Med Res 4 (Suppl 1):5-7

Bauernfeind RA, Swiryn S, Petropoulos AT, Coelho A, Gallastegui J, Kosen KM (1983) Concordance and discordance of drug reponses in atrioventricular reentrant tachycardia. J Am Coll Cardiol 2:345-350

Bauman JL, Bauernfeind RA, Hoff JV, Strasberg B, Swiryn S, Rosen KM (1984) Tor-

sade de pointes due to quinidine: observations in 31 patients. Am Heart J 107:425-430
Benditt DG, Woodrow-Benson P, Klein GT, Pritzker MR, Kriett JM, Anderson RW (1983) Prevention of recurrent sudden cardiac arrest: role of provocative electropharmacologic testing. J Am Coll Cardiol 2:418-425
Bennett DH (1978) Disopyramide in patients with the Wolff-Parkinson-White syndrome and atrial fibrillation. Chest 74:624-628
Bexton RS, Hellestrand KJ, Cory-Pearce R, Spurrell RAJ, English TAH, Camm AJ (1982) Direct electrophysiological effects of disopyramide—evaluated in denervated human heart (abstr). Br Heart J 48:87
Bigger JT (1985) Patients with malignant or potentially malignant ventricular arrhythmias: opportunities and limitations of drug therapy in prevention of sudden death. J Am Coll Cardiol 5 (Suppl):23B-26B
Bigger JT, Hoffman BF (1980) Antiarrhythmic drugs. In: Gilman AG, Goodman LS, Gilman A eds. The pharmacological basis of therapeutics, 6th ed. Macmillan, London, pp 761-792
Bigger JT, Rolnitzky LM (1985) The Evaluation of antiarrhythmic drug efficacy. In: Reiser HJ, Horowitz LN (eds) Mechanisms and treatment of cardiac arrhythmias: relevance of basic studies to clinical management. Urban and Schwarzenberg, Baltimore, pp 117-135
Bigger JT, Weld FM, Rolnitzky LM (1981) Prevalence, characteristics and significance of ventricular tachycardia (three or more complexes) detected with ambulatory electrocardiographic recording in the late hospital phase of acute myocardial infarction. Am J Cardiol 48:815-823
Bigger JT, Fleiss JL, Kleiger R, Miller JP, Rolnitzky LM, The Multicenter Post-Infarction Research Group (1984) The relationships among ventricular arrhythmias, left ventricular dysfunction, and mortality in the two years after myocardial infarction. Circulation 69:250-258
Birkhead JS, Vaughan Williams EM (1977) Dual effect of disopyramide on atrial and atrioventricular conduction and refractory periods. Br Heart J 39:657-660
Breithardt G, Seipel L, Abendroth RR (1981) Comparison of the antiarrhythmic efficacy of disopyramide and mexiletine against stimulus induced ventricular tachycardia. J Cardiovasc Pharmacol 3:1026-1037
Brugada P, Wellens HJJ (1984) Effects of intravenous and oral disopyramide on paroxysmal atrioventricular nodal tachycardia. Am J Cardiol 53:88-92
Califf RM, McKinnis RA, Burks J, Lee KL, Harrell FE, Behak VS, Pryor DB, Wagner GS, Kosati RA (1982) Prognostic implications of ventricular arrhythmias during 24 hours ambulatory monitoring in patients undergoing cardiac catheterization for coronary artery disease. Am J Cardiol 50:23-81
Camm J, Ward D, Spurrell R (1980) Response of atrial flutter to overdrive atrial pacing and intravenous disopyramide phosphate, singly and in combination. Br Heart J 44:240-247
Campbell TJ (1983) Kinetics of onset of rate-dependent effects of Class I antiarrhythmic drugs are important in determining their effects on refractoriness in guinea-pig ventricle, and provide a theoretical basis for their subclassification. Cardiovasc Res 17:344-352
Campbell TJ (1987) Differing electrophysiological effects of Class IA, IB & IC antiarrhythmic drugs on guinea-pig sinoatrial node. Br J Pharmacol 91:395-401
Campbell TJ, Morgan JJ (1980) Treatment of atrial arrhythmias after cardiac surgery with intravenous disopyramide. Aust NZ J Med 10:644-649
Campbell TJ, Gavaghan TP, Morgan JJ (1985) Intravenous sotalol for the treatment of

atrial fibrillation and flutter after cardiopulmonary bypass: comparison with disopyramide and digoxin in a randomised trial. Br Heart J 54:86-90
Cardiac Arrhythmia Pilot Study Investigators (1986) The cardiac arrhythmia pilot study. Am J Cardiol 57:91-95
Carliner NH, Fisher ML, Crouthamel WG, Narang PK, Plotnick GD (1980) Relation of ventricular premature beat suppression to serum quinidine concentration determined by a new and specific assay. Am Heart J 100:483-489
Caron JF, Libersa CC, Kher AR, Kacet S, Wanszelbaum H, Dupuis BA, Poirier JM, Lekieffre JP (1985) Comparative study of encainide and disopyramide in chronic ventricular arrhythmias: a double-blind placebo-controlled crossover study. J Am Coll Cardiol 5:1457-1463
Cooper MJ, Hunt LT, Waywood TA, Denniss AR, Richards DA, Uther JB, Ross DL (1985) Ambulatory monitoring in patients with sustained ventricular tachycardia based on coronary artery diease (abstr). Aust NZ J Med 15:506
Cooper MJ, Hunt LJ, Palmer KJ, Dennis AR, Richards DA, Uther JB, Ross DL (1986) Prediction of pro-arrhythmic drug effects at electrophysiologic study for ventricular tachycardia. Circulation 74 (Suppl II):482
David BM, Madsen BW, Ilett KF (1980) Plasma binding of disopyramide. Br J Clin Pharmacol 9:614-618
Deal BJ, Miller SM, Scagliotti D, Prechel D, Gallastegui JL, Hariman RJ (1986) Ventricular tachycardia in a young population without overt heart disease. Circulation 73:1111-1118
Deano DA, Wu D, Mautner RK, Sherman RH, Ehsani AE, Rosen KM (1977) The antiarrhythmic efficacy of intravenous therapy with disopyramide phosphate. Chest 71:597-606
Denes P, Gabstek A, Huang SK (1981) Clinical, electrocardiographic and follow-up observations in patients having ventricular fibrillation during Holter monitoring. Am J Cardiol 48:9-16
Denniss A, Richards DA, Cody DV, Russell PA, Young AA, Ross DL, Uther JB (1985) Comparable prognostic significance of delayed potentials and inducible ventricular tachycardia after myocardial infarction. Aust NZ Med J 15:520
Denniss AR, Ross DL, Cody DV, Richards DA, Russell PA, Young AA, Uther JB (1986) Randomized trial of antiarrhythmic drugs in patients with inducible ventricular tachyarrhythmias after recent myocardial infarction. Circulation 74 (Suppl II):213
Desai JM, Scheinman M, Peters RW, O'Young J (1979) Electrophysiological effects of disopyramide in patients with bundle branch block. Circulation 59:215-225
Dhingra RC, Rosen KM (1979) Procainamide and the sinus node. Chest 76:620-621
DiMarco JP, Garan H, Ruskin JN (1983) Quinidine for ventricular arrhythmias: value of electrophysiologic testing. Am J Cardiol 51:95
DiMarco JP, Lerman BB, Kron JL, Sellers TD (1985) Sustained ventricular tachyarrhythmias within 2 months of acute myocardial infarction: results of medical and surgical therapy in patients resuscitated from the initial episode. J Am Coll Cardiol 6:759-768
Doering W (1979) Quinidine-digoxin intraction. N Engl J Med 301:400-404
Drayer DE, Lowenthal DT, Restivo KM, Schwartz A, Cook CE, Reidenberg MM (1978) Steady-state serum levels of quinidine and active metabolites in cardiac patients with varying degrees of renal function. Clin Pharmacol Ther 24:31-39
Dreyfuss J, Bigger JT, Cohen AI, Schreiber EC (1972) Metabolism of procainamide in rhesus monkey and man, Clin Pharmacol Ther 13:366-371
Duff HJ, Roden DM, Primm RK, Carey EL, Oates JA, Woosley RL (1983) Mexiletine

for resistant ventricular tachycardia: comparison with lidocaine and enhancement of efficacy by combination with quinidine. Circulation 67:1124-1128

Editorial (1979) Ventricular fibrillation outside hospital. Lancet 2:508-509

Eisenberg MS, Hallstrom A, Bergner L (1982) Long-term survival after out-of-hospital cardiac arrest. N Engl J Med 306:1340-1343

Elison J, Strong JM, Lee WK, Atkinson AJ (1975) Antiarrhythmic potency of *N*-acetylprocainamide. Clin Pharmacol Ther 17:134-140

Fechter P, Ha H, Follath F, Nager F (1983) The antiarrhythmic effects of controlled release disopyramide phosphate and long acting propranolol in patients with ventricular arrhythmias. Eur J Clin Pharmacol 25:729-734

Fenster PE, Comess KA, Marsh R, Katzenberg C, Hager WD (1983) Conversion of atrial fibrillation to sinus rhythm by acute intravenous procainamide infusion. Am Heart J 106:501-504

Ferrick KJ, Bigger JT, Reifel JA, Livelli FD, Gang ES, Gliklich JI (1982) Congruence in efficacy of procainamide and quinidine for induced ventricular tachycardia (abstr). Circulation 66 (Suppl II):142

Flecainide-Quinidine Research Group (1983) Flecainide versus quinidine for treatment of chronic ventricular arrhythmias: a multicenter clinical trial. Circulation 67:1117-1123

Furberg CP (1983) Effect of antiarrhythmic drugs on mortality after myocardial infarction. Am J Cardiol 52:32C-36C

Gavaghan TP, Feneley MP, Campbell TJ, Morgan JJ (1985) Atrial tachyarrhythmias after cardiac surgery: results of disopyramide therapy. Aust NZ J Med 15:27-32

Giardina EGV, Dreyfuss J, Bigger JT, Shaw JM, Schreiber EC (1976) Metabolism of procainamide in normal and cardiac subjects. Clin Pharmacol Ther 19:339-351

Giardina EGV, Heissenbuttel RH, Bigger JT (1977) The relationship between the metabolism of procainamide and sulfamethazine. Circulation 55:388-394

Giardina E, Fenster P, Bigger JT, Mayersohn M, Perrier D, Marcus F (1980) Efficacy, plasma concentration and adverse effects of a new sustained-release procainamide preparation. Am J Cardiol 46:855-862

Graboys TB, Lowin B, Podrid PJ, DeSilva R (1982) Long term survival of patients with malignant ventricular arrhythmia treated with antiarrhythmic drugs. Am J Cardiol 50:437-443

Gradman AH, Batsford WP, Rieur EC, Leon L, Vanizetta AM (1985) Ambulatory electrocardiographic correlates of ventricular inducibility during programmed electrical stimulation. JAMA 5:1087-1093

Grande P, Sonne B, Pedersen A (1986) A controlled study of digoxin and quinidine in patients DC reverted from atrial fibrillation to sinus rhythm. Circulation 74 (Suppl II):101

Green NL, Werner JA, Trabaugh GB, Reid PR, Schaeffer AH, Kime GM (1981) Programmed pacing for the evaluation and therapy of ventricular tachycardia. In: Harrison DC (ed) Cardiac arrhythmias: a decade of progress. Hall, Boston, pp 617-629

Greenspan AM, Horowitz LN, Spielman SR, Josephson ME (1980) Large dose procainamide therapy for ventricular tachyarrhythmia. Am J Cardiol 46:453-462

Haffajee CI (1985) Clinical effects of class III antiarrhythmic agents. In: Reiser HJ, Horowitz LN (eds) Treatment of cardiac arrhythmias: relevance of basic studies to clinical management. Urban and Schwarzenberg, Baltimore, pp 382-294

Hartel G, Louhija A, Konttinen A (1974) Disopyramide in the prevention of recurrence of atrial fibrillation after electroversion. Clin Pharmaco Ther 15:551-555

Heel RC, Brogden TM, Speight TM (1978) Disopyramide: a review of its pharmacological properties and therapeutic use in treating cardiac arrhythmias. Drugs 15:331-368

Hellestrand KF, Nathan AW, Bexton RS, Camm AJ (1984) Electrophysiologic effects of flecainide acetate on sinus node function, anomalous atrioventricular connections, and pacemaker thresholds. Am J Cardiol 53:30B-38B

Henning B, Wit AL (1982) Multiple mechanisms for antiarrhythmic drug action on delayed after depolarizations and triggered activity in canine coronary sinus (abstr). Am J Cardiol 49:913

Hinderling PH, Garrett ER (1976) Pharmacokinetics of the antiarrhythmic disopyramide in healthy humans. J Pharmacokin et Biopharm 4:199-230

Hinkle LE, Carver ST, Stevens M (1969) The frequency of asymptomatic disturbances of cardiac rhythm and conduction in middle-aged men. Am J Cardiol 24:629-650

Hoffman A, Schutz E, White R, Follath F, Burckhardt O (1984) Suppression of high-grade ventricular ectopic activity by antiarrhythmic drug treatment as a marker for survival in patients with chronic coronary disease. Am Heart J 107:1103-1108

Holford NHG, Coates PE, Guentert TW, Riegelman S, Sheiner LB (1981) The effect of quinidine and its metabolites on the electrocardiogram and systolic time intervals: concentration-effect relationships. Br J Clin Pharmacol 11:187-195

Hondeghem L, Katzung BG (1980) Test of a model of antiarrhythmic drug action. Circulation 61:1217-1224

Horowitz LN (1985a) Clinical antiarrhythmic effects of calcium channel-blocking drugs. In: Reiser HJ, Horowitz LN (eds) Mechanisms and treatment of cardiac arrhythmias: relevance of basic studies to clinical management. Urban and Schwarzenberg, Baltimore, pp 329-335

Horowitz LN (1985b) Indices of antiarrhythmic drug efficacy using invasive techniques. In: Reiser HJ, Horowitz LN (eds) Mechanisms and treatment of cardiac arrhythmias: relevance of basic studies to clinical management. Urban and Schwarzenberg, Baltimore, pp 137-145

Horowitz LN, Josephson ME, Farshidi A, Spielman SR, Michelson EL, Greenspan AM (1978) Recurrent sustained ventricular tachycardia 3. Role of the electrophysiologic study in selection of antiarrhythmic regimens. Circulation 58:986-997

Hulting J, Jansson B (1977) Antiarrhythmic and electrocardiographic effect of single oral doses of disopyramide. Eur J Clin Pharmacol 11:91-99

Jaillon P, Rubenson D, Peters F, Mason JW, Winkle RA (1981) Electrophysiologic effects of *N*-acetylprocainamide in human beings. Am J Cardiol 47:1134-1140

Josephson ME, Caracta AR, Lau SH et al. (1973) Electrophysiological evaluation of disopyramide in man. Am Heart J 86:771-780

Josephson ME, Caracta AR, Ricciutti MA, Lau SM, Damato AN (1974a) Electrophysiologic properties of procainamide in man, Am J Cardiol 33:596-603

Josephson ME, Seides SF, Batsford WP, Weisfogel GM, Akhtar M, Caracta AR, Lau SH, Damato AN (1974b) The electrophysiological effects of intramuscular quinidine on the atrioventricular conducting system in man. Am Heart J 87:55-64

Kanovsky MS, Falcone RA, Dresden CA, Josephson ME, Simson MB (1984) Identification of patients with ventricular tachycardia after myocardial infarction: signal-averaged electrocardiogram, Holter monitoring and cardiac catheterization. Circulation 70:264-270

Karlson BW, Thorstensson I, Abjorn C (1986) Disopyramide in the maintenance of sinus rhythm after electroversion of atrial fibrillation—a placebo-controlled one year follow-up study. Circulation 74 (Suppl II):101

Katz MJ, Meyer CE, El-Etr A, Slooki SJ (1963) Clinical evaluation of a new antiarrhythmic agent, SC-7031. Curr Ther Res Clin Exp 5:343-350

Kennedy HL, Pescarmona JE, Bouchard RJ, Goldberg RJ (1980) Coronary artery status of apparently healthy subjects with frequent and complex ventricular ectopy. Ann Intern Med 92:179-185

Khan MM, Logan KR, McComb JM, Adgey AJ (1981) Management of recurrent ventricular tachyarrhythmias associated with Q-T prolongation. Am J Cardiol 47:1301-1308

Kim HG, Friedman HS (1979) Procainamide-induced sinus node dysfunction in patients with chronic renal failure. Chest 76:699-700

Kim SG, Seiden SW, Matos TA, Waspe LE, Fisher JD (1985) Discordance between ambulatory monitoring and programmed stimulation in assessing efficacy of class Ia antiarrhythmic agents in patients with ventricular tachycardia. J Am Coll Cardiol 6:539-544

Kjekshus J, Bathen J, Orning O, Storstein L (1984) A double-blind, crossover comparison of flecainide acetate and disopyramide phosphate in the treatment of ventricular premature complexes. Am J Cardiol 53:72B-78B

Kosowsky BD, Taylor J, Lown B, Ritchie RF (1973) Long term use of procainamide following acute myocardial infarction. Circulation 47:1204-1210

Kuchar DL, Thorburn CW, Sammel NL (1986) Late potentials detected after myocardial infarction: natural history and prognostic significance. Circulation 74:1280-1289

LaBarre A, Strauss HC, Scheinman MM, Evans GT, Bashore T, Tiedeman JS, Wallace AG (1979) Electrophysiologic effects of disopyramide phosphate on sinus node function in patients with sinus node dysfunction. Circulation 59:226-235

Lainal N, Griffiths JW, Carmichael DJS, Besterman EMM, Kidner PH, Gillham AP, Summers GD (1977) Oral disopyramide for the prevention of arrhythmias in patients with acute myocardial infarction admitted to open wards. Lancet 2:887-889

Lemery R, Brugada P, v. d. Dool A, Heijmeriks J, Bella PD, Dugernier T, Wellens HJJ (1986) Clinical course and long term follow-up of 60 patients with idiopathic ventricular tachycardia or ventricular fibrillation. Circulation 74(Suppl II):188

Lerman BB, Waxman HL, Buxton AE, Josephson ME (1983) Disophyramide: evaluation of electrophysiologic effects and clinical efficacy in patients with sustained ventricular tachycardia or ventricular fibrillation. Am J Cardiol 51:759-764

Lidell C, Rehnquist N, Sjogren A, Yli-Votika RJ, Ronnevik PK (1985) Comparative efficacy of oral sotalol and procainamide in patients with chronic ventricular arrhythmias: a multicenter study. Am Heart J 109:970-975

Lombardi F, Stein J, Podrid PJ, Graboys TB, Lown B (1986) Daily reproducibility of electrophysiologic test results in malignant ventricular arrhythmia. Am J Cardiol 57:96-101

Lown B (1979) Sudden cardiac death: the major challenge confronting contemporary cardiology. Am J Cardiol 43:313-328

Mason JW, Winkle RA (1978) Electrode-catheter arrhythmia induction in the selection and assessment of antiarrhythmic drug therapy for recurrent ventricular tachycardia. Circulation 58:971-985

Mason JW, Winkle RA, Rider AK, Stinson EB, Harrison DC (1977) The electrophysiologic effects of quinidine in the transplanted human heart. J Clin Invest 59:481-489

Mason JW, Winkle RA, Griffin JC (1981) Electrophysiologic-pharmacologic studies of ventricular arrhythmias: evaluation of methods. In: Harrison DC (ed) cardiac arrhythmias: a decade of progress. Hall, Boston, pp 631-642

Mark LC, Kayden HJ, Steele JM, Copper R, Berlin I, Rosenstine EA, Brodie BB (1951) The physiological disposition and cardiac effects of procaine amide. J Pharmacol Exp Ther 102:5-15

Mautz FR (1936) The reduction of cardiac irritability by the epicardial and systemic administration of drugs as a protection in cardiac surgery. J Thorac Surg 5:612-628

May GS, Eberlein KA, Furberg LO, Passamani ER, Demets PL (1982) Secondary prevention after myocardial infarction: a review of long-term trials. Prog Cardiovasc Dis 24:331-352

McHenry PL, Morris SN, Kavalier M, Jordan JW (1946) Comparative study of exercise-induced ventricular arrhythmias in normal subjects and patient with documented coronary artery disease. Am J Cardiol 37:609-616

Meffin PJ, Robert EW, Winkle RA, Harapat S, Peters FA, Harrison DC (1979) Role of concentration-dependent plasma protein binding in disopyramide disposition. J Pharmacokinet Biopharm 7:29-46

Millar Craig MW, Raftery EB (1977) A controlled trial of disopyramide in paroxysmal supraventricular arrhythmias. In: Ankier SI, Woodings DF (eds) Proceedings of the disopyramide (rythmodan) seminar. Viking, London, pp 55-60

Minardo JP (1986) Drug associated ventricular fibrillation: analysis of clinical features and QT_c prolongation (abst). J Am Coll Cardiol 7:158A

Minardo JP, Heger JJ, Zipes DP, Miles WM, Prystowsky EN (1986) Drug associated ventricular fibrillation: analysis of clinical features and QT_c prolongation. J Am Coll Cardiol (Abstract) 7:158A

Mirowski M, Reid PR, Mower MM, Watkins L, Gott VL, Schauble JF, Langer A, Heilman MS, Kolenik SA, Fischell RE, Weisfeldt ML (1980) Termination of malignant ventricular arrhythmias with an implanted automatic defibrillator in human beings. N Engl J Med 303:322-325

Mirowski M, Reid PR, Winkle RA, Mower MM, Watkins L, Stinson EB, Griffith LSC, Kallman CH, Weisfeldt ML (1983) Mortality in patients with implanted automatic defibrillators. Ann Intern Med 98:585-588

Mirro MJ, Manalan AS, Bailey JC, Watanabe AM (1980a) Anticholinergic effects of disopyramide and quinidine on guinea pig myocardium. Circ Res 47:855-865

Mirro MJ, Watanabe AM, Bailey JC (1980b) Electrophysiological effects of disopyramide and quinidine on guinea pig atria and canine cardiac Purkinje fibers. Circ Res 46:660-668

Mirro MJ, Watanabe AM, Bailey JC (1981) Electrophysiological effects of the optical isomers of disopyramide and quinidine in the dog. Circ Res 48:867-874

Mitchell LB, Duff HJ, Wyse DG (1986) Randomized comparison of non invasive and invasive approaches to drug therapy for sustained ventricular tachyarrhythmia. Circulation 74 (Suppl II):214

Mokler CM, Van Arman CG (1962) Pharmacology of a new antiarrhythmic agent, γ-disopropyl-amino-α-phenyl-α-(2-pyridyl)-butyramide (SC-7031). J Pharmacol Exp Ther 136:114-124

Morady F, Scheinman MM, Desai J, (1982) Disopyramide. Ann Intern Med 96:334-343

Morganroth J (1984) Computer recognition of cardiac arrhythmias and statistical approaches to arrhythmia analysis. Ann NY Acad Sci 432:117-128

Morganroth J, Somberg JC, Pool PE, Hsu P, Lee IK, Durkee J (1986) Comparative

study of encainide and quinidine in the treatment of ventricular arrhythmias. J Am Coll Cardiol 7:9-16
Moss AJ, Davis HT, DeCamilla J, Bayer LW (1979) Ventricular ectopic beats and their relation to sudden and nonsudden cardiac death after myocardial infarction. Circulation 60:998-1003
Mukharji J, Rude RE, Poole K, Gustavson N, Thomas LJ, Strauss HW, Jaffe AS, Muller JE, Roberts R, Raabe DS, Croft H, Passamani E, Braunwald E, Willerson JT, Milis Study Group (1984) Risk factors for sudden death after acute myocardial infarction: two-year follow-up. Am J Cardiol 54:31-36
Myerburg R, Kessler K, Kiem I, Pefkaros K, Condi C, Cooper D, Castellanos A (1981 a) Relationship between plasma levels of procainamide, suppression of premature ventricular complexes and prevention of recurrent ventricular tachycardia. Circulation 64:280-290
Myerburg RJ, Kessler KM, Pefkaros KC, Cooper D, Kiem I, Castellanos A (1981b) Effects of antiarrhythmic agents on premature ventricular contractions and on potentially lethal arrhythmias. In: Harrison DC (ed) Cardiac arrhythmias: a decade of progress. Hall, Boston, pp 571-584
Niarchos AP (1976) Disopyramide: serum level and arrhythmia conversion. Am Heart J 92:57-64
Nicholson WJ, Martin CE, Gracey JG, Knoch MR (1979) Disopyramide-induced ventricular fibrillation. Am J Cardiol 43:1053-1059
Platia EV (1986) Programmed stimulation (PES)—directed drug therapy for sustained ventricular tachyarrhythmias: long-term implications of altering therapy. Circulation 74(Suppl II):313
Platia EV, Reid PR (1984) Comparison of programmed electrical stimulation and ambulatory electrocardiographic (Holter) monitoring in the management of ventricular tachycardia and ventricular fibrillation. J Am Coll Cardiol 4:493-500
Podrid PJ (1984) Can antiarrhythmic drugs cause arrhythmia? J Clin Pharmacol 24:313-319
Podrid PJ (1985) Can antiarrhythmic drugs prevent sudden death? In: Reiser HJ, Horowitz LN (eds) Mechanism and treatment of cardiac arrhythmias; relevance of basic studies to clinical management. Urban and Schwarzenberg, Baltimore, pp 341-361
Podrid PJ, Schoeneberger A, Lown B, Lampert S, Matos T, Porterfiela J, Raeder E, Corrigan E (1983) Use of nonsustained ventricular tachycardia as a guide to antiarrhythmic drug therapy in patients with malignant ventricular arrhythmia. Am Heart J 105:181-188
Poser RF, Podrid PJ, Lombardi F, Lown B (1985) Aggravation of arrhythmia induced with antiarrhythmic drugs during electrophysiologic testing. Am Heart J 110:9-16
Pratt C, Francis M, Luck J, Wyndham C, Miller R, Quinones M (1983) Analysis of ambulatory electrocardiograms in 15 patients during spontaneous ventricular fibrillation with special reference to preceding arrhythmic events. J Am Coll Cardiol 2:789-797
Pratt CM, Young JB, Francis MJ, Taylor AA, Norton HJ, English L, Mann DE, Kopelen H, Quinones MA, Roberts R (1984) Comparative effect of disopyramide and ethmozine in suppressing complex ventricular arrhythmias by use of a double-blind, placebo-controlled, longitudinal crossover design. Circulation 69:288-297
Pratt CM, Hallstrom AP, Coromilas J, Romhilt DW, CAPS Investigators (1986) Apparent "arrhythmia reduction" by antiarrhythmic therapy in post-infarction patients: a natural history of the placebo group in the cardiac arrhythmia pilot study (CAPS). Circulation 74(Suppl II):214

Puech P (1970) Treatment of cardiac arrhythmias with disopyramide. Minerva Med 61:3763-3768
Rae AP, Sokoloff NM, Webb CR, Spielman SR, Greenspan AM, Horowitz LN (1985) Limitations of failure of procainamide during electrophysiologic testing to predict response to other medical therapy. J Am Coll Cardiol 6:410
Raftery EB, Cashman PMM (1976) Long-term recording of the electrocardiogram in a normal population. Postgrad Med J 52 (Suppl 7):32-37
Reidenberg MM, Drayer DE, Levy M, Warner H (1975) Polymorphic acetylation of procainamide in man. Clin Pharmacol Ther 17:722-730
Roden DM, Hoffman BF (1985) Action potential prolongation and induction of abnormal automaticity by low quinidine concentrations in canine Purkinje fibers. Circ Res 56:857-867
Roden MR, Gelband H, Hoffman BF (1972) Canine electrocardiographic and cardiac electrophysiologic changes induced by procainamide. Circulation 46:528-536
Roden DM, Reele SB, Higgins SB, Wilkinson GR, Smith RF, Oates JA, Woosley RC (1980) Antiarrhythmic efficacy, pharmacokinetics and safety of *N*-acetyloprocainamide in human subjects: comparison with procainamide. Am J Cardiol 46:463-468
Roden DM, Thompson KA, Hoffman BF, Woosley RL (1986) Clinical features and basic mechanisms of quinidine-induced arrhythmias. J Am Coll Cardiol 8:73A-78A
Rodstein M, Wolloch L, Gubner RS (1971) Mortality study of the significance of extrasystoles in an insured population. Circulation 44:617-625
Ross D, Vohra J, Cole P, Hunt D, Sloman G (1978) Electrophysiology of disopyramide in man. Aust NZ J Med 8:377-383
Routledge PA, Stargel WW, Wagner GS, Shand DG (1980) Increased alpha-1-acid glycoprotein and lidocaine disposition in myocardial infarction. Ann Intern Med 93:701-704
Ruskin JN, Dimarco JP, Garan H (1980) Electrophysiologic observations and selections of long-term antiarrhythmic therapy. N Engl J Med 303:607-13
Salerno DM, Hodges M, Granrud G, Sharkey (1983) Comparison of flecainide with quinidine for suppression of chronic stable ventricular ectopic depolarizations. Ann Intern Med 98:455-460
Salerno DM, Grandrud G, Dugbar D, Sharkey P, Krejci J, Fifield J, Riendl S, Maiden R, Hodges M (1986) Feasibility of long-term drug therapy for ventricular arrhythmia. Circulation 74(Suppl II):213
Sami M, Harrison DC, Kreamer H, Houston N, Shimasaki C, Debusk RF (1981) Antiarrhythmic efficacy of encainide and quinidine: validation of a model for drug assessment. Am J Cardiol 48:147-156
Schaeffer AH, Greene ML, Reid PR (1978) Suppression of the repetitive ventricular response: a index of long term antiarrhythmic effectiveness of aprindine for ventricular tachycardia in man. Am J Cardiol 42:1007-1012
Schaffer WA, Cobb LA (1975) Recurrent ventricular fibrillation and modes of death in survivors of out of hospital ventricular fibrillation. N Engl J Med 293:259-262
Schenck-Gustafsson K, von Bahr C, Dahlqvist R, Edhag O (1986) Long-term effects related to dose and concentration of disopyramide in atrial fibrillation. Circulation 74(Suppl II):101
Searle Laboratories (1977) Norpace (disopyramide phosphate), an antiarrhythmic drug. Investigational brochure, Searle Laboratories, Chicago
Sellers TD, Campbell WF, Bashore TM, Gallagher JJ (1977) Effects of procainamide and quinidine sulfate in the Wolff-Parkinson-White syndrome. Circulation 55:15-22

Selzer A (1982) Quinidine in perspective: the rise and fall of quinidine. Heart Lung 11:20-23

Senges J, Lengfelder W, Jauernig R, Czygan E, Brachmann J, Rizos I, Kubler W (1982) Comparative effects of sotalol, metoprolol and quinidine on sustained ventricular tachycardia. Circulation 66(Suppl II): II-142 (Abstr)

Singh JB, Rasul AM, Shah A, Adams E, Flessas A, Kocot SL (1984) Efficacy of mexiletine in chronic ventricular arrhythmias compared with quinidine: a single-blind, randomized trial. Am J Cardiol 53:84-87

Skale BT, Miles WM, Heger JJ, Zipes DP, Prystowsky EN (1986) Survivors of cardiac arrest: prevention of recurrence by drug therapy as predicted by electrophysiologic testing or electrocardiographic monitoring. Am J Cardiol 57:113-119

Slater W, Podrid P, Lampert S, Lown B (1986) Are there clinical predictors for arrhythmia aggravation? J Am Coll Cardiol 7:158A

Sondermark T, Jonsson B, Olsson A, Oro L, Wallin M, Edhag D, Sjogrew A, Danielsson M, Rosenhamer G (1975) Effect of quinidine on maintaining sinus rhythm after conversion of atrial fibrillation or flutter. A multicenter study from Stockholm. Br Heart J 37:486-492

Sokolow M, Ball RE (1956) Factors influencing conversion of chronic atrial fibrillation with special reference to serum quinidine concentration. Circulation 14:568-583

Spurrel RAJ, Thorburn CW, Camm J, Sowton E, Deuchar DC (1975) Effects of disopyramide on electrical physiological properties of the specialized conduction system in man and on the accessory atrioventricular pathway in Wolff-Parkinson-White syndrome. Br Med J 37:861-867

Strasberg B, Kanakis C, Dhingra RC, Rosen KM (1978) Myotonia dystrophica and mitral valve prolapse. Chest 78:845-848

Strasberg B, Sclarousky S, Erdberg A, Duffy CE, Lam W, Swiryn S, Agmon J, Rosen KM (1981) Procainamide-induced polymorphous ventricular tachycardia, Am J Cardiol 47:1309-1314

Steinbrecher VP, Fitchett DH (1980) Torsade de pointes, a cause of syncope with atrioventricular block. Arch Intern Med 140:1223-1226

Swerdlow CD, Yu JO, Jacobson E, Mann S, Winkle RA, Griffin JC, Ross DL, Mason JW (1983) Safety and efficacy of intravenous quinidine. Am J Med 75:36-41

Swiryn S, Bauernfeind RA, Wyndham CRC, Dhingra RC, Paliled E, Strasberg B, Rosen KM (1981) Effects of oral disopyramide phosphate on induction of paroxysmal supraventricular tachycardia. Circulation 64:169

Taggart W, Holyoak W (1983) Steady-state biovailability of two sustained-release quinidine preparations. Clin Ther 5:357

Tzivoni P, Keren A, Cohen AM, Loebel H, Zahavi I, Chanzbraun A, Stern S (1984) Magnesium therapy for torsades de pointes. Am J Cardiol 53:528-530

Vandenbosch R, Lisin N, Andriange M et al. (1975) Experimentation clinique du disopyramide administré par voie intraveineuse. Acta Cardiol (Brux) 30:267-278

Vaughan Williams EM (1980) Antiarrhythmic action and the puzzle of perhexiline. Academic, London

Velebit V, Podrid PJ, Lown B, Cohen BH, Graboys TB (1982) Aggravation and provocation of ventricular arrhythmias by antiarrhythmic drugs. Circulation 65:886

Vismara LA, Amsterdam EA, Mason DT (1975) Relation of ventricular arrhythmias in the late phase of acute myocardial infarction to sudden death after hospital discharge. Am J Med 59:6-12

Vismara LA, Vera Z, Miller RR, Mason DT (1977) Efficacy of disopyramide phosphate

in the treatment of refractory ventricular tachycardia. Am J Cardiol 39:1027-1034
Vlasses P, Rocci M, Porini K, Greenspan A, Ferguson R (1983) Immediate-release and sustained-release procainamide: bioavailability at steady state in cardiac patients. Ann Intern Med 98:613-614
Wald RW, Waxman MB, Colman JM (1981) Torsade de pointes ventricular tachycardia. A complication of disopyramide shared with quinidine. J Electrocardiol 14:301-308
Waxman HL, Buxton AE, Sadonski LM, Josephson ME (1983) The response to procainamide during electrophysiologic study for sustained ventricular tachyarrhythmias predicts the response to other medications. Circulation 67:30-37
Wellens HJJ, Bak FWM, Brugada P (1982) Ventricular tachycardia—the clinical problem. In: Josephson ME (ed) Ventricular tachycardia: Mechanisms and management. Futura, Mount Kisco, NY, pp 1-19
Wellens HJJ, Brugade P, Abdollah H (1984) Drug therapy of patients with arrhythmias associated with bypass tracts. Ann NY Acad Sci 432:272-278
Wenckebach KF (1914) Die unregelmässige Herztätigkeit und ihre klinische Bedeutung. Engelman, Leipzig
West TC, Amory DW (1960) Single fiber recording of the effects of quinidine at atrial and pacemaker sites in the isolated right atrium of the rabbit. J Pharmacol Exp Ther 130:183-193
Wilkinson PR, Desai J, Hollister J, Gonzalez R, Abbott JA, Scheinmann MM (1982) Electrophysiologic effects of disopyramide in patients with atrioventricular nodal dysfunction. Circulation 66:1211-1215
Willius FA, Keys TE (1942) A remarkably early reference to the use of cinchona in cardiac arrhythmia. Proc Staff Meet Mayo Clin 17:294-296
Winkle RA, Gradman AH, Fitzgerald JW (1978) Antiarrhythmic drug effect assessed from ventricular arrhythmia reduction in the ambulatory electrocardiogram and treadmill test: comparison of propranolol, procainamide and quinidine. Am J Cardiol 42:473
Woodcock BG, Rietbrock N (1982) Digitalis-quinidine interactions. Trends Pharmacol Sci 3:118-122
Woosley RL, Cerskus I, Roden DM (1984) Antiarrhythmic therapy: clinical pharmacology update. J Clin Pharmacol 24:295-305
Woske H, Belford J, Fastier FN, Brooks CMcC (1953) The effect of procaine amide on excitability, refractoriness and conduction in the mammalian heart. J Pharmacol Exp Ther 107:134-139
Yusuf S, Peto R, Lewis J, Collins R, Sleight P (1985) Beta blockade during and after myocardial infarction: an overview of the randomized trials. Prog Cardiovasc Dis 27:335-371
Zainal N, Griffith JW, Carmichael DJS, Besterman EM, Kidner PH, Gillham ED, Summers GD (1977) Oral disopyramide for the prevention of arrhythmias in patients with acute myocardial infarction admitted to open wards. Lancet 2:887-889
Zema M (1984) Serum drug concentrations and adverse effects in cardiac patients after administration of a new controlled-release disopyramide preparation. Ther Drug Monit 6:192-198

CHAPTER 10

Clinical Use of Class Ib Antiarrhythmic Drugs

D. W. G. HARRON and R. G. SHANKS

A. Introduction

Antiarrhythmic drugs have been classified into four classes depending on their cellular electrophysiology (VAUGHAN WILLIAMS 1974), as discussed in detail in Chap. 2.

Class I has been divided into three subgroups (VAUGHAN WILLIAMS 1984): class Ia drugs depress rapid depolarization and prolong the action potential duration, e.g. quinidine, procainamide, disopyramide; class Ib drugs slow rapid depolarization slightly and shorten the action potential duration, e.g. lignocaine, mexiletine, tocainide (and will be discussed in this review); and class Ic drugs markedly depress cellular depolarization but have little effect on action potential duration, e.g. encainide, flecainide, lorcainide.

Lignocaine is successfully used for the treatment of ventricular arrhythmias. However, it is only used intravenously because of its poor bioavailability and the formation of toxic metabolites when administered orally. Mexiletine and tocainide, whilst similar in structure (Fig. 1) and pharmacological activity, have a high oral bioavailability and long plasma elimination half-life, making them useful for long-term treatment of arrhythmias.

B. Electrophysiology

Studies in animal models indicating a reduction in membrane responsiveness, shortening of the effective refractory period and action potential duration with reduction in action potential upstroke velocity and amplitude, confirm the class Ib activity of these compounds. Small differences between the drugs occur, with lignocaine having little or no effect at therapeutic concentrations on effective refractory period compared with mexiletine, and tocainide causing small clinically unimportant increases in atrioventricular nodal conduction time, decreases in sinus node function and depression of ventricular activity (ALLEN and ARNOLD 1974; GRIFFIN et al. 1977; VAUGHAN WILLIAMS 1977; MOORE et al. 1978; OSHITA et al. 1980; ALMOTREFI and BAKER 1980). Mexiletine and tocainide, like lignocaine, shorten action potential duration to a greater extent than the shortening of absolute refractory period, resulting in an increase in the absolute refractory period/action potential duration ratio; this is thought to reduce the time during an action potential whereby an ectopic impulse may occur (MOORE et al. 1978; YAMAGUCHI et al. 1979).

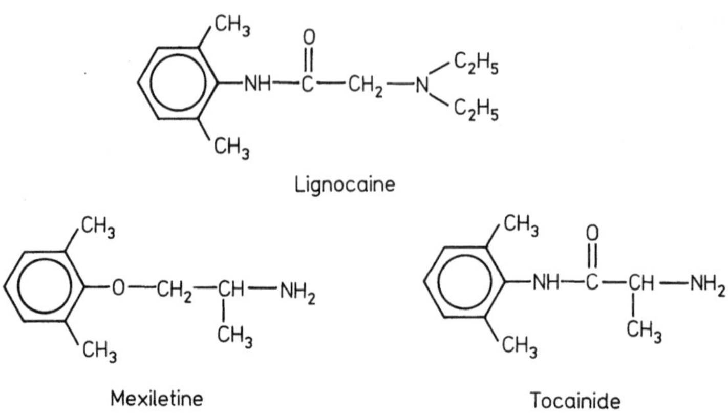

Fig. 1. Structures of class Ib antiarrhythmic drugs

Studies with mexiletine in patients indicate no effect on sinus node rate, sinus node recovery time, atrioventricular node and His-Purkinje conduction time, and atrial and atrioventricular refractory period; a shortening of the relative refractory period of the His-Purkinje system was reported by McCOMISH et al. (1977); however, Ross et al. (1977) reported an increase. These differences may result from differences in action in normal and diseased hearts. Tocainide 450 mg infusions in healthy volunteers had no significant effect on His bundle electrograms, sinus recovery time and ventricular effective refractory period (SWEDBERG et al. 1978). In patients with coronary artery disease, no significant effects were observed on sinus node function, intracardiac conduction time or effective refractory period (HOROWITZ et al. 1978). Small changes with higher concentrations of tocainide were reported by ANDERSON et al. (1978) in atrial (7%), ventricular (8%) and atrioventricular nodal (10%) effective refractory period.

The effects of lignocaine, mexiletine and tocainide in man on atrioventricular conduction time (PR interval), intraventricular conduction (QRS complex) and transnodal induction time indicate little or no change (Table 1).

In contrast, the class Ic agents encainide, flecainide and lorcainide (Table 1) prolong atrioventricular conduction, intraventricular conduction and transnodal conduction time and increase atrial and ventricular refractory periods. The effects on conduction and refractoriness with class Ia, class III and class IV agents are also shown in Table 1 for comparison.

C. Haemodynamic Effects

The haemodynamic effect of these drugs is dependent on the patient's clinical state and, in particular, left ventricular function. The majority of patients with arrhythmias have some left ventricular impairment. The effects of lignocaine, mexiletine and tocainide are similar.

Lignocaine causes dose-related effects on myocardial contractility, in that, at therapeutic doses (1-3 mg/kg), haemodynamic effects are mild and transient, whereas at higher doses (4-40 mg/kg) marked reductions occur in myocardial contractility, cardiac output and blood pressure (BINNION et al. 1969).

Table 1. Electrophysiological effects of lignocaine, mexiletine and tocainide compared with class Ia, Ic, III and IV antiarrhythmic drugs. (Zipes and Troup 1978; Camm 1984)

	Transnodal conduction time		AV node		Atria		Ventricle						
	A-H	H-V	ERP	FRP	ERP	FRP	ERP	FRP	PR	QRS	QT	QT_c	R-R
Lignocaine					↓		↓			→		↓→	
Mexiletine	↑→	↑→	↑→	↑→	→		→		→	→	→		→
Tocainide	↑→	→	↓	↓→	↓		↓		→	→	↓→		↓→
Procainamide					↑		↑			↑		↑	
Disopyramide	→	→↑	↓→	→	↑		↑		→	↑→	↑→	↑	↑→
Quinidine					↑		↑			↑		↑	
Encainide	↑→	↑	↑→		↑		↑		↑	↑	↑		
Flecainide	↑	↑	↑→		↑		↑		↑	↑	↑		
Lorcainide	↑	↑	↑→		↑		↑		↑	↑	↑		
Amiodarone	↑	→	↑	↑	↑		↑		→	→	↑		↓
Verapamil	↑	→	↑	↑	→		→		↑	→	→		↓

ERP, effective refractory period; FRP, functional refractory period; ↑, increase; ↓, decrease; →, no effect

Table 2. Haemodynamic effects of lignocaine, mexiletine and tocainide compared with other antiarrhythmic drugs

	Contractility	SVR	CO	BP
Lignocaine	↓→	→	→	→
Mexiletine	↓→	→	→	→
Tocainide	↓→	↑	→	→
Procainamide IV	↓	↓↓	→	↓↓
Procainamide PO	↓	↓	→	↓
Quinidine IV	↓↓	↓↓↓	↑→↓	↓↓↓
Quinidine PO	↓→	→↓	→	→↓
Amiodarone	↓	↓	↓→↑	↓
Verapamil	↓↓	↓↓	↓→↑	↓↓
Encainide	↓↓	↑	↓	→
Lorcainide	↓↓	↑→	↓→	→
Flecainide	↓↓	↑	↓	→
Disopyramide	↓↓↓	↑↑	↓↓↓	↑

↓ ↓↓ ↓↓↓, degree of depression; ↑, increase; →, no effect; SVR, systemic vascular resistance;
CO, cardiac output; BP, blood pressure

In patients with significant left ventricular dysfunction no significant alterations occurred in cardiac performance or blood pressure with lignocaine (BURTON et al. 1976).

Mexiletine does not affect ventricular performance in patients with normal left ventricular function following bolus injections up to 2 mg/kg or infusions up to 35 μg/kg per minute (POZENEL 1977). In patients with varying degrees of left ventricular dysfunction, an intravenous bolus up to 1.5 mg/kg produced a slight rise in left ventricular-end-diastolic pressure and a slight fall in cardiac output (BANIM et al. 1977). Mexiletine 300–400 mg administered orally three times daily had no significant haemodynamic effect in a group of patients with left ventricular ejection fractions of 48 % (STEIN et al. 1984).

Tocainide caused no detectable haemodynamic changes in healthy patients either at rest or with exercise after 450 mg intravenously (SWEDBERG et al. 1978). However, in patients with left ventricular dysfunction tocainide caused minor increases in left ventricular filling pressure and systemic vascular resistance, presumably compensating for a small reduction in myocardial contractility (WINKLE et al. 1978a).

It would appear therefore that mexiletine and tocainide in both the intravenous and oral forms are well tolerated haemodynamically, with lignocaine particularly suitable for patients with heart failure. The haemodynamic effects of class Ib drugs compare favourably with other antiarrhythmic agents, as shown in Table 2.

D. Pharmacokinetics

I. Lignocaine

The pharmacokinetics of intravenous lignocaine are best described by a two-compartment model (HARRISON et al. 1977); dose-independent pharmacokinetics of intravenous lignocaine after single doses (25–100 mg) have been described (OCHS et al. 1983), although this cannot be extrapolated to continuous infusion. The volume of distribution at steady state is 1.3 litres/kg, total body clearance 10 ml/min per kilogram, distribution half-life 8 min and plasma elimination half-life approximately 2 h. Lignocaine is reasonably well absorbed when administered orally; however, approximately 70 % undergoes first-pass metabolism to glycine xylidide and monoethylglycine xylidide, which has weak antiarrhythmic activity. Both metabolites can cause toxicity (STENSON et al. 1971; BLUMER et al. 1978). Approximately 40 %–80 % of lignocaine is bound to plasma proteins with a binding ratio related to the concentration of α_1 acid glycoprotein.

The therapeutic plasma concentration range varies from 1.5 to 6.0 µg/ml (HARRISON and ALDERMAN 1972). To achieve these plasma levels without toxicity due to peaks or subtherapeutic troughs occurring, various dosing regimes have been employed (see Sect. G).

II. Factors Affecting the Pharmacokinetics of Lignocaine

Old age, drugs and any disease state which influences hepatic blood flow or metabolism (O'MALLEY et al. 1971; WILLIAMS and MAMELOK 1980; PARK and BRECKENRIDGE 1981) may have significant effects on the pharmacokinetics of lignocaine.

In the presence of heart failure, plasma lignocaine concentrations increase by 50 % after long-term infusions (36–48 h). Metabolite accumulation may also contribute to clinical toxicity (HALKIN et al. 1975). With chronic alcoholic liver disease, clearance is reduced, volume of distribution at steady state increases and plasma elimination half-life can increase to 6 h (THOMSON et al. 1973); renal failure, whilst not influencing lignocaine kinetics, may affect metabolite elimination.

With long-term infusions of lignocaine an increased plasma elimination half-life occurs; this may be due to a change in hepatic extraction ratio or to an increase in α_1-acid glycoprotein; the latter rises significantly after acute myocardial infarction (SAWYER et al. 1981; BARCHOWSKY et al. 1982).

Drugs which may influence lignocaine pharmacokinetics include isoprenaline and glucagon, which increase hepatic blood flow, and propranolol and noradrenaline, which may decrease hepatic blood flow. Cimetidine has been shown to reduce lignocaine clearance by 25 %, with plasma levels increasing 50 % and five out of six subjects experiencing lignocaine toxicity (FEELY et al. 1982). Racial differences do not affect lignocaine kinetics (GOLDBERG et al. 1982).

III. Mexiletine

The pharmacokinetics of mexiletine following intravenous bolus are best represented by a three-compartment model (PRESCOTT et al. 1977; HASELBARTH et al. 1981) with myocardial uptake occurring rapidly (HOROWITZ et al. 1981). There is extensive uptake of mexiletine in the tissues reflected by its large apparent volume of distribution of 5-6.6 litres/kg. Plasma elimination half-life varies from to 6 to 12 h in normals (PRESCOTT et al. 1977; CAMPBELL et al. 1978a; HASELBARTH et al. 1981) and from 11 to 17 h in patients with arrhythmias (CAMPBELL et al. 1978b). Mexiletine is well absorbed after oral administration with peak plasma concentrations occurring within 2-4 h after administraion of the conventional preparation and 8-11 h after a sustained release preparation. Bioavailability is 79 %-88 % (PRESCOTT et al. 1977; HASELBARTH et al. 1981). Increasing doses of mexiletine from 100 to 600 mg in healthy volunteers give a progressive and linear increase in maximum plasma concentrations and the area under the plasma concentration time curve (AUC), indicating dose-independent pharmacokinetics (PRINGLE et al. 1986). In patients, administration of 720 mg/day sustained release mexiletine achieved steady state after three to four doses; therapeutic plasma concentrations occurred on day 2 (BOYLE et al. 1982).

The major metabolites of mexiletine, which have no pharmacological action, are parahydroxymexiletine, hydroxymethylmexiletine and their corresponding alcohols; they account for 20 % of eliminated mexiletine (BECKETT and CHIDOMERE 1977). Approximately 15 % of a dose of mexiletine is excreted unchanged in urine; changes in urine pH may alter renal clearance with 30 %-55 % excreted unchanged in acidic urine (BECKETT and CHIDOMERE 1977). Renal clearance of mexiletine ranged from 150 ml/min (pH 5.0) to 20 ml/min (pH 7.0), indicating active tubular secretion (renal clearance exceeds glomerular filtration rate) (PRESCOTT et al. 1977).

Studies with intramuscular mexiletine indicate that therapeutic plasma concentrations (> 0.75 μg/ml) are achieved after 400 mg in 7 of 9 patients, 15 min-2 h after administration (NICHOLAS et al. 1986).

IV. Factors Affecting the Pharmacokinetics of Mexiletine

Mexiletine disposition is influenced by urinary pH with increased plasma levels in association with a rise in urinary pH. In renal failure the plasma elimination half-life increased significantly in patients with a creatinine clearance of less than 10 ml/min (EL ALLAF et al. 1982). Ischaemic heart disease has no significant effect on mexiletine plasma elimination half-life (CAMPBELL et al. 1978b). However, following myocardial infarction, mexiletine absorption is decreased, especially in patients receiving narcotic analgesics (PRESCOTT et al. 1977).

V. Tocainide

Following intravenous administration of tocainide the pharmacokinetics are best described by a two-component model with a rapid distribution phase (7-13 min). The apparent V_d in healthy subjects is 1.46-3.2 litres/kg (LALKA et al. 1976; GRAFFNER et al. 1980), depending on dose, method of administration and whether to healthy volunteers or patients (3.2 litres/kg refers to patients with acute myocardial infarction given infusions of 0.5-0.75 mg/kg or 750 mg tocainide). Tocainide is well absorbed after oral administration, with peak plasma levels occurring after 1 h in fasting subjects. Food decreases peak levels and delays the rate of absorption, but overall bioavailability of approximately 100 % is not affected (LALKA et al. 1976; GRAFFNER et al. 1980; GRAFFNER 1981). The peak plasma level in volunteers is 1.82 µg/ml after a 400-mg dose. In patients with myocardial infarction receiving 400 mg three times a day, a plasma concentration of 5-7 µg/ml was recorded (BEATTIE et al. 1978; GRAFFNER et al. 1980; BOYES et al. 1981). The therapeutic range for tocainide is 4-10 µg/ml.

Protein binding ranges from 10 % to 15 % and is independent of concentration within the range 4-12 µg/ml (SEDMAN et al. 1982). Tocainide is metabolized 30 % to the glucuronide of *N*-carboxytocainide; this together with 39 %-52 % of unchanged drug is excreted in the urine within 72 h (ELVIN et al. 1980 a, b). The plasma elimination half-life ranges from 13.5 to 19.1 h following oral administration to patients (MEFFIN et al. 1977; WOOSLEY et al. 1977; BEATTIE et al. 1978; KLEIN et al. 1980). Following intravenous administration in both volunteers and patients half-life ranged from 11.3 to 14.3 h (LALKA et al. 1976; GRAFFNER et al. 1980).

VI. Factors Affecting the Pharmacokinetics of Tocainide

Although experimental studies have indicated that compounds which induce or inhibit hepatic enzymes influence the pharmacokinetics of tocainide, studies in man indicate little effect within the therapeutic plasma concentration range (ELVIN et al. 1980 a). In patients with decompensated liver failure, the plasma elimination half-life was prolonged to 27 h, the apparent volume of distribution increased and total body clearance was reduced (OLTMANS 1982). Plasma elimination half-life was prolonged (range 16.6-42.7 h) (WEIGERS et al. 1983), in patients with renal dysfunction (creatinine clearance less than 5 ml/min).

E. Therapeutic Use

I. Lignocaine

Lignocaine constitutes first-line therapy for the acute treatment of ventricular arrhythmias associated with acute myocardial infarction (AMI), digitalis toxicity, ventricular tachycardia (VT) and cardiac surgery, and together with cardioversion for ventricular fibrillation (VF).

Following acute myocardial infarction, frequent premature ventricular complexes (PVCs) are suppressed in 90% of patients (Gianelly et al. 1967; Lown and Vassauz 1968); continuous infusions of lignocaine can be administered to prevent recurrence of these arrhythmias although care must be taken to avoid the concentration rising to a toxic level. Lignocaine is less effective for arrhythmias occurring within the 1st h of acute myocardial infarction, early (R-on-T), PVC and more chronic ventricular arrhythmias (Pantridge and Geddes 1974; Lie et al. 1975; Zbarbaro et al. 1979).

For haemodynamically stable ventricular tachycardia, lignocaine is successful as initial therapy. However, in hypotensive patients with sustained VT, electrocardioversion, together with concomitant lignocaine to prevent reversion, is the therapy of choice (Lie et al. 1974). A comparison with bretylium showed both drugs to be equally effective in treating outpatient VF although lignocaine was better tolerated.

Lignocaine is also effective against digitalis-induced tachyarrhythmias (Castellanos et al. 1982) and may be considered for quinidine-induced (delayed repolarization) ventricular arrhythmias, although pacing, phenytoin or a β-adrenoceptor-blocking drug may be more effective (Smith and Gallagher 1980).

Supraventricular arrhythmias are less responsive to lignocaine when compared with procainamide or quinidine, with only 34% of arrhythmias responding favourably (Josephson et al. 1972).

The prophylactic use of lignocaine in patients admitted to hospital within 6 h of symptoms for acute myocardial infarction was studied by Lie et al. (1974) in a prospective controlled study. They showed that lignocaine provided effective prophylaxis against VF after AMI. No treated patient experienced VF and only two experienced VT, whilst in the control group nine patients experienced VF and 6 VT. However, toxicity occurred in 23% of patients aged 60–69 years and in 7% of the under 60 s; hospital mortality was not different between the two groups.

A double-blind placebo-controlled study in 269 patients treated with lignocaine in the pre-hospital phase of acute myocardial infarction indicated that significantly fewer ($P < 0.03$) patients receiving lignocaine therapy died (3/156) than placebo-treated patients (8/113) (Valentine et al. 1974). Similar results occurred in a study on 6024 patients given prophylactic intramuscular lignocaine when AMI was suspected outside hospital (Koster and Dunning 1985). However, a recent study (Dunn et al. 1985) in 420 patients with suspected myocardial infarction seen within 6 h of the onset of symptoms did not show any benefits and the authors were unable to advocate prophylactic lignocaine.

II. Mexiletine—Ventricular Arrhythmias in Patients with Acute Mocardial Infarction

Early open clinical trials investigating the effects of mexiletine on ventricular arrhythmias after acute myocardial infaction or surgery indicated that, with i.v. and oral administration, mexiletine suppressed 70%–80% of premature

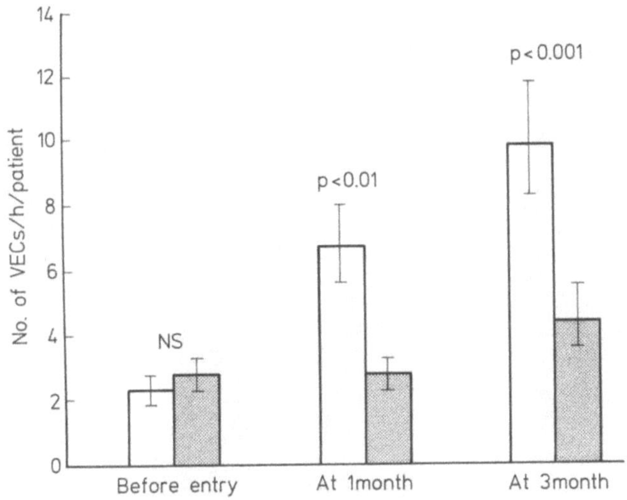

Fig. 2. Average number of ventricular ectopic complexes (VECs) per patient-hour (normalized by log transformation) with standard errors of mean and significance values. (CHAMBERLAIN et al. 1980)

ventricular complexes (PVCs) (TALBOT et al. 1973; CAMPBELL et al. 1973) (Tables 3-5). Chronic oral therapy produces similar reductions in PVCs (CAMPBELL et al. 1973; TALBOT et al. 1976). A placebo-controlled study in 344 patients administered 200-250 mg/8 h indicated a significant reduction in PVCs at 1 and 3 months (Fig. 2) (CHAMBERLAIN et al. 1980). METHA and CONTI (1982) also reported a significant effect, with mexiletine reducing PVCs by 66 % compared with 3 % for placebo. However, the IMPACT RESEARCH GROUP mexiletine trial (1984), which was double-blind placebo-controlled in 630 patients with recent AMI whilst showing a significant reduction in PVCs at 1 and 4 months, could not demonstrate it at 12 months; furthermore mortality was higher in the mexiletine group (7.6 %) compared with the placebo group (4.8 %). This result is in keeping with the mortality in other studies with antiarrhythmic drugs (MAY et al. 1983; FURBERG 1983), with β-adrenoceptor drugs being the only drugs at present to show decreased mortality after AMI.

III. Drug Refractory Ventricular Arrhythmias

Studies investigating the effects of mexiletine on drug refractory ventricular arrhythmias show variable response, with PODRID and LOWN (1981) and SAMI and LISBONA (1985) reporting a good response in 55 %-65 % of patients and HEGER et al. (1980) reporting 73 % of patients with less than 50 % reduction in PVCs. Long-term therapy was successful in 52 %-84 % of patients who were acutely controlled with mexiletine (PODRID and LOWN 1981; SAMI and LISBONA 1985; RUTLEDGE et al. 1985) (Tables 3-5). In patients with lignocaine-resistant ventricular arrhythmia, mexiletine suppressed the arrhythmias in 65 % of patients (CAMPBELL et al. 1973). Variable results have also been seen with mexiletine using serial electrophysiological testing, with mexiletine preventing induction of VT in 9 %-42 % (DIMARCO et al. 1981; PALILEO et al. 1982; WASPE et al. 1983). This variation is due not only to differences in patients but also to differences in pacing protocols.

Table 3. Effect of mexiletine on ventricular arrhythmias following acute myocardial infarction

Route/Dose	Study/patient details	No. of patients	Efficacy	Comments	References
Intravenous	AMI/surgery	32	22/32 ↓ 95 % PVC_s 7/32 ↓ 75 % PVC_s		Talbot et al. (1973)
Intravenous	Ventricular arrhythmias	27	20/27	Adverse effects—common	Campbell et al. (1973)
Intravenous	Lethal arrhythmias	10	8/10		Campbell et al. (1973)
Oral	Ventricular arrhythmias	17	14/17	Protected for 24 h	Campbell et al. (1973)
Chronic oral >1 200 mg/day <1 050 mg/day	Ventricular arrhythmias	18 51	15/18 32/51	Adverse Effects: 17 % 10 %	Campbell et al. (1973)
Chronic oral, loading 400–600 mg, maintenance 150–300 mg, 1–16 months	Ventricular arrhythmias	24	19/24	Three patients discontinued due to adverse effects	Talbot et al. (1976)
Intravenous followed by 200 mg/8 h	Ventricular arrhythmias	51	51/57	Ventricular tachycardia terminated in 9/11 3/57 reported gastric discomfort	Esser and Kikis (1978)
Mexiletine 100–400 mg × 8 h × 4 weeks	Placebo-controlled, mean 294 PVC/h	11	Placebo-reduced PVC_s, 3 % Mexiletine-reduced PVC_s, 66 %		Mehta and Conti (1982)

Table 3 (continued)

Route/Dose	Study/patient details	No. of patients	Efficacy	Comments	References
Oral, sustained release	Double-blind placebo-controlled patients with recent AMI	630	Significant reduction in PVC_s, at 1, 4 months,	Mortality—mexiletine, 7.6%; placebo, 4.8%	IMPACT RESEARCH GROUP (1984)
200–250 mg/8 h	Double-blind placebo-controlled patients with AMI	344	Significant reduction in PVC_s at 1, 3 month	Mexiletine: side effects 30/116, withdrawn; placebo: side effects, 6/125, withdrawn Mortality: 13% mexiletine 12% placebo	CHAMBERLAIN et al. (1980)

Table 4. Comparisons between mexiletine and other antiarrhythmic drugs on ventricular arrhythmias

Route/dose	Study/patient details	No. of patients	Efficacy	Comments	References
				Adverse effects	
Mexiletine 200 mg/8 h Procainamide 1.5 g/8 h Tolamolol 100 mg/8 h × 2 weeks	Ventricular arrhythmias associated with coronary artery disease	25	13/15 (52 %) 15/25 (60 %) 6/25 (24 %)	4/25 12/25 5/25	JEWITT et al. (1977)
Mexiletine 250 mg/8 h Procainamide 500 mg/4 h Placebo × 10 days	Ventricular arrhythmias after AMI	60 (20 each group)	68 % 65 % 23 %	2 withdrawals 1 withdrawals 3 withdrawals	CAMPBELL et al. (1975)
Mexiletine 3 × 250 mg/day Disopyramide 3 × 200 mg/day Atenolol 2 × 200 mg/day × 1 week	Chronic ventricular arrhythmias	20	Only mexiletine and disopyramide significantly reduced ectopic beats		VAN DURME et al. (1978)
Mexiletine i.v. 200 mg bolus inf 1 mg/min/5 mg/min Lignocaine i.v. 100 mg bolus inf 3 mg/min/2 mg/min	Randomized study period × 48 h Ventricular tachyarrhythmia 48 h after AMI	24	Mexiletine significantly better at reducing PVC_s + couplets than lignocaine		HOROWITZ et al. (1981)
Mexiletine 800 mg/daily/ 1 year Disopyramide 400 mg/day/ 1 year Procainamide 2 400 mg/day/ 1 year	Stable ventricular arrhythmia	45 (15 each group)	83.3 % 79.4 % 55.8 % } $P < 0.05$	Disopyramide and procainamide more marked—re: inotropic effect	TRIMARCO et al. (1983)

Table 4 (continued)

Route/dose	Study/patient details	No. of patients	Efficacy	Comments	References
Mexiletine 950 ± 202 mg/day (alone)	Ventricular rhythmias	21 17 17	3 controlled/ 62.5 ± 25% PVC_s suppression, 10 with VT	Limiting side effects in 12% patients	DUFF et al. (1983)
Quinidine 1042 ± 302 mg/day (alone)			59 ± 16% suppression (11 with VT)		
Mexiletine 800 ± 239 mg/day Quinidine 824 ± 298 mg/day	Combination		85.9 ± 26% PVC suppression (1 with VT)		
Mexiletine 200 mg/8 h Quinidine 200 mg/8 h	Chronic ventricular ectopy	13 13	5/13 effective 5/13 tolerated	Doses increased to >70% suppression of PVC_s or adverse effects	FENSTER and HANSON (1983)

Table 5. Effect of mexiletine on drug refractory ventricular arrhythmias

Route/dose	Study/patient details	No. of patients	Efficacy	Comments	References
Intravenous	Lignocaine-resistant ventricular arrhythmias	23	15/23		CAMPBELL et al. (1973)
300–400 mg × 8 h	Recurrent VT refractory to other drugs	15	2/15 > 90% ↓ PVC_s 11/15 < 50% ↓ PVC_s	Unsuccessful in abolishing or preventing VT in 11/25	HEGER et al. (1980)
Phase I 400 mg oral Phase II 2–400 mg/8 h × 2–3 days Phase III 8–32 months	Recurrent VT refractory to other drugs	88 78 44	65% 65% 84%		PODRID and LOWN. (1981)
250–1 500 mg/day 0.1–34.4 months	Ventricular arrhythmias resistant to other drugs	58	52% ↓ PVC_s 48% ↓ VT_s		RUTLEDGE et al. (1985)
100–300 mg 6–8 h	Ventricular arrhythmias resistant to other drugs	51	28/51 controlled	17/28 controlled > 1 year	SAMI and LISBONA (1985)
	Drug-resistant VT programmed electrical stimulation	41	17/41		DIMARCO et al. (1981)
1 073 ± 149 mg/day	Drug-resistant paroxysmal sustained VT Programmed electrical stimulation	17	1/11	5/17 possible proarrhythmic effects	PÁLILEO et al. (1982)

Table 5 (continued)

Route/dose	Study/patient details	No. of patients	Efficacy	Comments	References
250 ± 75 mg/8 h	Drug-resistant VT programmed electrical stimulation	33	3/33 mexiletine alone prevented VT 5/33 mexiletine and another drug prevented VT	Follow-up period 7.7 ± 4.1 months	WASPE et al. (1983)
800–1 000 mg/day	Drug-resistent ventricular ectopic activity	10	80 % good/excellent control 20 % partial control	6 month follow-up	ABINADER and COOPER (1979)
200–400 mg/8 h	Drug refractory ventricular arrhythmias	28	43 % effective 36 % ineffective 21 % drug intolerance		FENSTER and KERN (1983)
Arrhythmia following digitalis					
i.v.	arrhythmias following digitalis	7	5/7 ↓ 95 % PVC 2/7 ↓ 75 % PVC		TALBOT et al. (1973)

IV. Digitalis Arrhythmias

TALBOT et al. (1973) demonstrated that in five of seven patients with arrhythmias following digitalis administration, intravenous mexiletine suppressed PVCs by 95 % and in the other two patients PVCs were suppressed 75 %.

V. Comparisons with Other Antiarrhythmic Drugs

Comparisons with other antiarrhythmic drugs (Tables 3-5) indicate comparable efficacy between mexiletine and procainamide (JEWITT et al. 1977; CAMPBELL et al. 1977), although one group reported significantly more efficacy with mexiletine (83.3 %) than procainamide (55.8 %) (TRIMARCO et al. 1983). Studies comparing mexiletine with disopyramide indicate comparable efficacy (VAN DURME et al. 1978; TRIMARCO et al. 1983); similar results were seen with quinidine although quinidine and mexiletine in combination were more effective than either drug alone (DUFF et al. 1983). Improved efficacy (9 %-24 %) was also reported by WASPE et al. (1983) when mexiletine was combined with other unspecified antiarrhythmic drugs. Comparisons between mexiletine and the β-adrenoceptor-blocking agents tolamolol and atenolol indicate the latter's lack of acute effect in arrhythmia patients. Mexiletine has also been effective in patients refractory to oxprenolol (ABINADER and COOPER 1979).

VI. Tocainide—Ventricular Arrhythmias in Patients with Acute Myocardial Infarction

Tocainide is effective in the treatment of ventricular arrhythmias, reducing the frequency of PVCs by 53 %-100 % (Table 6). Doses ranged from 800 to 2 200 mg/day in divided doses, with treatment periods up to 27 months.

Placebo-controlled studies investigating the prophylaxis of tocainide on ventricular arrhythmias in patients with AMI (Table 6) indicated suppression of VT and frequent PVCs in all studies. Doses ranged from 400 to 800 mg 3 times a day for periods up to 6 months.

Combined data from the initial studies with tocainide (POTTAGE 1983) indicated that, out of 1 236 patients, 56 % had 90 % suppression of PVCs, 27 % had 75 % suppression and 17 % had no response or an increase in arrhythmia frequency. Out of 1 206 patients discharged from hospital taking tocainide, 87 were readmitted for VT/VF while receiving treatment with tocainide, 60 required cardioversion and 89 reported syncopal attacks. Whilst tocainide appears to decrease the frequency of PVC particularly after hospital discharge (CAMPBELL et al. 1979), it has not been demonstrated to lower the incidence of ventricular fibrillation during the course of a myocardial infarction (RYDEN et al. 1980).

Table 6. Effect of tocainide on ventricular arrhythmias following acute myocardial infarction

Route/dose	Study/patient details	No. of patients	Efficacy	References
Oral				
800/2 100 mg 5 days–27 months	Open study 5/17 PVC_s 17/17 VT_s	17	53 %	Winkle et al. (1978b)
Oral				
1 200/2 400 mg 1 months	Open study 7/17 PVC_s 10/17 VT_s	17	75 %	Ryan et al: (1979)
Oral				
1 200 mg 12 months	Open study 18/19 PVC_s 3/19 VT_s	19	58 %	Sonnhag (1980)
12 months	Open study 15/15 VT_s	15	85 %	Esterbrooks et al. (1983)
Oral				
1 200/1 800 mg 21 days	Double-blind study 12/12 PVC_s	12	12/12	Kuck et al. (1979)
Oral				
1 200/1 800 mg 8 days	Single-blind study 15/15 PVC_s	15	11/15	Winkle et al. (1976)
Oral				
800/2 200 mg 10 days	Single-blind study 20/20 PVC_s	20	8/10	Woosley et al. (1977)

Table 6 (continued)

Route/dose	Study/patient details	No. of patients	Efficacy	References
Oral 17.7 ± 4.9 mg/kg/day	Ventricular arrhythmias following AMI	18	7/18 Did not worsen heart failure in patients with ejection fraction <30%	Klein et al. (1980)
600 mg stat at 2 h 400 mg t.i.d. (24 h)	Double-blind study AMI	30 tocainide 38 placebo	Tocainide > placebo Two deaths on placebo	Campbell et al. (1979)
Oral 400 mg at 0, 4, 8 h (48 h)	Double-blind study AMI	18 tocainide 16 placebo	Tocainide > placebo	Swedberg and Holmberg (1981)
Oral 400 mg t.i.d. (6 months)	Double-blind study AMI	72 tocainide 74 placebo	Tocainide > placebo	Bastion et al. (1980)
Oral (2 weeks) 400 mg t.i.d. (2 weeks) 600 mg t.i.d. (2 weeks) 800 mg t.i.d.	Double-blind study AMI	10 5 4	80% reduction in PVC_s with tocainide	Le Winter et al. (1980)
750 mg i.v. 800 mg stat oral 400 mg t.i.d. (6 months)	Double-blind study AMI	31 tocainide 34 placebo	Tocainide > placebo	Ryden et al. (1980)

Table 6 (continued)

Route/dose	Study/patient details	No. of patients	Efficacy	Comments	References
300 mg × 8 h 800 mg × 6 h	Refractory ventricular arrhythmias	21	14/21	Two withdrawn due to CNS and GI adverse effects	HAFFAJEE et al. (1983)
Comparison between tocainide and other antiarrhythmic drugs on ventricular arrhythmias					
Tocainide 600 mg × 8 h	Placebo-controlled × 8 weeks	41	1/9 > 79 % PVC suppression (tocainide)	14/22 tocainide (13 discontinued)	WASENMILLER and ARONOW (1980)
Quinidine 300 mg × 6 h			6/13 > 75 % PVC suppression (quinidine)	9/19 quinidine (6 discontinued)	
Tocainide i.v. bolus/inf	Acute ventricular arrhythmias after open heart surgery	99	55 % > 80 % PVC suppression 74 % couplet suppression 87 % VT suppression	Tocainide 10 %—adverse effects Bolus 250, 250, 125 mg inf 104 mg/min, bolus 100, 50, 50 mg inf 2.08 mg/min	
Lignocaine i.v. bolus/inf			48 % > 80 % PVC suppression 68 % Couplet suppression 73 % VT suppression Tocainide vs lignocaine ($P < 0.004$)	Lignocaine 18 %—adverse effects	

Table 6 (continued)

Route/dose	Study/patient details	No. of patients	Efficacy	Comments	References
Tocainide i.v. inf	Prophylactic administration within 12 h of AMI symptoms	16	88 %	Adverse effects: Tocainide 6/16	KEEFE et al. (1986)
Lignocaine i.v. inf		13	46 %	Lignocaine 11/13	
Oral			*Holter*		MCLARAN et al. (1984)
Tocainide	Double-blind	10	Both reduced ($P < 0.05$) PVC frequency and grade		
			Exercise		
Dysopyramide	Crossover stable PVC_s		PVC grade reduced ($P < 0.01$) with disopyramide		

VII. Drug Refractory Ventricular Arrhythmias and Programmed Electrical Stimulation

Prevention of ventricular tachycardia produced by programmed electrical stimulation indicates a 10%-20% response rate with tocainide (EASLEY et al. 1985), comparable to that seen with mexiletine.

VIII. Comparisons with Other Antiarrhythmic Drugs

A comparison in 41 patients with PVCs between tocainide and quinidine indicated that in the doses given quinidine was more effective, benefited a larger number of patients treated and produced fewer adverse effects than tocainide (WASENMILLER and ARONOW 1980). Tocainide was significantly better ($P < 0.004$) than lignocaine in 99 patients with regard to suppression of PVC, couplets and VT (MORGANROTH et al. 1984). A comparison with disopyramide (MCLARAN et al. 1984) indicated similar efficacy at suppressing PVC frequency and grade. In patients with chronic stable PVCs and symptomatic ventricular arrhythmias an increased rate of response to tocainide was seen among those whose arrhythmia was responsive to lignocaine (MCDEVITT et al. 1976; WOOSLEY et al. 1977; RODEN et al. 1980; WINKLE et al. 1980; MALONEY et al. 1980; HAFFAJEE et al. 1983). A comparison of tocainide (16 patients) and lignocaine (13 patients) indicated comparative efficacy with less side effects associated with tocainide for the prophylaxis of ventricular arrhythmias associated with acute myocardial infarction (KEEFE et al. 1986).

F. Side Effects and Interactions

I. Lignocaine

Neurological side effects reported with lignocaine include: paraesthesias, drowsiness and agitation at plasma levels of 3-5 µg/ml; with higher plasma levels, nausea, vomiting, confusion, hearing impairment, dissociation, muscle twitching, convulsions and respiratory arrest occur (plasma level > 9 µg/ml). Psychotic reactions have also been reported (TURNER 1982). Whilst cardiovascular toxicity is not common, lignocaine can depress left ventricular function in patients with severe left ventricular dysfunction and induce heart block (GUPTA et al. 1974; CHENG and WADHWA 1973). However, in the majority of patients with left ventricular dysfunction, bundle branch block and sick sinus syndrome, the judicious use of lignocaine is usually well tolerated. The incidence of side effects is age related, occurring in 7% of patients < 60 years and 23% > 60 years (LIE et al. 1974). Lignocaine toxicity was assessed in 750 hospitalized patients (PFEIFER et al. 1976). Adverse effects associated with lignocaine therapy occurred in 6.3% of the group; 25% (or 1.6% of the total group) had a life-threatening event. In this study CNS disturbances accounted for 68% of all reactions. The authors indicate that CNS toxicity precedes significant cardiovascular toxicity and suggest careful monitoring of CNS function in patients receiving long-term (12–72 h) lignocaine infusions. Reports of proarrhythmic effects appear to be rare with lignocaine (ANDERSON 1984) although it has been implicated in patients with a long QT (BURKET et al. 1985).

Lignocaine has the potential for interacting with any drug that may affect hepatic blood flow or metabolism. Glucagon and isoprenaline may increase hepatic blood flow, conversely propranolol and noradrenaline decrease hepatic blood flow and lignocaine clearance (BRANCH et al. 1973). Lignocaine concentrations following loading infusions were shown to be 50% higher after cimetidine (enzyme inhibition), with the majority of subjects experiencing toxicity (FEELY et al. 1982).

II. Mexiletine

Gastrointestinal, neurological and cardiovascular side effects are associated with mexiletine therapy. These include nausea and vomiting, tremor [with long-term oral therapy (TALBOT et al. 1976)], dizziness, blurred vision, dysarthria, diplopia, ataxia, nystagmus, drowsiness and toxic confusional states, hypotension, sinus bradycardia and atrial fibrillation; widening of QRS complexes and atrioventricular dissociation may also occur.

Side effects tend to be less common with long-term oral therapy, but CAMPBELL et al. (1977) reported that 54% of 89 patients who received mexiletine as a bolus followed by infusion had side effects (44% severe). Side effects tend to occur at plasma concentrations >2.0 μg/ml. A recent study (RUTLEDGE et al. 1985) reported that in 58 drug refractory patients administered oral mexiletine, gastrointestinal side effects occurred in 12% of patients and neurological side effects in 5%; other adverse reactions included urinary retention and hair loss. No dose-related gastric intolerance was reported in 32% of 51 patients although gastric pain or burning occurred frequently after taking mexiletine (SAMI and LISBONA 1985). The same authors also reported a proarrhythmia effect in 10% of patients. DUFF et al. (1983) reported reduced side effects when mexiletine was given in combination with quinidine; the dose of mexiletine was reduced from 950 ± 202 to 800 ± 239 mg/day and quinidine from 1042 ± 362 to 824 ± 298 mg/day. A significantly greater antiarrhythmic response was also produced.

The interactions of mexiletine with other drugs have been reported for antacids, which delay absorption but do not affect the maximum plasma concentration or AUC (HERZOG et al. 1982). Atropine produces a similar effect, with metoclopramide enhancing the rate of mexiletine absorption but not affecting bioavailability (WING et al. 1980). Phenytoin and rifampicin enhance the metabolism of mexiletine and decrease plasma elimination half-life as a result of enzyme induction (BEGG et al. 1982; PENTIKAINEN et al. 1982). Digoxin concentrations are not altered by concurrent mexiletine therapy (SARIS et al. 1983). Mexiletine absorption is decreased in patients with myocardial infarction especially in those also receiving narcotics (PRESCOTT et al. 1977).

Table 7. Incidence of adverse effects in 369 patients treated with tocainide for periods up to 41.5 months (HORN et al. 1980)

Adverse effect	No. of patients (%)
Gastrointestinal	
Nausea	34
Vomiting	16
Anorexia	15
Central nervous system	
Dizziness	31
Lightheadedness	24
Tremors	22
Paraesthesias	16
Confusion	15
Nervousness	13
Other	
Palpitations	17
Shortness of breath	13
Rash	12

III. Tocainide

The side effects of tocainide were evaluated in 369 seriously ill patients up to a period of 41.5 months (HORN et al. 1980). Commonly occurring side effects are listed (Table 7). Therapy was discontinued in 16% of patients, 80% of whom had neurological and gastrointestinal effects. Other reasons for discontinuing therapy included: rash, hepatitis (LEVIN and FOX 1982), elevated liver enzymes, acute pulmonary oedema, interstitial pulmonary oedema, polyarthritis and pericarditis (the latter two were associated with elevated antinuclear antibody titre). HORN et al. (1980) also reported that tocainide aggravated preexisting heart failure in five patients, had a possible proarrhythmic effect in six patients, two reported convulsions and three developed a lupus erythematous-like illness, although a definite causal relationship with tocainide could not be established. Proarrhythmic effects of tocainide have been reported in other groups of patients (ENGLER and LEWINTER 1981; VELEBIT et al. 1982). Interstitial pneumonitis also occurred in other studies (PERLOW et al. 1981; BRAUDE et al. 1982).

Bradycardia was reported with both intravenous (SUTTON 1980; NYQUIST et al. 1980) and oral tocainide (MANDAL and DATTA 1983). The neurological effects include dizziness, light-headedness, tremor, paraesthesia, confusion, nervousness and frank psychosis (CURRIE and RAMSDALE 1984). The frequency of adverse effects increased at plasma concentrations >10 μg/ml or doses

>1 800 mg/kg. An important non-cardiovascular side effect not related to dose or concentration is agranulocytosis, which is estimated to occur in 0.18 % of patients (Volosin et al. 1985); this has restricted the use of the drug in the United Kingdom except for the treatment of life-threatening symptomatic ventricular arrhythmias associated with severely compromised left ventricular function in patients who did not respond to other therapy or in whom other therapies are contraindicated. In the United Kingdom, physicians have been advised to do frequent white cell counts especially during the first week of treatment. A cross-reactivity has been reported for lignocaine and tocainide with regard to rash and fever, which occur in 0 %-25 % of patients depending on the study; no cross-reactivity has been reported for tocainide and mexiletine (Duff et al. 1984). Tocainide does not prolong repolarization time (QT interval) and in fact may shorten the QT interval.

A lack of interaction has been reported between metoprolol and tocainide (Ikram 1980). Similarly no interaction has been reported between tocainide and digoxin (Ryan et al. 1979) or warfarin (Morganroth et al. 1985).

Concomitant administration of tocainide and propranolol to a patient resulted in severe paranoia which subsided on withdrawal of propranolol (Rubino and Jackson 1982). No significant interactions have been reported between tocainide, which undergoes hepatic metabolism, and enzyme inducers or inhibitors (Elvin et al. 1980a).

G. Dosage and Administration

I. Lignocaine

The loading dose of lignocaine is usually 1-2 mg/kg, given as an i.v. bolus over 1-2 min. Therapeutic plasma concentrations can be maintained for 90 min if a second i.v. bolus, half the concentration of the first, is given after 10-20 min. Other dosage regimens include: 100-mg i.v. bolus (over 1-2 min) followed 10 min later by another 100-mg i.v. bolus or 75-mg i.v. bolus followed by 50 mg at 5-min intervals up to 375 mg (Wyman et al. 1978). These regimens are then followed by a constant rate infusion of 2-5 mg/min to maintain plasma concentration within the therapeutic range. Dosage must be modified in the elderly and those with heart failure.

Modifications, including additional boluses or rapid infusions, and specific methods for achieving constant plasma concentrations have beeen developed (Wyman et al. 1978; Harrison 1978; Stargel et al. 1981; Pieper et al. 1982; Riddell et al. 1984). Guidelines as an approach to prevention of lignocaine toxicity include:

1. Lignocaine dose to be routinely calculated on the basis of body weight.
2. Patients with circulatory insufficiency should receive initially the lowest recommended infusion rate (10 μg/kg per minute).
3. In patients with pre-existing neurological dysfunction, where serum levels are not readily available, consider an alternative drug.
4. In prophylactic use monitor serum levels (impractical ?).

5. During prolonged infusions if serum levels rise, then reduce the infusion rate.
6. Any change in the haemodynamic or mental state of a patient is most likely to be due to early lignocaine toxicity (DAVIDSON et al. 1982).

II. Mexiletine

PRESCOTT et al. (1977) suggested that for intravenous therapy mexiletine should be administered in an initial bolus of 150-250 mg over 2-5 min followed by a loading infusion of 250 mg in 30 min, then 250 mg in 2.5 h and 500 mg in 8 h. This is followed by a maintenance infusion of 500-1 000 mg over 24 h. Oral mexiletine 200-300 mg given hourly produces a therapeutic steady-state plasma concentration of 1-2 µg/ml in 67 % of patients (POTTAGE 1977). Studies with slow-release mexiletine showed that 12-hourly administration of 360 mg SR mexiletine maintained therapeutic plasma concentrations (BOYLE et al. 1982). The same group investigated the effect of intramuscular mexiletine and showed 400 mg i.m. gave plasma concentration >0.75 µg/ml in seven of nine patients, with activity occurring at 15 min and lasting for at least 2 h (NICHOLAS et al. 1986). Mexiletine should be administered with caution in patients with sinus node dysfunction, AV conduction disease and advanced renal failure. Care must also be taken with patients with left ventricular dysfunction, hypotension and serious bradyarrhythmias.

III. Tocainide

Doses of tocainide required to achieve therapeutic plasma levels of 4-10 g/ml include the following regimens: 0.5-0.74 mg/kg per minute i.v. infusion for 15 min followed by an oral dose of 800 mg; and 500-750 mg infused i.v. over 15-30 min combined with an oral dose of 400-800 mg will achieve therapeutic plasma concentrations which are maintained for 8-12 h (NYQUIST et al. 1980; GRAFFNER et al. 1980; SUTTON 1980; BOYES et al. 1981).

H. Place in Therapy

Class 1b drugs are widely used in the therapy of cardiac arrhythmias. In the management of potentially lethal arrhythmias they are effective in 40 %-50 % of patients, disopyramide and quinidine are effective in 60 % and flecainide and encainide are effective in 80 %-90 % of patients. In electrophysiological testing class 1a, 1b and 1c drugs have a response rate of 15 %-40 %. However, marked differences exist with regard to side effects and to contraindications. The class 1a antiarrhythmic drug dysopyramide is markedly negatively inotropic, quinidine causes thrombocytopenia and procainamide causes systemic lupus. Moreover, the minor side effects such as anticholinergic activity with disopyramide and the gastrointestinal side effects of quinidine may play a more important part in limiting their use. They also prolong the QT interval and may be proarrhythmic. Whilst the class 1b agents commonly produce gastrointestinal and neurological side effects they are usually mild and well tolerated. Class 1b drugs have the added advantage of little or no negative inotropic effect and no effect on repolarization. The emergence of agranulocytosis with tocainide, however, has limited its use. Side effects with class 1c drugs tend to be neurological; arrhythmogenesis has also been reported, especially with higher doses.

In conclusion, the ideal antiarrhythmic drug should be effective against lethal arrhythmias; have dose-independent pharmacokinetics unaltered by disease state, renal function, etc.; and have an acceptable dosage regimen, no negative intropic effect, a wide therapeutic range, no active metabolites and limited side effects. In comparison with other class 1 drugs, judicial use of the class 1b group of drugs makes them acceptable antiarrhythmic agents using currently accepted criteria.

References

Abinader EG, Cooper M (1979) Mexiletine in control of chronic drug-resistant ventricular arrhythmia. J A M A 242:337-339

Allen JD, Arnold JMO (1974) Comparison of effects of mexiletine, lignocaine and procainamide on canine Purkinje fibres. J Physiol (Lond) 236:26P

Almotrefi AA, Baker JBE (1980) The antifibrillatory potency of aprindine, mexiletine, tocainide and lignocaine compared on Langendorff perfused hearts of rabbits and guinea pigs. J Pharm Pharmacol 32:746-751

Anderson JL (1984) Current understanding of lidocaine as an antiarrhythmic agent: a review. Clin Ther 6 (2):125-141

Anderson JL, Mason JW, Winkle RA et al. (1978) Clinical electrophysiologic effects of tocainide. Circulation 57:685-691

Banim SO, Stone D, Aniceto Da Silva, Balcon R (1977) Observations of the haemodynamics of mexiletine. Post Med J 53 (Suppl 1):74-76

Barchowsky A, Shand DG, Stargel WW, Wagner GS, Routledge FA (1982) On the role of α_1-acid glycoprotein in lignocaine accumulation following myocardial infarction. Br J Clin Pharmacol 13:411-415

Bastion BC, Macfarlane PW, McLaughlan JH, Ballantyne D, Clark R, Hillis WS, Rae AP, Hutton I (1980) A prospective randomized trial of tocainide in patients following myocardial infarction. Am Heart J 100:1017-1022

Beattie JM, Hutton I, McLauchlan J, Meyer MB (1978) Tocainide pharmacokinetics in patients with acute myocardial infarction. 51st scientific sessions. American Heart Association, Dallas, Texas

Beckett AH, Chidomere EC (1977) The distribution, metabolism and excretion of mexiletine in man. Postgrad Med J (Suppl 1) 53:60-66

Begg EJ, Chiwah PM, Webb C, Rodam DNW (1982) Enhanced metabolism of mexiletine after phenytoin administration. Br J Clin Pharmacol 14:219-223

Binnion PF, Murtagh G, Pollock AM, Fletcher E (1969) Relation between plasma lignocaine levels and induced haemodynamic changes. Br Med J 3:390

Blumer J, Strong JM, Atkinson AJ (1978) The convulsant potency of lidocaine and its *N*-dealkylated metabolites. J Pharmacol Exp Ther 180:31-36

Boyes RN, Meyer MB, Hutton I, Tweddel A (1981) Tocainide blood levels following intravenous and oral administration. In: Pottage A, Ryden L (eds) Workshop on

tocainide. Proceedings from a conference held in Copenhagen, 1979. Hassle, Mölndal, Sweden, pp 105-108
Boyle DM, Barber JM, Chapman C, Khalid I, Kinney C, McIlmoyle EL, Salathia K, Shanks RG (1982) Comparison of plasma concentrations and efficacy of mexiletine and a slow release preparation of mexiletine in patients admitted to a coronary care unit. J Cardiovasc Pharmacol 4:174-179
Branch RA, Shand DG, Wilkinson GR, Nies AS (1973) The reduction of lidocaine clearance by *dl*-propranolol: an example of hemodynamic drug interaction. J Pharmacol Exp Ther 184:515-519
Braude AC, Downar E, Chamberlain DW, Rebuck AS (1982) Tocainide associated interstitial pneumonitis. Thorax 37:309-310
Burket MW, Frrker TD, Temesy-Armos PN (1985) Polymorphous ventricular tachycardia provoked by lidocaine. Am J Cardiol 55:592-593
Burton JR, Mathew MT, Armstrong PW (1976) Comparative effects of lignocaine and procainamide on acutely impaired haemodynamics. Am J Med 61:215-220
Campbell NPS, Chaturvedi NC, Kelly JG, Strong JE, Shanks RG (1973) Mexiletine (Ko 1173) in the management of ventricular dysrhythmias. Lancet 2:404-407
Campbell RWF, Talbot RG, Dolder MA, Murray A, Prescott LF, Julian DG (1975) Comparison of procainamide and mexiletine in prevention of ventricular arrhythmias after acute myocardial infarction. Lancet 1:1257-1260
Campbell RWF, Dolder MA, Prescott LF, Talbot RG, Murray A, Julian DG (1977) Ventricular arrhythmias after acute myocardial infarction treated with procainamide or mexiletine. Postgrad Med J 53 (Suppl 1): 150-153
Campbell NPS, Kelly JG, Adgey AAJ, Shanks RG (1978a) Mexiletine in normal volunteers. Br J Clin Pharmacol 6:372-373
Campbell NPS, Kelly JG, Adgey AAJ, Shanks RG (1978b) The clinical pharmacology of mexiletine. Br J Clin Pharmacol 6:103-108
Campbell RWF, Bryson LG, Bailey BJ, Murray A, Julian DG (1979) Oral tocainide in suspected acute myocardial infarction. Circulation 60 (Suppl II):70
Camm AJ (1984) Electrophysiology of new anti-arrhythmic drugs. Eur Heart J 5 (Suppl B):75-79
Castellanos A, Ferreiro J, Pefkaros K et al. (1982) Effects of lidocaine on bidirectional tachycardia and on digitalis induced atrial tachycardia with block. Br Heart J 48:27-32
Chamberlain DA, Julian DG, Boyle D McC, Jewitt DE, Campbell RWF, Shanks RG (1980) Oral mexiletine in high risk patients after myocardial infarction. Lancet II:1324-1327
Cheng TO, Wadhwa K (1973) Sinus standstill following intravenous lidocaine administration. J A M A 223:790-792
Currie P, Ramsdale DR (1984) Paranoid psychosis induced by tocainide. Br Med J 288:606-607
Davidson R, Parker M, Atkinson AJ (1982) Excessive serum lidocaine levels during maintenance infusions: mechanism and prevention. Am Heart J 104:203-208
Dimarco JP, Garan H, Ruskin JN (1981) Mexiletine for refractory ventricular arrhythmias: results using serial electrophysiologic testing. Am J Cardiol 47:131-138
Duff HJ, Roden D, Kirby-Primm R, Oates JA, Woosley RL (1983) Mexiletine in the tratment of resistant ventricular arrhythmias: enhancement of efficacy and reduction in dose related side effects by combination with quinidine. Circulation 67 (5): 1124-1128
Duff HJ, Roden DM, Marney S et al. (1984) Molecular basis for the antigenicity of lidocaine analogs: tocainide and mexiletine. Am Heart J 107:585-589

Dunn HM, McComb JM, Kinney CD, Campbell NPS, Shanks RG, Mackenzie G, Adgey AAJ (1985) Prophylactic lidocaine in the early phase of suspected myocardial infarction. Am Heart J 110:353-362
Easley A, Higgins S, Thieme L, Lindenfeld J, Reiter MJ (1985) Tocainide in preventing inducible ventricular tachycardia in unselected patients (abstr). Clin Res 33 (1):6A
El Allaf D, Henrard L, Crochelet L, Delapierre D, Carlier J, Dresse A (1982) Pharmacokinetics of mexiletine in renal insufficiency. Br J Clin Pharmacol 14:431-435
Elvin AT, Lalka D, Stoeckel K, du Souich P, Axelson JC, Golden LH, McLean AJ (1980a) Tocainide kinetics and metabolism. Effects of phenobarbitone and substances for glucuranyl transferase. Clin Pharmacol Ther 28:652-658
Elvin AT, Keenaghan JB, Byrnes EW, Tenthorney Pa, McMaster PD, Takman BH, Lalka D, Meyer MB, Lenfeld RA (1980b) Tocainide conjugation in humans. Novel biotransformation pathway for a primary amine. J Pharm Sci 69:47-49
Engler RL, LeWinter M (1981) Tocainide-induced ventricular fibrillation. Am Heart J 101:494-496
Esser H, Kikis D (1978) Mexiletine in the suppression of ventricular ectopic activity: short and long term treatment. Proceedings of the symposium on management of ventricular tachycardia, role of mexiletine. Excerpta Medica, Amsterdam
Esterbrooks D, Aronow WS, Mohiuddin SM, Sketch MH, Mooss AN, Hee TT, Booth RW, Butler ML (1983) Effects of tocainide on ventricular arrhythmias refractory to standard oral anti-arrhythmic drugs but responsive to intravenous lignocaine. Curr Ther Res 33:272-278
Feely J, Wilkinson GR, McAllister CB, Wood AJ (1982) Increased toxicity and reduced clearance of lignocaine by cimetidine. Ann Intern Med 96:592-594
Fenster PE, Hanson CD (1983) Mexiletine and quinidine in ventricular ectopy. Clin Pharmacol Ther 34:136-142
Fenster PE, Kern KB (1983) Mexiletine in refractory ventricular arrhythmias. Clin Pharmacol Ther 34:777-784
Furberg CD (1983) Effect of anti-arrhythmic drugs on mortality after myocardial infarction. Am J Cardiol 52:320-326
Gianelly R, Vondergroeben JO, Spivack AP, Harrison DC (1967) Effect of lidocaine on ventricular arrhythmias in patients with coronary heart disease. N Engl J Med 277:1215-1219
Goldberg MJ, Spector R, Johnston GF (1982) Racial background and lidocaine pharmacokinetics. J Clin Pharmacol 22:391-394
Graffner C (1981) Tocainide pharmacokinetics: absorption metabolism, excretion; In: Pottage A, Ryden L, (eds) Workshop on tocainide. Proceedings from a conference held in Copenhagen, 1979. Hassle, Molndal, Sweden, pp 90-99
Graffner C, Conradson T, Hofoendahl S, Ryden L (1980) Tocainide kinetics after intravenous and oral administration in healthy subjects and in patients with acute myocardial infarction. Clin Pharmacol Ther 27:64-71
Griffin J, Schnittger F, Peters PJ, Meffin R, Hav R, Kernoff R, Winkle RA (1977) Effects of tocainide on ventricular fibrillation threshold. Circulation 56:158-162
Gupta, PK, Lichstein E, Chadda KD (1974) Lignocaine induced heart block in patients with bundle branch block. Am J Cardiol 33:487-492
Haffajee CI, Sacks CM, Alpert JS, Howe JP, Ockene IS, Paraskos JA, Dalen JE (1983) Chronic tocainide therapy for refractory high-grade ventricular arrhythmias. Clin Cardiol 6:72-78
Halkin H, Meffin P, Melman KL, Rowland M (1975) Influence of congestive heart failure on blood levels of lidocaine and its active monodeethylated metabolite. Clin Pharmacol Ther 17:669-675

Harrison DC (1978) Should lidocaine be administered routinely to all patients after acute myocardial infarction? Circulation 58:581-584

Harrison DC, Alderman EL (1972) The pharmacology and clinical use of lidocaine as an anti-arrhythmic drug. Mod Treat 9:139-175

Harrison DC, Meffin PJ, Winkle RA (1977) Clinical pharmacokinetics of anti-arrhythmic drugs. Prog Cardiocasc Dis 20:217-240

Haselbarth V, Doevendans Je, Wolf M (1981) Kinetics and bioavailability of mexiletine in healthy subjects. Clin Pharmacol Ther 29:729-736

Heger JJ, Nattel S, Rinkenberger RL, Zipes DP (1980) Mexiletine therapy in 15 patients with drug resistant ventricular tachycardia. Am J Cardiol 45:627-632

Herzog P, Holtermuller KH, Kasper W, Meinertz T, Trenk D, Jahnchen E (1982) Absorption of mexiletine after treatment with gastric antacids. Br J Clin Pharmacol 14:746-747

Horn HR, Hadidan Z, Johnson JL, Vassallo HG, Williams JH, Young MD (1980) Safety evaluation of tocainide in the American emergency use program. Am Heart J 100:1037-1040

Horowitz JD, Anavekar SN, Morris PM, Goble AJ, Doyle AE, Louis WJ (1981) Comparative trial of mexiletine and lignocaine in the treatment of early ventricular tachyarrhythmias after acute myocardial infarction. J Cardiovasc Pharmacol 3:409-419

Horowitz LN, Josephson ME, Farshidi A (1978) Human electropharmacology of tocainide, a lignocaine congener. Am J Cardiol 42:276-280

Ikram H (1980) Haemodynamic and electrophysiologic interactions between antiarrhythmic grugs and beta-blockers with special reference to tocainide. Am Heart J 100:1976

Impact Research Group (1984) International mexiletine and placebo anti-arrhythmic coronary trial: 1. Report on arrhythmia and other findings. J Am Coll Cardiol 4 (6): 1184-1163

Jewitt DE, Jackson G, McComish M (1977) Comparative anti-arrhythmic efficacy of mexiletine procainamide and tolamolol in patients with symptomatic ventricular arrhythmias. Postgrad Med J 53 (Suppl 1):158-161

Josephson ME, Caracta AR, Lau SH et al. (1972) Effects of lidocaine on refractory periods in man. Am Heart J 84:778-786

Keefe DL, Williams S, Torres V, Flowers D, Somberg JC (1986) Prophylactic tocainide or lidocaine in acute myocardial infarction. Am J Cardiol 57:527-531

Klein MD, Lavine PA, Ryan JJ (1980) Antiarrhythmic efficacy, pharmacokinetics and clinical safety of tocainide in convalescent myocardial infarction patients. Chest 7:726-730

Koster RW, Dunning AJ (1985) Intramuscular lidocaine for prevention of lethal arrhythmias in the pre-hospitalization phase of acute myocardial infarction. N Engl J Med 313:1105-10

Kuck KH, Hanrath P, Lubda J, Mathey D, Bleifeld W (1979) Antiarrhythmic effect of tocainide (lidocaine congener) on ventricular arrhythmias. Dtsch Med Wochenschr 104:1701-1705

Lalka D, Meyer MB, Duce BR, Elvin AT (1976) Kinetics of the oral anti-arrhythmic lignocaine congener tocainide. Clin Pharmacol Ther 19:757-777

Levin RI, Fox A (1982) Hepatitis in human subjects associated with tocainide administration. Clin Res 30:200a

LeWinter MM, Engler RL, Karliner JS (1980) Tocainide therapy for treatment of ventricular arrhythmias: assessment with ambulatory electrographic monitoring and treadmill exercise. Am J Cardiol 45:1045-1052

Lie KI, Wellens HJ, Van Capelle FJ, Durrer D (1974) Lidocaine in the prevention of primary ventricular fibrillation. N Engl J Med 291:1324-1326
Lie KI, Wellens HJJ, Downar E, Durrer D (1975) Observations on patients with primary ventricular fibrillation complicating acute myocardial infarction. Circulation 52:755-759
Lown B, Vassauz CC (1968) Lidocaine in acute myocardial infarction. Am Heart J 76:586-587
Maloney JD, Nissen RG, McColgan JM (1980) Open clinical studies at a referral centre: chronic maintenance tocainide therapy in patients with recurrent sustained ventricular tachycardia refractory to conventional anti-arrhythmic agent. Am Heart J 100:1023-1030
Mandal SK, Datta SK (1983) Nodal bradycardia induced by tocainide. Postgrad Med J 59:262-263
May GS, Furberg CD, Eberlein KA, Geraci BJ (1983) Secondary prevention after myocardial infarction: a review of short-term acute phase trials. Prog Cardiovasc Dis 25:335-359
McComish M, Kitson D, Robinson C, Jewitt DE (1977) Clinical electrophysiological effects of mexiletine, Postgrad Med J 53 (Suppl 1):85-91
McDevitt DG, Nies AS, Wilkinson GR, Smith RF, Woosley RL, Oates JA (1976) Antiarrhythmic effects of a lidocaine congener, tocainide, 2-amino-2′,6′-proprionoxylidide, in man. Clin Pharmacol Ther 19:396-402
McLaran CJ, Hossack KF, Nielson GH, Siskind V (1984) Oral tocainide versus disopyramide: a double-blind, randomized, crossover study of outpatients with stable ventricular premature beats. J Cardiovasc Pharmacol 6:657-662
Meffin PJ, Winkle RA, Blaschke TF, Fitzgerald J, Harrison DC, Hasapat SR, Bell PA (1977) Response optimization of drug dosage: antiarrhythmic studies with tocainide. Clin Pharmacol Ther 22:42-57
Mehta J, Conti CR (1982) Mexiletine, a new anti-arrhythmic agent for treatment of premature ventricular complexes. Am J Cardiol 49:455-460
Moore EN, Spear JF, Horowitz LN (1978) Electrophysiologic properties of a new antiarrhythmic drug—tocainide. Am J Cardiol 41:703-709
Morganroth J, Panidis IP, Harley S, Johnson J, Smith E, MacVaugh H (1984) Efficacy and safety of intravenous tocainide compared with intravenous lidocaine for acute ventricular arrhythmias immediately after cardiac surgery. Am J Cardiol 54:1253-1258
Morganroth J, Nestico PF, Horowitz LN (1985) A review of the uses and limitations of tocainide—a class 1B antiarrhythmic agent. Am Heart J 110:856-863
Nicholas J, Boyle DMcC, Kinney CD, Salathia K, Shanks RG (1986) Plasma concentrations and acceptability of mexiletine given by intramuscular injection in patients admitted to a coronary care unit. J Cardiovasc Pharmacol 8:21-28
Nyquist O, Forssell G, Norland RR, Schenck-Gustafsson K (1980) Haemodynamic and anti-arrhythmic effects of tocainide in patients with acute myocardial infarction. Am Heart J 100:1000-1005
Ochs HR, Knuchel M, Abernethy DR, Greenblatt DJ (1983) Dose-independent pharmacokinetics of intravenous lidocaine in humans. J Clin Pharmacol 23:286-188
Oltmanns D (1982) Pharmacokinetics of tocainide in patients with chronic liver disease. Naunyn Schmiedebergs Arch Pharmacol 321 (Suppl):R49
O'Malley K, Crooks J, Duke E, Stevenson IH (1971) Effect of age and sex on human drug metabolism. Br Med J 3:607-609
Oshita S, Sada H, Kojima M, Ban T (1980) Effects of tocainide and lignocaine on the transmembrane action potentials as related to external potassium and calcium

concentrations in guinea pig papillary muscles. Naunyn Schmiedebergs Arch Pharmacol 314:67-71

Palileo H, Hoff J, Swiryn S, Bauernfeind R, Strasberg B, Welch W, Rosen K (1982) Failure of mexiletine in drug-refractory paroxysmal sustained ventricular tachycardia. Am J Cardiol 49:1003

Pantridge JF, Geddes JS (1974) Primary ventricular fibrillation. Eur J Cardiol 1:335-337

Park BK, Breckenridge AM (1981) Clinical implications of enzyme induction and enzyme inhibition. Clin Pharmacokinet 6:1-24

Pentikainen PJ, Koivula IH, Hiltunen HA (1982) Effect of rifampicin treatment on the kinetics of mexiletine. Eur J Clin Pharmacol 23:261-266

Perlow GM, Jain BP, Pauker SG, Zarren HS, Wistran DC, Epstein RL (1981) Tocainide associated interstitial pneumonitis. Ann Intern Med 94:489-490

Pfeifer JH, Greenblatt DJ, Kockweser J (1976) Clinical use and toxicity of intravenous lidocaine: a report from the Boston collaborative drug surveillance program. Am Heart J 92:168-73

Pieper JA, Slaughter RL, Anderson GD et al. (1982) Lidocaine clinical pharmacokinetics. Drug Intell Clin Pharm 16:291-294

Podrid PJ, Lown B (1981) Mexiletine for ventricular arrhythmias. Am J Cardiol 47:895-902

Pottage A (1977) Oral dosage schedules for mexiletine. Postgrad Med J 53 (Suppl 1): 155-157

Pottage A (1983) Clinical profiles of newer class I antiarrhythmic agents—tocainide, mexiletine, encainide, flecainide and lorcainide. Am J Cardiol 52:24c-31c

Pozenel H (1977) Haemodynamic studies on mexiletine, a new anti-arrhythmic agent. Postgrad Med J 53 (Suppl 1):78-80

Prescott LF, Pottage A, Clements JA (1977) Adsorption, distribution and elimination of mexiletine. Postgrad Med J 53:50-55

Pringle T, Fox J, McNeill JA, Kinney CD, Liddle J, Harron DWG, Shanks RG (1986) Dose independent pharmacokinetics of mexiletine in healthy volunteers. Br J Clin Pharmacol 21:319-321

Riddell JG, McAllister CB, Wilkinson GR, Wood AJJ, Roden DM (1984) A new method for constant plasma drug concentrations: application to lidocaine. Ann Intern Med 100:25-28

Roden DM, Reele SB, Higgins SB et al. (1980) Tocainide therapy for refractory ventricular arrhythmias. Am Heart J 100:15-22

Roos JC, Paalman DCA, Dunning AJ (1977) Electrophysiological effects of mexiletine in man. Postgrad Med J 53 (Suppl 1):92-94

Rubino M, Jackson E (1982) Severe paranoia with concomitant tocainide and propranolol therapy. Clin Pharm 1:177-179

Rutledge JC, Harris F, Amsterdam EA, Skalsky E (1985) Clinical evaluation of oral mexiletine in the treatment of ventricular arrhythmias. J Am Coll Cardiol 6:780-784

Ryan W, Engler R, LeWinter M (1979) Efficacy of a new oral agent (tocainide) in the acute treatment of refractory ventricular arrhythmias. Am J Cardiol 43:285

Ryden L, Arnman K, Conradson T, Hofvendahl S, Mortensen O, Smedgard P (1980) Prophylaxis of ventricular tachyarrhythmias with intravenous and oral tocainide in patients with and recovering from acute myocardial infarction. Am Heart J 100:1006-1016

Sami M, Lisbona R (1985) Mexiletine: long-term efficacy and haemodynamic actions in patients with ventricular arrhythmias. Can J Cardiol 1:251-258

Saris SD, Lowenthal DT, Affrime MB (1983) Steady state digoxin concentration during oral mexiletine administration. Curr Ther Res 34:662-666
Sawyer DR, Ludden TM, Crawford MH (1981) Continuous infusion of lidocaine in patients with cardiac arrhytmias. Unpredictability of plasma concentrations. Arch Intern Med 141:43-45
Sedman AJ, Bloedow DC, Gal J (1982) Serum binding of tocainide and its enantiomers in human subjects. Res Commun Chem Pathol Pharmacol 38:165-168
Smith WM, Gallagher JJ (1980) Les torsades de pointes: an unusual ventricular arrhythmia. Ann Intern Med 93:578-584
Sonnhag C (1980) Efficacy and tolerance of tocainide during acute and long term treatment of chronic ventricular arrhythmias. Eur J Clin Pharmacol 18:301-310
Stargel WW, Shand DG, Routledge PA et al. (1981) Clinical comparison of rapid infusion and multiple injection methods for lidocaine loading. Am Heart J 102:872-876
Stein J, Podrid P, Lown B (1984) Effects of oral mexiletine on left and right ventricular function. Am J Cardiol 54:575-578
Stenson RE, Constantino RT, Harrison DC (1971) Interrelationships of hepatic blood flow, cardiac output, and blood levels of lidocaine in man. Circulation 43:205-211 Sutton PP (1980) Pharmacokinetics of tocainide in acute myocardial ischaemia. Br J Clin Pharmacol 10:530
Swedberg K, Holmberg S (1981) Antiarrhythmic effects and patient tolerance to tocainide in acute myocardial infarction—a randomized double blind study lasting 48 hours. In: Pottage A, Ryden L (eds) Workshop on tocainide. Proceedings from a conference in Copenhagen, 1977. Hassle, Molndal, Sweden, pp 190-193
Swedberg K, Pehrson J, Ryden L (1978) Electrocardiographic and haemodynamic effects of tocainide (W36095) in man. Eur J Clin Pharmacol 14:15-19
Talbot RG, Clark RA, Nimmo J, Neilson JMM, Julian DG, Prescott JF (1973) Treatment of ventricular arrhythmias with mexiletine (Ko 1173). Lancet 2:399-404
Talbot RG, Julian DG, Prescott LF (1976) Long-term treatment of ventricular arrhythmias with oral mexiletine. Am Heart J 91:58-65
Thomson PD, Melmon KL, Richardson JA et al. (1973) Lidocaine pharmacokinetics in advanced heart failure, liver disease and renal failure in humans. Ann Intern Med 78:499-508
Trimarco B, Ricciardelli B, DeLuca N, Volpe M, Sacca K, Rengo F, Condorelli M (1983) Disopyramide, mexiletine and procainamide in the long term and treatment of ventricular arrhythmias: antiarrhythmic efficacy and haemodynamic effects. Curr Ther Res 33 (3):472-487
Turner WM (1982) Lidocaine and psychotic reactions. Ann Intern Med 97:149-150
Valentine PA, Frew JL, Mashford ML, Sloman JG (1974) Lidocaine in the prevention of sudden death in the pre-hospital phase of acute infarction. N Engl J Med 291:1327-1331
Van Durme JP, Bogaert M, Bekaert I, DeClereq D, Moerman E (1978) Comparison of the anti-dysrhythmic efficacy of atenolol, disopyramide, mexiletine and placebo. Proceedings of the symposium on management of ventricular tachycardia—role of mexiletine. Excerpta Medica, Amsterdam, pp 21-26
Vaughan Williams EM (1974) Electrophysiological basis for a rational approach to anti-dysrhythmic drug therapy. Adv Drug Res 9:69-101
Vaughan Williams EM (1977) Mexiletine in isolated tissue models. Postgrad Med J 53 (Suppl 1):30-4
Vaughan Williams EM (1984) A classification of anti-arrhythmic drugs reassessed after a decade of new drugs. J Clin Pharmacol 24:129-147

Velebit V, Podrid P, Lown B, Cohen BH, Graboys TB (1982) Aggrevation and provocation of ventricular arrhythmias by anti-arrhythmic drugs. Circulation 65:886-893

Volosin K, Greenberg RM, Greenspon AJ (1985) Tocainide associated agranulocytosis. Am Heart J 109:1392-1393

Wasenmiller JE, Aronow WS (1980) Effect of tocainide and quinidine on premature ventricular contractions. Clin Pharmacol Ther 28:431-435

Waspe LE, Waxman HL, Buxton AE, Josephson ME (1983) Mexiletine for control of drug resistant ventricular tachycardia: clinical and electrophysiologic results in 44 patients. Am J Cardiol 51:1175-1181

Weigers U, Hanrath P, Kuck KH, Pottage A, Graffner C, Augustin J, Runge M (1983) Pharmacokinetics of tocainide in patients with renal dysfunction and during haemodialysis. Eur J Clin Pharmacol 24:503-507

Williams RL, Mamelok RD (1980) Hepatic disease and drug pharmacokinetics. Clin Pharmacokinet 5:528-547

Wing LMH, Meffin PJ, Grygeil JJ, Smith KJ, Birkett DJ (1980) The effect of metoclopramide and atropine on the absorption of orally administered mexiletine. Br J Clin Pharmacol 9:505-509

Winkle RA, Meffin PJ, Fitzgerald JW, Harrison DC (1976) Clinical efficacy and pharmacokinetics of a new orally effective anti-arrhythmic, tocainide. Circulation 54:884-889

Winkle RA, Anderson JL, Peters F, Meffin PJ, Fowles RE, Harrison DC (1978a) The haemodynamic effects of intravenous tocainide in patients with heart disease. Circulation 57:787-792

Winkle RA, Meffin PJ, Harrison DC (1978b) Long term tocainide therapy in ventricular arrhythmias. Circulation 57:1008-1016

Winkle RA, Mason JW, Harrison DC (1980) Tocainide for drug resistant ventricular arrhythmias: efficacy, side effects and lidocaine responsiveness for predicting tocainide success. Am Heart J 100:1031-1036

Woosley RL, McDevitt DG, Nies AS, Smith RF, Wilkinson GR, Oates JA (1977) Suppression of ventricular ectopic depolarizations by tocainide. Circulation 56:980-984

Wyman MG, Lalka D, Hammersmith L et al. (1978) Multiple bolus technique for lidocaine administration during the first hours of an acute myocardial infarction. Am J Cardiol 41:313-317

Yamaguchi I, Singh BN, Mandel MJ (1979) Electrophysiological actions of mexiletine on isolated rabbit atria and canine ventricular muscle and Purkinje fibres. Cardiovasc Res 13:288-96

Zbarbaro JA, Rawlings DA, Fozzard HA (1979) Suppression of ventricular arrhythmias with intravenous disopyramide and lidocaine efficacy compared in a randomized trial. Am J Cardiol 44:513-520

Zipes DP, Troup PJ (1978) New antiarrhythmic agents. Am J Cardiol 41:1005-1024

CHAPTER 11

Clinical Use of Class Ic Antiarrhythmic Drugs

J. C. SOMBERG

A. Introduction

In 1969, VAUGHAN WILLIAMS proposed a classification for antiarrhythmic drugs at an international meeting of cardiologists and electrophysiologists in Elsinor, Denmark (VAUGHAN WILLIAMS 1970). The majority of the antiarrhythmic agents, then and now, restrict the fast inward sodium current. However, distinct differences within this class I action have necessitated subdivision of class I drugs into a, b, and c groups. The most distinctive characteristic of the Ic agents is that then bind slowly and dissociate slowly from the Na channel (CAMPBELL and VAUGHAN WILLIAMS 1983). Thus, during tachycardias, not enough time will elapse between beats to permit the drug to dissociate from the Na channel, causing a population of channels to be "blocked" and thus slowing conduction markedly (VAUGHAN WILLIAMS 1984). These cellular electrophysiologic characteristics of the class Ic agents provide a rationale for both their potency and possible disadvantages. By binding to Na channels and remaining bound throughout the cardiac cycles at fast (though physiologic) heart rates, the drugs will decrease ectopy, the trigger for initiating ventricular tachycardia. Conduction will be slowed and the substrate critical for sustained ventricular tachycardia (VT) leading to sudden death may be favorably modified. The same electrophysiologic actions, however, in patients with an already diseased conducting system could delay conduction to the point of impairment with asystole, requiring insertion of a pacemaker.

Additionally, if an abnormal pathway exists conducting too rapidly for re-entry, a slowing of conduction could permit re-entry and sustained VT.

Several antiarrhythmic agents have recently been synthesized with properties assigning them to the Ic Class. Clinical experience has shown them to be highly effective in suppressing ventricular premature contraction (VPC) frequency and reducing the incidence of VT on Holter and at electrophysiologic testing, but the conduction disturbances and proarrhythmic actions predicted from their electrophysiologic properties have also been observed. It is necessary, therefore, to obtain a full understanding of the clinical pharmacology of these agents, especially with respect to potency, negative inotropism, and side effects, before they can be wisely and safely administered. This chapter reviews the clinical pharmacology of five class Ic agents, flecainide, lorcainide, encainide, propafenone, and indecainide.

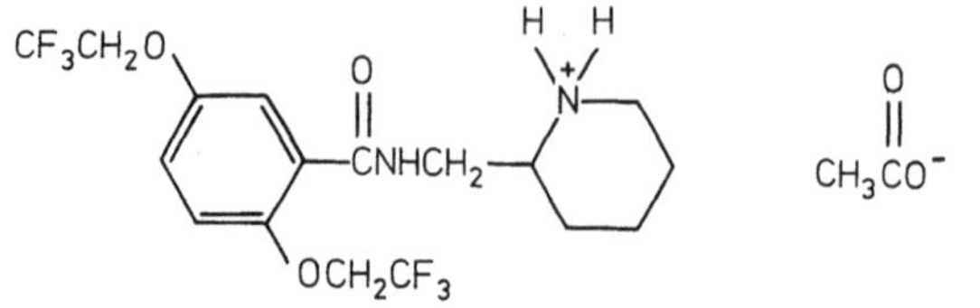

Fig. 1. Structure of flecainide acetate

B. Flecainide

Flecainide acetate is a newly approved antiarrhythmic agent developed at Riker Laboratories by Banitt et al. (1977) (Fig. 1). While investigating new fluoroalkylethoxybenzamides, they found some to have potent anesthetic and antiarrhythmic properties. Many modifications of the structure were tested on chloroform-induced ventricular fibrillation (VF) in the mouse (LAWSON 1968) and compared with quinidine, lidocaine, and procainamide. One compound, flecainide, was found to be 7–12 times more potent than the reference agents and possessed minimal toxicity.

I. Cellular Electrophysiology

CAMPBELL and VAUGHAN WILLIAMS (1983) have pointed out that "during the past decade a number of new antiarrhythmic drugs have been introduced, which share the property of restricting fast inward sodium current, but which exhibit some differences in clinical use. The major effect of both drugs (lorcainide and flecainide) is to depress V_{max} in a dose-dependent fashion without altering resting potential." In vivo both drugs produce a marked depression of His-Purkinje and intraventricular conduction with only minor effects on refractoriness (HELLESTRAND et al. 1982; KASPER et al. 1979; BAR et al. 1981). Additionally, the kinetics for onset and offset of rate-dependent depression of V_{max} have been described elsewhere in this volume.

II. Preclinical Studies

SCHMID et al. (1975) found flecainide to be extremely effective intravenously and orally against experimental arrhythmias in the dog produced by hydrocarbon-epinephrine, ouabain, aconitine, or coronary ligation. Studies by VERDOUW et al. (1979), performed in acutely ischemic pigs pretreated with flecainide, revealed that the incidence of ventricular fibrillation was reduced from 33% to 12%, VT incidence was reduced from 42% to 0%, and the number of premature ventricular contractions was reduced to 5% of the control frequency. These investigators also found a significant widening of QRS duration in the ECG and a decrease in myocardial contractility (*dp/dt*) associated with hypotension.

Flecainide caused a very significant dose-dependent decrease in the maximum velocity of depolarization (V_{max}) of dog Purkinje fibers, ventricular muscle, and guinea pig papillary muscle (BORCHARD and BOISTEN 1982; IKEDA et al. 1982). The action potential durations at 30% and 90% repolarization were decreased in Purkinje fibers but increased (slightly) in ventricular muscle (CAMPBELL 1983).

The electrophysiologic effects on the intact dog heart were studied by HODDESS et al. (1979). Low levels (0.4-0.7 μg/ml) of flecainide prolonged conduction in all parts of the conducting system, causing a maximal effect at 6.5 μg/ml. The tissues most sensitive to the lower flecainide concentrations were the bundle of His and the ventricle, with only a slight prolongation of the atrial refractory period and of the atrioventricular (AV) node relative and functional refractory periods. Ventricular responses in atrial fibrillation were slowed by 0.7 μg/ml flecainide, a combination with propranolol producing further slowing of ventricular response and increase in the AV nodal refractory period (HODDESS et al. 1979).

III. Clinical Electrophysiology

His bundle studies have shown that at low doses (1 mg/kg flecainide intravenously) there is a slight decrease in heart rate while the sinus node recovery time, calculated sinus node conduction time, and intraatrial conduction times increased (slightly). Intranodal, His-ventricular (HV), and ventricular-right ventricular apex conduction times were significantly prolonged from 13% to 33% (SEIPEL et al. 1980). Flecainide induced an average prolongation of the P-R interval by 26%, QRS interval by 29% and QT_c duration by 6%. Most of the QT prolongation was due to the increase in the QRS duration. QRS was prolonged on an average of 25%-50% by twice daily doses of 100 mg or 200 mg respectively. Since QRS prolongation was dose dependent, it could be used as a guide to warn of excessive flecainide serum concentrations.

IV. Pharmacokinetics

In dogs and rats the plasma half-life of flecainide was approximately 100 min (CONARD et al. 1975). In healthy young male volunteers, however, after a single intravenous or oral dose, the elimination half-time averaged 14 h (range 7-15 h). Considerably higher plasma levels and longer half-lives were observed in patients. Renal function, cardiac output, and the state of the patient's health must, therefore, be taken into account in dosage selection. Elimination half-life in a group with cardiac disease averages 10 h (HODGES et al. 1982).

After an oral dose, over 95% of the drug is absorbed and delivered to the systemic circulation unchanged, excluding any significant hepatic first-pass effect (CONARD et al. 1979). Although the drug is metabolized to active metabolites, these are far less potent than the parent drug and occur in the conjugated form in the plasma. GUEHLER et al. (1985) have concluded that the metabolites were unlikely to "potentiate flecainide's antiarrhythmic action or increase susceptibility to drug toxicity in the clinical setting." The effective serum concentration of flecainide ranges from 200 to 1200 ng/ml. The dose of 100 mg orally twice daily can produce this wide range in patients. Levels over 1000 ng/ml are potentially toxic and concentrations of 1600-2200 ng/ml have caused serious toxicity or proved fatal (LUI et al. 1982; SPIVACK et al.

1984). Some toxicity (CNS, gastrointestinal, arrhythmogenicity) may be noted at much lower doses within the therapeutic range (SELLERS and DIMARCO 1985). Determination of flecainide levels are clinically relevant, and when drug level monitoring was introduced in the experimental development program, the incidence of fatal arrhythmias markedly decreased. A number of chromotographic techniques are available for accurate determination of flecainide serum concentration (CHANG et al. 1984; BHAMRA et al. 1984).

Flecainide has been reported to cause a rise in serum digoxin concentration varying from 10% to 40% (LEWIS et al. 1984), attributed by TJANDRAMAGA et al. (1982) to a decrease in the volume of distribution. Other investigators have found no interaction. When an interaction has been observed, it has not been reported to be clinically significant.

V. Clinical Efficacy

Initial short-term trials, such as Somani's 1-day intravenous (1–2 mg/kg) study (SOMANI 1980), showed complete suppression of premature ventricular conctractions in nine of ten patients. ANDERSON et al. (1981) evaluated short-term flecainide therapy in 13 patients with frequent premature ventricular contractions (PVCs) (greater than 600 PVCs/12 h). Ten patients completed the trial; nine had a mean PVC suppression of 98% and all repetitive forms were reported to be eliminated. The tenth patient had a 68% suppression of PVCs. A 2-week outpatient trial showed a 95% suppression of PVCs in those patients. DUFF et al. (1981) showed similar results in a twice-daily dosing regimen at a mean effective total daily dose of 410 mg (range 200–600 mg/day). They reported a 97% mean suppression in PVCs and 100% suppression of VT. The effective response was dose dependent and was accompanied by a dose-related increase in PR and QRS intervals. The dose-related effects on ECG intervals and PVC suppression are also reported in the study of MEINERTZ et al. (1984). They reported that "a minimum effective therapeutic dose could be titrated in 9 of 14 patients at 100 mg twice dialy, in 3 of 14 patients at 150 mg twice daily and in 2 of 14 patients at 200 mg twice daily." Similar efficacy was also noted in the subsequent 48-week, long-term trial reported by these authors. SALERNO et al. (1983) compared flecainide to quinidine in a randomized placebo-controlled, double-blind trial in 19 patients. They reported "the mean suppression of total ventricular ectopic depolarization was 95% for flecainide and 56% for quinidine." In a larger study (280 patients) comparing flecainide with quinidine (Flecainide Quinidine Research Group 1983), 85% of the flecainide patients had at least an 80% suppression of PVCs, versus 57% suppression with quinidine. Sixty-eight percent of the flecainide patients had complete suppression of couplets and VT versus 33% with quinidine. Quinidine causes a significant QT prolongation, and flecainide causes QRS prolongation without JT prolongation. Flecainide's side effects (at 200 mg b.i.d.) included dizziness, blurred vision, headache, and nausea, while quinidine caused diarrhea, nausea, headache, and dizziness.

VI. Supraventricular Tachycardia

Pop et al. (1983) evaluated the effect of intravenous flecainide in 16 patients using programmed atrial stimulation at two pacing rates (100 and 120/min). All patients were reported to show an increased atrial vulnerabililty at both rates, abolished by flecainide (1 mg/kg) in most patients. Camm et al. (1985) found that intravenous flecainide terminated atrial fibrillation in 90%, atrial tachycardia in 100%, intra-AV nodal tachycardia in 89%, and atrioventricular reentrant tachycardia in 80%. Although they noted atrial flutter to slow, only a small proportion (20%) was terminated. Neuss et al. (1983), in 37 patients with paroxysmal supraventricular tachycardia (SVT) refractory to treatment with "conventional" antiarrhythmic drugs, found that over a mean treatment period of 14.2 months flecainide 200-400 mg daily completely suppressed the tachycardia. Kim et al. (1986) evaluated 15 patients with spontaneous or inducible sustained paroxysmal supraventricular tachcyardia. After infusion of oral flecainide SVT was noninducible in nine patients. Twelve patients continued oral flecainide treatment for a mean of 16 months (5- to 8-month range). Tachycardia recurred in three of four patients whose arrhythmias remained inducible after flecainide therapy and in one of eight patients whose SVT was suppressed. Hellestrand et al. (1983) evaluated 33 patients undergoing routine electrophysiology studies. Flecainide was given to 14 patients during sustained AV reentrant tachycardia and to 9 patients during sustained intra-AV nodal reentrant tachycardia. AV reentrant tachycardia was successfully terminated in 12 of 14 patients. In all 18 patients with accessory AV pathway, flecainide successfully increased both anterograde and retrograde accessory pathway effective refractory periods, with the anterograde pathway blocked in three patients and the retrograde in eight. Intra-AV nodal reentrant tachycardia was successfully terminated in eight of nine patients. After flecainide administration, re-initiation of intra-AV nodal reentrant tachycardia was not possible in four patients.

Goy et al. (1985) evaluated the effects of flecainide on the conversion of supraventricular arrhythmias to sinus rhythm in 50 patients [39 with atrial fibrillation, 6 with atrial flutter, 4 with SVT, and 1 with SVT and Wolff-Parkinson-White (WPW) syndrome]. Conversion was achieved in 36 patients (72%), 29 cases with AF, 4 with SVT, and 2 with atrial flutter. In 60 consecutive patients with AF (Borgeat et al. 1986), the overall conversion rate to sinus rhythm was 63%; AF converted in 18 patients (60%) treated with quinidine and 20 (67%) with flecainide. Adverse effects were more frequent in the quinidine group (27%) than in the flecainide group (7%), though less severe.

Ward et al. (1986) noted flecainide to be effective in four of five children who had tachycardias associated with the WPW syndrome. Fauchier et al. (1985) observed in patients with WPW syndrome (17 with spontaneous tachcyardias due to orthodromic reentry and 4 with arial arrhythmias) that intravenous flecainide, 2 mg/kg, over 5 min, terminated the SVT in 13 cases (blocking retrograde conduction in the accessory pathway) and terminated 3 of the 4 cases of atrial arrhythmias. Kappenberger et al. (1985) evaluated the action of flecainide in nine patients with severe symptomatic WPW syndrome. The shortest ventricular response during atrial fibrillation increased from 218 to 320 ms. In four patients sustained rapid atrial fibrillation converted to sinus rhythm. The rate of the circus movement tachycardia decreased from 166 to 130 beats/min after flecainide, due to a lengthening of retrograde ventricular atrial conduction time over the accessory pathway. While flecainide's efficacy in the WPW syndrome is based on increasing refractiveness especially in the retrograde part of the reentrant pathway (NEUSS et al. 1983), catecholamines may reverse this. BREMBILLA-PERROT et al. (1985) reported that "although anterograde conduction through a Kent Bundle with a short refractory period was suppressed by 300 mg flecainide acetate, the infusion of small amounts of isoproterenol caused the reappearance of WPW and permitted the induction of an atrial tachycardia with 1/1 conduction through the accessory pathway at a rate of 260 beats per minute." These observations suggest a need to evaluate therapy for the preexcitation syndrome by an exercise test on antiarrhythmic therapy as well as at electrophysiologic studies (possibly employing a catecholamine infusion).

VII. Programmed Stimulation Studies

Programmed electrical stimulation (PES) studies evaluate antiarrhythmic agents in patients with more severely deranged left ventricular function and with greater electrical instability than in sponsored PVC suppression studies. ANDERSON et al. (1983) reported that flecainide prevented VT induction in 9 of 15 patients (60%). Ten patients were placed on flecainide therapy for an average duration of 6.5 months; nine of the ten patients were effectively treated, with one patient developing VT. FLOWERS et al. (1985) studied 40 patients who initially received flecainide intravenously at electrophysiologic testing. Flecainide prevented VT induction in 26 (65%). Twenty-one patients were discharged on flecainide therapy, 100 mg twice daily (a reduced dosage based on the proarrhythmia incidence previously seen in this type of population). Sixteen patients have done well, one died suddenly, one developed VT, and one died of noncardiac cause over an 11-month period. The authors concluded "flecainide therapy guided by programmed electrical stimulation is effective at a reduced dose".

In a recent study in 93 patients long term efficacy (SOMBERG et al. 1987), 20 had a prior history of at least one cardiac arrest and 73 had sustained VT. The mean radionuclear ejection fraction was 32% ± 5%. After flecainide 44 patients no longer had VT (47%). Procainamide was evaluated in 69 patients, 24 had an adverse reaction to procainamide, and 28 were protected on procainamide (40% efficacy). The mean serum concentration of flecainide in the protected group was 298 ± 36 µg/ml and 4.3 µg/ml for procainamide. Both flecainide and procainamide significantly prolonged refractoriness and lengthened QRS duration, but only procainamide increased the JT interval. All 93 patients were discharged on antiarrhythmic therapy, 42 on flecainide, 27 on other antiarrhythmic drugs guided by electrophysiologic study (EPS), and 24 on amiodarone (when all other agents had failed). Six of the 42 patients on flecainide complained of adverse side effects, none severe enough to stop therapy; 4 (9%) died suddenly over 18 ± 4 months. Of the 27 patients on other therapy, 8 have died, 3 suddenly (11%), 4 with myocardial infarctions (MIs) and 1 with congestive heart failure. Of the 24 patients on amiodarone, 11 have died, 5 (21%) suddenly, 4 of congestive heart failure, 1 of pulmonary fibrosis, and 1 of MI. During acute intravenous administration of flecainide, seven patients developed spontaneous sustained VT (mean serum level, 357 ± 72 µm/ml); two terminated spontaneously, one required defibrillation, while four failed defibrillation, but responded to lidocaine. Although flecainide was well tolerated both intravenously and orally, a risk of proarrhythmia was significant. Evaluation of flecainide by intravenous administration at EPS predicted success with chronic oral therapy.

The Flecainide Ventricular Tachycardia Study Group (1986) reported on 49 patients with sustained, and 45 with nonsustained, VT, who had been refractory to, or intolerant of, other antiarrhythmic agents. Flecainide plasma levels were monitored and remained in the therapeutic range of 0.2–1.0 µg/ml. Minimum efficacy requirements included elimination of sustained VT and reduction of other ventricular arrhythmias as determined by one or more of the following: 24-h electrocardiographic monitoring, programmed electrical stimulation, exercise testing and/or hospital telemetry monitoring. "Sixty-eight patients (72%) were dischargd with flecainide therapy. After 8 months, 45 patients (48%) were still taking flecainide, 22 of 49 (45%) with sustained and 23 of 45 (51%) with nonsustained VT. Nine patients with sustained VT and 1 patient with nonsustained VT had aggravation of arrhythmias. Two patients had third-degree heart block. Nine patients died after discharge from the hospital: six from out of hospital sudden death and three from acute myocardial infarction."

To compare the predictive effectiveness of procainamide and flecainide Wynn et al. 1986, evaluated 153 patients who underwent PES studies because of either sustained or nonsustained VT. Procainamide prevented VT induction in 79 of 153 patients. Of the remaining 74 with inducible VT on procainamide, 55 were protected by another antiarrhythmic agent ($P < 0.001$). Twenty-nine of 55 patients were protected by flecainide at PES testing; 26 were protected by another agent also. Of the 26 VT inducible patients who received flecainide, 15 were protected by another agent ($P < 0.01$). Thus, if procainamide or flecainide prevented VT induction, other drugs were also likely to be effective, but when they did not prevent VT induction efficacy of other drugs was not excluded.

Lal et al. (1985) in another EP study, before and during flecainide treatment, concluded: "Sustained VT became noninducible in 5 patients, nonsustained in 5 and VT slowed in 13 patients (36% efficacy rate)." Platia et al. (1985), found that in only 2 of 17 patients was PES induced VT prevented by flecainide (12% efficacy). Livelli et al. (1984) reported a similar low efficacy in five patients with VT/VF. VT was prevented in one patient, but he died suddenly 2 months later while on flècainide; one patient was inducible for VT, and three had a proarrhythmic response to flecainide. Our own experience, however, was that flecainide had a high efficacy as well as a significant proarrhythmic effect.

VIII. Arrhythmogenicity

Antiarrhythmic agents may increase PVC frequency (VELEBIT 1982), facilitate induction of VT at EPS (TORRES et la. 1985), or exacerbate an arrhythmia to life-threatening severity (RUSKIN et al. 1983). The more potent agents were also the most proarrhythmic (WINKLE et al. 1981a). In the early studies with flecainide, an oral dose of 200 mg b.i.d. was given to patients with frequent PVCs only, but when patients with life-threatening arrhythmias were treated (sustained VT and history of cardiac arrest) this dosage caused a significant and unacceptable incidence of arrhythmogenicity. Halving the dosage reduced the induction of life-threatening arrhythmias. LUI et al. (1982) reported a case without previous VT who developed sustained VT at a high therapeutic concentration (1479 μg/ml). HOHNLOSER et al. (1983) reported a similar case with a cardiac arrest after 6 days of flecainide therapy (plasma level of 840 μg/ml). SPIVACK et al. (1984) reported the death of a patient recently started on flecainide who could not be resuscitated and had a serum concentration of 2200 μg/ml. NATHAN et al. (1984) reported proarrhythmia in 7 out of 152 patients treated with flecainide. Five patients developed VT or VF, only three of whom had preexisting ventricular arrhythmias. SELLERS and DIMARCO (1985) treated five patients with sustained VT or VF. They reported that "3 of these 5 consecutive patients developed a VT with an unusual electrocardiographic morphology shortly after initiation of flecainide therapy, despite the apparent suppression of isolated spontaneous ventricular premature beats in the hours preceding the occurrence of sustained tachycardia." The authors concluded that "the early reports of successful suppression of ventricular premature beats in stable patients may not be directly applicable to sudden death patients with recurrent sustained VT." They go on to point out that "life threatening arrhythmias may appear as the first manifestation of drug toxicity." Additional cases of flecainide toxicity have been reported in association with QT_c prolongation (WEHR et al. 1985). We have found, in three additional cases of severe VT provoked by flecainide that were unresponsive to defibrillation and resuscitation efforts, that lidocaine may be the most effective avaiable therapy to terminate ventricular tachycardia (WYNN et al. 1986).

Besides provoking VT, flecainide has been reported to make defibrillation more difficult. This effect on raising the defibrillation threshold may be similar to flecainide's effect on raising the ventricular threshold for pacing and thus causing pacing failure at times (WALKER et al. 1985).

IX. Hemodynamic Effects

SERRUYS et al. (1983) reported "that flecainide (2 mg/kg i.v.) has a negative inotropic effect not only under resting conditions but also (though less apparent) during pacing induced tachycardia." The effects were immediate and on average caused approximately a 10% decrease in contractility. HOLTZMAN et al. (1984) reported that "single and multiple doses of flecainide (200 mg twice a day) decreased the left ventricular ejection fraction (EF), but did not effect either the cardiac output or vascular resistance." JACKSON et al. (1985) found, in patients, after acute MI with (group I) and without (group 2) left ventricular decompensation, that flecainide induced significant cardiodepression in both groups: cardiac index fell 9% in group 1 and 18% in group 2. Other respective changes were: stroke volume index −13% and −23%; stroke work, −14% and −27%; pulmonary artery pressure, +27% and +21%; and systemic vascular resistance +11% and +18%. These effects are comparable to those of disopyramide in the compromised group, and similar caution is required in giving flecainide to patients with compromised left ventricular function.

In the Cardiac Arrhythmia Pilot Study (1986) on 502 patients with at least 10 ventricular premature depolarizations (VPD) (24 h HOLTER) between 6 and 60 days after MI, the incidence of congestive heart failure in the placebo group was not statistically different from that in the flecainide groups, both being higher than in patients on encainide. However, I Have administered flecainide both intravenously (2 mg/kg) as well as orally to patients with ejection fractions less than 20% without any clinical signs of congestive failure nor was proarrhythmia more probable in the patients with the most severe left ventricular dysfunction (WYNN et al. 1986; SOMBERG et al. 1987). Caution is appropriate but a low EF is not an absolute contraindication to flecainide therapy.

X. Clinical Use

The recent CARDIAC ARRHYTHMIA PILOT STUDY trial found only flecainide and encainide, the two Ic agents, to be effective in suppression of PVC frequency and complex ventricular ectopy (CARDIAC ARRHYTHMIA PILOT STUDY 1986). Flecainide should be administered to patients for the control of life-threatening VT, VF, or the suppression of PVCs when they are symptomatic or felt to be a potential trigger for VT or VF, and should be reserved for patients in whom less potentially arrhythmogenic agents (like beta-blockers, tocainide, mexiletine, and possibly procainamide) have failed. Flecainide should be started at 100 mg twice daily. Monitoring of serum concentrations is useful, because high concentrations (greater than 1 000 μg/ml) must be avoided, the first manifestation of toxicity being sustained VT or VF. CNS toxicity (blurred vision, dizziness, headache, nausea, paresthesias) is dose related, rarely seen at 100 mg b.i.d.

Permanent pacing in patients with sick sinus syndrome, may be needed even at the lower dose (FLOWERS et al. 1985). The proarrhythmic effects of flecainide are its greatest disadvantage. The incidence is not out of proportion to that of other drugs but the severity and difficulty at times in arrhythmia termination make flecainide proarrhythmia so troubling. Still flecainide is a useful addition to available antiarrhythmic therapy.

C. Lorcainide

Lorcainide is *n*-(4-chlorophenyl)*n*-(1-methylethyl)-4-piperidenyl benzene acetamide hydrochloride (JANSSEN; Fig. 2), possesses antiarrhythmic efficacy in animals and man and is categorized as class Ic although subtle differences do distinguish lorcainide from other agents of this class.

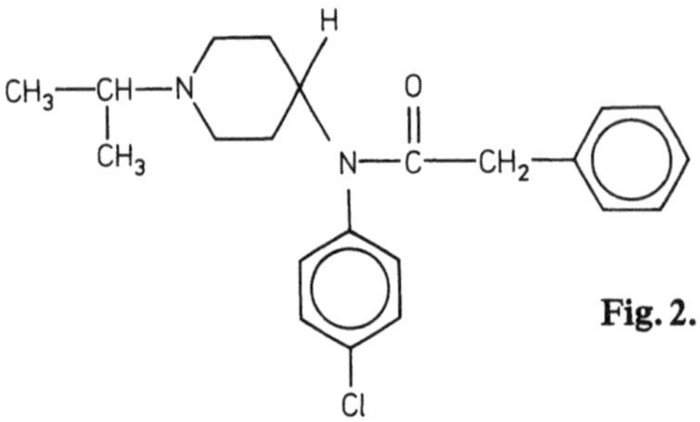

Fig. 2. Structure of lorcainide

I. Preclinical Studies

Lorcainide reduced the frequency and duration of arrhythmias induced by local application of either aconitine or acetylcholine (CARMELIET 1978) or by ouabain or ligature of a coronary artery (XHONNEUX et al. 1975); lorcainide 2.5 mg/kg i.v. reduced premature ventricular contractions (VPCs) by more than 50% in 14 of 29 animals after coronary ligation (CARMELIET 1978). After oral lorcainide of 20 mg/kg three of six animals showed greater than 50% reduction in PVC frequency. Arrhythmias developed on average in 90 min after ouabain infusion at 1 μg/min; intravenous lorcainide 0.63 mg/kg prevented arrhythmias from occurring in eight of the nineteen animals and reduced the PVC frequency by 75% in another two animals. Lorcainide reduced VF threshold (VFT) in ten anesthetized dogs, for 70 min after 5 mg/kg. In isolated guinea pig and rabbit hearts lorcainide increased VFT (ALMOTREFI and BAKER 1981), encainide being 7 times and lorcainide 14 times more potent than lidocaine. In 26 rats pretreated with lorcainide the arrhythmias induced by adenosine diphosphate were significantly reduced in duration in a dose-dependent fashion (dose range 1.25-5 mg/kg) (DECLERCK et al. 1978).

In unanesthetized dogs lorcainide 20 mg/kg orally caused a 15% increase in QRS duration; i.v. administration increased heart rate as well as PR and QRS durations (CARMELIET et al. 1978). At steady state in closed-chested anesthetized dogs, lorcainide 100-1307 μg/liter increased the PR, QRS, AH, and HV intervals. Wenchebach cycle length increased as well as atrial and right ventricular refractory period. Norlorcainide was an active metabilite, with, at similar drug levels, the same electrophysiologic effects (KEEFE et al. 1984). Studies using a digitalized dog model and programmed electrical stimulation methodology have also reported efficacy of acute intravenously administered lorcainide (SOMBERG et al. 1982).

II. Cellular Electrophysiology

In several studies lorcainide was shown to have the effects typical of a class Ic agent, depressing MRD, with slow onset/offset kinetics (CARMELIET et al. 1978, CARMELIET and ZAMAN 1979; SHINGENOBU et al. 1980).

Lorcainide is ten times more potent than lidocaine. At higher concentrations (9.8 mg/liter) lorcainide can cause an abnormal action potential in isolated canine ventricular tissue, causing abnormalities in conduction and the development of VT and VF (SENGES et al. 1982).

Lorcainide like flecainide depressed V_{max} without altering resting potential, but mildly prolonging APD (COWAN and VAUGHAN WILLIAMS 1981; CAMPBELL and VAUGHAN WILLIAMS 1983; and this volume, chap. 7).

III. Clinical Electrophysiology

Intravenous lorcainide, evaluated in 23 patients, did not affect the refractory period of the atrium, ventricle, atrioventricular node, or AH interval (BARR et al. 1981). SOLTI et al. (1986) observed in 30 patients an increase in atrial and ventricular ERP, H-V time, and ERP of the accessory pathway. STROOBANDT and KESTELOOR (1985) reported that lorcainide prolonged QRS in both patients with normal and patients with long QRS, but the group with a QRS of 110 ms showed the greatest increase. GSTOTTNER and GMEINER observed that lorcainide 1.25 or 2.5 mg/kg, over 2 or 4 min in 21 patients increased sinus rate and atrial ERP. The corrected sinus node recovery time rose after lorcainide 2.5 mg/kg, most markedly in patients with sinus node dysfunction. The PA interval remained unchanged, while the AH interval slightly increased. The functional and effective AV nodal refractory periods changed variably. HV intervals, pooled HV intervals, and QRS width lengthened in all patients, but especially in patients with preexisting His bundle delay. KASPER et al. (1979) noted similar findings and found that two patients with preexisting conduction abnormalities developed third-degree AV block. Thus, sinus node and AV nodal function is usually not impaired by lorcainide, while ventricular conduction (HV delay) is invariably observed. ECHT et al. 1983 reported AH interval prolongation after I week of lorcainide, but AH prolongation has not been regularly observed in other studies in animals and man.

IV. Pharmacokinetics

Absorption is rapid and extensive, approximately 100% at steady state with continuous oral dosing (KLOTZ et al. 1978). Initially lorcainide has poor bioavailability, which increases with chronic doses. JAHNCHEN et al. (1979) showed that the mean bioavailability increased from 27% ± 24% for an initial dose of 150 mg to 50% ± 30% for an initial dose of 300 mg in patients with PVCs. There is significant hepatic extraction of lorcainide, with a variable elimination half-life after IV administration ranging from 5 to 12 h (mean 7.8 h) (JAHNCHEN et al. 1979; KATES et al. 1983). KLOTZ et al. (1978b) reported a half-life in animals of 5.1 h and of 7.6 h in patients with PVCs prolonged in patients with alcoholic cirrhosis (KLOTZ et al. 1979 a). SOMANI (1981) reported the plasma half-life to be very variable, ranging from 6.2 to 23.1 h (mean 13.1 ± 5 h). Half-life increases with age ($r = 0.68$, $P < 0.01$) in parallel with a change in the volume of distribution ($r = 0.52$, $P < 0.05$). The volume of distribution (Vd) is approximately 6 liters/kg (KLOTZ et al. 1979 b). The $t_{1/2}$ for the initial alpha distribution phase is 0.3 h using a two-compartment model (KLOTZ et al. 1978) with a triexponential decay equation providing a better data fit (KATES et al. 1983). The free fraction is 0.24 to 0.29, mean 0.27. The active metabolite is not found in significant quantities following acute intravenous administration.

Following oral administration, half-lives of lorcainide and norlorcainide were 9.6 ± 2.8 and 26.8 ± 8.2 h respectively (Kates et al. 1983). Lorcainide and its active metabolite are most accurately measured by a liquid chromotographic technique (Yee et al. 1981). After 4 weeks at 100 mg b.i.d., mean plasma level concentration was 164 μg/liter (range 24.6–285 μg/liter). Clearance after multiple oral doses decreased from 988 ± 425 to 666 ± 27 ml/min.

Lorcainide, with rapid saturable first-pass metabolism, has two hydroylated metabolites and one dealkylated metabolite (Woestenborghs et al. 1979), the latter (norlorcainide) being active (Keefe et al. 1984) and accumulating to levels nearly twice those of the parent compound. Lorcainide in the myocardium accumulates to 15 times the plasma concentration.

Intravenous lorcainide reduces PVC frequency at a plasma level of 120–150 μg/ml and at between 200–500 μg/ml after oral therapy (Klotz et al. 1979 b, c). Meinertz et al. (1979) reported that QRS widening is dose related and can be used to monitor lorcainide therapy.

Somani et al. (1984) did not find renal disease to alter kinetics of lorcainide. The $t_{1/2}$ beta was 8.61 ± 6.3 h, not significantly different from 7.04 ± 4.12 h off dialysis.

Intravenous infusion of lorcainide frequently induces feelings of warmth, dizziness, or lightheadedness. Oral therapy is associated in 50% of cases initially with sleep disturbances, frequent waking, and vivid dreams, but many tolerate therapy and may not experience problems after 2–3 months of continued therapy, but gastrointestingal disturbance, a metallic taste in the mouth, and headaches may occur with chronic therapy. Hyponatremia was found in some patients (Somani et al. 1984), which normalized upon cessation of therapy.

V. Clinical Efficacy Intravenously

Anderson et al. (1985), in a randomized parallel study with a crossover option in 30 hospitalized patients with frequent (>1/min) complex ventricular arrhythmias, infused lorcainide 2 mg/kg at 2 mg/min, supplemented if needed with 100 mg in 1 maintenance (dose 8 mg/h); lidocaine loading was at 1 mg/kg (25 mg/min) supplemented if needed with 50 mg in 2 min with a maintenance dose of 2 or 3 mg/min (as needed). Arrhythmias were compared for 2 hours before and after loading. The median frequency of premature ventricular complexes decreased by 76% after lidocaine ($P < 0.05$) and by 93% after lorcainide ($P < 0.001$) (difference, $P = 0.06$). More than 95% arrhythmia suppression was achieved by lorcainide in 47% of patients, and but by lidocaine in only 13% ($P < 0.05$). Couplets were abolished after lorcainide, and reduced by 89% after lidocaine. Both drugs suppressed runs of premature beats. On crossover, lorcainide led to greater suppression of premature ventricular complexes in six of nine patients (67%) compared with none of seven patients ($P < 0.01$) crossing over to lidocaine. Adverse effects were minor with both drugs. Blevins et al. (1986) randomized 30 patients with greater than 30 PVCs/h unassociated with an acute MI to i.v. lorcainide or lidocaine. Nonresponders detected by bedside telemetry were crossed over. Clinical response was 6 of 25 (24%), including 2 of 9 crossovers with lorcainide, and 8 of 26 (31%), including 3 of 12 crossovers with lidocaine (NS). Lorcainide 1.9 mg/kg i.v. followed by constant infusion of 0.18 mg/kg per hour failed to suppress VPCs in five of seven patients (KLOTZ et al. 1979 c). VLAY and MALLIS (1986) compared i.v. lorcainide in 14 patients with 14 other patients who received oral lorcainide. Only 2 of 12 subjects receiving IV lorcainide experienced CNS side effects (headache, dizziness, sleep disturbance), while 12 of 14 subjects on oral use of lorcainide developed side effects, implying that the side effect may have been related to norlorcainide. Efficacy was similar in both groups. LLOYD et al. (1981) compared lidocaine with lorcainide in 50 post-MI patients, efficacy and side effects being similar with both.

VI. Clinical Efficacy: Oral Therapy

Lorcainide was evaluated initially by KESTELOOT and STROOBANDT (1977) in ten patients over 2 weeks with significant arrhythmia reduction. Cocco and STROZZI (1978) evaluated lorcainide in seven patients with malignant ventricular arrhythmias that were resistant to other antiarrhythmic agents, observing a complete disappearance of arrhythmias in six and 50% suppression in one. MORGANROTH (1981) evaluated lorcainide in 15 ambulatory patients with at least 30 PVCs/h (HOLTER). In eight (67%) of the twelve patients who completed the study 70% reduction in ventricular ectopy was observed. Eight of the ten patients with VT had a 91% mean reduction in arrhythmia frequency and seven of the eight had a 100% reduction. Nausea and insomnia induced three patients to withdraw from the study.

MEAD et al. (1985) showed that 12 of 14 patients achieved almost total suppression of ventricular ectopy throughout 1 year of treatment, at a dosage of between 200 and 400 mg/day. In a fixed-dose, randomized, crossover comparison, at 100 mg two or three times daily, lorcainide was shown to suppress ventricular ectopy in 7 of 12 patients (58%) (FALK et al. 1986), whereas quinidine was effective in 59% but a significant number of patients on lorcainide withdrew early from the study because of side effects.

VLAY et al. (1984) found that, although a significant number of patients receiving lorcainide reported CNS side effects, lorcainide appeared to be more efficacious than quinidine. The mean reduction in total PVCs was only 16% on quinidine, but 68% on lorcainide.

Lorcainide was studied in a double-blind, placebo-controlled, crossover trial of 39 patients with ventricular arrhythmias, at a dose of 100 mg bid (SINGH et al. 1984). Twenty-two of these patients had a significant reduction in total PVCs, 13 had no significant change, and 4 had a paradoxical increase in PVC frequency. After 6 months of treatment, 61% of the patients still had significant suppression in PVC frequency on lorcainide.

KEEFE et al. (1982) evaluated lorcainide in 10 patients with frequent (71/min) PVCs in a randomized double-blind, placebo-controlled crossover trial. All patients had a reduction in PVCs averaging 82% ± 20% ($P < 0.01$).

Lorcainide in MI patients, first i.v. for 24 h and then following initial i.v. therapy lorcainide 200 mg daily (orally) for 10 days was as effective as lidocaine (LLOYD et al. 1981).

VII. Programmed Electrical Stimulation Studies

In patients with VT, SOMBERG et al. (1984 a) employing PES, found lorcainide prevented induction of VT in 69 of 100 patients. Forty-six patients continued on lorcainide, 9 on procainamide, and 45 on other drugs. Eighty percent remained on lorcainide therapy, whereas only 47% have continued on other drugs, over 17.5 months (mean). Despite sleep disturbance and a need for sedation, lorcainide was tolerated and remained effective.

9 of 29 patients who had at least three beats of ventricular tachycardia (HOLTER) 7-12 days post MI, 28 were inducible into ventricular tachycardia. Lorcainide prevented induction of VT in 21 of 28; lidocaine protected 5 of 21; and procainamide protected 12 of 22 patients. After 12.5 ± 4.0 months, 19 patients were still being successfully treated with lorcainide. There were no sudden deaths or significant adverse effects.

SAKSENA et al. (1983) evaluated 12 patients; 1 patient could not tolerate therapy and 4 of the 11 patients were protected on lorcainide therapy. In 76 patients (CHESNIE et al. 1984), 60 studied by Holter and exercise text; 16 by invasive EPS, Lorcainide suppressed ventricular ectopic activity in 21 (38%) of 56 patients in the Holter group, and in 6 (40%) of 15 who had invasive testing.

ECHT et al. (1983) found that oral but not i.v. lorcainide prolonged the AH interval. Fifty patients with sustained VT or VF refractory to other drugs in 7 ± 2 trials, were treated with oral lorcainide (ECHT et al. 1985). Twenty-three underwent PES both before and after oral lorcainide, and all 23 remained inducible, but the VT cycle length increased from 273 ± 65 ms to 388 ± 97 ms ($P < 0.001$). These inducible patients were nevertheless started on chronic oral lorcainide, but had a poor outcome.

SOMBERG et al. (1986) evaluated by PES 38 patients with a prior history of cardiac arrest. Lorcainide prevented VT or VF induction in 14 patients and failed in 24 (efficacy 37%). Procainamide had failed (cardiac arrest or breakthrough VT) in 16 patients, 7 patients had progressively severe adverse side effects and thus only 15 patients were tested on procainamide with 7 protected. This result was better than that reported by Echt and associates. Fourteen patients were started on oral lorcainide and 24 on other drugs selected by PES. After 29 ± 7 months, three are alive on lorcainide, five discontinued therapy due to side effects, six died (three sudden deaths), two had MI, and there was 1 noncardiac death.

VIII. Arrhythmogenicity

Although lorcainide has caused a worsening of arrhythmias (PODRID and MORGANROTH 1985; TORRES et al. 1985), the incidence is not higher than with other drugs, and is lower than that reported for other class Ic agents.

IX. Hemodynamic Effects

Carmeliet et al. (1978) showed that, *in vitro*, lorcainide caused a slight depression of isometric contractions of isolated cat papillary muscle. In dogs a single 10 mg/kg oral dose of lorcainide reduced left ventricular and aortic pressure for 2 h, and caused a 27 % drop in the maximum left ventricular stroke work.

Meinertz et al. (1980) found that in patients with ejection fractions (EFs) of 13 %-30 % lorcainide 2 mg/kg infused in 5 min reduced EF by an average of 5 %, and the time for circumferential cardiac fiber shortening was prolonged but caused no significant changes in cardiac output or aortic pressure. Shita et al. (1981) found 150 mg lorcainide was well tolerated in patients following an acute MI. Although direct comparisons between lorcainide and other drugs were not made. Vanhalaweyk et al. (1984) concluded that encainide, flecainide, lorcainide, and tocainide have similar hemodynamic effects and the "data presented indicates that on the basis of hemodynamic action no one drug can be preferred above the other." The development of congestive heart failure on lorcainide is very rare and it appears clinically to be less negatively inotropic than flecainide.

X. Clinical Recommendations

Lorcainide decreases VPC frequency in patients with nonsustained and sustained VT, but is considerably less effective in patients with a prior cardiac arrest. The drug has a low incidence of proarrhythmia and causes minimal left ventricular depression. The main disadvantage is a frequent sleep-wake disturbance, often severe enough for withdarwal from therapy. With appropriate counseling and sedation, lorcainide may be tolerated well enough to be useful in selected patients. Lorcainide should be started at 100 mg twice daily. Failure to obtain a targeted reduction in arrhythmia frequency, or to achieve a blood level in the range of 200-800 µg/ml, may justify increasing the dosage after 5 to 7 days to 200 mg twice daily. Occasional patients require higher doses, but drug levels should be monitored to avoid a drug level greater than 800 µg/ml. For more rapid drug effect 200 mg may be given intravenously, 10 mg every 5 min by intermittent bolus, which has proven safe and convenient in our laboratory.

XI. Lorcainide and Supraventricular Tachycardia

Somani and diGiorgi (1980) reported that lorcainide was effective in one patient with supraventricular and ventricular arrhythmias. Kasper et al. (1983) found, in 20 patients with the Wolff-Parkinson-White syndrome, that i.v. lorcainide, 2 mg/kg, decreased sinus cycle length in all patients from 708 ± 117 to 636 ± 94 ms. The A-V conduction time, however, lengthened from 84 ± 22 to 94 ± 22 ms, and the atrial ERP increased from 230 ± 22 to 243 ± 35 ms. Retrograde conduction over the accessory pathway was blocked in 5 of 18 patients, and slowed in the remaining 13 patients (from 107 ± 32 to 162 ± 57 ms). Anterograde conduction over the accessory pathway was blocked in six patients and slowed in the others. Circus movement tachycardia could be induced in 14 patients before and in 10 patients after lorcainide. In 15 patients atral fibrillation could be induced and after lorcainide anterograde conduction during AF was blocked in 5 patients. The shortest R-R interval over the accessory pathway during induced atrial fibrillation increased from 228±35 to 304 ± 103 ms.

D. Encainide

Dykstra et al. (1973) selected encainide, 4-methoxy-2′-[2(1-methyl-2-piperidyl)ethyl] benzanilide (Fig. 3) from several active compounds as being the most potent antiarrhythmic. It is effective in man in both ventricular and supraventricular arrhythmias.

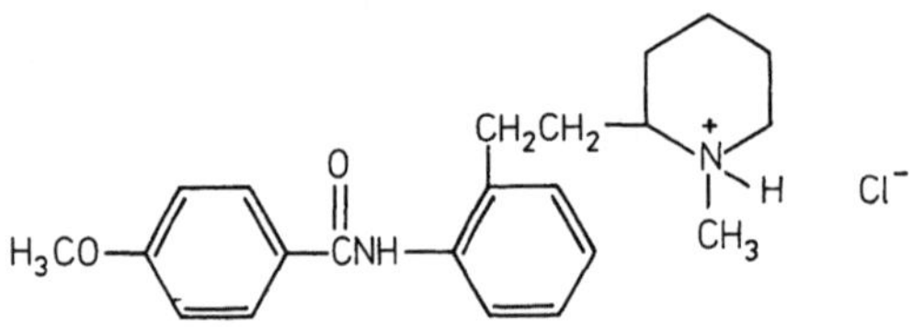

Fig. 3. Structure of encainide

I. Cellular Electrophysiology

Encainide decreased V_{max} (Gibson et al. 1978; Carmeliet 1980) at concentrations between 10^{-7} and 10^{-5} *M* and did so in a frequency dependent manner (Campbell and Vaughan Williams 1983) placing it in class Ic (Chap. 7). Encainide, as other Ic agents, has little effect on human action potential duration (Camm 1984). Clinical studies suggested that two metabolites, *O*-demethyl encainide (ODE) and 3-methoxy-*O*-demethylencainide (MODE), contribute to the antiarrhythmic effects of encainide. In Purkinje fibers the relative potency of ODE to MODE was 9 to 1, threshold concentrations of encainide, MODE, and ODE being 10^{-8}, 10^{-7}, and 10^{-8}, respectively (Elharrar and Zipes 1982). At a concentration of 10^{-6} *M* these compounds significantly decrease APD-90, V_{max} and propagation velocity. ODE significantly decreased action potential amplitude, while encainide and MODE did not.

In the isolated perfused guinea pig (Carmeliet 1980) and rabbit hearts (Dresel 1984) encainide (0.1-1.4 mg/liter) slowed conduction velocity more in the His Purkinje system and ventricle than in the atrium or AV node. Ventricular and AV nodal ERP_s at fixed basal cycle lengths were changed little, but were prolonged after extra systolic depolarizations.

II. Preclinical Studies

Acute administration of encainide intravenously at doses 0.9 or 2.7 mg/kg increased QRS duration by 15%-29% and HV by 31%-48% without changes in either AV nodal conduction or effective refractory period of the atria, AV node, or ventricle (SAMI et al. 1979 a). There was no significant change in HR, corrected sinus node recovery time, AH interval, or QT_c interval. Peak plasma concentrations averaged 450–4 000 μg/ml and were dose dependent. SAMUELSSON and HARRISON (1981) observed similar but larger effects. The changes, possibly due to the contribution of metabolites in the latter study. His bundle studies in anesthetized dogs given an infusion of ODE to attain steady state plasma levels comparable to those measured clinically during encainide therapy in man (149–230 μg/ml) support the interpretation that the difference between the electrophysiologic actions of intravenous and chronic oral encainide are due to the effects of the ODE metabolite (DUFF et al. 1983), which induced increases in PR (29 %), QRS (42 %), atrial ERP (62 %), ventricular ERP (13 %), AH (23 %), and HV (62 %). Thus, encainide when it is metabolized to ODE may have a profile more like quinidine (cf. the increases in ERP) than other class Ic agents.

BYRNE et al. (1977) studied encainide in dogs, cats, and monkeys with atrial fibrillation induced by aconitine or by electrial stimulation. Ventricular tachycardia was induced by ouabain or digoxin, or by ligature of a coronary artery. Encainide successfully converted the atrial fibrillation, and the application·ventricular tachycardias were slowed. Ventricular ectopic rate in conscious dogs 18–20 h after two-stage ligation of a coronary artery was also markedly reduced by encainide (2 mg/kg). In the perfused rabbit heart encainide (0.18, 0.35, and 0.71 mg/litre) increased the VF threshold by 50 %, 145 %, and 225 % respectively (ALMOTREFI and BAKER 1981). Single intravenous doses of 0.5, 1, 2, or 5 mg/kg encainide slowed overall heart rate and decreased ventricular activity by 23 %, 29 %, 38 %, and 86 % (BYRNE and GOMOLL 1982). Antiarrhythmic effects were seen within 2–4 min of injection and lasted up to 4 h. Injection of 0.25, 0.5, or 1 mg/kg ODE provoked a maximum decrease of 22 %, 47 %, and 72 % in the incidence of ectopic acitivity (GOMOLL et al. 1981). MODE had the least efficacy and the shortest duration of action. Doses of 0.18–0.76 mg /kg MODE caused a 25 %–55 % decrease in VPC frequency. At a dose of 10 mg/kg encainide, three of three dogs died; four of four given 2 mg/kg ODE and one of six given 1.52 mg/kg MODE died within 1–8 min following i.v. injection.

III. Clinical Electrophysiology

SAMI et al. (1979) observed that in five patients given 0.6 mg/kg and five given 0.9 mg/kg encainide i.v. over 15 min, the average peak plasma concentration was 0.49 ± 0.35 μg/ml. Encainide significantly prolonged HV and QRS intervals by 31 % ± 7 % and 18 % ± 9 % respectively. QT interval was hardly lengthened (2 % ± 9 %) and no significant changes were noted in HR, BP, and AH interval, corrected sinus node recovery time, Wenkebach cycle length, or ERPs of the atrium, AV node, or RV. In 15 patients (JACKMAN et al. 1982) EPSs were performed before and after 3 or more days of oral encainide therapy 100–300 mg/day. Encainide significantly lengthened the AH interval (7.45 ± 21.5 to 105.5 ± 39.1 ms), the shortest atrial pacing cycle length maintaining 1:1 AV nodal conduction (339 ± 71 to 417 ± 89 ms), HV interval

(48 ± 8 to 67 ± 13 ms), QRS interval 104 ± 31 to 132 ± 36 ms), right atrial ERP (234 ± 27 to 283 ± 39 ms), right ventricular ERP (236 ± 16 to 267 ± 37 ms) and QT interval (364 ± 38 to 417 ± 55 ms), but HR was unchanged. In four patients who had accessory AV pathways, encainide abolished anterograde conduction over the pathway in two patients, and increased the retrograde ERP and conduction time in all four. These effects show that in man, as in animals, metabolite effects on atrium, AV node, ventricular ERP and APO are similar to those of quinidine. LIBERSA et al. (1984) performed serial electrophysiologic studies over 1 h of intravenous administration. As the concentration of ODE increased the prolongation of conduction and refractoriness in the AV node became more pronounced. SCHWARTZ et al. (1984) studied the effects of encainide in six patients with bundle branch block and concluded that patients with preexisting infranodal conduction disease were more susceptible to depression of His Purkinje conduction.

IV. Pharmacokinetics

Encainide's pharmacokinetics are influenced by the formation of metabolites responsible for the effects of the drug in 90% of patients. Since ODE is nine times more potent than encainide, it is mainly responsible for the overall drug action (MASON and PETERS 1981; WINKLE et al. 1981b). In a study in 11 patients RODEN et al. (1980) noted that the pharmacokinetic effects of encainide persisted long after encainide itself was undetectable in plasma. They concluded that "the effects of encainide therapy were mediated by the variable generation of active metabolites." All ten of the responding patients had measurable quantitites of ODE, while the nonresponder failed to produce ODE in plasma. Patients unable to metabolize encainide to ODE were also poor metabolizers for debrisoquine (WOOSLEY et al. 1986). Two hundred other drugs also utilize this pathway for their metabolism and the specific cytochrome P450 isoenzyme that controls this oxidation has recently been characterized (GUENGERICH et al. 1986; DISTLERATH et al. 1985).

WINKLE et al. (1983) studied 12 patients with encainide-responsive frequent complex ventricular ectopic activity for a period of 1 year. Encainide reduced VPC frequency by 97%-99% with nearly total suppression of pairs and salvos. Administration b.i.d. or q.i.d. caused metabolite concentration to exceed that of encainide severalfold. The median time to arrhythmia return after drug withdrawal was 12-14 h. At the time of arrhythmia return, encainide was no longer detectable but the average concentration of ODE and MODE was 72 ± 49 and 172 ± 74 μg/ml respectively. The metabolites were eliminated, with an apparent elimination half-life of 11.4 ± 9.6 h for ODE and greater than 24 h for 3-MODE. RODEN et al. (1986) estimated the minimally effective plasma concentration to be 300 μg/ml for encainide, 35 μg/ml for ODE, 100 μg/ml for MODE. KATES et al. (1982) found in 13 patients on encainide for 6 months steady state concentration of encainide to be 56.3 μg/ml, 215 μg/ml for ODE, and 185 μg/ml for MODE after encainide doses ranging from 100 to 250 mg/day. Those patients found subsequently unable to metabolize encainide to ODE were also poor metabolizers for debrisoquine (WOOSLEY et al. 1986).

MCALLISTER et al. (1986) have noted that "the low plasma concentration of ODE and MODE in poor metabolizers would be expected to result in inefficacious therapy, but in such individuals, chronic oral therapy results in accumulation of unmetabolized encainide in far higher levels" adequate for an antiarrhythmic effect. These observations led ANTONACCIO and VERJEE (1986) to recommend dosing of encainide to start at 25 mg three times daily for 4-7 days and then to be titrated upward to 35 and 50 mg three times a day every 4-7 days.

BERGSTRAND et al. (1986 b) evaluated seven extensive metabolites of encainide in patients with renal failure and compared them with eight healthy normal subjects. After a single dose of encainide, the systemic and oral clearance was significantly lower, and the elimination $t_{1/2}$ longer, in renal failure. There was a significant reduction in the steady state volume of distribution in renal failure. After chronic dosing to steady state quantitatively similar changes were seen with 80% higher levels of ODE and 167% higher levels of MODE. The authors concluded that "patients with renal failure require lower doses of encainide because of both a reduced encainide clearance and increased accumulation of active metabolites." Six patients who were extensive metabolizers of encainide who also had cirrhosis were compared with eight normal subjects (BERGSTRAND et al. 1986 a). Patients with cirrhosis had lower systemic and oral clearance of encainide resulting in a threefold increase in oral bioavailability. Encainide concentration was significantly higher, but the concentrations of ODE and MODE were normal. It was concluded that "encainide in cirrhosis causes a three – to fourfold increase in parent drug concentration but because no change occurs in the levels of the metabolites, dosage adjustments are probably not required."

V. Clinical Efficacy

KESTELOOT and STROOBANDT (1979) found encainide effectively to suppress VPCs in 31 out of 33 cases treated. RODEN et al. (1980) found "total suppression of arrhythmias" in ten patients, which was "subsequently verified in a placebo-controlled crossover study." WINKLE et al. (1981 b), in nine patients with frequent and complex premature ventricular complexes, reported wide intersubject variation in bioavailability (mean 42% ± 24%, range 7.4%-82%), clearance (13.2 ± 5.6 ml/min/kg), and half-life (3.4 ± 1.7 h i.v. and 2.5 ± 0.8 h oral), but eight of the nine had more than 90% suppression of premature ventricular complexes. SAMI et al. 1981 found that in 20 patients with VPCs "the reduction in the average number of PVCs/hour on ambulatory ECG monitoring and on treadmill exercise test was greater with encainide than with quinidine ($P < 0.01$)." In a multicenter (eight-site) single-blind, placebo-controlled study in which 25-75 mg encainide was administered four times a day, described by MORGANROTH (1986), 80% of the patients demonstrated at least a 75% reduction in PVC frequency, with a 96% median decrease in VPCs. CNS adverse effects (asthenia, blurred vision, dizziness, and headache) were common (up to 25%) at the 200-mg daily dose. In a single-blind placebo controlled trial in three centers using lower doses of encainide (10, 20, and then 30 mg four times a day), there were 45, 65, and 84% median decreases in VPC frequency respectively, but only the results in the 20 and 30 mg groups were statistically significantly different from placebo. In a double-blind randomized, parallel, placebo-controlled trial in 125 patients (MORGANROTH et al. 1986 a), patients were randomized to 10 mg, 25 mg, or 50 mg *three* times a day, and VPC frequency on encainide compared with placebo was significantly decreased in the 25- and 50-mg groups. A multicenter comparative trial (double blind placebo controlled crossover trial) comparing encainide 25 and 50 mg q.i.d. with quinidine 200 and 400 mg q.i.d. was undertaken in 187 patients (MORGANROTH et al. 1986 c). Patients reporting side effects on the first drug could be prematurely advanced into the second placebo period. Encainide was more effective in terms of both percentage responders with VPC decrease and median VPC percentage decrease when comparing 25 mg encainide with 200 mg quinidine. However, the percentage of patients responding with a 75% VPC decrease was the same for the two agents. There was no difference in the incidence of proarrhythmia in terms of a VPC increase or incidence of VT events. Encainide was tolerated better at both doses than quinidine.

VI. Programmed Stimulation Studies

HOROWITZ (1986) observed that VT was not inducible after i.v. encainide in 13 of 62 (21%) patients, 80% with coronary artery disease, in whom several drugs (4.5 ± 1.4) had previously been found ineffective. In another 57 patients, 86% with coronary artery disease, in whom two of nine drugs had previously failed to control VT, VT was not inducible in 17 (30%) after oral encainide. In eight patients in whom i.v. encainide prevented initiation of VT, oral encainide prevented VT induction in 7 (88%). In six patients in whom VT was initiated after i.v. encainide administration, VT was also initiated during oral encainide treatment. Thus, the results with i.v. and oral encainide were concordant in 13 of 14 patients evaluated (93%) by PES.

Mason and Peters (1981) found in 38 patients with recurrent “strikingly drug refractory“ VT, that encainide 150–250 mg divided into four or six doses eliminated the VT (Holter) in 54% for 6 months of therapy, and in 29% for 18–30 months of therapy. Twelve patients (32%) had side effects that could be due to encainide and in four the arrhythmias may have been worsened by encainide. DUFF et al. (1982), BREN et al. (1981), and WINKLE et al. (1981 a) have noted worsening of ventricular arrhythmias, both spontaneous and induced at PES, by encainide, patients with malignant ventricular arrhythmias having the highest incidence of proarrhythmic events. DUFF et al. (1985) found that one of 11 patients with *sustained* VT repsonded to encainide (though 6 responded to other agents), but of 26 with *nonsustained* VT, 21 responded to encainide. Ventricular arrhythmias worsened in 4 of 8 patients with previous sustained VT, but in only 3 of 29 with *nonsustained* VT.

So far not enough patients have been studied to estimate the value of PES in selecting encainide as a suitable treatment for VT.

VII. Treatment of Supraventricular Tachycardia (SVT)

CASSAGENAU et al. (1985) found in ten patients with recurrent supraventricular tachycardia (SVT) that i.v. encainide interrupted the tachycardia in two. After 30 min, five patients could have the SVT reinitiated, but the cycle length was slowed from 326 ± 21 to 397 ± 51 ms. In four the SVT could still be initiated on chronic therapy and three still experienced "short events" of well-tolerated SVT. The electrophysiologic effects of encainide after acute i.v. (1 mg/kg over 6 min) and oral administration (75-150 mg/day for 48-72 h) were evaluated in ten patients with PSVT (Chimienti et al. 1985). The mechanism of the PSVT was related to a reentry through an accessory AV pathway in six cases while in the other four the reentry was confined to the AV node. PA, AH, and HV intervals lengthened from 43 ± 5 to 50 ± 14, 78 ± 20 to 92 ± 23, and 38 ± 7 to 49 ± 13 ms after intravenous encainide and to 48 ± 7, 94 ± 34 and 44 ± 9 ms after oral encainide, respectively. Atrial and ventricular ERPs were only slightly increased, but the Wenckebach point lengthened from 316 ± 28 to 354 ± 32 ms after i.v. to 359 ± 45 ms after oral, therapy. The tachycardia cycle length was 403 ± 48 increasing to 433 ± 85 ms in three patients; the other seven were no longer provokable. After 18 ± 8 months the incidence of tachycardia was still markedly reduced. Of 15 children with refractory SVT (STRASURGER et al. 1986), 10 had "incessant" tachycardia (>10% of the day) in 5 of whom SVT was controlled by encainide alone, and in 4 partially controlled by oral encainide in combination with other drugs. In five children with accessory pathway tachycardia, encainide was effective in three. BRUGADU et al. (1984) defined incessant SVT as the daily presence of the tachycardia for more than 50% of the day. In eight patients with incessant SVT with retrograde conduction over an accessory pathway, and in three with reccurrent atrial tachycardia, i.v. encainide terminated the SVT in seven of nine patients, and oral encainide (100-325 mg/day) completely suppressed the incessant SVT in eight patients over 5-20 months.

In 30 patients with AF and WPW, i.v. and oral encainide blocked anterograde accessory pathway conduction in approximately half of the patients and slowed the ventricular response in the majority of the remaining half (Naccarelli et al. 1984).

PRYSTOWSKY et al. (1984) treated 19 patients with accessory pathway with a mean daily dose of encainide of 197 mg, which caused complete block in the accessory pathway in 8 of 14 patients when conduction was anterograde, in 7 of 14 with retrograde conduction. In 19 patients in whom AV reentrant tachycardia was initiated at control, encainide prevented the tachycardia in 10 and the cycle length of the tachycardia increased from 314 ± 53 to 418 ± 81 ms in the other 9. Fifteen patients continued encainide treatment for a mean of 18 months and all but one remained asymptomatic. KUNZE et al. (1984) found in 8 patients with WPW syndrome and 4 with a accessory concealed pathway that encainide, given 1-1.5 mg/kg i.v. or for 4 weeks orally at 75-200 mg/day, prevented induction of the tachycardia (i.v. 4 of 11 and oral 2 of 7) or significantly prolonged the tachycardia cycle length (7 of 1 i.v. and 5 of 7 oral). Both i.v. and oral encainide were also evaluated in 13 patients with accessory AV pathways (7 overt, 1 intermittent, and 5 concealed). Intravenous encainide 1.5 mg/kg blocked the accessory AV pathway in four of seven patients, while oral encainide was effective in three of four patients (ABDOLLAH et al. 1984). In this study, i.v. and oral encainide had minimal effects on retrograde conduction over the accessory pathway. Five patients (out of 12) responded to oral encainide therapy over 10.5 months.

Encainide is successful in some cases of SVT, but ineffective in others, perhaps due to different concentrations of metabolites especially ODE, known to have a greater effect on the AV node than MODE or encainide.

VIII. Arrhythmogenicity

CHESNIE et al. (1983) reported that high encainide dosages and high plasma concentrations of ODE (307 µg/ml) were associated with drug toxicity in a heterogenous group of patients. DAWSON et al. (1984) evaluated the effects of ODE in induced and spontaneous arrhythmias in 25 conscious dogs with a "mottled" MI provoked by transient coronary occlusion. Plasma levels of the metabolite above 100 µg/ml suppressed (92 %) spontaneous ventricular ectopic activity 48 h after MI. In 15 dogs treated with ODE, the VF threshold decreased an average of 23 % from baseline level (23 ± 8 mA to 18 ± 9 mA), the effect being concentration dependent at plasma concentrations of ODE above 150 µg/ml. In 7 of the 17 dogs treated with ODE, VF could not be terminated by the countershock protocol (previously always effective). DAWSON et al. (1984) concluded that at plasma levels higher than 150 µg/ml ODE "lowers the VF threshold and at higher levels impedes defibrillation."

RINKENBERGER et al. (1982) reported on 7 patients in whom 3 to 32 repetitive ventricular responses were provoked on control PES; after encainide none were provoked. NICHOLSON et al. (1982) have noted that after encainide sustained VT may accelerate in rate with worsening of symptoms. The difficulty of terminating the VT has also been reported, along with patient deaths (WINKLE et al. 1981a). There is no correlation between encainide dose or plasma concentration and the proarrhythmic effect, which may be precipitated by the recommended oral dose of encainide, 25–50 mg three times a day. DUFF et al. (1985) stated that the proarrhythmic response to encainide "tended to be associated with higher encainide concentrations and greater prolongation of the QRS interval."

Anecdotal reports have appeared regarding the utility of $NaHCO_3$ in terminating arrhythmias thought due to encainide. BAJAJ et al. (1986) found that conduction slowing by ODE can be acutely reversed with either NaCl or $NaHCO_3$ with prolongation of repolarization. PENTEL et al. (1986) report a case of massive encainide toxicity in which "sodium bicarbonate was followed by prompt improvement in both cardiac rhythm and blood pressure." When the blood pH was normal, administration of hypertonic sodium bicarbonate was followed by modest but prompt reduction in QRS duration.

IX. Hemodynamic Effects and Adverse Side Effects

BRUTSAERT (1981) found that encainide (0.1–10 μg/ml) did not alter the contraction or relaxation of cat papillary muscle. In the rabbit heart, negative inotropic effects were noted with encainide at high doses (2×10^{-5} n*M*) (GOMOLL et al. 1986). GOMOLL and MAYOL (1982) reported that in open-chested chloralose-anesthetized dogs cardiac output was increased (3%–6%) by encainide, but decreased (4%–14%) by both ODE and MODE. Calculated total peripheral resistance was decreased (7%–10%) by encainide but significantly increased (5%–16%) by ODE. SAMI et al. (1983) using gated cardiac imaging in 19 patients, 63% of whom had a history of prior CHF, found that oral encainide 75–200 mg, did not significantly affect ejection fraction, averaging 22% ± 10% before and 25% ± 14% after encainide. Digoxin blood levels in ten patients averaged 1.04 ± 0,43 μg/ml and 1.22 ± 0.47 μg/ml during encainide therapy. Following routine diagnostic catheterization TUCKER et al. (1982) administered 0.9 mg/kg encainide over 15 min. Systemic vascular resistance and cardiac index were unchanged at the end of the infusion, but increased by 9% and decreased by 8% respectively at 30 min. LVEDP *decreased* by 15% while LV *dp/dt* remained unchanged. In the subgroup with a lowered cardiac output (CI < 2.4 litres/min/m^2), cardiac performance was significantly lower after encainide. DIBIANCO et al. (1982) found in 21 patients, with 16 undergoing radionuclide angiographic evaluation on placebo or encainide, that oral encainide did not reduce exercise capacity or left ventricular ejection fraction at rest or during supine exercise. In the Cardiac Arrhythmia Pilot Study (1986) incidence of congestive heart failure (CHF) was lower on encainide than on flecainide. Since, however, the *placebo* group had the same high incidence as the flecainide group, the results may not be reliable since the groups were not randomized by EF, permitting a possible sampling error.

X. Adverse Profile

SOYLA reviewed 1245 patients receiving oral encainide and reported an overall incidence of proarrhythmia of 9.2% (115 of the 1254 patients) rising to 16% in patients with a history of cardiomyopathy. There was one case of new and three cases of worsened CHF. Other adverse reactions included dizziness (26%), abnormal or blurred vision (19%), headache (15%), taste problems (4%), and tremors (3%).

Encainide does not change serum digoxin concentration. Plasma concentrations of encainide, ODE, and MODE significantly increased following cimetidine administration by 32%, 43% and 36%, respectively (QUART et al. 1986). SALERNO et al. (1986) reported that encainide induced hyperglycemia. Mean control serum glucose was 128 ± 50 mg/dl rising after 1 month of encainide to 168 ± 139 mg/dl. Furthermore on encainide five episodes of hyperglycemia needing therapy developed in four patients. In patients with previous diabetes, encainide may produce hyperglycemia requiring therapy but the effect disappears upon termination of encainide therapy.

XI. Clinical Use Recommendations

Encainide appears to have little negative inotropic properties but is significant ly proarrhythmic, especially in patients with severe arrhythmias and low ejection fractions. It is far safer in patients with PVCs and nonsustained VT. Therapy should be started at 25 mg every 8 h and increased only at 5- to 7-day intervals. It cannot, without further evaluation be routinely utilized in patients with sustained VT, prior cardiac arrest, and nonsustained VT and low EF. In patients with moderate to severe renal impairment, encainide should be startet at one-third the normal dose, i.e., 25 mg once a day. No adjustment in patients with hepatic disease is required.

E. Propafenone

Propafenone, (2′-3 propylamine)-2-(hydroxy)-propoxy-3 Propiophenone HCl, was originally developed as a beta-blocker (PHILIPSBORN et al. 1984; (Fig. 4); MCLEOD et al. 1984).

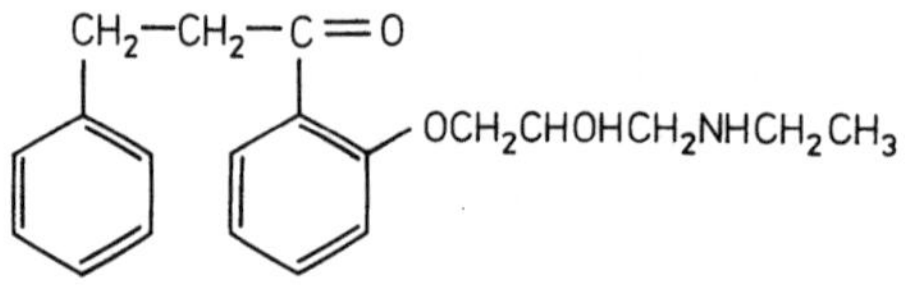

Fig. 4. Structure of propafenone

I. Cellular Electrophysiology

DUKES and VAUGHAN WILLIAMS (1984) evaluated the electrophysiologic effects of propafenone in the rabbit over a wide range of concentrations. Propafenone caused a dose-dependent decrease in the maximum rate of depolarization and conduction velocity in tissue depolarized by the fast sodium channel, with characteristics classifying it as a Ic agent. The effect on the beta receptor may be clinically significant but its potency as a calcium antagonist was determined to be relatively weak. Action potential duration and effective refractory period were lengthened in both atrium and ventricle and these effects were long lasting and persisted beyond the "wash out" period of the drug when other measurements had returned to baseline values. These observations distinguished propafenone from flecainide, lorcainide, and encainide and show similarities to the electrophysiologic effects of the encainide metabolite ODE.

In ischemic canine cardiac cells (ZEILER et al. 1984) propafenone reduced the amplitude and rate of rise of myocardial and Purkinje fiber action potentials more than in normal cells. SHEN et al. (1984) showed that propafenone i.v. prolonged the AH and HV intervals, and QRS duration in man, but not the sinus node recovery time. COUMEL et al. (1984), however, found that propafenone caused sinus node slowing and might cause SA block; it also prolonged PR interval and the corrected sinus node recovery time. PHILIPSBORN et al.

(1984) found in the isolated guinea pig atrium and rat aortic strips, that the metabolite 5-hydroxypropafenone has "a smaller effect on the maximum following frequency, a greater negative inotropic effect, a greater Ca^{2+} antagonist effect, and a very distinctly weaker beta-adrenoreceptor blocking effect than propafenone itself". In contrast to those in vitro observations-, 5-hydroxypropafenone showed a *stronger* antiarrhythmic potency in the aconitine and infarcted arrhythmia model in the rat. In His bundle studies, 5-hydroxypropafenone caused a "more marked prolongation of the conduction time in the atrium, AV node, and His Purkinje system," but the beta-blocking effect of the metabolite was smaller than that of propafenone. The authors suggested that the difference between "the in vivo and in vitro potency of 5-hydroxypropafenone may be explained by differences in pharmacokinetics, e.g., by a smaller distribution volume."

II. Preclinical Studies

Propafenone is active against arrhythmias induced by calcium chloride or left coronary occlusion (Hapke and Prigge 1976). Karagueuzian et al. (1982) found that propafenone 4 mg/kg i.v. over 2 min immediately and completely suppressed VT in seven conscious dogs with previously ligated left coronary arteries. Lidocaine was much less successful and was ineffective if the VT was greater than 160 beats/min. In anesthetized closed chest dogs (Karagueuzian et al. 1984) propafenone (4 mg/kg) increased heart rate in association with lowered aortic and pulmonary systolic pressure but had no effect on pulmonary and aortic diastolic pressure. Cardiac output decreased from 4.5 ± 1 to 3.8 ± 0.7 liters/min. Lidocaine also decreased aortic systolic pressure and increased heart rate. Propafenone, unlike lidocaine, significantly increased A-V nodal functional ERP, and slowed intraventricular conduction. Neither drug altered sinus node recovery time or ERP of the right ventricle.

III. Pharmacokinetics

Ninety-five percent of propafenone is absorbed after an oral dose (Hollmann et al. 1983a) with maximum plasma concentration achieved approximately 2–3 h after administration (Hollmann 1983b). Biovailability of the 300-mg film-coated tablets is approximately 12%, reflecting extensive presystemic clearance. A nonlinear increase in maximum plasma propafenone concentration after oral doses of the drug has been observed in normal volunteers with a six fold increase in plasma concentration after a three fold increase in dose from 150 to 450 mg (Connolly et al. 1983). Salerno et al. (1984) noted that with three fold increase in dosage from 300 to 900 mg/day there was a tenfold increase in mean plasma concentration. The elimination half-life was first reported to be between 3.6 and 4.6 h, but Connolly et al. (1984) found that range to be wider 2.4–11.8 h and Siddowy et al. (1983) observed a range of 1.8–32.2 h, mean 7.7 h. The drug is extensively metabolized by oxidative metabolism to yield 5-hydroxy and hydroxymethoxy metaboiltes. The 5-hydroxy derivative is active, and occurs in significant quantities (Philipsborn et al.

1984). Oxidation is via the debrisoquine pathway to some extent. KATES et al. (1985) have reported that the 5-hydroxy metabolite may also not decay linearly and that these metabolites may accumulate during chronic oral therapy. The concentrations of 5-hydroxypropafenone and *n*-depropylpropafenone did not decay in a log-linear manner during the sampling period and thus the determination of the $t_{1/2}$ was not possible. At 24 h after propafenone dosing, concentrations of 5-OHP and NDPP were 63% ± 37% and 50% ± 21% of their mean steady state levels.

SALERNO et al. (1984) reported minimum effective trough plasma concentrations from 91 to 3271 μg/ml, and CONNOLLY et al. (1984) reported effective concentrations between 64 and 1044 μg/ml.

IV. Clinical Efficacy

Propafenone was evaluated in 30 patients with frequent VPCs with 14 repsonding to therapy acutely (FURLANELLO et al. 1983), and was also reported to be effective in VPC suppresion by BAEDEKER et al. (1977). SOYZA et al. (1984), in 30 patients with "clinically significant" ventricular ectopy (30 VPCs/h), found that 20 responded with a greater than 85% reduction. Similar results were reported by RUTSCH (1978) and BECK and HOCHREIN (1978). SALERNO et al. (1984), using a single-blind, dose-ranging study, followed by a double-blind comparison with placebo, found that 8 of 12 patients achieved ≧80% suppression of total VPCs (mean 83%), paired VPCs suppression ≧90%, and VT was eliminated in 11. The effectiveness of propafenone was confirmed during the double-blind trial ($P < 0.05$ vs. placebo) and during treatment for 6 months ($P < 0.05$ *vs.* single-blind placebo). Propafenone prolonged the PR interval by 16%, and the QRS by 18%. Propafenone also increased serum digoxin concentration in five of five patients on combined therapy. Side effects included worsening CHF in one patient, and two patients developed conduction abnormlities. REHNQVIST et al. (1984) compared propafenone with i.v. lidocaine in patients post-MI, with a 75% (propafenone) and 73% (lidocaine) reduction in PVC frequency.

V. Programmed Electrical Stimulation Studies

In 16 patients with VT or nonfatal cardiac arrest (CONNOLLY et al. 1983), VT was rendered not inducible by propafenone in only one patient, was nonsustained in one, and was harder to induce in two patients. Propafenone increased the cycle length of the tachycardia from 307 ± 67 to 382 ± 107 ms. A proarrhythmic effect was noted in four patients. At a mean plasma concentration of 753 ± 428 μg/ml. CHILSON et al. (1985) observed that "Oral propafenone was given to a maximal dose of 300 μg every 8 hours. Ten of 25 patients developed side effects or had inadequate suppression of spontaneous arrhythmias." EPS studies were performed on the 15 patients who had an adequate response. Three patients were protected (no VT induced) but 12 patients were still inducible. Propafenone increased PR interval from 168 ± 46 to 188 ± 25 ms, the HV interval from 47 ± 10 to 65 ± 13 ms, the shortest atrial pacing cycle length from 385 ± 44 to 436 ± 42 ms, the ventricular ERP from 231 ± 17 to 255 ± 19 ms, and the ventricular functional ERP from 260 ± 15 to 278 ± 17 ms. Ten patients were "considered to have a satisfactory electrophysiologic response" and placed on propafenone orally, 8 doing well over 11 months. SHEN et al. (1984) evaluated 28 patients with recurrent VT. Propafenone 2 mg/kg was given i.v. and then followed by an i.v. drip 1 mg/min in the first 14, and then 2 mg/min in the second 14 patients. None in the first group were protected for VT induction, and only 3 out of 14 no longer had VT provokable in the second group. Intravenous propafenone was evaluated in an additional 14 patients with only one patient rendered noninducible (DOHERTY et al. 1984).

VI. Arrhythmogenicity and Side Effects

Common adverse effects included visual blurring, paresthesias, dizziness, and headache. Gastrointestinal symptoms, constipation, and nausea are also observed. Propafenone has significant proarrhythmic action (WINKLE et al. 1981; MUHIDDIN et al. 1982; NATHAN et al. 1984). One-third of patients undergoing electrophysiologic testing by WINKLE et al. (1983a) showed an aggravation of arrhythmias. NATHAN et al. (1984) reported one case of fatal VT due to propafenone. STAVENS et al (1985) administered propafenone to 16 patients with VT and 3 developed incessant VT, but these subsided after lidocaine. DOHERTY et al. 1984 found 2 of 14 patients developed spontaneous VT on propafenone and 5 patients with inducible VT were rendered more difficult to terminate.

VII. Hemodynamic Effects

SALERNO et al. (1983) noted a decrease in left ventricular perfusion as evaluated by 2-D echocardiography in 9 of 12 patients on propafenone. SHEN et al. (1984) found i.v. propafenone significantly increased intracardiac pressure (right atrial, pulmonary arterial, and pulmonary capillary wedge pressures), increased vascular resistance (pulmonary and systemic), and decreased cardiac index in 28 patients. PODRID and LOWN (1984) noted that propafenone caused a significant decrease in ejection fraction as measured by M-mode echocardiography in patients with preexisting left ventricular dysfunction. Patients without abnormalities in left ventricular function had no detectable change in LV function on propafenone. Two patients developed clinically symptomatic CHF and one required discontinuation of propafenone. BAKER et al. (1984) employing radionuclide angiography studied 22 patients with different degrees of left ventricular function. In the group with good LV function, EF was 52% ± 9% before drug and 48% ± 11% after propafenone. In eight patients with LV dysfunction the EF went from 42% ± 9% to 34% ± 6% ($P < 0.05$). Thus, a significant degree of LV functional impairment was noted in those patients with preexisting LV dysfunction. BRODSKY and ALLEN et al. (1985) reported that in ten patients on maintenance therapy with propafenone, while EF did not change, three had to stop the drug because of dyspnea.

VIII. Supraventricular Tachycardia

Thirty-two patients were treated with a mean dose of 900 mg/day propafenone for 6.6 months on average for a number of SVT conditions (COUMEL et al. 1984). Atrial flutter and fibrillation were less sensitive to propafenone than to quinidine or amiodarone, but eight cases of atrial tachycardia were more senstive to propafenone than to amiodarone or beta-blockers. Twelve cases of SVT showed an "intermediate response" and three patients with resistant junctional tachycardia were improved with propafenone. Electrophysiologic studies were perfomed in ten patients with AV nodal reentrant paroxysmal SVT before and after i.v. administration of propafenone (1.5 mg/kg) (GARCIA-CIVERA et al. 1984). All patients utilized an AV nodal slow pathway for anterograde conduction and an AV nodal fast pathway for retrograde conduction. Propafenone depressed retrograde fast pathway conduction. Off therapy nine of ten patients had induction of the tachycardias but following propafenone seven of the nine could no longer be provoked into VT. MANZ et al. (1985) evaluated 14 patients with WPW and 10 patients with AV nodal reentrant tachycardia. Both i.v. and oral propafenone increased the ERP of the right atrium and AV node, but the oral preparation was more effective than the i.v. in blocking the anterograde accessory pathway. The mean cycle length of SVT increased from 338 ± 60 ms to 387 ± 56 ms after i.v. propafenone and from 336 ± 65 to 367 ± 65 ms after oral therapy. The induction of SVT was prevented by i.v. drug in 10 of 20 patients and in 4 additional patients with oral propafenone. On oral therapy, six of seven patients whose SVT could not be initiated by EP testing remained free from recurrence, whereas five of seven patients with inducible SVT had a recurrence. FURLANELLO et al. (1984) found 8 out of 25 patients who had WPW syndrome, and 5 out of 8 with AV reentry tachycardia were protected by propafenone. Acute EPS predicted the clinical outcome in 100% of patients with WPW and 60% of patients with AV reentrant tachycardia.

BREITHARDT et al. (1984) carried out EPS initially in 23 patients (group 1, mean age, 38 years), in 19 of whom, and in 24 additional patients (group 2, mean age, 42 years), long-term therapy with oral propafenone was started. Both groups had a history of tachycardia for 12 years (mean); 14 patients had previously had attacks of syncope. During EPS in group 1, propafenone did not change spontaneous sinus rate, corrected sinus node recovery time, or the AH interval, but HV time, QRS duration, and ERP of atria and ventricles were significantly prolonged. The ERP of the accessory pathway increased from 238 to 322 ms. The maximum 1:1 conduction frequency of the accessory pathway decreased from 231 to 176 beats/min, with complete anterograde block in six patients. The shortest R-R interval during AF increased from 232 to 303 ms. The retrograde refractory period of the accessory pathway was prolonged from 245 to 295 ms. Complete, or 2:1, retrograde block during basic drive occurred in three patients and one patient, respectively. In 6 of 15 patients, propafenone made sustained SVT either no longer inducible or nonsustained. The cycle length of induced SVT increased from 324 to 395 ms. During long-term administration (follow-up duration 2–3 years), 17 of 43 patients did not report any episode of symptomatic tachycardia, but in 18 patients, the frequency and severity of attacks was unchanged. One patient with dilated cardiomyopathy died suddenly. Side effects necessitating discontinuation of medication were observed in only two patients.

Experience with propafenone in pediatric patients with SVT has been favorable. Propafenone, 300 mg/m^2 per day, administered three or four times a day was effective in 85.7% of the 35 patients (21 drugs) (DRESSLER et al. 1985). In children, average age 9.8 ± 4.7 years, propafenone (1.5 mg/kg i.v.) was effective in controlling SVT associated with the preexcitation syndrome (MUSTO et al. 1986).

IX. Clinical Recommendations

Propafenone is effective in reducing PVC frequency and in the treatment of SVT especially those associated with the preexcitation syndromes. The drug's efficacy and safety in patients with sustained VT, or prior cardiac arrest, who are usually referred for EPS is not established. The minimal dose required for efficacy, optimum serum levels to be targeted, the role of metabolites, and the drug's efficacy in "slow" metabolizers of debrisoquinine are still undefined. The significance of the digoxin and cimetidine interactions needs clinical evaluation as well as the disposition of the drug and its active metabolites in hepatic and renal disease.

F. Indecainide

I. Electrophysiology

Indecainide is 9-3-1-methylethyl amino propyl-9*H*-fluorene-9-carboxamide (Fig. 5). Its electrophysiologic effects were evaluated in rabbit atria, Purkinje fibers and papillary muscle in comparison with aprindine, which it resembles chemically (DENNIS and VAUGHAN WILLIAMS 1986). Both drugs depressed the maximum rate of depolarization in a dose-related manner, indecainide being ten times more potent than aprindine. Indecainide shortened action potential duration at 50% repolarization in ventricular muscle. The effects of indecainide were long lasting but had no effect on the effective refractory period measured by interpolated premature stimuli, and from its kinetics it was categorized as class Ic. STEINBERG and WIEST (1984) found in Purkinje fibers indecainide 10^{-6} and 3×10^{-6} *M* decreased the maximal rate or rise of phase 0 (V_{max}), conduction velocity, action potential duration, and effective refractory

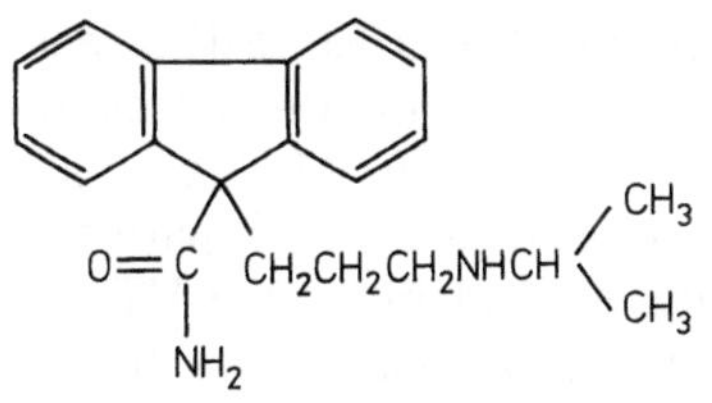

Fig. 5. Structure of indecainide

period and shifted the membrane response curve by 5 mV in a hyperpolarizing direction. In papillary muscle APD was unchanged, but V_{max} was decreased. "Recovery of V_{max} from maximum block was half completed in Purkinje and muscle fibers in 52 and 49 s respectively," thus placing indecainide in the same slow recovery group as flecainide and lorcainide (i.e., class Ic). Dennis and Vaughan Williams (1986) evaluated the action of indecainide on rabbit sinoatrial and atrioventricular node. Indecainide, at concentrations up to 2.9 mmol/liter in five animals, did not produce a sinus bradycardia while the drug slightly prolonged AV conduction time from 49 to 57 ms. These investigators concluded that "indecainide does not block channels carrying inward calcium current in nodal tissue."

II. Preclinical Studies

Studies in dogs have found indecainide to be at least 2.6 times more potent than aprindine and 6 times more potent than disopyramide in suppression of ouabain-induced arrhythmias (Holland and Lacefield 1983). The metabolism of indecainide has been studied in animals (Lindstrom et al. 1984) and man (Farid and Fasola 1984). The drug is metabolized by the liver and its metabolites are excreted in urine. Recently, a liquid chromotagraphic assay has been described (Farid et al. 1985). Sandusky and Meyers (1985) found in rats and dogs that indecainide was nontoxic on prolonged administration, but caused an increase in the PR, QRS, and QT intervals.

III. Hemodynamic Studies

The hemodynamic effects of indecainide have been compared with other drugs in patients with an acute myocardial infarction (Bernard et al. 1986). A significant decrease of more than 10% of the initial value was seen in systolic blood pressure after lorcainide, indecainide, and tocainide. Peripheral vascular resistance increased markedly, while pulmonary capillary pressure increased by more than 40% from baseline after mexiletine, indecainide, and tocainide.

IV. Clinical Studies

Horowitz et al. (1985) found that indecainide 60–90 mg/kg given i.v. at a rate of 12.5–15 µg/kg per minute had no effect on sinus node function or atrial and ventricular effective refractory period. The AH interval increased from 106 ± 13 to 130 ± 24 ms and the HV interval increased from 57 ± 7 to 73 ± 19 ms. The QRS duration increased from 102 ± 9 to 120 ± 13 ms, while no change in the JT interval was observed. Induction of VT at EP testing was prevented in only one of ten patients, but in the remaining nine the VT cycle length was significantly prolonged (248 ± 47 to 320 ± 71 ms). Miura et al. (1987) evaluated the antiarrhythmic efficacy of indecainide in 15 patients with a history of cardiac arrest or VT. The patients all had inducible VT off therapy and then were evaluated on procainamide and indecainide (1 mg/kg i.v.). Indecainide did not significantly change the baseline HR, BP, and QTc intervals from control. The PR and QRS intervals were significantly prolonged. Only 3 out of 15 patients were protected on indecainide; the VT rate was significantly slowed from 218 to 224 beats/min. One patient developed a proarrhythmic response on indecainide.

V. Final Thougths

The cornucopia of new drugs to treat arrhythmias reveals new possibilities and dangers. The class Ic grouping offers pharmacologic agents that markedly decrease automaticity and may modify conduction in the reentry pathway. The dispersion of repolarization is not affected by pure Ic agents like lorcainide and flecainide and this may have particular advantage in some patients. Although these new agents do possess their own side effects, they are mostly minor, resulting in a low discontinuation rate (especially with flecainide and encainide). The utility of blood level determinations has also been reemphasized, especially in relation to the toxicity of flecainide. Along with the increased potency of the class Ic agents is an associated risk of aggravation of arrhythmias in a small, but significant, segment of the population. This necessitates cautious administration, especially to patients with ventricular tachycardia, or cardiac arrest in association with poor ventricular function (EF < 30 %). In this important subset of patients, in hospital initiation of therapy is advised and "follow-up" Holter monitor and/or electrophysiology study on antiarrhythmic therapy is mandatory. Therapeutic drug level monitoring and observing the effects of these agents on QRS duration (avoiding greater than 50 % prolongation) may enhance the safety of the class Ic agents.

One can look at the arrhythmia problem as multifactorial. PVCs (trigger mechanism) and the vulnerability of the myocardium to sustain a life-threatening ventricular tachycardia (substrate) are two variables. Physicians treating patients at risk for sudden death should consider PVC frequency and vulnerability as interrelated but separate variables. Antiarrhythmic drug efficacy can be assessed in terms of a reduction in trigger mechanisms (PVCs), for which class Ic agents do so well. Drug efficacy can also be decreasing myocardial vulnerability (PES induction of VT); Ic agents are effective here as well, but are no panacea. Thus, agents are now available that may favorably affect both substrate and trigger and be tolerated during long-term therapy.

References

Abdollah H, Brugada P, Green M, Wehr M, Wellens H (1984) Clinical efficacy and electrophysiologic effects of intravenous and oral encainide in patients with accessory atrio-ventricular pathways and supraventricular arrhythmias. Am J Cardiol 54(6):544–549

Almotrefi AA, Baker JB (1981) Antifibrillatory efficacy of encainide, lorcainide, and ORG6001 compared with lignocaine in isolated hearts of rabbits and guinea pigs. Br J Pharmacol 73:373–377

Anderson J, Anastasiou-Nana M, Lutz J., Writer S (1985) Comparison of intravenous lorcainide with lidocaine for acute therapy of complex ventricular arrhythmias. Results of randomized study with cross-over option. J Am Coll Cardiol 5:333–341
Anderson J, Lutz J, Allison S (1983) Electrophysiologic and antiarrhythmic effects of oral flecainide in patients with inducible ventricular tachycardia. J Am Coll Cardiol 2:105–114
Anderson J, Stuart J, Perry B, Hamersveld D, Johnson T, Conard G, Chang S, Kvam D, Pitt B (1981) Oral flecainide acetate for the treatment of ventricular arrhythmias. N Engl J Med 305:473–477
Antonaccio M, Verjee S (1986) Dosing recommendations for encainide. Am J Cardiol 58:114–116c
Baedeker W, Wirtzfeld A, Sack D, Obersohl K (1977) The antiarrhythmic effect of propafenone in ventricular tachycardia. Herz/Kreisl 9:348–352
Baker BJ, Dinh H, Kroskey D, deSoyza N, Murphy M, Franciosa J (1984) Effect of propafenone on left ventricular ejection fraction. Am J Cardiol 54:20–22d.
Bannitt EH, Bronn WR, Coyne WE et al. (1977) Antiarrhythmics. 2. Synthesis and antiarrhythmic activity of *N*-(piperidylalkyl) trifluoroethoxybenzamides. J Med Chem 20:821–826
Bar F, Farre J, Ross D, Vanagt E, Gorgels A, Wellens H (1981) Electrophysiological effects of lorcainide, a new antiarrhythmic drug. Br Heart J 45:292–298
Bajaj A, Woosley R, Roden D (1986) Acute reversal of *O*-demethyl encainade induced conduction slowing. J Am Coll Cardiol 7 (Suppl 2):82a
Beck OA, Hochrein H (1978) Indications and risks of antiarrhythmic treatment with propafenone. Dtsch Med Wochenschr 103:1261–1265
Bergstrand R, Wang T, Roden D, Avant G, Sutton W, Siddoway L, Wolfenden H, Woosley R, Wilkinson G, Wood A (1986a) Encainide disposition of patients with chronic cirrhosis. Clin Pharmacol Ther 40:148–154
Bergstrand R, Wang T, Roden D, Stone W, Wolfenden H, Woosley R, Wilkinson G, Wood A (1986) Encainide disposition in patients with renal failure. Clin Pharmacol Ther 40:64–70
Bernard R, Renard M, Shita A, de-Hemptinne J, Gillet J, Lewinson H, Liebens I, Waterschoot P (1986) Hemodynamic repercussions and clinical tolerance of six Class I antiarrhythmic agents after acute myocardial infarction. Ann Cardiol Angeiol (Paris) 35(4):195–198
Bhamra RK, Flanagan RJ, Holt DW (1984) High performance liquid chromatographic method for the measurement of mexiletine and flecainide in blood plasma or serum. J Chromatogr 307:439–444.
Blevins R, Kerin N, Mathias P, Pesola D, Faitel K, Jarandilla R, Garfinkel C, Rubenfire M (1986) Intravenous lorcainide versus lidocaine in the treatment of frequent and complex ventricular arrhythmias. Am Heart J 111:447–451
Borchard U, Boisten M (1982) Effect of flecainide on action potentials in alternating current-induced arrhythmias in mammalian myocardium. J Cardiovasc Pharmacol 4:205–212
Borgeat A, Goy JJ, Maendly R, Kaufmann U, Grbic M, Sigwart U (1986) Flecainide versus quinidine for conversion of atrial fibrillation to sinus rhythm. Am J Cardiol 58(6):496–498
Breithardt G, Borggrefe M, Wiebringhaus E, Seipel LA (1984) Effect of propafenone in the Wolff-Parkinson-White syndrome: electrophysiologic findings and long term follow up. Am J Cardiol 54:29–39D
Brembilla-Perrot B, Admant P, Le-Helloco A, Pernot C (1985) Loss of efficacy of flecainide in the Wolff-Parkinson-White syndrome after isoproterenol administration. Eur Heart J 6(12):1074–1078

Bren G, Varghese PJ, Katzs R, Roth A (1981) Arrhythmogenicity of encainide, the role of QT interval. Am J Cardiol 47:498

Brodsky Ma, Allen BJ (1985) Propafenone. Chest 88(2):164-165

Brugada, P, Abdollah H, Hein JJ, Wellens H (1984) Suppression of incessant supraventricular tachycardia by intravenous and oral encainide. J Am Coll Cardiol 4(6):1255-1260

Brutsaert D (1981) Effect of encainide on myocardial contractility of cat papillary muscle. Eur J Pharamacol 76:267-269.

Byrne JE, Gomoll AW (1982) Antiarrhythmic action of encainide versus ventricular arrhythmias in the conscious dog following coronary artery ligation. Can J Physiol Pharmacol 60:369-375

Byrne J, Gomoll A, McKinney G (1977) Antiarrhythmic properties of MJ9067 in acute animal models. J Pharmacol Exp Ther 200(1):147-154

Camm A (1984) Cardiac electrophysiology of four new antiarrhythmic drugs: encainide, flecainide, lorcainide and tocainide. Eur Heart J 5(Suppl b):75-79

Camm A, Hellsterand K, Nathan A, Bexton R (1985) Clinical usefulness of flecainide acetate in the treatment of paroxysmal supraventricular arrhythmias. Drugs 29(Suppl 4):7-13

Campbell TJ (1983) Importance of physiochemical properties in determining the kinetics of the effects of Class I antiarrhythmic drugs on maximum rate of depolarization in guinea pig ventricle. Br J Pharmacol 80:33-40

Campbell TJ, Vaughan Williams M (1983) Voltage and time dependent depression of maximum rate of depolarization of guinea pig ventricular action potentials by two antiarrhythmic drugs, flecainide and lorcainide. Cardiovasc Res 17:251-258

Cardiac Arrhythmia Pilot Study (1986) Pro-arrhythmic and other adverse effects of encainide, flecainide, imipramine, moricizine, and placebo observed in the cardiac arrhythmia pilot study (abstr.) Circulation 74(Suppl II):313

Carmeliet E (1980) Electrophysiologic effects of encainide on isolated cardiac muscle and Purkinje fibers and on the Langendorff perfused guinea pig heart. Eur J Pharmacol 61:247-262

Carmeliet E, Zaman MY (1979) Comparative effects of lighocaine and lorcainide on conduction in the Langendorff-perfused guinea pig heart. Cardiovasc Res 13:439-449

Carmeliet E, Janssen P, Marsboom R, Van Nueten JM, Xhonneux R (1978) Antiarrhythmic electrophysiologic and hemodynamic effects of lorcainide. Arch Int Pharmacodyn 231:104-130

Cassagneau B, Miquel J, Puel J, Mordant B, Kher A, Fauvel J, Bounhoure J (1985) Antiarrhythmic and electrophysiological effects of encainide in supraventricular tachyarrhythmias. Arch Mal Coeur 78:113-119

Chang SF, Miller AM, Fox JM, Welscher TM (1984) Application of a bonded phase extraction column for rapid sample preparation of flecainide from human plasma for high performance liquid chromatographic analysis, fluorescence or ultraviolet detection. Ther Drug Monit 6:105-111

Chesnie B, Podrid P, Lown B, Raeder E (1983) Encainide for refractory ventricular tachyarrhythmia. Am J Cardiol 52:495-500

Chesnie B, Lampert S, Podrid P, Lown B (1984) Lorcainide in patients with refractory ventricular tachyarrhythmias. J Am Coll Cardiol 3(6):1531-1539

Chilson D, Heger J, Zipes D, Browne K, Prystowsky E (1985) Electrophysiologic effects and clinical efficacy of oral propafenone therapy in patients with ventricular tachycardia. J Am Coll Cardiol 5:1407-1413

Chimienti M, Salerno J, Moize M, Klersy C, Marangoni E, Previtali M, Bianchi P,

Montemartini C, Bobba P (1985) Intravenous and oral encainide, electrophysiologic effects in patients with paroxysmal reciprocating supraventricular tachycardia. G Ital Cardiol 15(5):553-542
Cocco G, Strozzi C (1978) Initial clinical experience of lorcainide (RO13-1042), a new antiarrhythmic agent. Eur J Clin Pharmacol 14:105-109
Conard G, Jernberg MJ, Carlson GL, Ober RE (1975) Metabolism of R-818, an antiarrhythmic candidate in rats. Pharmacologist 17:194
Conard GJ, Carlson GL, Frost JW, Ober RE (1979) Human plasma pharmacokinetics of flecainide acetate (R-818), a new antiarryhthmic, following single oral and intravenous doses (abstr.). Clin Pharmacol Ther 25:218
Connolly S, Kates R, Lebsack C, Harrison D, Winkle R (1983) Clinical pharmacology of propafenone, Circulation 68:589-596
Connolly S, Leback C, Winkle R, Harrison D, Kates R (1984) Propaferone disposition kinetics in cardiac arrhythmia. Clin Pharmacol Ther 36(2):163-168
Coumel P, Leclerq J, Assayag P (1984) European experience with the antiarrhythmic efficacy of propafenone for supraventricular and ventricular arrhythmias. Am J Cardiol 54(9):60-66D
Cowan JC, Vaughan Williams EM (1981) Characterization of a new oral antiarrhythmic drug, flecainide. Eur J Pharmacol 73:333-342
Dawson A, Roden D, Duff H, Woosley R, Smith R (1984) Differential effects of *O*-demethyl encainide on induced and spontaneous arrhythmias in the conscious dog. Am J Cardiol 54:654-658
Dennis P, Vaughan Williams EM (1985) Effects on rabbit cardiac potentials of aprindine, and indecainide, a new antiarrhythmic agent, in normoxia and hypoxia. Br J Pharmacol 85(1):11-19
Dennis P, Vaughan Williams EM (1986) The effects of indecainide, a new antidysrhythmic drug on nodal tissue on the isolated rabbit heart. J Cardiovasc Pharmacol 8(1):1-5
DeClerck F, van Gorp L, Xhonneux R (1978) Effects of lorcainide, a new antiarrhythmic compound, on the changes in heart rhythm induced by intravenous injection of adenosine 5-diphosphate in rats. Arch Int Pharmacodyn Ther 231:222-231
DiBianco R, Fletcher D, Cohen A, Gottdieher JS, Singh JN, Katz RJ, Bates HR, Saverbrunn B (1982) Treatment of frequent ventricular arrhythmias with encainide: assessment using serial ambulatory electrocartdiograms, intracardiac electrophysiologc studies, treadmill exercise tests, and radionucleide cineangiographic studies. Circulation 65:1134-1147
Distlerath L, Reilly P, Martin M, Davis G, Wilkinson G, Guengerich F (1985) Purification and characterization of the human liver cytochromes P450 involved in debrisoquine 4-hydroxylation and phenacetin *O*-deethylation, two prototypes for genetic polymorphism in oxidative drug metabolism. J Biol Chem 260(15):9057-9067
Doherty J, Waxman H, Kienzle M, Cassidy D, Marchlinsky F, Buxton A, Josephson M (1984) Limited role of intravenous propafenone hydrochloride in the treatment of sustained ventricular tachycardia: electrophysiologic effects and results of programmed ventricular stimulation. J Am Coll Cardiol 4(2):378-381
Dresel P (1984) Effect of encainide and its two major metabolites on cardiac conduction. J Pharmacol Exp Ther 228:180-186
Dressler F, Gravinghoff L, Grutte E, Jungst BK, Liersch R, Nomay H, Puls I, Rautenburg HW, Schmaltz A, Schumacher G (1985) The treatment of arrhythmias in infants and children using propafenone. Monatsschr Kinderheilkd 133(3):154-157
Duff HJ, Roden DM, Maffucci R, Vesper B, Conard G, Higgins S, Oates J, Smith R, Woosley R (1981) Suppression of resistant ventricular arrhythmias by twice daily dosing with flecainide. Am J Cardiol 48:1133-1140

Duff H, Roden D, Dawson A, Oates J, Smith R, Woosley R (1982) Comparison of effects of placebo and encainide on programmed electrical stimulation and ventricular arrhythmia frequency. Am J Cardiol 50:305-311

Duff HJ, Dawson AK, Roden DM, Oates JA, Smith RF, Woosley RL (1983) Electrophysiologic actions of *o*-demeltyl encainide, an active metabolite. Circulation 68:385-391

Duff H, Roden D, Carey E, Wang T, Primm R, Woosley R (1985) Spectrum of antiarrhythmic response to encainide. Am J Cardiol 56:887-891

Dukes I, Vaughan Williams EM (1984) The multiple modes of action of propafenone. Eur Heart J 5:115-125

Dykstra SJ, Minielli J, Lawson J, Furguson H, Duncan K (1973) Lysergic acid and quinidine analogs (2-*O*-acylaminophenethyl) piperidines. J Med Chem 16(9):1015-1020

Echt D, Mitchell L, Kates R, Winkle R (1983) Comparison of the electrophysiologic effects of intravenous and oral lorcainide in patients with recurrent ventricular tachycardia. Circulation 68(2):392-399

Echt D, Shapiro M, Trusso J, Mason J, Winkle R (1985) Treatment with oral lorcainide in patients with sustained ventricular tachycardia and fibrillation. Am Heart J 109:28-32

Elharrar V, Zipes D (1982) Effects of encainide and metabolites (MJ14030 and MJ9444) on canine cardiac Purkinje and ventricular fibers. J Pharmacol Exp Ther 220:440-447

Falk RH, O'Brien J (1986) Lorcainide: a comparative trial with quinidine gluconate in patients with previously untreated ventricular arrhythmias. Chest 4:537-540

Farid K, Fasola A (1984) The disposition and metabolism of C14 indecainide in man. Pharmacologist 26:171

Farid NA, Fasola A, Nash J (1985) Liquid chromatographic determination of indecainide, a new antiarrhythmic drug and its metabolite, desisopropylindecainide in biologic samples. J Chromatogr 337(2):329-340

Fauchier J, Cosnay P, Rouesnel P, Moquet B, Bonnet P, Scala P, Demeyer J (1985) Effect of oral and injectable flecainide in patients with accessory atrial ventricular pathway. Arch Mal Coeur 78:81-90

Flecainide Quinidine Research Group (1983) Flecainide versus quinidine for treatment of chronic ventricular arrhythmias. A multicenter clinical trial. Circulation 67:1117-1123

Flecainide Ventricular Tachycardia Study Group (1986) Treatment of resistant ventricular tachycardia with flecainide acetate. Am J Cardiol 57:1299-1304

Flowers D, O'Gallagher D, Torres V, Miura D, Somberg JC (1985) Flecainide: long term treatment using a reduced dosing schedule. Am J Cardiol 55:79-83

Furlanello F, Disertori M, Vergara G, Guarnerio M, Dal Forno P, Padrini R, Gobbato S, Ferrari M (1983) Clinical evaluation of new antiarrhythmic agents, an experience with proprafenone. Int J Clin Pharmacol Res 3(2):101-105

Furlanello F, Vergara G, Dal Forno P, Disertori M, Inama G (1984) Clinical experience with propafenone in paroxysmal supraventricular reciprocating tachycardia. G Ital Cardiol 14(5):379-380

Garcia-Civera R, Sanjuan R, Morell S, Ferrero J, Miralles L, Llavador J, Lopes-Merino V (1984) Effects of propafenone on induction and maintenance of atrioventricular nodal re-entrant tachycardia. PACE 7(4):649-655

Gibson J, Somani P, Bassett AL (1978) Electrophysiologic effects of encainide (MJ9067) on canine Purkinje fibers. Eur J Pharmacol 52:161-169

Gomoll AW, Mayol RF (1982) Comparative hemodynamic effects of encainide and its two major metabolites in anesthetized dogs. Pharmacologist 24:233

Gomoll AW, Byrne JE, Mayol RF (1981) Comparative antiarrhythmic and local anesthetic action of encainide and its two major metabolites (abstr). Pharmacologist 23:209
Gomoll A, Byrne J, Antonnaccio M (1986) Electrophysiology, hemodynamic and antiarrhythmic efficacy model studies on encainide. Am J Cardiol 58:10-17c
Goy J, Grbic M, Hurni M, Finci L, Maendly R, Duc J, Sigwart U (1985) Conversion of supraventricular arrhythmias to sinus rhythm using flecainide. Eur Heart J 6(6):518-524
Gstottner CK, Gmeiner R (1979) Intracardiac electrophysiological effects of lorcainide in man. Eur J Clin Pharmacol 15:241-247
Guehler J, Gornick C, Tobler G, Almquist A, Schmid J, Benson W, Benditt D (1985) Electrophysiological effects of flecainide acetate and its major metabolites in the canine heart. Am J Cardiol 55:807-812
Guengerich F, Distelrath L, Reilly P, Wolff T, Shimada T, Umbenhauer DA, Martin M (1986) Human liver cytochromes P450 involved in polymorphism of drug oxidation. Xenobiotica 16(5):367-378
Hapke H, Prigge E (1976) Zur Pharmakologie von 2′-[2-hydroxy-3-(propylamino)-propoxy-3-phenylpropi (propafenon, Sa 79)-hydrochlorid. Arzneimittelforschung 26:1849-1857
Hellestrand KJ, Bexton RS, Nathan AW, Spurrell RAJ, Camm AJ (1982) Acute electrophysiological effects of flecainide acetate on cardiac conduction and refractoriness in man. Br Heart J 48:140-148
Hellestrand K, Nathan A, Bexton R, Spurrell R, Camm AJ (1983) Cardiac electrophysiologic effects of flecainide acetate for paroxysmal reentrant junctional tachycardias. Am J Cardiol 51(5):770-776
Hoddess AB, Follansbee WP, Spear JF, Moore EN (1979) Electrophysiological effects of a new antiarrhythmic agent, flecainide, on the intact canine heart. J Cardiovasc Pharmacol 1:427-439
Hodges M, Haugland JM, Granrud G, Conard CJ, Asinger RW, Mikell FL, Krejci J (1982) Suppression of ventricular ectopic depolarizations by flecainide acetate, a new antiarrhythmic agent. Circulation 65:879-885
Hohnloser S, Zeiher A, Hust MH, Wollschlager H, Just H (1983) Flecainide induced aggravation of ventricular tachycardia. Clin Cardiol 6:130-135
Holland D, Lacefield W (1983) LY135837: a potent new antiarrhythmic agent. Fed Proc 42:1290
Hollmann M, Brode E, Holtz D, Kaumeiler S, Kehrhahn O (1983a) Investigations on the pharmacokinetics of propafenone in man. Arzneimittelforschung 33:763-770
Hollmann M, Hege H, Brode E, Buhler V, Holtz D, Kaumeier S, Kehrhahn O, Lietz H, Schwartz J, Stieren B, Weymann J (1983b) Pharmacokinetic and metabolic studies of propafenone in volunteers. In: Schlepper M, Molsson B (eds) Cardiac arrhythmias. Proceedings of the first international rythmonorm congress. Springer, Berlin Heidelberg New York, pp 125-132
Hotzman JL, Berry DA, Kvam DC, Mottonen L, Borrell G (1984) Application of second order polynomial equation to the study of pharmacodynamic interactions: the effect of flecainide acetate and propranolol on cardiac output and vascular resistance. J Pharmacol Exp Ther 231:286-290
Horowitz L (1986) Encainide in lethal ventricular arrhythmias evaluated by electrophysiologic testing and decrease in symptoms. Am J Cardiol 58:83-86c
Horowitz L, Spielman S, Webb C, Morganroth J, Greenspan A (1985) The clinical electrophysiology of intravenous indecainide. Am Heart J 110: 784-788
Ikeda N, Davis L, Hauswirth O, Singh BN (1982) Flecainide: electrophysiologic profile

in isolated cardiac muscle of an antiarrhythmic with different effects in ventricular muscle and Purkinje fibers (abstr). Circulation 66 (Suppl II):379

Jackman W, Zipes D, Naccarelli G, Rinkenberger R, Heger J, Prystowsky E (1982) Electrophysiology of oral encainide. Am J Cardiol 49:1270-1278

Jackson N, Verma SP, Frais MA, Silke B, Hafizullah M, Reynolds G, Taylor SH (1985) Hemodynamic dose response effects of flecainide in acute myocardial infarction, with and without left ventricular decompensation. Clin Pharmacol Ther 37:619-624

Jahnchen E, Bechtold H, Kasper W, Kersting F, Just H, Heykants J, Meinertz T (1979) Lorcainide I. Saturable presystemic elimination. Clin Pharmacol Ther 26:187-204

Kappenberger L, Fromer M, Shenasa M, Gloor H (1985) Evaluation of flecainide acetate in rapid atrial fibrillation complicating Wolff-Parkinson-White syndrome. Clin Cardiol 8(6):321-326

Karagueuzian H, Fujimoto T, Katoh T, Peter T, McCullen A, Mandell W (1982) Suppression of ventricular arrhythmias by propafenone, a new antiarrhythmic agent during acute myocardial infarction in the conscious dog, a comparison study with lidocaine. Circulation 66(6):1190-1198

Karagueuzian H, Katoh T, McCullen A, Mandell W, Peter T (1984) Electrophysiologic and hemodynamic effects of propafenone, a new antiarrhythmic agent, on the anesthetized, closed chest dog: a comparative study with lidocaine. Am Heart J 107:418-424

Kasper W, Meinertz T, Kersting F, Lollgen H, Lang K, Just H (1979) Electrophysiological actions of lorcainide in patients with cardiac disease. J Cardiovasc Pharmacol 1:343-352

Kasper W, Treese N, Meinertz T, Jahnchen E, Pop T (1983) Electrophysiologic effects of lorcainide on the accessory pathway in the Wolff-Parkinson-White syndrome. Am J Cardiol 51(10):1618-1622

Kates R, Harrison D, Winkle R (1982) Metabolite accumulation during long term oral encainide administration. Pharmacol Ther 40:427-432

Kates RE, Keefe DL, Winkle RA (1983) Lorcainide disposition. Kinetics in arrhythmia patients. Clin Pharmacol Ther 33:28-34

Kates R, Yee YG, Winkle R (1985) Metabolite accumulation during chronic propafenone dosing in arrhythmia. Clin Pharmacol Ther 37:610-614

Keefe D, Peters F, Winkle R (1982) Randomized double-blind placebo controlled cross-over trial, documenting oral lorcainide efficacy and suppression of symptomatic ventricular tachyarrhythmias. Am Heart J 103:511-518

Keefe D, Kates R, Winkle R (1984) Comparative electrophysiology of lorcainide and norlorcainide in the dog. J Cardiovasc Pharmacol 6:808-815

Kesteloot H, Stroobandt R (1977) Clinical experience with lorcainide (R15899), a new artiarrhythmic drug. Arch Int Pharmacodyn Ther 230(2):225-234

Kesteloot H, Stroobandt R (1979) Clinical experience of encainide (MJ9067): a new antiarrhythmic drug. Eur J Clin Pharmacol 16:323-326

Kim S, Lal R, Ruffy R (1986) Treatment of paroxysmal reentrant supraventricular tachycardia with flecainide acetate. Am J Cardiol 58(1):80-85

Klotz U, Muller-Seydlitz P, Heimburg P (1978) Pharmacokinetics of lorcainide in man, a new antiarrhythmic agent. Clin Pharmacokinet 3:407-418

Klotz U, Fischer C, Muller-Seydlitz P, Schultz J, Muller WA (1979a) Alterations in the disposition of differently cleared drugs in patients with cirrhosis. Clin Pharmacol Ther 26:221-227

Klotz U, Muller-Seydlitz P, Heimburg P (1979b) Disposition and antiarrhythmic effect of lorcainide. Int J Clin Pharmacol Biopharm 17:152-158

Klotz U, Muller-Seydlitz P, Heimburg P (1979c) Lorcainide infusion in the treatment of ventricular premature beats. Eur J Pharmacol 16:1-6
Kunze K, Kuck K, Schluter M, Kuch B, Bleifeld W (1984) Electrophysiologic and clinical effects of intravenous and oral encainide in accessory atrioventricular pathway. Am J Cardiol 54(3):323-329
Lal R, Chapman P, Naccarrelli G, Troup P, Rinkenberger R, Dougherty A, Ruffy R (1985) Short and long term experience with flecainide acetate in the management of refractory life-threatening ventricular arrhythmias. J Am Coll Cardiol 6:772-779
Lawson JW (1968) Antiarrhythmic activity of some isoquinoline derivatives determined by a rapid screening procedure in the mouse. J Pharmacol Exp Ther 160:22-31
Leinonen H, Heikkila J, Sundberg S, Gordin A (1984) Lorcainide in the prophylaxis of ventricular arrhythmias in acute myocardial infarction. Ann Clin Res 16:18-22
Lewis G, Holtzman J (1984) Interaction of flecainide with digoxin and propranolol. Am J Cardiol 53:52-57b
Libersa CC, Lekieffre J, Caron J, Poirier J, Plady S, Kacei S, Kher A (1984) Electrophysiologic effects of encainide and its metabolite in 11 patients. Eur Heart J 5(1):290
Lindstrom T, Lacefield W, Whitaker G (1984) Metabolism of the antiarrhythmic agent, indecainide, in rats, dogs, and monkeys. Drug Metab Dispos 12(6):691-697
Livelli F, Ferrick K, Bigger JT, Reiffel J, Gliklich J, Coromilas J (1984) Mixed response to flecainide acetate (abstr). J Am Coll Cardiol 3:583
Lloyd E, Workman L, Commerford P, Mabin T (1981) Evaluation of lorcainide, a new antiarrhythmic agent. S Afr Med J 60:929-931
Lui H, Lee G, Dietrich P, Low R, Mason D (1982) Flecainide induced QT prolongation and ventricular tachycardia. Am Heart J 103:567-569
Manz M, Steinbeck G, Luderitz B (1985) Usefulness of programmed stimulation in predicting efficacy of propafenone in long term antiarrhythmic therapy for paroxysmal supraventricular tachycardia. Am J Cardiol 56(10):593-597
Mason J, Peters F (1981) Antiarrhythmic efficacy of encainide in patients with refractory recurrent ventricular tachycardia. Circulation 63:670-675
McAllister C, Wolfenden H, Aslanian W, Woosley R, Wilkinson G (1986) Oxidative metabolism of encainide: polymorphism, pharmacokinetics and clinical considerations. Xenobiotica 16(5):483-490
McLeod A, Stiles G, Shand D (1984) Demonstration of beta adrenoreceptor blockade by propafenone hydrochloride: clinical pharmacologic, radioligand binding and adenylate cyclase activation studies. J Pharmacol Exp Ther 228(2):461-466
Mead R, Keefe D, Kates R, Winkle RA (1985) Chronic lorcainide therapy for symptomatic premature ventricular complexes: efficacy pharmacokinetics and evidence for nor-lorcainide antiarrhythmic effect. Am J Cardiol 55:72-78
Meinertz T, Kasper W, Kersting F, Just H, Bechtold H, Jahnchen E (1979) Lorcainide II. Plasma concentration-effect relationship. Clin Pharmacol Ther 26:196-204
Meinertz T, Kersting F, Kasper W, Just H, Bechtold H, Jahnchen E (1980) Hemodynamic effects of a single intravenous dose of lorcainide in patients with heart disease. Eur J Clin Pharmacol 18:461-465
Meinertz T, Zehender M, Geibel A, Treese N, Hofmann T, Kasper W, Pop T (1984) Long term antiarrhythmic therapy with flecainide. Am J Cardiol 54:91-96
Miura D, Wynn J, Keefe D, Laux B, Somberg J (1987) Antiarrhythmic effects of indecainide in drug resistant populations with ventricular tachycardia and programmed electrical stimulation. Am Heart J 113:65-69
Morganroth J (1981) The efficacy and safety of lorcainide in the treatment of ventricular arrhythmias as assessed by frequent Holter monitoring. Prognosis in pharma-

cotherapy of life threatening arrhythmias. In: Jahnchen E, Meinertz TH, Towse G (eds) Royal Society of Medicine. International congress and symposium, series no 49. Academic, London, The Royal Society of Medicine, London

Morganroth J (1986a) Encainide for ventricular arrhythmias: placebo controlled and standard comparison trials. Am J Cardiol 58:74-82C

Morganroth J, Pool P, Miller R, Hsu P, Lee I, Clark D (1986b) Dose response range of encainide for benign and potential lethal ventricular arrhythmias. Am J Cardiol 57:769-774

Morganroth J, Somberg JC, Pool P, Hsu P, Lee I, Durkee J (1986c) Comparative study of encainide and quinidine in the treatment of ventricular arrhythmias. J Am Coll Cardiol 7:9-16

Muhiddin K, Nathan AW, Hellestrand KJ, Banim SO, Camm AJ (1982) Ventricular tachycardia associated with flecainide. Lancet 2:1220-1221

Musto B, D'onofrio A, Musto A, Cavallaro C, Marsico F (1986) Electrophysiologic effects in clinical efficacy of propafenone in pediatric patients with paroxysmal supraventricular re-entry tachycardia. G Ital Cardiol 16(4): 336-343

Naccarelli G, Rinkenberg R, Dougherty A, Geibel R (1984) Use of encainide in treating atrial fibrillation in the Wolff-Parkinson-White syndrome. Circulation 70:II-445 (abstr.)

Nathan A, Hellestrand K, Bexton R, Banim S, Spurrell R, Camm AJ (1984) Proarrhythmic effects of the new antiarrhythmic agent, flecainide acetate. Am Heart J 107:222-228

Neuss H (1985) Long term use of flecainide in patients with supraventricular tachycardia. Drugs 29 (Suppl 4):21-25

Neuss H, Buss J, Schlepper M, Berthold R, Mitrovic V, Kramer A, Musial W (1983) Effect of flecainide on electrophysiological properties of accessory pathways in the Wolff-Parkinson-White syndrome. Eur Heart J 4(5):347-353

Nicholson M, Campbell RWF, Julian D (1982) Encainide in management of ventricular tachycardia. Aust NZ J Med 12:313

Pentel P, Goldsmith S, Salerno D, Nasraway S, Palmer D (1986) Effect of hypertonic sodium bicarbonate on encainide overdose. Am J Cardiol 57:878-880

Philipsborn G, Gries J, Hofmann H, Kreiskott H, Kretzschmar R, Muller D, Raschack M, Teschendorf H (1984) Pharmacology studies on propafenone and its main metabolite 5-hydroxy propafenone. Arzneimittelforschung 34(2):1489-1497

Platia E, Estes M, Heine D, Griffith L, Garan H, Ruskin J, Reid P (1985) Flecainide: electrophysiologic and antiarrhythmic properties in refractory ventricular tachycardia. Am J Cardiol 55:956-962

Podrid P, Lown B (1984) Propafenone, a new agent for ventricular arrhythmias. J Am Coll Cardiol 4:117-125

Podrid P, Morganroth J (1985) Aggravation of arrhythmias during drug therapy; experience with flecainide acetate. Prac Cardiol 11(12):55-70

Pop T, Treese N, Kana J, Meinertz T, Kasper W (1983) Effect of intravenous flecainide on atrial vulnerability in man. Klin Wochenschr 61(12):609-615

Prystowsky E, Klein G, Rinkenberger R, Heger J, Naccarelli G, Zipes D (1984) Clinical efficacy and electrophysiologic effects of encainide in patients with Wolff-Parkinson-White syndrome. Circulation 69(2):278-287

Quart B, Gallo D, Sami M, Wood A (1986) Drug interaction studies and encainide use in renal and hepatic impairment. Am J Cardiol 58:104-113c

Rehnqvist N, Ericsson C, Eriksson C, Olsson G, Svensson G (1984) Comparative investigation of the antiarrhythmic effect of propafenone (rytmonorn) and lidocaine in patients with ventricular arrhythmias during acute myocardial infarctions. Acta Med Scand 216:525-530

Rinkenberger RL, Prystowsky E, Jackman W, Naccarelli G, Heger J, Zipes D (1982) Drug conversion of non-sustained ventricular tachycardia to sustained ventricular tachycardia during serial electrophysiologic studies: identification of drugs that exacerbate tachycardia and potential mechanisms. Am Heart J 103:177-184

Roden D, Stots B, Reele S, Higgins S, Mayol R, Gammans R, Oates J, Woosley R (1980) Total suppression of ventricular arrhythmias by encainide. Pharmacokinetic and electrocardiographic characteristics. N Engl J Med 302:877-882

Roden D, Wood A, Wilkinson G, Woosley R (1986) Disposition kinetics of encainide and metabolites. Am J Cardiol 58:4-9c

Ruskin J, McGovern B, Garan H, DiMarco J, Kelly E (1983) Antiarrhythmic drugs: a possible cause of out-of-hospital cardiac arrest. N Engl J Med 309:1302-1306

Rutsch W (1978) Influencing ventricular extra systoles with propafenone. Herz/Kreisl 10:183-186

Saksena S, Rothbart S, Cappello G, Bernstein A, Somani P (1983) Clinical and electrophysiological effects of chronic lorcainide therapy in refractory ventricular tachycardia. J Am Coll Cardiol 2(3):538-544

Salerno D, Hodges M, Granrud G, Sharkey P (1983) Comparison of flecainide with quinidine for suppression of chronic stable ventricular ectopic depolarizations. Ann Int Med 98:455-460

Salerno D, Granrud G, Sharkey P, Asinger R, Hodges M (1984) A controlled trial of propafenone for treatment of frequent and repetitive ventricular premature complexes. Am J Cardiol 53:77-83

Salerno D, Fifield J, Krejci J, Larson T, Hodgess M (1986) Encainide induced hyperglycemia (abstr). Circulation 74 [suppl]:II:104

Sami M, Mason J, Oh G, Harrison D (1979a) Canine electrophysiology of encainide, a new antiarrhythmic drug. Am J Cardiol 43:1149-1154

Sami M, Mason J, Peters F, Harrison D (1979b) Clinical electrophysiologic effects of encainide a newly developed antiarrhythmic agent. Am J Cardiol 44:526-532

Sami M, Harrison D, Kreaemer H, Houston N, Shimasaki C, DeBusk R (1981) Antiarrhythmic efficacy of encainide and quinidine: validation of a model for drug assessment. Am J Cardiol 48:147-156

Sami M, Derbekyan V, Lisbona R (1983) Hemodynamic effects of encainide in patients with ventricular arrhythmia and poor ventricular function. Am J Cardiol 52:507-511

Samuelsson R, Harrison D (1981) Electrophysiologic evaluation of encainide with use of monophasic action potential recording. Am J Cardiol 48:871-876

Sandusky G, Meyers D (1985) Toxicology of indecainide hydrochloric after intravenous administration to rats and dogs. Toxicol Lett 26(2-3):107-110

Schmid JR, Seebeck BD, Henrie CL, Banitt EH, Kvam DC (1975) Some antiarrhythmic actions of a new compound, R-818, in dogs and mice (abstr). Fed Proc 34:775

Schwartz A, Shapiro W, Sauve M, Shen E, Bhandari A, Morady F, Scheinman M (1984) Adverse electrophysiologic effects of encainide in patients with bundle branch block (abstr). Clin Res 32:1

Seipel L, Abendroth R, Breithardt G (1980) Electrophysiologic effects of flecainide (R-818), in man (abstr). Circulation 62 (Suppl 3):III-153

Sellers TD, DiMarco JP (1985) Sinusoidal ventricular tachycardia associated with flecainide acetate. Chest 5:647-649

Senges J, Rizos I, Brachmann J, Tian-Li G, Lengfelder RW, Kubler W (1982) Arrhythmogenic effect of toxic concentrations of the antiarrhythmic drug lorcainide on the isolated canine ventricle. J Pharmacol Exp Ther 223:547-551

Serruys PW, Vanhaleweyk G, Van den brand M, Verdou P, Lubsen J, Hugenholtz PG

(1983) Hemodynamic effects of intravenous flecainide acetate in patients with coronary artery disease. Br J Clin Pharmacol 16:51-59

Shen EN, Sung RJ, Morady F, Schwartz AB, Scheinman MM, DiCarlo L, Shapiro W (1984) Electrophysiologic and hemodynamic effects of intravenous prapafenone in patients with recurrent ventricular tachycardia. J Am Coll Cardiol 3:1291-1297

Shigenobu K, Tatsuno-Atoda H, Asano T, Kasuya Y (1980) Electrophysiologic effects of a new antiarrhythmic agent, lorcainide, on the isolated cardiac muscle fibers as compared with disopyramide. J Pharmacobiodyn 3:677-685

Shita A, Bernard R, Mostinckx R, Debacker M (1981) Hemodynamic reactions after intravenous injections of lorcainide hydrochloride in acute myocardial infarction. Eur J Cardiol 12:237-242

Siddoway L, McAllister CB, Wang T, Bergstrand R, Roden D, Wilkinson GR, Woosley R (1983) Polymorphic oxidative metabolism of propafenone in man. Circulation 68(3):64

Singh SN, DiBianco R, Kostroff L, Fletcher RD, Cockrell JL (1984) Lorcainide for high frequency ventricular arrhythmia: preliminary results of a short term double blind and placebo controlled crossover study and long term follow-up. Am J Cardiol 54:22-28B

Solti F, Czako E, Szatmary L (1986) Electrophysiological characteristics of lorcainide, a new antiarrhythmic drug. Int J Clin Pharmacol Ther Toxicol 24:26-29

Somani P (1980) Antiarrhythmic effects of flecainide. Clin Pharmacol Ther 27:464-470

Somani P (1981) Pharmacokinetics of lorcainide, a new antiarrhythmic drug in patients with cardiac rhythm disorders. Am J Cardiol 48:157-163

Somani P, di Giorgi S (1980) Resistant ventricular arrhythmias treated with lorcainide, a new antiarrhythmic drug. Chest 78(4):658-660

Somani P, Simon V, Gupka K, King P, Shapira R, Stochard H (1984a) Lorcainide kinetics and protein binding in patients with end stage renal disease. :121-125

Somani P, Temesy-Armos P, Leighton R, Goodenday L, Fraker T (1984b) Hyponatremia in patients treated with lorcainide, a new antiarrhythmic drug. Am Heart J 108:1443-1448

Somberg JC, Wellens H, Keren J, Miura DS (1982) A new animal model for the electrophysiologic testing of antiarrhythmic drugs. Fed Proc 41(5):1711

Somberg J, Butler, Flowers D, Keefe D, Torres V, Miura D (1984a) Evaluation of lorcainide in patients with symptomatic ventricular tachycardia. Am J Cardiol 54:43-48B

Somberg J, Butler B, Torres V, Flowers D, Tepper D, Keefe D, Miura D (1984b) Lorcainide therapy for the high risk patient post myocardial infarction. Am J Cardiol 54:37-42B

Somberg JC, Laux B, Wynn J, Keefe D, Miura D (1986) Lorcainide therapy in cardiac arrest population. Am Heart J 111:648-653

Somberg JC, Tepper D, Sacker H, Schwartz J (1987) Chronic flecainide therapy selected by electrophysiology testing of intravenous flecainide. Am Heart I 114:18-25

Soyka L (1986) Safety of encainide for the treatment of ventricular arrhythmias. Am J Cardiol 58:96-103c

Soyza N, Terry L, Murphy M, Thompson C, Doherty J, Sakhaii M, Dinh H (1984) Effect of propafenone in patients with stable ventricular arrhythmias. Am Heart J 108:285-289

Spivack C, Gottlieb S, Miura D, Somberg JC (1984) Flecainide toxicity. Am J Cardiol 53:329-330

Stavens C, McGovern B, Garan H, Ruskin J (1985) Aggravation of electrically provoked ventricular tachycardia during treatment with propafenone. Am Heart J 110:24-29

Steinberg M, Wiest S (1984) Electrophysiological studies of indecainide hydrochloride, a new antiarrhythmic agent, in canine cardiac tissue. J Cardiovasc Pharmacol 6(4):614-621

Strasburger J, Moak J, Smith R, McVey-Duncan P, Armstrong K, Garson A (1986) Encainide for refractory supraventricular tachycardia in children. Am J Cardiol 58(5):49c-54c

Stroobandt R, Kesteloot H (1985) Efficacy and tolerance of intravenous lorcainide in patients with normal and impaired intra-ventricular conduction. Acta Cardiol (Brux) 40(6):637-648

Tjandramaga TB, Verbesselt R, Van-Hecken A, Mullie A, DeSchepper PJ (1982) Oral digoxin pharmacokinetics during multiple dose flecainide treatment. Arch Int Pharmacodyn 260:302-303

Torres V, Flowers D, Somberg J (1985) The arrhythmogenicity of antiarrhythmic agents. Am Heart J 109(5):1090-1097

Tucker C, Winkle R, Peters F, Harrison DC (1982) Acute hemodynamic effects of intravenous encainide in patients with heart disease. Am Heart J 104 (2/1):209-215

Vanhaleweyk G, Serruys P, Hugenholtz P (1984) Hemodynamic effects of encainide, flecainide, lorcainide and tocainide. Eur Heart J 5:67-74

Vaughan Williams EM (1970) Classification of antiarrhythmic drugs. In: Sandoe et al. (eds) Cardiac arrhythmias. Astra, Sodertalge, Sweden, pp 449-473

Vaughan Williams EM (1984) A classification of antiarrhythmic actions reassessed after a decade of new drugs. J Clin Pharmacol 24:129-147

Velebit V, Podrid P, Lown B, Cohen B, Graboys TB (1982) Aggravation and provocation of ventricular arrhythmias by antiarrhythmic drugs. Circulation 65:886-894

Verdouw, PD, Deckers JW, Conard JG (1979) Antiarrhythmic and hemodynamic actions of flecainide acetate (R-818) in ischemic porcine heart. J Cardiovasc Pharmacol 1:473-486

Vlay S, Mallis G (1986) Intravenous and oral lorcainide: assessment of central nervous system toxicity and antiarrhythmic efficacy. Am Heart J 111:452-455

Vlay SC, Mallis GI, Singh S, Cohn PF (1984) Comparison of lorcainide and quinidine in the treatment of ventricular ectopy. Chest 86:80-83

Vorgeat A, Goy J, Maendly R, Kaufmann U, Grbic M, Sigwart U (1986) Flecainide versus quinidine for conversion of atrial fibrillation to sinus rhythm. Am J Cardiol 58(6):496-498

Walker P, Papouchado M, James M, Clark L (1985) Pacing failure due to flecainide acetate. PACE 8:900-902

Ward D, Jones S, Shinebourne E (1986) Use of flecainide acetate for refractory junctional tachycardia in children with the Wolff-Parkinson-White syndrome. Am J Cardiol 57(10):787-790

Wehr M, Noll B, Krappe J (1985) Flecainide induced aggravation of ventricular arrhythmias. Am J Cardiol 55:1643-1644

Winkle R, Mason J, Griffin J, Ross D (1981a) Malignant ventricular tachyarrhythmias associated with the use of encainide. Am Heart J 102:857-864

Winkle R, Peters F, Kates R, Tucker C, Harrison D (1981b) Clinical pharmacology and antiarrhythmic efficacy of encainide in patients with chronic ventricular arrhythmias. 64(2):290-296

Winkle R, Peters F, Kates R, Harrison D (1983) Possible contribution of encainide metabolites to the long term antiarrhythmic efficacy of encainide. Am J Cardiol 51:1182-1188

Woestenborghs R, Michiels M, Heykants J (1979) Simultaneous gas chromotographic determination of lorcainide hydrochloride and three of its principle metabolites in biological samples. J Chromotogr 164:169–176

Woosley R, Roden D, Guizhu D, Wang T, Altenbern D, Oates J, Wilkinson G (1986) Co-inheritance of the polymorphic metabolism of encainide and debrisoquin. Clin Pharmacol Ther 39:282–289

Wynn J, Fingerhood M, Keefe D, Maza S, Miura D, Somberg JC (1986) Refractory ventricular tachycardia with flecainide. Am Heart J 112(1):174–175

Xhonneux R, Remeysm P (1975) Effectiveness of R-15889 (lorcainide) and ouabain induced ventricular arrhythmias in anesthetized dogs. Janssen Research Technical Report, Janssen Pharmaceutica, Belgium, N9969

Yee Y, Kates R (1981) High performance liquid chromotographic analysis of lorcainide and its active metabolite, nor-lorcainide in plasma. J Chromotogr 223:454–459

Zeiler R, Gough W, El-Sherif N (1984) Electrophysiologic effects of propefanone on canine ischemic cardiac cells. Am J Cardiol 54:424–429

Class II Agents

CHAPTER 12a

Arrhythmias in the Normal Human Heart

P. TAGGART

> ... Every affection of the mind that is attended with either pain or pleasure, hope or fear, is the cause of an agitation whose influence extends to the heart ... A strong man who, having received an injury and affront from one more powerful than himself, and upon whom he could not have his revenge, was so overcome with hatred and spite and passion, which he yet communicated to no one, that at last he fell into a strange distemper, suffering from extreme oppression and pain of the heart and breast ...
>
> *Exercitatio Anatomica De Motu*
> *Cordis Et Sanguinis in Animalibus.*
> *W. Harvey, Frankfurt-am-Main, 1628*

A. Introduction

This chapter is concerned with arrhythmias occurring in the normal heart particularly in response to stress. Various aspects are considered including the incidence of arrhythmia in people with apparently normal hearts as determined by electrocardiographic (ECG) recordings at rest, during 24-h ambulatory monitoring and during exercise; evidence that emotion may induce arrhythmia in normal people; the relationship of the sympathetic and parasympathetic nervous system; and prognostic importance and clinical implications. Problems arise in view of the inevitable difficulties of defining the "normal". Research interest has been largely directed towards the abnormal heart and there is a relative paucity of information on the "normal". Some reliance therefore has to be placed on control groups which in a number of instances are lacking in detail. There is a wide variation in the literature in the degree to which normality has been established, ranging from angiographic evidence in a few, through non-invasive studies in some to merely the absence of symptoms in the majority. The asymptomatic subjects tend to span a wide age range and it is inevitable that a proportion with significant cardiac disease will have been included. This may be important in view of the fact that the number exhibiting arrhythmia is relatively small.

A second problem concerns the difficulty in defining the "normal", the very concept of which becomes increasingly remote in the wake of advances in investigative techniques. Indeed it has often been said that the normal heart is merely an underinvestigated one. New methods of investigation enable subjects to be identified which, if they relate to any specific clinical picture, may qualify for the distinction of being called a syndrome. Consequently normality is a transient state not only with respect to the development of disease with age but also on account of ever increasing difficulties to satisfy the demands of investigative criteria.

Paradoxically as advances in diagnostic technique increase the problems in defining the normal so the need increases to set new standards in order to provide a yardstick for clinical interpretation. For example the increasing use of 24-h ambulatory ECG monitoring has revealed that arrhythmia is a more common finding in people with no apparent heart disease than had been appreciated from routine resting ECGs in which the recording time lasted a few seconds. Questions thereby arise such as: whether or not people showing such arrhythmia are truly "normal": what type and frequency of arrhythmia may be considered as physiological behaviour of the "normal" heart; will the development of new investigative methods reveal differences in "normal" hearts showing arrhythmia compared with those that do not, which will lead them to be reclassified as recognized normal variants or even pathological.

Should hearts that have undergone adaptive changes be considered normal? Physiological hypertrophy due to conditioning to high altitude has been shown to induce a prolongation of action potential duration and a reduced susceptibility to hypoxia-induced shortening, this being the opposite of the expected in view of the less favourable surface/volume ratio (VAUGHAN WILLIAMS 1986). Athletic training is another form of adaptation associated with a wide variety of ECG abnormalities some of which are indistinguishable from those of ischaemia (see Sect. C.V). Yet trained athletes could hardly be considered as having abnormal hearts in the accepted sense of the word. Perhaps hearts adapted to hypoxia and athletic training would be better classified as supernormal.

People with documented arrhythmia such as paroxysmal tachycardia, even in the absence of any known aetiology such as the Wolff-Parkinson-White syndrome or the long Q-T syndrome, tend to be classified separately. The complexities of the AV intranodal conduction pathways probably favour reentry more in some persons than others given the right circumstances. It is likely that these people occupy the extreme ends of a normal Gaussian curve and are in essence within the spectrum of normality. However, as difficulties of classification and definition are already sufficient, people with known arrhythmias have been excluded from the following text.

B. Incidence of Arrhythmia in Normal Subjects

This section examines the incidence and type of arrhythmia reported in "normal or asymptomatic" subjects in the resting ECG, during ambulatory ECG monitoring and during exercise. The studies include a wide variation in the ages and selection of subjects, the extent of screening for the presence of underlying disease, the circumstances of the study and the information provided.

I. Resting 12-Lead Electrocardiogram

Both supraventricular and ventricular ectopic beats are a frequent finding in asymptomatic people in the routine ECG and their incidence increases with age (His and Lamb 1962; Chiang et al. 1969). Whereas there is an association between the ventricular ectopic beats and coronary heart disease the same does not hold true for supraventricular ectopic beats.

The most comprehensive study is that of His and Lamb, who reported the electrocardiographic findings of 122 043 United States Air Force personnel. All were male, age 16 to over-50 years and asymptomatic. The vast majority were fliers and required to be in exceptionally good health in order to qualify. None of the abnormalities recorded were as a result of a clinical event or a new ECG abnormality in the presence of a previous normal baseline record. The incidence of abnormalities was therefore taken to be representative of an asymptomatic, apparently healthy population. The number of subjects showing specific rhythm abnormalities subdivided according to age is shown in Table 1.

II. Ambulatory ECG Monitoring

Hinkle et al. (1969) looked at a random sample of 301 men whose median age was 55 years. They excluded those they considered had definite or probable coronary heart disease. The remainder were divided into high-, medium- or low-risk groups according to risk factor profiles. The results are summarized in Table 2. The men were studied with 6-h recordings, during which time they performed a Valsalva manoeuvre, a Master's exercise test, drank 500 cc of iced water rapidly followed by hot coffee, performed paper and pencil psychological tests for 1 h and had a 1-h interview with a sociologist. A blood sample was then taken. A very large meal was eaten after which they walked briskly

Table 1. Resting 12-lead ECG in 122 043 asymptomatic men. Total number of subjects and ages (top). The number of subjects showing specific rhythm abnormalities are shown as percentages. When the numbers are small the number of individuals showing abnormalities are given in parentheses. (Recalculated from His and Lamb 1962)

Number Ages (years)	7072 16–19	58 840 20–29	34 994 30–39	20 403 40–49	734 50+
Atrial rhythm	14.5	7.4	2.3	2.7	1.6
Supraventricular premature beats	7.9	7.0	5.3	6.7	2.7
Ventricular premature beats	4.6	6.0	8.3	12.6	21.7
First AV block	6.0	7.0	5.6	6.9	13.6
Second AV block	(3.0)	(4.0)	(1.0)		
Third AV block	(1.0)	(2.0)			
Atrial flutter/fibrillation		(3.0)	(2.0)	(1.0)	
Nodal rhythm	(1.0)	(12.0)	(5.0)		
Ventricular tachycardia	(1.0)	(2.0)			

Table 2. Data from 264 six-hour ambulatory ECG recordings from 225 men classified according to risk profiles as high, medium and low risk. (HINKLE et al. 1969)

	High Risk	Medium Risk	Low Risk
Supraventricular premature beats			
0/1000	11	22	13
0.01-9/1000	38	99	28
> 10/1000	4	7	3
Bigeminy, Trigeminy, Pairs, Multiple consecutive	15	32	5
Ventricular premature beats			
0/1000	15	49	26
0.01-9/1000	30	70	17
> 10/1000	8	9	1
Bigeminy, Trigeminy, Pairs, Paroxysmal VT, VPC from three or four foci	34	47	5

up 13 stairs. They then underwent a 2-h psychological test period. At the end of the recording period the subjects claimed they felt tired.

RAFTERY and CASHMAN (1976) studied 29 men and 23 women aged between 20 and over-70 years. They noticed a low incidence of ectopic beats, the highest incidence being in the over-70 group. They suggested that higher counts therefore signified disease.

CLARKE et al. (1976) studied 41 male and 45 female company volunteers aged between 16 and 65 years with 48-h recordings. Ventricular ectopic beats were observed in 63 subjects, being multifocal in 13. Supraventricular tachycardia was observed in four subjects and ventricular tachycardia in two subjects. Ten had bradyarrhythmias and two had grade II heart block.

BRODSKY et al. (1977) studied 50 medical students. Supraventricular ectopic beats were noted in 28 and ventricular ectopic beats in 25. Marked sinus arrhythmia was observed in 25. Fourteen had sinus pauses of more than 1.75 s usually during sinus arrhythmia. Transient type 1 second-degree block was seen in three. They concluded that supraventricular and ventricular ectopic beats were unusual in a young healthy population but that sinus arrhythmia, sinus pauses and bradycardia were common.

SABOTKA et al. (1981) studied 50 white female volunteers aged between 22 and 28 years, selected from a group of 67. Thirty-two had atrial premature beats with only one having more than 100 in a 24-h period. Twenty-seven subjects had ventricular ectopic beats, only 3 having more than 50 in a 24-h period. One had a three-beat run of ventricular tachycardia. Two had transient type 1 second-degree AV block.

Kostis et al. (1981) studied 51 male and 50 female patients aged 16-68 years who were referred for differential diagnosis of chest pain, but who were subsequently shown to have a normal coronary arteriogram. They noted ventricular ectopic beats in 39 although only 4 had more than 100 in the 24-h period and less than 5 had more than 5 in any one hour.

Gomer et al. (1986) studied 147 actively employed healthy men aged 15-65 years. Ninety-five percent of the men aged 15-39 years had less than 2.9 ventricular extrasystoles/h and the same proportion of men aged 40 years and older had less than 36 ventricular extrasystoles/h.

None of these above studies showed any clear relation between the incidence of either supraventricular or ventricular ectopic beats with sex, blood pressure, cigarette smoking or the consumption of tea, coffee or alcohol. The incidence, however, did increase with age.

Several studies on adolescents, children and infants have been undertaken. Dickinson and Scott (1984) studied 100 healthy boys aged 14-16 years. Sinus arrhythmia was observed in all and was the only variation in 17%, 15% had changes compatible with sinus arrest or temporary complete sinoatrial block and one boy showed type 2 second-degree sinoatrial block. Escape rhythms were seen in 26%, first-degree AV block in 12% and second-degree Mobitz type 1 in 11%. Ventricular ectopic beats were observed in 41%, being unifocal in 75% and multifocal in 25%. Short runs of ventricular tachycardia were observed in 3%. Ventricular premature beats were absent at rates over 110/min, suggesting they were suppressed by even mild exercise. Supraventricular extrasystoles were seen in 44, varying from 1 to 243 during the 24-h period. No runs of supraventricular tachycardia were recorded.

Scott et al. (1980) studied 131 healthy boys aged 10-13 years, using 48-h recordings. Sinus arrhythmia was recorded in all and in 36 (25%) there was no other change. No Mobitz type 1 second-degree block was recorded but Mobitz type 2 occurred in two boys. Complete sinoatrial block was seen in 8.4% but never lasted more than one cycle and was always followed by a junctional beat. Premature beats were always single, being atrial in 13% and ventricular in 26%. Except in two boys they never occurred at a frequency of more than four in the 24-h period.

Southall et al. (1981) monitored 49 boys and 43 girls aged between 7 and 11 years. At the lowest rates 41 (45%) had junctional escape rhythms, the maximum duration of which was 25 min. Nine had P-R intervals in excess of 0.2 s including three with Mobitz type 1 block. Nineteen (21%) had isolated supraventricular ectopic beats or ventricular ectopic beats. Sixty (65%) had sinus pauses indistinguishable from sinus arrest or sinoatrial block.

Southall et al. (1980) monitored 134 infants under the age of 10 days. At the lowest rates 109 had sinus bradycardia and 25 had junctional escape rhythms. Atrial ectopic beats were observed in 19 but only 1 had more than 12 in 1 h. In a randomly selected subgroup of 71, sinus pauses were seen in 51 (72%). Sinoatrial exit block or sinus arrest was seen in 5 (7%). 2:1 sinoatrial block was seen in 8 (11%) and Wenkebach pauses in 23 (32%).

Table 3. Twenty-four-hour monitoring: reported incidence of supraventricular (SVE) and ventricular ectopic (VE) beats, complex ventricular ectopy (frequent, multiform, R on T, couplets) and ventricular tachycardia (VT) in "normal" subjects

	Subjects		SVE	VE	Complex	
	No.	Age			VE	VT
HINKLE and CASHMAN et al. (1969)	225	55				
RAFTERY et al. (1976)	52	20-70	10	9		
CLARKE et al. (1976)	86	16-65		63	13	2
PROBLETE et al. (1978)	30	47		10	4	
BRODSKY et al. (1977)	50		28	25		
SABOTKA et al. (1981)	50	22-28	32	27	3	1
KOSTIS et al. (1981)	101	49		39	4	
GOMER et al. (1986)	147	15-39	95% < 21/h	95% < 2.9/h	Rare	4
		> 40	95% < 19/h	95% < 36/h		
DICKINSON and SCOTT (1984)	100	14-16	26	4		3
SCOTT et al. (1980)	131	10-13	17	34		0
SOUTHALL et al. (1981)	92	7-11	19	1		
SOUTHALL et al. (1980)	134	< 10 days	19			

The incidence of supraventricular and ventricular ectopic beats, complex ventricular ectopy (frequent, multiform, R on T or couplets) and ventricular tachycardia in the above studies is summarized in Table 3.

III. Exercise

LAMB and HIS (1962) studied 1851 military flying personnel aged 25-45 years using a double Master's test: 290 (15.7%) developed ventricular ectopic beats and 141 (7.5%) developed supraventricular ectopic beats. They noted that in 431 who showed ectopic activity in the resting electrocardiogram before exercise, exercise either induced or aggravated arrhythmia in 39 (9%).

MCHENRY et al. (1972) studied 650 Indiana State policemen aged 25-54 years, of whom 561 were considered to have normal hearts; the results are tabulated according to age groups in Table 4. Fifty-one showed supraventricular ectopic beats, 183 showed ventricular ectopic beats. Complex ventricular ectopic activity was observed in 32 and ventricular tachycardia in 6.

One hundred and sixty-three normal subjects with a median age of 43 years were studied by JELINEK and LOWN (1974) using either treadmill or bicycle exercise. Ventricular ectopic activity was observed in 31 (19%) with complex ventricular ectopic beats in 7 (4.3%) and ventricular tachycardia in 3 (1.8%).

MCHENRY et al. (1976) studied 141 patients with chest pain but normal coronary arteriograms (group A) and 144 age-matched "controls" with no coronary angiographic evidence of abnormal anatomy (group B). The total incidence of ventricular ectopic beats in group A was 16 and in group B was 44,

Table 4. Exercise: reported incidence of supraventricular (SVE) and ventricular (VE) ectopic beats, complex ventricular ectopy (frequent, multiform, R on T, couplets) and ventricular tachycardia (VT) in "normal" subjects. Figures refer to the number of subjects

	Subjects		Exercise type	SVE	VE	Complex	
	No.	Age				VE	VT
LAMB and HIS et al. (1962)	1851	25-45	Masters	141	290		
MCHENRY et al. (1972)	266	25-34	Treadmill	17	77	9	1
	237	35-44		26	81	19	4
	58	45-54		8	25	4	1
JELINEK and LOWN et al. (1974)	163	43	Treadmill or bike		31	7	3
MCHENRY et al. (1976)	141 (A)	31-59	Treadmill		22	1	0
	144 (B)	33-57			61	9	1
FARIS et al. (1976)	462	25-54	Treadmill		144		
PROBLETE et al. (1978)	30	47	Treadmill		2	1	0
NORTHCOTE et al. (1983)	21	23-43	Squash	13	8		2
FLEG and LAKATTA et al. (1984)	922	21-96	Treadmill				10
TOFF et al. (1986)	920	16-75	Treadmill	36	35	67	17

with complex ventricular ectopic beats in 1 and 9 respectively and ventricular tachycardia in 1 of the group B.

Two treadmill exercise tests were performed on Indiana policemen at just under 3 years apart (FARIS et al. 1976). These authors found the reproducibility disappointing, being little better than that expected by chance. PROBLETE et al. (1978) compared exercise testing with 24-h monitoring and the combination of the two. They observed ventricular ectopic beats in 2 of 30 normal subjects during exercise testing, in 10 of the 30 during 24-h tape recording and in 12 when both tests were used. They concluded that the 24-h monitor was a more useful test in demonstrating arrhythmic activity than the exercise test but the combination of both was better still.

NORTHCOTE et al. (1983) studied 21 squash players aged 23-43 years. Seven had one ventricular premature beat during play but only three had more than five during 40 min play. Two of these had one episode of ventricular tachycardia, one a four-beat salvo and the other a nine-beat salvo. The latter subject, aged 40 years, had coupled ventricular ectopic beats. In the first 30 min after play, seven had one ventricular premature beat and three had more than five. Most of these occurred in the first 9 min. One subject had a five-beat run of ventricular tachycardia. Thirteen had one supraventricular ectopic beat during play, four having more than five. Six had occasional supraventricular ectopic beats after exercise. Of the eight subjects who had ventricular ectopic beats at some stage, five had at least one risk factor although X^2 tests did not show any difference in the relationship to risk factors. Each subject who developed ventricular ectopic beats during or immediately post exercise had maximal treadmill exercise testing. All had excellent exercise tolerance and only one had ventricular premature beats in the form of three isolated beats.

Fleg and Lakatta (1984) studied 597 men and 325 women aged 21-96 years from the Baltimore Longitudinal Study on Ageing. None had any clinically evident heart disease. Ventricular tachycardia was observed in ten. Only one of these subjects was younger than 65 years of age. All episodes were asymptomatic and unsustained, the longest run being a six-beat salvo. Multiple salvos occurred in four subjects. After 2 years follow-up all subjects were reported to be well.

Toff et al. (1986) performed a total of 1346 exercise tests on 920 asymptomatic aircrew/allied personnel, aged 16-75 years. Exercise induced atrial fibrillation in 5, supraventricular tachycardia in 1 and ventricular ectopic beats in 35, with multiform beats in 16, couplets in 44 and salvos of ventricular tachycardia in 17. Conduction disturbance was observed in 16 (sinoatrial block in 3, left bundle branch block in 8, right bundle branch block in 4, left anterior hemiblock in 1 and left posterior hemiblock in 1). The authors commented that although the incidence of important rhythm conduction disturbances on exercise in this population was low these abnormalities accounted for 8.7 of the 104 professional aircrew licences withdrawn for cardiovascular reasons during the 7-year period of the study.

A summary of the reported incidence of supraventricular and ventricular ectopy and ventricular tachycardia in these studies is presented in Table 4.

C. Emotionally Induced Arrhythmia

I. ECG Evidence

For over half a century it has been recognized that emotion may influence the ECG by affecting heart rate, conduction or the ST/T wave configuration. In 1918 Hyde and Scalapino observed the effect of music on a person who was particularly fond of music. They noticed changes in heart rate and R wave amplitude in response to Tchaikovsky's Death Symphony, The Toreador's Song and The Souza National Emblem.

Ten normal subjects and 27 with mental illness, all with apparently normal healthy hearts, were subjected to an unexpected loud noise. Many showed extrasystoles immediately following the stimulus (Landis and Slight 1929).

Loftus et al. (1945) described a 26-year-old female who developed a wandering pacemaker at times of pathological sexual excitement associated with obscene behaviour and shouting her desire for intercourse.

Electrocardiograms were recorded on 193 patients with psychiatric illness by Blom (1951). He noted sinus bradycardia in four; marked sinus arrhythmia in ten; a wandering pacemaker in one; ventricular, atrial and nodal ectopics in three, three and one respectively and a P-R interval of >0.2 s in four.

An apparently healthy young aviator developed left bundle branch block within a few seconds of the shock of an unexpected pistol shot being fired

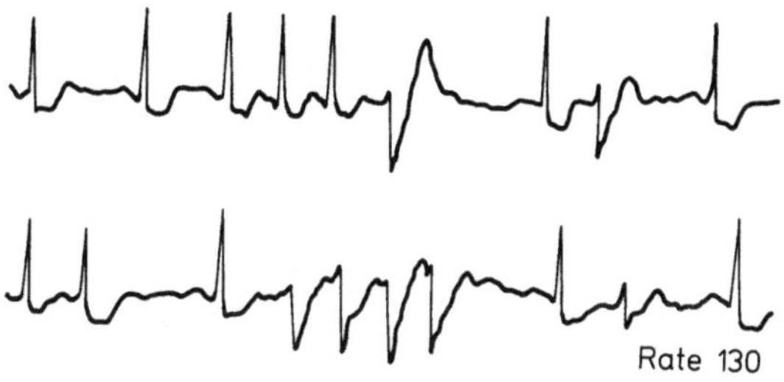

Fig. 1. Male aged 56 years with coronary heart disease monitored while speaking to an audience, showing ST depression, supraventricular and ventricular ectopic beats, together with a short run of ventricular tachycardia. The control trace was normal

close to him (GRAYBIEL et al. 1944). The conduction disturbance subsequently returned to normal.

BENEDICT and EVANS (1952) described a 27-year-old Negro who developed first- and second-degree heart block which fluctuated according to his emotional state. Five of 24 persons with coronary heart disease developed ventricular arrhythmia while driving in city traffic. Two had runs of ventricular tachycardia. No arrhythmia was observed in 32 normal subjects in the same circumstances (TAGGART et al. 1969). Seven people with coronary heart disease and 23 with apparently normal hearts were monitored during the more intense emotion of public speaking. Multiple and often multifocal ventricular and supraventricular ectopic beats were observed in five of the seven people with coronary heart disease (Fig. 1). Ventricular ectopic beats at a rate of more than 6/min were noted in 5 of the 23 normal subjects. In one subject they

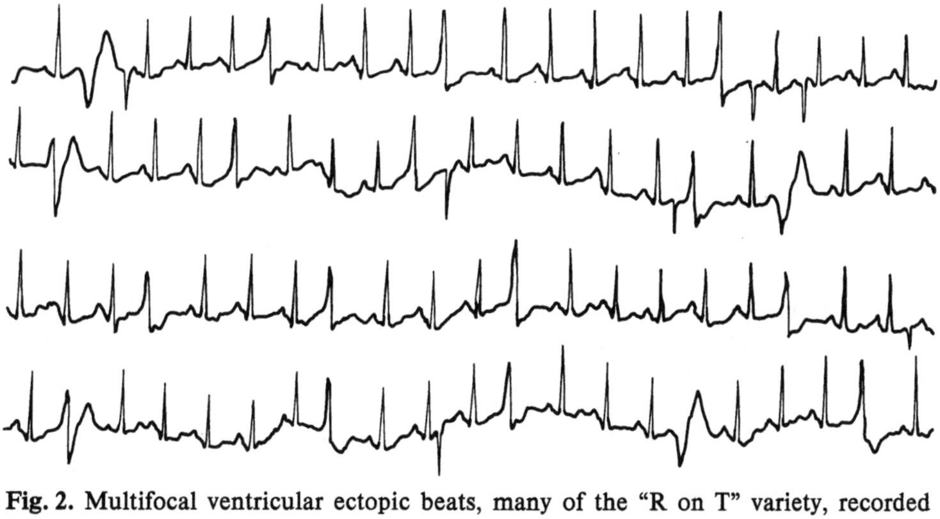

Fig. 2. Multifocal ventricular ectopic beats, many of the "R on T" variety, recorded from a male doctor aged 46 years with no apparent heart disease while speaking to an audience

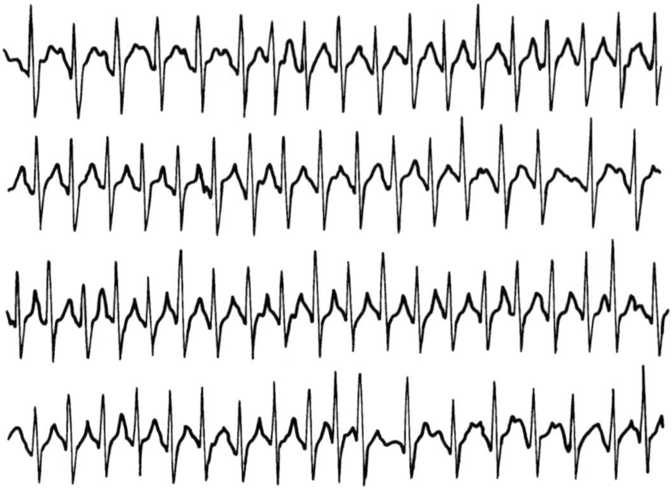

Fig. 3. Continuous strip recorded from a male parachutist, age 36 years. Irregular tachycardia during the free fall period (see text)

were, at times, prolific and often multifocal (Fig. 2) (TAGGART et al. 1973). Ten racing drivers were monitored during competition. Heart rates were in the region of 180/min throughout. No arrhythmia was observed during the event or in the slowing down period afterwards (TAGGART et al. 1969).

Fifteen subjects of varying experience were monitored during a parachute descent. Ectopic beats at a rate of more than six per minute developd in four, although these were not prolific, multifocal or of R on T type. Sinus arrhythmia was prominent on landing but never during the anticipatory phase. A complex rhythm developed during the anticipatory phase in one subject (Fig. 3). The rhythm is irregular and the heart rate 180/min. The longest R-R intervals are equivalent to a heart rate of 135/min (second strip) and the shortest to a rate of 300/min (bottom strip). This is probably a chaotic atrial tachycardia with a least three separate foci in competition. No arrhythmia was observed in any of the other subjects (TAGGART et al. 1983).

II. Emotion and Sympathetic Activity

Many emotional stress situations are associated with a tachycardia ranging from a little over 100/min during the everyday stress of driving; through about 120-150/min during the more intense emotion of public speaking; and to 165-180/min during the more extreme challenges of racing car driving and parachute jumping (TAGGART et al. 1983).

Many studies have shown increased catecholamine release. VON EULER and LUNDBERG (1954) were the first to demonstrate an increased catecholamine excretion in response to anxiety. They noticed a fourfold increase in urinary noradrenaline and adrenaline excretion in pilots during flight compared with the same pilots during ground activities. Since then a number of studies have been undertaken demonstrating increased catecholamine excretion response to emotions of different types and intensities, amongst which will be mentioned a few.

A considerable volume of work has emanated from the Karolinska Institute in Stockholm under the direction of LEVI (1971). Twelve female clerks had their pay basis changed from their usual salary system to an accelerated piece rate system. The urinary excretion of adrenaline increased by 40% and noradrenaline by 20% (LEVI 1964). Twenty female clerks were shown four films portraying different moods. A tranquil or boring film was associated with no change in the urinary catecholamine excretion whereas films portraying excitement or humour were associated with increases (LEVI 1965).

Young military subjects performed on a shooting range for 75 h without being aware of the time of day or whether or not it was light or dark. Urinary adrenaline excretion showed a smooth sinusoidal curve with pronounced diurnal variation, the highest levels occurring in the afternoon and the lowest after midnight. The lowest levels coincided with the worst performance and the greatest feelings of fatigue (LEVI 1966).

PATKAI et al. (1967) studied 52 students in response to stressful psychological tests and noted that those with a higher adrenaline excretion rate performed better.

The mere process of admission to hospital has been shown to be associated with an increase in catecholamine excretion (TOLSON et al. 1965).

Several workers have studied the effect of examinations, showing increased catecholamine production (BOGDONOFF et al. 1960; JONES et al. 1968). We have observed increased plasma catecholamine concentrations in normal subjects driving cars in city traffic, speaking in public, racing car driving and parachute jumping (TAGGART et al. 1983). The plasma concentrations roughly parallel the intensity of the arousal, as judged by subjective and objective assessment and the degree of associated tachycardia.

III. Arrhythmogenic Action of Catecholamines and Enhanced Sympathetic Activity

Catecholamines and sympathetic stimulation are well known to encourage arrhythmia formation (WIT et al. 1985; Chap. 14). Following experimental coronary artery occlusion arrhythmias may be substantially reduced by thoracic sympathectomy (COX and ROBERTSON 1936) and cardiac denervation (SCHAAL et al. 1969; EBERT et al. 1970) whereas an increased sensitivity to arrhythmias has been shown with stellate ganglion stimulation (HARRIS 1966). The association of palpitation, tachycardia and arrhythmias in patients with phaeochromocytoma is well known. These occur irrespective of whether the tumour secretes predominantly noradrenaline (SAYER et al. 1954) or adrenaline (PAGE et al. 1969). Many arrhythmias may be suppressed by beta-adrenergic blockade, as discussed in Chap. 2. Oxprenolol has been shown to be effective in suppressing emotionally induced arrhythmic activity in normal subjects monitored while speaking before an audience (TAGGART et al. 1973). Not all types of emotional stress are associated with a tachycardia. Some situations are associated with parasympathetic overactivity.

IV. Emotion and Parasympathetic Activity

Not all forms of emotional arousal are associated with a tachycardia. Certain situations involving the witness, experience, or even the anticipation of pain may produce a bradycardia. The medical student watching his/her first operation who develops vasovagal syncope provides a spectacular example. A study of young healthy people watching violent films showed a reduction of heart rate below resting levels in spite of a greatly increased urinary excretion of catecholamines (CARRUTHERS and TAGGART 1973).

Dentistry has been shown to be associated with an absence of increase in heart rate in the presence of greatly increased plasma adrenaline levels (Figs. 4, 5) (TAGGART et al. 1976).

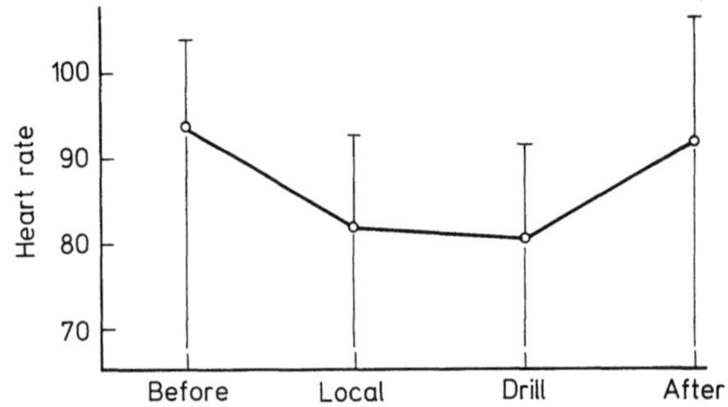

Fig. 4. Mean heart rate (± SD) in 11 young women undergoing dental surgery at 10 min before local anaesthetic (*left*); 1 min before local anaesthetic (*left centre*); during drilling (*right centre*); and after

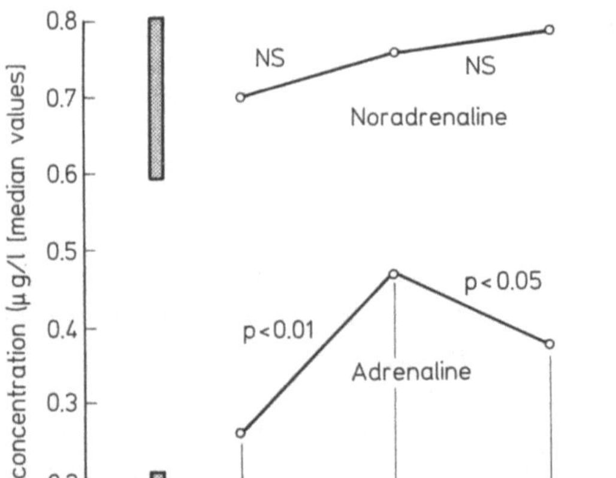

Fig. 5. Plasma noradrenaline and adrenaline concentrations in 11 young women undergoing dental procedures. *Shaded areas* represent normal ranges. NS, not significant

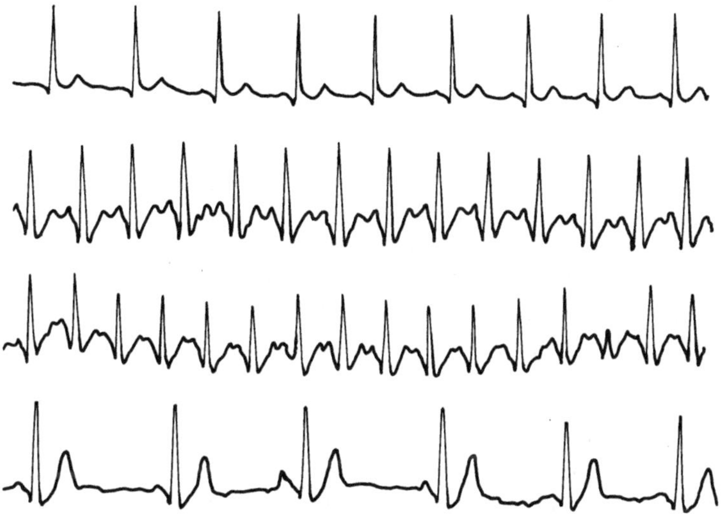

Fig. 6a–d. Male novice parachutist, aged 22 years. *A* **a** Control; *B* **b** shortly before leaving the aircraft; **c** *C* during descent; and *D* **d′** a few seconds after landing and fracturing his ankle

These observations would suggest that vagal activity was dominant and overriding greatly enhanced sympathetic activity. In this context one anecdotal observation may be mentioned. A free-fall parachutist was monitored with a tape-recorded ECG. During the fall his heart rate increased to 185/min. This rate persisted until he landed. Immediately upon landing he fractured his ankle and the rate dropped instantaneously to 60/min. A blood sample was taken within seconds of landing and showed greatly increased plasma adrenaline concentration (Fig. 6). He had been free of pain at the time of the precipitous fall in heart rate and when the blood sample was taken. This might suggest that the anticipation of pain and the realization of his injury produced vagal activity which again dominated the greatly increased sympathetic activity as evidenced by the elevated plasma adrenaline concentration. The above examples illustrate the principle of accentuated antagonism whereby the higher the level of sympathetic activity the greater is the depressant effect of a given level of vagal interaction (LEVY 1977.

V. Individual Susceptibility and Variation

The question arises as to why arrhythmias occur in some people and not in others. Several factors are probably involved. Different situations may produce different autonomic (sympathetic/parasympathetic) responses. Some personalities seek out stress situations whereas others dread them. What is pleasurable to one person may be terrifying to another, eg. parachute jumping. Some people are more easily affected by stress ("sensitive") than others ("thick-skinned").

Attempts have been made to separate the physiology of specific emotions, in particular those of fear, anger and aggression, and to equate these with different secretion patterns of adrenaline and noradrenaline. VON EULER (1964), who reviewed the available evidence, considered that the passive emotion of fear equated more closely with adrenaline and the active aggressive emotion, such as anger, equated more with noradrenaline.

LEPESCHKIN et al. (1960) infused adrenaline and noradrenaline separately in 60 young women and 40 men, all with apparently normal hearts. Noradrenaline infusion was associated with ectopic rhythm only when the blood pressure level was high and the heart rate was low, whereas they occurred at all heart rates following adrenaline infusion. These differences in the action of the two hormones was considered to be due to the vagal reflex effects accompanying the hypertension induced by noradrenaline. Some people may be electrophysiologically more labile and thereby more susceptible than others. For example, electrocardiographic ST and T wave changes indistinguishable from those of myocardial ischaemia occur in a substantial number of asymptomatic people both at rest and on exercise. These people are often young, athletic, with normal echocardiograms and angiographic coronary artery anatomy (TAGGART et al. 1979, 1982; MARCOMICHELAKIS et al. 1980). Characteristic of ambulatory ECG recordings in these subjects is that the ECG abnormalities come and go. They have been shown to be catecholamine related since they may be readily manipulated by adrenaline infusion and beta-blockade. In order for the ST-T configuration to vacillate between normal and abnormal patterns there must be coincidental changes in the usual regional differences in repolarization times of action potentials in different parts of the heart which summate to form the clinical ECG. This suggest that in a proportion of people variations in regional repolarization may occur readily in response to fluctuations in autonomic balance. Such regional differences would be expected to favour re-entry.

D. Prognosis of Ectopy and Arrhythmia in Apparently Healthy People

I. Historical Note

Roughly 2000 years ago HEROPHILUS described irregularities which were almost certainly extrasystoles (ARCIERI 1945). Subsequent writers referred to "the intermittent pulse". In the second century A. D. GALEN instilled the belief that it was an indicator of disease and implied a poor prognosis (SCHERF and SCOTT 1973). This view prevailed until the seventeenth century.

It is likely that this belief was strengthened by the inclusion of a number of more malignant arrhythmias amongst accounts of the "intermittent pulse" and also patients exhibiting arrhythmia due to severe systemic illness. Although the majority of writers had adhered to Galen's original doctrine, several authors disagreed and by the early twentieth century the pendulum of medical opinion was swinging towards classifying extrasystoles as innocent phenomena which were of little significance in the absence of disease (HORAN and KENNEDY 1984).

In 1906 the development of the string galvanometer by EINTHOVEN provided a means of recording and analyzing different types of arrhythmic activity. This facility, combined with data from actuarial studies, supported the benign view of extrasystoles and by the mid-1960's it was considered that ventricular ectopy was not necessarily a manifestation of disease, but occurred in normal people (HORAN and KENNEDY 1984).

II. Ectopy and Arrhythmia in Normal People

Assessment of the importance of arrhythmia occurring in apparently healthy people is handicapped by uncertainties as to the "normality" of the study population as mentioned earlier. Ventricular ectopy has commanded the greatest attention.

The configuration of ventricular ectopic beats is not specifically related to the presence or absence of disease (BODENHEIMER et al. 1977). In general, normal subjects have a tendency to right ventricular ectopic beats (ROSENBAUM 1969; MANNING et al. 1968) although they are not confined solely to the right ventricle (KOSTIS et al. 1981). The reasons are not clear. The configuration recorded on the electrocardiogram depends not only on the site of the focus of origin but also on the site of exit, which may be in either the right or left ventricle.

Considerable day-to-day variation of frequency of ventricular ectopic beats occurs in both normal subjects and those with coronary heart disease (KOSTIS et al. 1981; MORGANROTH et al. 1978). ECG monitoring for a short period suggests a bimodal distribution of ventricular ectopic beats, one group with a low incidence probably composed of normal subjects and another group with a high incidence, probably composed of people with abnormal hearts. When monitoring is continued for up to 3 days, however, the trends tend to even out, suggesting that ventricular ectopy of low frequency occurs in the majority of people.

Infrequent uniform and non-repetitive ventricular ectopic beats are well established as having a good prognosis in healthy persons (RUSKIN 1985; HORAN and KENNEDY 1984; ROSE et al. 1978). Whereas frequent or complex ventricular ectopy in persons with coronary heart disease or myocardial disease identifies a subset who are at greater risk of sudden death, subjects who are asymptomatic and apparently healthy, as determined by non-invasive techniques have a relatively good long-term prognosis (BARRETT et al. 1981). KENNEDY et al. (1985) followed up 73 asymptomatic subjects with frequent or complex ventricular ectopic beats for a mean period of 6.5 years. The subjects have been shown on 24-h monitoring to have multiform ectopy in 63%, couplets in 60% and ventricular tachycardia in 26%. They concluded that the long-term prognosis in asymptomatic healthy subjects with frequent and complex ventricular ectopy was similar to that of the healthy American population and that there was no increased risk of death. However, RUSKIN (1985) pointed out that the subjects involved were a highly select subgroup of patients which might have induced bias towards the favourable outcome, although even taking this into account the overall prognosis is nevertheless excellent.

III. Sport

1. Athletes

Sudden death in competitive athletes is rare and most commonly associated with cardiomyopathy or anomalous coronary arteries (MARON et al. 1980). A wide variety of abnormalities in the electrocardiogram has been reported (for references see VIITASALO et al. 1982). These authors describe ambulatory ECG recordings on 35 highly trained athletes and 35 control subjects. Sinus pauses exceeeding 2 s were seen in 37% of the athletes compared with 5.7% of the control group; first-degree AV block (type I) occurred in 23% compared with 6% of controls; second-degree AV block (type II) occurred in 8.6% of the athletes but in no control subjects. Fewer ventricular extrasystoles were observed in the athletes as compared with the control subjects. Two control subjects had ventricular tachycardia. The authors noted that blocks invariably disappeared as heart rate increased and considered that the findings, particularly as regards the AV blocks, represented a physiological phenomenon.

2. Non-athletes

NORTHCOTE et al. (1986) described 60 sudden cardiac deaths associated with squash. Coronary heart disease was found to be present in 51. In two, no pathological abnormality was found on autopsy and death was considered to be due to an arrhythmia. A number of other studies have included accounts of persons dying of no obvious cause and which have been ascribed to rhythm disturbances. Arrhythmias may occur in response to thermal stress (TAGGART et al. 1972) and many sports are associated with an excessive increase in body temperature. Metabolic factors related to exercise may encourage arrhythmia formation. Exercise increases both free fatty acid and catecholamine concentrations, which are known to be associated with arrhythmia (JOHNSON et al. 1969; KURIEN et al. 1971). In addition, exercise-induced hyperkalaemia (LINTON et al. 1984) and lactic acidosis (BOUHUYS et al. 1966) in combination with elevated catecholamines may also encourage arrhythmic activity (TAGGART et al., in preparation).

Arrhythmias are common in the post-exercise period (ADAMS 1972). Immediately after ceasing exercise venous return is reduced as a result of pooling. Continued enhanced sympathetic activity may perpetuate an oxygen debt and also increase circulating free fatty acid concentrations. Smoking also increases circulating free fatty acids (KERSHBAUM et al. 1961). The hot bath or shower and cigarette after the game may therefore be a time of more mental than electrophysiological relaxation.

E. Emotion and Sudden Death

The possibility that emotion may cause sudden cardiac death as a result of the development of a malignant arrhythmia has been the subject of speculation for many years. Evidence, however, is lacking in view of the suddenness of the occurrence and the unavailability of the only knowledgeable witness, the victim. The majority of the victims of sudden cardiac death have known coronary artery disease although a few have apparently normal hearts at postmortem. It is highly probably that emotion is responsible for a number of sudden deaths amongst those with coronary disease. Emotion is well known to induce angina in susceptible subjects at which time they would be at risk of arrhythmia. Emotion has been shown to induce ST/T segment changes and arrhythmias in persons with coronary disease in response to stressful emotions and even relatively mild emotions (TAGGART et al. 1973). To what extent emotion may induce lethal arrhythmias in people with normal hearts must remain a matter of conjecture. However, there is a vast amount of anecdotal evidence, some of which is presented below.

CANNON (1942) described a number of reports from the anthropological literature of healthy people dying without any obvious explanation after being intimidated or by terrifying augury or prediction of death by a "medicine man". An example is quoted, taken from the writings of Dr. Herbert BASEDOW from his book, *The Australian Aboriginal*, in which he describes vivid pictures of the horrifying effect of bone pointing on the ignorant, superstitious natives:

> The man who discovers that he is being boned by an enemy is, indeed, a pitiable sight. He stands aghast, with his eyes staring at the treacherous pointer, and with his hands lifted as though to ward off the lethal medium, which he imagines is pouring into his body. His cheeks blanch and his eyes become glassy and the expression of his face becomes horribly distorted ... he attempts to shriek but usually the sound chokes in his throat, and all that one might see is froth at his mouth. His body begins to tremble and the muscles twitch involuntarily. He sways backwards and falls to the ground, and after a short time appears to be in a swoon; but soon after he rises as if in mortal agony, and, covering his face with his hands, begins to moan. After a while he becomes very composed and crawls to his wirley. From this time onwards he sickens and frets refusing to eat and keeping aloof from the daily affairs of the tribe. Unless help is forthcoming in the shape of a counter-charm administered by the hand of the Nangarri, or medicine man, his death is only a matter of a comparatively short time. If the coming of the medicine man is opportune he might be saved.

Dr. BASEDOW goes on to say that if the medicine man arrives the recovery is amazing. The victim, who until the moment of his arrival was close to death, raises his head and gazes in wonderment at the medicine man, and his recovery is speedy. Many such examples are quoted in CANNON's text. The possibility was raised that poison may be involved and administered by the medicine man perhaps, in order to further his reputation of exercising supernatural power. However, this was considered overall to be unlikely since many such episodes were observed by medical men and other reliable witnesses who considered this question. It was CANNON's considered opinion that these deaths were due to extreme overactivity of the sympathetic nervous system.

ENGEL (1971) described reports of 170 sudden deaths from newspaper clippings and classified them according to the type of emotion and circumstance surrounding the death. Many of the victims were in an older age group and had established disease, either cardiac or other. It could therefore be argued that in many of the instances quoted any association with emotion or a particular emotion was fortuitous. However, a number of deaths were reported in apparently healthy people in whom the relation to fear was clear. The youngest reported cases were a 3-year-old child who died when caught in a severe downpour and a terrified 4-year-old girl who died while having some milk teeth extracted. A 35-year-old man was described who was accused of robbery. He told his lawyer in the courtroom "I'm scared to death"! Whereupon he immediately collapsed and died.

ENGEL (1971) quotes numerous examples of sudden death occurring among animals in response to psychological stress. These include trapped animals dying when escape is impossible; deaths after fights without injury and after the deaths of mates; the death of a llama within minutes of seeing her mate of 13 years shot and killed; and in animals subjected to restraint. He comments that whereas most writers tend to stress the sympathetic response to stress, that vagal influences are frequently observed and suggests that some of the lethal arrhythmias may be induced by rapid shifts between sympathetic and parasympathetic. This relationship he illustrates by vasodepressor syncope which characteristically occurs in settings in which the impulse to flee is inhibited, when tachycardia is then replaced by bradycardia. REICH (1985) studied the circumstances surrounding the onset of life-threatening ventricular arrhythmias. They found a high association with intense or unusual emotional states, often involving anger. Such an association was present in approximately 20% of cases. Their patients included a higher proportion of subjects with no apparent organic heart disease. Reviewing the literature he concluded that an acute emotional disturbance played a role in the development of arrhythmias but this role was limited. It was considered likely that subgroups could be identified with specific vulnerabilities to psychological stress and that persons with arrhythmias in the absence of apparent structural heart disease were such a subgroup. ELIOTT and BUELL (1983) considered that some individuals whom they described as "hot reactors" have an excessive cardiac response to stress. Patients with the long Q-T syndrome are an example of a population who are highly prone to arrhythmia during sympathetic arousal. However, theirs are not strictly speaking normal hearts.

It has been known for many years that neural mechanisms are involved in the genesis of arrhythmias (MALLIANI et al. 1980; LOWN et al. 1977). Stimulation of cardiac sympathetic nerves and stellate ganglia may increase the vulnerability to ventricular fibrillation. A number of animal studies have shown that stress may substantially reduce the ventricular fibrillation threshold (LOWN 1982). WELLENS et al. (1972) recorded the onset of ventricular fibrillation in a young girl when she was woken by thunder and subsequently on each occasion when she was woken by an alarm clock.

F. Clinical Considerations

It is apparent that the precise incidence and prognosis of arrhythmias in normal persons remains uncertain, although in general the incidence is low and the prognosis good. This holds true for complex ventricular ectopy for which treatment is not necessarily indicated (KENNEDY et al. 1985; RUSKIN 1985).

However, emotional responses to stress are regarded by many as disturbing and unpleasant. Palpitation, which is a common symptom, may be distracting and alarming and possibly in certain predisposed persons dangerous.

It has been shown that beta-blockade may be useful in preventing stress-induced ectopic activity and palpitation. Oxprenolol administered in a single oral dose of 40 mg was effective in abolishing the tachycardia, palpitation and arrhythmia observed in people monitored while speaking to an audience (TAGGART et al. 1973). The relative absence of side effects of beta-blockade and the low dose required suggest that this may be a useful therapeutic manoeuvre in selected people. Unquestionably the overenthusiastic administration of antiarrhythmic drugs is not to be recommended, particularly in view of the significant incidence of cardiac and other side effects associated with them (RUSKIN et al. 1983; VELEBIT et al. 1982).

Since VAUGHAN WILLIAMS's (1970) original classification the current trend has been to develop antiarrhythmic regimes according to specific electrophysiological requirements. The preceding sections have drawn attention to an interplay between sympathetic and parasympathetic which can be quite profound. It has been shown that rapid swings in dominance of one or other component may occur. Sudden changes of this type in the presence of already greatly elevated catecholamines would be an ideal theoretical setting for arrhythmogenesis. It may be appropriate for future development to take into account the influence of swings in autonomic balance on the electrophysiological behaviour of antiarrhythmic agents.

References

Adams CW (1972) Exercise and the heart. Introduction. Am Cardiol 30:713–715

Arcieri GP (1945) The circulation of the blood; and Andrea Cesalpino of Arezzo. SF Vanni, New York

Barrett BA, Peter CT, Swan JJC, Singh BN, Mandel WJ (1981) The frequency and prognostic significance of electrocardiographic abnormalities in clinically normal individuals. Prog Cardiovasc Dis 22:299–319

Benedict RV, Evans JM (1952) Second degree heart block and Wenkebach phenomenon associated with anxiety. Am Heart J 43:626–633

Blom GE (1951) Review of electrocardiographic changes in emotional states. J Nerv Ment Dis 113:283–300

Bodenheimer MM, Banka VS, Helfant RH (1977) Relation between the site of origin of ventricular premature complexes and the presence and severity of coronary artery disease. Am J. Cardiol 40:865

Bogdonoff M, Harlan W, Estes E, Kirshner N (1959) Changes in urinary catecholamine excretion accompanying carbohydrate and lipid responses to oral examination (Abstr.) Circulation 20:674

Bouhuys A, Poole J, Binhorsl RA (1966) Metabolic acidosis of exercise in healthy males. J Appl Physiol 21:1040–1046

Brodsky M, Denes P, Kanakis C, Rosen KM, Wu D (1977) Arrhythmias documented by 24-hour continuous electrocardiographic monitoring in 15 male medical students without apparent heart disease. Am J Cardiol 39:390–395

Cannon WB (1942) "Voodoo" death. Am Anthropol 44:169-181
Carruthers M, Taggart P (1973) Vagotonicity of violence: biochemical and cardiac responses to violent films and television programmes. Br Med J 3:384-389
Chiang BN, Pearlman LV, Ostrander LD, Epstein FH (1969) Relationship of premature systoles to coronary heart disease and sudden death in the Tecumseh epidemiologic study. Ann Intern Med 70:1159-1166
Clarke JM, Hamer J, Shelton JR, Taylor S, Venning GR (1976) The rhythm of the normal human heart. Lancet 2:508-512
Cox WV, Robertson HF (1936) Effect of stellate ganglionectomy on cardiac function of intact dogs and its effect on extent of myocardial infarction and on cardiac function following coronary artery occlusion. Am Heart J 12:285-300
Dickinson DF, Scott O (1984) Ambulatory electrocardiographic monitoring in 100 healthy teenage boys. Br Heart J 51:179-183
Ebert PA, Vanderbeek RB, Allgood RJ, Sabiston DC Jr (1970) Effect of chronic cardiac denervation on arrhythmias after coronary artery ligation. Cardiovasc Res 4:141-147
Einthoven W (1906) L'éléctrocardiogramme. Arch Intern Physiol 4:132
Elliot RS, Buell JC (1983) The role of the CNS in cardiovascular disorders. Hosp Pract 18:189-199
Engel GL (1971) Sudden and rapid death during psychological stress. Folklore or folk wisdom? Ann Intern Med 74:771-782
Faris JV, Jordan JW, McHenry PL, Morris SN (1976) Prevalence and reproducibility of exercise-induced ventricular arrhythmias during maximal exercise testing in normal men. Am J Cardiol 37:617-621
Fleg JL, Lakatta EG (1984) Prevalence and prognosis of exercise-induced non-sustained ventricular tachycardia in apparently healthy volunteers. Am J Cardiol 54:762-764
Gomer KO, Hogstedt C, Bodin L, Soderholm B (1966) Frequency of extrasystoles in healthy male employees. Br Heart J 55:259-264
Graybiel A, McFarland RA, Gates DC, Webster FA (1944) Analysis of electrocardiograms obtained from 1000 young healthy aviators. Am Heart J 27:524-549
Harris AS (1966) Mechanism and therapy of cardiac arrhythmias. In: Dreifus LS, Likoff W (eds) Genesis of ventricular tachycardia and fibrillation following coronary occlusion. New York, Grune and Stratton, pp 293-301
Hinkle LE, Carver ST, Stevens M (1969) The frequency of asymptomatic disturbances of cardiac rhythm and conduction in middle-aged men. Am J Cardiol 24:629-650
His RH, Lamb LE (1962) Electrocardiographic findings in 122043 individuals. Circulation 25:947-961
Horan MJ, Kennedy HL (1984) Ventricular ectopy. History, epidemiology and clinical implications. JAMA 251:380-386
Hyde IH, Scalapino W (1918) The influence of music upon electrocardiograms and blood pressure. Am J Physiol 14:35-38
Jelinek MV, Lown B (1974) Exercise stress testing for exposure of cardiac arrhythmia. Prog Cardiovasc Dis 16:497-522
Johnson RH, Walton JL, Kerbs HA, Williamson DH (1969) Metabolic fuels during and after severe exercise in athletes and non-athletes. Lancet 2:452-455
Jones MT, Bridges PK, Leak D (1968) Relationship between the cardiovascular and sympathetic responses to the psychological stress of an examination. Clin Sci 35:73-79
Kennedy HL, Buckingham TA, Goldberg RA, Kennedy LJ, Sprague MK, Whitlock JA (1985) Long-term-follow-up of asymptomatic healthy subjects with frequent and complex ventricular ectopy. N Engl J Med 312:193-197

Kershbaum A, Bellet S, Dickstein ER, Feinberg LJ (1961) Effect of cigarette smoking and nicotine on serum free fatty acids: based on a study in the human subject and the experimental animal. Circ Res 9:631-638

Kostis JB, Juo PT, McCrone K, Moreyra AE, Aglitz NM, Gotzoyannis S, Natarajan N (1981) Premature ventricular complexes in the absence of identifiable heart disease. Circulation 63:1351-1356

Kurien BA, Yates PA, Oliver MF (1971) The role of free fatty acids in the production of ventricular arrhythmia after acute coronary artery occlusion. Eur J Clin Invest 1:225-241

Lamb LE, His RH (1962) Influence of exercise on premature contractions. Am J Cardiol 9:209-216

Landis C, Slight D (1929) Studies of emotional reactions; cardiac responses. J Gen Psychol 2:413-420

Lepeschkin E, Marchet H, Schroeder G, Wagner R, Paula D, de Silva P, Raab W (1969) Effect of epinephrine and norepinephrine on the electrocardiogram of 100 normal subjects. Am J Cardiol 5:594-603

Levi L (1964) The stress of everyday work as reflected in productiveness, subjective feelings and urinary output of adrenaline and noradrenaline under salaried and piece-work conditions. J Psychosom Res 8:199-202

Levi L (1965) The urinary output of adrenaline and noradrenaline during pleasant and unpleasant emotional states. Psychosom Med 27:80-85

Levi L (1966) Physical and mental stress reactions during experimental conditions simulating combat. Forsvarsmedicin 2:3-8

Levi L (ed) (1971) Society, stress and disease. Oxford University Press, London

Levy MN (1977) Parasympathetic control of the heart. In: Randall WC (ed) Neural regulation of the heart. Oxford University Press, New York, pp 95-129

Linton RAF, Lim M, Wolff CB, Wilmhurst P, Band DM (1984) Arterial plasma potassium measured continuously during exercise in man. Clin Sci 67:427-431

Loftus TA, Gold H, Diethelm O (1945) Cardiac changes in presence of intense emotion. Am J Psychiatry 101:697-698

Lown B (1982) Mental stress, arrhythmias and sudden death. Am J Med 72:177-180

Lown B, Verrier RL, Rabinowitz SH (1977) Neural and psychologic mechanisms and the problem of sudden cardiac death. Am J Cardiol 39:890-902

Malliani A, Schwartz PJ, Zanchetti A (1980) Neural mechanism in life threatening arrhythmias. Am Heart J 100:705-715

Manning GW, Ahuja SP, Gutierrez MR (1968) Electrocardiographic differentiation between ventricular ectopic beats from subjects who have normal and diseased hearts. Acta Cardiol (Brux) 23:462

Marcomichelakis J, Donaldson R, Green J, Joseph S, Kelly HB, Taggart P, Somerville W (1980) Exercise testing after beta-blockade: improved specificity and predictive value in detecting coronary heart disease. Br Heart J 43:252-261

Maron BJ, Roberts WC, McAllister HA, Rosing DR, Epstein SE (1980) Sudden death in young athletes. Circulation 60:218-229

McHenry PL, Fish C, Jordan JW (1972) Cardiac arrhythmias during maximal treadmill exercise testing in clinically normal men. Am J Cardiol 29:331-336

McHenry PL, Jordan JW, Kavalier M, Morris SN (1976) Comparative study of exercise induced ventricular arrhythmias in normal subjects and patients with documented coronary artery disease. Am J Cardiol 37:609-616

Morganroth J, Dunkman WB, Horowitz LN, Josephson ME, Michelson EL, Pearlman AS (1978) Limitations of routine long-term electrocardiographic monitoring to assess ventricular ectopic frequency. Circulation 58:408

Northcote RJ, Ballantyne D, MacFarlane (1983) Ambulatory electrocardiography in squash players. Br Heart J 50:372-377

Northcote RJ, Flannigan C, Ballantyne D (1986) Sudden death and vigorous exercise—a study of 60 deaths associated with squash. Br Heart J 55:198-203

Page LB, Riker JW, Berberich FR (1969) Phaeochromocytoma with predominant epinephrine secretion. Am J Med 47:648-652

Patkai P, Frankenhaeuser M, Rissler A, Bjorkvall C (1967) Catecholamine excretion, performance and subjective stress. Scand J Psychol 8:113-122

Problete PF, Kennedy HL, Caralis DG (1978) Detection of ventricular ectopy in patients with coronary heart disease and normal subjects by exercise testing and ambulatory electrocardiography. Chest 74: 402-407

Raftery EB, Cashman PMM (1976) Long term recording of the electrocardiogram in a normal population. Postgrad Med J [Suppl 17] 52:32-37

Reich P (1985) Psychological predisposition to life threatening arrhythmias, Annu Rev Med 36:397-405

Rose G, Baxter PJ, Reid DD, McCartney P (1978) Prevalence and prognosis of electrocardiographic findings in middle-aged men. Br Heart J 40:636-643

Rosenbaum MB (1969) Classification of ventricular extrasystoles according to form. J Electrocardiogr 2:289

Ruskin JN (1985) Ventricular extrasystoles in healthy subjects. Engl J Med 312:238-239

Ruskin JN, McGovern B, Garen H, DiMarco JP, Kelly E (1983) Anti-arrhythmic drugs: a possible cause of out-of-hospital cardiac arrest. N Engl Med 309:1302-1306

Sabotka PA, Mayer JH, Bauernfeind RA, Kanakis C, Rosen KM (1981) Arrhythmias documented by 24-hour continuous ambulatory electrocardiographic monitoring in young women without apparent heart disease. Am Heart J 101:753-759

Sayer WJ, Moser M, Mattingly TW (1954) Phaeochromocytoma and abnormal electrocardiograms. Am Heart J 48:42-53

Schaal SF, Wallis AG, Sealy WC (1969) Protective influence of cardiac denervation against arrhythmias of myocardial infarction. Cardiovasc Res 3:241-244

Scherf D, Schott A (1973) Extrasystoles and allied arrhythmias. 2nd edn. Year Book Medical Publishers, Chicago, p 2

Scott O, Williams GJ, Fiddler GI (1980) Results of 24-hour ambulatory monitoring of electrocardiogram in 131 healthy boys aged 10-13 years. Br Heart J 44:304-308

Southall DP, Richard J, Mitchell P, Brown DJ, Johnston PGB, Shinebourne EA (1980) Study of cardiac rhythm in healthy newborn infants. Br Heart J 43:14-20

Southall DP, Johnston FG, Shinebourne EA, Johnston PGB (1981) 24-hour electrocardiographic study of heart rate and rhythm patterns in population of healthy children. Br Heart 45:281-291

Taggart P, Gibbons D, Somerville W (1969) Some effects of motor car driving on the normal and abnormal heart. Br Med J 4:130-134

Taggart P, Carruthers M, Parkinson P (1972) Cardiac responses to thermal physical and emotional stress. Br. Med J 3:71-76

Taggart P, Carruthers M, Somerville W (1973) Electrocardiograms, plasma catecholamines and lipids, and their modification by oxprenolol when speaking before an audience. Lancet 2:341-346

Taggart P, Hedworth-Whitty RB, Carruthers M, Gordon PD (1976) Observations on the electrocardiogram and plasma catecholamines during dental procedures; the forgotten vagus. Br Med J 2:787-789

Taggart P, Carruthers M, Joseph S, Kelly HB, Marcomichelakis J, Noble D, O'Neill G, Somerville W (1979) Electrocardiographic changes resembling myocardial ischaemia in asymptomatic men with normal coronary arteriogram. Br Heart J 41:214-225

Taggart P, Donaldson R, Green J, Joseph S, Kelly HB, Marcomichelakis J, Noble D, White J (1982) The inter-relationship of heart rate and autonomic activity in asymptomatic men with unobstructed coronary arteries: studies by atrial pacing, adrenaline infusion and autonomic blockade. Br Heart J 47:19-25

Taggart P, Carruthers M, Somerville W (1983) Some effects of emotion on the normal and abnormal heart. In: Proctor Harvey W (ed) Recurrent problems in cardiology, vol 7, no 12. Year Book Medical Publishers, Chicago, pp 1-29

Toff WD, Bennett G, Joy M (1986) Exercise-induced arrhythmia and conduction disturbance in an asymptomatic population (Abstr 2953). American Heart Association, World Congress of Cardiology, September, Washington

Tolson WW, Mason JW, Sachar EJ, Hamburg DA, Handlon JH, Fisherman JR (1965) Urinary catecholamine responses associated with hospital admission in normal human subjects. J Psychosom Res 8:365-372

Vaughan Williams EM (1970) Classification of anti-arrhythmic drugs. In: Sandoe E, Flensted-Jensen, E, Olesen KH (eds) Symposium on cardiac arrhythmia. Astra, Södertälje, pp 449-472

Vaughan Williams EM (1986) Ventricular hypertrophy—physiological mechanisms. J Cardiovasc Pharmacol [Suppl 3] 8:512-516

Velebit V, Podrid JP, Lown B, Cohen BH, Graboys TB (1982) Aggravation and provocation of ventricular arrhythmias by anti-arrhythmic drugs. Circulation 65:886-894

Viitasalo MT, Carla R, Eisalo A (1982) Ambulatory electrocardiographic recording in endurance athletes. Br Heart 47:213-220

Von Euler US (1964) Quantification of stress by catecholamine analysis. Clin Pharmacol Ther 5:398-404

Von Euler US, Lundberg U (1954) Effect of flying on the epinephrine excretion in Air Force personnel. J Appl Physiol 6:551-555

Wellens HJJ, Vermeulen A, Durrer D (1972) Ventricular fibrillation on arousal from sleep by auditory stimuli. Circulation 46:661-665

Wit AL, Hoffman BF, Rosen MR (1975) Electrophysiology and pharmacology of cardiac arrhythmias. IX. Cardiac electrophysiologic effects of beta-adrenergic receptor stimulation and blockade. Part A. Am Heart J 90:521-533

CHAPTER 12b

Adrenergic Arrhythmogenicity

E. M. Vaughan Williams

A. Sympathetic Excitation and Arrythmias

An association between stress or emotion and a "heart attack" has long been recognized. A reasonable explanation would be that an increase in blood pressure, or pulse pressure, in a coronary artery already containing an atheromatous plaque could cause the latter to split and expose underlying collagen to platelets, which in turn would initiate thrombosis. The protection offerred by long-term beta-blockade to postmyocardial infarction (MI) patients against reinfarction and sudden death has been claimed to be due to a blunting of hypertensive surges, thus obviating the wall-stress which could split a plaque. If this explanation is correct in attributing the prophylactic benefits of beta-blockers solely to protection of coronary arteries, the alpha-adrenoceptor blockers or other antihypertensive agents not acting on adrenergic receptors would be expected to provide similar prophylaxis.

On the other hand there is direct evidence that extreme exertion or emotional stress is arrhythmogenic, as detailed in the first part of this chapter. The fact that the abnormalities do not occur after beta-blockade implies that they are caused by stimulation of myocardial beta-adrenoceptors. In the long-QT syndrome (LQTS) arrhythmias are precipitated by exercise or emotion, but the heart rate of such patients actually does not increase as much as would be expected in normal people. This may be due (Chap. 23) to a deficit of innervation from the right stellate ganglion, from which the sinoatrial node (SAN) normally has its predominant sympathetic control. There is a compensatory excess of activity from the left stellate, possibly related to loss of an inhibitory influence from afferent nerves normally accompanying the right sympathetic. It may be concluded, therefore, that apart from vascular effects, high sympathetic stimulation, especially if heterogeneously distributed, is directly arrhythmogenic, and it is important to investigate the mechanisms by which this effect is induced.

One obvious possibility is that accessory pacemakers are activated and provide an out-of-phase depolarization from an ectopic site. It has already been noted (Chaps. 1, 2) that specialized pacemaker-type cells occur in and around the AV node and bundle branches as well as in the SAN, and occasional extrasystoles are comparatively common, but usually innocuous. The question arises whether sympathetic stimulation has further electrophysiological effects, apart from the initiation of ectopic beats, which could increase the probability of progression to a re-entry tachycardia or ventricular fibrillation (VF).

B. Adrenoceptors in the Heart

Although beta$_1$-adrenoceptors predominate in the heart, there is evidence that beta$_2$-adrenoceptors are also present. PRICHARD (1971) observed that in many patients it was impossible to block the tachycardia induced by isoprenaline or by the Valsalva manoeuvre with the beta$_1$-selective blocker practolol, even when administered in very high dosage. Recent studies with the non-selective 125I-labelled cyanopindolol and various selective adrenoceptor agonists and antagonists (BRODDE et al. 1983; HEITZ et al. 1983) indicated the presence of a mixed population of binding sites, the authors concluding "that beta$_1$- and beta$_2$-adrenoceptors coexist in the left ventricle and the left atrium of the human heart". Binding studies do not, however, establish the functional significance of the sites. There is evidence that in rat hearts beta$_1$-adrenoceptors predominate (BRYAN et al. 1981) and that both beta$_1$- and beta$_2$-receptors coexist in cat hearts (CARLSSON et al. 1972), but conflicting conclusions have been reached about the existence of single or mixed populations in guinea pigs (JOHANSSON and PERSSON 1983) and rabbits. Functional evidence was obtained for the mediation of increases in heart rate by beta$_2$-adrenoceptors in conscious rabbits (VAUGHAN WILLIAMS et al. 1975), and the presence of beta$_2$-adrenoceptors has recently been demonstrated in rabbit atria by binding studies (BRODDE et al. 1982). In contrast, COSTIN et al. (1983) maintained that rabbit atria contained beta$_1$-adrenoceptors only.

Evidence for the existence and functional role of myocardial alpha-adrenoceptors was obtained long ago (GOVIER 1968; GIOTTI et al. 1973), and the possibility that stimulation of alpha-receptors may be involved in the genesis of arrhythmias induced by ischaemia and reperfusion has been suggested more recently (SHERIDAN et al. 1980; Chap. 22). The myocardial receptors are probably of the alpha-1 subtype, because binding sites for alpha$_2$-receptors have been looked for but not found, and selective stimulation of alpha$_2$-adrenoceptors had no electrophysiological effects (DUKES and VAUGHAN WILLIAMS 1984). During the past few years selective agonists and antagonists for all subtypes of adrenoceptors have been developed, making it possible for the first time to study the electrophysiological and other effects of stimulation of the individual types of receptor (DUKES and VAUGHAN WILLIAMS 1984). The agonists and antagonists used are listed in Table 1. If a selective agonist for, say,

Table 1. Selective adrenoceptor agonists and antagonists

	Alpha$_1$	Alpha$_2$	Beta$_1$	Beta$_2$
Agonists	St 587 (Phenylepherine) Noradrenaline + β_1 and β_2 antagonists	BHT 933	Isoprenaline + α_1, α_2 and β_2 antagonists	Pirbuterol
Antagonists	Prazosin	WY 25309	Atenolol	ICI 118 551

an alpha$_1$-adrenoceptor had the same effect as a non-selective agonist (noradrenaline) in the presence of selective beta$_1$- and beta$_2$-receptor antagonists, then it was reasonable to assume that the effect was indeed mediated by alpha$_1$-adrenoceptors.

C. Effects of Stimulation of Individual Receptor Types

It was found that both beta$_1$- and beta$_2$-adrenoceptors mediated an increase in frequency of the rabbit SAN, but the mechanisms by which the frequency was raised were not the same. Beta$_1$ stimulation accelerated the rate of rise of the upstroke of the SAN action potential, from which it could be deduced that more current was flowing through calcium-selective ion channels, since various studies (see Chap. 1) have shown that calcium current, not the fast inward sodium current, is responsible for this phase of depolarization in the SA and AV nodes (the maximum diastolic potential being insufficiently negative in these tissues for sodium channels to recover from inactivation). An increased inward calcium current is consistent with the very large positive inotropic effect of beta$_1$ stimulation in myocardial cells. In contrast, highly selective beta$_2$ stimulation did not increase the rate of rise of the action potential of SAN cells, and had no significant positive inotropic action in the myocardium. Both beta$_1$ and beta$_2$ stimulation shortened the action potential duration (APD); indeed the shortening of diastole by accelerated repolarization made a major contribution to the increase in frequency. Here again, however, the mechanisms were different. Beta$_1$-receptor stimulation shortens APD by increasing potassium conductance (*g*K), an effect which persists in the presence of cardiac glycosides (GADSBY 1983). Beta$_2$-receptor stimulation, on the other hand, activates the Na/K pump (SMITH and KENDALL 1984), and since the pump is electrogenic, adding one net negative charge to the interior of the cell per cycle of the pump, repolarization is accelerated. The Na/K pump activation occurs in skeletal as well as cardiac muscle, causing a lowering of plasma potassiumm, which is itself an arrhythmogenic factor.

Selective alpha$_2$-receptor stimulation had no cardiac electrophysioloical effect, which is consistent with the lack of evidence for the existence of alpha$_2$-receptors on myocardial cells. Selective alpha$_1$-receptor stimulation caused a sinus *bradycardia*, due to delayed repolarization of SAN cells. In the myocardium there was a positive inotropic response to alpha$_1$-receptor agonists, but its magnitude was much smaller than the response to beta$_1$ agonists. Whereas the latter increased the rate of development of tension, and raised peak tension threefold, the alpha agonists caused no change in the rate of development of tension, and raised peak tension by not more than 50%. The duration of contraction was prolonged, in association with delayed repolarization, so that the area-under-the-curve of contraction was substantially increased. Unlike beta$_1$ agonists, alpha$_1$ agonists do not increase the level of intracellular cyclic adenosine monophosphate (cAMP). Perhaps the alpha-receptors provide a reserve mechanism for increasing contractile force in situations in which beta$_1$ stimulation is ineffective because of failure of the adenyl-

Table 2. Effects of stimulation of individual adrenoceptor subgroups

Receptor	$Beta_1$	$Beta_2$	$Alpha_1$	$Alpha_2$
Contraction	+++	0 ?+	+	0 (None on myocardium)
Action potential duration	Shortened by increased gk	Shortened by Na/K pump stimulation	Lengthened	0
Ca-selective channels in nodes	Opened	0	0	0
Arteries		Relaxed	Large vessels constricted	Small vessels constricted

cyclase system (e.g. if ischaemia creates an unfavourable ATP/ADP ratio for the GMP complex to exist in its active form). Thus an adrenergic source of heterogeneity is added to that created by the ischaemia itself. Hypoxia shortens APD, but low pH lengthens it (VAUGHAN WILLIAMS and WHYTE 1967). Since both hypoxia and low pH are part of the ischaemic environment, wide dispersion of repolarization times is to be expected according to which of the factors predominates locally.

The effects of selective adrenoceptor stimulation have been summarized in Table 2. Many of these effects would increase the probability of re-entry arrhythmias, especially the shortening of APD and the fall in plasma potassium.

D. Distribution of Sympathetic Innervation

Long-term beta-blockade (or sympathetic denervation) causes an adaptational response in cardiac cells which delays repolarization. The effect takes a week to develop, reaches a plateau at 3 weeks, and persists for many days after cessation of treatment (VAUGHAN WILLIAMS 1985; Chap. 2). It is not caused by the bradycardia associated with beta-blockade, because prolonged administration of alinidine, which produces a more profound bradycardia, does not induce APD prolongation (VAUGHAN WILLIAMS et al. 1986). Since the response occurs in all cardiac tissues there is a uniform prolongation of refractory period, an antiarrhythmic action (class III) which might be relevant to the prophylactic effect of long-term beta-blockade.

If, as a consequence, for example, of local destruction of sympathetic nerves by ischaemia or infarction, there is a heterogeneous distribution of sympathetic innervation, the denervated regions will adapt and have cells with delayed repolarization. When sympathetic drive is increased during exercise or emotion, APD will be shortened in the innervated areas but not in the denervated, so that at the border between the two there will be great heterogeneity of repolarization times, a highly arrhythmogenic situation. An extreme example of this occurs in the LQTS. The right stellate normally supplies the SAN and RV free wall, and the left stellate supplies the AVN and LV. In consequence of the deficient right-sided innervation APD will be longer (hence the long QT) in the RV, and will not shorten in response to exercise. On the left, compensatory sympathetic hyperactivity will shorten APD in the LV and at the right/left junction there will be maximum dispersion of repolarization times, leading to re-entry tachycardia or even VF. Left-sided sympathetic cardiac denervation or prolonged beta-blockade can reduce the dispersion by causing an adaptational prolongation of APD in the left ventricle and by making it less responsive to sympathetic drive. The restored balance of APD on the two sides may lead to mutual cancellation of late repolarization vectors on the ECG, so that a *lengthening* of APD in the LV may paradoxically result in a shortening of the QT interval.

In conclusion, therefore, in addition to the possible arrhythmogenic effect of activation of ectopic pacemaking, we must add several other factors to adrenergic arrhythmogenesis. Alpha$_1$-receptor stimulation lengthens APD, but beta-receptor stimulation shortens it, so that dispersion of repolarization times may occur if the distribution of alpha- and beta-receptors is heterogeneous. Since beta-receptor activation fails in ischaemia, but alpha-receptor stimulation does not, the uneven distribution will be exaggereated in areas of local ischaemia. Dispersion will also be greatly increased if sympathetically denervated and innervated regions are juxtaposed. Finally the fall in plasma potassium induced by beta$_2$-receptor stimulation is an additional arrhythmogenic factor.

E. Reperfusion Arrhythmias

All the above phenomena are concerned with the myocardial effects of adrenergic stimulation, but vascular effects may also be of importance. Reperfusion of underperfused regions causes a massive release of noradrenaline, and localized vasoconstriction may induce heterogeneity in the distribution of flow in the microcirculation during the postocclusion hyperaemic period. In consequence there could be uneven recovery from hypoxia-induced shortening of APD, again leading to dispersion of repolarization times in neighbouring areas, with a high probability of re-entry. Incidentally it is of interest that vasoconstriction in poststenotic regions is mediated by alpha-2 adrenoceptors (Heusch et al. 1985), so that although these are absent from myocardial cells they may yet play a role in adrenergic arrhythmogenesis.

References

Brodde O-E, Leifert F-J, Krehl H-J (1982) Co-existence of beta-1 and beta-2-adrenoceptors in the rabbit heart: quantitative analysis of the regional distribution by (−)-3H-dihydroalprenolol binding. J Cardiovasc Pharmacol 4:34–43

Brodde O-E, Karad K, Zerkowski H-R, Rohm N, Reidemeister JC (1983) Coexistence of beta-1 and beta-2-adrenoceptors in human heart and lung. Br J Pharmacol 78:72P

Bryan LJ, Cole JJ, O'Donnell SR, Wanstall JC (1981) A study designed to explore the hypothesis that beta-1-adrenoceptors are "innervated" receptors and beta-2-adrenoceptors are "hormonal" receptors. J Pharmacol Exp Ther 216:395-400

Carlsson E, Ablad B, Brandstrom A, Carlsson B (1972) Differentiated blockade of the chronotropic effects of various adrenergic stimuli in the cat heart. Life Sci 11:953-958

Costin BI, O'Donnell SR, Wanstall JC (1983) Chronotropic responses of rabbit isolated atria to beta adrenoceptor agonists are mediated by only beta-1-adrenoceptors. J Pharm Pharmacol 35:752-754

Dukes ID, Vaughan Williams EM (1984) Effects of selective alpha-1, alpha-2, beta-1 and beta-2-adrenoceptor stimulation on potentials and contractions in the rabbit heart. J Physiol (Lond) 355:523-546

Gadsby DC (1983) Beta-adrenoceptor agonists increase membrane K^+-conductance in cardiac Purkinje fibres. Nature 306:691-693

Giotti A, Ledda F, Mannaioni PF (1973) Effects of noradrenaline and isoprenaline in combination with alpha and beta receptor blocking substances on the action potential of cardiac Purkinje fibres. J Physiol (Lond) 229:99-113

Govier WC (1968) Myocardial alpha adrenergic receptors and their role in the production of a positive inotropic effect by sympathomimetic agents. J Pharmacol Exp Ther 159:82-90

Heitz A, Schwartz J, Velly J (1983) Beta-adrenoceptors of the human myocardium: determination of beta-1 and beta-2 subtypes by radioligand binding. Br J Pharmacol 80:711-717

Heusch G, Deussen A, Schipke J, Vogelsang H, Hoffmann V, Thämer V (1985) Role of cardiac sympathetic nerves in the genesis of myocardial ischemia distal to coronary stenosis. J Cardiovasc Pharmacol [Suppl 5] 7:S13-S18

Johansson L-H, Persson H (1983) Beta-2-adrenoceptors in guinea-pig atria. J Pharm Pharmacol 35:804-807

Prichard BNC (1971) Panel discussion. Postgrad Med J [Suppl] 47:112

Sheridan DJ, Penkoske PA, Sobel BE, Corr PB (1980) Alpha adrenergic contributions to dysrhythmia during myocardial ischemia and reperfusion in cats. J Clin Invest 65:161-171

Smith SR, Kendall MJ (1984) Metabolic responses to beta-2-stimulants. J R Coll Physicians Lond 18:190-194

Vaughan Williams EM (1985) Delayed ventricular repolarization as an anti-arrhythmic principle. Eur Heart J [Suppl D] 6:145-149

Vaughan Williams EM, Whyte JM (1967) Chemosensitivity of cardiac muscle. J Physiol (Lond) 189:119-137

Vaughan Williams EM, Raine AEG, Cabrera AA, Whyte JM (1975) The effects of prolonged beta-adrenoceptor blockade on heart weight and cardiac intracellular potentials in rabbits. Cardiovasc Res 9:579-592

Vaughan Williams EM, Dennis PD, Garnham C (1986) Circadian rhythm of heart rate in the rabbit; prolongation of action potential duration by sustained beta adrenoceptor blockade is not due to associated bradycardia. Cardiovasc Res 20:528-535

CHAPTER 13

Antiarrhythmic Properties of Beta-Adrenoceptor Blockade During and After Myocardial Infarction

P. Sleight and P. Rossi

A. Introduction

Acute myocardial infarction is generally caused by a fresh thrombus which occludes a major coronary artery. It carries a mortality of about 50% and many patients die before receiving any medical or paramedical help (Armstrong et al. 1972; Kinlen 1973). Studies from Chamberlain's Group in Brighton (O'Doherty et al. 1983) have shown that many of the early deaths occur because of the development of serious ventricular arrhythmia which is most prevalent during the first 1-3 h after the onset; thereafter the incidence decays very rapidly. Several studies have shown the superiority of continuous ECG recording on magnetic tape, which records many arrhythmias not noticed on monitor screens (Mogensen 1970; Vetter and Julian 1975). Most of the trials of antiarrhythmic drugs used in the treatment of myocardial infarction have taken place after this early very lethal phase. They have generally concentrated on reductions in less lethal arrhythmia such as ventricular ectopic beats; these have been shown to bear little relation to lethal arrhythmia, unless care is taken to identify the "R on T" ectopic beat (Fig. 1).

Many drugs can be shown to reduce the numbers of ectopic beats on Holter tapes, but it has been a source of considerable disappointment that, apart from amiodarone and beta-blocking agents, they have failed to reduce mortality. Indeed the majority of drugs used have been shown to have proarrhythmic effects in some patients. (see Chaps. 5, 6, 9, 11). Too many trials of the effects of antiarrhythmic drug treatment have failed to distinguish the type of arrhythmia. Ventricular extrasystoles are often "lumped" as one class of arrhythmia. This fails to take note of the increase in late diastolic extrasystoles (escape beats) which may often increase in number with beta-blockade, because of the bradycardia caused by the treatment. These have no therapeutic importance but may mask a simultaneous decrease in potentially lethal R on T ventricular extrasystoles. There has also been a failure to distinguish both the types of patient entered (e.g. "definite" MI versus "threatened" MI) and the time from the onset of pain.

Thus the treatment of arrhythmia in the setting of acute myocardial infarction remains poorly understood and poorly based scientifically.

Another source of confusion has been the plethora of trials which are individually too small to yield reliable results. The resulting mixture of positive, negative or inconclusive trials has confused the profession.

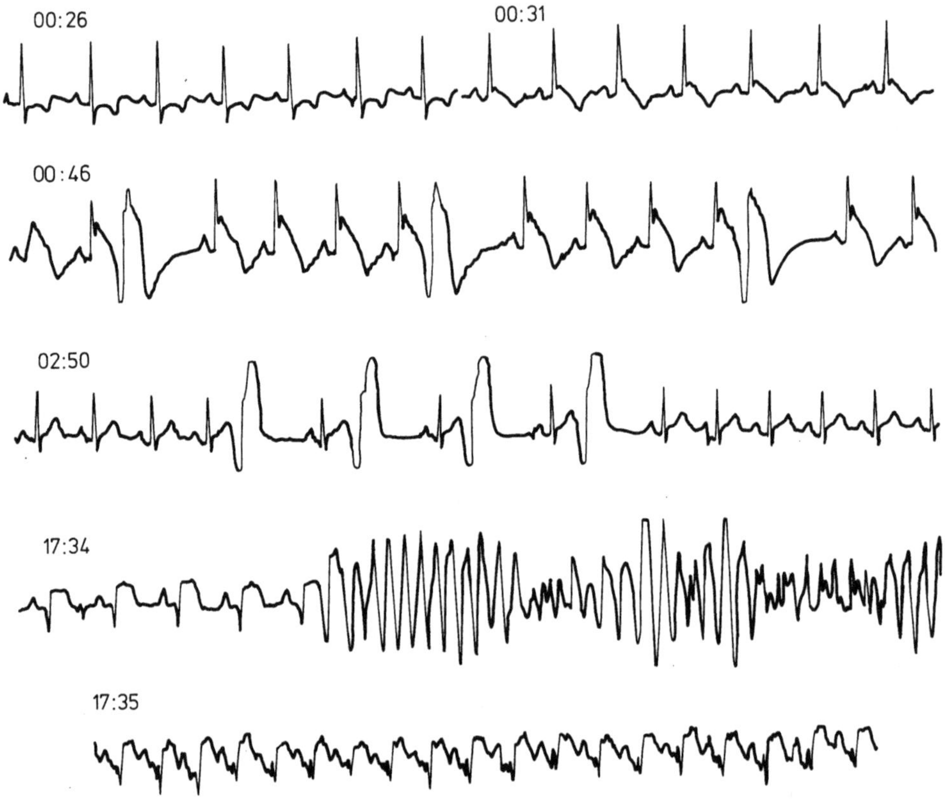

Fig. 1. Successive ECG strips from a patient in CCU (time indicated on each strip). Note the development of ST segment elevation, followed by R on T ventricular extrasystoles, culminating in ventricular fibrillation (torsades de pointes). The final strip taken after DC shock reversion to sinus rhythm. Note also the long interval between the first R on T ectopic and the development of ventricular fibrillation-17 h. ROSSI 1983

A further important consideration is the accuracy of recording of the arrhythmia (see above) and, even more important, the accuracy of analysis. Good-quality Holter monitoring is essential. Many commercial arrhythmia analysis systems perform very poorly compared with manual counts from paper records (ROSSI 1983).

The need for more reliable detection of moderate reductions in a serious arrhythmia or in mortality has led to the development of large multicentre international trials which involve the cooperation of hundreds of coronary care units. We have developed one such group coordinated from Oxford which now comprises over 400 collaborating units in Europe, North America and Australasia. ISIS-1 (International Studies of Infarct Survival) randomized over 16000 patients to receive intravenous atenolol or control treatment on admission to hospital (ISIS-1 1986). ISIS-2 (1988) already gave a preliminary result of a 35% reduction in mortality by streptokinase in patients treated within 4 h (ISIS-2 1987). So far we have recruited over 12000 patients to a 2×2 factorial double-blind randomized placebo-controlled trial of high-dose intravenous streptokinase and/or aspirin. These very large trials have been developed in conjunction with the MRC clinical trials service unit in Oxford (R. PETO) and with the coordinating efforts of S. Yusuf and R. Collins.

I have recently reviewed the more general aspects of beta-blockade during and following myocardial infarction (SLEIGHT 1986). This review deals particularly with arrhythmia; it will be based on an overview of all the available evidence from these and other trials (YUSUF et al. 1985; ISIS-1 1986), and particularly on the work of one of my postgraduate students, Dr. Paulo Rossi (ROSSI et al. 1983; ROSSI 1983).

B. Mechanisms of Arrhythmia in Acute Myocardial Infarction

Hypoxia shortens action potential duration (APD), but low pH lengthens APD. In ischaemia both factors are present, which leads to heterogeneity in repolarization times compared with normal tissue and hence sets the scene for serious disorders of conduction, and hence arrhythmia. This heterogeneity leads to late afterpotentials in the signal-averaged ECG (see COBBE, Chap. 16).

In addition, the rise in catecholamines which accompanies infarction (BENEDICT and GRAHAME-SMITH 1979) and which is greatest with larger infarctions (JEWITT et al. 1969), also exacerbates heterogeneity of repolarization times (Chap. 12) and has been shown to lower the electrical threshold for ventricular fibrillation in animals. Beta-adrenoceptor blockade has been shown to reduce the incidence, frequency and severity of ventricular arrhythmia after experimental coronary occlusion (FEARON 1967; KHAN et al. 1973; MENKEN et al. 1979).

Other factors which may be important are the leakage of potassium from cells, caused by hypoxia (HARRIS et al. 1954), and the rise in free fatty acids (FFAs) (OLIVER et al. 1968; OPIE et al. 1977). There is little evidence bearing on the effect of beta-blockade on this rise in FFAs (OPIE 1987, personal communication). It is clear that the mechanisms of arrhythmia in acute infarction are complex and probably differ during the different time phases during and after infarction (Chap. 26).

ROSSI (1983) found a striking difference in the natural history and timing of ventricular extrasystoles between those patients with initial ST segment elevation in the ECG ("definite" infarction) compared with those with initial ECGs which were normal or showed only ST depression and/or T wave changes ("threatened" infarction). The "definite" infarctions showed an initial early peak in ectopic beats, whereas the patients with "threatened" infarction gradually built up a slower later peak, presumably as infarction developed (Fig. 2). All the patients who developed ventricular fibrillation belonged to the group with "definite" infarction.

The sinus rates of the non -beta-blocked control groups of these two groups of patients were also different, with greater tachycardia in those with "definite" infarction (Fig. 3).

Supraventricular and some ventricular arrhythmia appears to be related to the amount of myocardial damage (NIELSEN 1973) but ventricular fibrillation does not (ADGEY et al. 1969; COBB et al. 1975; ROSSI 1983).

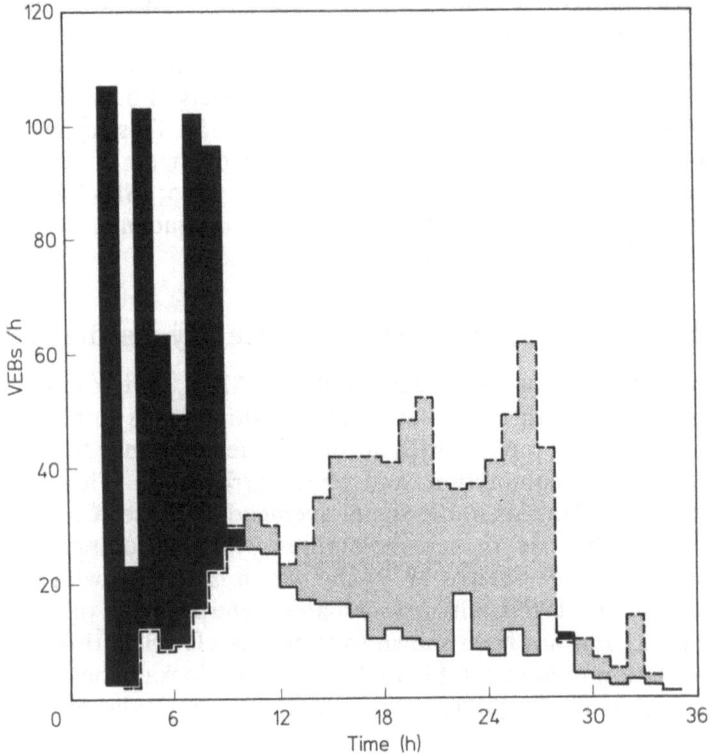

120
100
80
60
40
20
0
VEBs /h
6
12
18
24
30
36
Time (h)

Fig. 2.

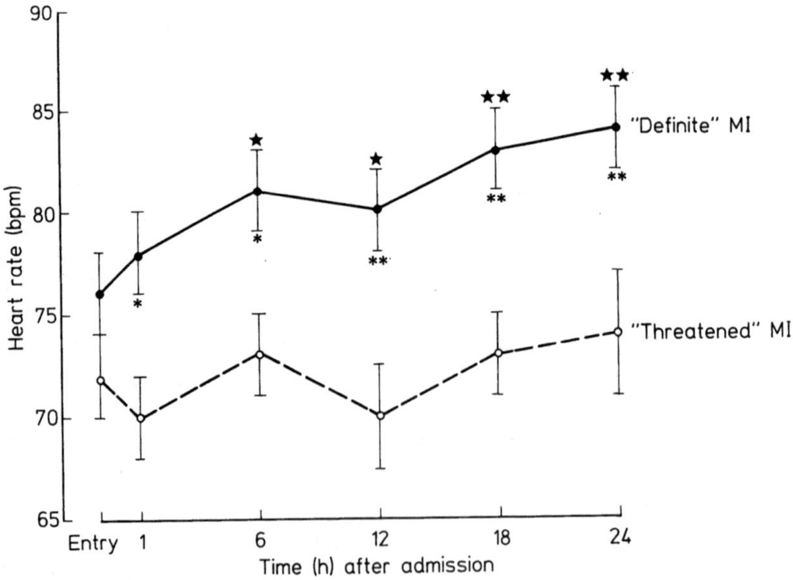

90
85
80
75
70
65
Heart rate (bpm)
"Definite" MI
"Threatened" MI
Entry
1
6
12
18
24
Time (h) after admission

Fig. 3.

C. Use of Beta-Blockade in the Acute Phase of Acute Myocardial Infarction

There have been relatively few randomized controlled studies using Holter monitoring to assess the effect of beta-blockade on the arrhythmia of acute myocardial infarction. In Oxford (Rossi et al. 1983) we studied 182 patients recorded for 24 h after a mean time of slightly over 5 h from the onset of pain. The patients were an unselected subgroup of the ISIS-1 Pilot Study (Yusuf et al. 1983). Patients with suspected myocardial infarction were admitted to the trial if they were in a suitable haemodynamic state to tolerate beta-blockade (heart rate > 60 beats, systolic blood pressure > 90 mm Hg, no second-or third-degree heart block, no severe heart failure and no contraindications to, or indications for, beta-blockade.) Patients already taking beta-blockers or verapamil were also excluded. Between October 1978 and August 1981, 477 patients were randomized and 24-h tapes were obtained on 224 of these. Forty-two were discarded because of technical faults in recording, leaving 182 for analysis. Patients were randomly allocated to control or treatment groups. Treatment group patients received 5 mg atenolol slowly intravenously over 5 min, followed by an oral dose of 50 mg at 10 min if the injection was well tolerated. Control patients did not receive routine antiarrhythmic prophylaxis, in accordance with our unit policy.

Holter recordings using the Oxford Medilog I or II recorders were made from admission to the trial and for the next 24 h. Atenolol caused a highly significant reduction in sinus rate (Fig. 4). There were also highly significant reductions in arrhythmia. Because the types of arrhythmia were not normally distributed, individuals' average counts were expressed as logarithms. I will deal first with supraventricular and then with the more serious ventricular arrhythmias.

I. Supraventricular Arrhythmia

These disturbances are not usually lethal but can be a considerable therapeutic problem in the setting of acute myocardial infarction. Fast atrial fibrillation may be very poorly tolerated when the circulation is already compromised; the resultant circulatory embarrassment may be sufficient to produce acute pulmonary oedema and/or shock, and recurrence or increase of chest pain.

◀

Fig. 2. Differences in patterns of ventricular ectopic beat frequency in patients with "definite" infarction (i. e. ST segment elevation on the initial ECG) and those with "threatened" infarction (normal initial ECG, or ST or T wave depression). "Definite" infarction = early peak, *solid line*; "threatened" infarction = later peak, *dotted line*. (Rossi 1983)

Fig. 3. Difference in heart rates (8 beats/min, $2P = < 0.025$) over the first 24 h following admission between control patients with "definite" and "threatened" infarction. (Rossi 1983)

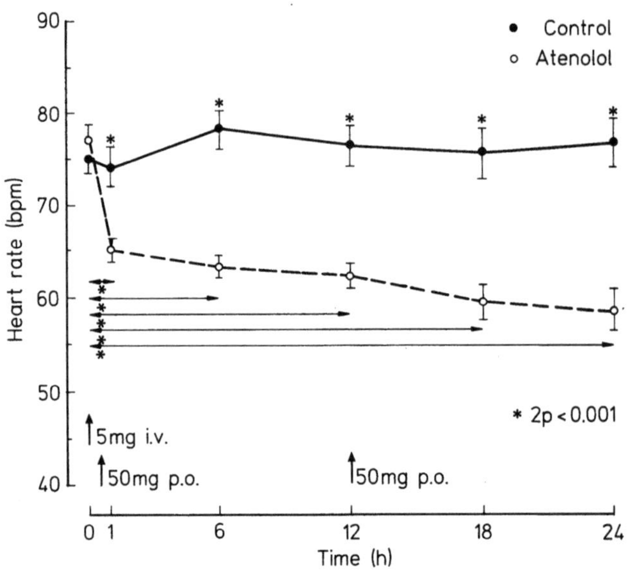

Fig. 4. Heart rate changes after admission in control (●) and atenololtreated patients (○). Atenolol rapidly reduced heart rate from 77 ± 1 to 65 ± 1/min over the 1st h ($2P < 0.001$). (ROSSI 1983)

It may be difficult to decide whether a particular arrhythmia has any prognostic implications per se. This difficulty arises because larger infarctions themselves lead both to a poor prognosis and also to the development of arrhythmia. Patients who develop atrial fibrillation certainly have a higher mortality (JULIAN et al. 1964; LOWN et al. 1967; NORRIS 1969; MOGENSEN 1970; HENNING et al. 1979). It is possible that the arrhythmia itself might lead to an increase in myocardial infarct size.

Beta-blockade leads to an increase in the refractory period of the atria and of the AV node (ROBINSON et al. 1978; EDVARDSSON and OLSSON 1981).

EVEMY and PENTECOST (1978) and ROSSI (1983) both found significant reductions in supraventricular arrhythmia using intravenous practolol and atenolol respectively. In the latter study supraventricular arrhythmia as a whole was reduced from 19.3 % of control patients to 16.5 % of atenolol-treated patients ($2P = < 0.025$). Atrial fibrillation was more strikingly reduced, from 12 % to 4.9 % ($2P < 0.005$). This is particularly useful in view of the difficulty in clinical management of this arrhythmia. Digoxin may not reduce heart rate for several hours; DC cardioversion may not be appropriate because of (a) the strong possibility of the recurrence of atrial fibrillation; (b) the consequent need for repeat DC shock with the subsequent risk of further myocardial damage.

My present policy, if beta-blockade cannot be used or has been unsuccessful in preventing fast supraventricular arrhythmia, is to use intravenous amiodarone.

II. Ventricular Arrhythmia

In our study R on T ventricular extrasystoles were seen in about two-thirds of the control patients, compared with one-fourth of the beta-blocked patients ($2P = < 0.0001$). They occurred on average during 2.9 h of recording in the controls, compared with 0.8 h in the treated group (Fig. 5).

These reductions were seen whether the initial ECG was relatively normal ("threatened" infarction) or whether it was compatible with "definite" infarction.

There was also a highly significant reduction in repetitive ventricular arrhythmia (couplets, triplets and ventricular tachycardia) from $\log_{10}$ 0.601 ± 0.06 in the controls to 0.327 ± 0.04 in the treated group (Fig. 6). The reduction in ventricular fibrillation was substantial (five controls vs. one treated patient) but probably because of small numbers this did not achieve significance ($P = < 0.10$). In hospital, mortality was similarly non-significantly reduced (one vs. five). Six patients in the atenolol group and 15 in the control group either died or had ventricular fibrillation and were discharged alive

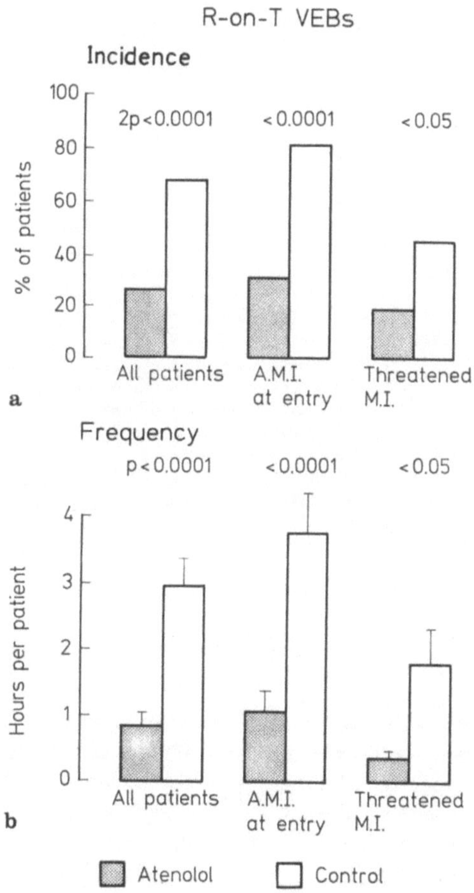

Fig. 5. a Incidence of R on T ventricular extrasystoles in beta-blocker (atenolol) and control groups. **b** Effect of treatment on the frequency of R on T extrasystoles (number of hours/patient in first 24 h after acute myocardial infarction). All *P* values are two-tailed (2 *P*). (Rossi et al. 1983)

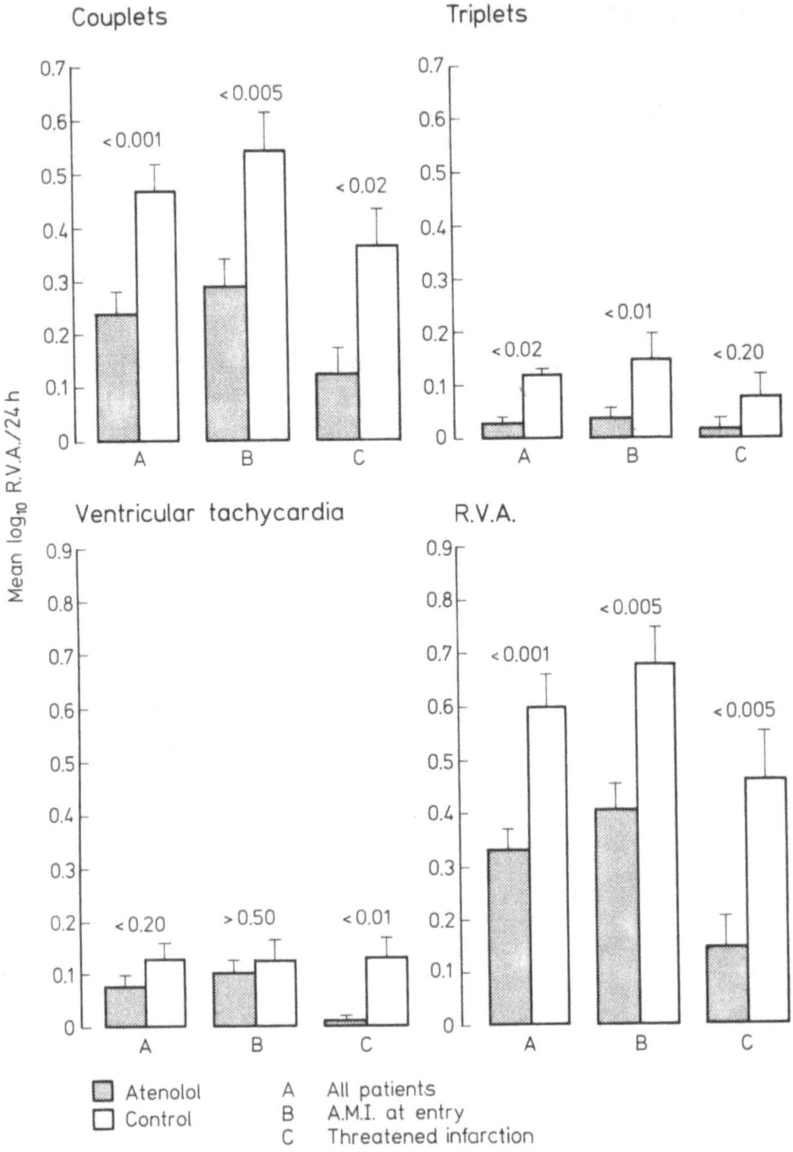

Fig. 6. Frequency of repetitive ventricular arrhythmia in first 24 h after acute myocardial infarction. Patients randomized to beta-blocker group had fewer episodes of repetitive arrhythmia irrespective of subgroup analysis (see text). Note that the greatest effect is shown when these arrhythmias are analyzed as groups (couplets, triplets, and ventricular tachycardia). Values expressed as $\log_{10}$ mean ± SEM. All P values are two-tailed ($2P$). (Rossi et al. 1983)

($\chi^2 = 5.31$, $2P = < 0.025$). The patients treated with atenolol needed significantly less other antiarrhythmic drugs.

This trial therefore strongly suggested a useful antiarrhythmic action of beta-blockade on repetitive ventricular arrhythmia and on the R on T ventricular extrasystole—a potentially lethal event (CAMPBELL et al. 1981).

In contrast, an earlier study from Nottingham (WILCOX et al. 1980) had shown no reduction of arrhythmia by oxprenolol in acute myocardial infarction.

This may have been due to either the late start of treatment (mean of 17 h), inadequate early effect (oral dosing was used) or the use of a beta-blocker with intrinsic sympathetic activity (our review of all trials suggested less benefit in those trials which used beta-blockers such as oxprenolol and pindolol which have ISA) (see below, YUSUF et al. 1985).

III. Ventricular Fibrillation

I have stated earlier that reduction of ventricular ectopic activity might be irrelevant to the problem of prevention of ventricular fibrillation (VF), although the study of CAMPBELL et al. (1981) suggests that reduction in R on T ectopics might be useful.

RYDEN et al. (1983) in one of the earlier metoprolol trials found a significant reduction in VF (17/698 placebo vs. 6/697 metoprolol-treated patients). NORRIS et al. (1984) also found a reduction of VF in the PREMIS study (14/371 vs. 2/364). However, this study may be flawed, since patients already on beta-blockers before the infarction occurred had these stopped. The difference might therefore represent more "rebound VF" in the control group, rather than a reduction in the treated group. Norris has argued that non-selective blocking agents might have more benefit in reducing ventricular fibrillation than selective drugs. This argument has some plausibility since catecholamines can be shown to lower serum potassium by a β_2-effect (BROWN et al. 1983). Lower serum potassium predisposes to ventricular fibrillation. Blocking this effect should protect against VF.

In practice, however, there is considerable evidence that there is no difference between selective and non-selective beta-blockers in protection against sudden death in the postinfarction trials (YUSUF et al. 1985).

More recent data suggest also that initial serum potassium levels in patients admitted with myocardial infarction are not different between patients already on cardioselective or non-selective beta-blockers, or on no treatment (RAJ et al. 1985).

Despite the promising results with both atenolol and metoprolol in earlier trials, there was surprisingly no significant reduction in ventricular fibrillation (or cardiac arrest) in either the much larger ISIS or MIAMI studies (about 16000 and 6000 patients respectively), although both trials showed non-significant reductions in these events.

When an overview of all the available evidence is taken (ISIS-1 1986), there is a 15% reduction in ventricular fibrillation ($P = < 0.05$) and this overview is overwhelmingly based on β_1-receptor blockade.

D. Postinfarction Beta-Blockade

I. Prevention of Arrhythmia in Patients Who Have Survived a Myocardial Infarction

There is now overwhelming evidence for a worthwhile 20 % reduction in mortality by beta-blockade from the large trials of beta-blockade *after* infarction (timolol, Norwegian study; propranolol, Beta-blocker Heart Attack Trial; metoprolol, Gothenberg; for review and references see YUSUF et al. 1985).

The Norwegian Multicentre Study Group (PEDERSEN 1985) have recently reported the 6-year follow-up of the original postinfarction timolol trial. They concluded that the previous observed benefit was maintained for at least 6 years, despite the confounding effect of crossover to active treatment of patients originally on placebo.

It is of considerable interest that the majority of the lives saved occurred through a reduction in sudden death. Some of these deaths (about one-third of the control group in the metoprolol study) are due to cardiac rupture but the remainder can be presumed to be arrhythmic. A later analysis of ISIS-1 data strongly suggested that reduction of cardiac rupture was a major factor in the benefit of early β-blockade (JULIAN et al. 1988).

The mechanism by which beta-blocking agents reduce sudden death and arrhythmia postinfarction is not clear (Chap. 2). It is likely that it is indeed due to $beta_1$-receptor blockade since this is the one common property of all the above three drugs.

1. Ancillary Properties

a. Membrane-Stabilitizing Activity

Some beta-blockers reduce the maximum rate of rise of phase 0 of the cardiac action potential, without changing the membrane potential (DAVIS and TEMTE 1968). This so-called membrane-stabilizing activity (MSA) is antiarrhythmic, but it needs far higher doses than are used clinically (LUCCHESI et al. 1967). It seems clear from an overview of the ancillary properties of the drugs used in the trials (YUSUF et al. 1985) that MSA has no clinical advantages.

b. Intrinsic Sympathetic Activity

Drugs with intrinsic sympathetic activity (ISA) (pindolol and oxprenolol) seem less effective (YUSUF et al. 1985). When all the large trials postinfarction are pooled, they can be divided into about 8500 patients randomized to trials using drugs with ISA, and about 12 000 patients to drugs without ISA. The non-ISA drugs reduce mortality by about 30 %, the ISA drugs by only 10 %. It should be noted that this analysis is "post hoc" and therefore not entirely reliable; nevertheless the numbers involved are substantial and consistent.

The mechanism of arrhythmia reduction by $beta_1$-blockade is possibly partly by blockade of the catechol-induced heterogeneity of repolarization, partly by reduction in threshold for VF (see Chap. 28) and probably partly by reduction in myocardial ischaemia as a consequence of bradycardia (with better diastolic coronary flow) and reduction in cardiac work (rate pressure product).

c. Cardioselectivity

Pooling all types of postinfarction beta-blocker trials, the reduction in sudden death is about 30% (95% confidence limits, 20%–40%, $P = < 0.0001$). For non-sudden death the reduction in mortality is only 12% and non-significant.

It thus seems that the majority of the effect on survival is by the reduction in serious arrhythmia achieved by beta_1-blockade.

E. Adverse Effects of Beta-Blockade

I. Acute Phase of Myocardial Infarction

There have been no unexpected adverse effects of intravenous beta-blockade in the early phase of myocardial infarction. In the ISIS-1 (1986) study, there was a small but highly significant increase (from 3.4% control to 5.0% atenolol) in the need for inotropic drugs such as isoprenaline or dobutamine. It was reassuring that this adverse experience occurred at the very time of the major benefit in mortality, namely the first 24–36 h. In other words, the adverse effects were largely or wholly reversible. My own experience has been that some or most of the intervention with inotropes was unnecessary or misguided. It is quite common for the resident to be unduly alarmed about the blood pressure or pulse rate of a patient, when the patient himself is in a manifestly adequate circulatory state, with warm periphery. It is necessary to emphasize that asymptomatic hypotension and bradycardia is the therapeutic aim.

We also saw a trend towards heart block (atenolol 2.2% vs. 1.9% control). This was more likely in those with first-degree block on the initial ECG, so perhaps beta-blockade should be used with caution in this group.

There was no indication of an increase in heart failure or an excess requirement for diuretics in those given beta-blockade.

II. Post Myocardial Infarction Use of Beta-Blockade: Long-term Adverse Effects

Many of the effects attributed to beta-blockade, e.g. lassitude, cold extremities, dizziness and even shortness of breath, are common results of the infarction itself. Our review of the published trials (Yusuf et al. 1985), where placebo treatment was used, makes this clear. Nevertheless there is a small excess of all these symptoms in the beta-blocker group. Some of the symptoms may be alleviated by either changing the drug (e.g. from a non-selective to a selective blocker) or by adding ancillary treatment to increase vasodilation and cardiac output.

I have remarked above on the favourable experience of, and need to continue with, beta-blockade in patients who have large hearts and poor left ventricular function.

F. Conclusions

Beta-blockade by intravenous injection represents the best proven means of reducing both supraventricular and ventricular arrhythmia in acute myocardial infaction, but only about half the patients admitted to coronary care will be considered suitable because of fears of shock, heart failure or heart block. These patients are also the lower risk part of the spectrum, which again diminishes the potential for this treatment.

For those patients who have recovered from infarction, a greater proportion can be treated. Fears of the precipitation of heart failure by beta-blockade have some foundation. But HANSTEEN et al. (1982) showed that patients considered at high risk because of left ventricular failure were those who benefited most from beta-blockade, particularly in a significant reduction in sudden death, if the heart failure was controllable by concomitant diuretic treatment. More recently OLSSON and REHNQVIST (1986) have shown in another randomized placebo-controlled trial that postinfarction treatment with metoprolol is particularly beneficial for those patients with signs of left ventricular failure in the CCU, and those with larger hearts on predischarge chest X-ray. Mortality over the 3-year follow-up period was halved (placebo, 33 %; metoprolol, 16 %; $P = < 0.05$). Similarly, in our large ISIS-1 study of acute infarction, our fears of side effects from intravenous atenolol proved largely unfounded. It seems likely that in future we will be able to extend this treatment to somewhat higher-risk patients than were treated in the trials, since these patients will show higher absolute benefit.

References

Adgey AAJ, Nelson PG, Scott ME, Geddes JS, Allen JD, Zaida SA, Pantridge JF (1969) Management of ventricular fibrillation outside hospital. Lancet 1:1169–1171

Armstrong A, Duncan B, Oliver MF, Julian DG, Donald KW, Fulton M, Lutz W, Morrison SL (1972) Natural history of acute coronary heart attacks: a community study. Br Heart J 34:67–80

Benedict CR, Grahame-Smith DG (1979) Plasma adrenaline and noradrenaline concentrations and dopamine-β-hydroxylase activity in myocardial infarction with and without cardiogenic shock. Br Heart J 42:214–220

Brown MJ, Brown DC, Murphy MB (1983) Hypokalemia from beta 2-receptor stimulation by circulating epinephrine. N Engl J Med 309-1414-1419

Campbell RWF, Murray A, Julian DG (1981) Ventricular arrhythmias in the first 12 hours of acute myocardial infarction: natural history study. Br Heart J 46:351–357

Cobb LA, Baum RS, Alvarez H III, Schaffer WA (1975) Resuscitation from out-of-hospital ventricular fibrillation: 4 years follow-up. Circulation (Suppl 3) 51 52:223–228

Davis LD, Temte JV (1968) Effects of propranolol on the transmembrane potentials of ventricular muscle and Purkinje fibers of the dog. Circ Res 22:661–677

Edvardsson N, Olsson SB (1981) Effects of acute and chronic beta-receptor blockade on ventricular repolarisation in man. Br Heart J 45:628–636

Evemy KL, Pentecost BL (1978) Intravenous and oral practolol in the acute stages of myocardial infarction. Eur J Cardiol 7(5-6):391-398

Fearon RE (1967) Propranolol in the prevention of ventricular fibrillation due to experimental coronary artery occlusion. Am J Cardiol 20:222-228

Hansteen V, Møinichen E, Lorentsen E, Andersen A, Strøm O, Søiland K, Dyrbekk D, et al. (1982) One year's treatment with propranolol after myocardial infarction: preliminary report of Norwegian Multicentre Trial. Br Med J 284:155-160

Harris AS, Bisteni A, Russell RA, Brigham JC, Firestone JE (1954) Excitatory factors in ventricular tachycardia resulting from myocardial ischemia: potassium a major excitant. Science 119:200-203

Henning H, Gilpin EA, Covell JW, Swan EA, O'Rourke RA, Ross J Jr (1979) Prognosis after acute myocardial infarction: a multivariate analysis of mortality and survival. Circulation 59:1124-1136

ISIS-1 (1986) Randomised trial of intravenous atenolol among 16027 cases of suspected acute myocardial infarction. Lancet 2:57-66

ISIS-2 (1987) Intravenous streptokinase given within 0-4 hours of the onset of myocardial infarction reduced mortality in ISIS-2. Lancet 1, 502

ISIS-2 (1988) Randomized trial of intravenous streptokinase, oral aspirin, both, or neither, among 17, 184 cases of suspected acute myocardial infarction. Lancet 2:349-360

Jewitt DE, Mercer CJ, Reid D, Valori C, Thomas M, Shillingford JP (1969) Free noradrenaline and adrenaline excretion in relation to the development of cardiac arrhythmias and heart failure in patients with acute myocardial infarction. Lancet 1:635-641

Julian DG, Valentine PA, Miller GG (1964) Disturbances of rate, rhythm and conduction in acute myocardial infarction: a prospective study of 100 consecutive unselected patients with the aid of electrocardiographic monitoring. Am J Med 37:915-927

Julian D, Chamberlain D, Sandoe E, Kahrs J, Kala R, Henning R, Sleight P, Collins R (1988) Mechanisms for the early mortality reduction produced by beta blockade started early in acute myocardial infarction. Lancet 1:921-923

Khan MI, Hamilton JT, Manning GW (1973) Early arrhythmias following experimental coronary occlusion in conscious dogs and their modification by beta-adrenoceptor blocking drugs. Am Heart J 86:347-358

Kinlen LJ (1973) Incidence and presentation of myocardial infarction in an English community. Br Heart J 35:616-622

Lown B, Vassaux C, Hood WB Jr, Fakhro AM, Kaplinsky E, Roberge G (1967) Unresolved problems in coronary care. Am J Cardiol 20:494-508

Lucchesi BR, Whitsitt LS, Stickney JL (1967) Antiarrhythmic effects of beta-adrenergic blocking agents. Ann NY Acad Sci 139:940-951

Menken U, Wiegand V, Bucher P, Meesmann W (1979) Prophylaxis of ventricular fibrillation after acute experimental coronary occlusion by chronic beta-adrenoceptor blockade with atenolol. Cardiovasc Res 13:588-594

Mogensen L (1970) Ventricular tachyarrhythmias and lignocaine prophylaxis in acute myocardial infarction: a clinical and therapeutic study. Acta Med Scand [Suppl] 513:1-80

Nielsen BL (1973) ST-segment elevation in acute myocardial infarction: prognostic importance. Circulation 48:338-345

Norris RM (1969) Bradyarrhythmia after myocardial infarction (Letter). Lancet 1:313-314

Norris RM, Brown MA, Clarke ED, Barnaby PF, Geary GG, Logan RL, Sharpe DN

(1984) Prevention of ventricular fibrillation during acute myocardial infarction by intravenous propanolol. Lancet 2:883-886

O'Doherty M, Tayler DI, Quinn E, Vincent R, Chamberlain DA (1983) Five hundred patients with myocardial infarction monitored within one hour of symptoms. Br Med J 286:1405-1408

Oliver MF, Kurien VA, Greenwood TW (1968) Relation between serum-free-fattyacids and arrhythmias and death after acute myocardial infarction. Lancet 1:710-715

Olsson G, Rehnqvist N (1986) Effect of metoprolol in postinfarction patients with increased heart size. Eur Heart J 7:468-474

Opie LH, Tansey M, Kennelly BM (1977) Proposed metabolic vicious circle in patients with large myocardial infarcts and high plasma-free-fatty-acid concentrations. Lancet 2:890-892

Pedersen TR (1985) Six-year follow-up of the Norwegian Multicenter study on timolol after acute myocardial infarction. N Engl J Med 313:1055-1058

Raj SM, Simpson E, Rodger JC (1985) Pre-treatment with beta-blockers and the incidence of hypokalaemia in patients with acute chest pain. Eur Heart J 6(1):125

Robinson C, Birkhead J, Crook B, Jennings K, Jewitt D (1978) Clinical electrophysiological effect of atenolol—a new cardioselective beta-blocking agent. Br Heart J 40:14-21

Rossi PRF (1983) The effect of beta-blockade on ventricular arrhythmias in acute myocardial infarction. PhD thesis, University of Oxford

Rossi PRF, Yusuf S, Ramsdale D, Furze L, Sleight P (1983) Reduction of ventricular arrhythmias by early intravenous atenolol in suspected acute myocardial infarction. Br Med J 286:506-510

Rydén L, Arniego R, Arnman K, Herlitz J, Hjalmarson Å, Holmberg S, Reyes C, et al. (1983) A double-blind trial of metoprolol in acute myocardial infarction: effects on ventricular tachyarrhythmias. N Engl J Med 308:614-618

Sleight P (1986) Use of beta adrenoceptor blockade during and after acute myocardial infarction. Annu Rev Med 37:415-425

Vetter NJ, Julian DG (1975) Comparison of arrhythmia computer and conventional monitoring in coronary care unit. Lancet 1:1151-1154

Wilcox RG, Rowley JM, Hampton JR, Mitchell JRA, Roland JM, Banks DC (1980) Randomised placebo-controlled trial comparing oxprenolol with disopyramide phosphate in immediate treatment of suspected myocardial infarction. Lancet 2:765-769

Yusuf S, Sleight P, Rossi PRF, Ramsdale D, Peto R, Furze L, Sterry H, et al. (1983) Reduction in infarct size, arrhythmias and chest pain by early intravenous beta blockade in suspected acute myocardial infarction. Circulation 67:32-45

Yusuf S, Peto R, Lewis J, Collins R, Sleight P (1985) Beta blockade during and after myocardial infarction: on overview of the randomized trials. Prog Cardiovasc Dis 27 (5):335-371

Class III Agents

CHAPTER 14

Class III Antiarrhythmic Action

S. B. Olsson

A. Introduction

Class III antiarrhythmic action, e.g. the principle of delayed myocardial repolarization as an antiarrhythmic action, was described in 1970. This mode of action has become increasingly interesting with the increasing knowledge about mechanisms involved in initiation and perpetuation of cardiac arrhythmias and with the increasing number of antiarrhythmic drugs belonging to this class. Consequently, the possible class III mode of action of different antiarrhythmic drugs has been analyzed also in the intact human heart. For this purpose, several different methods have been applied. The present paper aims to describe some basic facts of importance for the demonstration and interpretation of delayed repolarization as an antiarrhythmic action in the clinical setting and to give some examples of this mechanism.

B. Ionic Fluxes Affecting Repolarization

It is well recognized that several mechanisms are involved in the regulation of myocardial repolarization. The actual ionic and metabolic mechanisms regulating the repolarization process of the myocardium are far from well understood. Under physiological conditions, phase 1 of the repolarization is presumably a result not only of inactivation of the sodium inward current, but also of activation of an early outward current, a fall in potassium conductance and the existence of a slow inward current (Carmeliet 1977). The end of the plateau phase of the action potential is explained by the activation of the delayed outward current mainly carried by potassium ions, the inactivation of the slow inward current and the increase in outward background currents especially involving calcium. The final part of repolarization, phase 3 of the action potential, is essentially due to the charging of the membrane by an outward current almost exclusively dependent on membrane potential.

In the clinical situation, myocardial repolarization may be affected by a number of factors other than antiarrhythmic drugs. For instance, electrocardiographic T wave changes of different types, but all compatible with delayed ventricular repolarization, are found as a consequence of a decrease of heart rate and of a decreased plasma concentration of calcium or potassium. Apart from such modulations of myocardial repolarization through mechanisms of physiological nature, a delay of repolarization may reflect a pathological condition. Thus, a delayed repolarization may be associated with some types of ventricular tachyarrhythmia, e.g. the long QT syndrome (LQTS) (Schwartz 1970).

C. Evaluation of Myocardial Repolarization in Humans

Although the microelectrode recording technique can be applied in a "floating manner" in order to record action potentials from the intact heart (Bromberger-Barnea et al. 1959), the technique is limited to use during cardiac surgery and is seriously compromised by the mechanical activity of the heart. Using percutaneous cardiac catheterization with suction or contact electrodes, monophasic signals may be recorded from the endocardial surface of the right or left heart (Olsson et al. 1985; Franz et al. 1986). The monophasic action

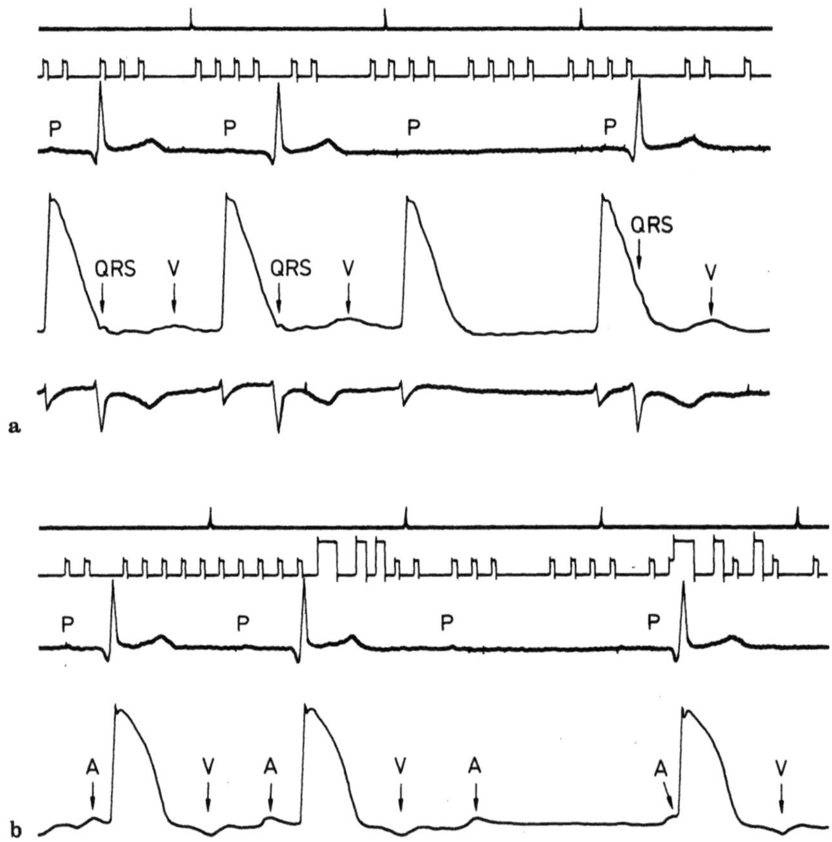

Fig. 1. Example of electrical and mechanical disturbances in MAP recordings. **a** This illustrates an atrial MAP recording in a patient with total AV block. Note the effect of electrical (QRS) and mechanical *(V)* ventricular activity upon the atrial MAP signal. **b** This illustrates ventricular MAP recording in the same patient. Note the effect of electrical (P) and mechanical (A) atrial activity as well as the effect of mechanical ventricular activity *(V)* upon ventricular MAP recordings. (Olsson et al. 1985)

potential (MAP) recorded in this way exhibits a marked resemblance during the entire course of repolarization with that of a transmembrane action potential recorded intracellularly from the immediate vicinity of the MAP recording site (FRANZ et al. 1986). The mechanical activity of the heart does, however, influence the recorded signal, often resulting in "afterpotentials" which may be artefacts making interpretation of the final part of the repolarization course difficult (OLSSON et al. 1985) Nevertheless, during this period of the cardiac cycle when movement artefacts occur, there may be true electrocardiographic repolarization abnormalities, as in patients with LQTS and ventricular arrhythmias. MAP recordings performed in such patients have documented pronounced disturbances of the normal ventricular repolarization during this part of the cardiac cycle (BONATTI et al. 1985). Unfortunately, however, it is not at present possible to distinguish unequivocally whether these findings are due to artefacts or are real pathoelectrophysiological phenomena. *Figure 1* illustrates how electrical and mechanical atrial and ventricular activities may influence MAP recordings.

Although the MAP recording method is the only readily applicable technique for assessing the entire myocardial cellular repolarization course, measurements by simpler techniques have documented a parallelism with the duration of the MAP. Thus, the duration of the local pacing evoked response (DONALDSON and RICKARDS 1982) as well as that of the paced QT interval (lead V2) (EDVARDSSON and OLSSON 1981a) may be used in order to study changes of ventricular myocardial repolarization.

By programmed electrical stimulation, the effective refractory period, e.g. the refractoriness, of ventricular (VERP) or atrial (AERP) myocardium may be explored (OLSSON et al. 1977). The method utilizes percutaneously introduced electrode catheters and external pacemakers capable of delivering ex-

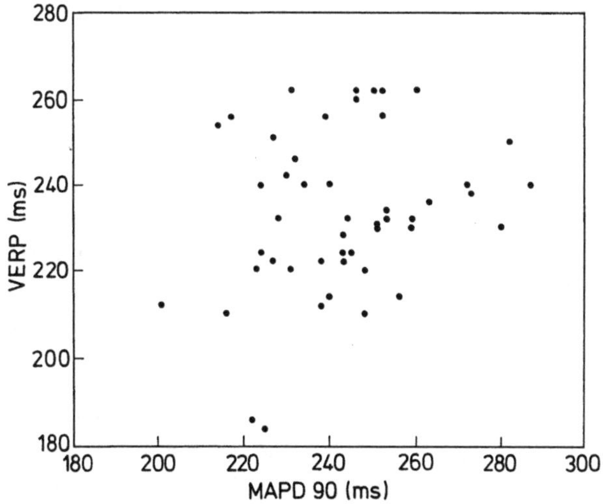

Fig. 2. Relation between right ventricular MAP duration, measured at 90% repolarization and ventricular effective refractory period estimated at the site of MAP recording in presumably healthy hearts. The values have been extracted from EDVARDSSON et al. (1984a)

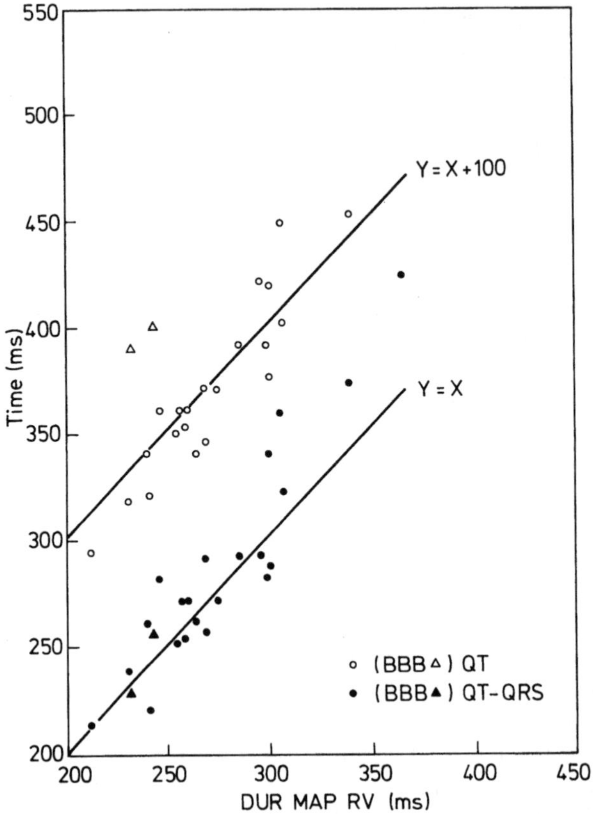

Fig. 3. Relation between QT time, QT minus QRS time and duration of right ventricular MAP during spontaneous regular rhythm. In two cases there is an obvious prolongation of the QT time compared with the duration of the MAP. These patients had bundle branch blocks and, when correction for the QRS duration is made, the relationship fits with that found in other patients. Line of identity and line Y = X + 100 are included for comparison. (Olsson 1972)

tra stimuli with increments of a few milliseconds. Normal values concerning ventricular repolarization and refractoriness have been obtained from studies in healthy male volunteers using recording of MAPs and programmed ventricular stimulation technique (Edvardsson et al. 1984a). A further analysis of the data presented in that study revealed, however, a poor correlation between the duration of the right ventricular MAP and the VERP, estimated at the MAP recording site (Fig. 2).

The effect of several class III antiarrhythmic drugs devoid of concommitant class I mode of antiarrhythmic action upon almost simultaneously estimated VERP and MAP duration from the same site has been analyzed. Although both variables were prolonged, the MAP duration was sometimes more prolonged than the VERP (see Table 1). Differences between individuals, however in prolongation of these variables were quite marked. Consequently, although it is true that VERP is affected by class III antiarrhythmic action, the observed prolongation of this variable is less valid as a measure of delay of repolarization than that of the MAP recording method. Furthermore, prolongation of refractoriness occurs after class I antiarrhythmic drugs without concomitant prolongation of the MAP.

It should be pointed out that some antiarrhythmic drugs with a class I mode of action also prolong the final part of the phase 3 of the myocardial action potential. This action may be detected by measuring the MAP duration of the 90% repolarization level. In order to illustrate more clearly a genuine prolongation of the MAP (class III action), measurements should be made not only at the 90% repolarization level, but also at the beginning of phase 3, for instance at the 50% level of repolarization.

Although QT interval has often been used as a general measure of ventricular myocardial repolarization, it includes the QRS complex. Moreover, there are well-recognized problems involved in delineation of the QT interval which limit the use of this method to monitor subtle changes of repolarization. As the T-wave is an expression of ventricular repolarization, one would expect a relation between the QT time and the duration of the ventricular MAP. Indeed, an overall relationship does exist between the duration of the right ventricular MAP and the QT time, but the MAP duration may differ by as much as 50 ms from any given JT time (QT-QRS) (OLSSON 1972) (Fig. 3). *Changes* of MAP duration and QT in any individual can be followed with much higher precision, however (EDVARDSSON and OLSSON 1981a).

D. Class III Antiarrhythmic Action In Vitro

In 1970, VAUGHAN WILLIAMS suggested a classification of antiarrhythmic drugs, which has since been widely accepted. The third class of action of antiarrhythmic mechanisms was thus the prolongation of the myocardial action potential. The first drug shown to have predominantly this action was amiodarone (SINGH and VAUGHAN WILLIAMS 1970) (Fig. 4). Since then, several other drugs have been found to prolong APD acutely in isolated myocardial preparations (BEXTON and CAMM 1982; PLATOU 1982). In addition, chronic administration of beta-blockers may also induce a prolongation of myocardial cellular repolarization (VAUGHAN WILLIAMS et al. 1975 EDVARDSSON and OLSSON 1981a). Drugs suggested to possess class III antiarrhythmic mode of action include:

1. Amiodarone; 2. Bretylium; 3. Sotalol; 4. Meobentine; 5. Clofilium; 6. Melperone; 7. Inpea; 8. Bunaphtine; 9. Prolonged beta-blockade;

It should be stressed that the class III mode of antiarrhythmic action is by definition only a *myocardial* action. The effects of the class III drugs upon the action potentials of different components of the specialized conductive system may therefore differ within this class of drugs. Available information does not permit detailed analyses of the effect of all the drugs presented in Table 1 upon the repolarization of all components of the specialized conductive system. It should be pointed out, however, that amiodarone in fact diminishes the normal difference between the duration of the action potential of the His-Purkinje system and the ventricular myocardium (VAUGHAN WILLIAMS 1985).

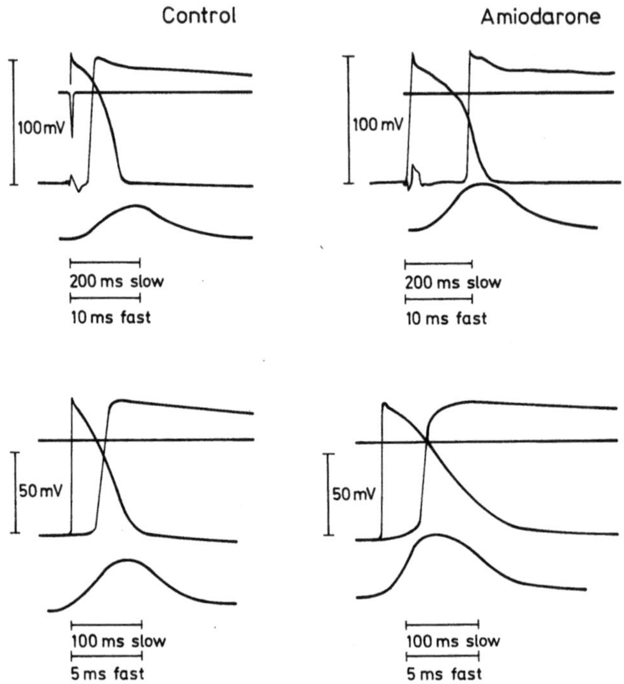

Fig. 4. Effect of chronic treatment with intraperitoneally administered amiodarone upon atrial and ventricular myocardial action potentials of guinea pigs. Note the marked prolongation of repolarization of both atrial and ventricular tissue. (VAUGHAN WILLIAMS 1970)

E. Class III Antiarrhythmic Action in Man

The effect of amiodarone upon human myocardial repolarization is well documented although no published study has hitherto illustrated the chronic effect of the drug upon myocardial cellular repolarization as well as refractoriness. Thus, QT prolongation, often combined with the development of a U-like wave, was identified shortly after the drug had been recognized as an antiarrhythmic agent. (ARMENTA et al. 1970) In fact, the QT interval may be used as a predictor of the plasma and myocardial concentrations of amiodarone (DEBBIAS et al. 1984). The degree of QT prolongation has also been suggested to reflect the antiarrhythmic efficacy of this drug (TORRES et al. 1986). The prolongation of the QT interval seen after chronic administration of amiodarone parallels the prolongation of right ventricular **MAP**, whilst after acute administration of the drug only a minimal prolongation of the right ventricular MAP is observed (BLOMSTRÖM et al. 1987).

Body surface recordings cannot reveal the desired prolongation of atrial repolarization after amiodarone. Using programmed stimulation or MAP recording technique, marked prolongation of atrial repolarization has been verified after chronic administration of amiodarone in man (OLSSON et al. 1973).

Fig. 5 illustrates this mechanism in a patient who had taken 200 mg amiodarone daily for 6 weeks.

Acute administration of sotalol leads to a prolongation of the QT interval (BRACHMANN et al. 1985). Using invasive techniques, e.g. programmed right ventricular stimulation and right ventricular MAP recording, a prolongation of ventricular refractoriness and repolarization has also been demonstrated after intravenous injection of sotalol, as illustrated in Fig. 6 (EDVARDSSON et al. 1980). Interestingly, the D-isomer of sotalol has no beta-blocking activity but a class III mechanism of the same magnitude as the racemic form of the drug (EDVARDSSON et al. 1984b).

Chronic administration of metoprolol, a beta-blocker devoid of acute class III mode of action, leads to a slight prolongation of repolarization (EDVARDSSON and OLSSON 1981a). This was documented by MAP recording technique in eight healthy male volunteers, who were given 400 mg metoprolol daily for 5 weeks. In contrast, acute administration of the drug to these and eight additional volunteers, was not followed by any significant prolongation of right ventricular MAP. The study used a pacing rate of 120/min and resulted in an average MAP prolongation from 252 to 268 ms. This prolongation of repolarization seen after chronic metoprolol treatment is much less pronounced than that seen after chronic administration of an equipotent beta-blocking dose of sotalol (BLOMSTRÖM-LUNDQVIST et al. 1986).

The results of the invasive studies of repolarization effects of metoprolol and sotalol are not, however, directly comparable because healthy volunteers took part in the metoprolol study, whilst in the sotalol study patients with different cardiac diseases and chronic digitalis treatment were investigated. Furthermore, a pacing rate of 100/min was used in the sotalol study. In spite of these limitations, the results do, however, suggest that acute sotalol administration is followed by a more pronounced class III effect than chronic

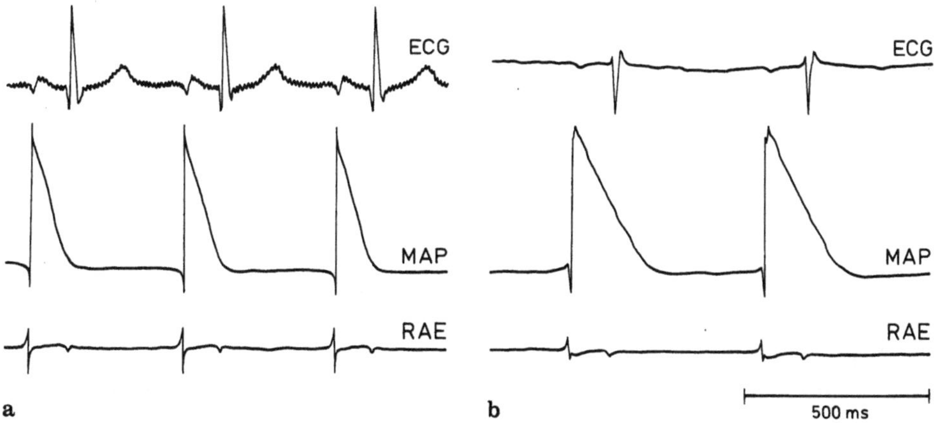

Fig. 5. Electrocardiogram, right atrial monophasic action potential and unipolar right atrial electrogram before *(a)* and 4 weeks after *(b)* treatment with amiodarone. The latter recording shows a decreased heart rate and pronounced prolongation of atrial monophasic action potential. (OLSSON et al. 1973)

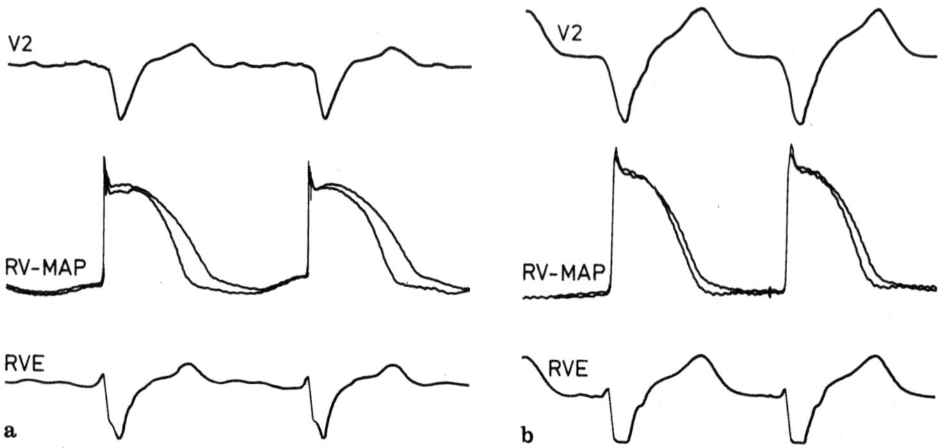

Fig. 6. Effect of acute administration of 100 mg sotalol intravenously (a) and chronic administration of 400 mg metoprolol daily for 4 weeks (b) upon duration of right ventricular monophasic action potential. Note that the delay of repolarization is more marked after intravenous administration of sotalol. In the sotalol study, a pacing interval of 600 ms has been used, whilst in the metoprolol study stimulation is performed using an interval of 500 ms. (EDVARDSSON and OLSSON 1981a; EDVARDSSON et al. 1980)

Table 1. Effect of different antiarrhythmic drugs upon invasively recorded measures of human myocardial repolarization and refractoriness

Drug type	Position	No. of cases	Rate	Duration			
				MAP	(ms)	ERP	(ms)
Ia Procainamide (EDVARDSSON et al. 1984c)	RV	8	500	+2	NS	−5	NS[a]
Ib Lidocaine (EDVARDSSON et al. 1984c	RV	8	500	−7	$P<0.01$	+3	NS
Mexiletine (HARPER et al. 1979)	RV	9	500/600	−7	NS	−8	NS
Ic Flecainide (OLSSON and EDVARDSSON 1981)	RV	9	500/600	+24	$P<0.01$	+22	$P<0.05$
II Metoprolol acute (EDVARDSSON and OLSSON 1981a)	RV	16	500	−1	NS	−1	NS
III Amiodarone chronic (OLSSON et al. 1973)	RA	8	NC	+74	NP	NM	m
Amiodarone acute (BLOMSTRÖM et al. 1987)	RV	13	controlled	+6	$P<0.03$	+7	$P<0.04$

	Sotalol (EDVARDSSON et al. 1980)	RV	7	600	+42	$P<0.01$	+30	$P<0.01$
	D-Sotalol (BLOMSTRÖM-LUNDQUIST et al. 1986)	RV	6	500	+26	$P<0.025$	+20	$P<0.01$
	Melperone (EDVARDSSON and OLSSON 1981b)	RA	7	600	+34	$P<0.05$	NM	-
	Bunaphtine (FENICI et al. 1977)	RA	11	NC	+33	$P<0.0025$	NM	-
	Bunaphtine (BONIATTI et al. 1976)	RV	3	NC	+60[b]	—	NM	-
	Metoprolol chronic (EDVARDSSON and OLSSON 1981a)	RV	8	500	+16	$P<0.01$	+24	$P<0.01$
IV	Verapamil (EDVARDSSON et al. 1985)	RA	10	600	−58	$P<0.025$	+10	NS
	Verapamil (EDVARDSSON et al. 1985)	RV	10	600	+4	NS	+6	NS
M	Digoxin (EDVARDSSON et al. 1984c)	RV	8	500	−2	NS	+13	$P<0.05$
	Atropine (EDVARDSSON et al. 1984c)	RV	8	500	−7	NS	−18	$P<0.05$

Drug types refer to the classification of VAUGHAN WILLIAMS. M, miscellaneous; RV, right ventricle; RA, right atrium; MAP, monophasic action potentail; ERP, effective refractory period; NC, not controlled; NM, not measured; NP, not presented. The duration of the MAP is measured at 90 % repolarization

[a] VERP of right ventricular apex significantly increased

[b] "Total duration"

administration of metoprolol. Fig. 6 illustrates the class III actions of i.v. sotalol and oral metoprolol in these studies.

Following intravenous administration of bunaphtine, atrial MAP duration is lengthened by 18.4 % (FENICI et al. 1976) and right ventricular MAP duration also (BONATTI et al. 1976); an average prolongation from 276.6 to 336.6 ms was noted in three patients, receiving 1.5–2.0 mg/kg body weight of the drug. The limited patient material does not permit comparison of the magnitude of class III effect of this drug with those of other drugs studied with this technique. Melperone, a drug used in psychiatric disorders, has a well-documented class III effect in vitro and in animal models (PLATOU 1982). This effect has also been demonstrated in man by atrial MAP recording technique (EDVARDSSON and OLSSON 1981b). No estimation of refractoriness was, however, made in this study.

As mentioned earlier, several other drugs are capable of prolonging myocardial repolarization in vitro and have therefore been considered to possess class III antiarrhythmic mode of action. Evidence that these drugs prolong atrial and ventricular repolarization in the intact human heart is, however, lacking or very limited. Table 1 summarizes the available literature regarding class III activity demonstrated by invasive techniques in man. For comparison, data obtained with the same technique in studies of some antiarrhythmic drugs with other modes of action are also presented.

F. Class III Antiarrhythmic Action Related to Arrhythmia Mechanism

Since delay of repolarization may exert an antiarrhythmic action, it is reasonable to suggest that class III action might be especially effective in arrhythmias which are initiated or maintained by an acceleration of repolarization.

It is well recognized that patients with hyperthyroidism are likely to develop atrial fibrillation. In this situation, accelerated atrial repolarization seems to be an important factor. Thus, Freedberg et al. (1970) showed that atria of rabbits made hyperthyroid had a marked shortening of their action potentials. Also in other forms of atrial fibrillation, acceleration of myocardial repolarization is an important arrhythmogenic factor. Several studies have revealed an accelerated atrial myocardial repolarization in patients unable to maintain permanent sinus rhythm after DC conversion (Olsson et al. 1971; Cotoi et al. 1972). Ventricular arrhythmias occuring in the acute phase of myocardial infarction are examples of other arrhythmias which at least partly may be dependent upon rapid repolarization. Acceleration of repolarization of the ischemic myocardial cell is a characteristic electrophysiological finding in this situation (Janse et al. 1979).

Surprisingly, only very limited information is available concerning the effect of different antiarrhythmic drugs upon various arrhythmias associated with acceleration of repolarization. In only one of these arrhythmias has a class III antiarrhythmic drug demonstrated a superior efficacy compared with antiarrhythmic drugs with other modes of action. Thus, amiodarone is the most powerful antiarrhythmic treatment today in atrial fibrillation. It may thus not only convert atrial fibrillation to sinus rhythm, but may also be used as a maintenance therapy in recurrent paroxysmal atrial fibrillation and after DC conversion of chronic atrial fibrillation.

References

Armenta J, Zambrano A, Duclós F (1970) Effectos de la amiodarona sobra la repolarizacion ventricular. Nota clinica. Rev Esp Cardiol 23:409–419

Bexton RS, Camm AJ (1982) Drugs with a class III antiarrhythmic action. Pharmacol Ther 17:315–355

Blomström P, Bodnar J, Edvardsson N, B-Lundqvist C, Olsson SB (1987) Acute effect of intravenous amiodarone on ventricular refractoriness and repolarization. Pace 10 (40:II):1004

Blomström-Lundqvist C, Dohnal M, Hirsch I, Lindblad A, Hjalmarsson Å, Olsson SB, Edvardsson N (1986) Effect of long term treatment with metoprolol and sotalol on ventricular repolarization measured by use of transoesophageal atrial pacing. Br Heart J 55:181-186

Bonatti V, Finardi A, Cabasson J, Botti G (1976) L'effetto della bunaftine sui potenziali d'azione monofascici del miocardio atriale e ventricolare destro nell'uomo. G Ital Cardiol 6:440-449

Bonatti V, Rolli A, Botti G (1985) Monophasic action potential studies in human subjects with prolonged ventricular repolarization and long QT syndromes. Eur Heart J [Suppl D]6:131-143

Brachmann J, Senges J, Lengfelder W, Jauernig R, Rizos I, Czygan E, Cobbe SM, Kubler W (1985) Contribution of delayed ventricular repolarization to the anti-arrhythmic efficacy of sotalol. Eur Heart J [Suppl D]6:171-174

Bromberger-Barnea B, Caldini P, Wittenstein GJ (1959) Transmembrane potentials of the normal and hypothermic human heart. Circ Res 7:138-140

Carmeliet E (1977) Repolarisation and frequency in cardiac cells. J Physiol (Paris) 73:903-923

Cotoi S, Gavrilescu S, Pop T, Vicas E (1972) The prognostic value of right atrium monophasic action potential after conversion of atrial fibrillation. Eur J Clin Invest 2:472-474

Debbas NMG, duCailar C, Bexton RS, Demaille JG, Camm AJ, Puech P (1984) The QT interval: a predictor of the plasma and myocardial concentrations of amiodarone. Br Heart J 51: 316-320

Donaldson RM, Rickards A (1982) Evaluation of drug-induced changes in myocardial repolarization using the paced evoked response. Br Heart J 48:381-387

Edvardsson N, Olsson SB (1981a) Effects of acute and chronic beta receptor blockade on ventricular repolarisation in man. Br Heart J 45:628-636

Edvardsson N, Olsson SB (1981b) Effect of intravenous melperone on atrial repolarization in man. Scand J Clin Lab Invest 41:87-90

Edvardsson N, Hirsch I, Emanuelsson H, Pontén J, Olsson SB (1980) Sotalol-induced delayed ventricular repolarization in man. Eur Heart J 1:335-343

Edvardsson N, Hirsch I, Olsson SB (1984a) Right ventricular monophasic action potentials in healthy young men. Pace 7:813-821

Edvardsson N, Hirsch I, Amlie J, Olsson SB (1984b) Class III antiarrhythmic activity of D-sotalol on human ventricular refractoriness and repolarization (Abstr). Circulation [Suppl 2]70:443

Edvardsson N, Hirsch I, Olsson SB (1984c) Acute effects of lignocaine, procainamide, metoprolol, digoxin and atropine on human myocardial refractoriness. Cardiovasc Res 17:463-470

Edvardsson N, Talwar KK, Hirsch I, Olsson SB (1985) Acute effects of verapamil on the human right atrium and ventricle. Cardiovascular Pharmacotherapy International Symposium, April 22-25, Geneva

Fenici R, Marchei M, Bollocci F, Zecchi P (1977) Effect of bunaphtine on right atrial repolarisation in man. Br Heart J 39:787-794

Franz MR, Burkhoff D, Spurgeon H, Weisfeldt ML, Lakatta EG (1986) In vitro validation of a new cardiac catheter technique for recording monophasic action potentials. Eur Heart J 7:34-41

Freedberg As, Papp JG, Vaughan Williams EM (1970) The effect of altered thyroid state on atrial intracellular potentials. J Physiol 207:357-369

Harper RW, Olsson SB, Varnauskas E (1979) Effect of mexiletine on monophasic action potentials recorded from the right ventricle in man. Cardiovasc Res 13: 303-310

Janse MJ, Cinca J, Moréna H, Fiolet JW, Kléber AG, de Vries PG, Becker AE, Durrer D (1979) The "border zone" in myocardial ischemia. An electrophysiological, metabolic and histochemical correlation in the pig heart. Circ Res 44: 576-588

Olsson SB (1972) Right ventricular monophasic action potentials during regular rhythm. A heart catheterization in man. Acta Med Scand 151:145-157

Olsson SB, Edvardsson N (1981) Clinical electrophysiologic study of antiarrhythmic properties of flecainide: acute intraventricular delayed conduction and prolonged repolarization in regular paced and premature beats using intracardiac monophasic action potentials with programmed stimulation. Am Heart J 102:864-871

Olsson SB, Cotoi S, Varnauskas E (1971) Monophasic action potential and sinus rhythm stability after conversion of atrial fibrillation. Acta Med Scand 190:381-387

Olsson SB, Brorson L, Varnauskas E (1973) Class 3 anti-arrhythmic action in man. Observations from monophasic action potential recordings and amiodarone treatment. Br Heart J 35: 1255-1259

Olsson B, Brorson L, Harper R, Rydén L (1977) Estimation of ventricular refractoriness in man by the extra stimulus method. Cardiovasc Res 2:31-38

Olsson SB, Brorson L, Edvardsson N, Varnauskas E (1985) Estimation of ventricular repolarization in man by monophasic action potential recording technique. Eur Heart J [Suppl D]6:71-79

Platou ES (1982) Class III antiarrhythmic action with special reference to the electrophysiological and haemodynamic effects of melperone. Thesis, University of Tromsö

Schwartz PJ (1980) The long QT syndrome. In: Kulbertus HE, Wellens HJJ (eds) Sudden death. Nijhoff, The Hague, pp 358-378

Singh BN, Vaughan Williams EM (1970) The effect of amiodarone, a new antianginal drug, on intracellular potentials and other properties of isolated cardiac muscle. Br J Pharmacol 39:657-667

Torres V, Tepper D, Flowers BAD, Wynn J, Lam S, Keefe D, Miura DS, Somberg JC (1986) QT Prolongation and the antiarrhythmic efficacy of amiodarone. J Am Coll Cardiol 7:142-147

Vaughan Williams EM (1970) Classification of anti-arrhythmic drugs. In: Sandoe E, Flensted-Jensen E, Olesen KH (eds) Symposium on cardiac arrhythmias. Astra, Södertälje, pp 449-466

Vaughan Williams EM (1985) Delayed ventricular repolarization as an anti-arrhythmic principle. Eur Heart J [Suppl D]6:145-149

Vaughan Williams EM, Raine AEG, Cabrera AA, Whyte JN (1975) The effects of prolonged betaadrenoceptor blockade on heart weight and cardiac intracellular potentials in rabbits. Cardiovasc Res 9:579-592

CHAPTER 15

Amiodarone: Electropharmacologic Properties

B. N. SINGH

Developed specifically as a coronary vasodilator and an antianginal compound in 1962, in the past 10 years or so amiodarone hydrochloride (Fig. 1) has recently attracted much experimental and clinical interest as an antiarrhythmic agent. Few, if any, other antidysrhythmic compounds have stimulated as much interest as has amiodarone in relation to the control of refractory arhythmias. Its extreme potency in the prophylactic control of most supraventricular and ventricular arrhythmias is now well established (ROSENBAUM et al. 1976, 1983; HEGER et al. 1981, 1984; NADEMANEE et al. 1981, 1982a, 1983; GRABOYS et al. 1983; ZIPES et al. 1984; SINGH et al. 1980). However, the fundamenal mechanism whereby it induces its salutary effects is far from being certain. For this reason, the effects of the compound on cardiac electrophysiology relative to its associated pharmacologic properties are of much theoretical as well as practical importance.

As in the case of numerous antiarrhythmic agents, the overall effects of amiodarone on the cardiac action potentials may result from its direct as well indirect actions. Barring its intrinsic effects, the compound has the propensity

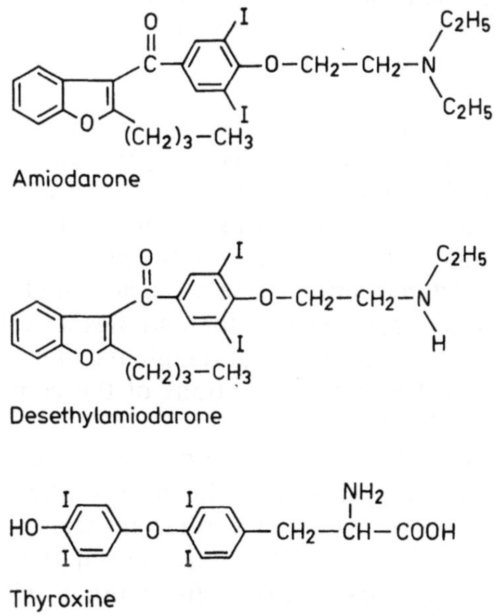

Fig. 1. Structural formulae of amiodarone, desethylamiodarone, and thyroxine. Note the presence of iodine in amiodarone and desethylamiodarone

to antagonize noncompetitively alpha- and beta-adrenergic receptors (POLSTER and BROEKHUYSEN 1976; CHARLIER et al. 1968) with a poorly understood and complex interrelationship with thyroid hormone metabolism (SINGH 1983). The purpose of this chapter is to discuss the overall pharmacodynamic actions of the drug with particular reference to its electrophysiologic properties.

A. Development of Amiodarone

Amiodarone was synthesized by Labaz labaratories in Belgium as an antianginal agent during a systematic search for potent coronary vasodilators (CHARLIER et al. 1962). The drug was one of a series of derivatives that were synthesized on the basis of the benzofuran moiety of the khellin molecule and its natural congeners all of which were reasonably potent coronary vasodilators. The very first compound was benziodarone, in which the presence of two iodine atoms was believed to augment the overall pharmacologic properties compared with those of its precursor, benzarone. Benziodarone underwent brief clinical evaluation but was abandoned when it was discovered that it induced jaundice and hepatotoxicity in man. It was superseded by amiodarone, which was found to be a more potent coronary vasodilator. In a series of comprehensive pharmacologic studies, CHARLIER et al. (1968) in a variety of isolated tissue preparations and in anesthetized and in intact conscious dogs clearly demonstrated somewhat unusual properties of the compound. Their studies indicated a slow onset and offset of action of amiodarone. For example, when it was given orally to instrumented dogs, the reduction in heart rate, tension-time index, and systemic pressure were gradual and did not appear to reach a steady state at a constant daily dose for about 5–6 weeks. Similarly, the regression of the observed changes was not complete even 5 weeks after drug withdrawal.

When intravenous amiodarone was given there was an increase in coronary blood flow and reduced myocardial oxygen consumption. Intravenous amiodarone also tended to attenuate the tachycardia and enhanced contractility produced by isoproterenol, suggesting an interaction with the autonomic nervous system. Although the differences between the effects of the parenterally and orally administered amiodarone were not emphasized by CHARLIER et al. (1968), the overall effects noted in their studies were construed as representing a "new biological profile" for an antianginal compound. The first report documenting the clinical antianginal actions of the compound appeared in 1967 (VASTESAEGER et al. 1967).

The antiarrhythmic effects of amiodarone in experimental animals appeared in 1969 (CHARLIER et al. 1969). The earliest attempt to delineate the fundamental mechanism of action in cardiac muscle was reported in 1970 (SINGH and VAUGHAN WILLIAMS 1970a) and again in 1971 as an integral part of a doctoral dissertation (SINGH 1971). In common with the drug sotalol (SINGH and VAUGHAN WILLIAMS 1970b), it was suggested that amiodarone might be a potent antiarrhythmic compound. The first clinical report with the drug as an antiarrhythmic agent was with the *intravenous* amiodarone in 1970 (VAN SCHEPDAEL and SOLVAY 1970) and after oral therapy in 1974 (ROSENBAUM et al. 1974). The effects of the compound have now been studied in a variety of animal and clinical arrhythmias. The emergence of amiodarone has been a major landmark in the development of antiarrhythmic therapy (SINGH and ZIPES 1983) but the precise cellular mode of action of the compound remains incompletely understood. In the section that follows, the relevant data are briefly discussed. The discussion of the electrophysiologic actions of the compound will be preceded by a description of the compound's interaction with the autonomic nervous system. The thesis will be developed that the precise understanding of the action of this unusually potent compound may provide ideas about both the development of similar but safer compounds, and the fundamental mechanisms of arrhythmias themselves.

B. Amiodarone and Antiadrenergic Antagonism

The acute antiadrenergic actions of amiodarone have previously been established both in vitro (POLSTER and BROEKHUYSEN 1976) as well as in vivo (CHARLIER 1970; CHARLIER et al. 1972; KOBAYASHI et al. 1983) experimental models. POLSTER and BROEKHUYSEN (1976) compared the effects of the competitive beta-antagonist propranolol in isolated rabbit atria with those of amiodarone. The pA_2 value for propranolol was 8.33. Amiodarone acted as a noncompetitive beta-antagonist with a pD_2 value of 4.17 with isoproterenol as an agonist. The effects of amiodarone on alpha-receptor blockade were inves-

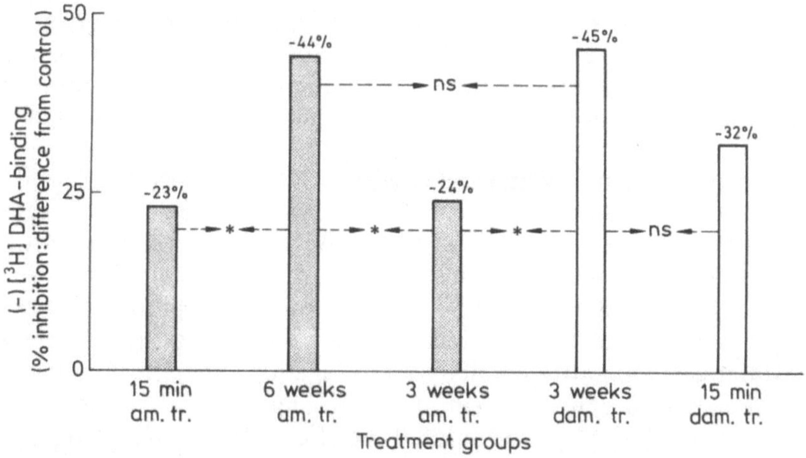

Fig. 2. Effects of acute and chronic treatment with amiodarone and desethylamiodarone on beta-receptor density (B_{max}) in the rabbit ventricular myocardium. Both agents depress B_{max}, with amiodarone exhibiting the trend to produce a greater reduction following chronic than after acute drug administration. The data raise the possibility that the chronic effect may be due to the summated effects of the parent compound and those of the metabolite. The greater chronic effect may also result from an added effect of selective hypothyroidism (see text). (VENKATESH et al. 1986).

tigated in isolated rat-aortic strips induced to contract by norepinephrine. The pA_2 value for phentolamine was found to be 8.69 whereas the pD_2 value for amiodarone was 4.06, the drug having no effect on calcium permeability in this preparation. Subsequently, CHARLIER (1970) found that in anesthetized dogs amiodarone produced bradycardia independently of beta-receptor blockade as the effect persisted after the administration of propranolol. It is of interest that although amiodarine has been found to decrease cholinergic receptors in the rat heart and brain (COHEN-ARMON et al. 1984), some studies have failed to demonstrate a significant interaction with the cholinergic component of the autonomic nervous system (SINGH and VAUGHAN WILLIAMS 1970a; KOBAYASHI et al. 1983).

It is known that bradycardia develops in a gradual manner as a function of time on a constant dose of amiodarone (CHARLIER et al. 1968; SINGH 1971), raising the possibility of progressive decrease in myocardial beta-adrenoceptors. This possibility has recently drawn increasing attention (NOKIN et al. 1983; SHARMA et al. 1983). It is now confirmed that amiodarone does antagonize beta-receptormediated actions in a noncompetitive fashion and exerts a significant effect on beta-receptor density following acute as well as chronic administration. For example, VENKATESH et al. (1986a) have shown that when amiodarone and its principal metabolite desethylamiodarone were given acutely and chronically to rabbits, there was a significant reduction in myocardial beta-receptor density (B_{max}) without an effect on receptor affinity (K_D). The effect of chronic amiodarone administration (Fig. 2) over 6 weeks was more pronounced (−45%) than that (−25%) following acute intravenous drug injection. However, this difference was unrelated to dose since the effects of 20 mg/kg and 40 mg/kg chronic dosing regimens on B_{max} were statistically indistinguishable. Nor was the greater effect following chronic therapy attributable to serum and tissue levels following chronic dose drug administration, since both the serum and myocardial levels 15 min after acute intravenous amiodarone administration were considerably higher than the corresponding levels following 6 weeks of chronic drug dosing.

The data confirm and extend the results of NOKIN et al (1983), who also performed direct ligand-binding assays on rat myocardial beta-adrenoceptors, but differ in providing evidence for a greater effect following chronic drug administration than after acute. The observations of NOKIN et al. (1983) are also of interest in that they showed that both pretreatment with propranolol as well as with amiodarone abolished the increases in beta-receptor density induced by myocardial ischemia after coronary artery ligation. The effects of amiodarone on adrenergic receptors are similar in different animal species. For example, SHARMA et al. (1983), who studied the effects of chronic (6 weeks) oral amiodarone administration to cats on beta- and alpha-receptor density in ventricular muscle, found no effect on alpha-receptor density but observed a significant reduction in beta-receptor density without a change in receptor affinity. The reason for the dissociated effect of amiodarone on alpha- and beta-receptors as determined by radioligand binding is at present unclear especially in light of the fact that in in vitro systems noncompetitive effects against alpha- and beta-catecholamine receptors have been demonstrated

(POLSTER and BROEKHUYSEN 1976). GAGNOl et al. (1985) have also found that in rat heart membrane preparations amiodarone noncompetitively antagonized the activation of adenylate cyclase by isoproterenol, glucagon, and secretin but not by sodium fluoride. The authors suggested that noncompetitive beta-antagonistic properties of amiodarone might be due to the inhibition of the coupling of beta-receptors with the regulatory unit of the adenylate cyclase complex and/or to a decrease in the number of functional beta-receptors at the surface of the myocardial cell. The net result in vivo is the attenuation of the positive chronotropic actions of catecholamines (GAGNOL et al. 1985), a property of possible significance in mediating the antiarrhythmic and anti-ischemic effects of the compound.

The fact that bradycardia during chronic amiodarone therapy develops as a function of time is consistent with the data of VENKATESH et al. (1986a), indicating a gradual decrease in the number of beta-receptors. In part this may be due to the additive effect of the metabolite during chronic drug administration. This may also be explained on the basis of a secondary consequence of selective hypothyroidism induced by amiodarone (SINGH 1983; VENKATESH et al. 1986a). It is known that in hypothyroidism a significant decrease in the density of myocardial beta-adrenoceptors occurs, the converse in hyperthyroidism (WILLIAMS et al. 1977). The changes in cardiac rate and rhythm during altered thyroid state may therefore in part be due to the associated alteration in the adrenergic state (VENKATESH et al. 1986a; WILLIAMS et al. 1977). Thus, if the principal mechanism of action of amiodarone were a selective inhibition of T_3 action on cardiac muscle, the marked reduction in beta-receptor density in cardiac muscle following chronic amiodarone treatment may in part be due to the drug-induced myocardial hypothyroid state (see below). The observations of BACQ et al. (1976) on the effects of amiodarone on neurotransmitter overflow from the spleen induced by electrical stimulation of the splenic nerve suggest that the drug might also exert a significant adrenergic blocking action at high drug concentration.

C. Pharmacokinetic and Hemodynamic Effects

A detailed discussion of the pharmacokinetic and hemodynamic properties of the compound are beyond the scope of this chapter. However, salient features will be summarized briefly (see HOLT et al. 1983; STOREY et al. 1983; KANNAN et al. 1982; LATIN et al. 1984). Amiodarone is well absorbed with a bioavailability between 22 % and 86 %, is highly protein bound, is metabolized extensively in the liver and at least one active metabolite (desethylamiodarine) has been pharmacologically characterized. The major pathway of excretion is the biliary tract with little or no renal excretion. The elimination half-life of the drug varies between 15 and 110 days and that of the metabolite is even longer. Amiodarone is an amphiphilic compound. There is a reasonable linearity between plasma drug concentrations and dose of amiodarone. Its main pharmacokinetic features are listed in Table 1.

Table 1. Clinical pharmacokinetic profile of amiodarone

Absorption rate	T_{max}: 2–12 h (lag time, 0.4–3 h)
Extent of absorption	Poor and slow
Bioavailability	Variable (22 %–86 %)
Protein binding	96.3 % ± 0.6 %
Volume of distribution	1.3–65.8 liters/kg (acute)
Elimination	Negligible renal excretion
Biotransformation	Hepatic and intestinal
Elimination half-life	3.2–20.7 h (acute)[a]; 13.7–52.6 days (chronic)
Total body clearance	0.10–0.77 liters/min
Pattern of elimination kinetics	First order
Metabolites	Major: mono *N*-desethylamiodarone; Minor: bis-*N*-desethylamiodarone, deiodinated metabolites
Therapeutic plasma range	1–2.5 μg/ml
Dose schedule	Once daily
Special factors	Slow onset and offset of action

[a] A terminal elimination half-life of 24.8 ± 11.7 days was reported by Holt et al, after a 400 mg i.v. dose in six healthy volunteers.

The hemodynamic effects of amiodarone represent a complex interplay of simultaneous alterations in preload, afterload, ventricular contractility, heart rate, and coronary blood flow, the overall effect being dependent upon dose of the drug, rate of infusion, the diluent in which the drug is suspended, and the presence or absence of myocardial ischemia as well as the degree of ventricular dysfunction. In 5% aqueous solutions, the drug (2.5–10 mg/kg) reduced heart rate and blood pressure, depressed systemic vascular resistance, increased coronary blood flow, but did not reduce cardiac output; at the highest dose used it increased the left ventricular end diastolic pressure and reduced the *LV* d*p*/d*t*. These findings in open chest-anesthetized dogs (SINGH et al. 1976; PETTA and ZACCHEO 1971) have also been found in man, in which the intravenous amiodarone suspended in Tween 80 exerted a slightly depressant effect on cardiac performance, but such an effect was short lived (SCHWARTZ et al. 1983). The drug increased cardiac output and decreased the left ventricular filling pressures while decreasing systemic and coronary vascular resistance (COTÉ et al. 1979). The data are consistent with the finding that during chronic therapy with the drug there is no significant fall in the left ventricular ejection fraction determined by radionuclide ventriculography (SINGH 1983) even in patients with severely depressed baseline ejection fractions. The data harmonize with the observations that rarely does the compound aggravate cardiac failure in patients with markedly reduced myocardial function or even in those with manifest ventricular decompensation (ROSENBAUM et al. 1983; HEGER et al. 1981, 1984; NADEMANEE et al. 1981, 1982a, 1983).

D. Electrophysiologic Effects of Amiodarone

When the action of amiodarone on cardiac muscle is considered, a number of features of its pharmacology appears to be of importance. First, the drug is not soluble in the usual physiologic media; thus, superfusion studies in isolated cardiac tissues can only be undertaken in homologous plasma or blood or in an especially modified extracellular environment (HAUSWIRTH and SINGH 1978; YABEK et al. 1986). Second, when the drug is given intravenously to experimental animals and in man, the electrophysiologic effects are much less striking than those noted after the chronic administration at a constant dose over long periods (WELLENS et al. 1984; MORADY et al. 1986; IKEDA et al. 1984). An explanation for the delayed onset of the action of the drug thus appears crucial to a better understanding of its actions in the control of cardiac arrhythmias. Third, the elimination half-life of amiodarone is extremely long and variable (HOLT et al. 1983; STOREY et al. 1983; KANNAN et al. 1982) and the steady-state effect of the compound cannot be predicted simply on the basis of the plasma and tissue drug levels of the parent compound or its active metablite, desethylamiodarone (LAMBERT et al. 1986; KATO et al. 1986). For these reasons, the acute and the chronic effects of amiodarone and its metabolite need to be differentiated. Fourth, the interpretation of its effects must allow for the fact that the compound is an iodinated molecule with a complex interrelationship with the metabolism of the thyroid hormone to which in part it bears a structural resemblance (Fig. 1). There are features of the drug's electrophysiologic actions which closely resemble those of hypothyroid cardiac muscle and are prevented by the administration of relatively small amounts of thyroxine. Finally, amiodarone is a coronary vasodilator (SINGH et al. 1976) exhibiting a significant interaction with the autonomic nervous system which may be of importance in mediating the overall pharmacologic and electrophysiologic effects of the compound in controlling cardiac arrhythmias.

I. Early Electrophysiologic Observations

In the first reported studies (SINGH and VAUGHAN WILLIAMS 1970a), amiodarone was given 20 mg/kg intraperitoneally (as a 5% aqueous solution) for 1-6 weeks to rabbits, and atria and ventricular tissues were removed and studied electrophysiologically by the microelectrode technique. The drug had no significant effect on the resting membrane potential, or the action potential amplitude in either tissue; it produced about a 10% reduction only in the V_{max} at a stimulus frequency 10% above the spontaneous frequency of the sinus node in the case of the atria and at 1 Hz in the case of the ventricular fibers. The major effect was a considerable prolongation of the action potential duration in both atrial and ventricular tissue (Fig. 3). Thus, by inference, the effective refractory period was also prolonged. As far as the effects of amiodarone in the atrial muscle were concerned, they bore a striking resemblance to those induced by thyroid gland ablation in the rabbit (FREEDBERG et al. 1970). Two other features of importance were noted. First, the effect of the drug on the action potential duration was slow in onset and continued to increase in a gra-

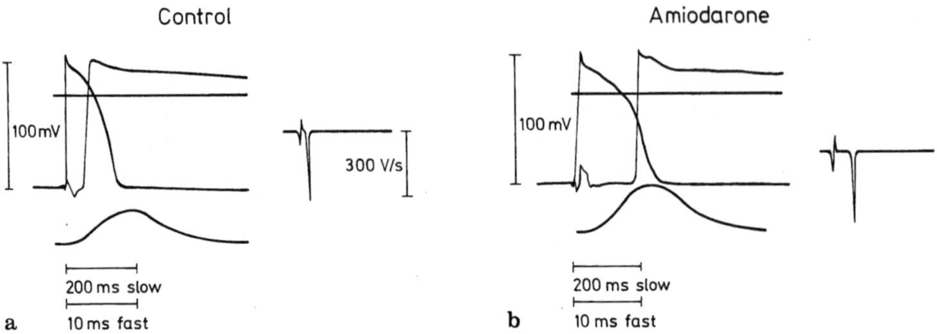

a b

Fig. 3. The effects of chronic amiodarone administration on the characteristics of transmembrane potentials in rabbit ventricular myocardium. **a** A typical recording from the ventricular muscle of a control rabbit, and **b** from one treated with amiodarone 20 mg/kg intraperitoneally for 6 weeks. In each panel, the *upper trace* represents zero potential; the *middle trace,* the transmembrane potential at slow and fast sweep speeds; the lower trace, contraction. On the *right trace* is shown the extracellular electrogram and the differentiated signal of the rate of rise of phase 0 of the action potential. Note that the major effect of the drug was to increase the time course of repolarization (and, by inference, the effective refractory period) with a minimal effect on the upstroke velocity of phase 0. (SINGH and VAUGHAN WILLIAMS 1970a)

dual manner as a function of time on a *constant* daily dose. For example, at a daily intraperitoneal dose of 20 mg/kg amiodarone in the rabbit, it was found that the ventricular action potential increased by 11% at 1 week, 23% at 3 weeks, and 30% after 6 weeks of drug administration. Second, 5 μg thyroxine given daily at the same time as amiodarone for the last 3 weeks of the 6-week drug administration precluded the expected development of the action potential lengthening effected by the drug alone. Moreover, the administration of iodine as potassium iodide in amounts equivalent to that contained in the daily amiodarone dose used in these experiments did not produce changes similar to those induced by the drug. It was therefore postulated that the actions of chronically administered amiodarone could be counteracted by stimulation of thyroxine-dependent pathways (SINGH and VAUGHAN WILLIAMS 1970a). Amiodarone had no effect on the weight of the thyroid gland, however, implying the absence of a direct antagonism to thyroid hormones, which still exerted their negative feedback action on the production of TSH.

A number of studies (MASON et al. 1983, 1984; AOMINE et al. 1984; GLOOR et al. 1983) in recent years have further enlarged our understanding of the actions of amiodarone in various cardiac tissues.

II. Amiodarone and Slow-Channel Potentials

In the sinus node preparations of the isolated rabbit atria superfused with amiodarone, GOUPIL and LENFANT (1976) reported significant drug-induced decrease in the action potential amplitude, the maximal diastolic potential, and the slope of phase 4 of pacemaker potentials. That amiodarone does de-

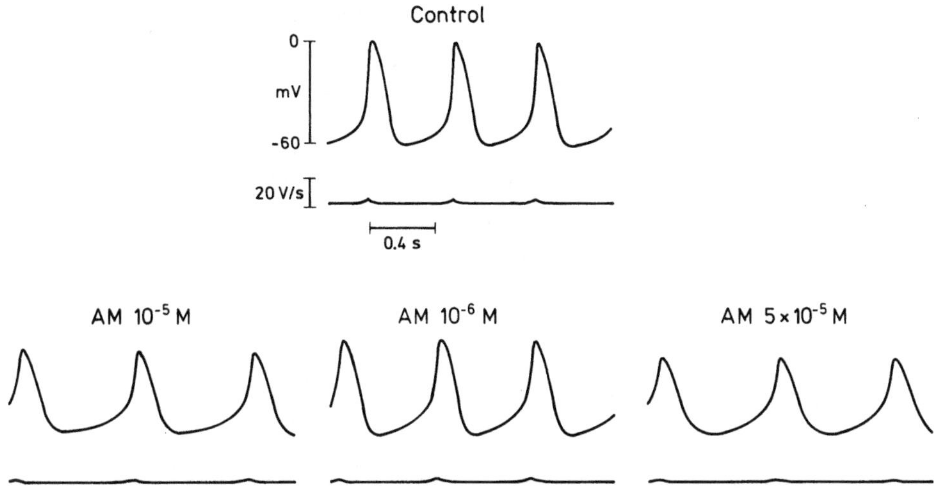

Fig. 4. The effects of increasing amiodarone concentrations on a single dominant sinus node pacemaker cell action potential. The preparation is discharging spontaneously. In this and all subsequent illustrations of intracellular recordings the *lower trace* shows the phase 0 (upstroke) V_{max}. Voltage and time calibrations are the same for each illustration in the figure. (YABEK et al. 1986)

crease phase 4 depolarization in the SA node and reduce the amplitude of the pacemaker potentials in the rabbit sinoatrial node (Fig. 4) was recently confirmed by YABEK et al. (1986). In their study the sinus cycle length was significantly lengthened by amiodarone as well as by desethylamiodarone.

A recent study (GLOOR et al. 1983) in which amiodarone was injected directly into the sinus and AV nodal arteries has raised the possibility that the drug might also act by inhibiting the slow channel in nodal tissues. It is thus conceivable that the drug's acute effects (WELLENS et al. 1984) to a significant extent may be mediated via its antiadrenergic (see below) and calcium antagonistic actions in the SA and AV nodes. These effects are consistent with the marked decreases in sinus cycle length by the depression of phase 4 depolarization noted in the studies reported by YABEK et al. (1986) as well as in those of GOUPIL and LENFANT (1976). It is also conceivable that the slow-channel blocking action may be the basis for impaired myocardial contractility evident with the intravenous drug in high doses.

III. Amiodarone Effects on Fast-Channel Potentials

The studies of YABEK et al. (1986) have also focused on defining the effects of amiodarone and its metabolite on the electrophysiologic parameters of other cardiac tissues during acute superfusion studies. Over a range of concentrations, 10^{-6} to 5×10^{-5} M — (0.68–34 μg/ml), both amiodarone and desethylamiodarone dissolved in appropriate superfusion media exerted distinct but quantitatively and qualitatively similar electrophysiologic actions in isolated canine and rabbit cardiac muscle. At 1.0 Hz stimulus frequency, neither drug

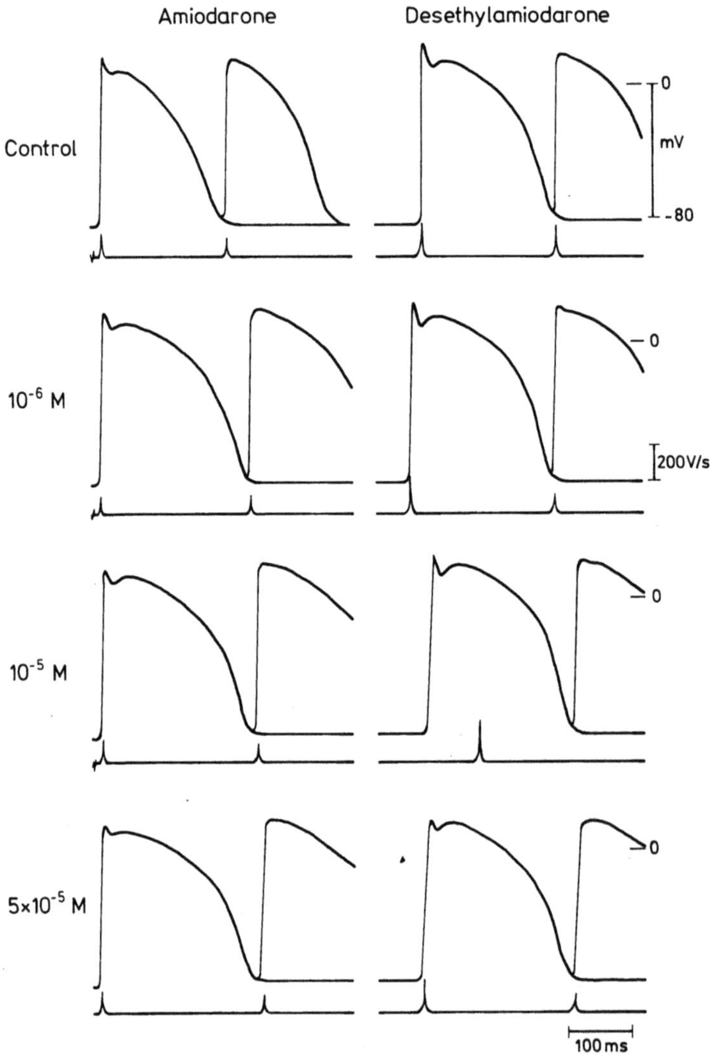

Fig. 5. Effects of increasing concentrations of amiodarone and desethylamiodarone on ventricular myocardial action potentials. For each drug, the various action potentials were obtained from a single myocardial cell. Voltage and time calibrations are the same for each illustration. (Yabek et al. 1986)

had a significant effect on action potential amplitude, overshoot, upstroke velocity of phase 0, or the resting membrane potential of rabbit atria, canine ventricular muscle, or Purkinje fibers even at the highest drug concentration (34 μg/ml). Modest increases in the action potential duration at 50% repolarization time (APD_{50}) and at 90% (APD_{90}) and in the effective refractory period (ERP), however, occurred in the ventricle (Fig. 5); in the atria these changes were less marked. A lack of change in the ratio of APD_{90}/ERP indicated that the change in the ERP was due essentially to the delayed repolarization. An

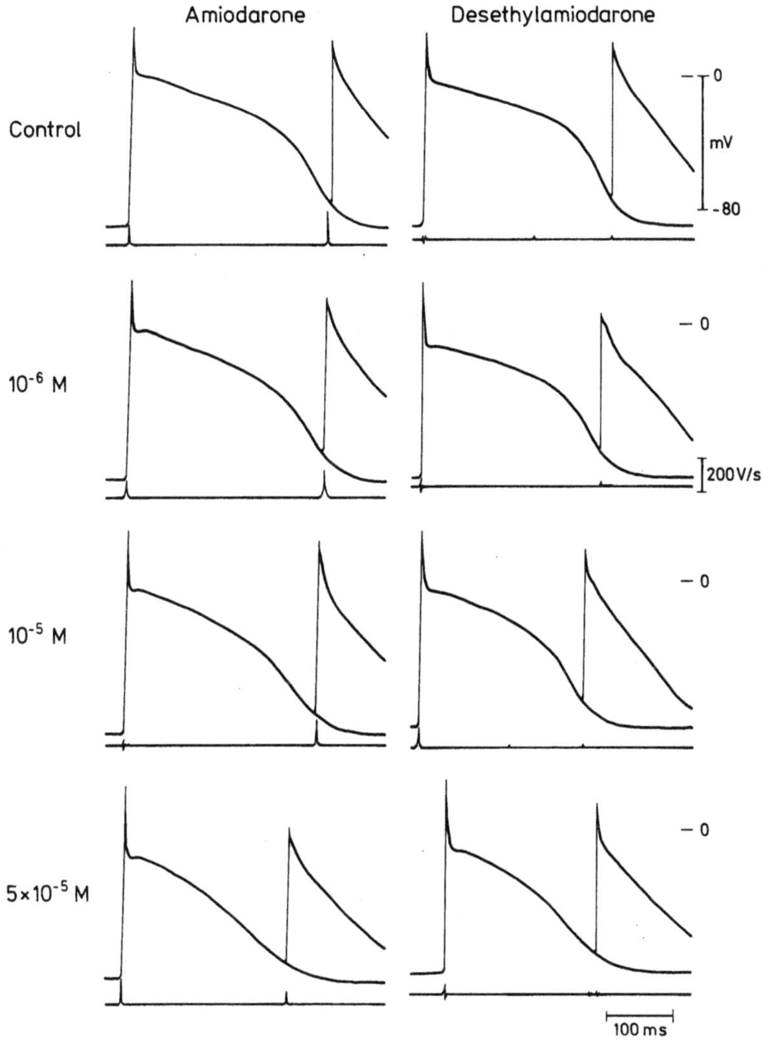

Fig. 6. Results from a typical experiment showing the effects of increasing amiodarone and desethylamiodarone concentrations on Purkinje fiber action potentials. For each drug, the four action potentials were obtained from a single cell. Voltage and time calibrations are the same for each panel. (YABEK et al. 1986)

unexpected finding reported by YABEK et al. (1986) was that in Purkinje fibers both amiodarone and its metabolite significantly *decreased* APD_{90} and *shortened* the ERP especially at the higher concentrations (Fig. 6). In the case of amiodarone, similar observations have previously been made by AOMINE et al. (1984). It is noteworthy that the shortening of the Purkinje fibre ERP was accompanied by a lengthening of the ERP in ventricular muscle. The ERP in the Purkinje fibre is normally longer than that in the ventricular muscle. Therefore, the observed differential effect of amiodarone and its metabolite on the voltage-dependent changes in the ERP may contribute to an overall electrical stability in the heart and constitute an antiarrhythmic mechanism (YABEK et al. 1986). The possibility must be considered that such an effect might mediate at least in part the acute antiarrhythmic actions of the compound (but see discussion of this point in Chap. 2). However, it must be emphasized that it is not known whether such a differential effect on repolarization is also induced by the drug during chronic administration.

IV. Effects on Fast-Sodium Channel Kinetics and Use Dependency

SINGH and VAUGHAN WILLIAMS (1970a) reported that in rabbits treated chronically with amiodarone V_{max} was reduced only by about 10%. Although such an effect was not felt to be significant, recent data have suggested that the overall magnitude of the effect may vary with animals species used, the concentratons of the drug and its metabolite that are tested, as well as with the route of drug administration. The depressant effect of amiodarone on V_{max} has been most apparent at high stimulation frequency (MASON et al. 1983) and in depolarized or ischemic myocardial fibers (Chap. 16). During superfusion studies involving the left atrial free wall, amiodarone produced a slow development of concentration-dependent prolongation of the action potential duration; this was accompanied by a significant decrease in V_{max}, reflecting an inhibitory effect on the fast sodium channel as has been reported in the case of nerve fibers (COURTNEY 1975). YABEK et al. (1986) found a significant rate-dependent block of V_{max} in canine Purkinje fibers and ventricular muscle at a concentration of 34 µg/ml amiodarone and desethylamiodarone.

Of uncertain significance (to its clinical action) are the findings of MASON et al. (1983, 1984), who, during voltage clamp studies, showed that extremely high concentrations of amiodarone (over 50 µg/ml) exerted a rate-dependent depressant effect on the inactivated fast-Na channels both in guinea pig myocardial fibers acutely superfused with amiodarone and in similar fibers removed from animals chronically treated with the drug over a period of 28 days. Amiodarone had a particular affinity for inactivated sodium-channels and had a marked effect on the kinetics of recovery from inactivation. For example, under drug-free conditions and at normal resting membrane potentials, recovery from inactivation was usually complete in 10 ms (MASON et al. 1984). Under the influence of amiodarone, the partial recovery of V_{max} occurred rapidly, but a considerable fraction tended to recover slowly. This slow component of recovery at −80 to −90 mV had a time constant of about 163 ms. Thus, their data indicated that the use-dependent block of V_{max} induced by amiodarone developed during inactivation and recovered during rest. The data from MASON et al. (1984) are also of interest insofar as they showed that the drug lengthened the action potential duration both acutely as well as after chronic drug administration; it also reduced or prevented the occurrence of depolarization-induced automaticity.

V. Ionic Correlates of the Electrophysiologic Effects of Amiodarone

While the effects of amiodarone on the gross parameters of cardiac action potential are now reasonably well delineated in acute superfusion studies and following chronic drug administration, there remains paucity of data on the ionic correlates of these changes. Using the double sucrose gap technique in frog and ferret ventricular fibers, NELIAT et al. (1982a) provided further confirmation of the drug's effect in lengthening the action potential duration, in reducing the slope of diastolic depolarization in the atrium, and in inhibiting the pacemaker activity in the Purkinje fibers and of the repetitive activity in the atrium. In a further study, NELIAT et al. (1982b) demonstrated that amiodarone decreased outward potassium current consistent with the prolongation of the action potential duration. They also found that high concentrations of the drug depressed inward currents, retarding the kinetics of reactivation of the fast sodium and the slow calcium currents. Further data are, however, needed to define the significance of these changes in altering conduction and refractoriness in cardiac muscle in relation to the control of cardiac arrhythmias.

VI. Electrophysiologic Effects of Amiodarone Following Chronic Administration

The acute superfusion effects with amiodarone and its metabolite, while being modest with respect to repolarization, have been found to be no greater at the higher drug concentrations compared with those at the lower ones. It appears that, despite very high concentrations of amiodarone, a finite duration of exposure of the myocardium to the drug on a chronic basis is necessary for the full expression of the electrophysiologic effect. The most striking and consistent electrophysiologic effects of amiodarone occur when the drug is administered chronically as originally reported (SINGH and VAUGHAN WILLIAMS 1970a). These effects have now been shown (IKEDA et al. 1984) to occur in atria, sinus node fibers, AV nodal fibers, and ventricular tissues (Fig. 7). The chronic effects in Purkinje fibers have not been clearly defined. The reason for the gradual increase in the intensity of the effect as a function of time remains uncertain. Investigations have been undertaken to determine whether the phenomenon of the delayed onset of drug action is due to the formation of the active metabolite or to the slow build-up of amiodarone in the tissues or the myocardial membranes. Neither possibility is well supported by experimental data.

VII. Significance of the Activity of Desethylamiodarone

It has been found that although the metabolite (desethylamiodarone) is active, having qualitatively the same pharmacodynamic profile, its action is not immediate (see below). The drug has the propensity to reduce responses to sti-

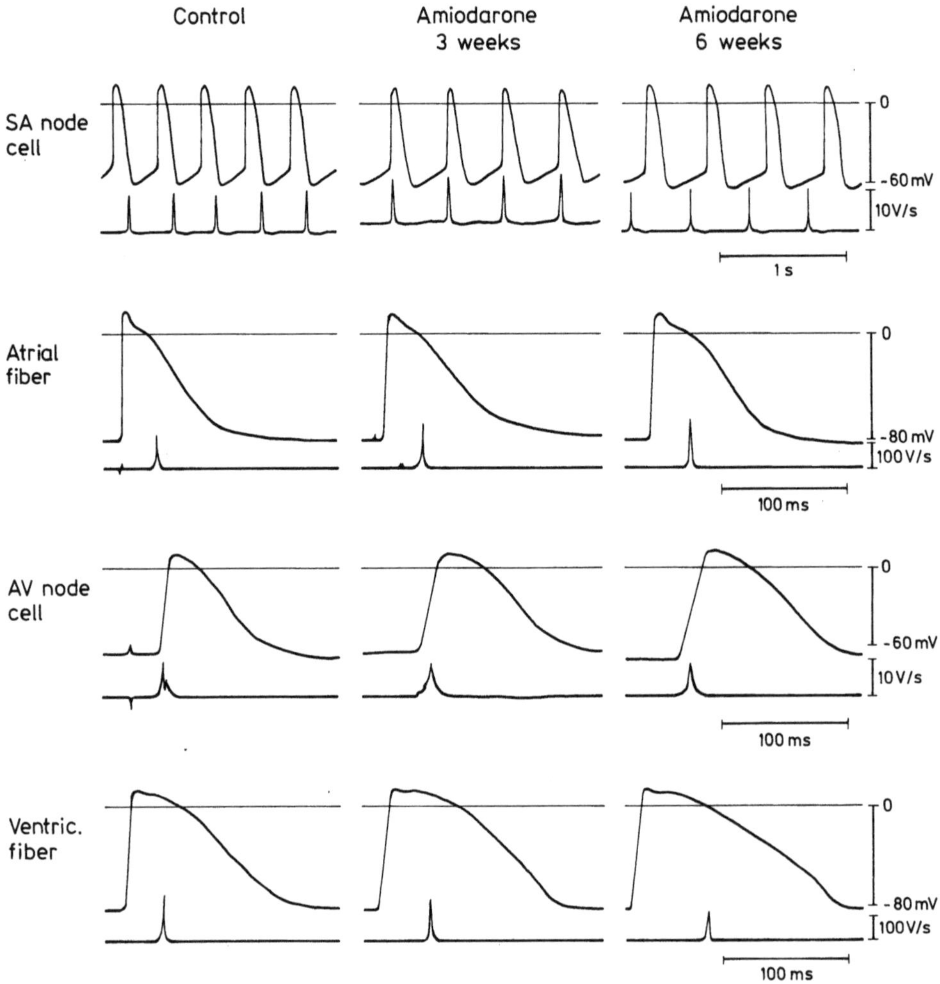

Fig. 7. Effects of chronic administration of amiodarone on various types of cardiac action potentials in the rabbit heart compared with representative control recordings. Note the progressive increase in repolarization time as a function of time on a constant daily dose. (IKEDA et al. 1984)

mulation of beta-adrenoceptors after acute as well as chronic administration (VENKATESH et al. 1986a), to alter thyroid hormone metabolism after chronic administration (VENKATESH et al. 1986b), and to interact pharmacokinetically with cardiac glycosides (VENKATESH et al. 1986c). As in the case of the parent compound, the metabolite appears to influence electrophysiologic parameters as a function of time, and its elimination half-life is longer than that of amiodarone (KANNAN et al. 1984). Thus, while the activity of the metabolite will be additive to that of the parent compound during chronic administration, the delay in the onset of amiodarone action can not be attributed solely to the effects of desethylamiodarone.

VIII. Significance of Myocardial and Sarcolemmal Amiodarone Concentrations

The more pronounced pharmacologic efficacy of amiodarone following chronic administration despite low plasma drug concentrations and the lesser effects of the drug after acute intravenous administration when drug levels are maximum has not been explained on the basis of the pharmacokinetic behavior of the drug. The recent studies of VENKATESH et al. (1986d) have shown that the gradual increase in cardiac repolarization as a function of time is not related to the rate of accumulation of amiodarone in myocardial tissues or in the sarcolemmal preparations (Fig. 8). The effects of amiodarone on transmembrane action potential recordings from rabbit ventricular myocardium were studied in relation to the drug concentrations in the serum, myocardium, and myocardial sarcolemma both after acute intravenous drug administrations and after 4 weeks oral administration of 20 mg/kg per day amiodarone. Following the 15 min acute drug administration when amiodarine concentrations were maximal in the serum (4.72 ± 1.23 μg/ml), cardiac muscle (34.5 ± 7.6 μg/g), and sarcolemma (1.94 mg/g protein), the electrophysiologic changes were *insignificant*. However, following chronic treatment when the levels of amiodarone were *low* in the serum [0.05 ± 0.01 μg/ml(Am),

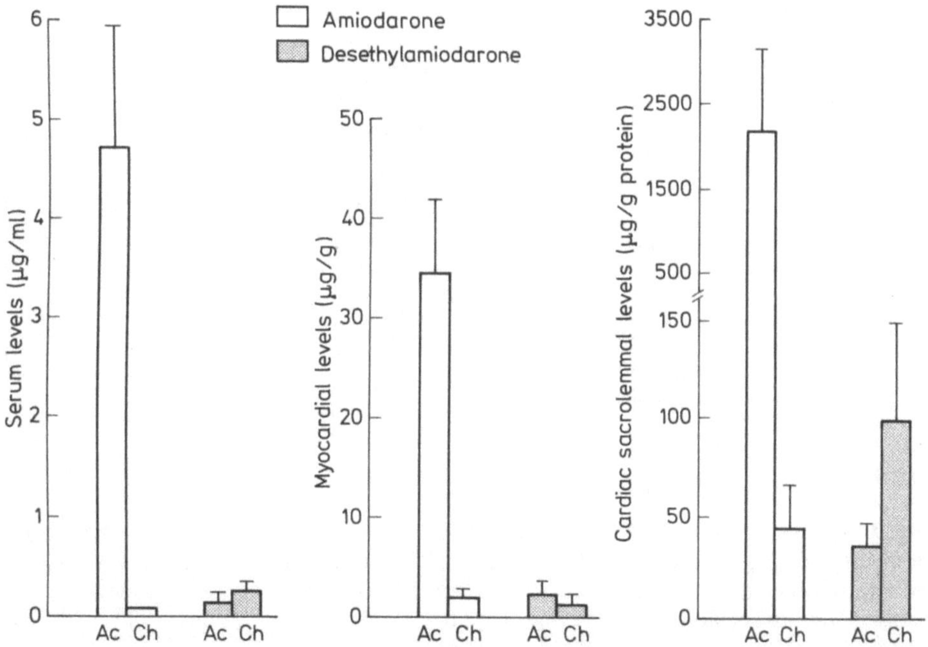

Fig. 8. Concentrations of amiodarone, desethylamiodarone in the serum, myocardium, and ventricular sarcolemmal preparation are shown. Note the significant reduction in the concentrations of amiodarone following chronic (Ch) treatment in all of the tissues analyzed, and the levels of desethylamiodarone in the serum and sarcolemma following chronic therapy; $\beta < 0.0001$. *Ac*, acute. (VENKATESH et al. 1986d)

0.25 ± 0.08 μg/ml (desethylamiodarone)], cardiac muscle [1.91 ± 0.9 μg/g (amiodarone), 1.35 ± 1.33 μg/g (desethylamiodarone)] and the myocardial membranes [0.043 mg/g protein (amiodarone), 0.097 mg/g protein (desethylamiodarone)], there was a 54.3% increase in action potential duration at 90% repolarization ($P < 0.01$) and 65% increase in the effective refractory period ($P < 0.01$) of rabbit ventricular myocardium. These data are supported by the observations of LAMBERT et al. (1986), who found that the increases in the ventricular effective refractory period as a function of time were dissociated from the tissue or serum concentrations of amiodarone as well as from the myocardial disposition of the metabolite. It is also noteworthy that PATTERSON et al. (1983) recently showed (Fig. 10), that such differences between the acute and chronic dosing of amiodarone in their experimental canine model of sudden death could not be accounted for by lower serum and tissue levels of amiodarone following acute drug administration. Thus, the magnitude of the electrophysiologic and antiarrhythmic effects induced by amiodarone are not explained by the pharmacokinetics of the drug, but are probably related to drug-induced adaptational changes in cellular metabolism (compare the effects of long-term beta-blockade on APD, Chap. 2, 12b). A finite duration of exposure of the myocardium to the drug on a chronic basis appears necessary for the maximal steady state electrophysiologic effect to become established.

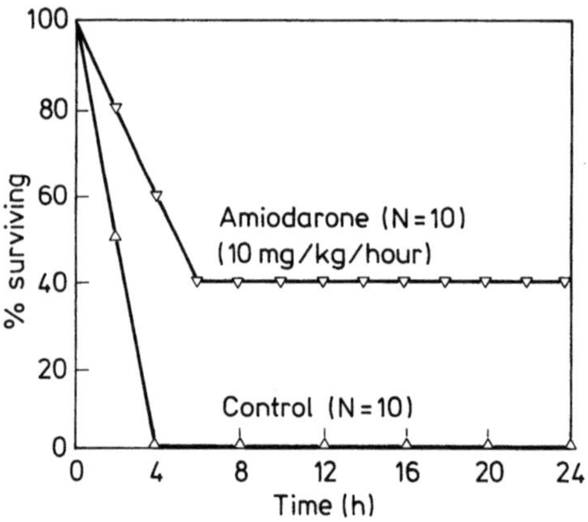

Fig. 9

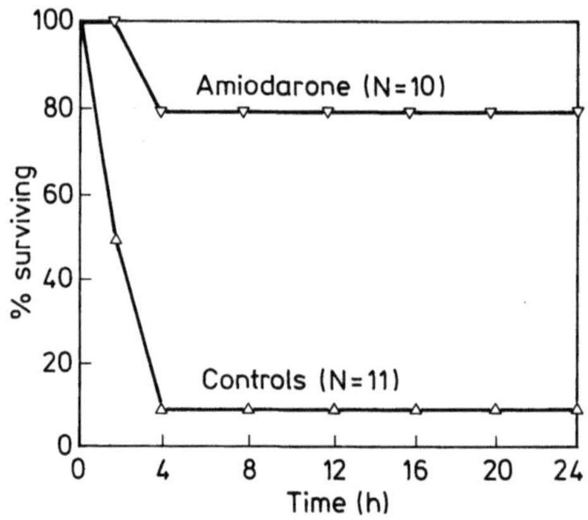

Fig. 10

Fig. 9. Survival in conscious canine preparation of sudden coronary death after short-term intravenous amiodarone administration. The cumulative survival curves for control and intravenous amiodarone groups are shown. Survival was significantly increased by intravenous amiodarone ($\beta < 0.05$). (PATTERSON et al. 1983 with the permission of the authors and of the *American Herart Association*)

Fig. 10. Survival in conscious canine preparation of sudden coronary death after short-term chronic oral amiodarone administration. The cumulative survival curve for control and chronically administered amiodarone groups are shown. Survival was significantly increased by oral amiodarone ($\beta < 0.002$). The effects of the intravenous drug were much less striking. (PATTERSON et al. 1983)

IX. Amiodarone Action and Metabolism of Thyroid Hormones

It has been repeatedly emphasized that the electrophysiologic changes induced by amiodarone following chronic drug administration closely resemble those produced by thyroid gland ablation (SINGH 1983; JOHNSON et al. 1973; GAVRILESCU et al. 1976; SHARP et al. 1985). Such an effect is not due to the iodine contained in the amiodarone molecule since the administration of iodine alone in doses equivalent to those contained in the effective dose of amiodarone had no significant effect on the time course of atrial action potentials (SINGH and VAUGHAN WILLIAMS 1970a). On the other hand, the concomitant administration of amiodarone and thyroid hormone prevented the development of repolarization changes evident after amiodarone alone (SINGH and VAUGHAN WILLIAMS 1970a). These observations raised the possibility that the fundamental electrophysiologic effect of amiodarone may at least in part be related to a reduction in the action of T_3 on cardiac muscle, since there is an inhibition of the peripheral conversion of T_4 to T_3 (PRITCHARD et al. 1975; BURGER et al. 1976; MELMED et al. 1981; HERSHMAN et al. 1986), resulting in a decrease in T_3, an increase in rT_3 and a small increase in T_4 in the plasma (VENKATESH et al. 1986b) due to the blockade of 5′-monodeiodinase (SOGOL et al. 1983). A direct inhibition of T_3 nuclear binding by amiodarone and/or its metabolite desethylamiodarone (WIERSINGA and BROENIK 1983; LATHAM et al. 1985) has been postulated to result in a hypothyroid state at a cellular level (SINGH and NADEMANEE 1983; NADEMANEE et al. 1982b). Since the electrophysiologic effects of hypothyroidism (SINGH 1983; JOHNSON et al. 1973; GAVRILESCU et al. 1976; SHARP et al. 1985) resemble those observed following long-term amiodarone treatment, this phenomenon has been thought to exhibit some cardiospecificity (SINGH and VAUGHAN WILLIAMS 1970a; SINGH and NADEMANEE 1983; NADEMANEE et al. 1982b).

X. Amiodarone-Membrane Lipid Interactions and Effects on Membrane Fluidity

Amiodarone is a complex molecule with a charge, nonpolar hydrophobic moiety, and a small hydrophilic side chain. This ampiphilic nature confers on the drug the propensity to alter lipid metabolism of the myocardium intracellularly as well as in the membrane. For example, GROSS and SOMANI (1986) have shown that amiodarone alters lipid metabolism of the cardiac cell as evidenced by the development of lysosomal and myelinoid inclusion bodies. From the standpoint of the drug's electropharmacologic actions, its effects on the membrane lipid dynamics are of more direct relevance. CHATELAIN (1985a) determined the in vitro effects of amiodarone on lipid dynamics using the fluorescent probe DPH in the erythrocyte ghost, the brain synaptic membrane (CHATELAIN 1985b) and the multilamellar vesicles synthesized from neutral phospholipids (CHATELAIN et al. 1986). In both preparations, incubation with increasing concentrations of amiodarone led to significant decreases in membrane fluidity. Whether this is the basis for the known electrophysiologic actions of amiodarone is unknown, but it would clearly be of interest to determine whether a similar decrease in membrane fluidity occurs in hypothyroid tissues.

Other experimental data indicate that amiodarone has effects at the level of membrane proteins. For example, it has been found that the drug selectively inhibits the Na^+/K^+ ATPase of the guinea-pig myocardial particulate fraction (Broekhuysen et al. 1972); its interaction with the effects of stimulation of beta-adrenergic receptors has been well defined as has been its inhibitory effects on the adenylate cyclase activity (Nokin et al. 1983). However, the available data on the alterations produced by amiodarone in biological membranes must be regarded as preliminary and further work is necessary to delineate the significance of the observed effects in mediating the compound's electrophysiologic and antiarrhythmic actions.

E. In Vivo Electrophysiologic Effects of Amiodarone

I. Experimental Observations

Chronic therapy with excellent control of arrhythmias (Heger et al. 1984; Mostow et al. 1984) in man has been associated with plasma amiodarone levels of 1-3 µg/ml with slightly lower levels of desethylamiodarone. As indicated above, in acute in vitro experiments, concentrations up to 34 µg/ml amiodarone and of the metabolite had considerably less effect on repolarization or refractoriness than when the drug was administered on a chronic basis. This is consistent with little or no effect on ventricular refractoriness in unanesthetized patients following acute intravenous drug (5 mg/kg) administration (Wellens et al. 1984; Ikeda et al. 1984; Gomes et al. 1984; Hariman et al. 1984) despite plasma levels often exceeding 10 µg/ml (Ikeda et al. 1984).

There are now increasing data which suggest that the overall effects in anesthetized animals may differ from those observed in man because in the latter extracardiac or other associated pharmacologic actions of the compound may influence the net effects. For example, in *anesthetized* animals intravenous amiodarone in doses up to 10 mg/kg has been reported to increase the ventricular effective refractory period by 30% (Platou and Refsum 1982) in the absence of a significant increase in the QT_c interval as noted in another study (Singh et al. 1976). Jaillon et al. (1982) showed that in pentobarbital-anesthetized animals intravenous amiodarone (1.25-10 mg/kg) produced a dose-related decrease in heart rate and prolonged the sinus node recovery time while having no effect on His-Purkinje conduction time. Modest increases in atrial and ventricular effective refractory periods were associated with a marked decrease in atrioventricular nodal conduction. Similar results have been reported by others (Platou and Refsum 1982; Pickoff et al. 1983) in adult as well as neonatal dogs (Yabek et al. 1985). In the latter, the effects of amiodarone have been foound to be less striking, consistent with the findings of Yabek et al. (1985) in acute superfusion studies. The overall observations suggest that the reported variable and modest antiarrhythmic actions of acutely administered amiodarone (GOMES et al. 1984; HARIMAN et al. 1984) may in part be due to the noncompetitive alpha- and beta-adrenergic receptor antagonism (POLSTER and BROEKHUYSEN 1976; CHARLIER et al. 1968). The inhibitory actions on alpha- and beta-receptors are known to be associated with distinct electrophysiologic effects (ROSEN et al. 1971; GIOTTI et al. 1973; Chap. 12), which may contribute to the observed effects of amiodarone during intravenous injections.

II. Clinical Electrophysiologic Effects

The experimental and clinical data emphasize that the maximal or steady state effects with amiodarone in cardiac muscle do not become apparent acutely despite extremely high drug concentrations. This is reflected in the marked differences found between the effects of acute intravenous versus chronic oral drug therapy. For example, following intravenous amiodarone administration the electrophysiologic effects are much less striking (WELLENS et al. 1984; MORADY et al. 1986; IKEDA et al. 1984). The main acute effect is the lengthening of AV nodal refractoriness and intranodal conduction (AH interval) time (WELLENS et al. 1984; IKEDA et al. 1984; HARIMAN et al. 1984; PLATOU and REFSUM 1982) with minimal effect on the effective refractory periods of the atrial, ventricular, the bypass tract, or the His-Purkinje tissue when the drug is administered in a dose of 5 mg/kg body weight. There is no significant effect on the HV or QRS intervals nor the QTc duration. The acute effect on the atrioventricular node may be due to the blockade of the slow-channels and/or the noncompetitive adrenergic antagonism exerted by the drug (POLSTER and BROEKHUYSEN 1976; CHARLIER et al. 1968; VENKATESH et al. 1986a). At somewhat higher doses (10 mg/kg) intravenous amiodarone has been shown to increase the ventricular effective refractory period by 20–30 ms and some lengthening of the QRS duration at fast stimulus frequencies consistent with a use-dependent effect on fast-sodium channels (WELLENS et al. 1984). Despite the well-documented in vitro depressant effects of amiodarone on sinus node automaticity, intravenous amiodarone in conscious man does not produce the expected reduction in heart rate (WELLENS et al. 1984; MORADU et al. 1986; IKEDA et al. 1984), presumably due to the opposing effects of sympathetic activation resulting from the peripheral vasodilator actions of the drug (SINGH 1983).

In contrast, when administered chronically, amiodarone predictably lengthens repolarization (QT_c) and refractoriness in most cardiac tissues as a function of time with little or no change in the QRS duration and a modest increase in the HV interval with a significant prolongation of the AH interval (HEGER et al. 1984; NADEMANEE et al. 1982a; WAXMAN et al. 1982; FINERMAN et al. 1982). As far as the effects on repolarization are concerned, they are consistent with those previously noted with studies in rabbits chronically treated with amiodarone. For example, 20 mg/kg amiodarone (SINGH and VAUGHAN WILLIAMS 1970a) increased the ventricular action potential by 11% at 1 week, 23% at 3 weeks, and 30% following 6 weeks of drug administration. This is consistent with the prolongation of the monophasic action potentials in experimental animals (CABASSON 1978). In man (OLSSON et al. 1973) after 6 weeks of oral treatment with amiodarone, the duration of the monophasic action potential recorded by suction electrode in atria was also increased by about 30%. These observations are concordant with the observation that the QT_c interval in man increased progressively on a constant dose of amiodarone, reaching what appeared to be a steady state effect at 6 weeks (PRITCHARD et al. 1975).

After chronic treatment in man there is marked increase in the effective and the functional refractory periods in most cardiac tissues (atria, ventricles, AV node, His-Purkinje system, accessory tracts of the heart) as a function of time with little or no change in the QRS duration and a modest increase in the HV interval (WELLENS et al. 1984; MORADY et al. 1986). An increase in the QRS duration does, however, occur (WELLENS et al. 1984) as do increases in the infranodal conduction (anterograde or retrograde) following fast-stimulation frequencies, again reflecting effects on the fast-sodium channel (SHENASA et al. 1984). It is clear that the overall electrophysiologic changes, which are accompanied by a progressive decrease in heart rate, are significantly greater during chronic therapy when compared with those found after acute intravenous administration. There is now substantial evidence that such a difference, also noted in animals, is not accounted for by differences in serum drug concentrations.

These compounds exert significant but quantitatively and qualitatively similar acute electrophysiologic effects in isolated cardiac muscle. However, unlike the effects during chronic therapy, which are dominated by a marked lengthening of the action potential duration, those after acute superfusion with these drugs are associated with less striking alterations in repolarization and refractoriness despite extremely high drug concentrations. These differences are consistent with the observations that the overall electrophysiologic effects of intravenously administered amiodarone in man differ from those found after long-term chronic drug administration. On the other hand, since the potency of desethylamiodarone was no greater than that of amiodarone in terms of the electrophysiologic effects, the known latency of the onset of the antiarrhythmic actions of amiodarone is unlikely to be due solely to the formation of the metabolite. The depressant effect of amiodarone on the characteristics of the sinus node potentials suggests that the drug might exert an acute calcium antagonistic effect and antiadrenergic effects, the summation of which which might be of particular significance in the AV node. Finally, the fact that amiodarone exerted a marked use-dependent effect on V_{max} in isolated cardiac muscle suggests the possibility of a significant beneficial acute effect on conduction and refractoriness during rapid tachyarrhythmias.

F. Effects of Amiodarone in Experimental Arrhythmias

The antiarrhythmic effect of amiodarone has now been demonstrated in a wide variety of experimentally induced cardiac arrhythmias. As might be predicted from the electrophysiologic actions, the antiarrhythmic effects of acute intravenous doses versus chronic oral dosing of amiodarone differ significantly. For example, in ventricular fibrillation produced by chloroform inhalation or calcium chloride administration in mice or rats, the effects of intravenous amiodarone were found to be weak (CHARLIER et al. 1969). However, high doses of the drug were found to be effective in suppressing ventricular tachyarrhythmias produced by aconitine hydrochloride in the rat and dog (CHARLIER et al. 1969; SINGH and VAUGHAN WILLIAMS 1970a). It has also been reported that the intravenous drug may be effective in the suppression of multifocal premature ventricular ectopic beats produced by the injection of epinephrine in anesthetized dogs and rabbits and by the injection of barium chloride in the anesthetized dogs and rabbits. Pretreatment with amiodarone of anesthetized guinea pigs has also been reported to increase the dose of intravenous infusion of ouabain required to produce ventricular firbrillation due to glycoside intoxication (SINGH and VAUGHAN WILLIAMS 1970a).

A number of studies have emphasized the antifibrillatory effects of acutely and chronically administered amiodarone in a variety of experimental models of arrhythmia. For instance, LUBBE et al. (1978) reported that pretreatment of rats with amiodarone for 2 min to 3 weeks before the hearts were removed and studied as a Langendorff preparation induced a dose-related decrease in the spontaneous heart rate with an increase in the ventricular fibrillation threshold both before as well as after coronary artery ligation. The drug also reduced the numbers of premature ventricular ectopic beats as well as ventricular tachycardia and fibrillation following coronary artery ligation and after reperfusion. The study provided convincing evidence for the protective effects of the compound against the increases in ventricular vulnerability in the early phases following coronary artery occlusion and against reperfusion ventricular fibrillation. These data are consistent with those reported by SCHOENFELD (1982), who found 20–50 mg/kg orally administered amiodarone had a markedly protective effect against both early as well as late fatal ventricular fibrillation following coronary artery ligation in rats.

The antifibrillatory effects of amiodarone have also been established in a number of experimentally induced models of ischemic arrhythmias in conscious as well as anesthetized animals (ROSENBAUM et al. 1976; VENKATESH et al. 1986b; OLSSON et al. 1973). Following coronary ligation in the dog, ROSENBAUM et al. (1976) found that amiodarone prevented the occurrence of ventricular fibrillation in all ten pretreated animals given 40 mg/kg oral amiodarone for 1–4 weeks. In contrast, ventricular fibrillation occurred in seven of eight untreated animals. The drug also exerted a potent effect on the occurrence of premature ventricular contractions. CHEW et al. (1982) found that chronic pretreatment with amiodarone for a period of 4 weeks also markedly attenuated the frequency of ventricular arrhythmias following coronary artery ligation in conscious instrumented dogs.

Particularly noteworthy are the data reported by PATTERSON et al. (1983) in a sudden death canine model produced by sequential ligation of coronary arteries. In this model both short-term as well as long-term administration of amiodarone reduced the incidence of ventricular fibrillation. In the control series, there was a 91%–100% (two series) incidence of ventricular fibrillation,

60% following short-term intravenous drug administration and only 20% following chronic drug administration. The differences between the effects of the acutely and chronically administered drug (which were significant) were not accounted for on the basis of differences in the plasma or myocardial tissue levels of amiodarone. The data emphasize the greater efficacy of chronically administered amiodarone as an antifibrillatory agent and indicate its particular role in the prevention of sudden death especially in patients with ischemic heart disease.

G. Clinical Antiarrhythmic Effects of Amiodarone

The electropharmacologic properties of amiodarone suggest that the compound should exhibit a broad spectrum of antiarrhythmic actions. The available clinical data are in line with the experimental observations. The extensive and growing literature will not be discussed herein; only the major findings will be summarized.

Intravenously administered amiodarone (5 mg/kg) given over 5-10 min has been found to slow the ventricular response in atrial flutter and fibrillation in most patients. In long-standing cases the compound effects a conversion only rarely, although this may occur frequently in cases of recent onset of the arrhythmia. The intravenous drug may also slow the tachycardia cycle length in cases of reentrant supraventricular tachycardia and terminate it in a significant number by producing conduction block in the antegrade limb of the reentrant circuit (Gomes et al. 1984). These effects in atrial flutter/fibrillation and in paroxysmal tachycardia are paralleled in experimental arrhythmias by a reduction in the number of premature ventricular ectopic beats as well as termination of ventricular tachycardia and fibrillation following coronary artery ligation and after reperfusion. The study provided convincing evidence for the protective effects of the compound against the increases in ventricular vulnerability in the early phases following coronary artery occlusion and against reperfusion ventricular fibrillation. These data are consistent with those reported by Schoenfeld et al. (1982), who found 25-50 mg/kg orally administered amiodarone had a markedly protective effect against both early as well as late fatal ventricular fibrillation following coronary artery ligation in rats.

The major clinical utility of amiodarone as an antiarrhythmic compound is in the prophylactic control of supraventricular as well as ventricular arrhythmias. The beneficial effect in these cases may relate to the intrinsic action of the drug in atrial and ventricular tissue as well as in the AV node and the accessory tracts of the heart in terms of changes in conduction and refractoriness but also in terms of suppression of supraventricular and ventricular ectopic beats which provide the trigger mechanism for the initiation of the tachyarrhythmias.

Given orally, amiodarone (200-400 mg/day, often in lower doses) slows the ventricular response in atrial flutter and fibrillation at rest and with exercise; it produces conversion after a variable period in over 30% of cases and may maintain stability of sinus rhythm after chemical or electrical conversion in over 50% of cases, a figure significantly higher than that produced by most class IA compounds. In the cases of atrial flutter, the flutter rate is often markedly reduced if conversion to sinus rhythm has not occurred but unlike the situation in the case of quinidine and disopyramide (anticholinergic effect), acceleration of the ventricular response does not occur, because A-V conduction is depressed (anti-sympathetic effect). Numerous studies have shown that amiodarone is perhaps the most effective agent in the control of paroxysmal atrial fibrillation (GRABOYS et al. 1983).

Although amiodarone has been used in the case of most types of reentrant supraventricular tachycardias, including those due to AV nodal reentry and those using accessory tracts of the heart, the drug is of particular value in cases with antidromic tachycardias and in the atrial flutter and fibrillation complicating the Wolff-Parkinson-White syndrome. A high rate of success can be achieved with relatively modest doses of the drug in this setting. However, long-term drug usage here may lead to significant side effects and an increasing trend for the use of surgical ablation is now developing especially in centers able to undertake detailed electrophysiologic evaluation to localize the accessory tracts (Chap. 21).

The advent of amiodarone has had a major impact on the treatment of recurrent life-threatening ventricular tachyarrhythmias (SINGH and ZIPES 1983). Numerous studies have shown that, following the initial loading dose, a maintenance regimen of 300-400 mg/day of the drug is effective in controlling malignant arrhythmias (including those with cardiac arrest) in 60%-75% of cases in which conventional agents (especially class I compounds) have failed. There is, however, controversy whether long-term success of therapy can be predicted more accurately by the effects of the drug on the suppression of spontaneously occurring arrhythmias on Holter recordings or on the basis of inducible tachycardia induced by programmed electrical stimulation of the heart. The issue is not clearly resolved but the two approaches do not appear to be mutually exclusive. It is also generally agreed that the failure of the drug to prevent ventricular induction does not necessarily preclude an excellent clinical outcome. There is emerging data which, by life-table analysis, suggest that amiodarone may enhance survival in patients resuscitated from sudden cardiac deaths (NADEMANEE et al 1983; ZIPES et al. 1984). It is not certain which components of the various actions of the drug—antiectopic, antifibrillatory, anti-ischemic, antiadrenergic—are the crucial determinants of its salutary effects in this regard.

The available data indicate that the observed beneficial effects of the compound should be balanced against the well-known potentially serious deleterious actions. Particularly important is the development of pulmonary toxcity (in 5%-13%, although a lower incidence appears to be associated with lower doses) and hepatoxicity (1%). Troublesome but less serious side effects include altered thyroid state, peripheral neuropathy, CNS and gastrointestinal disturbances, photosensitivity, corneal microdeposits, myopathy, and especially bluish skin pigmentation the incidence of which is unrelated to dose but increases as a function of duration of therapy. Numerous minor side effects of the drug have been reported.

It should also be emphasized that amiodarone exhibits electrophysiologic and hemodynamic interactions with a number of other cardioactive compounds but the drug's propensity to induce pharmacokinetic interactions (e.g., digoxin, warfarin, quinidine, procainamide, NAPA, propafenone, flecainide, phenytoin sodium among others) may be associated with serious effects. Clearly, in the case of amiodarone an appropriate caution needs to be exercised that takes account of an extremely potent action and a complex side effect profile which is not readily predictable on the basis of plasma drug monitoring.

H. Conclusions on the Antiarrhythmic Mechanisms of Actions of Amiodarone

There is now little doubt about the potency of amiodarone as a broad-spectrum antiarrhythmic agent for the prophylactic control of most supraventricular and ventricular tachyarrhythmias. While the most readily measurable correlate of its antiarrhythmic action is its property to lengthen the action potential duration in most cardiac tissues (the class III action), this alone is unlikely to be the sole basis of the drug's extraordinary potency as an antiarrhythmic compound. There are several compounds which also induce a lengthening of the action potential duration (SINGH et al. 1980; HAUSWIRTH and SINGH 1978; Chap. 16, 17) but which are conserably less potent than amiodarone in the control of arrhythmias. Thus, the question arises as to which of the various electrophysiologic properties discussed in this chapter are the major determinants of the drug's antiarrhythmic potency, as discussed in Chapt. 2. It has been indicated that there are major differences between the actions of the acutely administered drug and those following protracted oral administration, the difference not being accountable in terms of serum, tissue, or membrane concentrations of the compound or its active metabolite. The dominant action of the drug is its ability to prolong the cardiac action potential duration which probably forms the basis of its antifibrillatory effects undoubtedly modulated by the complex array of its associated pharmacologic properties and its anti-ischemic potential. The relative significance of the drug's associated effects on the slow channel, the fast channel, alpha- and beta-receptor-mediated effects, cholinergic receptors, and adrenergic neurons remains to be determined.

Acknowledgments. The original work cited in this chapter has been shared with a number of colleagues whose contributions are gratefully acknowledged. My indebtedness is also expressed to Lawrence Kimble for his help with the preparation of this chapter. The studies were supported by grants from the Medical Research Service of the Veterans Administration and the American Heart Association, the Greater Los Angeles Affiliate, Los Angeles, California.

References

Aomine M, McCullough J, Mayuga R, Morrone W, Singer D (1984) Cellular electrophysiologic effects of acute exposure to amiodarone on guinea pig heart. Fed Proc 43:961

Bacq ZM, Blakeley AGH, Summers RJ (1976) The effects of amiodarone, an alpha and beta receptor antagonist, on adrenergic transmission in the cat spleen. Biochem Pharmacol 25:1195-1199

Broekhuysen J, Charlier R, Ghislain J (1972) Action of amiodarone on guinea-pig heart sodium and potassium activated adenosine triphosphatase. Biochem Pharmacol 21:2951-2960

Burger A, Dinichert C, Nicod P, Jenny M, Lemarchard BP, Valloton MB (1976) Effect of amiodarone on serum triiodothyronine, thyroxine and thyrotropin: a drug influencing peripheral metabolism of thyroid hormones. J Clin Invest 58:255-264

Cabasson J (1978) Analyse des effects électrophysiologiques de l'amiodarone, perhexiline and bepridil on the cardiac rhythms of the anaesthetized dog in chronic heart block. Arch Int Pharmacodyn 233:65-75

Charlier R (1970) Cardiac actions in the dog of a new antagonist of adrenergic excitation which does not produce competitive blockade of adrenoceptors. Br J Pharmacol 39:668

Charlier R, Deltour G, Tondeur R, Binon F (1962) Recherche dans la série des benzofurannes. VII. Etude pharmacologique préliminaire du butyl-2(diiodo 3′5′-beta-*N*-diethylamino-ethoxy-4′-benzoyl)-3benzofurannes. Arch Int Pharmacodyn 139:255-262

Charlier R, Deltour G, Baudine A, Chaillet F (1968) Pharmacology of amiodarone, an antianginal drug with a new biological profile. Arzneimittelforschung 18:1408-1430

Charlier R, Delaunois G, Bauthier J, Deltour G (1969) Recherche dans la série des benzofurannes. XL. Propriété anti-arrhythmiques de l'amiodarone. Cardiologia 54:83-89

Charlier R, Delaunois G, Bauthier J (1972) Opposite effects of amiodarone and beta-blocking agents on cardiac functions under adrenergic stimulation. Arzneimittelforschung 22:545-552

Chatelain P (1985a) Effects of amiodarone on lipid dynamics in erythrocyte membrane in vitro and after chronic treatment. Arch Int Pharmacodyn 276:327-328

Chatelain P (1985b) Modulation by amiodarone of membrane fluidity and Na^+/K^+ ATPase activity in rat brain synaptosomes. Biochem Biophys Res Commun 129:148-154

Chatelain P, Ferriera J, Laruel R, Ruysschaert JM (1986) Amiodarone-induced modifications of the phospholipid physical state. A fluorescence polarization study. Biochem Pharmacol 35(18):3007-3013

Chew CYC, Collet JT, Campbell C, Kannan R, Singh BN (1982) Beneficial effects of amiodarone pretreatment on early ischemic ventricular arrhythmias relative to infarct size and regional myocardial blood flow in the conscious dog. J Cardiovasc Pharmacol 4:1028-1036

Cohen-Armon M, Schreiber G, Sokolovsky M (1984) Interaction of the anti-arrhythmic drug amiodarone with the muscarinic receptor in rat heart and brain. J Cardiovasc Pharmacol 6:1148-1155

Coté P, Bourassa MG, Delaze, et al. (1979) Effects of amiodarone on cardiac and coronary hemodynamics and on myocardial metabolism in patients with coronary artery disease. Circulation 59:1165-1172

Courtney KR (1975) Mechanism of frequency-dependent inhibition of sodium currents in myelinated nerve by the quaternary lidocaine derivative GEA 968. J Pharmacol Exp Ther 195:225-236
Finerman WB Jr, Hamer A, Peter T (1982) Electrophysiologic effects of chronic amiodarone therapy in patients with ventricular arrhythmias. Am Heart J 104:987-993
Freedberg AS, Papp GJ, Vaughan Williams EM (1970) The effects of altered thyroid state on atrial intracellular potentials. J Physiol 207:357-370
Gagnol JP, Devos C, Clinet M, Nokin P (1985) Amiodarone: biochemical aspects and hemodynamic effects. Drugs [Suppl 3] 29:1-10
Gavrilescu S, Luca C, Streian C, Lungu G, Deutsch G (1976) Monophasic action potential of right atrium and electrophysiologic properties of AV conducting system in patients with hypothyroidism. Br Heart J 38:1350-1354
Giotti A, Ledda F, Mannaioni DF (1973) Effects of noradrenaline and isoprenaline, in combination with alpha- and beta-receptor blocking substances on the action potential of cardiac Purkinje fibers. J Physiol 229:99P
Gloor HO, Urthaler F, James TN (1983) Acute effects of amiodarone upon the canine sinus node and atrioventricular junctional region. J Clin Invest 71:1457
Gomes JAC, Kang PS, Hariman RJ, El-Sherif N, Lyons J (1984) Electrophysiologic effects and mechanisms of termination of supraventricular tachycardia by intravenous amiodarone. Am Heart J 107:214-222
Goupil N, Lenfant J (1976) The effects of amiodarone on the sinus node activity of the rabbit heart. Eur J Pharmacol 39:23
Graboys TB, Podrid PJ, Lown B (1983) Efficacy of amiodarone for refractory supraventricular tachyarrhythmias Am Heart J 106:870-879
Gross SA, Somani P (1986) Amiodarone-induced ultrastructural changes in the canine myocardial fibers. Am Heart J 112:771-779
Hariman RJ, Gomes AC, Kang KS, El-Sherif N (1984) Effects of intravenous amiodarone in patients with inducible repetitive ventricular responses and ventricular tachycardia. Am Heart J 107:1109-1117
Hauswirth O, Singh BN (1978) Ionic mechanisms in heart muscle in relation to the genesis and the pharmacologic therapy of cardiac arrhythmias. Pharmacol Rev 30:5-63
Heger JJ, Prystowky EN, Jackman WN, Naccarelli GV, Warfel KA, Rinkenberger RL, Zipes DP (1981) Amiodarone: clinical efficacy and electrophysiology during long term therapy for recurrent ventricular tachycardia or ventricular fibrillation. N Engl J Med 305:539-545
Heger JJ, Prystowsky EN, Miles WM, Zipes DP (1984) Clinical use and pharmacology of amiodarone. Med Clin North Am 68(5):1339-1367
Hershman JW, Nademanee K, Masahiro S, Pekary AE, Ross R, Singh BN, DiStefano JJ III (1986) Thyroxine and triiodothyronine kinetics in cardiac patients taking amiodarone. Acta Endocrin ol (Copenh) 111:193-199
Holt DW, Tucker GT, Jackson PR, Storey GCA (1983) Amiodarone pharmacokinetics. Am Heart J 106:840
Ikeda N, Nademanee K, Kannan R, Singh BN (1984) Electrophysiologic effects of amiodarone:experimental and clinical observations relative to serum and tissue concentrations. Am Heart J 108:890-899
Jaillon P, Heckle J, Jais J-M, Povedra-Sierra J, Cheymol G (1982) Acute effects of intravenous prifuroline and amiodarone on canine automaticity, conduction and refractoriness. J Cardiovasc Pharmacol 4: 486-492
Johnson PN, Freedberg AS, Marshall JM (1973) Action of thyroid hormone on the

transmembrane potentials from sino-atrial nodal cells and atrial muscle cells in isolated atria of rabbits. Cardiology 58:273–289

Kannan R, Nademanee K, Hendrickson JA, Rostami HJ, Singh BN (1982) Amiodarone kinetics after oral doses. Clin Pharmacol Ther 31:438–444

Kannan R, Ikeda N, Drachenberg M, Wagner R, Ooktens M, Singh BN (1984) Serum and myocardial kinetics of amiodarone and its major metabolite desethylamiodarone in rabbits. J Pharm Sci 73:1208–1211

Kato R, Venkatesh N, Yabek S, Kannan R, Singh BN (1988) The comparative electrophysiologic effects of desethylamiodarone and amiodarone after chronic dosing in rabbits Amer Heart J 115:351–359

Kobayashi M, Godin D, Nadeau R (1983) Acute effects of amiodarone in the isolated dog heart. Can J Physiol Pharmacol 61:308–314

Lambert C, Vermeulen M, Cardinal R, Nadeau R (1986) Effect of induction of amiodarone biotransformation on ventricular refractory periods in rats. J Pharmacol Exp Ther 238:307–312

Latham KR, Sellittie DF, Goldstein RE (1985) Interaction of amiodarone and desethylamiodarone with nuclear thyroid hormone receptors (Abstr). J Am Coll Cardiol 5:466

Latini R, Togoni G, Kates KE (1984) Clinical pharmacokinetics of amiodarone. Clin Pharmacokin et 9:136–156

Lubbe WF, McFadyen ML, Muller CA, Worthington M, Opie LH (1978) Protective action of amiodarone against ventricular fibrillation in the isolated perfused rat heart. Am J Cardiol 43:533–540

Mason JW, Hondeghem LM, Katzung BG (1983) Amiodarone blocks inactivated cardiac sodium channels. Pflugers Arch 396:79–85

Mason JW, Hondeghem LM, Katzung BG (1984) Block of inactivated sodium channels and of depolarization-induced automaticity in guinea-pig papillary muscle by amiodarone. Circ Res 55:277–285

Melmed S, Nademanee K, Reed AW, Hendrickson JA, Singh BN, Hershman JM (1981) Hyperthyroxinemia with bradycardia and normal thyrotropin secretion after chronic amiodarone administration. J Clin Endocrine Metab 53:997–302

Morady F, Di Carlo LA, Krol RB, Berman JM, Buitleir M (1986) Acute and chronic effects of amiodarone on ventricular refractoriness, intraventricular conduction and ventricular tachycardia induction. J Am Coll Cardiol 7(1):148–157

Mostow ND, Rakita L, Vrobel TR, Noon DL, Blumer J (1984) Amiodarone: correlation of serum concentration with suppression of complex ventricular ectopic activity. Am J Cardiol 54:569–574

Nademanee K, Hendrickson JA, Cannom DS, Goldreyer BN, Singh BN (1981) Control of refractory life-threatening ventricular arrhythmias by amiodarone. Am Heart J 101:759

Nademanee K, Hendrickson J, Kannan R, Singh BN (1982a) Antiarrhythmic efficacy and electrophysiologic actions of amiodarone in patients with life-threatening arrhythmias. Am Heart J 103:950–959

Nademanee K, Singh BN, Hendrickson JA, Reed AW, Melmed S, Hershman JM (1982b) Pharmacokinetic significance of reverse T_3 levels during amiodarone treatment:a potential method for monitoring chronic drug therapy. Circulation 66:202–214

Nademanee K, Singh BN, Hendrickson JA, Lopez B, Weiss JN, Feld G (1983) Amiodarone in refractory life-threatening ventricular arrhythmias. Ann Intern Med 98:577–584

Neliat G (1982a) Electrophysiologic effects of butoprozine on isolated heart preparations. Comparison with amiodarone and verapamil. Arch Int Pharmacodyn 255:220–236

Neliat G (1982b) Effects of butoprozine on ionic currents in frog atrial and ferret ventricular fibers. Comparison with amiodarone and verapamil. Arch Int Pharmacodyn 255:237–255

Nokin P, Clinet M, Schoenfeld P (1983) Cardiac beta-adrenoceptor modulation by amiodarone. Bioch Pharmacol 32(17):2473–2477

Olsson B, Brorson L, Varnauskas E (1973) Antiarrhythmic action in man: observations from monophasic action potential recordings and amiodarone treatment. Br Heart J 35:1255–1262

Patterson E, Eller BT, Abrams GD, Vasilades J, Lucchesi BR (1983) Ventricular fibrillation in conscious canine preparation of sudden coronary death. Prevention by short and long-term amiodarone administration. Circulation 68:857–864

Petta JM, Zaccheo VJ (1971) Comparative profile of L3428 and other anti-anginal agents on cardiac hemodynamics. J Pharmacol Exp Ther 176:328–338

Pickoff AS, Singh S, Flinn CJ, Torres E, Ezrin AM, Gelband H (1983) Dose-dependent electrophysiologic effects of amiodarone in the immature canine heart. Am J Cardiol 52:621–625

Platou ES, Refsum H (1982) Class III antiarrhythmic action in experimental atrial fibrillation and flutter in dogs. J Cardiovasc Pharmacol 4:839–846

Polster P, Broekhuysen J (1976) The adrenergic antagonism of amiodarone. Biochem Pharmacol 25:131–136

Pritchard DA, Singh BN, Hurley PJ (1975) Effects of amiodarone on thyroid function in patients with ischaemic heart disease. Br Heart J 37:856–863

Rosen MR, Gelband H, Hoffman BF (1971) Effects of phentolamine on electrophysiologic properties of isolated canine Purkinje fibers. J Pharmacol Exp Ther 179:586–595

Rosenbaum MB, Chiale PA, Ryba D, Elizari MV (1974) Control of tachyarrhythmias associated with Wolff-Parkinson-White syndrome by amiodarone hydrochloride. Am J Cardiol 34:215–222

Rosenbaum MB, Chiale PA, Halpern MS, Nau GJ, Przybylski J, Levi RJ, Lazzari JO, Elizarri MV (1976) Clinical efficacy of amiodarone as an antiarrhythmic agent. Am J Cardiol 38:934–942

Rosenbaum MB, Chiale PA, Haedo A, Lazzari JO, Elizarri MV (1983) Ten years of experience with amiodarone. Am Heart J 106:957–964

Schoenfeld P (1982) Comparison des effects de l'amiodarone et du propranolol sur l'incidence et la sévérité des arrhythmies ventriculaires après ligature de l'artère coronarienne chez le rat anesthésié. J Cardiol 11:499–506

Schwartz A, Shen E, Morady F, et al. (1983) Hemodynamic effects of intravenous amiodarone in patients with depressed left ventricular function and recurrent ventricular tachycardia. Am Heart J 106:848–856

Sharma AD, Corr PB, Sobel BE (1983) Modulation by amiodarone of cardiac adrenergic receptors and their electrophysiologic responsiveness to catecholamines (Abstr) Circulation 68(2):393

Sharp NA, Neel DS, Parsons RL (1985) Influence of thyroid hormone levels on the electrical and mechanical properties of rabbit papillary muscle. J Mol Cell Cardiol 17:119–132

Shenasa M, Denker S, Mahmud R, Lehmann M, Estrada A, Akhtar M (1984) Effect of amiodarone on conduction and refractoriness of the His-Purkinje system in the human heart. J Am Coll Cardiol 4:105–110

Singh BN (1971) A study of the pharmacological actions of certain drugs and hor-

mones with a particular reference to cardiac muscle. PhD Thesis, University of Oxford
Singh BN (1983) Amiodarone: historical development and pharmacologic profile. Am Heart J 106:788-797
Singh BN, Nademanee K (1983) Amiodarone and thyroid function: clinical implications during antiarrhythmic therapy. Am Heart J 106(4):857-869
Singh BN, Vaughan Williams EM (1970a) The effect of amiodarone, a new anti-anginal drug, on cardiac muscle. Br J Pharmacol 39:657-667
Singh BN, Vaughan Williams EM (1970b) A third class of antiarrhythmic action. Effects on atrial and ventricular intracellular potentials and other pharmacologic actions on cardiac muscle of MJ1999 and AH3474. Br J Pharmacol 39:675-686
Singh BN, Zipes DP (eds) (1983) Amiodarone: basic concepts and clinical applications. Am Heart J 106:787-797
Singh BN, Jewitt DE, Downey JM, Kirk ES, Sonnenblick EH (1976) Effects of amiodarone and L8040, novel antianginal and antiarrhythmic drugs, on cardiac and coronary hemodynamics and on cardiac intracellular potentials. Clin Exp Pharmacol Physiol 3:426-436
Singh BN, Collett JT, Chew CYC (1980) New perspectives in the pharmacologic therapy of cardiac arrhythmias. Prog Cardiovasc Dis 22:243-301
Sogol PB, Hershman JM, Reed AW (1983) The effects of amiodarone on serum thyroid hormones and hepatic thyroxine 5'—monodeiodination in rats. Endocrinology 113:1464-1469
Storey GCA, Adams PC, Campbell RWF, Holt HW (1983) High performance liquid chromatographic measurement of amiodarone and desethylamiodarone in small tissue samples after enzymatic digestion. J Clin Pathol 36:785-789
Van Schepdael J, Solvay H (1970) Etude clinique de l'amiodarone dans les troubles du rhythme cardiaque. Presse Med 78:1849-1855
Vastesaeger M, Gillot P, Rasson G (1967) Etude clinique d'une nouvelle médication anti-angoreuse. Acta Cardiol(Brux) 22:483-490
Venkatesh N, Padbury JF, Singh BN (1986a) Effects of amiodarone and desethylamiodarone on rabbit myocardial beta-adrenoceptors and serum thyroid hormones—absence of relationship to serum and myocardial drug concentrations. J Cardiovasc Pharmacol 8:989-997
Venkatesh N, Al-Sarraf L, Hershman JM, Singh BN (1986b) Effects of desethylamiodarone on thyroid hormone metabolism in rats: comparison with the effects of amiodarone. Proc Soc Exp Biol Med 181:233-236
Venkatesh N, Al-Sarraf L, Singh BN (1986c) Digoxin-desethylamiodarone interaction in rats. Comparison with that of amiodarone. J Cardiovasc Pharmacol 8:309-313
Venkatesh N, Somani P, Bersohn M, Phair R, Kato R, Singh BN (1986d) Electropharmacology of amiodarone: absence of relationship to serum myocardial and cardiac sarcolemmal membrane drug concentrations. Am Heart J 112:916-922
Waxman HL, Groh WC, Marchlinski FE, Buxton AE, Sadowski LW, Horowitz LN, Josephson ME, Kastor JA (1982) Amiodarone for control of sustained ventricular tachyarrhythmias: clinical and electrophysiological effects in 51 patients. Am J Cardiol 50:1066
Wellens HJJ, Brugada P, Abdollah H, Dassen WR (1984) A comparison of the electrophysiologic effects of intravenous and oral amiodarone in the same patient. Circulation 69:120-127
Wiersinga WM, Broenik MM (1983) In vitro inhibition of nuclear thyroid hormone binding by amiodarone and desethylamiodarone in rat liver and cardiac muscle (Abstr). 59th American Thyroid Association, Oct 5th-8th, New Orleans

Williams LT, Lefkowitz RJ, Hathaway DR, Watanabe AM, Besch HR (1977) Thyroid hormone regulation of beta-adrenergic receptor number. J Biol Chem 252:2787
Yabek S, Kato R, Singh BN (1985) Acute effects of amiodarone on the electrophysiologic properties of isolated neonatal and adult cardiac fibers. J Am Coll Cardiol 15:1109-1115
Yabek S, Kato R, Singh BN (1986) Acute electrophysiologic effects of amiodarone and desethylamiodarone in isolated cardiac muscle. J Cardiovasc Pharmacol 8:197-207
Zipes DP, Prystowsky EN, Heger JJ (1984) Amiodarone: electrophysiologic actions, pharmacokinetics and clinical effects. J Am Coll Cardiol 3:1059

CHAPTER 16

Sotalol

S. M. COBBE

A. Basic Pharmacology

I. Structure

The chemical entity 4-(2-isopropylamino-1-hydroxyethyl) methane sulphonanilide hydrochloride (MJ 1999—sotalol) was first described in 1964 by DUNGAN and LISH. The chemical structure is illustrated in Fig. 1. Sotalol is a white substance, molecular wt. 309, which is readily soluble in water an in physiological solutions at room temperature. Sotalol was initially synthesized as a beta-adrenergic receptor blocker, but it has subsequently been shown to possess additional effects on the action potential which make it unique among clinically available beta-receptor blockers, and of particular interest as a class III antiarrhythmic drug.

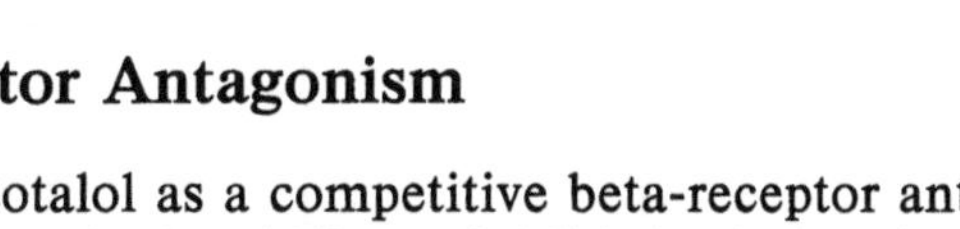

Fig. 1. Chemical structure of sotalol

II. Beta-Receptor Antagonism

The properties of sotalol as a competitive beta-receptor antagonist have been demonstrated in vitro by its ability to inhibit both the chronotropic effects of isoprenaline on rabbit atria (BLINKS 1967; STRAUSS et al. 1970) and its inotropic effect in rabbit atrial and ventricular muscle (LEVY and RICHARDS 1965; COBBE and MANLEY 1985). These studies have shown sotalol to be devoid of intrinsic sympathomimetic activity. Studies on guinea pig trachea (PATIL 1968) indicate that sotalol blocks beta$_2$- as well as beta$_1$-receptors. Sotalol is a relatively weak beta-receptor antagonist, with a pA_2 of 6.4 (BLINKS 1967) as compared with a value of 8.7 for propranolol. There is some evidence that sotalol may block transmitter release from sympathetic nerve endings in addition to its competitive activity at cardiac and smooth muscle beta-receptors (ARAMENDIA and KAUMANN 1967; BARTLET and HASSAN 1969). In common with other agents in this class, the principal beta-receptor-blocking activity resides in the laevo isomer of sotalol, which is approximately 44 times more active then *d*-sotalol (Patil 1968) and 14 times more active than racemic sotalol (Manley et al. 1986).

III. Effects on Action Potential Duration

Kaumann and Olson observed in 1968 that sotalol lengthened action potential duration and produced aftercontractions in cat papillary muscle, but it was not until 1970 that systematic intracellular microelectrode studies explored the unique action of sotalol in delaying repolarization (Singh and Vaughan Williams 1970; Strauss et al. 1970). A dose-dependent lengthening was demonstrated in atrial and ventricular myocardial cells and Purkinje fibres from various species (Fig. 2). Sotalol was found to have 1/300th of the local anaesthetic potency of propranolol at equimolar concentrations. The possibility that sotalol may have some class I action at high concentration has been the source of some disagreement in various studies. There is no evidence of inhibition of action potential upstroke velocity at concentrations as high as 10^{-3} mol/litre in canine Purkinje fibres and ventricular myocardium (Strauss et al. 1970). Concentrations up to 5×10^{-4} mol/litre had no effect on the rapid inward sodium channel in rabbit atria in one report (Strauss et al. 1970), but in another study on the same tissue, a concentration of 3.2×10^{-4} mol/litre produced a decrease of 25 % in conduction velocity and of 36 % in action potential upstroke velocity (Singh and Vaughan Williams 1970). Sotalol has no class I effect on rabbit ventricular muscle at 1×10^{-4} mol/litre in either isolated papillary muscle or in the perfused interventricular septum (Cobbe and Manley 1985; Cobbe et al. 1985a). There is no reduction of upstroke velocity in cat papillary muscle at 1.6×10^{-4} mol/litre, but a 29 % decrease occurs at 3.2×10^{-4} mol/litre (Singh and Vaughan Williams 1970). Carmeliet (1985)

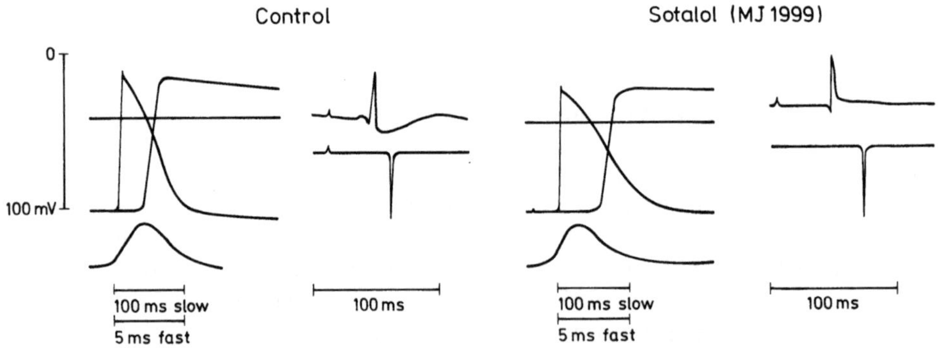

Fig. 2. Acute effect of sotalol 1.62×10^{-4} mol/litre on atrial transmembrane action potentials. *Left Horizontal trace,* zero potential; *Middle traces,* action potential at slow and fast sweep speeds; *bottom trace,* contraction. *Right Upper trace,* extracellular electrogram for measurement of conduction delay. *Lower trace,* differential of action potential. Depth of spike represents upstroke velocity. Note increase of action potential duration with no effect on upstroke velocity or conduction. (Courtesy of Dr. E. M. Vaughan Williams)

found no evidence of class I action in guinea pig papillary muscle in concentrations up to 1×10^{-4} mol/litre, but CULLING et al. (1984) noted that sotalol 1×10^{-4} mol/litre reduced action potential amplitude and upstroke velocity by 7% and 12% respectively in Langendorff-perfused guinea pig ventricle. It is clear that these discrepancies are most likely to be attributable to species differences, or variations in experimental conditions or heart rate. Given that the therapeutic plasma concentration of sotalol in man is of the order of 1×10^{-5} mol/litre, however, it is reasonable to assume that sotalol has no class I effect in clinical usage.

The lengthening of action potential duration induced by sotalol involves a prolongation of phases 2 and 3. The magnitude of the effect varies between tissues. For example, the increase in action potential duration produced by 5×10^{-4} mol/litre sotalol was 15% in canine ventricular muscle compared with 37% in Purkinje fibres in the same preparation (STRAUSS et al. 1970). Sotalol has a direct effect on the sinus node, causing an increase in sinus cycle length in isolated preparations which is attributable to a delay in repolarization of sinoatrial cells rather than a slowing of phase 4 depolarization (KATO et al. 1986). In other words, sotalol causes a specific bradycardia as a result of its class III action in addition to its effect on heart rate mediated by beta-receptor antagonism. The basis of the effects of sotalol on action potential duration has been investigated using voltage-clamp techniques (CARMELIET 1985). There was a substantial reduction in the amplitude of the time-dependent outward K^+ current activated during the plateau of the action potential, and an additional small reduction in background K^+ current. The ED_{50} for the reduction of i_K was 1×10^{-5} mol/litre. Sotalol had no effect on the slow inward current in the absence of exogenous catecholamines.

SINGH and VAUGHAN WILLIAMS (1970) concluded in their original study on sotalol that the ability of the drug to cause uniform lengthening of action potential duration represented a new potential antiarrhythmic action. Indeed, the term "class III" had its origins in their report, and in the recognition that the properties of sotalol were different from those of the conventional class I agents, and were not attibutable to competitive beta-receptor antagonism alone. It is implicit in the concept of class III action that the effective refractory period (ERP) of the cell should be increased in parallel with the action potential duration (APD_{90}). This prediction has been confirmed in studies in canine and rabbit papillary muscle, which indicate that the ratio ERP/APD_{90} remains close to unity despite large increases in APD_{90} (STRAUSS et al. 1970; COBBE and MANLEY 1985).

IV. Differential Effects of Optical Isomers of Sotalol

The clinical interest in sotalol as an antiarrhythmic drug has led to an evaluation of the differential effects of the dextro- and laevo-isomers. As mentioned above, the *l*-isomer of sotalol is approximately 44 times more potent a beta-receptor antagonist. than the *d*-isomer (PATIL 1968). *d*-Sotalol could therefore find use as a relatively "pure" class III agent in cases where beta-receptor antagonism is undesirable. Studies of the relative potency of the class III action of the two isomers have reached differing conclusions. CARMELIET (1985) found *d*- and *l*-sotalol to have identical actions on action potential duration and voltage clamp currents in guinea-pig ventricle, and no difference between the effects of the isomers was found in canine Purkinje fibres and ventricular muscle (KATO et al. 1986). The effects of *d*-sotalol on monophasic action potential duration in the dog ventricle were identical to those of the racemic compound (TAGGART et al. 1985). In contrast to these results, MANLEY et al. (1986) found *d*-sotalol to have a greater effect than *l*-sotalol on rabbit papillary muscle, although the difference was only significant at high concentrations.

V. Subsidiary Class III Actions of Other Beta-blockers?

Increasing interest in class III properties of sotalol has led to re-evaluation of other beta-receptor antagonists with reference to possible acute class III effects. TAGGART et al. (1984) studied the effects of several beta-receptor antagonists on the ventricular monophasic action potential duration (MAPD) in open-chested dogs. The dogs were anaesthetized with chloralose and urethane, and pretreated with propranolol or pindolol. Under these conditions, oxprenolol, nadolol and sotalol caused an increase in MAPD, while atenolol, pindolol, propranolol and timolol did not. Although this result was interpreted as indicating that oxprenolol and nadolol had class III action, a more likely explanation is that the dogs were inadequately beta-blocked by the condition-

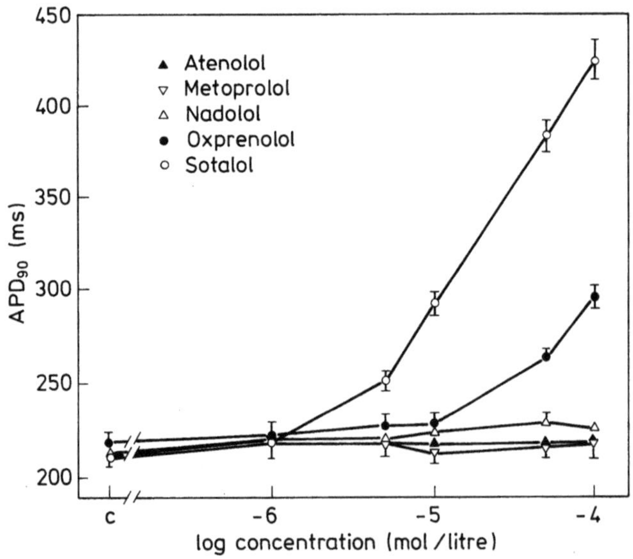

Fig. 3. Comparison of the effects of various beta-blockers on action potential duration recorded in vitro from rabbit papillary muscles. Recordings were made under control conditions (*c*) and in the presence of drug concentrations from 10^{-6} to 10^{-4} mol/litre. See text for details.

ing dose of propranolol, and that the effect of the additional dose of nadolol and oxprenolol was to counteract a catecholamine-induced shortening of action potential duration. Such an effect has been demonstrated in a similar open-chested canine preparation (AMLIE et al. 1982), in which atenolol caused significant increases in MAPD of 5 %-6 %, despite its lack of class III action in vitro (COBBE and MANLEY 1985). A comparative in vitro study of the effects of atenolol, metoprolol, nadolol, oxprenolol and sotalol on action potential duration in rabbit papillary muscle has recently been performed (MANLEY et al. 1986). Atenolol, metoprolol and nadolol had no effect on action potential duration or effective refractory period in concentrations up to 1×10^{-4} mol/litre (Fig. 3). Oxprenolol caused an increase in action potential duration, but this was only significant at concentrations above 5×10^{-5} mol/litre, and was associated with a substantial class I effect. At clinically relevant concentrations, only sotalol caused a significant lengthening of the action potential.

It should be emphasized that the above observations relate only to the *acute* effects of beta-receptor antagonists on action potential duration. The phenomenon of the chronic adaptational increase in action potential duration during prolonged beta-receptor blockade (RAINE and VAUGHAN WILLIAMS 1981) will be discussed in Sect. CIII.

B. Antiarrhythmic Activity of Sotalol in Experimental Models

I. Simple Arrhythmia Models

Preliminary studies of the antiarrhythmic properties of sotalol confirmed its efficacy against catecholamine-induced arrhythmias (SOMANI et al. 1966; SHARMA 1967) as would be predicted from its beta-receptor antagonist properties. Sotalol 1.5 mg/kg protects against ouabain-induced ventricular fibrillation in guinea pigs (SINGH and VAUGHAN WILLIAMS 1970). The differences between *d*- and *l*-sotalol were investigated in dogs by SOMANI and WATSON (1968). As expected, *l*-sotalol was approximately 20 times more effective than *d*-sotalol in antagonizing adrenergically induced ventricular tachycardia. Neither isomer was effective in suppressing ouabain-induced ventricular tachycardia.

II. Effects of Changes in Experimental Conditions

Before considering the effects of sotalol in ischaemic and infarcted myocardium (see the following), it is instructive to consider the way in which the actions of sotalol are changed under other experimental conditions. STRAUSS et al. (1970) noted that the effect of sotalol on action potential duration was greatest at lower heart rates, and diminished with shortening of the cycle length of stimulation. LEVY and RICHARDS (1965) noted that sotalol 1×10^{-5} mol/litre failed to reduce the maximum following frequency in electrically stimulated rabbit left atria. These important observations are in marked contrast to the use-dependent properties of class I agents, which result in a greater suppression of the sodium channel at shorter cycle lengths (CAMPBELL 1983). Thus whereas drugs such as lignocaine will exert a greater suppressant effect on premature beats or during tachycardia than at slow rates, the opposite effect would be predicted for sotalol. However, the effect of sotalol in increasing sinus cycle length will tend to maximize its class III activity.

The observations on the rate dependency of the class III effect of sotalol help to explain some apparent inconsistencies in the literature. Thus while CARMELIET (1985) found that a concentration of 1×10^{-4} mol/litre sotalol caused a lengthening of 30% in action potential duration in guinea pig papillary muscle stimulated at a cycle length of 1000 ms, the same concentration resulted in an increase of only 5% in Langendorff-perfused guinea pig ventricular muscle at a cycle length of 300 ms (CULLING et al. 1984).
The effect of alterations in extracellular potassium concentrations $(K^+)_e$ on the actions of sotalol are of interest. A reduction of $(K^+)_e$ from 3.5 to 2.5 mmol/litre led to a decrease in the extent of action potential lengthening produced by sotalol 2×10^{-4} mol/litre from 36% to 25% in endocardial fibres and from 17% to 6% in epicardial fibres of dog left ventricle (BRACHMANN et al. 1984). The effects of increases of $(K^+)_e$ to 8 and 12 mmol/litre were studied by COBBE and MANLEY (1985). These levels of $(K^+)_e$ were chosen to simulate those occurring in acutely ischaemic myocardium (WEISS and SHINE 1981). Increases in $(K^+)_e$ led to parallel shortening in action potential duration in both control and sotalol-treated papillary muscles. Even when $(K^+)_e$ reached 12 mmol/litre, however, APD_{90} was 62% longer in the presence of sotalol 1×10^{-4} mol/litre than in controls. There was no evidence of a "hidden" class I effect brought out by hyperkalaemia in this study. The duration of postrepolarization refractoriness was identical in the control and sotalol groups, indicating that sotalol had not interfered with the process of recovery of excitation after full repolarization (GETTES and REUTER 1974).

The effects of hypoxia with and without substrate depletion on the class III action of sotalol were studied in rabbit papillary muscles (COBBE et al. 1985b). The ability of sotalol to maintain an increase in APD_{90} over controls was preserved under "mild" hypoxic conditions of superfusion with buffer, equilibrated with 95% N_2, 5% CO_2 and containing 5 mmol/litre glucose. The extent of action potential shortening was small under these conditions, consistent with the known ability of the cell to maintain the action potential plateau by virtue of glycolytically derived ATP production (McDONALD et al. 1971). A contrasting effect was noted under "severe" hypoxic conditions (95% N_2, 5% CO_2, zero glucose). The extent of shortening of APD_{90} in the control fibres was much greater, and the class III effect of sotalol was lost within 5 min of exposure to hypoxia. Even under these conditions, however, the ERP/APD_{90} ration remained close to unity, and there were only minor effects on sodium channel activity.

III. Acute Ischaemia

The interactions between the electrophysiological effects of sotalol and the changes brought about by acute myocardial ischaemia have been studied by

CULLING et al. (1984) and COBBE et al. (1985a). At first sight, the results obtained in the two studies appear inconsistent despite the use of the same concentration of sotalol (1×10^{-4} mol/litre). Most of the apparent discrepancies can be explained on the basis of differences in experimental design. CULLING et al. (1984) used Langendorff-perfused guinea pig hearts stimulated at a cycle length of 300 ms. Under these conditions, as mentioned previously, the pre-ischaemic values indicated that sotalol exerted minimal class III activity with a significant class I effect. A period of 30 min low-flow ischaemia was used. The incidence of ventricular tachycardia was significantly reduced by sotalol during ischaemia, and on reperfusion. There was a non-significant reduction in the incidence of ventricular fibrillation from 20% to 10% during ischaemia, and a significant reduction from 80% to 43% on reperfusion. The mean action potential upstroke velocity ($\dot{V}_{max}$) was consistently lower during ischaemia in the sotalol-treated hearts, while stimulation threshold, conduction time and QRS duration were greater in the sotalol-treated group during ischaemia than in the controls. The increase in APD_{90} induced by sotalol prior to ischaemia disappeared within 5 min of the onset of ischaemia. The ERP in the sotalol group was consistently greater than in controls but the magnitude of the difference diminished from +39% at 4 min to +6% at 30 min. Sotalol prevented the shortening in APD_{90} which occured in the control hearts on reperfusion. These results indicate that sotalol did indeed exert an antiarrhythmic effect, but that only the effect on reperfusion arrhythmias could conceivably be attributed to its influence on action potential duration.

The study by COBBE et al. (1985a) used the isolated interventricular septum of the rabbit heart, and thus recorded changes in endocardial as opposed to epicardial cells. The cycle length was 1000 ms, and under control conditions sotalol exerted a substantial class III effect, causing in increase of 52% in APD_{90} and 31% in ERP. There was no significant change in action potential amplitude or $\dot{V}_{max}$. During 30 min zero-flow ischaemia, the shortening of APD_{90} in the sotalol group was much faster than in controls, with the disappearance of any significant effect on mean APD_{90} after 24 min. The loss of the class III effect on ERP was even more dramatic, so that the value in the sotalol group was significantly *shorter* than controls by 30 min. In contrast to the findings of Culling et al., there was a trend towards greater preservation of $\dot{V}_{max}$ in the sotalol group. This observation, coupled with the absence of development of postrepolarization refractoriness in the sotalol group, suggested a relative preservation of the sodium channel in the presence of sotalol. The most likely explanation of this phenomenon is that the rate of increase of $(K^+)_e$ was slowed by sotalol. This suggestion is supported by the finding that under conditions of "simulated ischaemia" (hypoxia, acidosis and a constant $(K^+)_e$ of 12 mmol/litre) there was no difference in the degree of postrepolarization refractoriness or in other indices of sodium channel activity between the control and sotalol groups. (COBBE et al. 1985b). The mechanism of this postulated effect of sotalol on $(K^+)_e$ during ischaemia is unclear. One explanation is that sotalol lessened the severity of the metabolic insult during ischaemia by virtue of its beta-blocking action, and that this led to a slowing in the rate of leakage of potassium into the extracellular fluid (WIEGAND et al. 1979). Alternatively, it may be that the inhibition of potassium efflux was a direct consequence of the action of sotalol on outward potassium current (Carmeliet 1985). In both instances the lack of washout of the extracellular space during zero flow ischaemia would allow local factors to influence $(K^+)_e$.

IV. Comparison of Effects of Sotalol and Amiodarone in Ischaemic Myocardium

The electrophysiological effects of sotalol and amiodarone in ischaemic myocardium have recently been compared (Cobbe et al. 1985a; Cobbe and Manley 1988). Hearts from animals chronically treated with amiodarone were used. The effect of amiodarone in the interventricular septum was studied under identical experimental conditions to those described for sotalol. Amiodarone caused a smaller increase in APD_{90} (13% vs. 52%) and ERP (13% vs. 31%). Unlike sotalol there was clear evidence of class I action in the amiodarone-treated muscles, with a reduction of $\dot{V}_{max}$ of 14%. During ischaemia, the

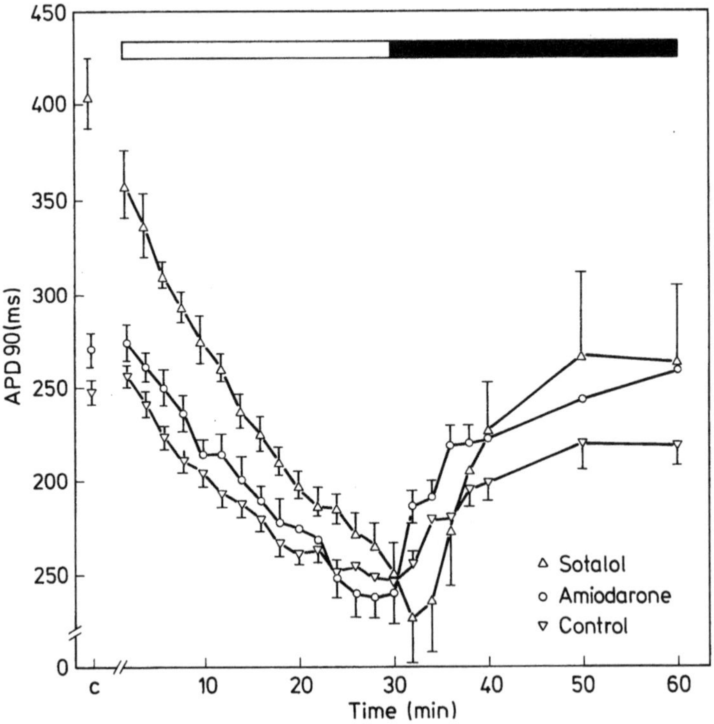

Fig. 4. Comparison of the effects of sotalol and amiodarone on action potential duration (APD_{90}) before, during and after 30 min zero flow ischaemia. Both drugs caused significant lengthening in APD_{90} prior to ischaemia. There was a shortening in APD_{90} in all groups during ischaemia, with loss of the class III effect of amiodarone and sotalol in relation to controls. The class III effects were re-established on reperfusion. (Adapted from Cobbe et al. 1985a; 1988) △ sotalol; ▽ control; ○ amiodarone

class III effect of amiodarone in lengthening APD_{90} in relation to controls disappeared within 10 min (Fig. 4), while by 30 min the mean APD_{90} tended to be shorter in the amiodarone group than controls. The most striking difference between the properties of sotalol and amiodarone in ischaemia, however, was in the response of the sodium channel. In contrast to sotalol, amiodarone caused a more rapid decrease in $\dot{V}_{max}$ than in controls. There was a rapid increase in stimulation threshold, conduction delay, ERP and postrepolarization refractoriness, resulting in the majority of the muscles failing to maintain 1:1 responses to stimulation. These changes, indicative of profound depression of the rapid sodium channel, disappeared within 2 min of reperfusion.

In summary, therefore, these experiments on sotalol and amiodarone in acute ischaemia indicate that neither drug was able to maintain its effect on APD_{90} in the face of severe ischaemia. Sotalol exerted a protective effect on the sodium channel, while the modest class I properties of amiodarone in normal myocardium were greatly enhanced during ischaemia—a property common to other class I agents (CARDINAL et al. 1981).

V. Postinfarction

The effects of sotalol have been studied not only in acutely ischaemic myocardium, as described above, but also in established experimental myocardial infarction. Studies have used the Harris technique for two-stage ligation of the left anterior descending artery of the dog (HARRIS 1950). At 18–24 h after ligation, sustained ventricular tachycardia is normally seen, and is considered to be automatic in origin (DAVIS et al. 1982). Neither *d*- nor *l*-sotalol was effective in suppressing this arrhythmia, even at high doses (SOMANI and WATSON 1968). Spontaneous arrhythmias have subsided by 48–72 h (EL-SHERIF et al. 1977), but sustained ventricular tachycardia may be initiated by programmed ventricular stimulation at 3–7 days (COBBE et al. 1985c); and isochronal mapping studies have confirmed epicardial re-entry as the basis for this arrhythmia (MEHRA et al. 1983).

BRACHMANN et al. (1984) studied epicardial and endocardial preparations in vitro from 4-day-old canine myocardial infarctions. They found that sotalol 2×10^{-4} mol/litre caused an increase of 17% and 12% respectively in APD_{90} from cells in the non-infarcted and infarcted areas of the epicardium, with comparable effects on ERP. Studies on endocardial cells revealed a greater percentage increase in APD_{90} and ERP than on the epicardial surface. Sotalol lengthened APD_{90} by 36% and ERP by 27% in the normal zone. The most marked effect of sotalol was in the endocardial infarcted zone, where sotalol increased ADP_{90} by 31% and ERP by 42%. No effect of sotalol on action potential amplitude, conduction velocity or V_{max} was demonstrated in either non-infarcted or infarcted areas.

The ability of sotalol to prevent initiation of ventricular tachycardia in conscious dogs 3–7 days after experimental myocardial infarction was demonstrated by COBBE et al. (1983). Sotalol was completely or partially effective in 11 of 19 studies (58%), as compared with 1 of 14 (7%) with metoprolol. Similar rates of efficacy ranging from 54% to 67% have been reported from subsequent studies in similar experimental models using either racemic or *d*-sotalol (PATTERSON et al. 1984; LYNCH et al. 1984, 1985). These results suggest that the efficacy of sotalol is attributable to its class III action rather than to beta-receptor blockade alone. COBBE et al. compared the effect of sotalol on the ERP in the non-infarcted ventricle with that on the refractory period in the infarcted area, measured from implanted "composite" electrodes. At a mean plasma concentration of 2.13 µg/ml, sotalol increased QT interval by 13%, ERP in the normal zone by 14%, and refractory period in the infarcted zone by 41%. Metoprolol had no significant effect in the infarcted area.

In addition to its ability to suppress the inducibility of ventricular tachycardia in the postinfarction canine model, *d*-sotalol has been shown to be effective in preventing the development of ventricular fibrillation in a conscious canine model of sudden cardiac death (LYNCH et al. 1985), in which acute occlusion of the left circumflex artery is produced in the presence of prior anterior infarction. The protective effect of *d*-sotalol was achieved using a large dose (8 mg/kg every 8 h), while a single pretreatment with 8 mg/kg of either *dl* or *d*-sotalol did not achieve a statistically significant improvement in survival (PATTERSON et al. 1984; LYNCH et al. 1984).

The results of the in vitro and in vivo studies of sotalol in the postinfarction period described by BRACHMANN et al. (1984) and COBBE et al. (1983) respectively are consistent in suggesting that sotalol has an effect in the infarcted area which is greater than in the normal myocardium. The exact nature of this effect is unclear, since it results in a proportionately greater increase in refractory period than in action potential duration. Nevertheless, the clear implication is that the loss of the class III effect in acute ischaemia may be a transient phenomenon, possibly related to the particular conditions experienced in the myocardium at that time, and that the class III properties of sotalol are re-established in the early postinfarction period.

C. Clinical Pharmacology

I. Beta-Blocking Activity

As predicted from its profile in experimental studies (LEVY and RICHARDS 1965), sotalol is effective in man as a non-selective beta-receptor antagonist without intrinsic sympathomimetic activity (SHANKS et al. 1974; FRISHMAN 1979). FRANKL and SOLOFF (1968) showed significant blockade of the chronotropic response to isoprenaline 20 µg, using intravenous doses of sotalol as low as 0.06 mg/kg. The effects of doses of 0.2, 0.4 and 0.6 mg/kg intravenously were investigated by BROOKS et al. (1970) in a group of patients with various cardiac diseases undergoing catheterization. A maximal reduction in resting heart rate and cardiac index was achieved with a dose of 0.4 mg/kg. There was no effect on stroke index or left ventricular end-diastolic pressure in this group, nor any adverse clinical effects. Subsequent experience with sotalol in patients with poor left ventricular function, however, has confirmed its ability to exacerbate congestive cardiac failure in a proportion of cases (SENGES et al. 1984; NADEMANEE et al. 1985). The latter study, however, did not demonstrate any reduction in mean left ventricular ejection fraction.

II. Clinical Evidence of Acute Class III Activity

Despite the characterization of the class III properties of sotalol in 1970 by SINGH and VAUGHAN WILLIAMS and STRAUSS et al., a considerable period elapsed before clinical studies were undertaken to establish whether this effect was seen in man. WARD et al. (1979) studied the effect of intravenous sotalol 0.4 mg/kg on intracardiac conduction and refractory periods. They demonstrated significant increases in sinus cycle length, AV nodal conduction time (AH interval) and AV nodal effective and functional refractory periods. These changes were considered to be attributable to beta-blockade. The increase in atrial ERP was small (4%) although significant, while the change in ventricular ERP was not considered significant. The conclusion from this study was that sotalol had no substantially different electrophysiological properties from other beta-blockers.

A different result was obtained by EDVARDSSON et al. (1980), using the larger intravenous dose of 100 mg (approximately 1.5 mg/kg). They demonstrated an increase of 13% in ERP in the right ventricular outflow tract, corresponding to a lengthening of 17% in monophasic action potential duration (MAPD) recorded at the same site. There was no overall change in the ERP/MAPD ratio. Data from three patients treated for at least 12 weeks suggested that the class III effects of sotalol persisted during chronic therapy. Subsequent intravenous studies have used doses ranging from 0.4 to 1.5 $mg \cdot kg^{-1}$, and confirmed that sotalol increases atrial, AV nodal and ventricular ERP by 10%-20%, with comparable changes in surface QT interval (BENNETT 1982; NATHAN et al. 1982; SENGES et al. 1984; TOUBOUL et al. 1984; NADEMANEE et al. 1985). The mean plasma concentrations of sotalol measured 10-20 min after injection have been in the range 2-3 µg/ml. Sotalol has been shown to increase the ERP of both fast and slow limbs in patients with dual AV nodal pathways, and of extranodal accessory pathways (BENNETT 1982; NATHAN et al. 1982; RIZOS et al. 1984). Although the studies mentioned above have demonstrated changes in ventricular MAPD and ERP which would not be predicted from the known effects of conventional beta-receptor blockers, they did not directly compare sotalol with another beta-blocker. Such a study was performed by ECHT et al. (1982), comparing sotalol 0.3-0.6 mg/kg with propranolol 0.15-0.2 mg/kg in a double-blind protocol. Despite the small dose of sotalol employed, they demonstrated highly significant increases of 10% in ventricular MAPD and ERP, while no significant changes occurred with propranolol. Another randomized comparison of intravenous sotalol and propranolol on the QT-interval in patients with implanted ventricular pacemakers reached similar conclusions (CREAMER et al. 1986).

Preliminary reports of the clinical electrophysiological properties of *d*-sotalol have confirmed that the isomer causes acute increases in QT_c and ERP, as would be predicted. In keeping with its weak beta-receptor-blocking properties, there is little effect on sinus cycle length (McComb et al. 1985).

III. Clinical Evidence of Chronic Class III Activity

The studies described above confirm that racemic sotalol has acute class III electrophysiological effects in man when present at a plasma concentration of 2–3 μg/ml. The effects of sotalol on ventricular repolarization during long-term therapy are of additional importance if the drug is to be used prophylactically. It is important to establish that tachyphylaxis does not occur, but a question of more interest has been the relationship between the acute class III action of sotalol and the chronic adaptational increase in APD which occurs during prolonged beta-receptor blockade. The latter phenomenon was demonstrated in rabbits by Raine and Vaughan Williams (1981), who found an increase in 20%–23% in ventricular APD_{90} and ERP. This "chronic class III effect" was postulated as one of the mechanisms whereby beta-receptor blockers reduce the incidence of sudden death during follow-up after myocardial infarction (Norwegian Multicentre Study Group 1981; Beta-Blocker Heart Attack Trial Research Group 1982). The presence of a chronic adaptational increase in ventricular MAPD and ERP in man was confirmed by Edvardsson and Olsson (1981) in a study of the acute and chronic effects of metoprolol in volunteers. They demonstrated an increase of 6% in both MAPD and ERP after therapy with 400 mg daily for 5 weeks.

The question relevant to sotalol therapy, therefore, is whether the chronic adaptational response to beta-receptor blockade will enhance the acute effects of sotalol on APD and ERP, or whether it will result in other beta-blockers achieving the same effects during chronic therapy that sotalol demonstrates acutely. This question has recently been addressed by several groups. A common feature of these studies has been the direct comparison of QT intervals or MAPD between patients on oral sotalol and another beta-blocker, using identical paced or comparable sinus cycle lengths to avoid the necessity for rate correction, which is known to be unreliable during beta-receptor blockade (Milne et al. 1980).

Creamer et al. (1986) have compared the effects of chronic oral therapy with sotalol 160 mg b.d. and propranolol 80 mg b.d. on the QT and JT intervals in a crossover study on patients with implanted ventricular pacemakers. The mean increase in QT during chronic oral sotalol therapy was 11.5%, in contrast to the 6.5% lengthening which occurred acutely in response to intravenous infusion of 1.5 mg/kg sotalol. This effect on QT interval occurred despite a lower mean plasma level of sotalol on oral therapy, and suggests that the class III effect of sotalol is enhanced during oral therapy. In contrast to this finding, however, was the failure to demonstrate a significant adaptive lengthening of QT interval during oral propranolol therapy, despite an apparently equal degree of beta-receptor blockade to that occurring on sotalol. This finding is inconsistent with the observations of Edvardsson and Olsson (1981).

The probable explanation is that propranolol causes a small acute shortening in APD both experimentally (RAINE and VAUGHAN WILLIAMS 1980) and clinically (ECHT et al. 1982; CREAMER et al. 1986). This acute shortening will tend to mask the adaptative lengthening of APD in response to chronic beta-blockade unless measurement is made after the drug has been excreted (RAINE and VAUGHAN WILLIAMS 1980). Comparison between the chronic effects of sotalol and other beta-blockers may be made more easily if agents such as atenolol and metoprolol are used, since neither of these causes any acute change in APD or ERP (EDVARDSSON and OLSSON 1981; MANLEY et al. 1986; WAY et al. 1986).

The effects of 1 month's therapy with atenolol 50 mg b.d. or sotalol 160 mg b.d. on MAPD and ERP were compared by WAY et al. (1986) in a crossover study of patients with chronic stable angina. Chronic oral sotalol produced a lengthening of 8% in right ventricular MAPD and 9% in ERP compared with chronic atenolol. Intravenous sotalol increased MAPD and ERP by 19% and 17% respectively in patients receiving oral atenolol, while intravenous atenolol caused no further increases in patients receiving oral sotalol.

Two studies have compared oral sotalol with either atenolol or metoprolol in patients after myocardial infarction. COBBE et al. (1988) compared sotalol 160 mg b.d. and atenolol 50 mg b.d. in 104 patients, starting 7-10 days after infarction. Patients received either sotalol or atenolol for 6 months and the other drug for a further 6 months. The mean heart rate was not significantly different between the groups receiving sotalol or atenolol at any stage. Sotalol caused a lengthening in QT interval of 9% over atenolol by 6 days, and this increase remained stable for the duration of the study, including the crossover period. There was an increase in absolute but not corrected QT interval in response to atenolol therapy. BLOMSTRÖM-LUNDQVIST et al. (1986) compared sotalol 160 mg b.d. and metoprolol 100 mg b.d. in a crossover study of 20 patients performed 12 or more months after infarction. QT intervals were measured during transoesophageal atrial pacing at constant rate. Under these conditions, mean QT interval during chronic sotalol therapy was 5% longer than on metoprolol. The mean plasma levels of sotalol during chronic oral therapy in the studies discussed above was 1.4-3.1 μg/ml.

In summary, therefore, there is now clear clinical evidence that sotalol has an acute and long-term class III electrophysiological effect which is not attributable to beta-receptor blockade, although the class III effect is seen at higher doses than are required for beta-blockade alone. Still further evidence for this conclusion is now available from preliminary clinical studies on *d*-sotalol (BARBEY et al. 1985; McCOMB et al. 1985).

IV. Clinical Pharmacokinetics

The clinical pharmacokinetics of sotalol were reviewed by SHANKS et al. (1974), JOHNSSON and REGARDH (1976) and ANTTILA et al. (1976). The overall bioavailability of sotalol after an oral dose is >60%, and the drug does not undergo any first-pass, hepatic metabolism. The effective plasma concentration for beta-receptor blockade ranges from 0.5 to 4 μg/ml, although, as discussed in previous sections, the effective concentration for a class III effect appears to be 2–4 μg/ml. Approximately 50% of sotalol in the plasma is bound to serum albumin (JOHNSSON and REGARDH 1976). Sotalol is weakly lipid soluble, and thus does not readily pass the blood-brain barrier.

Excretion of the drug occurs principally via the kidneys, with approximately 60% of the dose recoverable in the urine. There are no active metabolites of clinical importance. The elimination half-life of sotalol is 5–13 h (JOHNSSON and REGARDH 1976), and the half-life for reversal of the effects on QT-interval was found to parallel the plasma half-life of sotalol in one study on the effect of sotalol overdosage (ELONEN et al. 1979). As would be anticipated from its mode of excretion, the elimination half-life of sotalol is considerably lengthened in renal failure (TJANDRAMAGA et al. 1976), and toxic effects have been reported when appropriate dosage reduction was not instituted (KONTOPOULOS et al. 1981).

V. Metabolic Effects

A possible adverse effect of sotalol on plasma lipoprotein levels was suggested by LEHTONEN and VIIKARI (1979) in an open study on 12 hypertensive subjects. In common with other non-selective beta-receptor blockers, sotalol inhibited lipolysis, resulting in a significant reduction in plasma free fatty acid levels, and a increase in plasma triglycerides of 66% over 12 months. This effect is commonly seen with other beta-receptor antagonists (VAN BRUMMELEN 1983). The unusual feature of the study of Lehtonen and Viikari was the finding of a significant increase of 17% in total cholesterol and 32% in low-density lipoprotein (LDL)-cholesterol, with a 28% reduction in high-density lipoprotein (HDL)-cholesterol concentration. These potentially adverse effects on plasma cholesterol were not seen with other beta-receptor blockers. It should be noted, however, that the study by Lehtonen and Viikari was open and uncontrolled, and that no data on concomitant medication, diet or changes in body weight were reported. A repeat study in hypertensives utilizing a similar protocol (NORTHCOTE et al. 1986) failed to demonstrate any adverse effect on plasma cholesterol or its subfractions.

D. Antiarrhythmic Efficacy in Clinical Practice

I. Supraventricular Arrhythmias

Despite the considerable interest in sotalol as a class III antiarrhythmic agent, there are relatively few detailed reports of the drug's efficacy. The vast majority of studies have been open and uncontrolled, and there remains considerable scope for adequate trials to assess the clinical value of sotalol as an antiarrhythmic drug, and its efficacy in relation to other available agents.

Several open studies have confirmed the ability of intravenous sotalol to terminate paroxysmal supraventricular tachycardia in approximately 75% of episodes (FOGELMAN et al. 1972; PRAKASH et al. 1972; LATOUR et al. 1977; TEO et al. 1985). The doses used in these series ranged from 20 to 100 mg, and thus it is likely that in many instances the drug was acting as a beta-receptor blocker rather than as a class III agent. RIZOS et al. (1984) found that sotalol 1.5 mg/kg prevented reinduction of supraventricular tachycardia at electrophysiological testing in 59% of patients, as compared with a 28% response to metoprolol. The efficacy of intravenous sotalol in converting atrial flutter to sinus rhythm was only of the order of 0%-33% when doses of 20-60 mg were used (FOGELMAN et al. 1972; PRAKASH et al. 1972; LATOUR et al. 1977; TEO et al. 1985). The larger dose of 100 mg, however, was effective in six of seven patients (TEO et al. 1985). Intravenous sotalol was sometimes effective in converting atrial fibrillation of recent onset to sinus rhythm, and slowed ventricular rate even when fibrillation persisted (FOGELMAN et al. 1972; PRAKASH et al. 1980; TEO et al. 1985). The drug was only effective in restoring sinus rhythm in 1/17 patients with chronic atrial fibrillation in the studies of FOGELMAN et al. (1972) and EDVARDSSON et al. (1980).

CAMPBELL et al. (1985) compared intravenous sotalol with digoxin and disopyramide in a randomized trial of 40 patients with atrial fibrillation or flutter after cardiopulmonary bypass. Although the overall success in restoring sinus rhythm was equal at 85% in the groups, sotalol was more rapidly effective, and was associated with fewer relapses.

There have been no controlled trials of the ability of sotalol to prevent recurrent supraventricular arrhythmias. PRAKASH et al. (1972) successfully prevented relapse in three of four patients with paroxysmal supraventricular tachycardia, while SIMON and BERMAN (1979) reported successful prophylaxis in 13/13 patients treated with oral doses ranging from 80 to 320 mg daily.

II. Ventricular Arrhythmias

The effect of sotalol on ventricular arrhythmias has been investigated both in relation to suppression of ventricular premature beats, and to the prophylaxis of recurrent sustained tachycardia and sudden death. Preliminary uncontrolled studies suggested that intravenous sotalol was effective in suppressing lignocaine-resistant ventricular ectopic activity in acute myocardial infarction (LATOUR et al. 1977). A randomized study in 30 patients with acute infarction confirmed a significant reduction in the incidence of ventricular arrhythmia and creatine kinase release in response to intravenous sotalol (LLOYD et al. 1982). No large-scale trials of the impact of sotalol therapy on mortality or incidence of ventricular fibrillation in acute myocardial infarction have been performed.

A large-scale multicentre trial of oral sotalol in patients discharged after infarction was reported by JULIAN et al. (1982). Patients were not recruited on the basis of the presence of ventricular arrhythmias. The overall results of the trial showed an 18% reduction of mortality in the sotalol group, which did not reach statistical significance although it was comparable in magnitude to the reduction seen in other secondary prevention trials (YUSUF et al. 1985). Analysis of the mode of death in the sotalol and control groups showed no suggestion of a trend towards reduction of instantaneous deaths, as might have been hoped for in view of the class III antiarrhythmic effect of sotalol.

COBBE et al. (1985d) compared the effects of sotalol and atenolol on patients up to 12 months postinfarction. As in other studies of unselected postinfarction patients (LICHSTEIN et al. 1983), the incidence of spontaneous ventricular arrhythmia was low. There was no difference in the incidence of significant arrhythmias or in the overall ectopic frequency between the groups receiving sotalol and atenolol, despite the clear demonstration of a persistent effect of sotalol on Q-T interval.

These studies may have failed to demonstrate a clear-cut antiarrhythmic action of sotalol in postinfarction patients because the incidence of arrhythmia in the study population was too low. A trial was initiated in which selected high-risk patients were recruited on the basis of frequent ventricular premature beats and depressed ejection fraction shortly after myocardial infarction (the TEST study). The drugs used were encainide, sotalol and timolol. Unfortunately, this trial was abandoned as a result of a higher mortality rate than predicted in the sotalol group, although the data were not significant.

Studies of sotalol have, however, been performed in patients with stable, frequent ventricular premature beats, at least 6 months after myocardial infarction. BURCKHARDT et al. (1983) performed a double-blind placebo-controlled trial of sotalol 320 mg daily in 14 patients. There was a 42 % reduction in overall ectopic frequency which was not significant, but a significant reduction in high-grade arrhythmias occured. MYBURGH et al. (1979) reported a significant reduction of 68 % in ventricular premature beat frequency in a placebo-controlled study, using 320 mg sotalol daily. A reduction of 89 % was achieved during an open dose-titration phase of the study. The frequency of high-grade arrhythmias on placebo and sotalol was not reported. BURKHARDT et al. (1984) evaluated the use of sotalol as a substitute for amiodarone in six patients with high-grade ventricular ectopic activity controlled on therapy. It was possible to substitute sotalol in five of six patients without loss of control of the arrhythmias. A randomized crossover study comparing sotalol and procainamide in chronic ventricular arrhythmias was reported by LIDELL et al. (1985). Adequate control of ventricular premature complexes (>75 % suppression) was achieved in 22/33 patients on sotalol, and 13/33 on procainamide.

The role of sotalol in the prophylaxis of recurrent ventricular tachycardia or fibrillation has been described in two reports (SENGES et al. 1984; NADEMANEE et al. 1985). The acute efficacy of sotalol in preventing ventricular tachycardia or fibrillation initiated by programmed stimulation in the two studies was 12/18 (67 %) and 15/33 (46 %) respectively. NADEMANEE et al. (1985) also studied the effects of oral sotalol on spontaneous arrhythmias using ambulatory monitoring. Sotalol reduced the frequency of ventricular premature beats by 73 %, doublets by 89 % and ventricular tachycardia beats by 95 %. There was a correlation between the demonstration of drug efficacy by programmed stimulation and on ambulatory monitoring. Oral therapy with sotalol was instituted in a total of nine sotalol responders by SENGES et al. (1984), and follow-up continued for a mean of 16 months. Recurrent ventricular tachycardia occurred in only one patient, in whom the drug had been withdrawn. NADEMANEE et al. (1985) treated 25 patients with oral sotalol, of whom 15 were considered to be acute responders to invasive electrophysiological testing. Eight of the 15 responders were able to tolerate treatment and remained free from recurrent ventricular tachycardia or fibrillation, in comparison with 3/10 long-term successes from the non-responders.

The two studies described above were open and uncontrolled. No systematic studies have been published comparing the efficacy of sotalol with other antiarrhythmic drugs in the prevention of sustained ventricular tachyarrhythmias. A total of 15 of the patients in the study by SENGES et al. (1984) were also tested with quinidine (10 mg/kg). The overall acute efficacy of quinidine was 2/15 (13 %) as compared with 9/15 (60 %) for sotalol.

E. Adverse Effects

Sotalol therapy may be associated with any of the characteristic adverse effects of beta-receptor antagonism such as bradycardia, hypotension, bronchospasm or congestive cardiac failure. In addition to these class-specific problems, however, sotalol therapy has been shown to predispose to the development of polymorphic ventricular tachycardia ("torsade de pointes") associated with gross lengthening of the QT interval. The underlying mechanism is considered to be regional inhomogeneity of ventricular repolarization, which may permit re-excitation of areas with short APD, and the creation of re-entry. Such a mechanism may be initiated experimentally by perfusing areas of the heart at different temperatures to induce imhomogeneity of APD (KUO et al. 1983). Another possible mechanism for sotalol-related torsade de pointes is the development of early afterdepolarizations, which may initiate triggered activity (CRANEFIELD 1977). Early afterdepolarizations have been noted in vitro in the presence of high concentrations of sotalol (KAUMANN and OLSON 1968; STRAUSS et al. 1970).

The first clinical reports of torsade de pointes induced by sotalol were attributable to very high plasma concentrations (7.5–16 μg/ml) as a result of self-poisoning (ELONEN et al. 1979; NEUVONEN et al. 1981). Gross lengthening of QT_c interval up to 180 % of control was reported. Another case occurred in a patient with renal failure treated with 480 mg sotalol daily (KONTOPOULOS et al. 1981). A series of 13 patients with torsade de pointes was reported by McKIBBIN et al. (1984) in association with daily doses of 80–480 mg. Twelve of these patients were taking a fixed sotalol/hydrochlorothiazide combination tablet. The majority of these patients had one or more predisposing factors for torsade de pointes, such as a plasma potassium level <3.5 mmol/litre (8/13), impaired renal function or concurrent therapy with other agents known to increase Q-T interval such as tricyclic antidepressants and disopyramide. Three patients, however, had potassium concentrations in the range 3.6–3.8 mmol/litre with no other predisposing factors. Plasma sotalol levels were not recorded in the study by McKIBBIN et al. (1984), nor in a case report by KUCK et al. (1984), where lengthening of the Q-T_c interval to 590 ms was associated with torsade de pointes despite normal electrolyte levels. KRAPF and GERTSCH (1985) reported an episode of torsade de pointes in a patient with a normal plasma potassium (4.0 mmol/litre) and magnesium concentration. The plasma sotalol concentration was 2.45 µg/ml and the Q-T_c 530 ms. The patient was, however, receiving maprotilene, a tetracyclic antidepressant which has itself been reported to cause lengthening of Q-T_c and torsade de pointes (HERRMAN et al. 1983).

The overall incidence of torsade de pointes is unknown, and although the manufacturers reported 25 cases in 1984 (0.002% of prescriptions; ANTONACCIO et al. 1984), the true figure may be much higher. It seems clear, however, that the principal predisposing factors to excessive Q-T_c prolongation and torsade de pointes are high plasma sotalol levels, potassium depletion and the coadministration of other drugs or presence of other factors known to cause Q-T_c lengthening. The presence of moderade Q-T_c lengthening by sotalol is not arrhythmogenic, and in a recent study of 29 episodes of Q-T_c >500 ms in postinfarction patients, ambulatory recording revealed no episodes of torsade de pointes (COBBE et al., unpublished observations). It is noteworthy that the plasma potassium concentration was >4.0 mmol/litre in 84% of these patients, and only one had a value <3.8 mmol/litre.

F. Conclusions

The foregoing discussion on sotalol has described the unique class III properties of this beta-receptor blocker, and reviewed clinical experience of its use as an antiarrhythmic agent. There is still a lack of reliable clinical data on its efficacy, but it is hoped that this shortcoming will be rectified. The clinical evaluation of *d*-sotalol will be a subject of great interest in the next few years.

References

Amlie JP, Refsum H, Landmark KH (1982) Prolonged ventricular refractoriness and action potential duration during β-adrenoreceptor blockade in the dog heart in situ. J Cardiovasc Pharmacol 4:157–162

Antonaccio MJ, Lessem JN, Soyka LF (1984) Sotalol, hypokalaemia, syncope and torsade de pointes (Letter). Br Heart J 52:358

Anttila M, Arstila M, Pfeffer M, Tikkannen R, Vallinkoski V, Sundqvist H (1976) Human pharmacokinetics of sotalol. Acta Pharmacol Toxicol (Copenh) 39:118–128

Aramendia P, Kaumann AI (1967) Inhibition of sympathomimetic effects on the cardiovascular system by 4-(2-isopropylamino-1-hydroxyethyl) methanesulfonanilide hydrochloride (MJ1999). J Pharmacol Exp Ther 155:259–266

Barbey JT, Echt DS, Thompson KA, Roden DM, Woosley RL (1985) Effect of *d*-sotalol on ventricular arrhythmias in man. Circulation [Suppl 3] 72:170

Bartlet AL, Hassan T (1969) A nerve blocking action of MJ1999. Br J Pharmacol 36:176P

Bennett DH (1982) Acute prolongation of myocardial refractoriness by sotalol. Br Heart J 47:521–526

Beta-Blocker Heart Attack Trial Research Group (1982) A randomised trial of propranolol in patients with acute myocardial infarction. JAMA 247:1707-1714
Blinks JR (1967) Evaluation of the effects of several beta-adrenergic blocking agents. Ann NY Acad Sci 139:679-685
Blomstrom-Lundqvist C, Dohnal M, Hirsch I, Lindblad A, Hjalmarson A, Olsson SB, Edvardsson N (1986) Effect of long term treatment with metoprolol and sotalol on ventricular repolarization measured by use of transoesophageal atrial pacing. Br Heart J 55:181-186
Brachmann J, Senges J, Gao T-L, Schols W, Aidonidis I, Jauernig R, Kubler W (1984) Class III effects of sotalol are dependent on the extracellular potassium concentration in normal and depressed fibers following myocardial infarction. J Am Coll Cardiol 2:617
Brooks H, Banas J, Meister S, Szucs M, Dalen J, Dexter L (1970) Sotalol-induced beta blockade in cardiac patients. Circulation 42:99-110
Burckhardt D, Pfisterer M, Hoffmann A, Burkart F, Emmenegger H, Jost M, Bolli P, Buehler F (1983) Effects of the beta-adrenoceptor blocking agent sotalol on ventricular arrhythmias in patients with chronic ischemic heart diesease. Cardiology [Suppl 1] 70:114-121
Burckhardt D, White RA, Hoffman A (1984) Replacement of amiodarone by sotalol for repetitive ventricular premature beats. Am Heart J 107:167-168
Campbell TJ (1983) Kinetics of onset of rate-dependent effects of class I antiarrhythmic drugs are important in determining their effects on refractoriness in guinea-pig ventricle, and provide a theoretical basis for their subclassification. Cardiovasc Res 17:344-352
Campbell TJ, Gavaghan TP, Morgan JJ (1985) Intravenous sotalol for the treatment of atrial fibrillation and flutter after cardiopulmonary bypass. Comparison with disopyramide and digoxin in a randomized trial. Br Heart J 54:86-90
Cardinal R, Janse MJ, van Eeden I, Werner G, Naumann d'Alnoncourt C, Durrer D (1981) The effects of lidocaine on intracellular and extracellular potentials, activation, and ventricular arrhythmias during acute regional ischemia in the isolated porcine heart. Circ Res 49:792-806
Carmeliet E (1985) An electrophysiologic and voltage clamp analysis of the effects of sotalol on isolated cardiac muscle and Purkinje fibers. J Pharmacol Exp Ther 232:817-825
Cobbe SM, Manley BS (1985) Effects of elevated extracellular potassium concentrations on the class III antiarrhythmic action of sotalol. Cardiovasc Res 19:69-75
Cobbe SM, Manley BS (1987) The influence of ischaemia on the electrophysiological properties of amiodarone in chronically treated rabbit heads. Eur Heart J 11:1241-8
Cobbe SM, Hoffman E, Ritzenhoff A, Brachmann J, Kubler W, Senges J (1983) Action of sotalol on potential re-entrant pathways and ventricular tachyarrhythmias in conscious dogs in the late post-myocardial infarction phase. Circulation 68:865-871
Cobbe SM, Manley BS, Alexopoulos D (1985a) The influence of acute myocardial ischaemia on the class III antiarrhythmic action of sotalol. Cardiovasc Res 19:661-667
Cobbe SM, Manley BS, Alexopoulos D (1985b) Interaction of the effects of hypoxia, substrate depletion, acidosis and hyperkalaemia on the class III antiarrhythmic properties of sotalol. Cardiovasc Res 19:668-673
Cobbe SM, Hoffmann E, Ritzenhoff A, Brachmann J, Kubler W, Senges J (1985c) Day to day variations in inducibility of ventricular tachyarrhythmias during the late post-myocardial infarction phase in conscious dogs. Circulation 72:200-204

Cobbe SM, Alexopoulos D, Winner SJ, McCaie CP, Cobbe PF, Johnston J (1988) a comparison of the long-term effects of sotalol and atenolol on QT interval and arrhythmias after myocardial infarction. Eur Heart J 9:24-31
Cranefield PF (1977) Action potentials, after potentials and arrhythmias. Circ Res 41:415-423
Creamer JE, Nathan AW, Shennan A, Camm AJ (1986) Acute and chronic effects of sotalol and propranolol on ventricular repolarization using constant-rate pacing. Am J Cardiol 57:1092-1096
Culling W, Penny WJ, Sheridan DJ (1984) Effects of sotalol on arrhythmias and electrophysiology during myocardial ischaemia and reperfusion. Cardiovasc Res 18:397-404
Davis J, Glassman R, Wit AL (1982) Method for evaluating the effects of antiarrhythmic drugs on ventricular tachycardias with different electrophysiologic characteristics and different mechanisms in the infarcted canine heart. Am J Cardiol 49:1176-1184
Dungan KW, Lish PW (1964) Potency and specificity of new adrenergic beta-receptor blocking agents. Fed Proc 23:124
Echt DS, Berte LE, Clusin WT, Samuelsson RG, Harrison DC, Mason JW (1982) Prolongation of the human cardiac monophasic action potential by sotalol. Am J Cardiol 50:1082-1086
Edvardsson N, Olsson SB (1981) Effects of acute and chronic beta-receptor blockade on ventricular repolarization in man. Br Heart J 45:628-636
Edvardsson N, Hirsch I, Emanuelsson H, Ponten J, Olsson SB (1980) Sotalol-induced delayed ventricular repolarization in man. Eur Heart J 1:335-343
Elonen E, Neuvonen PJ, Tarssanen L, Kalar R (1979) Sotalol intoxication with prolonged Q-T interval and severe tachyarrhythmias. Br Med J 1:1184
El-Sherif N, Scherlag BJ, Lazzara R, Hope RR (1977) Reentrant ventricular arrhythmias in the late myocardial infarction period. I. Conduction characteristics in the infarction zone. Circulation 55:686-702
Fogelman F, Lightman SL, Sillett RW, McNicol MW (1972) The treatment of cardiac arrhythmias with sotalol. Eur J Clin Pharmacol 5:72-76
Frankl WS, Soloff LA (1968) Sotalol: A safe beta adrenergic receptor blocking agent. Am J Cardiol 22:266-272
Frishman W (1979) Appraisal and reappraisal of cardiac therapy. Clinical pharmacology of the new beta-adrenergic blocking drugs. Part 1. Pharmacodynamic and pharmacokinetic properties. Am Heart J 97:663-670
Gettes LS, Reuter H (1974) Slow recovery from inactivation of inward currents in mammalian myocardial fibres. J Physiol (Lond) 240:703-724
Harris AS (1950) Delayed development of ventricular ectopic rhythms following experimental coronary occlusion. Circulation 1:1318-1328
Herrmann HC, Kaplan LM, Bierer BE (1983) Q-T prolongation and torsade de pointes ventricular tachycardia produced by the tetracyclic antidepressant agent maprotilene. Am J Cardiol 51:904-906
Johnsson G, Regardh CG (1976) Clinical pharmacokinetics of β-adrenoreceptor blocking drugs. Clin Pharmacokinet 1:233-263
Julian DG, Prescott RJ, Jackson FS, Szekely P (1982) Controlled trial of sotalol for one year after myocardial infarction. Lancet 1:1142-1147
Kato R, Ikeda N, Yabek SM, Kannan R, Singh BN (1986) Electrophysiologic effects of the levo- and dextrorotatory isomers of sotalol in isolated cardiac muscle and their in vivo pharmacokinetics. J Am Coll Cardiol 7:116-125
Kaumann AJ, Olson CB (1968) Temporal relation between long-lasting aftercontractions and action potentials in cat papillary muscles. Science 161:293-295

Kontopoulos A, Filindris A, Manoudis F, Metaxas P (1981) Sotalol-induced torsade de pointes. Postgrad Med J 57:321-323
Krapf R, Gertsch M (1985) Torsade de pointes induced by sotalol despite therapeutic plasma concentrations. Br Med J 290:1784-1785
Kuck KH, Kunze KP, Roewer N, Bleifeld W (1984) Sotalol-induced torsade de pointes. Am Heart J 107:179-180
Kuo CS, Munakata K, Reddy CP, Surawicz B (1983) Characteristics and possible mechanism of ventricular arrhythmia dependent on the dispersion of action potential durations. Circulation 67:1356-1367
Laasko M, Pentikainen PJ, Pyorala K, Neuvonen PJ (1981) Prolongation of the Q-T interval caused by sotalol—possible association with ventricular tachyarrhythmias. Eur Heart J 2:353-358
Latour Y, Dumont G, Brosseau A, Lelorier J (1977) Effects of sotalol in twenty patients with cardiac arrhythmias. Int J Pharmacol Biopharm 15:275-278
Lehtonen A, Viikari J (1979) Long-term effect of sotalol on plasma lipids. Clin Sci Mol Med 57:405-407S
Levy JV, Richards V (1965) Inotropic and metabolic effects of three beta-adrenergic receptor blocking drugs on isolated rabbit left atria. J Pharmacol Exp Ther 150:361-369
Lichstein E, Morganroth J, Harris R, Hubble E (1983) Effect of propranolol on ventricular arrhythmias; the Beta-Blocker Heart Attack Trial experience. Circulation [Suppl 1] 67:I-5-10
Lidell C, Rehnqvist N, Sjogren A, Yli Uotila RJ, Ronnevik PK (1985) Comparative efficacy of oral sotalol and procainamide in patients with chronic ventricular arrhythmias: a multicenter study. Am Heart J 109:970-976
Lloyd EA, Gordon GD, Mabin TA, Charles RG, Commerford PJ, Opie LH (1982) Intravenous sotalol in acute myocardial infarction. Circulation [Suppl 2] 66:3
Lynch JJ, Wilber DJ, Montgomery DG, Hsieh TM, Patterson E, Lucchesi BR (1984) Antiarrhythmic and antifibrillatory actions of the levo- and dextrorotatory isomers of sotalol. J Cardiovasc Pharmacol 6:1132-1141
Lynch JJ, Coskey LA, Montgomery DG, Lucchesi BR (1985) Prevention of ventricular fibrillation by dextrorotatory sotalol in a conscious canine model of sudden coronary death. Am Heart J 109:949-958
Manley BS, Alexopoulos D, Robinson GJ, Cobbe SM (1986) Subsidiary class III effects of beta-blockers? A comparison of atenolol, metoprolol, nadolol, oxprenolol and sotalol. Cadiovasc Res 20:705-709
McComb JM, McGovern BA, Garan H, Ruskin JN (1985) d-Sotalol: electrophysiologic properties in relation to dose and to plasma concentration. Circulation [Suppl 3] 72:169
McDonald TF, Hunter EG, MacLeod DP (1971) Adenosinetriphosphate partition in cardiac muscle with respect to transmembrane electrical activity. Pflugers Arch 322:95-108
McKibbin JK, Pocock WA, Barlow JB, Scott Millar RN, Obel IWP (1984) Sotalol, hypokalaemia, syncope and torsade de pointes. Br Heart J 51:157-163
Mehra R, Zeiler RH, Gough WB, El-Sherif N (1983) Reentrant ventricular arrhythmias in the late myocardial infarction period. 9. Electrophysiological-anatomical correlation of reentry circuits. Circulation 67:11-23
Milne JR, Camm AJ, Ward DE, Spurrell RAJ (1980) Effect of intravenous propranolol on Q-T interval. Br Heart J 43:1-6
Myburgh DP, Goldman AP, Cartoon J, Schamroth JM (1979) The efficacy of sotalol in suppressing ventricular ectopic beats. S Afr Med J 56:295-298
Nademanee K, Feld G, Hendrickson JA, Singh PN, Singh BN (1985) Electrophysio-

logic and antiarrhythmic effects of sotalol in patients with life-treatening ventricular tachyarrhythmias. Circulation 72:555-564
Nathan AW, Hellestrand KJ, Bexton RS, Ward DE, Spurrell RAJ, Camm AJ (1982) Electrophysiological effects of sotalol—just another beta blocker? Br Heart J 47:515-520
Neuvonen PJ, Elonen E, Vuorenmaa T, Laasko M (1981) Prolonged Q-T interval and severe tachyarrhythmias, common features of sotalol intoxication. Eur J Clin Pharmacol 20:85-89
Northcote RJ, Packard CJ, Ballantyne D (1986) The effect of sotalol on plasma lipoproteins. Clin Chim Acta 158:187-192
Norwegian Multicentre Study Group (1981) Timolol-induced reduction in mortality and reinfarction in patients surviving acute myocardial infarction. N Engl J Med 304:801-807
Patil PN (1968) Steric aspects of adrenergic drugs. VIII. Optical isomers of beta adrenergic receptor antagonists. J Pharmacol Exp Ther 160:308-314
Patterson E, Lynch JJ, Lucchesi BR (1984) The antiarrhythmic and antifibrillatory actions of the beta-adrenergic receptor antagonist *d,l*-sotalol. J Pharmacol Exp Ther 230:519-526
Prakash R, Parmley WW, Allen HN, Matloff JM (1972) Effect of sotalol on clinical arrhythmias. Am J Cardiol 29:397-400
Raine AEG, Vaughan Williams EM (1980) Adaptational responses to prolonged beta-adrenoceptor blockade in adult rabbits. Br J Pharmacol 70:205-218
Raine AEG, Vaughan Williams EM (1981) Adaptation to prolonged β-blockade of rabbit atrial, Purkinje and ventricular potentials, and of papillary muscle contraction. Circ Res 46:804-812
Rizos I, Senges J, Jauernig R, Lengfelder W, Czygan E, Brachmann J, Kubler W (1984) Differential effects of sotalol and metoprolol on induction of paroxysmal supraventricular tachycardia. Am J Cardiol 53:1022-1027
Senges J, Lengfelder W, Jauernig R, Czygan E, Brachmann J, Rizos I, Cobbe S, Kubler W (1984) Electrophysiological testing in assessment of sotalol for sustained ventricular tachycardia. Circulation 69:577-584
Shanks RG, Brown HO, Carruthers SG, Kelly JG (1974) Clinical pharmacology of sotalol. Excerpta Med Int Congr Ser 341
Sharma PL (1967) Mechanism of action of a new adrenergic beta-receptor antagonist, MJ 1999, in the prevention of halothane-adrenaline-induced and ouabain-induced ventricular tachycardia in the dog. Ind J Med Res 55:1357-1365
Simon A, Berman E (1979) Long term sotalol in patients with arrhythmias. J Clin Pharmacol 19:547-556
Singh BN, Vaughan Williams EM (1970) A third class of anti-arrhythmic action. Effects on atrial and ventricular intracellular potentials, and other pharmacological actions on cardiac muscle, of MJ1999 and AH3474. Br J Pharmacol 39:675-687
Somani P, Fleming JG, Chan GK et al. (1966) Antagonism of epinephrine-induced cardiac arrhythmias by 4-(2-isopropylamino-1-hydroxyethyl) methane sulfonanilide (MJ1999). J Pharmacol Exp Ther 151:32-37
Somani P, Watson DL (1968) Antiarrhythmic activity of the dextro- and levo-rotatory isomers of 4-(2-isopropylamino-1-hydroxyethyl) methane sulphonanilide (MJ1999). J Pharmacol Exp Ther 164:317-325
Strauss HC, Bigger JT, Hoffman BF (1970) Electrophysiological and beta-blocking effects of MJ1999 on dog and rabbit cardiac tissue. Circ Res 26:661-678
Taggart P, Donaldson R, Abed J, Nashat F (1984) Class III action of β-blocking agents. Cardiovasc Res 18:683-689

Taggart P, Sutton P, Donaldson R (1985) *d*-Sotalol: a new potent class III antiarrhythmic agent. Clin Sci 69:631-636
Teo KK, Harte M, Horgan JH (1985) Sotalol infusion in the treatment of supraventricular tachyarrhythmias. Chest 87:113-118
Tjandramaga TB, Thomas J, Verbeeck R, Verbessbelt R, Verbenchmaes R, de Schepper PJ (1976) The effect of end-stage renal failure and haemodialysis on the elimination kinetics of sotalol. Br J Clin Pharmacol 3:259-265
Touboul P, Atallah G, Kirkorian G, Lamaud M, Moleur P (1984) Clinical electrophysiology of intravenous sotalol, a beta-blocking drug with class III antiarrhythmic properties. Am Heart J 107:888-895
Van Brummelen P (1983) The relevance of intrinsic sympathomimetic activity for β-blocker-induced changes in plasma lipids. J Cardiovasc Pharmacol [Suppl 1] 5:551-555
Ward DE, Camm AJ, Spurrell RAJ (1979) The acute cardiac electrophysiological effects of intravenous sotalol hydrochloride. Clin Cardiol 2:185-191
Way BPJ, Forfar JC, Cobbe SM (1986) Acute and chronic effects of sotalol and atenolol on monophasic action potential duration and effective refractory period. Clin Sci [Suppl 13] 70:5P
Weiss J, Shine KI (1981) Extracellular potassium accumulation during myocardial ischemia: implications for arrhythmogenesis. J Mol Cell Cardiol 13:699-704
Wiegand V, Guggi M, Meesmann M, Kessler M, Greitschus F (1979) Extracellular potassium activity changes in the canine myocardium after acute coronary occlusion and the influence of beta-blockade. Cardiovasc Res 13:297-302
Yusuf S, Peto R, Lewis J, Collins R, Sleight P (1985) Beta blockade during and after myocardial infarction: an overview of the randomised trials. Prog Cardiovasc Dis 27:335-371

CHAPTER 17

Clofilium and Other Class III Agents

M. I. Steinberg and J. K. Smallwood

A. Introduction

If the wavelength of a reentrant impulse exceeds the path length of the circuit, reentry cannot be maintained (see Cranefield 1975). Since wavelength is the product of conduction velocity and the longest refractory period in the circuit, prolonging the time course of repolarization without slowing conduction could be expected to abolish (Wit et al. 1974a,b) or at least slow (Allessie et al. 1977) reentrant arrhythmias. This concept, which serves as the electrophysiological basis for the antiarrhythmic activity of class III agents, has been reviewed in this volume and elsewhere (Wellens et al. 1984; Singh and Nademanee 1985; Steinberg and Michelson 1985; Vaughan Williams 1985a). In theory, a class III drug should have little or no effect on automaticity but possess specific activity against reentrant rhythms that are believed to underlie most potentially life-threatening arrhythmias (see Zipes 1975; Josephson et al. 1978). The distinction between the antiectopic activity characteristic of class I antiarrhythmic drugs and the specific antifibrillatory effects of class III agents has been emphasized previously (Bacaner 1983a; Anderson 1984; Lucchesi 1984; Sasyniuk 1984; Lucchesi and Lynch 1986), although there is no evidence that drugs called "antifibrillatory" on the basis of an experimental model are actually more efficacious in protecting man against VF.

The class III concept has gained widespread interest recently as reports on the validity of this approach to treating cardiac arrhythmias have appeared. Recent studies have stressed the basic and clinical electrophysiological characteristics of agents that possess class III activity, e.g., *dl* and *d*-sotalol (Carmeliet 1985; Lynch et al. 1985a), bretylium (Patterson and Lucchesi 1983) and *N*-acetylprocainamide (Dangman and Hoffman 1981). Class III compounds are also extremely effective in prolonging atrial refractoriness and abolishing experimentally induced *atrial flutter* of the circus movement (Feld et al. 1986) or leading circle type (Boyden 1986). Finally, the undisputed efficacy of amiodarone in treating a wide variety of arrhythmias (Naccarelli et al. 1985) has provided impetus to the development of newer class III agents that lack the side effects associated with amiodarone (Thomis and Tenthorey 1983).

At least two major issues, however, limit the general therapeutic application of the class III concept. First, it is unclear whether the currently available agents favorably influence cardiac arrhythmias solely by prolonging cardiac repolarization since most agents described to date also possess a diversity of electrophysiological and pharmacological properties (see STEINBERG and MICHELSON 1985). Secondly, drugs that delay repolarization will necessarily cause some Q-T interval prolongation (RODEN and WOOSLEY 1985), and a prolonged Q-T interval, whether idiopathic or drug related, is known to be associated with the development of severe ventricular arrhythmias (SCHWARTZ et al. 1975; SURAWICZ and KNOEBEL 1984). Q-T interval, however, can be increased by widening of QRS or by heterogeneity of conduction or action potential duration, but there is no evidence that uniform prolongation of APD is arrhythmogenic.

The availability of highly selective class III agents permits a more rigorous study of the possible efficacy of delayed repolarization as an antiarrhythmic effect. First, the demonstration of potent antiarrhythmic activity in man and experimental animals in doses that selectively increase refractoriness would strengthen the basic tenet of the class III concept (VAUGHAN WILLIAMS 1982). Secondly, the availability of selective class III drugs should help clarify the role of delayed repolarization per se as an antiarrhythmic or possibly proarrhythmic factor. For example, the relative importance of Q-T interval prolongation could be studied in isolation from other electrophysiological effects associated with the proarrhythmic liability of antiarrhythmic drugs (TORRES et al. 1985).

Clofilium (p-chloro-*N*,*N*-diethyl-*N*-heptylbenzenebutanaminium tosylate) is a highly selective and potent class III agent (see STEINBERG et al. 1984). This quaternary ammonium agent prolongs cardiac action potential duration in vitro (STEINBERG and MOLLOY 1979; STEINBERG et al. 1981) and ventricular (STEINBERG and MICHELSON 1985) and atrial (BOYDEN 1986) refractoriness in vivo without affecting conduction velocity in normal or ischemic tissues. Clofilium has antiarrhythmic activity against electrical- or drug-induced ventricular arrhythmias in experimental animals (KOPIA et al. 1985; STEINBERG and MICHELSON 1985; JOHNSON et al. 1986), increases ventricular fibrillation threshold (STEINBERG and MOLLOY 1979; KOPIA et al. 1985; KOWEY et al. 1985), and decreases the defibrillation threshold (TACKER et al. 1980; DAWSON et al. 1985) in nonischemic models. However, spontaneous arrhythmias associated with acute (KOWEY et al. 1985) or chronic ischemia (KOPIA et al. 1985) are somewhat less responsive to clofilium. Recent studies suggest that clofilium prolongs the action potential duration by selectively decreasing potassium ion conductance (SNYDERS and KATZUNG 1985).

In this chapter, we will describe studies emphasizing some of the factors that influence the expression of the class III activity of clofilium in vivo. Since clofilium is a quaternary ammonium salt, its oral bioavailability is to some extent limited by carrier-mediated gastrointestinal absorption mechanisms for charged amines (PATTERSON et al. 1980; TURNHEIM and LAUTERBACH 1980). Therefore, we will also discuss the electrophysiological properties of a nonquaternary amine analog of clofilium, *LY190147*. This new class III agent does not cause excessive increases in the rate-corrected Q-T interval (Q-T_c) even though cardiac refractoriness can be increased to a similar extent as with clofilium. A possible cellular basis for this effect will be discussed. Finally, we will review briefly experimental and clinical results obtained with compounds described in the recent literature that possess varying degrees of class III activity.

B. Clofilium

I. Specificity

The concentration of clofilium needed to prolong the duration of Purkinje fiber action potentials in vitro ranges between 10 and 100 nM (STEINBERG and MOLLOY 1979). However, even micromolar concentrations are without direct effects on the sympathetic nervous system. For example, the response of the isolated rat vas deferens to field stimulation (Fig. 1a) is unchanged in the presence of clofilium (10^{-5} M) while bretylium produces its characteristic inhibition of adrenergic neuronal function (BOURA and GREEN 1959). In addition, the chronotropic responses to isoproterenol in isolated guinea pig atria are unaffected by high concentrations of clofilium (Fig. 1b). Thus, clofilium lacks effects on alpha-receptors, beta$_1$-receptors, or norepinephrine release mechanisms. Although clofilium lacks effects on blood pressure or cardiac

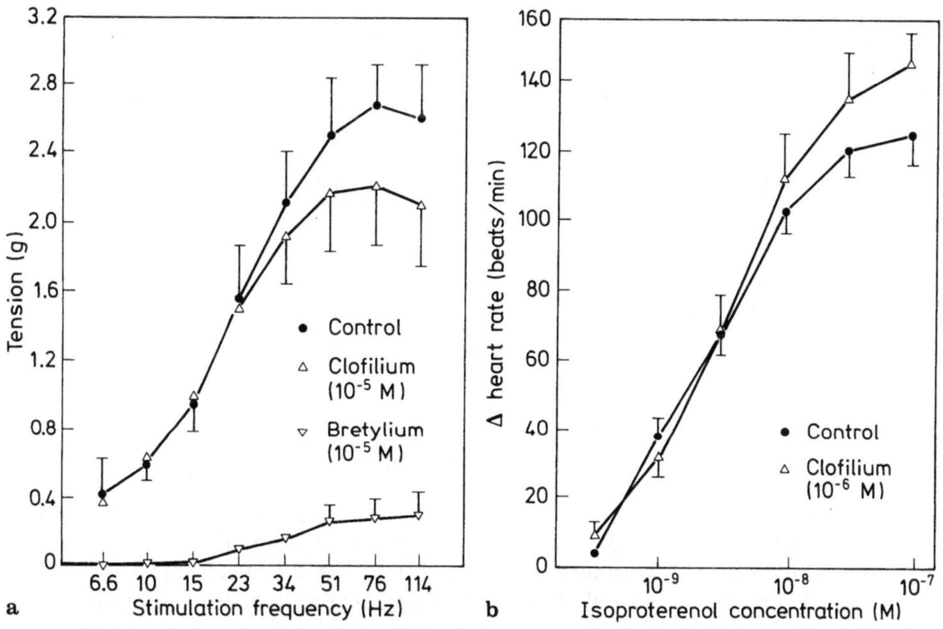

Fig. 1a. Effect of clofilium and bretylium on the twitch response to field stimulation in the guinea pig vas deferens. Guinea pig vas deferentia were suspended in a Ringer's bicarbonate buffer at 35 °C and stimulated every 2 min with a 2-s train of square wave pulses (100-150 V) at frequencies from 7 to 114 Hz. The resting tension was 0.5 g. The data are expressed in grams of tension and represent the mean ± SE of eight tissues in each treatment group. **b.** The effect of clofilium (10^{-6} M) on the positive chronotropic response of spontaneously beating guinea pig atria to *l*-isoproterenol. Guinea pig atria were isolated and suspended in Ringer's bicarbonate buffer at 35 °C under a resting tension of 0.5 g. Each point represents the mean ± SE of six to eight atria for each group

conduction parameters, modest increases in cardiac contractility have been noted in anesthetized dogs (Steinberg and Michelson 1985) and conscious human subjects (Steinberg et al. 1984).

II. Effects on Refractoriness in Conscious and Anesthetized Dogs

Intravenously administered clofilium increases the ventricular effective refractory period (VERP) and prolongs the Q-T_c interval in barbiturate- and halothane-anesthetized dogs (Sullivan and Steinberg 1981). Of note, the Q-T_c interval is especially prolonged under halothane anesthesia. However, the maximal effects of clofilium are reduced considerably if the drug is administered to conscious animals and refractoriness is determined in the awake state (Fig. 2). These observations suggest that the expression of the class III activity of this agent in vivo may be modified by the activity of the CNS in general and the level of autonomic tone in particular.

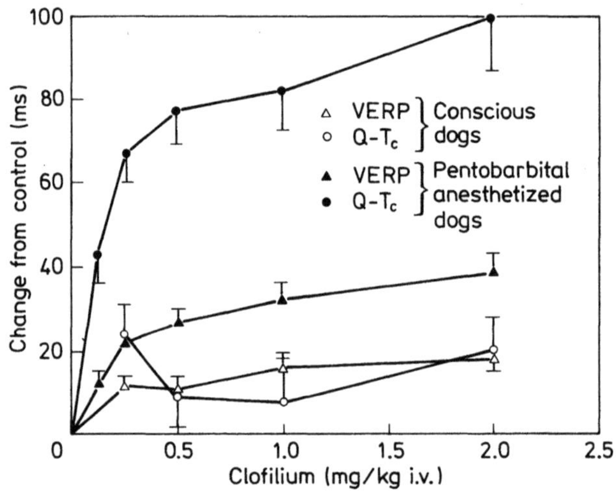

Fig. 2. Effect of clofilium on the ventricular effective refractory period (VERP) and rate-corrected Q-T interval (Q-T_c) of conscious and pentobarbital-anesthetized dogs. The Q-T_c interval (Bazett's formula) was derived from a lead II electrocardiogram. The VERP was determined in closed chest dogs by applying a premature stimulus (S_2) at varying intervals after every 20 pacing (150 beats/min) stimuli (S_1) through a 5F bipolar catheter in the right ventricular apex. The VERP was defined as the longest S_1-S_2 interval which failed to elicit a propagated impulse. After determining control values, clofilium was given cumulatively by intravenous bolus, and measurements were repeated 15 min after each dose. Each point represents the mean ± SE of ten experiments. The control Q-T_c intervals for conscious and pentobarbital-anesthetized dogs were 359 ± 5 and 368 ± 9 ms, respectively. The control VERPs for conscious and pentobarbital-anesthetized dogs were 174 ± 6 and 174 ± 4 ms, respectively

III. Autonomic Nerve Stimulation

It is well known that stimulation of the sympathetic nervous system can decrease basal refractory period (HAN et al. 1964; MARTINS and ZIPES 1980; LEVY 1984) and facilitate development of ventricular fibrillation (see LOWN and VERRIER 1976), but much less is known regarding the effects of autonomic nerve stimulation on drug-induced changes in refractoriness. To determine whether differences in autonomic tone might provide a basis for the differential effect of clofilium on refractoriness in conscious and anesthetized animals (see Sect. B.II), we studied the effect of sympathetic nerve stimulation on the ventricular refractory period before and after clofilium in pentobarbital-anesthetized dogs (Fig. 3). Before clofilium, stimulation of the dorsal and ventral branches of the decentralized right ansa subclavia caused a significant decrease in right endocardial VERP in saline-infused dogs (Fig. 3, left, control). Bilateral vagal stimulation in the absence of sympathetic stimulation did not influence refractoriness, but concurrent vagal stimulation antagonized the decrease in VERP caused by sympathetic stimulation (Fig. 3, left). Clofilium

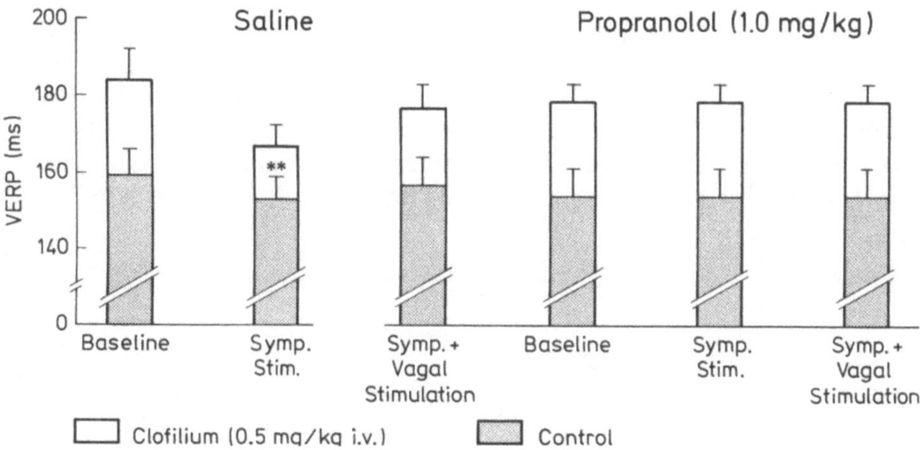

Fig. 3. Effect of autonomic nerve stimulation and clofilium on the ventricular effective refractory period (VERP) in the absense or presence of propranolol pretreatment in pentobarbital-anesthetized dogs. *d,l*-Propranolol was administered intravenously 10–15 min prior to the measurement of VERP in the propranolol-pretreated group ($n = 3$). The VERP was determined as in Fig. 2. The right stellate ganglion was stimulated with a train of impulses (10 Hz, 5 ms duration) of sufficient strength (2–10 V) to increase the heart rate to about 200 beats/min. The vagus was stimulated bilaterally with a train of impulses (30 Hz, 2 ms duration) of sufficient strength (2–6 V) to cause complete sinoatrial arrest. In the saline group ($n = 4$), the predrug VERP was less during sympathetic stimulation than during either baseline (no stimulation) or concurrent vagal and sympathetic stimulation ($p < 0.05$, one-way ANOVA). The clofilium-induced increase in the VERP was significantly less during sympathetic stimulation than during either baseline or concurrent sympathetic and vagal stimulation ($P < 0.05$, one-way ANOVA)

(0.5 mg/kg) increased the refractory period at baseline in the absence of autonomic stimulation (25 ± 2 ms), but during sympathetic stimulation, clofilium increased VERP only about half as much (14 ± 3 ms). Concurrent vagal stimulation antagonized the sympathetically mediated inhibition of clofilium's effect on refractoriness. In the presence of propranolol (1.0 mg/kg, Fig. 3, right), clofilium increased baseline VERP to the same extent as in saline-pretreated dogs (25 ± 3 ms); however, the ability of sympathetic stimulation to antagonize clofilium's effect was abolished. In agreement with Jaillon et al. (1980), propranolol did not increase basal VERP and thus it is unlikely that basal sympathetic tone to the ventricle is prominent in the barbiturate-anesthetized dog.

These results demonstrate that elevated sympathetic tone can antagonize the ability of clofilium to increase ventricular refractoriness and that vagal tone acts by modulating the effect of sympathetic stimulation. The permissive, rather than primary, role of vagal tone in modifying cardiac refractoriness has been emphasized previously, both in isolated tissue studies (Bailey et al. 1979) and in intact animals (Kolman et al. 1976; Martins and Zipes 1980). In conscious animals, sympathetic tone should be relatively high in comparison with barbiturate-anesthetized animals (see Peiss and Manning 1964). The marked attenuation of clofilium's effect in conscious dogs may indicate that sympathetic tone was elevated in these dogs to a greater extent than vagal tone. The idea that ventricular sympathetic tone is elevated in the conscious dog is consistent with studies showing a small but reproducible reduction in both VFT (Dawson et al. 1980) and ventricular refractoriness (Laforet et al. 1957) in conscious dogs compared with barbiturate-anesthetized dogs. The fact that propranolol potentiated the effect of clofilium on refractor-

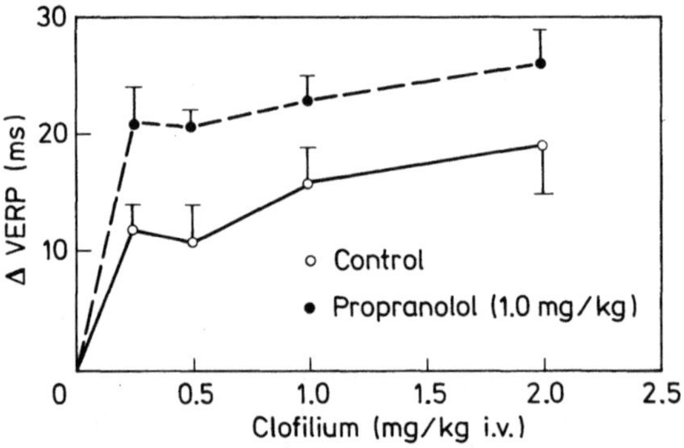

Fig. 4. Effect of clofilium on the ventricular effective refractory period (VERP) of conscious dogs in the absence or presence of propranolol pretreatment. VERP was determined as described in Fig. 2. Clofilium was given cumulatively as an intravenous bolus to ten control dogs and six dogs pretreated for 10–15 min with *d,l*-propranolol (1.0 mg/kg). The baseline values of VERP in control and propranolol-pretreated dogs prior to clofilium administration were 174 ± 6 and 171 ± 8 ms, respectively. The propranolol-pretreated group was significantly different from the control group at 0.5 mg/kg clofilium ($P < 0.05$, Student's t test)

iness in conscious dogs (Fig. 4) is further evidence that ventricular sympathetic tone is elevated in the conscious dog. It may be that stresses accompanying the experimental determination of cardiac refractoriness in conscious animals preferentially elevate sympathetic tone. The mechanism by which sympathetic tone influences the effect of clofilium is unknown, but beta-adrenoceptor stimulation shortens APD and it is noteworthy that the EC_{50} of clofilium in prolonging action potential duration in isolated Purkinje fibers is increased about seven fold in fibers exposed to low concentrations of norepinephrine (see STEINBERG et al. 1984).

IV. Arrhythmogenicity

The marked potentiation of the electrophysiological effects of clofilium in halothane-anesthetized dogs referred to in Sect. B.II provided an opportunity to study the relationship between isolated Q-T_c interval prolongation and the proarrhythmic potential of a pure class III agent. In preliminary studies we reported that serious ventricular arrhythmias could be induced in halothane-anesthetized dogs following acute intravenous administration of clofilium (STEINBERG et al. 1984). In an expanded study, we administered clofilium (0.5 mg/kg) to 50 halothane-anesthetized dogs by intravenous bolus and monitored the animals for 15 min, during which time the increase in the Q-T_c interval and VERP reached a steady state. No other significant cardiovascular or electrophysiological effects were noted. Table 1 shows that within 15 min after dosing, 26% of the animals developed spontaneous ventricular arrhythmias. No difference was noted in the baseline Q-T_c interval between the group developing arrhythmias and the group that was arrhythmia free. However, the predrug heart rate was lower in dogs that developed arrhythmias after clofilium (Table 1), and the Q-T_c interval increased more in response to clofilium

Table 1. The heart rate (HR) and Q-T_c interval response to clofilium in halothane-anesthetized dogs; relationship to arrhythmogenicity[a]

	Baseline		Post clofilium	
	Q-T_c (ms)	HR (beats/min)	ΔQ-T_c (ms)	ΔHR (beats/min)
Dogs developing arrhythmias ($n = 13$)[b]	287 ± 9	78 ± 3	150 ± 18[c]	-15 ± 3[c]
Dogs without arrhythmias ($n = 37$)	272 ± 5	96 ± 2	108 ± 8	-21 ± 1

[a] Dogs were anesthetized with halothane (1.5%–2.0% v/v in O_2) and were administered 0.5 mg/kg clofilium as an intravenous bolus. Over the next 15 min the dogs were observed for the appearance of arrhythmias.
[b] A proarrhythmic response was defined as the occurrence of more than five nonsinus beats during the 15-min period after clofilium administration.
[c] Significantly different from the arrhythmia-free group $P < 0.05$, Student's t-test.

in dogs with arrhythmias. The bradycardic response to clofilium was similar in both groups. Thus, clofilium-halothane arrhythmias were associated with a drug-induced increase in the Q-T_c interval of about 150 ms and were more likely to occur in dogs with initially low heart rates. No proarrhythmic effects were noted in normal, conscious dogs (in which, as shown in Fig. 2, the Q-T_c interval increased only 10–30 ms), in animals anesthetized with barbiturate (in which the Q-T_c interval increased about 80 ms and the baseline heart rate was high) or under chloralose anesthesia (in which the baseline heart rate was low but the Q-T_c interval increased even less than under barbiturate anesthesia). Others (KOPIA et al. 1985) have reported proarrhythmic responses in conscious or barbiturate-anesthetized dogs in the presence of chronic myocardial ischemia following rather large doses of clofilium (2 mg/kg, i.v.). Such doses are considerably higher than the 0.1–0.5 mg/kg doses that produce nearly maximal increases in ventricular refractoriness in both dogs and humans (see Figs. 2, 5).

The clofilium-halothane arrhythmias appear similar to the bradycardia-dependent ventricular arrhythmias that occur after the administration of agents known markedly to delay ventricular repolarization and increase the Q-T interval. For example, ventricular tachycardia resembling the torsade de pointes pattern was noted after administration of cesium salts to normal dogs (BRACHMANN et al. 1983; LEVINE et al. 1985) or *N*-acetylprocainamide to dogs with chemically induced AV nodal block (DANGMAN and HOFFMAN 1981). These arrhythmias were also induced in isolated tissues and appeared to result from triggered automaticity due to drug-induced early afterdepolarizations. The excessive prolongation of the Q-T_c interval associated with these agents may be indicative of *nonuniform* ventricular repolarization (see KENNY and SUTTON 1985), which (especially in the presence of bradycardia) can predispose to reentry, delayed afterdepolarizations, or both (HAN et al. 1966; LEVINE et al. 1985). The nonuniformity may be more important than the absolute delay of repolarization, but the precise nature of clofilium-mediated arrhythmias is unknown. It is difficult, for example, to differentiate between abnormal automaticity arising from depolarized cells and early afterdepolarizations arising from plateau potentials (DAMIANO and ROSEN 1984). However, in unpublished studies, we have noted that overdrive ventricular pacing can abolish clofilium arrhythmias as would be expected for triggered mechanisms arising from more negative potentials (LEVINE et al. 1985).

V. Programmed Electrical Stimulation in Patients

Clinical reports have described the class III properties of intravenously administered clofilium and its efficacy in programmed stimulation protocols (GREENE et al. 1983a; PLATIA and REID 1984). The electrophysiological effects of clofilium have been studied in a total of 75 patients in various clinical centers. Figure 5 shows the dose-response relationship for clofilium on the ventricular effective refractory period and Q-T_c interval in patients undergoing programmed stimulation protocols. Since there were no changes in cardiac conduction as evidenced by a lack of effect on the QRS duration, P-R interval, or intracardiac H-V and A-H intervals, a prolongation of ventricular recovery times most likely explains the increase in ventricular refractory period. In 19 patients with sustained ventricular tachycardia (VT) induced before clofilium, 8 patients (42 %) were either noninducible after clofilium or were converted to nonsustained VT (mean dose of 160 µg/kg). On the other hand, 6 of 75 patients (8 %) experienced proarrhythmic responses at a mean dose of 197 µg/kg. Clofilium's effect on both refractory period and Q-T_c interval in man is more analogous to the effect of the drug in pentobarbital-anesthetized dogs than in conscious dogs (cf. Figs. 2, 5). The basis for the differential response to clofilium in conscious dogs and humans is unclear. If basal ventricular sympathetic tone were in fact low in humans compared with the dog, then the antagonism of clofilium's effects by sympathetic stimulation (Fig. 3) could provide an explanation for the reduced effectiveness of clofilium in conscious dogs compared with conscious humans. Programmed stimulation studies in humans have in fact demonstrated that vagal tone predominates over sympathetic tone in determining background right ventricular refractoriness (PRYSTOWSKY 1981) and directly increases the resting Q-T interval (BROWNE et al. 1982). Differences in the level of basal autonomic tone among patients might also explain why clofilium precipitates arrhythmias in some patients. In these patients, the response to the drug may be more analogous to the effects of clofilium in halothane-anesthetized dogs (i.e., large increases in the Q-T_c interval relative to refractoriness). Heterogeneity of repolarization may widen as both sympathetic background and drug concentration rise. Un-

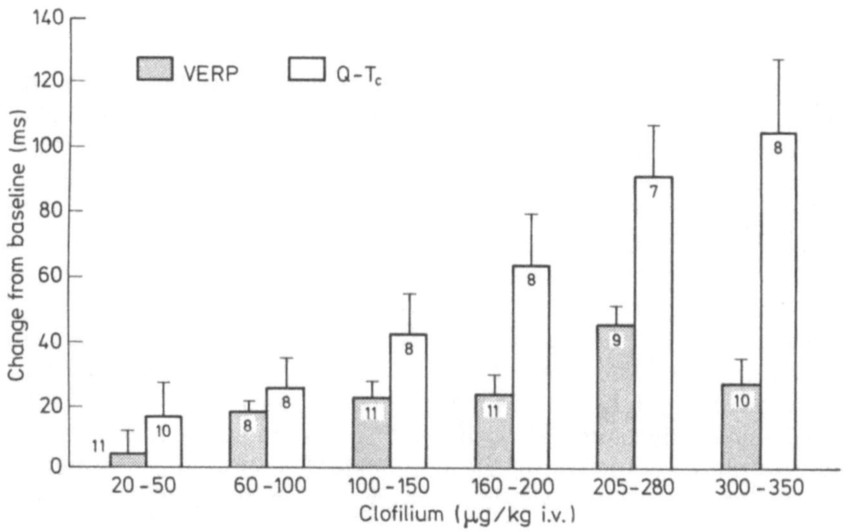

Fig. 5. Effect of clofilium on the ventricular effective refractory period (VERP) and the Q-T_c interval of patients undergoing programmed electrical stimulation testing. VERP was determined by the extrastimulus technique and Q-T_c by Bazett's formula. Each patient received only one dose of clofilium. The number of patients in each dosage interval is indicated in the figure

fortunately, not enough patients had both proarrhythmic effects and complete Q-T_c interval and refractory period measurements to investigate this point further. In any case, the proarrhythmic risk in this relatively small group of patients treated with clofilium appears to be similar to that described for other antiarrhythmic drugs (PODRID 1985; TORRES et al. 1985). The different responses to clofilium in conscious dogs and in humans suggest that conscious dogs may not be the most appropriate model for evaluation of the electrophysiological or antifibrillatory effects of class III drugs as has been proposed (see LUCCHESI and LYNCH 1986). Similar caution regarding the suitability of the conscious dog as a model for VT induction by programmed electrical stimulation was recently noted (ECHT et al. 1983).

C. LY190147, a Tertiary Amine Class III Agent

I. Basic Electrophysiology

LY190147 is a newly developed tertiary amine class III agent whose molecular structure is indicated in Fig. 6. This agent is similar in potency to clofilium in prolonging the action potential duration and refractory period of canine Pur-

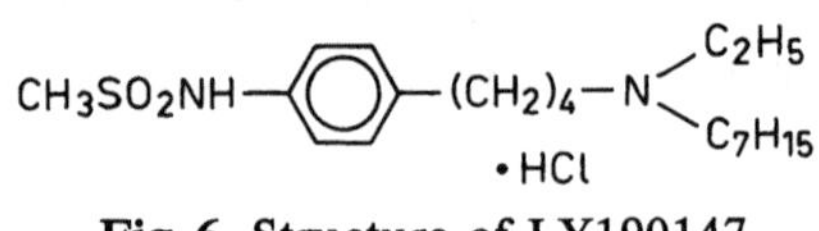

Fig. 6. Structure of LY190147

kinje fibers (SMALLWOOD et al. 1986). Following administration of LY190147 to halothane-anesthetized dogs by intravenous bolus, both the ventricular refractory period and the Q-T_c interval increased (Fig. 7) without effects on blood pressure, heart rate, or ventricular conduction time. Like clofilium, there were no effects on paced or unpaced His-ventricular (HV) conduction intervals as determined by intracardiac recording techniques (Table 2). Under halothane anesthesia, the maximum increase in the Q-T_c interval (approximately 100 ms) occurred after a dose of 1.0 mg/kg; further increases in the dose caused a marked fall in the Q-T_c interval toward control values and a small decrease in refractory period (Fig. 7). In contrast, the maximal Q-T_c interval increase after clofilium was approximately 150 ms (see Table 1). Thus, LY190147 selectively increases cardiac refractoriness but at high doses does not increase the Q-T_c interval to the same extent as clofilium. Furthermore, unlike clofilium, proarrhythmic effects were not observed in the halothane model.

A possible cellular basis for this apparent biphasic effect of LY190147 was investigated in isolated tissues. Low concentrations of the drug (up to 3×10^{-7} *M*) selectively increased action potential duration without affecting the resting potential, conduction time, or maximal upstroke velocity (Fig. 8, middle). At higher concentrations (10^{-6} *M* and 3×10^{-6} *M*), action potential

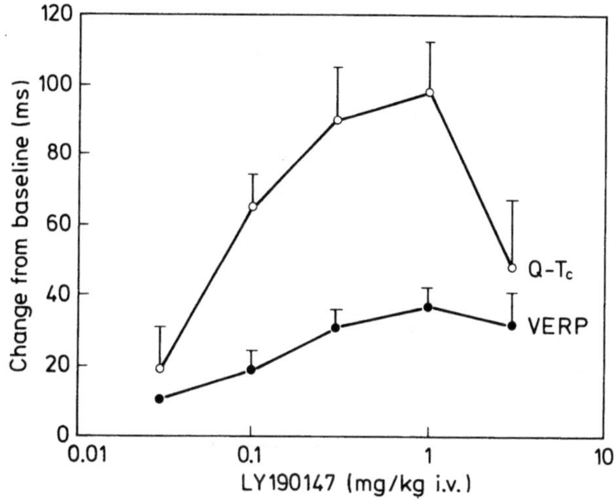

Fig. 7. Effect of LY190147 on the ventricular effective refractory period (VERP) and Q-T_c interval of halothane-anesthetized dogs. Dogs were anesthetized with 1.5% halothane in oxygen. VERP and Q-T_c were determined as described in Fig. 2. LY190147 was administered cumulatively by intravenous bolus, and 15 min after each dose measurements were repeated. The data represent the mean change from baseline ±SE of from two to seven experiments at each dose. The mean baseline values for the VERP and Q-T_c interval were 175 ± 3 and 358 ± 10 ms, respectively. Values were significantly greater than baseline values at all doeses except at 0.03 mg/kg (paired *t*-test, $P < 0.05$)

Table 2. The lack of effect of saline, clofilium, or LY190147 on infra-His conduction in chloralose-anesthetized dogs

	H-V interval[a] (Percent change from control)	
	Unpaced	Paced
Saline ($n = 8$)	1.3 ± 0.8	0.3 ± 1.0
Clofilium[b] 1 mg/kg ($n = 8$)	0.0 ± 1.0	1.6 ± 0.9
LY190147[b] 1 mg/kg ($n = 8$)	−1.1 ± 0.9	−1.3 ± 1.7

[a] The His-bundle electrogram was recorded by placing a bipolar catheter into the aorta, near the aortic valve. The H-V interval was defined as the interval from the beginning of the His-bundle depolarization to the beginning of the ventricular depolarization, and represents the conduction time through the His-Purkinje system. The baseline value for both the unpaced and paced H-V intervals in the saline group was 27 ± 2 ms. The baseline values were not significantly different among groups.

[b] The dose indicated was infused over a 60-min period.

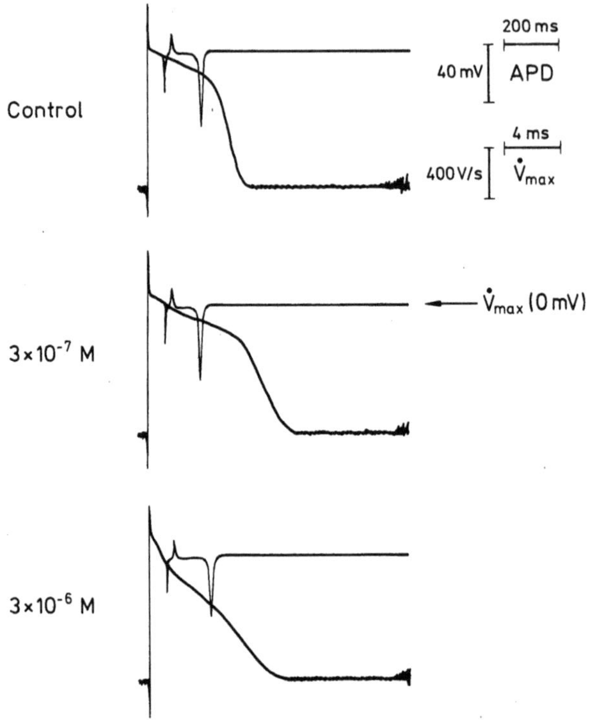

Fig. 8. Effect of LY190147 on the action potential of a canine Purkinje fiber cell. The downward deflection from the 0 mV baseline is the derivative of the upstroke velocity (V/s). The horizontal and vertical scales for the action potential duration (APD) and maximal upstroke velocity ($\dot{V}_{max}$) are shown in the figure. Note that at 3×10^{-7} *M (middle)* there is a symmetrical prolongation of the action potential duration with no significant change in the upstroke velocity or conduction time from the stimulus artifact. At 3×10^{-6} *M (bottom)* the action potential duration shortens and the upstroke velocity decreases along with an increase in the conduction time

duration decreased toward control, maximal upstroke velocity decreased and conduction time increased (Fig. 8, bottom). In eight experiments at 3×10^{-6} *M*, LY190147 significantly decreased upstroke velocity by 18% ± 3% and increased conduction time by 12% ± 5% without changing the action potential duration at 70% or 95% of full repolarization. Judging from its effects on upstroke velocity and conduction, the return of the action potential duration toward control at higher concentrations of LY190147 likely results from inhibition of the slowly inactivating sodium current present during plateau potentials (CARMELIET and SAIKAWA 1982; COLATSKY 1982).

These in vitro results parallel the biphasic effects of the drug on Q-T_c interval in vivo, suggesting that the biphasic effects of LY190147 may result from typical class III effects at low concentrations and from an inhibition of sodium conductance (class I effect) at high concentrations. Of course, further work on the effects of LY190147 in isolated ventricular muscle is required before concluding that a causal relation exists between action potential duration measured in vitro and the Q-T interval. In fact, the relation between the Q-T interval and ventricular action potential duration is rather complex (see VAUGHAN WILLIAMS 1985b). The Q-T interval may be a better indicator of differences in recovery times in various parts of the heart (homogeneity of repolarization) than of individual action potential durations (SURAWICZ and KNOEBEL 1984; RODEN and WOOSLEY 1985). Regardless of the precise relation between the Q-T interval and action potential duration, however, excessively large increases in the Q-T interval or in the QT/VERP ratio by class III drugs should be avoided. Thus, to the extent that an abnormally prolonged Q-T_c interval reflects the proarrhythmic potential of class III agents, the presence of some class I activity may be desirable to limit excessive delays in ventricular repolarization. Slowing conduction within the reentrant loop might appear to negate the desired class III electrophysiological effect (see WELLENS et al. 1984). However, it may be possible to antagonize excess prolongation of action potential duration without affecting maximal rate of depolarization and conduction since only a slight inhibition of plateau sodium current is required to shorten action potential duration due to the nonlinear relation between current and voltage at plateau potentials (see CARMELIET and VEREECKE 1979).

II. Ventricular Fibrillation Threshold

If refractoriness can be increased without excessively increasing the Q-T_c interval, the therapeutic index of LY190147 might be improved relative to other

Table 3. The effect of various class III agents on the ventricular fibrillation threshold (VFT), the Q-T_c interval, and the VFT/Q-T_c ratio in pentobarbital-anesthetized dogs[a]

	Percentage change from baseline		
	VFT[b]	Q-T_c	VFT/Q-T_c
Saline	−7.1 ± 6.6	−0.2 ± 1.7	−7.0 ± 6.4
LY190147 (1 mg/kg, i.v.)	46.9 ± 21.8[c]	9.9 ± 3.1[c]	34.7 ± 21.3[c]
d-Sotalol (5 mg/kg, i.v.)	20.2 ± 9.2	13.8 ± 1.3[c]	5.4 ± 7.7
Bretylium (10 mg/kg, i.v.)	28.8 ± 11.5	6.4 ± 2.1[c]	22.0 ± 12.7

[a] Each agent was administered as a 15-min infusion. The values shown were taken 3 h after the end of the infusion and represent the mean percentage change from baseline ± SE of six experiments. The control values for VFT, Q-T_c, and VFT/Q-T_c in the saline group were 21 mA, 351 ms, and 60 mA/s, respectively. There were no differences in baseline values among groups

[b] To determine the VFT, the vulnerable period of the T wave was scanned at 2-ms increments by a premature stimulus (S_2) of 10-ms pulse width delivered after every 15 paced beats. The vulnerable period was defined as the range of S_1-S_2 intervals in which an extrasystole occurred following the delivery of S_2. Each time the vulnerable period was scanned, the current of the S_2 was increased by 1 mA until fibrillation occurred

[c] Significantly different from the saline group. $P < 0.05$, Fisher's LSD

class III agents. To explore this possibility further, we examined the effects of the class III agents *d*-sotalol, bretylium, and LY190147 on the ventricular fibrillation threshold (VFT) using the single extrastimulus method (Table 3). The doses of *d*-sotalol and bretylium were chosen to approximate those previously reported to elicit near maximal antiarrhythmic or antifibrillatory activity in dogs (ANDERSON et al. 1980; LYNCH et al. 1984). Ventricular fibrillation threshold measurements were made at the time of maximum effect (3 h after infusion of each drug). LY190147 (1.0 mg/kg) increased the VFT by nearly 47% and the Q-T_c interval by only about 10%, causing a significant increase in the VFT/Q-T_c ratio. Bretylium (10 mg/kg) also increased the VFT and Q-T_c intervals but the VFT/Q-T_c ratio did not increase as much as after LY190147. *d*-Sotalol (5 mg/kg) increased both parameters to about the same extent, resulting in little change in the VFT/Q-T_c ratio. Thus, in viewing the VFT/Q-T_c ratio as a "therapeutic index" for class III drugs, LY190147, with its potent class III activity and self-limiting effects on Q-T_c interval, may possess some therapeutic advantages. However, especially in the presence of myocardial ischemia, there is not necessarily a correlation among Q-T_c interval, refractoriness, and efficacy for class III agents (KOPIA et al. 1985; KOWEY et al. 1985).

D. Other Class III Agents

Numerous substances with diverse molecular structures possess class III activity. The pharmacological and electrophysiological properties of some of these have been reviewed (BEXTON and CAMM 1982; THOMIS and TENTHOREY 1983; STEINBERG and MICHELSON 1985; STEINBERG et al. 1986). The class III effects of amiodarone and sotalol were discussed in the previous two chapters.

I. Bretylium and Congeners

The mechanism(s) for the antiarrhythmic and antifibrillatory activity of bretylium continues to generate considerable debate. Bretylium is known to possess direct class III membrane effects in the ventricle (see PATTERSON and LUCCHESI 1983; LUCCHESI 1984; SASYNIUK 1984; ANDERSON 1985); however, the ionic mechanism(s) for these effects is largely unknown. Recent studies indicated that very high concentrations in vitro (1 m*M*) blocked both sodium and potassium channels in squid axon or isolated chicken heart cells. Moreover, a direct relationship between antifibrillatory effects and potassium channel block has been suggested (BACANER et al. 1986). Conversely, recent studies in vivo have concluded that the acute antifibrillatory effect of bretylium in dogs (FRAME and WANG 1986) and cats (KOWEY et al. 1985) is dependent mainly on the block of sympathetic transmission and not on effects on refractory period per se. The role of newly synthesized catecholamines in the occurrence of VF due to acute coronary artery occlusion was also recently emphasized (ABRAHAMSSON et al. 1985). Studies in vitro have demonstrated that the depolarization and conduction block associated with acute hypoxia were reversible by bretylium acting through a catecholamine release-dependent mechanism (NISHIMURA and WATANABE 1983). In clinical studies, bretylium has been shown to elevate plasma catecholamine levels after slow intravenous doses of 2-10 mg/kg without changing the Q-T interval (DUFF et al. 1985). Likewise, in patients undergoing programmed stimulation, neither acute (BAUERNFEIND et al. 1983) nor chronic (GREENE et al. 1983b) bretylium suppressed induced tachycardia or altered ventricular refractoriness. The lack of effect on ventricular refractoriness or QT interval after acute dosing in clinical studies (see ANDERSON 1985) suggests that the acute antifibrillatory effects of bretylium in man are probably only partly related to direct class III activity.

Other recent experimental reports on the ability of bretylium to reduce electrophysiological disparity are also at least partly explainable by the antisympathetic actions of the drug (which follow the brief immediate release of transmitter). For example, bretylium increased homogeneity of both conduction velocity and excitability across the boundary between normal and ischemic myocardium by slowing conduction in normal tissue and increasing excitability in ischemic tissue (FUJIMOTO et al. 1983). Bretylium reduced the disparity of refractory periods after acute coronary artery occlusion in cats (KOWEY et al. 1985) and reduced the disparity of refractory periods caused by quinidine in dogs (INOUE et al. 1985a). Both the decrease in refractory period accompanying acute coronary artery occlusion and the overshoot upon reperfusion were reduced if bretylium was administered for 24 h prior to the occlusion (GIBSON et al. 1983). Chronic bretylium therapy prevented reperfusion arrhythmias in this model, but acute therapy (1 h before occlusion) was ineffective. Pretreatment with bretylium for 90 min before coronary artery occlusion reduced reperfusion arrhythmias only if the area of myocardium at risk was considered in detail (WENGER et al. 1984). *Acute* administration of bretylium (phase of transmitter release) also failed to reduce the defibrillation threshold in anesthetized dogs (KOO et al. 1984; DAWSON et al. 1985; CHO et al. 1986). But chronic bretylium dosing schedules (10 mg/kg daily for 2 days) prevented VT induction by programmed stimulation in dogs with established ischemia, and markedly decreased mortality in dogs with infarcts subjected to acute thrombus formation (HOLLAND et al. 1983). Thus, the antifibrillatory and antiarrhythmic activity of bretylium is probably due to a mixture of both direct (class III) and indirect effects, with the greatest efficacy apparent in those models in which the sympathetic nervous system is involved in the maintenance or initiation of arrhythmia. The apparently greater efficacy of chronically administered bretylium is consistent with this notion.

Bethanidine is an adrenergic blocking drug which, because of improved oral absorption, has been proposed as an alternative to bretylium (BACANER 1983b). Class III activity was shown in guinea pig papillary muscle in vitro (INOUE 1985b) and in rat myocardium in vivo and in vitro (NORTHOVER 1985). However, intravenous bethanidine failed to increase refractoriness and was only marginally effective against programmed stimulation arrhythmias or sudden death in conscious dogs with chronic ischemia (PATTERSON et al. 1984). Likewise, the drug failed to prolong refractoriness and prevent electrically induced tachycardia in patients with a history of ventricular fibrillation (DI-

Marco et al. 1984). Another study in a similar patient group demonstrated the effectiveness of bethanidine against programmed stimulation arrhythmias in the absence of any change in refractory periods (Benditt et al. 1984). A high incidence of hypotension, however, was reported in both clinical studies. Meobentine is a bethanidine congener with less likelihood of interacting with the sympathetic nervous system (Wastila et al. 1981). Meobentine elicited class III activity in rats (Northover 1985) and markedly increased the ventricular fibrillation threshold in normal and ischemic canine myocardium (Zimmerman et al. 1984). However, meobentine, like bethanidine, failed to increase refractory periods or to control arrhythmias after ischemic injury in dogs (Zimmerman et al. 1984). In patients, meobentine only slightly increased refractoriness and showed marginal effectiveness against electrically induced arrhythmias (Anderson et al. 1985). In a group of patients with frequent ectopy, no antiarrhythmic effects were noted after oral doses of up to 20 mg/kg (Duff et al. 1984). These clinical studies both reported a high incidence of gastrointestinal and hypotensive side effects.

UM 301, a quaternary ammonium agent with some structural similarity to bretylium, was recently described (Gibson et al. 1986). This agent increased atrial and ventricular refractoriness and elevated ventricular fibrillatory threshold at doses (5–10 mg/kg) that lacked negative hemodynamic effects and caused only little or no depression of myocardial conduction velocity in normal tissues. UM 301 was also effective in vivo against ouabain arrhythmias and arrhythmias arising spontaneously in the 2-day postinfarct dog. The effectiveness of UM 301, a relatively selective class III agent, against these automatic arrhythmias is of interest and requires further investigation.

II. Agents with Mixed Electrophysiological Activity

Bepridil is a calcium entry blocking agent with potent antiarrhythmic activity in animals and man (Flaim and Cummings 1986). In dogs with chronic infarcts, intravenous bepridil slowed intramyocardial conduction velocity and increased the Q-T_c interval and refractoriness (Lynch et al. 1985b). In patients, bepridil was effective against programmed stimulation induced arrhythmias in doses that produced relatively selective class III activity (Prystowsky 1985; Somberg et al. 1985). However, bepridil had little effect on the action potential duration of isolated Purkinje fibers, decreased the upstroke velocity of ouabain-intoxicated fibers (Kane et al. 1983), and in isolated cardiac myocytes blocked both sodium and calcium but not potassium currents (Yatani et al. 1986). Thus, the electrophysiological basis for its apparent class III effects in vivo remains obscure. AQ-A39 is a verapamil analog with calcium antagonistic activity that also possesses pronounced class III effects in vitro (Hohnloser et al. 1982).

The electrophysiological effects of the selective alpha$_1$-antagonist, indoramin, were studied in isolated guinea pig myocardial preparations and in anesthetized cats (Harron 1986). Indoramin produced bradycardia, and increased the S-T interval and refractory period without altering QRS duration in isolated beating hearts. Similar effects, as well as hypotension, were seen in anesthetized cats (6 mg/kg). However, 10 mg/kg also increased QRS duration, P-Q interval, and diastolic threshold. Only the higher dose significantly reduced the incidence of fibrillation in rats subjected to coronary artery occlusion. In one study, indoramin also lowered blood pressure and increased ventricular refractory period in patients (BUTROUS and CAMM 1986); however, others have reported no important effects on either Q-T interval or ventricular refractoriness (MANZ et al. 1986).

Melperone is a neuroleptic that has been studied extensively in regard to its class III electrophysiological effects (see PLATOU et al. 1982). Recently this agent has also been found to possess marked class I activity in isolated rabbit cardiac tissues in concentrations that also show class III effects (MILLAR and VAUGHAN WILLIAMS 1983). The psychotomimetic and neuronal potassium channel antagonist phencyclidine was reported to possess fairly selective class III effects in amphibian isolated cardiac tissues (D'AMICO et al. 1983). However, recent studies using guinea pig and rat tissues have also reported decreases in upstroke velocity at concentrations that increase the action potential duration (TEMMA et al. 1985). The latter investigators also postulated that class III activity may be at least partly due to an increase in the slow inward current. The class Ia antiarrhythmic agents, pirmenol and quinidine, also demonstrated pronounced class III activity in vitro. Pirmenol increased action potential durations of rabbit conducting tissues in concentrations that also depressed the maximum upstroke velocity (DUKES et al. 1986). Relatively low concentrations of quinidine (1×10^{-6} *M*) selectively increased action potential duration in isolated canine Purkinje fibers (RODEN and HOFFMAN 1985). The class III effects were potentiated at long cycle lengths and low extracellular potassium concentration; early afterdepolarizations were also noted in some preparations. Studies with somewhat higher concentrations of quinidine in rabbit Purkinje fibers had earlier shown the increase in action potential duration to be due to a block of the delayed rectifier current (COLATSKY 1982). To what extent, if any, the class III activity of pirmenol or quinidine contribute to their antiarrhythmic efficacy or potential proarrhythmic liability requires additional study. The N-acetyl metabolite of procainamide (NAPA) is well known to possess class III activity both in vitro and in vivo (see KEEFE et al. 1981) More recent clinical reports have demonstrated relatively selective class III effects of intravenous NAPA in programmed stimulation studies (SUNG et al. 1983) and antiarrhythmic efficacy in patients treated orally for periods up to 4 years (ATKINSON et al. 1983). Like procainamide, however, NAPA possesses significant ganglionic blocking activity (REYNOLDS and GORCZYNSKI 1980; PEARLE et al. 1983).

E. Summary and Conclusions

Data summarized here and elsewhere demonstrate that selective class III agents possess potent antiarrhythmic properties in animal models as well as in man. However, these agents also possess a proarrhythmic liability possibly related to their propensity to increase heterogeneity of repolarization. Whether antiarrhythmic or proarrhythmic effects are manifest depends on the abnormal cardiac substrate that is responsible for initiating or maintaining the arrhythmia (WIT and ROSEN 1983; ROSEN and WIT 1983). So far the electrophysiological properties of only a few selective class III drugs have been described in sufficient detail to suggest the pharmacological basis for their antiarrhythmic or proarrhythmic activity. New drugs with selective class III activity are under active investigation (REISER and SULLIVAN 1986), and the trend toward electrophysiological selectivity in new drug design will doubtless continue.

In this chapter, we discussed some of the factors that might determine the net effect of a selective class III agent administered in vivo. The results outlined in Sect. B.III demonstrate that sympathetic nerve stimulation can antagonize the effect of clofilium on ventricular refractoriness. Thus, the differential effect of clofilium on the Q-T_c interval and VERP in conscious and anesthetized dogs (Fig. 2), and the differing responses to clofilium in conscious humans versus conscious dogs (Figs. 2, 5), could involve differences in the balance between parasympathetic and sympathetic tone. The intensity of the Q-T interval effect may also be influenced by pharmacokinetic variables. For example, although ventricular refractoriness can be increased by clofilium to the same extent after oral or intravenous dosing, less of an increase in the Q-T_c interval occurs after the former (STEINBERG et al. 1984). One could hypothesize that in the case of clofilium, with its extremely high myocardial affinity (LINDSTROM et al. 1982), rapid intravenous dosing could result in marked disparities in regional distribution and hence a proarrhythmic heterogeneity of myocardial repolarization. Dosing methods that provide for more gradual delivery of drug to the myocardium (e.g., oral dosing or prolonged intravenous infusion) may increase the therapeutic index of the compound by increasing homogeneity of repolarization. Finally, as discussed in Sect. C, the presence of some class I activity inherent in LY190147 provides two classes of antiarrhythmic action. Blockade of plateau sodium current limits action potential prolongation as concentration is raised, giving this particular class III drug a "built-in" limit to Q-T interval changes.

References

Abrahamsson T, Almgren O, Carlsson L, Svensson L (1985) Antiarrhythmic effect of reducing myocardial noradrenaline stores with α-methyl-metatyrosine. J Cardiovasc Pharmacol 7:S81–S85

Allessie MA, Bonke FIM, Schopman FJG (1977) Circus movement in rabbit atrial muscle as a mechanism of tachycardia. III. The "leading circle" concept: a new model of circus movement in cardiac tissue without the involvement of an anatomical obstacle. Circ Res 41:9–18

Anderson JL (1984) Antifibrillatory versus antiectopic therapy. Am J Cardiol 54:7A–13A

Anderson JL (1985) Bretylium tosylate: profile of the only available class III antiarrhythmic agent. Clin Ther 7:205–224

Anderson JL, Patterson S, Conlon M, Pasyk S, Pitt B, Lucchesi BR (1980) Kinetics of antifibrillatory effects of bretylium: correlation with myocardial drug concentration. Am J Cardiol 46:583-592

Anderson JL, Reid PR, Platia EV, Akhtar M, Ruskin JN, Schaal SF, Jueng P, Long RA, Wenger TL (1985) Meobentine sulfate: antiarrhythmic and electrophysiologic effects assessed by programmed electrical stimulation and ambulatory monitoring in patients with complex ventricular tachyarrhythmia. Am Heart J 110:774-784

Atkinson AJ, Lertora JJL, Kushner W, Chao GC, Nevin MJ (1983) Efficacy and safety of N-acetylprocainamide in long-term treatment of ventricular arrhythmias. Clin Pharmacol Ther 33:565-576

Bacaner MB (1983a) Prophylaxis and therapy of ventricular fibrillation: bretylium reviewed and lidocaine refuted. Int J Cardiol 4:133-152

Bacaner MB (1983b) Bethanidine sulfate as an antifibrillatory drug. In: Scriabine A (ed) Cardiovascular drugs. Raven, New York p 69 (New drugs annual, vol I)

Bacaner MB, Clay JR, Shrier A, Brochu RM (1986) Potassium channel blockade: a mechanism for suppressing ventricular fibrillation. Proc Natl Acad Sci 83:2223-2227

Bailey JC, Watanabe AM, Besch HR, Lathrop DA (1979) Acetylcholine antagonism of the electrophysiological effects of isoproterenol on canine cardiac Purkinje fibers. Circ Res 44:378-383

Bauernfeind RA, Hoff JV, Swiryn S, Palileo E, Strasberg B, Scagliotti D, Rosen K (1983) Electrophysiologic testing of bretylium tosylate in sustained ventricular tachycardia. Am Heart J 105:973-980

Benditt DG, Benson DW, Dunnigan A, Kriett JM, Pritzker MR, Bacaner MB (1984) Antiarrhythmic and electrophysiologic actions of bethanidine sulfate in primary ventricular fibrillation or life-threatening ventricular tachycardia. Am J Cardiol 53:1268-1274

Bexton RS, Camm AJ (1982) Drugs with class III antiarrhythmic action. Pharmacol Ther 17:315-355

Boura ALA, Green AE (1959) The actions of bretylium: adrenergic neurone blocking and other effects. Br J Pharmacol 14:536-548

Boyden PA (1986) Effects of pharmacologic agents on induced atrial flutter in dogs with right atrial enlargement. J Cardiovasc Pharmacol 8:170-177

Brachmann J, Scherlag BJ, Rosenshtraukh LV, Lazzara R (1983) Bradycardia dependent triggered activity: relevance to drug-induced multiform ventricular tachycardia. Circulation 68:846-856

Browne KF, Zipes DP, Heger JJ, Prystowsky EN (1982) Influence of the autonomic nervous system on the Q-T interval in man. Am J Cardiol 50:1099-1103

Browne KF, Prystowsky E, Heger JJ, Zipes DP (1983) Modulation of the Q-T interval by the autonomic nervous system. PACE 6:1050-1056

Butrous GS, Camm AJ (1986) Clinical cardiac electrophysiological assessment of indoramin. J Cardiovasc Pharmacol 8:S137-S143

Carmeliet E (1985) Electrophysiologic and voltage clamp analysis of the effects of sotalol on isolated cardiac muscle and Purkinje fibers. J Pharmacol Exp Ther 232:817-825

Carmeliet E, Saikawa T (1982) Shortening of the action potential and reduction of pacemaker activity by lidocaine, quinidine and procainamide in sheep cardiac Purkinje fibers. An effect of Na or K currents? Circ Res 50:257-272

Carmeliet E, Vereecke J (1979) Electrogenesis of the action potential and automaticity. In: Berne RM, Sperelakis N, Geiger SR (eds) The cardiovascular system. American Physiological Society, Bethesda p 269 (Handbook of physiology, vol I)

Chow MSS, Kluger J, Lawrence R, Feldman A (1986) The effect of lidocaine and bretylium on the defibrillation threshold during cardiac arrest and cardiopulmonary resuscitation. Proc Soc Exp Biol Med 182:63-67

Colatsky TJ (1982) Mechanisms of action of lidocaine and quinidine on action potential duration in rabbit cardiac Purkinje fibers. An effect on steady state sodium currents. Circ Res 50:17-27

Cranefield PF (1975) The conduction of the cardiac impulse. Futura, Mt Kisco, pp 31-46

Damiano BP, Rosen MR (1984) Effects of pacing on triggered activity induced by early afterdepolarizations. Circulation 69:1013-1025

D'Amico GA, Kline RP, Maayani S, Weinstein H, Kupersmith J (1983) Effects of phencyclidine on cardiac action potential: pH dependence and structure activity relationships. Eur J Pharmacol 88:283-290

Dangman KH, Hoffman BF (1981) In vivo and in vitro antiarrhythmic and arrhythmogenic effects of N-acetyl procainamide. J Pharmacol Exp Ther 217:851-862

Dawson AK, Leon AS, Taylor HL (1980) Effect of pentobarbital anesthesia on vulnerability to ventricular fibrillation. Am J Physiol 237:H427-H431

Dawson AK, Steinberg MI, Shapland JE (1985) Effect of class I and class III drugs on current and energy required for internal defibrillation. Circulation 72 (Suppl III):384

DiMarco JP, Sellers TD, Shipe JR, Savory J, Spyker DH (1984) Acute electrophysiologic effects of bethanedine sulfate in patients with ventricular tachycardia or fibrillation. Am Heart J 108:1244-1249

Duff HJ, Oates JA, Roden DM, Woosley RL (1984) The antiarrhythmic activity of meobentine sulfate in man. J Cardiovasc Pharmacol 6:650-656

Duff HJ, Roden DM, Yacobi A, Robertson D, Wang T, Maffucci RJ, Oates JA, Woosley RL (1985) Bretylium: relations between plasma concentrations and pharmacologic actions in high-frequency ventricular arrhythmias. Am J Cardiol 55:395-401

Dukes ID, Vaughan Williams EM, Dennis PD (1986) Electrophysiological and cardiovascular effects of pirmenol, a new class 1 antiarrhythmic drug. J Cardiovasc Pharmacol 8:227-234

Echt DS, Griffin JC, Ford AJ, Knutti JW, Feldman RC, Mason JW (1983) Nature of inducible ventricular tachyarrhythmias in a canine chronic myocardial infarction model. Am J Cardiol 52:1127-1132

Feld GK, Venkatesh N, Singh BN (1986) Pharmacologic conversion and suppression of experimental canine atrial flutter: differing effects of *d*-sotalol, quinidine and lidocaine and significance of changes in refractoriness and conduction. Circulation 74:197-204

Flaim SF, Cummings DM (1986) Bepridil hydrochloride: a review of its pharmacologic properties. Curr Ther Res 39:568-597

Frame VB, Wang H-H (1986) Importance of interaction with adrenergic neurons for antifibrillatory action of bretylium in the dog. J Cardiovasc Pharmacol 8:336-345

Fujimoto T, Hamamoto H, Peter T, McCullen A, Melvin N, Mandel WJ (1983) Electrophysiologic effects of bretylium on canine ventricular muscle during acute ischemia and reperfusion. Am Heart J 105:966-972

Gibson JK, Stewart JR, Li Y-P, Lucchesi BR (1983) Electrophysiologic effects of bretylium tosylate on the canine heart during coronary artery occlusion and reperfusion. J Cardiovasc Pharmacol 5:517-524

Gibson JK, Patterson E, Lucchesi BR (1986) Electrophysiologic, antiarrhythmic and cardiovascular actions of UM 301, a quaternary ammonium compound. J Pharmacol Exp Ther 237:318-325

Greene HL, Werner JA, Gross BW, Sears JK, Trobaugh GB, Cobb LA (1983a) Prolongation of cardiac refractory times in man by clofilium phosphate, a new antiarrhythmic agent. Am Heart J 106:492-501

Greene HL, Werner JA, Gross BW, Sears GK (1983b) Failure of bretylium to suppress inducible ventricular tachycardia. Am Heart J 105:717-721

Han J, Garcia de Jalon P, Moe GK (1964) Adrenergic effects on ventricular vulnerability. Circ Res 14:516-524

Han J, Millet D, Chizzonitti B, Moe GK (1966) Temporal dispersion of recovery of excitability in atrium and ventricle as a function of heart rate. Am Heart J 71:481-487

Harron DWG (1986) Experimental evidence for the antiarrhythmic action of indoramin. J Cardiovasc Pharmacol 8:S131-S136

Hohnloser S, Weirich J, Homberger H, Antoni H (1982) Electrophysiological studies on effects of AQ-A39 in the isolated guinea pig heart and myocardial preparations. Arzneim-Forsch Drug Res 32:730-734

Holland K, Patterson E, Lucchesi BR (1983) Prevention of ventricular fibrillation by bretylium in a conscious canine model of sudden coronary death. Am Heart J 105:711-717

Inoue H, Toda I, Nozaki A, Matsuo H, Sugimoto T (1985a) Effects of bretylium tosylate on inhomogeneity of refractoriness and ventricular fibrillation threshold in canine hearts with quinidine-induced long QT interval. Cardiovasc Res 19:655-660

Inoue D, Nakanishi T, Asayama J, Katsume H, Ijichi H (1985b) Electrophysiological effects of bethanidine sulfate on guinea-pig papillary muscle. Eur J Pharmacol 108:301-303

Jaillon P, Heckle J, Weissenburger J, Cheymol G (1980) Cardiac electrophysiologic properties of *dl*-propranolol, *d*-propranolol, *l*-propranolol and *dl*-pindolol in anesthetized dogs. J Pharmacol Exp Ther 212:347-353

Johnson GL, Ehrreich SJ, El-Hage AN, Balazs T (1986) Effects of antiarrhythmic agents on isoproterenol-induced ventricular fibrillation in heavy rats: a possible model of sudden cardiac death. Res Commun Chem Pathol Pharmacol 51:351-364

Josephson ME, Horowitz LN, Farshidi A, Kastor JA (1978) Recurrent sustained ventricular tachycardia. 1. Mechanisms. Circulation 57:431-440

Kane KA, Berdeja GY, Sanchez-Perez S, Pastelin G (1983) Electrophysiological effects of lidocaine, *l*-chlorpheniramine, and bepridil on normal and ouabain-intoxicated canine Purkinje fibers. J Cardiovasc Pharmacol 5:109-115

Keefe DLD, Kates RE, Harrison DC (1981) New antiarrhythmic drugs: their place in therapy. Drugs 22:363-400

Kenny RA, Sutton R (1985) The prolonged QT interval-a frequently unrecognized abnormality. Postgrad Med J 61:379-386

Kolman BS, Verrier RL, Lown B (1976) The effect of vagus nerve stimulation upon excitability of the canine ventricle. Am J Cardiol 37:1041-1045

Koo CC, Allen JD, Pantridge JF (1984) Lack of effect of bretylium tosylate on electrical ventricular defibrillation in a controlled study. Cardiovasc Res 18:762-767

Kopia GA, Eller BT, Patterson E, Shea MJ, Lucchesi BR (1985) Antiarrhythmic and electrophysiologic actions of clofilium in experimental canine models. Eur J Pharmacol 116:49-61

Kowey PR, Friehling TD, O'Connor KM, Wetstein L, Kelliher GJ (1985) The effect of bretylium and clofilium on dispersion of refractoriness and vulnerability to ventricular fibrillation in the ischemic feline heart. Am Heart J 110:363-370

Laforet EG, Angelakos ET, Hegnauer AH (1957) Ventricular excitability in the unanesthetized dog and its modification by pentobarbital anesthesia. Am J Physiol 189:596-598

Levine JH, Spear JF, Guarnieri T, Weisfeldt, ML, DeLangen CDJ, Becker LC,

Moore EN (1985) Cesium chloride-induced long QT syndrome: demonstration of after depolarizations and triggered activity in vivo. Circulation 72:1092-1103
Levy MN (1984) Cardiac sympathetic-parasympathetic interactions. Fed Proc 43:2598-2602
Linstrom TD, Murphy PJ, Smallwood JK, Wiest SA, Steinberg MI (1982) Correlation between the disposition of [^{14}C]-clofilium and its cardiac electrophysiological effect. J Pharmacol Exp Ther 221:584-589
Lown B, Verrier RL (1976) Neural activity and ventricular fibrillation. N Engl J Med 294:1165-1170
Lucchesi BR (1984) Rationale of therapy in the patient with acute myocardial infarction and life-threatening arrhythmias: a focus on bretylium. Am J Cardiol 54:14A-19A
Lucchesi BR, Lynch JJ (1986) Preclinical assessment of antiarrhythmic drugs. Fed Proc 45:2197-2205
Lynch JJ, Wilber DJ, Montgomery DG, Hsieh TM, Patterson E, Lucchesi BR (1984) Antiarrhythmic and antifibrillatory actions of the levo- and dextrorotatory isomers of sotalol. J Cardiovasc Pharmacol 6:1132-1141
Lynch JJ, Coskey LA, Montgomery DG, Lucchesi BR (1985a) Prevention of ventricular fibrillation by dextrorotatory sotalol in a conscious canine model of sudden coronary death. Am Heart J 109:949-958
Lynch JJ, Montgomery DG, Ventura A, Lucchesi BR (1985b) Antiarrhythmic and electrophysiologic effects of bepridil in chronically infarcted conscious dogs. J Pharmacol Exp Ther 234:72-80
Manz M, Wagner W-L, Grube E, Lüderitz B (1986) Current concepts in the treatment of cardiac arrhythmias. J Cardiovasc Pharmacol 8:S124-S130
Martins JB, Zipes DP (1980) Effects of sympathetic and vagal nerves on recovery properties of the endocardium and epicardium of the canine left ventricle. Circ Res 46:100-110
Millar JS, Vaughan Williams EM (1983) Pharmacological mapping of regional effects in the rabbit heart of some new antiarrhythmic drugs. Br J Pharmacol 79:701-709
Naccarelli GV, Rinkenberger RL, Dougherty AH, Giebel RA (1985) Amiodarone: pharmacology and antiarrhythmic and adverse effects. Pharmacotherapy 5:298-313
Nishimura M, Watanabe Y (1983) Membrane action and catecholamine release action of bretylium tosylate in normoxic and hypoxic canine Purkinje fibers. J Am Coll Cardiol 2:287-295
Northover BJ (1985) A comparison of the effects of bethanidine, meobentine and quinidine on the electrical activity of rat hearts in vivo and in vitro. Br J Pharmacol 84:755-763
Patterson E, Lucchesi BR (1983) Bretylium: a prototype for future development of antidysrhythmic agents. Am Heart J 106:426-431
Patterson E, Stetson P, Lucchesi BR (1980) Plasma and myocardial tissue concentrations of UM-272 (N,N-dimethylpropranolol) after oral administration in dogs. J Pharmacol Exp Ther 214:449-453
Patterson E, Amalfitano DJ, Lucchesi BR (1984) Development of ventricular tachyarrhythmias in the conscious canine during the recovery phase of experimental ischemic injury: effect of bethanidine administration. J Cardiovasc Pharmacol 6:470-475
Pearle DL, Souza JD, Gillis R (1983) Comparative vagolytic effects of procainamide and N-acetylprocainamide in the dog. J Cardiovasc Pharmacol 5:450-453
Peiss CN, Manning JW (1964) Effects of sodium pentobarbital on electrical and reflex activation of the cardiovascular system. Circ Res 14:228-235

Platia EV, Reid PR (1984) Dose ranging studies of clofilium, an antiarrhythmic quaternary ammonium. Clin Pharmacol Ther 35:193-202
Platou ES, Refsum H, Amlie JP, Landmark K (1982) Plasma levels and cardiac electrophysiological effects of melperone in the dog. Eur J Pharmacol 82:1-7
Podrid PJ (1985) Antiarrhythmic drug therapy (part 1). Benefits and hazards. Chest 88:452-460
Prystowsky EN (1985) Electrophysiologic and antiarrhythmic properties of bepridil. Am J Cardiol 55:59C-62C
Prystowsky EN, Jackman WM, Rinkenberger RL, Heger JJ, Zipes DP (1981) Effect of autonomic blockade on ventricular refractoriness and atrioventricular nodal conduction in humans. Circ Res 49:511-518
Reiser HJ, Sullivan ME (1986) Antiarrhythmic drug therapy: new drugs and changing concepts. Fed Proc 45:2206-2212
Reynolds RD, Gorczynski RJ (1980) Comparison of the autonomic effects of procainamide and *N*-acetylprocainamide in the dog. J Pharmacol Exp Ther 212:579-583
Roden DM, Hoffman BF (1985) Action potential prolongation and induction of abnormal automaticity by low quinidine concentrations in canine Purkinje fibers. Circ Res 56:857-867
Roden DM, Woosley RL (1985) QT prolongation and arrhythmia suppression. Am Heart J 109:411-415
Rosen DM, Wit AL (1983) Electropharmacology of antiarrhythmic drugs. Am Heart J 106:829-839
Sasyniuk B (1984) Concept of reentry versus automaticity. Am J Cardiol 54:1A-6A
Schwartz PJ, Periti M, Malliani A (1975) The long Q-T syndrome. Am Heart J 89:378-390
Singh BN, Nademanee K (1985) Control of cardiac arrhythmias by selective lengthening of repolarization: theoretic considerations and clinical observations. Am Heart J 109:421-430
Smallwood JK, Wiest SA, Molloy BB, Steinberg MI (1986) Class III antiarrhythmic properties of LY190147 [*N*-(4-(4-(ethylheptylamino)-butyl)phenyl)methane sulfonamide hydrochloride]. Fed Proc 45:804
Snyders DJ, Katzung BG (1985) Clofilium reduces the plateau potassium current in isolated cardiac myocytes. Circulation 72:III-233
Somberg J, Torres V, Flowers D, Miura D, Butler B, Gottleib S (1985) Prolongation of QT interval and antiarrhythmic action of bepridil. Am Heart J 109:19-27
Steinberg MI, Michelson EL (1985) Cardiac electrophysiological effects of specific class III substances. In: Reiser HJ, Horowitz LN (eds) Mechanisms and treatment of cardiac arrhythmias: relevance of basic studies to clinical management. Urban and Schwarzenberg, Baltimore, p 263
Steinberg MI, Molloy BB (1979) Clofilium—a new antifibrillatory agent that selectively increases cellular refractoriness. Life Sci 25:1397-1406
Steinberg MI, Sullivan ME, Wiest SA, Rockhold FW, Molloy BB (1981) Cellular electrophysiology of clofilium, a new antifibrillatory agent, in normal and ischemic canine Purkinje fibers. J Cardiovasc Pharmacol 3:881-895
Steinberg MI, Lindstrom TD, Fasola AF (1984) Clofilium: a class III antiarrhythmic agent. In: Scriabine A (ed) Cardiovascular drugs. Raven Press, New York, p 103 (New drugs annual, vol 2)
Steinberg MI, Lacefield WB, Robertson DW (1986) Class I and III antiarrhythmic drugs. In: Bailey DM (ed) Annual reports in medicinal chemistry, vol 21. Academic, Orlando, p 95
Sullivan ME, Steinberg MI (1981) Cardiovascular and electrocardiographic effects of clofilium in conscious or anesthetized dogs. Pharmacologist 23:209

Sung RJ, Juma Z, Saksena S (1983) Electrophysiological properties and antiarrhythmic mechanisms of intravenous *N*-acetylprocainamide in patients with ventricular dysrhythmias. Am Heart J 5:811-819

Surawicz B, Knoebel SB (1984) Long QT: good, bad or indifferent? J Am Coll Cardiol 4:398-413

Tacker WA, Niebauer MJ, Babbs CF, Combs WJ, Hahn BM, Barker MA, Seipel JF, Bourland JD, Geddes LA (1980) The effect of newer antiarrhythmic drugs on defibrillation threshold. Crit Care Med 8:177-180

Temma K, Akera T, Ng Y-C (1985) Cardiac actions of phencyclidine in isolated guinea pig and rat heart: possible involvement of slow channels. J Cardiovasc Pharmacol 7:297-306

Thomis JT, Tenthorey P (1983) Prolonged ventricular repolarization-a prevention of severe arrhythmias? In: Hess H-J (ed) Annual reports in medicinal chemistry, vol 18. Academic, New York, p 99

Torres V, Flowers D, Somberg JC (1985) The arrhythmogenicity of antiarrhythmic agents. Am Heart J 109:1090-1097

Turnheim K, Lauterbach F (1980) Interaction between intestinal absorption and secretion of monoquaternary ammonium compounds in guinea pigs—a concept for the absorption kinetics of organic cations. J Pharmacol Exp Ther 212:418-424

Vaughan Williams EM (1982) QT and action potential duration. Br Heart J 47:513-514

Vaughan Williams EM (1985a) Delayed ventricular repolarization as an antiarrhythmic principle. Eur Heart J [Suppl D] 6:145-149

Vaughan Williams EM (1985b) Class III antiarrhythmic action and the QT interval. In: Reiser JH, Horowitz LN (eds) Mechanism and treatment of cardiac arrhythmias; relevance of basic studies to clinical management. Urban and Schwarzenberg, Baltimore, p 295

Wastila WB, Copp FC, Walton E, Ellis CH, Maxwell RA (1981) Meobentine sulphate (bis[*N*-4-methoxybenzyl-*N'N''*-dimethylguanidine]sulphate): a new antidysrhythmic agent. J Pharm Pharmacol 33:594-596

Wellens HJJ, Brugada P, Farre J (1984) Ventricular arrhythmias: mechanisms and actions of antiarrhythmic drugs. Am Heart J 107:1053-1057

Wenger TL, Lederman S, Starmer CF, Brown T, Strauss HC (1984) A method for quantitating antifibrillatory effects of drugs after coronary reperfusion in dogs: improved outcome with bretylium. Circulation 69:142-148

Wit AL, Rosen MR (1983) Pathophysiologic mechanisms of cardiac arrhythmias. Am Heart J 106:798-811

Wit AL, Rosen MR, Hoffman BF (1974a) Electrophysiology and pharmacology of cardiac arrhythmias. II. Relationship of normal and abnormal electrical activity of cardiac fibers to the genesis of arrhythmias. B. Re-entry. Section I. Am Heart J 88:664-670

Wit AL, Rosen MR, Hoffman BF (1974b) Electrophysiology and pharmacology of cardiac arrhythmias. II. Relationships of normal and abnormal electrical activity of cardiac fibers to the genesis of arrhythmias. B. Re-entry. Section II. Am Heart J 88:798-806

Yatani A, Brown AM, Schwartz A (1986) Bepridil block of cardiac calcium and sodium channels. J Pharmacol Exp Ther 237:9-17

Zimmerman JM, Patterson E, Pitt B, Lucchesi BR (1984) Antidysrhythmic actions of meobentine sulfate. Am Heart J 107:1117-1124

Zipes DP (1975) Electrophysiological mechanisms involved in ventricular fibrillation. Circulation [Suppl III] 51/52:120-130

Class IV Agents

CHAPTER 18

Class IV Antiarrhythmic Agents: Utility in Supraventricular Arrhythmias and Their Proarrhythmic Potential

J. MORGANROTH and L. N. HOROWITZ

A. Introduction

Over 100 years ago calcium was identified as a necessary element for cardiac muscular contraction (RINGER 1883). It was only about 20 years ago (KAUFMAN and FLECKENSTEIN 1965) that calcium antagonists that inhibited the excitation contraction coupling process were identified. There are many different mechanisms by which calcium interacts within cellular processes and some drugs specifically act by blocking calcium channels and those are more appropriately termed the "calcium channel blockers" (KATZ and REUTER 1979). A complete classification system for calcium active agents is still evolving based on their pharmacologic and electrophysiologic actions.

B. Therapeutic Basis

The first therapeutic application of calcium channel blockers was as coronary vasodilators (GRUEN and FLECKENSTEIN 1972). Over the next decade, these agents were studied in patients with coronary spasm and chronic angina pectoris. In the early 1970s, the electrophysiologic effects of calcium channel blockers were actively studied both in the United States and Japan (ZIPES and FISCHER 1974; WATANABE and BESCH 1974). Unlike class IA antiarrhythmic drugs which predominantly affect the fast sodium channel, the calcium channel blocking agents depress sinoatrial automoticity and atrioventricular nodal depolarization predominantly by their effect on the second inward current that is carried predominantly by calcium (SINGH et al. 1978). Blockade of the calcium channel produces a slowing of conduction and prolongation of refractoriness within the sinus and atrioventricular nodes. Clinically, these electrophysiologic actions manifest themselves as a slowing of the sinus rate, prolongation of the PR interval on the electrocardiogram, and a reduction in the ventricular response to atrial arrhythmias (ROWLAND et al. 1979). Arrhythmias that involve the atrioventricular node can also be interrupted by calcium channel blockers.

C. Effect on the Ventricle

Little, if any, significant effect occurs in normal Purkinje fibers or cardiac muscle cells. In vitro studies have shown, however, that abnormal Purkinje fibers, particularly in which the fast sodium current has been inactivated, may have their electrophysiologic properties altered by calcium channel blockers (WIT et al. 1974). Thus, it is possible that calcium channel blockers may have an effect on ventricular arrhythmias. In ischemic settings such as experimental myocardial infarction, verapamil has been shown to improve conduction in the ischemic zone and thus may have an effect on ventricular arrhythmias (EL-SHERIF and LAZARRA 1979). In the experimental model, verapamil has also been shown to increase ventricular fibrillation threshold (FONDACARO et al. 1978).

D. Indications

Calcium channel blockers belong to the class IV group of antiarrhythmic drugs and are principally indicated for treatment of supraventricular arrhythmias. Because of their effect on the atrioventricular nodal conduction they may be useful in reducing the ventricular response during atrial fibrillation and in the chronic prevention of paroxysmal supraventricular reentrant tachycardias.

E. Atrial Fibrillation

In atrial fibrillation, calcium antagonists are expected to decrease ventricular response. In one study (WAXMAN et al. 1981) mean ventricular rate decreased from 146 to 114/min in two out of three of the patients studied. Overall about 70 % of patients with atrial fibrillation should experience about a 20 % reduction in ventricular response during verapamil therapy. Intravenous verapamil may achieve this effect within minutes (SCHAMROTH 1971) and thus in respect of rapidity of onset of action it is more advantageous than digitalis. Exercise-induced tachycardia in patients with atrial fibrillation can be markedly attenuated by the additions of verapamil in digitalized patients (MORGANROTH et al. 1982). While digitalis may control the ventricular response to atrial fibrillation at rest, marked increase in ventricular response is usually seen in such patients during exercise, which limits their exercise tolerance. Verapamil has also been used for the management of ventricular response in patients, with atrial fibrillation complicating acute myocardial infarction (HAGEMEIJER 1978).

Because of its negative inotropic effect, however, caution must be used in giving intravenous verapamil to any patient who has compromised left ventricular function. In addition, the use of verapamil is potentially hazardous in patients with atrial fibrillation and preexcitation syndrome if the agent should slow conduction over the normal atrioventricular pathway, thus allowing for an increased conduction rate over the accessory pathway and a paradoxical increase in the ventricular response rate (GULAMHUSSEIN et al. 1981).

F. Paroxsymal Supraventricular Tachyarrhythmias

Paroxsymal supraventricular tachyarrhythmias usually have a reentrant mechanism involving the atrioventricular node. Calcium channel blockers alter conduction and refractoriness within the atrioventricular node and terminate these reentrant rhythms by disrupting the reentrant circuit. The efficacy of verapamil and diltiazem in terminating such rhythm disturbances is extremely high and in general over 80 % of such episodes convert to normal sinus rhythm within minutes. In controlled trials in the United States, about 60 % of such patients reverted to normal sinus rhythm within 10 min after verapamil treatment. In patients in whom sinus rhythm is not restored the rate of the tachycardia is usually significantly reduced. In one study, in patients with different types of reentrant supraventricular tachycardia, 79 % of episodes were converted to normal sinus rhythm by an intravenous dose of verapamil of 0.075-0.15 mg/kg infused over 1-3 min (SUNG et al. 1980). As in patients with atrial fibrillation, the rapid action of intravenous calcium channel blockers is a marked advantage over other agents with slower onset of effect. Hypotension or a bradyarrhythmia are the principal concerns and avoidance of other concomitant negative inotropic agents such as beta-blockers is mandatory.

The use of calcium channel blockers in the chronic prophylactic treatment of supraventricular tachycardias has not been well studied but reports exist that attest to their effectiveness in this setting (WU et al. 1983; LIE et al. 1983; YEH et al. 1983).

G. Pharmacology and Adverse Clinical Effects

The principal side effect from calcium channel blockers such as verapamil, diltiazem, and nifedipine is hypotension. With verapamil, approximately 5 %-10 % of patients will have a reduction in blood pressure to levels of 90/60 mmHg and symptomatic hypotension may occur in 1 %-2 % of patients. Hypotension may be less prevalent with diltiazem but is more commonly seen in patients with nifedipine, which has more potent peripheral vasodilatory effects. Concomitant use of calcium channel blockers with other agents which have marked negative inotropic effects such as beta-blockers or disopyramide may be particularly dangerous when patients have the combination of a low left ventricular ejection fraction (< 30 %) and an elevated pulmonary artery wedge pressure (> 20 mmHg).

Significant sinus bradycardia or induced second- or third-degree atrioventricular block may be seen in approximately 0.5 %-2 % of patients with verapamil or diltiazem. Less impressive effects on sinoatrial and atrioventricular

Table 1. Differential pharmocology of verapamil, diltiazem, and nifedipine

	% Absolute oral bioavailability	Elimination half-life (h)	Effective plasma levels (ng/ml)	Principal site of metabolism	Dose (mg)
Oral verapamil	25	2-5	125-400	Hepatic and Renal	80 tid-120 qid
Intravenous verapamil	—	12	—		5-10 mg over 2 min
Diltiazem	40	3.5	50-200	Hepatic	30 tid-60 qid
Nifedipine	100	2	—	Renal	10 tid-30 qid

nodal function are seen with nifedipine. General adverse effects include dizziness, headache, nausea, edema, and rash. Abnormal liver function tests and the possibility of liver toxicity, though uncommon, have been seen. Drug interactions have not been entirely clarified but verapamil and nifedipine do increase serum digoxin concentration when coadministered with digoxin. The clinical importance of this change is yet to be determined. Table 1 details the differential pharmacology of verapamil, diltiazem, and nifedipine, which are the commonly used calcium channel blockers. The following sections will detail the effect of calcium channel blockers on ventricular arrhythmias and point out another potential side effect: the induction of proarrhythmia or arrhythmogenesis.

H. Calcium Blockers: Effect on Ventricular Arrhythmias

While in vitro data (reviewed above) suggest that calcium channel blockers may have an effect on abnormal ventricular Purkinje fibers, verapamil has not shown any ability to suppress chronic ventricular premature complexes (MULLER et al. 1983) or sustained ventricular tachycardia (WELLENS et al. 1977). Further data, however, are required to define more precisely the effectiveness of calcium channel blockers in patients with ventricular arrhythmias before any definite conclusions can be drawn. However, since calcium channel blockers do effect the electrophysiologic processes of abnormal fibers the potential for such agents in causing proarrhythmia (arrhythmogenesis) must be considered.

J. A Definition for Proarrhythmia

The widespread use of quantitative ambulatory electrocardiographic monitoring (Holter monitoring) has provided a means to determine the efficacy of antiarrhythmic agents as well as their ability to enhance ventricular arrhythmias (the so-called proarrhythmic effect) (MORGANROTH 1985; MORGANROTH et al.

1978; MORGANROTH and HOROWITZ 1984). In patients with benign or potentially lethal ventricular arrhythmias (MORGANROTH 1984) (thus excluding those patients with sustained ventricular tachycardia or cardiac arrest) the determination of whether an antiarrhythmic drug is inefficacious or producing a proarrhythmic event requires considerable thought. Previous definitions of proarrhythmia (MORGANROTH and HOROWITZ 1984) have used an arbitrarily defined algorithm to differentiate lack of efficacy from proarrhythmia in terms of a change in ventricular arrhythmia frequency (Table 2). This algorithm has been validated recently in 495 patients in whom frequent Holter monitoring was performed while the patient was receiving placebo. In this population five patients (1 %) would have been classified as having proarrhythmia using the algorithm in Table 1 despite the fact that only placebo was given. A tenfold or greater increase in the frequency of the beats, in the form of ventricular tachycardia, on placebo occurred in 9/274 (3 %) of patients. Thus, the proarrhythmic definition that we have adopted (Table 2) provides for a very low false positive rate of allocating patients to the proarrhythmic category.

Other important aspects of the proarrhythmia definition include the new onset of a ventricular arrhythmia that was not previously identified. For example, if an antiarrhythmic drug such as a calcium channel blocker is given to

Table 2. A definition of proarrhythmia

1. The *new* onset of:
   a) Ventricular premature complexes > five per hour
   b) Nonsustained ventricular tachycardia
   c) Sustained ventricular tachycardia
   d) Torsades de pointes (polymorphic ventricular tachycardia)
   e) Ventricular flutter /fibrillation
   f) A supraventricular tachyarrhythmia
2. Change in the frequency of a previously documented ventricular arrhythmia:
   a) Increase in the frequency of ventricular premature complexes (VPCs)

Mean VPC/h at baseline	Increase required for prorrhythmia
10– 50	Tenfold
51–100	Fivefold
101–300	Fourfold
> 300	Threefold

   b) A tenfold or greater increase in the mean hourly frequency of nonsustained ventricular tachycardia
3. A significantly more difficult cardioversion or termination of ventricular tachycardia or ventricular flutter /fibrillation as defined by the investigator
4. Occurrence of sudden cardiac death (defined as an unexpected, nontraumatic, non-self-inflicted fatality in patients who die within 1 h of the onset of the terminal event)

a patient without ventricular arrhythmias and a subsequent quantitative Holter monitor demonstrates > five ventricular premature complexes per hour (KOSTIS et al. 1981), a significant new ventricular arrhythmia has thus occurred. This response would be considered a proarrhythmic event. Patients without previous nonsustained ventricular tachycardia who now have sustained ventricular tachycardia on the antiarrhythmic drug are another example of a new onset of ventricular arrhythmia.

The occurrence of events such as a new onset or change in frequency or a type of ventricular arrhythmia (as defined in Table 2) should not be considered proarrhythmia if another clinical factor which may be the cause of the event can be identified. Such examples would include the occurrence of an event within 72 h of a myocardial infarction, in relation to hypokalemia, discontinuation of antiarrhythmic therapy, etc. In addition, proarrhythmia is not assigned if the ventricular arrhythmia event occurs after the patient has been on the same daily dose of the antiarrhythmic drug for ≧ 30 consecutive days. Since proarrhythmic events directly due to the drug should occur when the antiarrhythmic drug or its metabolites have reached steady state, a change in ventricular arrhythmia status over 30 days from its initiation (except for drugs with extremely long half-lives such as amiodarone) are more likely to be due to a change in the underlying clinical state than to a late proarrhythmia.

Since some of these proarrhythmic events may have a more important effect on the patient's well-being or clinical course than others it is important to segregate proarrhythmic events into various types. There are those that cause death. There are those that are "serious." Serious is defined as a worsened ventricular arrhythmia that required immediate termination with drugs, overdrive pacing, or cardioversion or if the proarrhythmic event was associated with hypotensive symptoms. All other proarrhythmic events, particularly those that include just an increase in ventricular arrhythmia frequency without symptoms, are considered "other proarrhythmic events." These may not have produced worsening of the cardiovascular symptoms or immediately affected the patient's well-being. The segregation of proarrhythmic events by this classification has been used to evaluate the proarrhythmic potential of flecainide (MORGANROTH 1986). This segregation has been extremely useful in demonstrating that proarrhythmic deaths and serious proarrhythmia are extremely rare occurrences from flecainide in patients with benign or potentially lethal ventricular arrhythmias (MORGANROTH 1984), whereas, in patients with sustained ventricular tachycardia or lethal ventricular arrhythmias (MORGANROTH and HOROWITZ 1984), a high rate of proarrhythmia can occur which is either fatal or serious particularly if an incorrect dosage of flecainide is used in such compromised patients.

K. Ventricular Proarrhythmia Form Calcium Channel Blockers

While there have been rare case reports of probable torsade de pointes from calcium channel blockers, to date the only available data directed at determining the proarrhythmic potential of calcium channel blockers has been performed in a study comparing bepridil with diltiazem. The purpose of this study was to define the relative safety of bepridil versus diltiazem in patients with coronary artery disease and stable chronic ventricular arrhythmias. Patients had to have at least 10 ventricular premature complexes/h on a 2-week placebo period using quantitative Holter monitoring. This was a multicenter study conducted at 25 sites within the United States in which patients, after the placebo qualification period, were randomized to bepridil or diltiazem therapy. Bepridil was used from 200 to 400 mg/day and diltiazem from 90 to 240 mg/day. None of these patients had previous sustained ventricular tachycardia or uncontrolled angina, hypertension, or congestive heart failure. Approximately half of the patients were on previous antiarrhythmic therapy for their ventricular arrhythmias and approximately half were receiving concomitant beta-blocker therapy for coronary artery disease. The definition of efficacy for suppression of ventricular arrhythmias was $\geqq 75\,\%$ reduction in the frequency of ventricular premature complexes from baseline and $> 95\,\%$ reduction in beats in ventricular tachycardia form. In this study, 23/85 (27 %) and 3/70 (4.3 %) of patients reached both of these efficacy end points on bepridil and diltiazem respectively ($P < 0.001$). Thus, this study demonstrated that bepridil has, as is well known, a combined class III effect in addition to its calcium channel blocking effect. Of interest was that of 52 patients with non-sustained ventricular tachycardia at baseline in the diltiazem arm, 16 (31 %) had $\geqq 95\,\%$ reduction in ventricular tachycardia beats, which was insignificantly different from bepridil [21/54 (39 %)]. Thus, this randomized study did demonstrate that diltiazem may have had an effect on repetitive ventricular arrhythmia forms but not on total ventricular arrhythmia frequency compared with bepridil. Eleven patients on diltiazem had proarrhythmia. This included two patients with the new onset of a supraventricular tachyarrhythmia, four patients with an asymptomatic increase in the frequency of ventricular premature complex, three patients with an asymptomatic increase in ventricular tachycardia, two patients with an increase in ventricular arrhythmia which was symptomatic, and one with the new onset of sustained ventricular tachycardia. No patients suffered sudden cardiac death while on diltiazem. This proarrhythmic rate was not significantly different from the bepridil group. Clinical trials in the United States of another calcium channel blocker, fostedil, were prematurely discontinued when an unexpectedly high incidence of sudden death and new supraventricular tachycardias were noted in its study in hypertensive patients. Fostedil had a very similar pharmacology profile to that of diltiazem but a longer half-life permitting once a day dosing. Thus, a proarrhythmic response must be considered as a possible adverse drug effect of calcium blockers and an enhanced recognition of this possibility may further define the prevalence and clinical significance of this phenomenon in the future.

L. Conclusion

Class IV antiarrhythmic agents—the calcium channel blockers—are particularly useful because of their unique electrophysiological effects. The change in sinoatrial and atrioventricular nodal function allows for easier control of the ventricular response in patients with atrial fibrillation. The conversion of reentrant supraventricular tachyarrhythmias and possible prophylaxis of paroxysmal episodes are important clinical uses. Cardiac adverse reactions are generally thought to be limited to sinoatrial bradycardia, atrioventricular block, or hypotension. The potential for their use in the treatment of ventricular arrhythmias has not been well studied and further data are needed. Other adverse cardiac effects such as proarrhythmia must be added to the list of adverse reactions and in further studies must be carefully considered. Calcium channel blockers have an extremely important role to play in the management of patients with supraventricular tachyarrhythmias.

References

El-Sherif, Lazarra R (1979) Reentrant ventricular arrhythmias in the late myocardial infarction period. 7. Effect of verapamil and D-600 and the role of the "slow channel". Circulation 60:605–615

Fondacaro ID, Han J, Yoon MS (1978) Effects of verapamil on ventricular rhythm during acute coronary occlusion. Am Heart J 96:81–86

Gruen G, Fleckenstein A (1972) Die electromechanische Entkoppelung der glatten Gefassmuskulatur als Grundprinzip der Coronardilation durch 4-(2-Nitrophenyl)-2,6-dimethyl-1,4-dihydropyridin-3,5-dicarbonsaure-dimethylester (Bay a 1040, Nifedipin). Arzneimittelforsch. 22:334

Gulamhussein S, Ko P, Carruthers S, Klein GJ (1981) Acceleration of the ventricular responses during atrial fibrillation in the Wolff-Parkinson-White syndrome after verapamil. Circulation 65:348–354

Hagemeijer F (1978) Verapamil in the management of supraventricular tachyarrhythmias occurring after a recent myocardial infarction. Circulation 57:751–755

Katz AM, Reuter H (1979) Cellular calcium and cardiac cell death. Am J Cardiol 44:188–190

Kaufmann R, Fleckenstein A (1965) Ca^{2+}-competitive electromechanische Entkoppelung durch Ni^{2+}- und Co^{2+}-Ionen am Warmblutermyokard. Pflugers Arch Ges Physiol 282:290

Kostis JB, McCrone K, Moreya AE Gotzoyannis S, Aglitz NM, Natarajan N, Kuo PT (1981) Premature ventricular complexes in the absence of identifiable heart disease. Circulation 63:1351–1356

Lie KI, Duren DR, Manger Cats D, David GK, Durrer D (1983) Long-term efficacy of verapamil in the treatment of paroxysmal supraventricular tachycardias. Am Heart J 105:668

Morganroth J (1984) Premature ventricular complexes: diagnosis and indication for therapy. JAMA 252:673–676

Morganroth J (1985) Ambulatory Holter electrocardiography: choice of technologies and clinical uses. Ann Intern Med 102:73–81

Morganroth J, Borland M, Chao G (1987) Application of a frequency definition of ventricular proarrhythmia. Am J Cardiol 59:97-99
Morganroth J, Horowitz LN (1984) Flecainide: its proarrhythmic effect and expected changes on the surface electrocardiogram. Am J Cardiol 53:89B-94B
Morganroth J, Michelson EL, Horowitz LN, Josephson ME, Pearlman AS, Dunkman B (1978) Limitations of routine long-term electrocardiographic monitoring to assess ventricular ectopic frequency. Circulation 58:408-414
Morganroth J, Chen CC, Sturm S, Dreifus LS (1982) Oral verapamil in the treatment of atrial fibrillation/flutter. Am J Cardiol 49:981
Morganroth J, Anderson JL, Gentzkow GD (1986) Classification by type of ventricular arrhythmia predicts frequency of adverse cardiac events from flecainide. J Am Coll Cardiol 28:607-615
Muller J, Morrison J, Stone P, Rude R, Rosner B, Roberts R, Pearle D, Turi Z, Schneider J, Serfas D, Hennekens C, Braunwald E (1983) Nifedipine therapy for threatened and acute myocardial infarction: a randomized double blind comparison. Circulation 68:III-120
Ringer S (1883) A further contribution regarding the influence of the different constituents of the blood on the contraction of the heart. J Physiol (London) 4:29-42
Rowland E, Evans T, Krickler D (1979) Effect of nifedipine on atrioventricular conduction as compared with verapamil: intracardiac electrophysiology study. Br Heart J 42:124-127
Schamroth L (1971) Immediate effects of intravenous verapamil on atrial fibrillation. Cardiovasc Res 5:419-424
Singh BN, Ellrodt G, Peters CT (1978) Verapamil: a review of its pharmacological properties and therapeutic use. Drugs 15:169-197
Sung RJ, Elser B, McAllister RG (1980) Intravenous verapamil for termination of reentrant supraventricular tachycardias. Ann Intern Med 93:682-689
Watanabe AM, Besch HR (1974) Subcellular myocardial effects of verapamil and D-600:comparison with propranolol. J Pharmacol Exp Ther 191:241-251
Waxman HL, Myerburg RJ, Appel R, Sung RJ (1981) Verapamil for control of ventricular rate in paroxysmal supraventricular tachycardia and atrial fibrillation or flutter. Ann Intern Med 94:1-6
Wellens HJJ, Bar FW, Lie KI, Duren DR, Dohmen HJ (1977) Effects of procainamide, propranolol and verapamil on mechanism of tachycardia in patients with chronic recurrent ventricular tachycardia. Am J Cardiol 40:579-585
Wit AL, Rosen MR, Hoffman BF (1974) Relationship of normal and abnormal electrical activity of cardiac fibers to the genesis of arrhythmias. II. Reentry section II. Am Heart J 88:798-807
Wu D, Kou H, Yeh S, Lin F, Hung J (1983) Effects of oral verapamil in patients with atrioventricular reentrant tachycardia incorporating an accessory pathway. Circulation 67:426-433
Yeh S, Kou H, Lin F, Hung J, Wu D (1983) Effects of oral diltiazem in paroxysmal supraventricular tachycardia. Am J Cardiol 52:271-278
Zipes DP, Fischer JC (1974) Effects of agents which inhibit the slow channel on sinus node automaticity and atrioventricular conduction in the dog. Circ Res 34:184-192

Class V Agents

CHAPTER 19

Specific Bradycardic Agents

W. KOBINGER

A. Introduction and Definition

Alinidine (St 567, *N*-allyl-clonidine), the prototype of the novel class of drugs discussed in this section, has been classified as an antiarrhythmic agent by MILLAR and VAUGHAN WILLIAMS (1981a) on the basis of its selective electrophysiological effects upon certain myocardial cells. Unlike other agents discussed in this volume, the therapeutic goal searched for was not primarily to interfere with abnormal heart beats, but simply to reduce the sinus node controlled heart rate. In patients where a decrease in myocardial oxygen consumption is a desired therapeutic measure, this might be achieved by a decrease in heart rate (SONNENBLICK and SKELTON 1971). The expected benefit of a retardation in heart rate for a patient with rigid narrowing of coronary arteries is even more obvious, if the bradycardia is mainly due to a prolongation of the diastolic period, the time span which allows perfusion of the myocardium.

Pharmacologically, a sinus node controlled heart rate can be reduced indirectly by interfering with activities of the autonomic nervous system (e.g. by beta-adrenoceptor-blocking drugs, certain veratrum alkaloids) or directly at the pacemaker cells of the atrial sinus node.

The term "specific bradycardic agents" has been proposed for drugs which decrease the heart rate by a direct effect on sinus node pacemaker cells and in concentrations or doses which are smaller than those which elicit other cardiovascular or pharmacological effects (KOBINGER 1985). Different chemical structures have been found to fit this definition, and fig. 1 reveals the two main groups deriving from alinidine (St 567) and from falipamil (AQ-A39) respectively (reviews: KOBINGER and LILLIE 1984a; HARRON and SHANKS 1985; KOBINGER 1985). [After completion of this chapter a symposium took place in Rotterdam in May 1987: "Heart rate reduction, a reinvented approach to cardiac physiology and therapy" (SIMOONS and HUGENHOLTZ 1987)].

From published data it appears that substances such as mixidine fumarate (TAKEDA et al. 1980; SIEGL et al. 1984) and some veratrum and harmala alkaloids have certain features of these specific bradycardic agents (KRAYER 1949; ZETLER et al. 1968).

Compound	MW	log P	pK_a
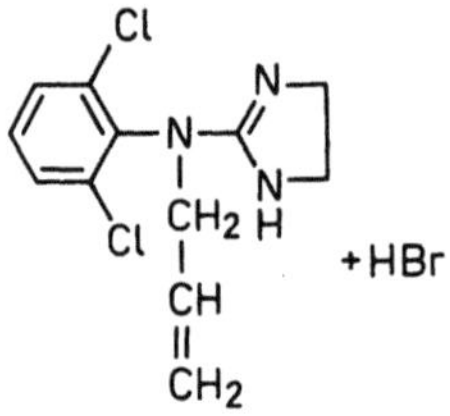 Alinidine (St 567)	351 (Br)	-1.66	10.4
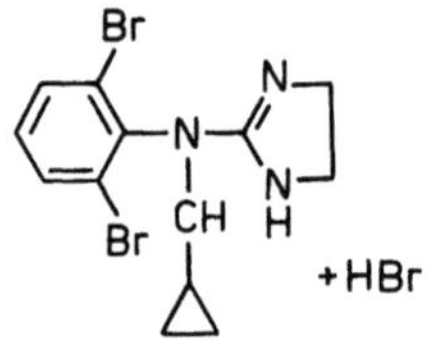 STH 2148	454 (Br)	-1.12	10.9
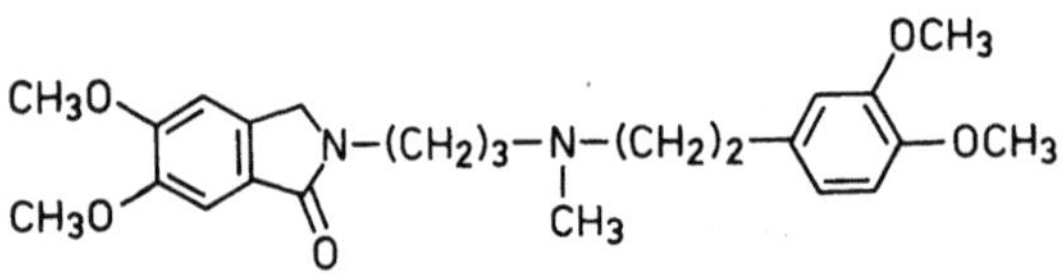 Falipamil (AQ-A 39)	465 (Cl)	0.43	8.58
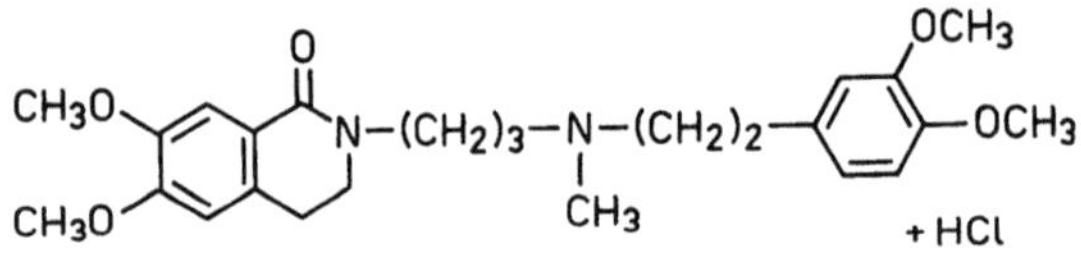 AQ-AH 208	479 (Cl)	0.96	8.77
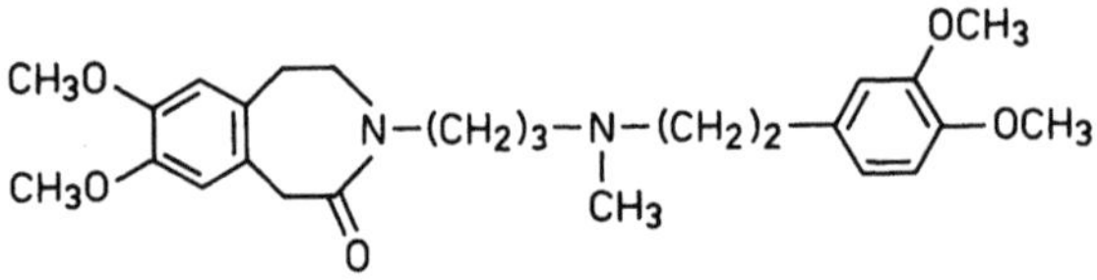 UL-FS 49	493 (Cl)	0.66	8.73

Fig. 1. Chemical structures of specific bradycardic agents. *MW*, molecular weight; log *P*, logarithm of the partition coefficient octanol/buffer pH 7.4; pK_a, dissociation constant

B. Alinidine

I. Pharmacology

1. Isolated Cardiac Preparations

Figure 2 illustrates the predominance of the bradycardic effect of alinidine in isolated guinea pig atria. In three different experimental setups, the following parameters were determined: spontaneous sinus rate, maximal driving frequency, as a test for effective refractory period, and contractile force in electrically driven left atria (Kobinger et al. 1979 a). As can be seen from the concentration response curves, alinidine decreased sinus rate in much smaller concentrations than maximal driving frequency and contractility. The concentrations which reduced each parameter by 30 % (EC_{30}) were determined graphically, and are given in Table 1 for alinidine, as well as for other drugs. It appears that, with the exception of clonidine, all reference substances exert a bradycardic action. However, comparison of the ratios between the three activities reveals for alinidine the predominance of the bradycardic effect in relation to the negative inotropic action and the decrease in the maximal driving frequency. The class I antiarrhythmics (classification according to Vaughan Williams 1984), quinidine and lidocaine, are distinguished by the predominance of the reduction of the maximal driving frequency.

Verapamil, a "calcium-channel blocking agent" (class IV antiarrhythmic), also had a prominent bradycardic effect, when compared with its action on the effective refractory period, but proved to be markedly negative inotropic. Mil-

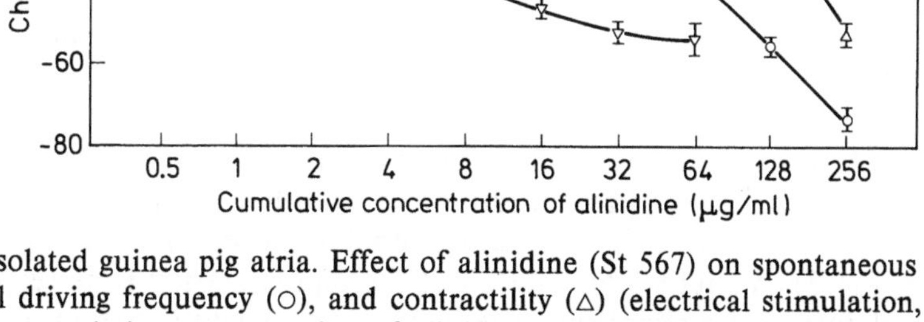

Fig. 2. Isolated guinea pig atria. Effect of alinidine (St 567) on spontaneous rate (▽), maximal driving frequency (○), and contractility (△) (electrical stimulation, 2.5 Hz). *Abscissa*, cumulative concentration of alinidine in the organ bath; *ordinate*, drug induced changes in percentage of control values. Means, SEM is indicated when it exceeds ± 2 %. Control values: atrial rate, 208 ± 7.6 beats/min ($n = 10$); contractility, 0.93 ± 0.056 g ($n = 10$); maximal driving frequency, 15.7 ± 0.376 beats/s ($n = 11$). (Kobinger et al. 1979a)

Table 1. Effects of substances on spontaneous rate, contractility and maximal driving frequency in isolated guinea pig atria. EC_{30} is the concentration which reduced predrug value by 30 %, evaluated from curves as in Fig. 2. (KOBINGER et al. 1979a)

Substance	Decrease in			Ratio of EC_{30}	
	Atrial rate (EC_{30} µg/ml)	Contractility (EC_{30} µg/ml)	Maximal driving frequency (EC_{30} µg/ml)	Contractility / Atrial rate	Maximal driving frequency / Atrial rate
Alinidine	2.9[a]	155[b]	40	53	14
Clonidine	440	96	43	0.22	0.10
Lidocaine	37	56	11	1.5	0.30
Verapamil	0.20	0.24	6.5	1.2	32
Carbachol	0.029	0.0065	Increase	0.22	—

[a] 8 µmol/litre
[b] 440 µmol/litre

LAR and VAUGHAN WILLIAMS (1981b) reported the preponderance of the bradycardic effect of alinidine over its effects on contractility, measured as peak force and rate of tension increase, maximal driving frequency as well as the electrical stimulation threshold in rabbit atria.

The direct effect of alinidine upon the sinus node and the profile different from the main classes of antiarrhythmic drugs were verified by intracellular recording techniques (Fig. 3). In sinus node cells of guinea pigs TRITTHART et al. (1981) reported the decrease in diastolic depolarization rate to cause the concentration-dependent decrease in discharge rate at 0.28–85 µmol/litre i.e. ca. 0.1–30 µg/ml. In these concentrations no changes in the action potential and tension development of electrically driven atrial fibres were observed; in concentrations of 285 µmol/litre a mild decrease in maximal rate of rise of action potential, and decreases in repolarization velocity and in contractility were seen. MILLAR and VAUGHAN WILLIAMS (1981b), BOUMAN et al. (1984) and HABERL and STEINBECK (1985) also reported the decrease in diastolic depolarization velocity by recording pacemaker potentials in rabbits and guinea pigs; in electrically driven dominant pacemaker fibres of rabbits the slope of phase 4 depolarization decreased from 64 ± 10 mV/s to 50 ± 15 mV/s ($n = 6$) after exposure to 2.9 µmol/litre alinidine and no other changes of the action potential were seen (phase 0, amplitude, repolarization, HABERL and STEINBECK 1985, Fig. 3). A slight prolongation of the repolarization time (especially the terminal part) and a slight increase in maximal diastolic potential was reported by MILLAR and VAUGHAN WILLIAMS (1981b, 1983) for atrial and ventricular fibres but not for Purkinje cells in the concentration of 6.25 µmol/litre alinidine. The increase in repolarization time is more pronounced in peripheral than in central, i.e. primary pacemaker areas of the sinus node (OP-

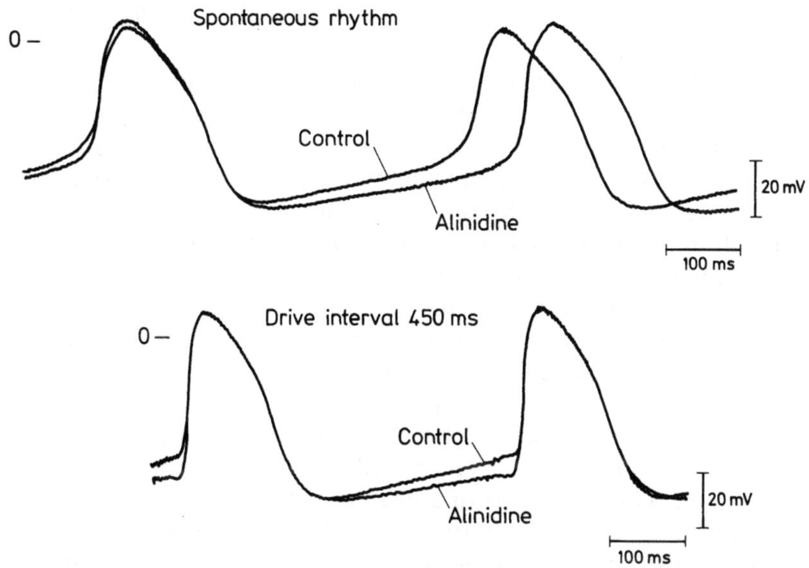

Fig. 3. Effect of alinidine (2.9 μ*M*) on the transmembrane potential of a dominant pacemaker fibre of the isolated sinus node of a rabbit heart during spontaneous rhythm *(upper panel)* and constant atrial drive *(lower panel)*. The recordings during control and after alinidine exposure are superimposed. Note the selective slowing of phase 4 depolarization when the effect of different basic cycle length is eliminated by atrial drive. (HABERL and STEINBECK 1985)

THOF et al. 1986). In rabbit "true" sinus node cells, the decrease in mean rate of repolarization amounted to 7 % (1 μmol/litre), 17 % (2.5 μmol/litre) and 30 % (6.25 μmol/litre), and in these experiments a decrease in maximal rate of depolarization (22 %, 28 %, 45 %) and no change in take off potential up to 2.5 μmol/litre was described (MILLAR and VAUGHAN WILLIAMS 1981b). In electrically driven atrial fibres 6.25 μmol/litre alinidine did not change the resting potential or overshoot potential and did not significantly reduce the maximal rate of depolarization (although there was a trend towards reduction) whereas 90 % repolarization time was increased ca. 10 % (MILLAR and VAUGHAN WILLIAMS 1981b).

The (corrected) sinus node recovery time, i.e. the interval between an electrical stimulus and the first spontaneous electrogram of sinus origin, was increased from 83 ± 47 ms to 126 ± 80 ms, $n = 9$, by 2.9 μmol/litre alinidine (HABERL and STEINBECK 1985). Alinidine, in concentrations lower than 114 μmol/litre, did not change the pacemaker hierarchy as measured by multiple intracellular impalements in rabbit atria; thus the effect cannot be explained by a change from a faster to a slower discharging sinus node cell (BOUMAN et al. 1984, 1985; HABERL and STEINBECK 1985; OPTHOF et al. 1986).

From these investigations it appears that in relevant concentrations (< 30 μmol/litre = < 10 μg/ml, Fig. 1) the deceleration of diastolic depolarization is the direct drug effect and the main determinant of the fall in discharge rate (MILLAR and VAUGHAN WILLIAMS 1983; BOUMAN et al. 1984; HABERL and STEINBECK 1985).

An increase in atrial refractory period (HABERL and STEINBECK 1985) and in atrial-Hisian conduction time but not in Hisian-Purkinje conduction time was observed in isolated preparations of rabbit cardiac tissue (MILLAR and VAUGHAN WILLIAMS 1983). The latter workers compared alinidine with two antiarrhythmics in concentrations which were approximately equipotent in reducing heart rate; all three drugs reached this goal by affecting quite different electrophysiological parameters in sinoatrial cells: melperone by prolongation of action potential duration (class III antiarrhythmic property), cibenzoline by a slowing of the depolarization of the sinoatrial pacemaker potential as well as prolongation of its action potential (class IV and class III antiarrhythmic properties) and alinidine mainly by retardation of the diastolic depolarization.

Alinidine did also reduce discharge rates in other rhythmically beating cardiac cells. Purkinje fibres were retarded in a concentration-dependent manner (2.85–85.5 μmol/litre) with a decrease in diastolic depolarization and standstill in the highest concentration. The standstill was in contrast to sinus node preparations, where alinidine only reduced frequency (TRITTHART et al. 1981; LILLIE and KOBINGER 1983a, 1984). In cultivated myocytes of neonatal rats the rate of automaticity was suppressed by alinidine 0.1–100 μmol/litre concentration dependently, without changing inotropic functions of the single cells (WARBANOW et al. 1984).

The drug exerted little changes in the action potential of stimulated ventricular papillary muscle of guinea pigs in concentrations up to 85.5 μmol/litre alinidine. In the very high concentration of 285 μmol/litre, the rate of rise was reduced and the last third of repolarization retarded. These changes were independent of stimulation frequency (1 and 5 Hz).

Contrary to atrial preparations (Table 1, EC_{30} = 440 μmol/litre), in papillary muscles a negative inotropic effect was observed already with 28.5 and 85.5 μmol/litre (peak tension development decreased by 48 % and 63 % respectively; TRITTHART et al. 1981).

In isolated blood-perfused dog heart preparations KAWADA et al. (1984a) demonstrated the selectivity of alinidine for sinus node automaticity versus other cardiac activities as AV nodal and intraventricular conduction, and the contractility of paced papillary muscle. HAGEMAN et al. (1985) infused alinidine into the sinus node artery of anaesthetized dogs. This reduced spontaneous sinus rate (5–25 μg/ml) but infusion into the AV node artery (1–100 μg/ml) had no effect on the A-H interval of the bundle electrogram.

2. Mode of Action

A beta-adrenoceptor-blocking effect of alinidine at cardiac sites has been excluded: in isolated, spontaneously beating guinea pig atria, alinidine shifted the dose-reponse curve of isoproterenol down to lower frequency values along the ordinate. No shift along the abscissa was observed, indicating a functional, but excluding a competitive, antagonism (KOBINGER et al. 1979a). A difference between alinidine and beta-adrenoceptor-blocking agents that might be of clinical importance in situations of stress has been demonstrated with isoprenaline in isolated guinea pig atria. Isoprenaline (0.1 μg/ml) increased both rate in sponteneously beating atria and contractility in electrically driven preparations (1 Hz). Propranolol (0.3 μg/ml) abolished the positive chronotropic and inotropic effect of isoprenaline, alinidine (3 μg/ml) abolished the chronotropic without reducing the inotropic effect (LILLIE and KOBINGER 1983b).

The bradycardic effect of alinidine in isolated guinea pig atria was not changed by phentolamine (1 μg/ml) or atropine (0.05 μg/ml), a finding that excluded the involvement of cardiac alpha-adrenoceptors or muscarinic receptors in the bradycardic action. On the other hand, an anticholinergic action of the drug was reported by an antagonism against the negative chronotropic action of carbachol on isolated guinea pig atria (KOBINGER and LILLIE 1984a). This observation was also made by JAHNEL and NAWRATH (1985), who explained this antagonism by opposite effects on the action potential and K^+ efflux rather than by an interaction at muscarinic receptors.

Obviously the slowing of pacemaker activity by alinidine is caused by changes of ionic membrane currents. It is important to emphasize that the main effect, bradycardia and slowing of diastolic depolarization in pacemaker cells in isolated tissues was observed at concentrations 10 μmol/litre (ca. 3 μg/ml). In humans the effective plasma concentration was even lower, reaching 100 ng/ml only, after an oral dose of 40 mg. It was already obvious on the basis of pharmacological results and on recordings of action potentials that a decrease in fast inward current (Na^+), slow inward current (Ca^{2+}/Na^+) or an increase in K^+ permeability cannot be responsible for the pacemaker retardation (see Sect. B.I.1). Based on this "diagnosis by exclusion" MILLAR and VAUGHAN WILLIAMS (1981a,b) presented the hypothesis that the action of alinidine might due to inhibition of anion-selective (Cl^-) membrane channels and proposed the substance as a prototype of "class V antiarrhythmic drugs". Inhibition of this Cl^- current by alinidine would explain the slowing of the depolarization as well as the repolarization of the action potential, changes which have been described in the foregoing section. The authors supported their hypothesis by experiments on rabbit atria, in which Cl^- was substituted by either Br^- or methylsulphate, i.e. anions with either higher or smaller permeability through anion channels than Cl^-. In accordance with their hypothesis, alinidine (1-6.25 μmol/litre) was more effective in the Br^- and less effective in the methylsulphate solution (MILLAR and VAUGHAN WILLIAMS 1981a,b).

Recently another possible mechanism was proposed, the blockade of "i_f" (or "i_h"), a mixed Na^+/K^+ voltage- and time-dependent inward current with a reversal potential between −30 and −40 mV, activated by hyperpolarization in Purkinje fibres and sinus node cells and eliminated by low concentrations of caesium (review: DI FRANCESCO 1985). Mentioned first on pharmacological reasoning (LILLIE and KOBINGER 1984), SNYDERS and VAN BOGAERT (1985; 1987) reported evidence that in Purkinje fibres alinidine inhibited i_f in a voltage-dependent manner. Thus alinidine 10 μg/ml (= 28 μmol/litre) shifted the voltage dependence of i_f activation in hyperpolarizing direction (7.8 ± 0.6 mV, $n = 18$, $P < 0.001$) and the conductance of the fully activated i_f

channel was reduced to 73 ± 2 % ($P < 0.001$) of the control value. These effects were concentration dependent and in Tyrode's solution buffered with 5 % CO_2—HCO_3 a half maximal shift of the current voltage curve was measured with 14 µmol/litre alinidine (SNYDERS and VAN BOGAERT 1987). In the presence of 2 mmol/litre caesium chloride (a known inhibitor of the i_f) alinidine no longer changed the steady state current-voltage relationship. There is therefore little doubt that the reduction of diastolic depolarization rate by alinidine in Purkinje cells is due to the inhibition of i_f.

This question is more delicate in isolated sinus node cells (maximum diastolic potential −58 mV), where the activation range of i_f (−60 to −90 mV) is so negative that it is difficult to see how it could contribute to pacemaking. BOUMAN et al. (1985) showed the voltage-dependent inhibition of i_f in voltage clamp experiments on small clusters (20–50 µm) of SA nodal cells of rabbits. Alinidine, 3–80 µmol/litre, shifted the negative part of the current-voltage relationship in a hyperpolarizing direction and parallel to the control curve. From their curves it appears that the greatest change (in percentage) by the drug occurs at relatively "positive" potentials (−60 mV). Providing arguments that in situ pacemaker cells are under hyperpolarizing influence, the authors concluded that i_f did modulate pacemaking in the sinus node and that blockade of i_f is the main rate-lowering mechanism of alinidine.

Recently SATOH and HASHIMOTO (1986) clamped rabbit sinoatrial node cells and found a suppression in the slow inward current (i_{si}), the outward current i_k as well as i_h ($=i_f$) in concentrations of 10, 30, 100 µg/ml (ca. 30–300 µmol/litre) alinidine. From their data it appears that i_{si} and i_k were inhibited dose dependently, while inhibition of i_h was the same for all three concentrations used, i.e. possibly supramaximal even in the lowest concentration. It would be of interest to show whether lower concentrations would reveal the specificity of alinidine at i_h. It might be mentioned that in similar experiments the "Ca-channel blocker" diltiazem did not affect i_h ($= i_f$) in concentrations (10 µmol/litre) much higher than those which nearly completely suppressed i_{si} (2.2 µmol/litre; MIYAMOTO et al. 1986).

The i_f hypothesis for the bradycardic action of alinidine, however, has recently been doubted by DENNIS and VAUGHAN WILLIAMS (1986): alinidine was shown to decrease spontaneous heart rate of rabbit atria in the presence of caesium.

3. Experiments in Intact Animals

As in isolated heart preparations, the decrease in sinus rate was the prominent cardiovascular action of alinidine in intact animals: slowing of the heart rate was observed in doses that only slightly or not at all affected other cardiovascular and pharmacological parameters. The bradycardic effect was seen in various animal species and dose dependently between 0.4 and 5 mg/kg i. v. (review: KOBINGER and LILLIE 1984a). Moreover, the heart-retarding potency of alinidine—as of the other specific bradycardic agents—is dependent on the heart rate before drug administration: a positive correlation exists between the control heart rate and the bradycardic effect of a given dose of the drug. This

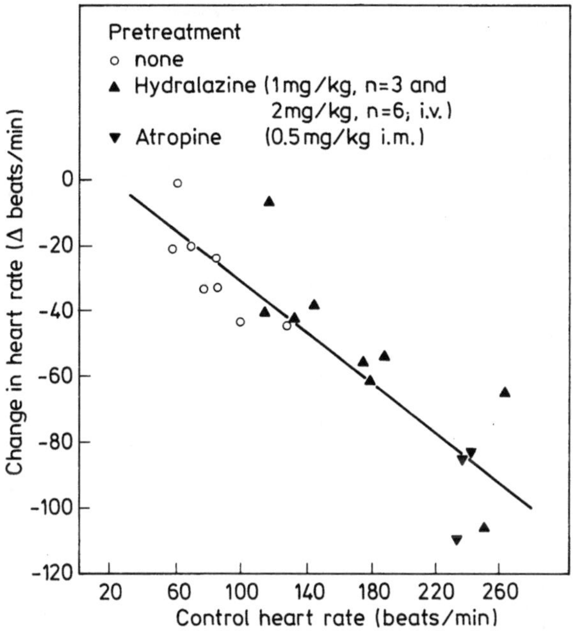

Fig. 4. Effect of alinidine on heart rate in conscious dogs. Correlation between control rate and bradycardic effect. *Abscissa*, control heart rate in beats/min; *ordinate*, change in heart rate (△ beats/min) induced by 2.5 mg/kg St 567 i.v. The dogs had either a genuine sinus rhythm (no pretreatment) or a sinus rhythm accelerated by injection of hydralazine or atropine. Pretreatment: ○, none; ▲, hydralazine (1 mg/kg, $n = 3$ and 2 mg/kg, $n = 6$ i.v.); ▼, atropine (0.5 mg/kg i.m.) (KOBINGER et al. (1979b)

is shown in Fig. 4 for conscious dogs that were either untreated or had an increased control rate induced by treatment with either hydralazine or atropine (KOBINGER et al. 1979b). The same correlation was shown later in humans during neurolept-anaesthesia and during controlled hypotension with sodium nitroprusside (FITZAL and ZIMPFER 1982). This can be described as a "self-limiting" mechanism and might explain why standstill or extreme bradycardia has not been reported and never been observed in this laboratory.

The ECG of cats and rats revealed that the bradycardic action of alinidine was due to a prolongation of the P-P interval; practically no changes were seen in the P-Q and QRS durations (KOBINGER et al. 1979a, b). There was a small increase in Q-T interval (Fig. 5) that can be explained by the bradycardia. In this respect, alinidine contrasts with antiarrhythmics, such as quinidine, which prolongs the QRS duration, verapamil, which markedly prolongs the P-Q interval, or *d*-sotalol, which increases the Q-T interval.

The ejection time, i.e. the time interval from the beginning of the carotid artery pressure rise until the first incisure of the pulse wave, was only slightly increased in cats. Therefore the main part of the increase in heart period (i.e. reciprocal value of heart rate) was calculated to be due to an increase in the diastolic period (interval from first incisure of pulse wave until next R peak in

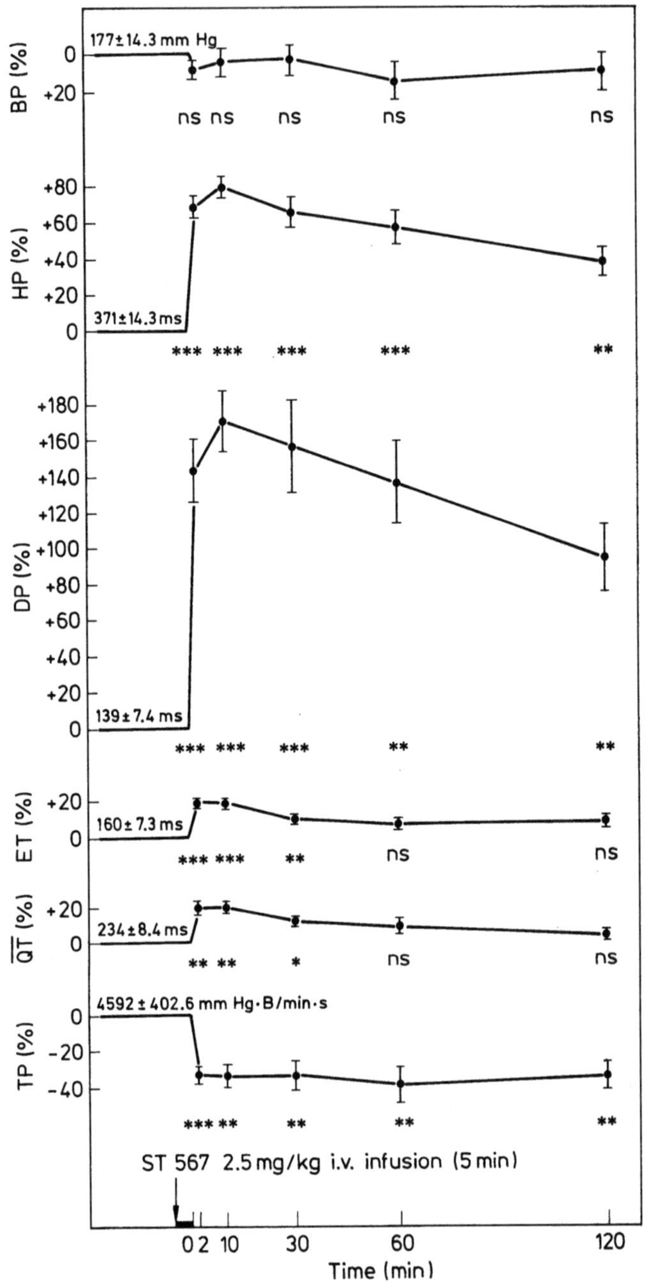

Fig. 5. Effect of alinidine (ST 567) on cardiovascular parameters in anaesthetized cats (chloralose). BP, systemic mean blood pressure; HP, heart period; DP, diastolic period; ET, ejection time; QT, QT interval from ECG; *TP*, triple product = systolic blood pressure × heart rate × ejection time. Each value is the mean of seven experiments. *Abscissa*, time after termination of infusion of 2.5 mg/kg St 567 for 5 min. *Ordinate*, changes in percentage of control values. *Numbers* at the beginning of each curve indicate the mean control values (± SE). Significance of differences between control and drug values, calculated by means of the confidence limits; n.s., not significant ($P > 0.05$); $*P \leqq 0.05$; $**P \leqq 0.01$; $***P \leqq 0.001$ (KOBINGER et al. (1979b).

the ECG; KOBINGER et al. 1979b; Fig. 5). Only minor changes or slight decreases of blood pressure were described following intravenous injection of alinidine in cats, rats, dogs and pigs (KOBINGER et al. 1979a, b; HARRON et al. 1982a; STRUYKER-BOUDIER et al. 1981; TRAUNECKER and WALLAND 1980; VERDOUW et al. 1980). Cardiac output (cardiac index) decreased in parallel with the heart rate. Pacing at a constant heart rate reduced the decrease in cardiac output but did not abolish it (2.5 mg/kg i.v.; SIEGL et al. 1984). After injection of alinidine in doses of 0.2 up to 6 mg/kg, there was little change, or a slight increase, in stroke volume and an increase in total peripheral vascular resistance (KOBINGER et a. 1979b; STRUYKER-BOUDIER et al. 1981; TRAUNECKER and WALLAND 1980; VERDOUW et al. 1980; SIEGL et al. 1984). In debuffered cats, blood pressure decreased more markedly than in normal animals, indicating that the baroreceptor-mediated pressor reflexes might be, at least partly, responsible for the increase in peripheral resistance (PICHLER 1982).

TRAUNECKER and WALLAND (1980) and VERDOUW et al. (1980) reported that alinidine caused a decrease in femoral, and no change in renal and cerebral, blood flow. Alinidine decreased myocardial contractility, as measured by the maximal rate of rise in left ventricular pressure (dp/dt_{max}) in cats, dogs and pigs (KOBINGER et al. 1979b; TRAUNECKER and WALLAND 1980; VERDOUW et al. 1980; SIEGL et al. 1984; KRUMPL et al. 1986b). The decreased contractility was due in part to a decrease in heart rate (staircase effect), as indicated by a significantly smaller effect of the drug on animals whose heart rate was kept constant by electrical pacing. At doses above 1 mg/kg i.v., however, alinidine had a direct negative inotropic effect (KOBINGER et al. 1979b; VERDOUW et al. 1980). In these experiments, where doses were carefully stepped up, the bradycardic effect was always observed with alinidine at smaller doses than necessary to achieve the negative inotropic action. It also appears that the negative inotropic effect of alinidine is of shorter duration than the negative chronotropic effect (TRAUNECKER and WALLAND 1980; SIEGL et al. 1984).

The product of systolic blood pressure × heart rate × ejection time—the "triple product"—has been shown to parallel myocardial oxygen consumption both in animal experiments and in angina patients (AMSTERDAM et al. 1974; NEILL et al. 1963). Alinidine decreased the triple product in cats (Fig. 5; KOBINGER et al. 1979b). In experiments in anaesthetized dogs, where myocardial blood flow as well as arterial and venous oxygen concentration were measured, a decrease in myocardial oxygen consumption (−44%) was reported following 3 mg/kg i.v. alinidine. This decrease was paralleled by a decrease in heart rate and coronary blood flow. Coronary venous oxygen concentration either remained unchanged or increased (TRAUNECKER and WALLAND 1980), suggesting the autoregulatory character of the decreased coronary flow and not a primary vasoconstriction by the drug. Similar results were reported for pigs (VERDOUW et al. 1980).

4. Investigations in Experimental Myocardial Ischaemia

In anaesthetized cats with open chests, occlusion of the anterior descending branch of the left coronary artery for 60 s leads to a ST segment elevation in a unipolar epicardial electrogram. This disturbance is reproducible at 15-min intervals for several hours (Kobinger et al. 1979b), and its degree has been reported to parallel myocardial tissue injury (Maroko et al. 1971; Sayen et al. 1958). For the duration of the experiments (60 min), the elevation of the electrogram elicited by the occlusion was significantly diminished by alinidine (1 mg/kg i.v.), an effect that paralleled the bradycardia (Kobinger et al. 1979b).

Schamhardt et al. (1981) measured regional myocardial blood flow in anaesthetized pigs by the microsphere technique. The partial occlusion of the proximal part of the left anterior descending coronary artery led to a 53% decrease in transmural blood flow and a pronounced decrease in the endocardial/epicardial blood flow ratio. Alinidine, at doses that reduced the heart rate by about 25% (0.4-0.9 mg/kg i.v.), given 15 min later increased blood flow to the endocardial layers of the ischaemic area by approximately 55% but did not affect flow to the epicardium. Consequently, the endocardial/epicardial blood flow ratio returned to control values, while the transmural blood flow in the ischaemic area remained unchanged. In contrast, in the non-ischaemic areas, the transmural flow was reduced, and the endocardial/epicardial flow ratio was not affected by the drug. Thus, the drug led to a redistribution of blood flow in favour of the endocardial layers of the myocardium in the ischaemic area. This shift was explained by the prolongation of the diastolic period (see Fig. 5), which is known preferentially to enhance endocardial flow. The diminished flow in the non-ischaemic parts of the heart is obviously caused by the reduced oxygen demand, which decreased the regional flow as an autoregulatory response.

In these experiments, myocardial wall thickness was monitored by an ultrasonic transducer. Following coronary artery stenosis, the mean velocity and magnitude of systolic wall thickening was reduced in the ischaemic myocardial segment. Alinidine improved these parameters parallel to the bradycardia and to the shifts in regional blood flow (Schamhardt et al. 1981). Krumpl et al. (1986a,b) used a model in instrumented, conscious dogs which combined partial coronary artery occlusion (circumflex branch of left coronary artery) and standardized exercise, imitating the main features of exertional angina pectoris in man. Under these conditions myocardial dysfunction appeared as a decrease in the systolic segment shortening (measured by the distance of piezoelectric crystals implanted subendocardially). The dysfunction was completely abolished in the exercise period following (20-35 min) the infusion of 1 mg/kg alinidine during 5 min. During peak exercise a reduction of heart rate (−18%) and left ventricular dp/dt_{max} (−24%) was measured and both changes may contribute to the beneficial effect on contractility in the ischaemic region.

5. Antiarrhythmic Properties

As pointed out in Sects. B.I.1 and B.I.3, alinidine, in sinus rate lowering concentrations or doses, did not show increases in refractory period or cell membrane effects such as other, known antiarrhythmics. In dogs 0.5-1.0 mg/kg i.v.

caused pronounced reduction in sinus rate; cumulative injection of higher doses, 2.9 and 15.5 mg/kg respectively, were required to abolish ventricular tachycardias induced by epinephrine and halothane or by toxic doses of ouabain. Alinidine up to 7.5 mg/kg was practically ineffective against ventricular tachycardias induced by coronary artery ligation (24 h thereafter) and 15.5 mg/kg reduced ectopic beats by 36% (HARRON et al. 1982a). The authors concluded that the sinus node was more responsive to the rate-lowering actions than were abnormal ventricular pacemaker sites.

The lack of antiarrhythmic properties in ventricular tachycardias illustrates the different effect of alinidine upon rhythmically discharging pacemaker activity and pathological occurring discharges. Class III antiarrhythmic properties were excluded in rabbits treated for 2 weeks with alinidine (VAUGHAN WILLIAMS et al. 1986). In contrast to these reports seems to be the beneficial effect upon ventricular fibrillation and death in anaesthetized rats with acute coronary occlusion (30 min): pretreatment with alinidine (1-6 mg/kg i.v.) reduced both parameters (combined as a factor "risk of death"). This effect was dependent on the dose and, thereby, on the degree of heart rate reduction immediately before coronary occlusion; it was discussed as secondary to the reduction in myocardial oxygen demand (HARRON et al. 1985).

II. Pharmacokinetics

After intravenous injection of 10 mg ^{14}C-labelled alinidine (specific activity, 10 μCi/mg) in humans, the plasma curves followed a multiexponential decline, representing a rapid distribution phase in tissues ($t/2\ \alpha = 35$ s) followed by two elimination phases ($t/2\ \beta = 44$ min; $t/2\ \gamma = 210$ min). After oral administration of 40 mg the absorption was calculated as 95%, the main route of elimination occurred via urine (12 h, 82%; ARNDTS et al. 1981; WIEGAND et al. 1982); the plasma concentration curve is depicted in Fig. 6. With the development of highly sensitive radioimmunoassays clonidine was detected as the *N*-dealkylated metabolite of alinidine, appearing in plasma and urine in amounts approximately 2-3 orders of magnitude smaller than that of the parent compound. In contrast to alinidine (see above), the plasma concentration curve of appearing clonidine reached a maximum not before 3-8 h after a single oral administration of alinidine and remained at the plateau level for up to 12 h (ARNDTS and FORSTER 1981; HARRON et al. 1982b). The time course of both substances alinidine and clonidine after a single oral dose of alinidine is depicted in Fig. 6 (lower part).

Data of plasma levels following repeated administration of alinidine have been reviewed by KOBINGER and LILLIE (1984a) and HARRON and SHANKS (1985). A 40 mg t.i.d. treatment schedule resulted in steady plasma levels of alinidine after 3 days, with trough and peak levels of ca. 42 and 180 ng/ml respectively (i.e. 0.12-0.51 μmol/litre). This treatment also resulted in clonidine plasma levels (ca. 0.3-0.8 ng/ml) nearly as high as following treatment with a moderate therapeutic dose of clonidine (0.1 mg t.i.d; HARRON et al. 1983).

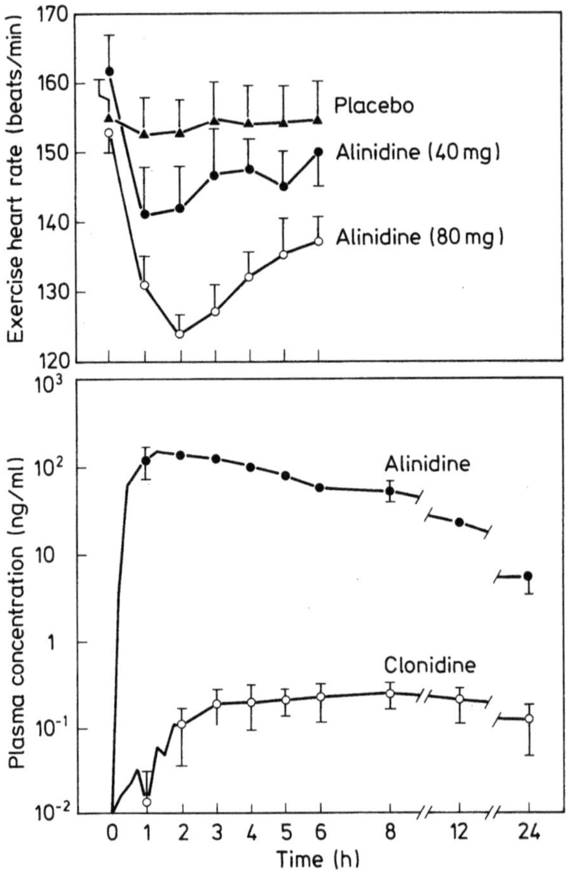

Fig. 6. Heart rate and plasma concentrations in human volunteers after a single oral administration of alinidine. *Upper*: exercise heart rate before and after several doses of alinidine and placebo respectively. Mean (± SEM, $n = 5$) (adapted from HARRON et al. 1981). *Lower*: plasma concentration of alinidine as well as clonidine following administration of 40 mg alinidine. Mean (± SD, $n = 5$) (*adapted from* HARRON et al. 1982b)

III. Clinical Pharmacology

In contrast to the known antiarrhythmics, alinidine showed little or no effects in electrocardiographic parameters in humans. WIEGAND et al. (1982) reported no changes in surface electrograms (duration of P, PQ time, QRS duration, $Q\text{-}T_c$) after intravenous injection of 40 mg alinidine to healthy volunteers, which led to a maximal decrease in heart rate of 19.2% ± 7.7% (SD, $n = 5$). A detailed study of the electrophysiological properties of alinidine (40 mg i.v.) by means of His-bundle electrograms on patients with sinus rhythm but various disturbances of the cardiac conduction system was reported by KASPER et al (1981). There was no change in SA conduction time, the effective refractory period of the right atrium, the AV node, and the right ventricle, the functional refractory period of the AV node, intra-atrial, AV, and intraventricular conduction as well as sinus node recovery time when corrected for the bradycardic effect (sinus cycle length increased by 23 %).

In humans with normal cardiac function, the decrease in heart rate was reported as the predominant action of alinidine (Fig. 6). Exercise tachycardia was more reduced than resting heart rate (HARRON et al. 1981, 1982c) as was the tachycardia increased by emotional stress (AUBÖCK et al. 1982); a positive correlation was found between predrug heart rates and the effect of alinidine (FITZAL and ZIMPFER 1982) as described for dogs (Fig. 4, Sect. B.I.3). Dose response relations between 20 and 80 mg i.v. and per os were demonstrated by HARRON et al. (1981) and SIMOONS and HUGENHOLTZ (1984). Blood pressure was not influenced or slightly decreased.

Detailed acute haemodynamic studies were reported from patients with suspected coronary artery diseases undergoing cardiac catheterization (JASKI and SERRUYS 1985). Alinidine (0.6 mg/kg i. v. over 3 min) resulted in a decrease in sinus rate (−13 %) but also in negative inotropic effects, which were still observed after atrial pacing and therefore the authors queried the specificity of the drug upon heart rate. Haemodynamic parameters were determined a few minutes after i.v. administration of the drug and a plasma concentration of 790 ± 112 ng/ml ($n = 12$) was shown during this study. As has been reviewed by HARRON and SHANKS (1985) a distinct heart rate lowering can be achieved by plasma concentrations between ca. 43 and 160 ng/ml alinidine thus negative inotropic effects would be expected only at considerably higher concentrations (Sects. B.I.1, B.I.3).

Alinidine was administered repeatedly for 8 days (40 mg orally once, twice or three times daily; HARRON et al. 1982c, 1983). Heart rate was reduced in the supine and standing position and, most pronounced, after exercise. Maximal effects were reached between 2 and 4 days; they did not substantially change during further treatment. The treatments also reduced systolic and diastolic blood pressure. In these studies, tiredness was reported on days 1 and 2 but not during the remainder of the study. This side effect, as well as dryness of the mouth as reported at 8–9 h after a single oral dose of alinidine (HARRON et al. 1982c), is probably caused by the metabolic formation of clonidine (see above). Clonidine, by means of its central sympathoinhibition, reduces blood pressure and heart rate (review: KOBINGER 1978). The question may be asked, therefore, whether the cardiovascular effect after chronic treatment with alinidine are attributable to this metabolite. Four arguments contradict this hypothesis:

1. The time course of the bradycardic effect after a single oral dose of alinidine follows closely the plasma concentration profile of alinidine (peak, 1–2 h) and not that of clonidine (plateau after 3–12 h, Fig. 6; HARRON et al. 1982b).
2. Following administration of single oral doses, which are equipotent in reducing blood pressure, alinidine, but not clonidine, decreased exercise heart rate (HARRON et al. 1981).
3. Fifteen minutes after intravenous injection of a single dose of alinidine, exercise-induced tachycardia was significantly reduced, but no decrease in

plasma norepinephrine level was measured, as would be expected for cardiovascular active levels of clonidine (STANEK et al. 1983).
4. Following chronic administration of clonidine (0.1 mg t.i.d) or alinidine (40 mg t. i. d) for 7 days, comparable plasma levels of clonidine were achieved. During alinidine administration, however, the exercise heart rate was decreased more markedly than during clonidine treatment, wheras the effects on blood pressure were similar (HARRON et al. 1983).
5. Bradycardia is observed at low concentrations in *isolated* tissues, in which clonidine would not, of course, be produced.

Data on clinical pharmacology were recently summarized by KOBINGER and LILLIE (1984 a) and HARRON and SHANKS (1985).

IV. Clinical Results

Patients with hyperkinetic heart syndrome received 0.5 mg/kg i.v. alinidine, which reduced heart rate at rest and during exercise (REITERER and STANEK 1980; STANEK et al. 1983). SIMOONS and HUGENHOLTZ (1982, 1984) investigated haemodynamic effects in patients with tachycardia, following myocardial infarction partly under treatment for ventricular failure and shock. Cumulative bolus injections of 10 mg alinidine each (maximal 40 mg) reduced heart rate in most patients without clinically significant negative inotropic actions. In particular, no elevation of left ventricular filling pressure was observed although some of these patients suffered from left ventricular failure. There were two exceptions with elevation of the pulmonary capillary wedge pressure (SIMOONS 1985).

In patients with coronary artery disease, alinidine (20 mg i.v.) decreased heart rate and blood pressure during rest and exercise (LÖLLGEN et al. 1981); it also decreased pulmonary artery pressure (12 % and 18 %) and pulmonary capillary wedge pressure (19 % and 28 %). There was an insignificant increase in cardiac output. The improvement of myocardial performance was explained by an increase of myocardial oxygen utilization and reduced oxygen consumption in the state of left ventricular dysfunction (LÖLLGEN et al. 1981). This observation is supported by recent animal data showing alinidine to improve contractility during exercise in ischaemic myocardial regions (KRUMPL et al. 1986 a, b; Sect. B.I.4). A decrease in left ventricular filling pressure in coronary patients was also reported by REITERER (1981).

Six patients with coronary artery disease were subjected to a stepwise increasing exercise test. During control exercise in five patients, the test was limited due to severe angina pectoris; after 20 mg i.v. alinidine, none of the patients complained of angina at the same workload. ST segment depression was significantly less pronounced after alinidine (LÖLLGEN et al. 1981).

In a double-blind, randomized study with 16 anginal patients, alinidine (40 mg per os) was compared with propranolol (40 mg) and placebo (SCHURMANS et al. 1982). Both active drugs reduced heart rate, blood pressure and ST depression at a given workload and increased maximal workload capacity; alinidine was more effective at 2 than 5 h after treatment, a course of activity which again follows the plasma level (and bradycardia) of alinidine and not that of clonidine (see Fig. 6). SIMOONS et al. (1982) also reported improved exercise tolerance in patients with stable angina at 2 h after 40 and 60 mg p.o. alinidine; the ST segment depression was smaller after 40 mg alinidine at each level of exercise (SIMOONS and BALAKUMARAN 1981).

MEINERTZ et al (1980, 1981) reported 12 patients with stable angina in a double-blind, randomized, placebo-controlled, crossover study for 10 weeks, where 40 mg alinidine was given t.i.d. orally. Heart rate and rate-pressure product during workload was reduced, and exercise tolerance was clearly improved in six, slightly in three and remained constant in three patients (MEINERTZ et al. 1981). In another long-term study in 32 patients with stable angina, alinidine (30–40 mg t.i.d.) was compared with metoprolol (50–100 mg t.i.d.) in a double-blind crossover study, where each drug was given for 5 weeks. Both drugs were equally effective in reducing the number of anginal attacks (from eight to three/week) and in increasing the physical work load performed without angina (from 87 to 99 and 92 W respectively; BALAKUMARAN et al. 1984; SIMOONS 1985).

V. Adverse Effects of Alinidine

As recently summarized by HARRON and SHANKS (1985) side effects are mainly due to the release of the metabolite clonidine: drowsiness, sedation and dry mouth. These effects parallel the plasma concentration of clonidine (maximum 6–8 h after a single dose), disappear after the first days of treatment and seem to be more prominent in healthy volunteers because they are rarely described in patients (review: HARRON and SHANKS 1985). There were no reports of severe haemodynamic effects. In healthy volunteers and occasionally in patients visual side effects occurred such as flashing lights, flickering, double and multiple contours. These observations were of short duration and, with chronic administration, generally lasted for 2–3 days and rarely persisted throughout the entire treatment period (review: SHANKS 1987).

C. Substances Chemically Related to Alinidine

Structure-activity relationships were published by STÄHLE et al. (1980). The alkylation of the exocyclic nitrogen atom of the clonidine molecule seems the essential step to change the pharmacological properties from a "clonidine-like" substance to a "specific bradycardic agent". HABERL and STEINBECK (1985) reported that clonidine exerted electrophysiological properties similar to those of alinidine (2.9 μmol/litre; slowing of phase 4-depolarization, negative chronotropic effect). This drug, however, has additional properties including stimulation of alpha$_2$-adrenoceptors and histamine H_2-receptors, in doses 1/100–1/1000 that of alinidine, changing essentially the character of the molecule (reviews: KOBINGER 1978, 1986).

Highest bradycardic potency was reached with STH 2148 (Fig. 1) which was about five times more potent than alinidine. This drug cannot metabolize to clonidine (the expected metabolite 2-(2,6-dibromphenylimino)-2-imidazolidine is less potent as a clonidine-like agent than clonidine; W. KOBINGER, L. PICHLER 1980, unpublished data); it therefore might be a more specific tool and therapeutic agent than alinidine.

D. Falipamil (AQ-A 39) and Congeners (AQ-AH 208, UL-FS 49)

These substances have some chemical resemblance to verapamil (Fig. 1), but their cardiovascular profile differs from that of the so-called calcium-channel blockers and much more resembles that of alinidine.

I. Pharmacology

1. Isolated Preparations

The profile of these drugs in comparison with that of calcium-channel blockers is illustrated in Table 2. Results were obtained from isolated guinea pig atria with dose response curves similar to those presented for alinidine in Fig. 2. There was, however, a much longer equilibration time (60 min) because of the slower onset of action with some of these drugs (KOBINGER and LILLIE 1984b). In addition, vascular relaxation was tested in isolated aortic strips of rabbits contracted in a Tyrode's solution with 42.7 mmol/litre KCl and 1.8 mmol/litre $CaCl_2$. The specificity of the bradycardic effect in comparison with the other parameters is revealed from the high "ratios of EC_{30}". The difference between falipamil and congeners on one side and the three calcium-channel blockers on the other is illustrated by the relative absence of vasorelaxing properties (aortic contracture) and of negative inotropic effects (atrial contractility) of the falipamil group: within the three drugs of the latter group UL-FS 49 exerted the highest, falipamil the least and AQ-AH 208 an intermediate specificity. The same ranking order can be seen for the bradycardic potency (atrial rate, EC_{30} in µmol/litre: falipamil, 1.3; AQ-AH 208, 0.29; UL-FS 49, 0.06). As mentioned for alinidine these drugs also exerted a cardiac anticholinergic activity, reducing the bradycardic effect of carbachol. The concentration of falipamil which shifted the concentration response curve of carbachol by a factor of 10 to the right was 1.7 µg/ml (3.5 µmmol/litre), that for UL-FS 49 was 11.3 µg/ml (23 µmol/l) showing again the greater specificity of the latter drug towards bradycardia (KOBINGER and LILLIE 1981, 1984b).

Intracellular recordings in electrically driven atrial preparations of guinea pigs showed no change in resting potential or in maximal upstroke velocity and amplitude for falipamil 4.3 µmol/litre (KRÄHENMANN und HEISTRACHER 1980; HOHNLOSER et al. 1982). Unlike alinidine, falipamil markedly increased the 50% and 90% repolarization time in this preparation as well as in isolated guinea pig papillary muscles and Purkinje fibres. Similar results were described for UL-FS 49: repolarization time increased with 0.2 µmol/litre but not with 0.04 µmol/litre; the latter concentration, however, effectively lowered sinus

Table 2. Effects of substances on spontaneous rate, contractility and maximal driving frequency in isolated guinea pig atria and on K^+-induced contracture in isolated aortic strips of rabbits. EC_{30} is the concentration which decreased predrug value by 30 %, evaluated from curves as in Fig. 2. (KOBINGER and LILLIE 1984b; AQ-AH 208 unpublished results, gained under the same experimental conditions)

Substance	Decrease in				Ratio of EC_{30}		
	Atrial rate EC_{30} (µg/ml)	Atrial contractility EC_{30} (µg/ml)	Atrial maximal driving frequency EC_{30} (µg/ml)	Aortic contracture EC_{30} (µg/ml)	Atrial contractility / Atrial rate	Atrial maximal driving frequency / Atrial rate	Aortic contracture / Atrial rate
Falipamil	0.608[a]	93.6	9.63	21.3	154	15.8	35.0
AQ-AH 208	0.137[b]	26.6	4.37	16.33	194	34.5	119
UL-FS 49	0.0300[c]	108	11.3	15.0	3 600	377	500
Verapamil	0.0694	0.0611	0.214	0.0284	0.880	3.08	0.409
Nifedipine	0.11[d]	0.0149	>0.8[e]	0.00397	0.135	>7.3	0.036
Diltiazem	0.14[f]	2.73	4.40	0.0741	19.5	31.6	0.53

[a] 1.3 µmol/litre
[b] 0.29 µmol/litre
[c] 0.06 µmol/litre
[d] Extrapolated flat concentration-response curve up to 0.1 µg/ml ($r = -0.64$), with 0.3µg/ml standstill in two out of two preparations
[e] Higher concentrations not tested because of negative inotropy
[f] Flat curve up to 0.3 µg/ml ($r = -0.65$), with 1 µg/ml arrhythmias in three out of four preparations

rate and the slope of the slow diastolic depolarization on spontaneously beating sinus node preparations, indicating the greater specificity of UL-FS 49 towards bradycardia (GRUBER et al. 1983). In rabbit sinus nodal preparations, falipamil (10 µmol/litre) reduced spontaneous discharge rate, rate of diastolic depolarization, slowed the upstroke and prolonged the duration of the action potential. Overshoot and maximal diastolic potential were decreased (OSTERRIEDER et al. 1981). Falipamil also reduced the rate of spontaneous discharges in Ba^{2+}-induced pacemakers of ventricular myocardium (43 µmol/litre) and in Purkinje fibres (25 µmol/litre; HOHNLOSER et al. 1982).

The decrease in contractility with high concentrations of falipamil in guinea pig atria was shown in Table 2 (EC_{30}=94 µg/ml=210 µmol/litre). It was also reported with lower concentrations and was frequency dependent in rabbit ventricular fibres (10 µmol/litre) and papillary muscles of guinea pigs (21.5 µmol/litre; TRAUTWEIN et al. 1981; HOHNLOSER et al. 1982). On the other hand, low concentrations of falipamil were observed to increase contractile force (KOBINGER and LILLIE 1981). URTHALER and WALKER (1984), using canine ventricular trabeculae with low-frequency stimulation (12/min), found the positive inotropic effect of falipamil (10–20 µmol/litre) abolished by pretreatment with reserpine and atropine and therefore due to sympathetic stimulation as well as to an antimuscarinic effect. In isolated guinea pig atria falipamil and UL-FS 49 reduced the rate, but not the contractile force as increased by isoprenaline (LILLIE and KOBINGER 1983b, 1986).

2. Mode of Action

Falipamil has been shown to prolong the action potential in driven and spontaneously beating cardiac cells (see above). In voltage clamp experiments on SA nodes of rabbits (20 µmol/litre) falipamil reduced the time- and voltage-dependent outward current i_K, explaining the prolongation of the action potential (OSTERRIEDER et al. 1981). However, this has not been considered to cause the bradycardia, as the concomitant shift of the maximal diastolic potential towards positive values counteracts this effect. Moreover an essential connection between bradycardia and prolonged action potential is unlikely, as UL-FS 49 is more potent as a bradycardic agent with less effect on action potential duration than falipamil (GRUBER et al. 1983; foregoing section). Therefore, the prolongation of the action potential by falipamil may be considered as an effect additional to the bradycardia, a property proposed for class III antiarrhythmic agents. As with alinidine, the reason for the bradycardia may be attributed to slowing of the diastolic depolarization rate.

An inhibition of the slow inward current system (i_{si}) by falipamil (20 µmol/litre) was observed in voltage clamp experiments in small preparations from rabbit hearts, either from ventricular or sinoatrial sites (TRAUTWEIN et al. 1981; OSTERRIEDER et al. 1981). This inhibition was strongly frequency and voltage dependent, the drug acting mainly at lower (i.e. depolarized) potentials and at higher rates of activity. From the results of TRAUTWEIN et al. (1981) it appears that falipamil reaches the blocking site only during depolarization and is removed during diastole. The i_{si} blockade in relatively positive voltage regions (> −45 mV) differentiates the drug from other calcium-channel blockers and might explain the specific action on the sinus node, which operates at a relatively depolarized diastolic level (PELZER et al. 1982; TRAUTWEIN et al. 1981, 1983).

There are, however, other results where blockade of i_{si} by falipamil does not explain the bradycardia. As reported above, 25 μmol/litre falipamil reduced discharge rate in Purkinje fibres where i_{si} has no influence on pacemaker activity (HOHNLOSER et al. 1982). The rate-lowering effect of verapamil was compared with that of falipamil in intact sinus node preparations of guinea pigs. Low Ca^{2+} increased and low Na^+ as well as high K^+ (depolarization!) decreased the bradycardic effect of falipamil (6.5 μmol/litre), but the action of verapamil was influenced in the opposite direction. Alinidine was also tested under these conditions and behaved like falipamil (LILLIE and KOBINGER 1983a, 1984) as did UL-FS 49 (0.2 μmol/litre) in high K^+ concentration (LILLIE and KOBINGER 1986).

When sinus rate was decreased by a supramaximal dose of falipamil (65 μmol/litre), addition of verapamil (0.2 μmol/litre) resulted in a significant further lowering of the rate. The same result was gained using the alinidine derivative STH 2148 (66 μmol/litre) instead of falipamil. When the rate was lowered by a supramaximal dose of STH 2148, addition of falipamil was ineffective (LILLIE and KOBINGER 1987). These results suggest that in intact, isolated guinea pig atria the bradycardic effects of alinidine (and alinidine-like drugs) and of falipamil are mediated by the same mechanism, which is different from that of "calcium-channel blockers". Recently, it has been published that in voltage clamp experiments with sheep Purkinje fibres AQ-A 39 and UL-FS 49 blocked the i_f current, i. e. the same current as is blocked by alinidine (see Sect. B.I.2). VAN BOGAERT and GOTHALS (1987) analysed this block with UL-FS 49, which mainly reduced the amplitude of the i_f activation curve, without shifting this curve. This distinguishes UL-FS 49 from alinidine, which mainly shifts the voltage-current (i_f) curve in a hyperpolarized direction, but only slightly reduces the amplitude. Moreover, these authors described the use dependence of the i_f block by UL-FS 49.

3. Experiments in Intact Animals

Falipamil and UL-FS 49 reduced heart rate in doses which do not or only little change other cardiovascular parameters. In anaesthetized cats prolongation of the diastolic period was the most prominent change (KOBINGER and LILLIE 1981, 1984b). For falipamil, reduction in contractility and blood pressure was reported in higher doses (ED 20%, mg/kg i. v.: dp/dt_{max}, 1.8; mean BP, 2.9; rate, 0.5; DÄMMGEN et al. 1981). The drug reduced cardiac output, but less than heart rate, increased stroke volume and did not change systemic vascular resistance in dogs and pigs (DÄMMGEN et al. 1981; VERDOUW et al. 1983). The triple product (heart rate x ejection time x blood pressure; see Sect. B.I.3) was reduced, indicating reduction of myocardial oxygen consumption (KOBINGER and LILLIE 1981). This was established by O_2 measurements in coronary blood and discussed to be primarily due to the heart rate reduction; in higher doses the negative inotropic action might contribute, as shown by electrical pacing in pigs. The absence of an increased vascular resistance indicated a slight vasodilatory effect on coronary arteries (VERDOUW et al. 1983).

In ECG studies in cats falipamil (0.1-10 mg/kg) and UL-FS 49 (0.3 mg/kg) showed little or no effect on P-Q and QRS intervals, contrary to verapamil, which markedly increased P-Q interval (KOBINGER and LILLIE 1981, 1984b; KOBINGER et al. 1979a, b). Falipamil increased Q-T interval and in parallel the effective refractory period as determined by R—triggered extra stimuli in the right ventricle. This effect is explained by the prolongation of the repolarization phase of the action potential, as shown in recordings from single cells (class III antiarrhythmic action, see Sect. D.I.1). In accordance with less "class III" activity, UL-FS 49 is also much less effective than falipamil in prolonging Q-T and effective refractory period, and this can be seen from Fig. 7, where effects on refractory period are plotted against bradycardic effects (cycle length). Sotalol is given as a reference for a class III antiarrhythmic agent. The higher specificity towards bradycardia is obvious for UL-FS 49, which approaches the values for alinidine. AQ-AH 208 closely resembled falipamil (Fig. 7).

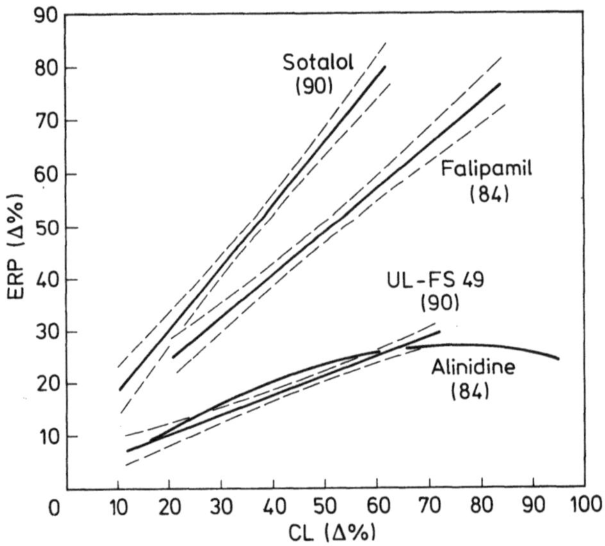

Fig. 7. Correlation between changes (△%) in effective refractory period (ERP) and in cycle length (CL) of cats (chloralose anaesthesia) as induced by i.v. injection or infusion of various drugs. ERP was determined by test pulses (voltage: double diastolic threshold) triggered by the R wave of the ECG delivered from a bipolar stimulating electrode introduced in the right ventricle. CL: RR′ interval of ECG. Regression lines are given and their 95% confidence limits. Number of data pairs used for calculations in parentheses. Sotalol, 15 mg/kg; falipamil, 10 mg/kg; UL-FS 49, 0.3 mg/kg (KOBINGER and LILLIE (1984b). Alinidine (10 mg/kg) did not fit into a linear but into a polynomial regression. The linear regression for AQ-AH 208 3.0 mg/kg (not depicted) did not differ significantly from that of falipamil (C. LILLIE 1986, unpublished results)

The preponderance of falipamil to decrease sinus rate in comparison to other cardiovascular parameters was further described by KAWADA et al. (1984b) and SIEGL et al. (1984).

4. Investigations in Experimental Myocardial Ischaemia

Decreases in signs of myocardial ischaemia by pretreatment with falipamil were already demonstrated in the first report on the drug: in cats with repeated acute occlusion (1 min) of a coronary artery branch falipamil (5 mg/kg i.v.) diminished the elevation of the ST segment of the epicardial electrogram (KOBINGER and LILLIE 1981). Similar results were gained with UL-FS 49 (0.5 mg/kg i.v.; KOBINGER and LILLIE 1987).

GROSS et al. (1985), DÄMMGEN and GROSS (1985) and GROSS and DÄMMGEN (1986) tested AQ-AH 208 i.v. (0.3 mg/kg; 0.5 mg/kg + 25 μg/kg/min) in anaesthetized, open chest dogs using various protocols of acute coronary artery branch occlusion and reperfusion. Pre- or postocclusion treatment with AQ-AH 208 resulted in increased coronary collateral blood flow mainly in subendocardial regions (radioactive microsphere technique), reduction of infarct area and improvement of myocardial function in the ischaemic region (systolic segment shortening, sonomicrometric technique). The authors concluded the beneficial effects of the drug as partly due to the increase in duration of diastole, and partly due to a vascular action, as electrical pacing did not completely reverse the effects (DÄMMGEN and GROSS 1985). In one report UL-FS 49 has also been used with similar, but much smaller, beneficial effects than AQ-AH 208, which might be explained by the very small dose of UL-FS 49 (i.v. bolus + infusion 105 μg/kg; DÄMMGEN et al. 1985). Using reasonable doses (0.5 mg/kg i.v.) KRUMPL et al. (1986b) demonstrated that UL-FS 49 prevented the regional contractile dysfunction in left ventricles of conscious dogs as evoked by coronary artery ischaemia and exercise. In these experiments UL-FS 49 produced a reduction of the heart rate without reduction in positive dp/dt_{max}, pointing out the selective effect of the bradycardia in an angina pectoris model (KRUMPL et al. 1986b—same experiments as referred for alinidine in Sect. B.I.4). GUTH et al. (1987) gained similar results (UL-FS 49 1 mg/kg i.v.) with respect to contractility in normal and ischaemic myocardial regions and, in addition, reported an increase of subendocardial and transmural blood flow in ischaemic myocardial regions.

II. Investigations with Falipamil and UL-FS 49 in Healthy Humans and Patients

Falipamil reduced heart rate in humans during physical work load dose dependently between 25 and 100 mg i.v. (BENEDIKTER et al. 1980; HILAIRE et al. 1983). In patients with sinus tachycardia this effect was dependent on the heart rate before medication; there was no effect in patients with atrial fibrillation (BENDER and GÜLKER 1982; HERZOG and SIMON 1983). During infusion of isoprenaline, falipamil was also administered (2 mg/kg i.v.) and decreased heart rate without reducing cardiac index, which was increased by isoprenaline. Stroke volume index and stroke work were increased by the test drug (Khosropour et al. 1983). Electrophysiological studies in patients with high sinus rate showed a decrease in heart rate, in improvement of AV conduction and an increase of the effective refractory period of the right atrium, right ventricle, in the H-V, Q-T and Q-T_c interval (Clementy and Hilaire 1984; Gülker et al. 1986). Patients with coronary heart disease were investigated in exercise tests in a placebo-controlled study. A single dose of falipamil 200 mg i.v. reduced ST-segment depression and duration of anginal pain with reduction in heart rate but without significant changes in blood pressure. No major side effects were reported in this study (Jost et al. 1985).

In healthy volunteers UL-FS 49 was given orally in single doses from 5 to 20 mg and t.i.d from 2.5 to 10 mg. The drug reduced heart rate at rest and at exercise up to 20% of control values without major side effects. With both falipamil and UL-FS 49 following repeated administration, occasionally a visual side effect occurred such as light flashes, light stripes, slow motion pictures, flickering and afterimages (Hellner and Gauri 1982; Franke 1987, personal communication).

E. Conclusion

Data have been provided which suggest a number of chemically different agents to be considered as a pharmacological class of its own: specific bradycardic agents. They are characterized by a slowing of the sinus rate within physiological limits as the dominating cardiovascular effect. Intracellular recordings of action potentials in pacemaker cells showed that retardation of the diastolic depolarization phase (phase 4) is the prominent effect of these drugs. So far all experiments performed to reveal the mode of action of specific bradycardic agents indicate that they have a different action upon ionic movements across myocardial cell membranes from that of other known antiarrhythmic agents. Therefore these drugs have been considered as another class of antiarrhythmics (class V). Evidence has been presented that these agents block the pacemaker current i_f (i_h), i.e. an inward current activated by hyperpolarization in Purkinje fibres and sinus node pacemaker cells. They will be applied to reduce sinus tachycardias, whenever this seems therapeutically indicated. Results from experimental models of myocardial ischaemia and from clinical trials indicate the therapeutic value of these drugs in ischaemic heart disease.

References

Amsterdam EA, Hughes JL, de Maria AN, Zelis R, Mason DT (1974) Indirect assessment of myocardial oxygen consumption in the evaluation of mechanisms and therapy of angina pectoris. Am J Cardiol 33:737–743

Arndts D, Forster HJ (1981) New aspects in the metabolism of alinidine in man. Eur J Metab Pharmacokinet 6:313–315

Arndts D, Leb G, Förster HJ (1981) Pharmacokinetics and metabolism of C-labelled alinidine in man and dog. Eur J Drug Metabol Pharmacol 6:225-236

Auböck, J, Konzett H, Olbrich E (1982) The effect of alinidine (St 567) on emotionally induced tachycardia in man. Eur J Clin Pharmacol 21:467-471

Balakumaran K, Lubsen J, Simoons ML, Jovanovic A, ten Cate FJ, van Es GA, Pieterse H (1984) Anti-anginal effect of the clonidine derivative alinidine: a comparison with metoprolol. Eur Heart J 5 [Suppl 1]:271

Bender F, Gülker H (1982) Ein neues Antiarrhythmicum (AQ-A 39) zur Behandlung von Sinustachykardien. Muench Med Wochenschr 124:97-98

Benedikter L, Trouvain H, Zimmer A (1980) Pharmacodynamics and pharmacokinetics of a new aryl-alkylamine with negative chronotropic effects in man. In: Rietbrock N, Woodcock BG, Neuhaus G (eds) Methods in clinical pharmacol, the proceedings of an international symposium held in Frankfurt/M. 6-8 May 1979, Vieweg, Braunschweig, pp 44-45

Bouman LN, Duivenvoorden JJ, Op't Hof T, Treytel BW (1984) Electrophysiological effects of alinidine on nodal and atrial fibres in the guinea-pig heart. J Pharmacol Exp Ther 229:551-556

Bouman LN, Jongsma HJ, Opthof T, van Ginneken ACG (1985) Does i_f contribute to pace-making in the rabbit sinoatrial node? J Physiol 358:51P

Clementy J, Hilaire A (1984) Electrophysiological effects of the bradycardic agent falipamil (AQ-A 39) Eur Heart J 5 [Suppl 1]:287 (abstr)

Dämmgen JW, Gross GJ (1985) AQ-AH 208, a new bradycardic agent, increases coronary collateral blood flow to ischemic myocardium. J Cardiovasc Pharmacol 7:1048-1054

Dämmgen J, Kadatz R, Diederen W (1981) Cardiovascular actions of 5,6-dimethoxy-2-[3[[(α(3,4-dimethoxy)-phenylethyl]methylamino]propyl]phthalimidine (AQ-A 39), a specific bradycardic agent. Arzneim Forsch Drug Res 31:666-670

Dämmgen JW, Lamping KA, Gross G (1985) Actions of two new bradycardic agents, AQ-AH 208 and UL-FS 49, on ischemic myocardial perfusion and function. J Cardiovasc Pharmacol 7:71-79

Dennis PD, Vaughan Williams EM (1986) Further studies of alinidine induced bradycardia in the presence of caesium. Cardiovasc Res 20:375-378

Di Francesco D (1985) The cardiac hyperpolarizing-activated current, i_f, origins and developments. Prog Biophys Mol Biol 46:163-183

Fitzal S, Zimpfer M (1982) Cardiovascular effects of *N*-allyl-clonidine (St 567, alinidine): a substance with specific bradycardic action during neuroleptanesthesia in humans. Int J Clin Pharmacol Ther Toxicol 20:404-407

Gross GJ, Dämmgen JW (1986) Beneficial effects of two specific bradycardic agents AQ-A 39 (falipamil) and AQ-AH 208 on reversible myocardial reperfusion damage in anesthetized dogs. J Pharmacol Exp Ther 238: 422-428

Gross GJ, Warltier DC, Dämmgen JW (1985) Effects of AQ-AH 208, a new specific bradycardic agent on myocardial ischemia-reperfusion injury in anesthetized dogs. J Cardiovasc Pharmacol 7:929-936

Gruber R, Lumper G, Zilberszac A, Heistracher P (1983) Effects of 1,3,4,5-tetrahydro-7,8-dimethoxy-3-[3-[[2-(3,4-dimethoxyphenyl)-ethyl]methylimino]propyl]-2H-3-benzazepin-2-on-hydrochloride (UL-FS 49 Cl) on the action potential of the guinea-pig heart. Naunyn Schmiedebergs Arch Pharmacol 324:R 32

Gülker H, Holtvogt J, Specker E, Thale J, Heuer H, Bender F (1986) Elektrophysiologische und haemodynamische Wirkungen der neuen bradykardisierenden Substanz AQ-A 39. Z Kardiol 75:47-51

Guth BD, Heusch G, Ross J (1987) Elimination of exercise-induced regional myocard-

ial dysfunction by a bradycardic agent in dogs with chronic coronary stenosis. Circulation 75:661-669
Haberl R, Steinbeck G (1985) Chronotropic action of alinidine on the isolated sinus node of the rabbit heart. Eur Heart J 6:730-736
Hageman GR, Neely BH, Urthaler F, James TH (1985) Negative chronotropic and parasympatholytic effects of alinidine on canine sinus node and AV junction. Am J Physiol 248:H324-H330
Harron DWG, Ridell JG, Shanks RG (1981) Alinidine reduces heart-rate without blockade of beta-adrenoceptors. Lancet i:351-353
Harron DWG, Shanks RG (1985) Pharmacology, clinical pharmacology and potential therapeutic uses of the specific bradycardic agent alinidine. Eur Heart J 6:722-729
Harron DWG Allen J, Wilson R, Shanks RG (1982a) Effect of alinidine on experimental cardial arrhythmias. J Cardiovasc Pharmacol 4:221-225
Harron DWG, Arndts D, Shanks RG (1982b) Alinidine pharmacokinetics following acute and chronic dosing. Br J Clin Pharmacol 13:821-827
Harron DWG, Jady K, Riddell JG, Shanks RG (1982c) Effects of alinidine, a novel bradycardic agents, on heart rate and blood pressure in man. J Cardiovasc Pharmacol 4: 213-220
Harron DWG, Arndts D, Finch M, Shanks RG (1983) An assessment of the contribution of clonidine metabolised from alinidine to the cardiovasular effects of alinidine. Br J Clin Pharmacol 16:451-455
Harron DWG, Brezina, M, Lillie C, Kobinger W (1985) Antifibrillatory properties of alinidine after coronary artery occlusion in rats. Eur J Pharmacol 110:301-308
Hellner KA, Gauri KK (1982) Über Akkomodationsphosphene. Fortschr Ophthalmol 79:169-170
Herzog H, Simon H (1983) Untersuchungen zur Frequenzsenkung bei Sinustachykardie und tachykardem Vorhofflimmern unterschiedlicher Ätiologie mit einer neuen bradykardisierenden Substanz (AQ-A 39). Herz/Kreislauf 7:354-357
Hilaire J, Broustet JP, Colle JP, Theron M (1983) Cardiovascular effects of AQ-A 39 in healthy volunteers. Br J Clin Pharmacol 16:627-631
Hohnloser SJ, Weirich J, Homburger H, Antoni H (1982) Electrophysiological studies on effects of AQ-A 39 in the isolated guinea-pig heart and myocardial preparations. Arzneim Forsch/Drug Res 32:730-734
Jähnel C, Nawrath M (1985) Interactions of alinidine and acetylcholine in guinea pig atria. Naunyn Schmiedebergs Arch Pharmacol 329:R54
Jaski BE, Serruys PW (1985) Anion-channel blockade with alinidine: a specific bradycardic drug for coronary heart disease without negative inotropic activity? Am J Cardiol 56:270-275
Jost S, Schulz W, Kober G (1985) Wirkung des neuen Kalziumantagonisten 5,6-Dimethoxy-2-[3[[α-(3,4-dimethoxy)-phenylethyl]methylamino]propyl]-phthalimidin-hydrochlorid (AQ-A 39 Cl) auf Hämodynamik und Ischämieparameter in Belastungs-EKG von Patienten mit koronarer Herzkrankheit. Arzneim Forsch/Drug Res 35:1279-1282
Kasper W, Meinertz T, Treese N, Kersting F, Pop T, Jähnchen E (1981) Clinical electrophysiological properties of *N*-allyl-clonidine (St 567) in man. J Cardiovasc Pharmacol 3:39-47
Kawada M, Satoh K, Taira N (1984a) Selectivity of alinidine, a bradycardic agent, for SA nodal automaticity versus other cardiac activities in isolated, blood-perfused dog-heart preparations. Arch Int Pharmacodyn Ther 272:88-102
Kawada M, Satoh K, Taira N (1984b) Analyses of the cardiac action of the bradycardic agent, AQ-A 39, by use of isolated, blood perfused dog-heart preparations. J Pharmacol Exp Ther 228:484-490

Khosropour R, Zimpfer M, Lackner F (1983) Slow channel calcium blockade to antagonize the chronotropic effects of isoproterenol. Anesthesiology 59:A43
Kobinger W (1978) Central α-adrenergic systems as targets for hypotensive drugs. Rev. Physiol Biochem Pharmacol 81:39-100
Kobinger W (1985) Specific bradycardic agents, a new approach to therapy in angina pectoris? Prog Pharmacol 5/4:89-100
Kobinger W (1986) Drugs as tools in research on adrenoceptors. Naunyn Schmiedebergs Arch Pharmacol 332:113-123
Kobinger W, Lillie C (1981) AQ-A 39 (5,6-dimethoxy-2-[3-[[α-(3,4-dimethoxy)-phenylethyl]methylamino]propyl]phthalimidine), a specific bradycardic agent with direct action on the heart. Eur J Pharmacol 72:153-164
Kobinger W, Lillie C (1984a) Alinidine. In: Scriabine A (ed) Cardiovascular drugs. Raven, New York, pp 193-210 (New drugs annual, vol 2)
Kobinger W, Lillie C (1984b) Cardiovascular characterization of UL-FS 49, 1,3,4,5-tetrahydro-7,8-dimethoxy-3-[3-[[2-(3,4-dimethoxyphenyl)-ethyl]methylimino]-propyl]-2H-3-benzazepin-2-on-hydrochloride, a new "specific bradycardic agent". Eur J Pharmacol 104:9-18
Kobinger W, Lillie C (1987) Specific bradycardic agents – a novel pharmacological class? Eur Heart J. 8 (Suppl L): 7-15
Kobinger W, Lillie C, Pichler L (1979a) *N*-Allyl-derivative of clonidine, a substance with specific bradycardic action at a cardiac site. Naunyn Schmiedebergs Arch Pharmacol 306:255-262
Kobinger W, Lillie C, Pichler L (1979b) Cardiovascular actions of *N*-allyl-clonidine (St 567), a substance with specific bradycardic action. Eur J Pharmacol 58:141-150
Krähenmann R, Heistracher P (1980) Electrophysiological studies of the effects of 5,6-dimethoxy-2-[3-[[α-(3,4-dimethoxy)phenylethyl]methylamino]propyl]phthalimidine (AQ-A 39) in isolated cardiac muscle. Naunyn Schmiedebergs Arch Pharmacol 311:R36
Krayer O (1949) Studies on veratrum alkaloids. VIII. Veratramine, an antagonist to the cardioaccelerator action of epinephrine. J Pharmacol Exp Ther 96:422-437
Krumpl G, Mayer N, Schneider W, Raberger G (1986a) Effects of alinidine on exercise-induced regional contractile dysfunction in dogs. Eur J Pharmacol 130:37-46
Krumpl G, Schneider W, Raberger G (1986b) Can exercise-induced regional contractile dysfunction be prevented by selective bradycardic agents? Naunyn Schmiedebergs Arch Pharmacol 334:540-543
Lillie C, Kobinger W (1983a) Comparison of the bradycardic effects of alinidine (St 567), AQ-A 39 and verapamil on guinea-pig sinoatrial node superfused with different Ca^{2+} and NaCl solutions. Eur J Pharmacol 87:25-33
Lillie C, Kobinger W (1983b) Actions of alinidine and AQ-A 39 on rate and contractility of guinea pig atria during β-adrenocoptor stimulation. J Cardiovasc Pharmacol 5:1048-1051
Lillie C, Kobinger W (1984) Decrease in bradycardic effect of AQ-A 39 and alinidine in guinea-pig sinoatrial node depolarized by high external K^+-concentration. Naunyn Schmiedebergs Arch Pharmacol 328:210-213
Lillie C, Kobinger W (1986) Investigations into the bradycardic effects of UL-FS 49 (1,3,4,5-tetrathydro-7,8-dimethoxy-3-[3-[[2-(3,4-dimethoxyphenyl)-ethyl]methylimino]propyl]-2-3-benzazepin-2-on hydrochloride) in isolated guinea pig atria. J Cardiovasc Pharmacol 8:791-797
Lillie C, Kobinger W (1987) Investigations differentiating the mechanism of specific bradycardic agents from that of calcium channel blockers. Naunyn Schmiedebergs Arch Pharmacol 335:331-333

Löllgen H, Just H, Wollschläger H, Kersting F (1981) Hemodynamic actions of alinidine during exercise in patients with coronary artery disease. Z Kardiol 70:425-428
Maroko PR, Kjekshus JK, Sobel BE, Watanabe T, Covell JW, Ross J Jr, Braunwald E (1971) Factors influencing infarct size following experimental coronary artery occlusions. Circulation 43:67-82
Meinertz T, Kasper W, Kersting F, Wiegand U, Jähnchen E, Pop T (1980) Wirkung von Alinidine bei koronarer Herzkrankheit. Z Kardiol 69:720
Meinertz T, Kasper W, Meier R, Wiegand V, Jähnchen E (1981) Beneficial effects of alinidine in patients with angina. Circulation [Suppl IV] 64:294
Millar JS, Vaughan Williams EM (1981a) Anion antagonism—a fifth class of antiarrhythmic action? Lancet i:1291-1293
Millar JS, Vaughan Williams EM (1981b) Pacemaker selectivity: influence on rabbit atria of ionic environment and of alinidine, a possible anion antagonist. Cardiovasc Res 15:335-350
Millar JS, Vaughan Williams EM (1983) Pharmacological mapping of regional effects in the rabbit heart of some new antiarrhythmic drugs. Br J Pharmacol 79:701-709
Miyamoto J, Kotake H, Mashiba H (1986) Study on bradycardia induced by diltiazem in the rabbit sinoatrial node. Arzneim Forsch/Drug Res 36:808-810
Neill Wa, Levine HJ, Wagman RJ, Gorlin R (1963) Left ventricular oxygen utilization in intact dogs: effect of systemic hemodynamic factors. Circ Res 12:163-169
Opthof T, Duivenvoorden JJ, van Ginneken ACG, Jongsma HJ, Bouman L (1986) Electrophysiological effects of alinidine (St 567) on sinoatrial node fibres in the rabbit heart. Cardiovasc Res 20:727-739
Osterrieder W, Pelzer D, Yang Q-F, Trautwein W (1981) The electrophysiological basis of the bradycardic action of AQ-A 39 on the sinoatrial node. Naunyn Schmiedebergs Arch Pharmacol 317:233-237
Pelzer D, Trautwein W, McDonald TF (1982) Calcium channel block and recovery from block in mammalian ventricular muscle treated with organic channel inhibitors. Pflügers Arch 394:97-105
Pichler L (1982) Effect of alinidine (St 567) on sympathetic and vagal activities. Arch Int Pharmacodyn 255:162-167
Reiterer W (1981) Belastungshämodynamik Koronarkranker nach selektiver Frequenzsenkung (Alinidine). Acta Med Austriaca 8:182-183
Reiterer W, Stanek B (1980) Einfluß von Alinidine auf die Herzfrequenz unter Ergometerarbeit. Z Kardiol 69:707
Satoh H, Hashimoto K (1986) Electrophysiological study of alinidine in voltage clamped rabbit sino-atrial node cells. Eur J Pharmacol 121:211-219
Sayen JJ, Sheldon WF, Peirce G, Kuo PT (1958) Polarographic oxygen, the epicardial electrocardiogram and muscle contraction in experimental acute regional ischemia of the left ventricle. Circ Res 6:779-798
Schamhardt HC, Verdouw PD, Saxena PR (1981) Improvement of perfusion and function of ischemic porcine myocardium after reduction of heart rate by alinidine. J Cardiovasc Pharmacol 3:728-738
Schurmans J, Piessens J, Kesteloot H, de Geest H (1982) Comparative effect of alinidine and propranolol in ischemic heart disease. Eur J Clin Pharmacol 23:389-396
Shanks RG (1987) The clinical pharmacology of alinidine and its side effects. Eur Heart J 8 (Suppl L):83-90
Siegl PKS, Wenger HC, Sweet CS (1984) Comparison of cardiovascular responses to the bradycardic drugs, alinidine, AQ-A 39, and mixidine, in the anesthetized dog. J Cardiovasc Pharmacol 6:565-574

Simoons ML (1985) Clinical evaluation of alinidine, a specific sinus node inhibitor. Prog Pharmacol 5[4]101–108
Simoons ML, Balakumaran K (1981) The effects of drugs on the exercise electrocardiogram. Cardiology [Suppl 2] 63:124–132
Simoons ML, Hugenholtz PG (1982) Alinidine, a new drug for treatment of sinus tachycardia in myocardial infarction of cardiogenic shock. Am J Cardiol 49:980
Simoons ML, Hugenholtz PG (1984) Haemodynamic effects of alinidine, a specific sinus node inhibitor, in patients with unstable angina or myocardial infarction. Eur Heart J 5:227–232
Simoons ML, Hugenholtz PG (eds) (1987) Heart rate reduction. A reinvented approach to cardiac physiology and therapy. Eur Heart J 8 (Suppl L)
Simoons ML, Tummers J, van Meurs-van Woezik H, van Domburg R (1982) Alinidine, a new agent which lowers heart rate in patients with angina pectoris. Eur Heart J 3:542–545
Snyders DJ, van Bogaert PP (1985) Mode of action of alinidine, a new bradycardic agent: a voltage-clamp study. J Am Coll Cardiol 5:494
Snyders DJ, van Bogaert PP (1987) Alinidine modifies the pacemaker current in sheep Purkinje fibers. Pflügers Arch 410:83–91
Sonnenblick EH, Skelton CL (1971) Oxygen consumption of the heart: physiological principles and clinical implications. Mod Concepts Cardiovasc Dis 40:9–16
Stähle H, Daniel H, Kobinger W, Lillie C, Pichler L (1980) Chemistry, pharmacology and structure-activity relationship with a new type of imidazolines exerting a specific bradycardic action at a cardiac site. J Med Chem 23:1217–1222
Stanek B, Reiterer W, Placheta P, Raberger G (1983) Acute effects of alinidine on heart rate and blood pressure in healthy subjects and patients with hyperkinetic heart syndrome. Eur J Clin Pharmacol 24:31–34
Struyker-Boudier H, Smits J, van Essen H (1981) Haemodynamic effects of alinidine (St 567), a specific bradycardic agent, in the conscious spontaneously hypertensive rat. Naunyn Schmiedebergs Arch Pharmacol 316 [Suppl]:R42
Takeda K, Akera T, Brody TM (1980) Cardiovascular actions of mixidine fumarate. Eur J Pharmacol 68:129–137
Traunecker W, Walland A (1980) Haemodynamic and electrophysiologic actions of alinidine in the dog. Arch Int Pharmacodyn 244:58–72
Trautwein W, Pelzer D, McDonald TF, Osterrieder W (1981) AQA 39, a new bradycardic agent which blocks myocardial calcium (Ca) channels in a frequency and voltage-dependent manner. Naunyn Schmiedebergs Arch Pharmacol 317:228–232
Trautwein W, Pelzer D, McDonald TF (1983) Interval- and voltage-dependent effects of the calcium channel-blocking agents D 600 and AQA 39 on mammalian ventricular muscle. Circ Res 52 [Suppl I]:60–68
Tritthart HA, Windisch H, Heuberger S (1981) The effects of the bradycardia-producing compound alinidine on action potentials and tension development in cardiac fibres. Naunyn Schmiedebergs Arch Pharmacol 316:172–177
Urthaler F, Walker AA (1984) Indirect stimulatory action of the calcium channel blocker AQ-A 39. J Pharmacol Exp Ther 230:336–340
Van Bogaert P-P, Gothals M (1987) Pharmacological influence on the pacemaker current of sheep cardiac Purkinje fibres. A comparison between three different molecules. Eur Heart J 8 (Suppl L):35–42
Vaughan Williams EM (1984) A classification of antiarrhythmic actions reassessed after a decade of new drugs. J Clin Pharmacol 24:129–147
Vaughan Williams EM, Dennis PD, Garnham C (1986) Circadian rhythm of heart rate in the rabbit: prolongation of action potential duration by sustained beta adrenoceptor blockade is not due to associated bradycardia. Cardiovasc Res 20:528–535

Verdouw PD, Saxena PR, Schamhardt HC, van der Hoek TM, Rutterman AM (1980) The effects of alinidine, an *N*-allyl-derivative of clonidine, on regional myocardial perfusion and performance in the pig with or without artrial pacing. Eur J Pharmacol 64:209-220

Verdouw PD, Gom HPA, Bijleveld RE (1983) Cardiovascular responses to increasing plasma concentrations of AQ-A 39 Cl, a new compound with negative chronotropic effects. Arzneim Forsch/Drug Res 33 [1]:702-706

Warbanow W, Wallukat G, Will-Shahab L (1984) Effects of the bradycardic substance alinidine: (St 567) on cultivated heart myocytes. In: Proc 12 congress of the Ges für Kardiol und Angiol der DDR, Rostock 6-10 November 1984, p 143

Wiegand UW, Kasper W, Meinertz T, Stützle U, Jähnchen E (1982) Pharmacokinetic and pharmacodynamic properties of alinidine in man. J Cardiovasc Pharmacol 4:59-62

Zetler G, Lenschow E, Prenger-Berninghoff W (1968) Die Wirkung von 11 Indol-Alkaloiden auf das Meerschweinchenherz in vivo und in vitro, verglichen mit 2 synthetischen Azepinoindolen, Chinidin und Quindonium. Naunyn Schmiedebergs Arch Pharmakol Exp Pathol 260:26-49

Other Therapies

CHAPTER 20

Use of Adenosine as an Antiarrhythmic Agent

P. PUECH, A. MUNOZ, A. SASSINE and J. BRUGADA

A. Introduction

Since the classical observation of DRURY and SZENT-GYORGYI (1929), it has been known that adenosine and adenosine triphosphate (ATP) exert pronounced electrophysiological effects on the mammalian heart. Subsequently, the properties of both compounds were widely investigated in vitro or in animals (BELLARDINELLI et al. 1982; CARDENAS 1964; JAMES 1965; JOHNSON and MCKINNON 1956; MUNOZ et al. 1984a; PELLEG 1984, 1985a; RUBIO et al. 1979; SCHRADER et al. 1975). The pronounced and transient depressant effects on atrioventricular conduction have attracted growing attention due to its clinical implication. In fact, these properties explain the high efficacy of purinergic compounds in terminating paroxysmal supraventricular tachycardia with a re-entrant mechanism involving the AV node (BELHASSEN and PELLEG 1984; GRECO et al. 1982; KOMOR and GARAS 1955; LATOUR et al. 1968; SOMLO 1955). ATP is routinely used in Europe for acute treatment of paroxysmal supraventricular tachycardia (injectable preparation: Striadyne, Laboratories Auclair, ATP-hormonotherapia, Richter). Adenosine has been investigated more recently in man, both in the United States and in Europe (DI MARCO et al. 1983, 1985; MUNOZ et al. 1984b; PUECH et al. 1985); differences in the mechanism of action demonstrated in preclinical work explain the interest of cardiologists for its clinical evaluation. In fact, while pharmacological properties and effectiveness of adenosine appear to be very similar to those of ATP, the possible differences in the mechanism of action could lead to fewer side effects and it can be thought that adenosine would probably be an alternative to other available compounds as a drug of first choice for acute management of supraventricular tachycardia.

B. Metabolism

Adenosine is cleared from the circulation by cellular uptake and metabolized within a few seconds after intravenous injection (ONTYD and SCHRADER 1984). This property explains the very short duration of action of adenosine and permits the compound to be administered without concern for long-lasting or cumulative adverse reaction.

In these conditions sampling for adenosine blood levels assay appears to be useless and, furthermore, complicated by the need for a specific inhibitor.

C. Electrophysiological Properties and Mechanism of Antiarrhythmic Action of Adenosine

Both ATP and adenosine have strong depressing effects on the sinus node and AV node when injected as a bolus. Many preclinical works have demonstrated that adenosine depresses automaticity of the sinus node and Purkinje fibres, hyperpolarizes the membranes and shortens the plateau phase of the action potential in the atrium and depresses the AV nodal conduction. These actions of adenosine are thought to be related to a direct effect on the potassium and calcium conductance, and also to an indirect antagonism of the electrophysiological properties of intracellular cyclic adenosine 3′5′-monophosphate (PELLEG 1985). The direct effects of adenosine are probably related to its interaction with a specific extracellular receptor. According to BURNSTOCK (1980), purinergic receptors are divided into P1 and P2 on the basis of several criteria: potency of purinergic effects, action of antagonists, influence on cyclic AMP and induction of prostaglandin synthesis. Methylxanthine and mainly aminophylline appear to exert a specific inhibitory effect on P1 cardiac receptors, not related to a phosphodiesterase inhibition.

Due to the specific metabolism of purinergic compounds, it has been suggested that the effects of ATP result from its rapid degradation to adenosine. However, differences between the nucleoside and the nucleotide exist regarding their interaction with the autonomic nervous system (MUNOZ et al. 1983). Involvement of the vagus in the mechanism of action of ATP has been well established both in preclinical work and in patients (HUGUES et al. 1980), whereas it has not been demonstrated for adenosine (FAVALE et al. 1985).

In patients, an intravenous bolus of adenosine or ATP is followed by an immediate sinusal bradycardia, in some cases followed by a secondary tachycardia. The effect is prevented by atropine after ATP while the adenosine effect is not modified. Both compounds depress AV conduction by lengthening the atrial to His bundle interval, leading to atrioventricular block. This effect, which is dose dependent, occurs within a delay of 10–20 s after injection, is short-lasting and returns to normal values within 1 min. Electrophysiological studies have demonstrated that this effect is not antagonized by atropine whereas aminophylline completely prevents the effects of adenosine. Dipyridanole, catecholamine blood levels and other factors interfering with the status of the AV node conduction are likely to modify the effect of adenosine on nodal conduction. On the same basis, it can be thought that the observed individual sensitivity to purinergic compounds probably depends on the neurovegatative balance of the patient.

D. Drug Regimen

The effective concentration of adenosine that reaches the heart in each patient depends on several factors: injected dose, speed of injection, circulation time from the periperal vein to the heart and site of injection. Furthermore, as discussed previously, the neurovegetative balance of the patient has to be taken into account. These observations explain the wide range of doses required to achieve effectiveness in patients. Therefore, the need for individualized dosage for each patient appears to be an important factor preventing occurrence of side effects (Di Marco et al. 1985). Rapid uptake and metabolism of adenosine which allow repeated administration enable the adequate dosage for each patient by starting at a low dose and then increasing the dose at 1-min intervals until the desired effect is produced. For a standard body weight (adult) a 5-mg first dose is proposed, then increased to 10 mg in case of failure. Due to the shorter circulation time, lower doses should be used in case of central injection (Leclercq and Coumel 1978). Injection of adenosine is to be performed after patient information (general side effects) and during continuous ECG monitoring within an intensive care unit (theoretical possible prolonged pauses or ventricular arrhythmia related to escaped ectopic beats). The injection is to be performed as quickly as possible, preferably followed by a bolus of saline serum: "the quicker the injection, the higher the effectiveness".

Lower dosages are to be used in elderly patients or children.

E. Indications

Adenosine appears to be a very effective treatment of supraventricular tachycardia, and also a useful tool for diagnostic purposes.

I. Therapeutic Use

The transient impairment of the AV nodal conduction provides efficacy for termination of tachycardia in which the re-entry loop includes the AV node. Tachycardia termination occurs 10–30 s after the injection and is achieved in

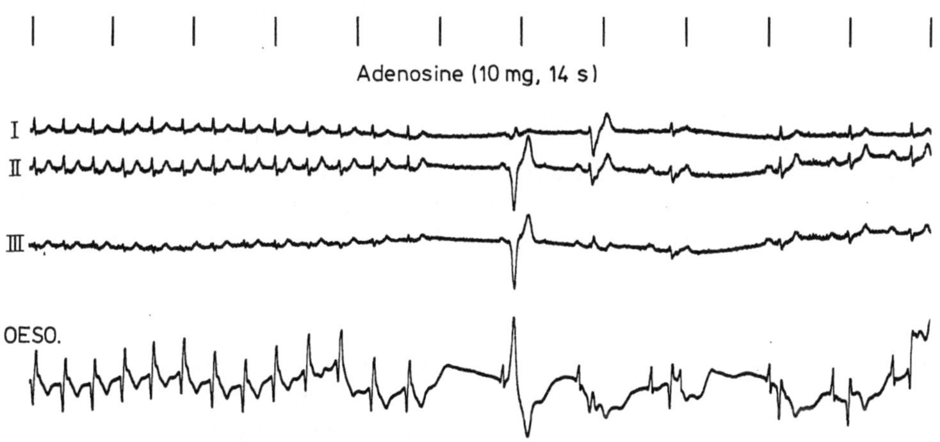

Fig. 1. Arrest of a paroxysmal junctional tachycardia after a bolus of adenosine. Intraoesophageal recording shows the atrial activity during tachycardia

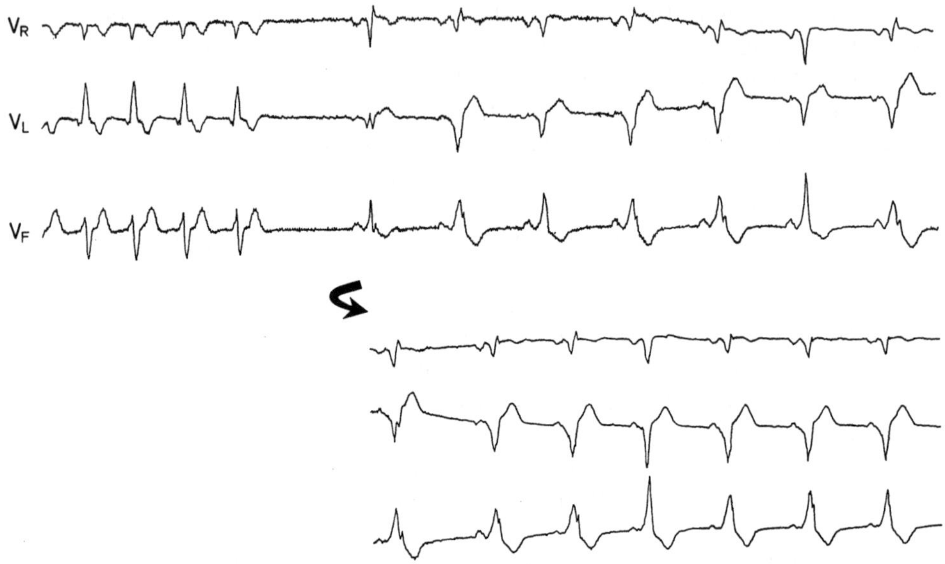

Fig. 2. In a patient with Wolff-Parkinson-White syndrome the depression of the AV-node conduction leads to the arrest of the reentrant tachycardia; before return to basal ECG a transient aspect of "super Wolff" is documented while the transmission to ventricles uses only the accessory pathway

more than 90 % of cases. Electrophysiological studies show that termination of the tachycardia results from a block in the slow anterograde pathway in patients with AV nodal re-entry and in the AV node in patients with re-entrant tachycardias using an accessory pathway. Figure 1 illustrates the arrest of a paroxysmal tachycardia after adenosine injection. In patients with the Wolff-Parkinson-White syndrome, adenosine increases the degree of pre-excitation (Fig. 2), the normal pathway being depressed while usually the accessory pathway remains unaffected (it is not known whether adenosine could, as does ATP in some specific cases, depress accessory pathways with long effective refractory periods) (PERROT and FAIVRE 1982).

A recent report on adenosine- and verapamil-sensitive ventricular tachycardias suggests that in a very selected population (four patients with structurally normal hearts who had exertionally related sustained ventricular tachycardia exhibiting a left bundle branch block morphology) adenosine could be used for the treatment of ventricular tachycardia (LERMAN et al. 1986).

II. Diagnostic Use

The inhibition of intranodal conduction by adenosine constitutes an aid for the diagnosis of ectopic atrial rhythm when the atrial activity is masked. The inhibition of the conduction to the ventricle allows recognition of abnormal atrial activity such as flutter and atrial tachycardia. The pharmacological test is also very useful to differentiate ventricular from supraventricular tachycardia from pre-existing or rate-dependent ventricular aberration. ATP is used during electrophysiological exploration for sick sinus syndrome and "vagal tone" evaluation (CASTILLO-FENOY et al. 1980). Whether adenosine could be used in the same way has not yet been established.

F. Tolerance and Side Effects

Adenosine apperars to be a well-tolerated compound. Minor side effects have occasionally been reported. Rhythm disturbances may consist of prolonged atrial pauses, sinus arrhythmia, occurrence of ventricular ectopic beats or transient atrial fibrillation. In all cases, they are transient and never lead to critical consequences. A search for an individual optimal dosage should decrease these observed adverse events. Despite the known vascular effects of purinergic compounds (DEAN et al 1951; GABA et al. 1986), no change in blood pressure has been reported after adenosine bolus. Flushing and transient dyspnoea can be encountered after adenosine injection; they are especially reported by patients at the time of the cessation of tachycardia and are always qualified as transient, minor, but unpleasant.

G. Drug Interaction

Due to their specific properties, xanthines which are direct antagonists to purinergic compounds lead to a non-effectiveness of adenosine injection which limits its therapeutic use in patients with chronic pulmonary disease. Conversely, in patients treated with dipyridamole, which is known to have a synergistic effect with adenosine, extreme caution must be observed to avoid cardiac adverse events such as prolonged AV block or atrial fibrillation. In fact, precautions are also to be taken when injecting adenosine in patients receiving pharmacological agents depressing the AV node conduction, such as calcium blockers or beta-blockers. The short-lasting effect of adenosine and its lack of strong haemodynamic effects should not lead to side effects in patients with impaired ventricular function or in patients receiving negative inotropic agents.

H. Contraindications

In patients with sinus node dysfunction adenosine is expected to have pronounced effects with long sinus pauses but these events could also occur when other modes of arrhythmia cessation are applied. Chronic treatment with dipyridamole carries a much greater risk, and a systematic check must be undertaken to ensure that this drug has not been administered before adenosine can be injected safely.

Bronchial asthma could be a contraindication for adenosine use. Due to its bronchoconstrictive effect documented during inhalation (Cushley et al. 1985), further work is needed to define the conditions for safe use of intravenous adenosine in asthmatic patients.

Compared with other drugs available for acute treatment of supraventicular tachycardia, purinergic compounds appear to be of greater interest because of their rapid inactivation which separates them from other products like diltiazem, verapamil, and digoxin, which have much longer half-lives. Thus, treatment can be started with low doses and increased or repeated without any risk of cumulative effects. Furthermore, this property is very useful if the diagnosis is not obvious.

In fact, ATP or adenosine can be proposed safely to terminate tachycardia when other compounds have failed or are contraindicated, mainly in the case of poor left ventricular function, congestive heart failure, marked hypotension or pretreatment with beta-blocking agents. However, growing clinical experience with adenosine suggests that it is as effective as ATP but probably produces fewer side effects, and therefore it may well become a first-choice drug for the treatment of paroxysmal junctional tachycardia.

References

Belhassen B, Pelleg A (1984) Acute management of paroxysmal supraventricular tachycardia: verapamil, adenosine triphosphate or adenosine. Am J Cardiol 54:225–227

Bellardinelli L, Fenton RA, West A et al. (1982) Extracellular action of adenosine and the antagonism by aminophylline on the atrioventricular conduction of isolated perfused guinea pig and rat hearts. Circ Res 51:569–579

Burnstock G (1980) Purinergic receptors in the heart. Circ Res 46:175–182

Cardenas M, Aceves J, Alarçon G (1964) Efecto del acido adenosin-trifosforico sobre las propriedades fisiologicas del corazon. Arch Inst Cardiol Mex 34:485–494

Castillo-Fenoy A, Thebaut JF, Achard F et al. (1980) Etude du mécanisme de la pause sinusale par l'enregistrement endocavitaire du potentiel sinusal chez l'homme. Arch Mal Coeur 73:805–815

Cushley MG, Tattersfield AE, Holgate ST (1985) Inhaled adenosine and guanosine on airway resistance in normal and asthmatic subjects. Br J Clin Pharmacol 15:161–165

Dean F, Gropper AL, Schroeder HA (1951) Circulatory and respiratory effects of adenosine triphosphate in man. Circulation 3:543–552

Di Marco JP, Sellers TD, Berne RM, West GA, Bellardinelli L (1983) Adenosine: electrophysiologic effects and therapeutic use for terminating paroxysmal supraventricular tachycardia. Circulation 68:1254–1263

Di Marco JP, Sellers TD, Lerman BB, Greenberg LM, Berne RM, Belardinelli L (1985) Diagnostic and therapeutic use of adenosine in patients with supraventricular tachycardias. JACC 6 (2):417–425

Drury AN, Szent-Gyorgyi A (1929) The physiological activity of adenine compounds with special reference to their action upon the mammalian heart. J Physiol 68:213–237

Favale S, Biase SV, Rizzo V, Bellardinelli R, Rizzon R (1985) Effects of adenosine and adenosine-5'-triphosphate on atrioventricular conduction in patients. JACC 5:1212-1219

Gaba S, Trigui F, Dujols P, Godard P, Michel FB, Prefaut C (1986) Compared effects of ATP versus adenosine on pulmonary circulation of COPD. Eur J Respir Dis 69:515-522

Greco R, Musto B, Arienzo V, Alborino A, Garefalo S, Marsico F (1982) Treatment of paroxysmal supraventricular tachycardia in infancy with digitalis, adenosine-5'-triphosphate and verapamil. A comparative study. Circulation 66:504-508

Hugues FC, Jan Y, Bors V (1980) Etude de l'action chronotrope de l'ATP chez l'homme. Coeur Med Interne 19:227-234

James TN (1965) The chronotropic action of ATP and related compounds studied by direct perfusion of the sinus node. J Pharmacol Exp Ther 149:233-247

Johnson EA, McKinnon MG (1956) Effect of acetylcholine and adenosine on cardiac cellular protentials. Nature 179:1174-1176

Komor K, Garas Z (1955) Adenosine triphosphate in paroxysmal tachycardia. Lancet ii:93-94

Latour H, Puech P, Grolleau R, Sat M, Balmes P (1968) L'utilisation de l'adénosine-5'-triphosphorique dans le diagnostic et le traitement des tachycardies. Arch Mal Coeur 61:293

Leclercq JF, Coumel P (1978) Les effets de l'adenosine triphosphate (ATP) sur le noeud-sinusal et le noeud auriculoventriculaire chez l'homme. Variations selon le lieu d'injection. Coeur Med Interne 17:541-546

Lerman B, Bellardini L, West A, Berne R, DiMarco JP (1986) Adenosine-sensitive ventricular tachycardia: evidence suggesting cyclic AMP-mediated triggered activity. Circulation 74:270-280

Munoz A, Sassine A, Carabantes G, Lehujeur C, Koliopoulos M, Puech P (1983) Effects of purinergic compounds on AV conduction: experimental studies in anesthetized dogs. In: Levy S, Gérard R (eds) Recent advances in Cardiac Arrhythmias. Libbey, London pp 35-42

Munoz A, Sassine A, Lehujeur C, Koliopoulos N, Puech P (1984) Mode d'action des substances purinergiques dérivées de l'adénine sur la conduction auriculo-ventriculaire. Etude expérimentale chez le chien. Arch Mal Coeur 1:143-150

Munoz A, Leenhardt A, Sassine A, Gallay P, Puech P (1984b) Therapeutic use of adenosine for terminating spontaneous paroxysmal supraventricular tachycardia. Eur Heart J 5:735-738

Ontyd J, Schrader J (1984) Measurement of adenosine, xanthine and hypoxanthine in human plasma. J Chromatogr 9:307-404

Pelleg A (1985) Cardiac cellular electrophysiologic actions of adenosine and adenosine triphosphate. Am Heart J 110:688-692

Pelleg A, Belhassen B, Reuben I, Lamado I (1984) Comparative electrophysiologic effects of adenosine triphosphate and adenosine in the canine heart: influence of atropine, propanolol, vagotomy, dipyridamole and aminophylline. Am J Cardiol 54:225-227

Perrot B, Faivre G (1982) Action de l'adénosine triphosphorique (ATP) sur les faisceaux accessoires de conduction. Arch Mal Coeur 75:593-601

Puech P, Sassine A, Munoz A, Masse C, Zettelmaier F, Leenhardt A, Yoshimura H (1985) In: Zipes P, Talife J (eds) Electrophysiologic effects of purines: clinical applications in cardiac electrophysiology and arrhythmias. Grune and Stratton, pp 443-450

Rubio R, Bellardini L, Thompson CI et al. (1979) Cardiac adenosine electrophysiological effects, possible significance in cell function and mechanism controlling its release. In: Bear HB, Drummed GI (eds) Physiological and regulatory functions of adenosine and adenine nucleotides. Raven, New York, pp 167–182

Schrader J, Rubino R, Berne RM (1975) Inhibition of slow action potentials of guinea pig atrial muscle by adenosine: a possible effect on Ca^{2+} influx. J Mol Cell Cardiol 7:427–433

Somlo E (1955) Adenosine triphosphate in paroxysmal tachycardia. Lancet i:1125

CHAPTER 21

Physical and Surgical Treatment of Cardiac Arrhythmias

A. J. CAMM and D. W. DAVIES

A. Introduction

Close scrutiny of the long-term response to antiarrhythmic drugs usually leads to the conclusion that although there are many currently available drugs which may be used alone or in combination, they are rarely completely effective and may be responsible for side effects. Therefore, alternative, non-pharmacological means of treatment are necessary for many patients. Over the past 20 years, a variety of techniques have been developed and evaluated. Implantable electronic devices and methods of ablation have been particularly effective. It is the purpose of this report to review these techniques in the context of antiarrhythmic therapy in general (Table 1).

Table 1. Suitability of various therapies for common arrhythmias

Arrhythmia	Drugs	Implanted device	Catheter ablation	Surgery
Atrial fibrillation	+	−	+	±
Atrial flutter	±	±	+	−
Atrial tachycardia	±	±	+	+
AVNRT	+	+	+	+
AVRT	+	±	+	+
VT	+	+	±	+
VF	±	+++	−	±

AVNRT, atrioventricular nodal re-entry tachycardia; AVRT, atrioventricular re-entry tachycardia; VF, ventricular fibrillation; VT, ventricular tachycardia.
Whilst it may seem possible to apply, eg. any therapy to a patient with AVNRT, careful individual consideration and assessment is mandatory before choice of appropriate therapy for each case

B. Implantable Electronic Devices

Since their advent in the therapy of cardiac tachyarrhythmias over 15 years ago, implantable pulse generators have grown in sophistication whilst, with the exception of the implantable defibrillator, they have decreased in size.

Table 2. Characteristics of "low- and high-energy" implantable devices for tachycardia termination

	Energy	Efficacy	Safety	Speed	Comfort
Pacemakers	25 μJ	+	+/−	+	++
Cardioverter/ defibrillators	25 J	++	+	−	−

They now range from ordinary demand pacemakers which prevent escape tachyarrhythmias by avoiding the precipitating bradycardia (Zoll et al. 1960; Sowton et al. 1964) or by preventing re-entry (Coumel 1970; Spurrell and Sowton 1976) through devices which recognize pathological tachycardias and then deliver pacing stimuli in attempts to terminate them (Spurrell et al. 1982; Sowton 1984; Nathan et al. 1986) to the automatic implantable cardioverter-defibrillator, which has resulted in a dramatic improvement in the expected survival of the patients with life-threatening ventricular arrhythmias in whom it has been implanted (Mirowski et al. 1985; Table 2).

Prevention of tachyarrhythmias by pacemakers is most easily achieved when the tachyarrhythmias occur as escape rhythms during periods of bradycardia (Zoll et al. 1960; Sowton et al. 1964) especially when the QT interval is abnormally prolonged. The prophylaxis of re-entry with dual-chamber pacemakers (Coumel 1970; Spurrell and Sowton 1976) has not been widely applied because the systems available are bradycardia support pacemakers not designed to be used in the prevention of tachycardia. The variety of complex unpredictable ways in which re-entry may be initiated (Olson and Bardy 1986) requires that, to be successful, such systems need to be capable of equally complex recognition and pacing algorithms (Malik et al. 1986).

Interruption of established tachyarrhythmias is a rapidly increasing application of implantable pulse generators. The mechanism of most pathological tachycardias is re-entry (Wellens 1975; Wellens et al. 1976; Denes et al. 1975; Coumel et al. 1979) (v.i.). Since the depolarization wavefront utilizes the same anatomical circuit for each successive beat, the part of the circuit immediately ahead of the wavefront must be excitable or the circus movement would be extinguished. It is this dependence upon the excitability of part of the circuit which offers the opportunity of termination of tachycardia by pacing stimuli. The aim is to achieve premature depolarization of the excitable gap by pacing so that the circulating wavefront encounters inexcitable tissue and tachycardia is terminated. In order to maximize the chance of success, the pacing stimulus should be delivered at or close to the circuit (Ward et al. 1979). However, this is rarely possible because of the distance between conventional sites for pacing electrodes and the circuits responsible for the tachycardias. Different termination algorithms exist, designed partly to overcome this problem.

The simplest method is underdrive pacing, usually triggered as a dual-demand pacing mode. Here, a pacemaker is programmed to pace at a bradycardia support rate when the heart rate either falls below that rate or exceeds another, programmable rate (Krikler et al. 1976). Thus, during tachycardia, pacing stimuli are delivered randomly and relatively slowly. Termination of tachycardia by this method thus occurs by chance which is increased if the pacing electrode is on circuit or the tachycardia is relatively slow. If neither of these obtains then the chances of termination are low.

It is more likely that a pathological tachycardia will be terminated if the pacing stimuli are delivered more rapidly than the tachycardia and preferably synchronized to the tachycardia. Such methods are broadly divided into overdrive pacing and extrastimulus pacing algorithms. Overdrive pacing consists of the delivery of a burst of (typically four to eight) pacing stimuli more rapidly than the tachycardia rate (Camm and Ward 1983). Various refinements to this simple principle have been described (Camm et al. 1980; Nathan et al. 1982) but the aim in all is to enter the circuit by means of the initial stimulus and to achieve termination, ideally only as a result of the last stimulus delivered. The extrastimulus method consists of the synchronized delivery of one or two premature pacing stimuli during tachycardia aiming to breach the circuit by virtue of their premature timing (Massumi et al. 1967; Durrer et al. 1967). Both of these principles have been applied with considerable success, predominantly to the management of re-entrant junctional tachycardias (Spurrell 1975, 1976; Kapperberger and Sowton 1981). However, problems remain. One is the unpredictable tendency for an attempt at termination to result in degeneration into atrial fibrillation (Table 3). As a result, although devices have successfully reduced symptoms by shortening the duration of abnormal tachycardias in many patients with the Wolff-Parkinson-White syndrome, they are no longer recommended for such patients because of the risk of dangerously fast ventricular rates should atrial fibrillation be produced. A similar problem exists with paced termination of ventricular tachycardia, potentially causing a faster tachycardia or ventricular fibrillation so that such devices have rarely been used to treat ventricular tachycardia. An important part of the assessment of patients for tachycardia reversion pacemakers (Table 4) is to examine for the tendency to provocation of unwanted arrhythmias by the device.

Table 3. Factors increasing the likelihood of provocation of unwanted arrhythmias by tachycardia-terminating devices

Closely coupled stimulation	
Higher energy stimulation	
Electrolyte abnormality	(e.g. hypokalaemia)
Ventricular stimulation	(more than atrial)
Drug toxicity	(e.g. digoxin)
Structural heart disease	(e.g. cardiomyopathy, CHD, WPW)

CHD, coronary heart disease; WPW, Wolff-Parkinson-White abnormality

Table 4. Requirements prior to prescribing tachycardia-terminating devices

1. Unsatisfactory drug control of—symptoms —risk	
2. Surgery difficult inappropriate or unsuccesful	
3. Electrophysiology study demonstration of	
– Responsive nature of tachycardia	multiple trials
– Absence of easily provoked unwanted arrhythmias	multiple trials
– Satisfactory tachycardia recognition	
– Absence of unwanted spontaneous arrhythmias	

Another problem is the difficulty in automatic diagnosis of a tachycardia by an implantable pulse generator. Presently available systems use heart rate analysis to distinguish a pathological tachycardia and problems occur when there is overlap in rates between sinus and abnormal tachycardia. Methods capable of rate-independent diagnosis of tachycardia will shortly be implanted and should overcome this problem (Davies et al. 1986a).

Higher energies than pacing stimuli can also be delivered by implantable devices in a attempt to terminate tachycardias. The implantable cardioverter delivers energies ranging from 0.1 to 5 J ("microshocks") and was intended mainly for treatment of ventricular tachycardia (Zipes et al. 1982). This has been withdrawn from use because of an unacceptably high incidence of acceleration of the arrhythmias (Zipes et al. 1984; Vergara et al. 1985). The highest energies are delivered by the implantable cardioverter-defibrillator (Mirowski et al. 1980). Usually between two epicardial patch electrodes, shocks of between 25 and 30 J are delivered when ventricular tachycardia or fibrillation is detected. Used in a population of patients already resuscitated from a "sudden cardiac death", the device has dramatically reduced their expected mortality way beyond that which could be expected using alternative therapies (Mirowski et al. 1985; Echt et al. 1985). Its limitations are related to reli-

Table 5. Future developments of implantable cardioverter-defibrillator

1. Improved programmability
2. Combination with pacing techniques
3. Improved defferentiation of tachycardias
4. Data storage and telemetry
5. Non-invasive EP testing facility
6. Implantation avoiding thoracotomy
7. Automatic maintenance (ęg. capacitor forming)
8. Reduced size and weight
9. Increased longevity
10. Lower cost

EP, electrophysiology

able battery life (estimated at only 18 months), its cost and its lack of flexibility, and discomfort to the patient if a shock is delivered when conscious during ventricular tachycardia. Many of these problems will be overcome as newer devices combining pacing and defibrillation shortly become available (Table 5).

C. Ablation of Arrhythmia Substrates

I. Basic Concepts

The principle of surgical and physical therapy for tachyarrhythmias is to identify and then accurately localize before resecting, destroying or isolating the tissue responsible for the arrhythmia whilst preserving as much as possible of surrounding normal tissue. Modern electrophysiological techniques have provided the clinician with the knowledge of mechanisms and substrates of cardiac arrhythmias which first led to the application of surgical techniques for their modification (Buchell et al. 1967; Dreifus et al. 1968; Cobb et al. 1968; Moss and McDonald 1971; Spurrell et al. 1973; Camm et al. 1979; Harken et al. 1980; Marchlinski and Josephson 1984).

II. Mapping Techniques

Tachyarrhythmias may be considered to arise from an abnormal localized focus or to originate as a result of the repeated circulation of a depolarization wavefront—re-entry. Focal tachycardia may be due to increased local automaticity or micro-re-entry. Automaticity of myocardial cells is enhanced by a number of factors such as localized hypoxia, hypokalaemia, hypercalcaemia and drugs such as digoxin and catecholamines. This often manifests as isolated ectopic depolarizations or non-sustained tachycardia. Such arrhythmias are not usually suitable for ablative therapy as they are difficult to provoke and, unless permanent, may not be present at the time of the procedure so that localization to guide ablation would be impossible. Re-entry tachycardias are arrhythmias which utilize a definable circuit such as one composed of atrial myocardium, the AV node, ventricular myocardium and an accessory pathway returning to the atria in atrioventricular re-entry tachycardia. Other examples include atrial flutter and most examples of ventricular tachycardia where re-entry is centred around an area of functional block such as a scar.

Prior to attempting ablation of an arrhythmia an accurate localization or mapping of its source is vital. The origins of an arrhythmia are usually localized by intracardiac mapping using catheter electrodes prior to attempting ablation of all or part of the substrate essential for the arrhythmia. There are three methods of obtaining this information and all are repeatable at the time of surgery. The most useful is to obtain information regarding the sequence of depolarization during tachycardia—so-called activation mapping (Josephson et al. 1982a). This involves moving one catheter electrode (the mapping electrode) to a variety of intracardiac sites whilst keeping others stationary as references. The electrograms obtained from the mapping electrode in its various positions are recorded simultaneously with cine-radiographic recordings of the positions. The purpose of moving the mapping electrode varies with the arrhythmia being studied. In the case of a ventricular or an atrial tachycardia, the intention is to place the mapping electrode as close as possible to the site of origin of the tachycardia identified by electrograms which precede the onset of respectively the surface QRS or P wave. Although this provides a reasonable guide for more accurate intraoperative mapping to be performed, the earliest activation may not always indicate the source of the problem when dealing with re-entry tachycardia since, with a circuit, there is always activity at some point which is essential for the perpetuation of the arrhythmia (Josephson et al. 1978). With ventricular tachycardia, it can be argued that the tissue which should be resected to prevent such arrhythmias is that which depolarizes *later* than the bulk of the myocardium and is responsible for continuation of depolarization throughout diastole. This identifies an area of slow conduction where depolarization persists whilst surrounding tissue recovers to become excitable by a depolarization wavefront emerging from the slow area. Thus, activation which spans diastole indicates the tissue which allows re-entry to persist, rather than the activation which occurs just before depolarization of the main bulk of myocardium. For example, with atrioventricular re-entry tachycardia, the purpose of mapping is to localize the accessory pathway which forms one of the links between atria and ventricles. This is achieved, usually during orthodromic atrioventricular re-entry tachycardia, by positioning the mapping electrode to record the earliest atrial activity during tachycardia. Since the atria are activated by the accessory pathway in such a circuit, this indirectly localizes the atrial insertion of the accessory pathway. Some workers can now record direct accessory pathway potentials which are of considerable academic interest and have been used as guides to ablation therapy (Jackman et al. 1983).

If tachycardia cannot be induced, two other methods of mapping may be employed, both of which are predominantly useful in ventricular tachycardia. One is pace mapping which consists of placing the mapping electrode at the site from which, using electrocardiographic criteria, the clinical tachycardia is suspected to have arisen. This site is then paced at the rate of the clinical tachycardia and the morphology of the resultant 12-lead ECG compared with the clinical arrhythmia (Josephson et al. 1982b; Holt et al. 1985a). The electrode is repositioned if necessary and the process repeated until a good match of the paced ECG with the clinical tachyarrhythmia is obtained. The electrode is then assumed to be at the site of origin of the arrhythmia. The last method of mapping is to record late or fragmented potentials during sinus rhythm (Cassidy et al. 1984). These occur at or around areas of slow conduction or block which may therefore predispose to re-entry. However, the finding lacks specificity for identifying the substrate of a clinical arrhythmia.

Cryomapping is a confirmatory mapping technique performed during surgery. Once other methods have localized the tissue to be destroyed then a cryoprobe is briefly applied to that site during tachycardia and then, after a short recovery period in the case of an accessory pathway, during atrial pacing to

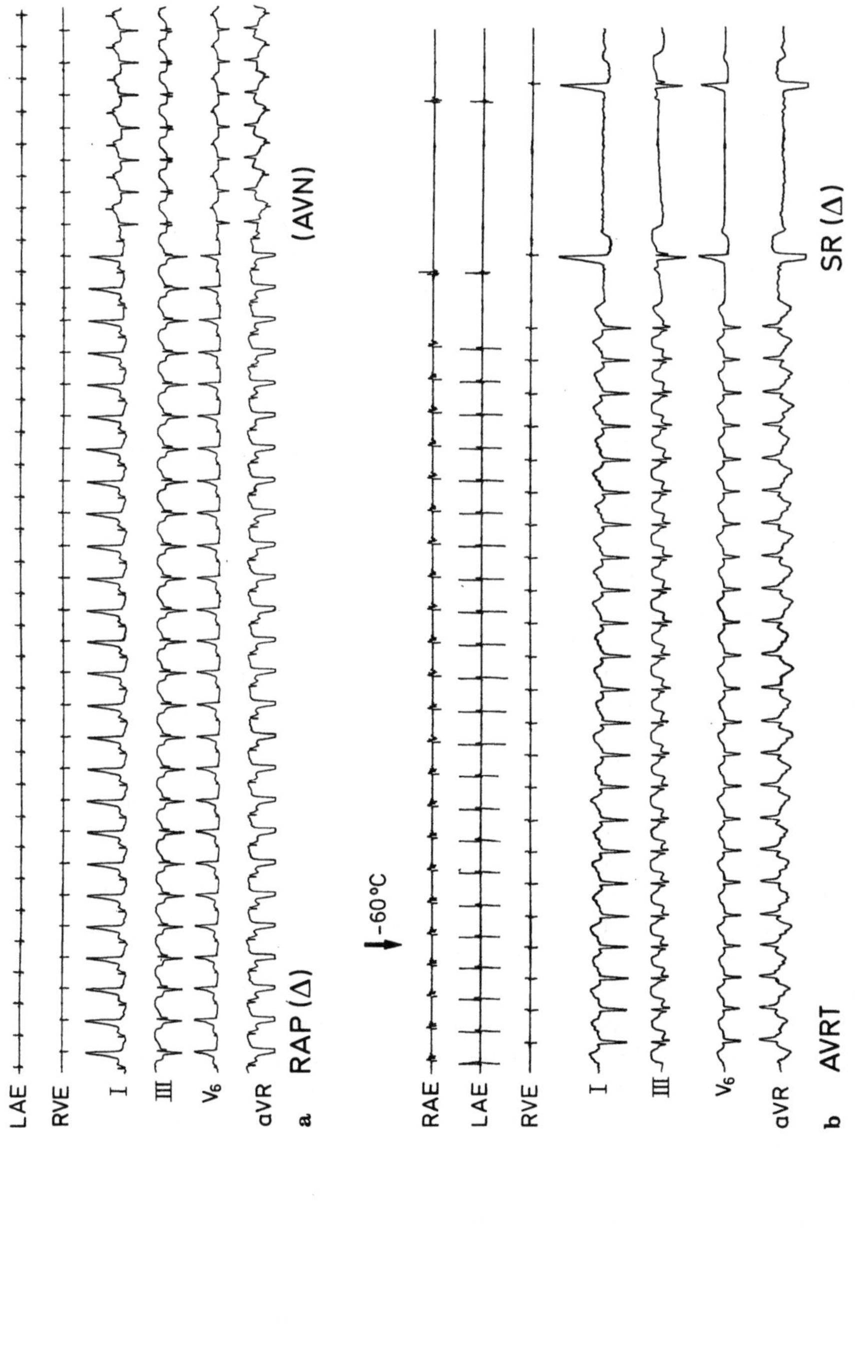

Fig. 1 a. An example of cryomapping; abolition of pre-excitation within seconds by application of cryothermy (↓ −60 °C). Pre-excitation is maximized by right atrial pacing (RAP) and abolished within 10 s of cryoprobe application unmasking AV nodal conduction (AVN). LAE, left atrial electrogram; RVE right ventricular electrogram. Surface ECG leads labelled conventionally. **b** Confirmation of cryomapping seen in *A*. Cryothermy (↓ −60 °C) is repeated at the same site but during orthodromic atrioventricular re-entry tachycardia (AVRT). AVRT is terminated within 6 s, abolishing retrograde accessory pathway conduction but interestingly leaving anterograde conduction present (note pre-excited sinus rhythm after tachycardia – SRΔ). RAE right atrial electrogram maximize pre-excitation. If the probe is on the correct site then, respectively, tachycardia will be terminated (or pre-excitation abolished) within seconds (Fig. 1 a, b). Such a short period of cryotherapy does not produce permanent damage so that the test can be repeated several times in the same patient if required.

III. Ablation Techniques

1. Catheter Ablation

This technique was originally reported by Vedel et al. (1979) as an accidental discovery. An external DC cardioversion performed by them during an electrophysiology study resulted in a "shorting" between one of the paddles and a catheter electrode positioned at the His bundle position. Following the cardioversion, the patient was in complete heart block. This was first applied deliberately to ablate normal AV conduction in man by Scheinman et al. (1982) and has become the chosen method for achieving AV nodal/His bundle ablation in patients with refractory supraventricular arrhythmias for which there exists a world registry. Other groups have pioneered its application to destruction of accessory pathways (Fisher et al. 1984; Morady et al. 1987; Davies et al. 1987a), for direct ablation of a right atrial tachycardia focus (Silka et al. 1985) and for ablating ventricular tachycardia (Hartzler 1983; Fontaine et al. 1987) after careful mapping in all three situations.

Catheter ablation performed by the usual method reproduces the original accident by using standard defibrillators connected to temporary pacing electrodes. To guarantee ablation, high energies of at least 50 J and usually over 200 J are delivered. Such energies result in an explosion at the catheter tip associated with transient but dramatic elevation of pressure and local temperature (Boyd and Holt 1985). Adverse effects including subsequent dyspnoea (Schofield et al. 1986), ventricular tachycardia (Davies et al. 1987b) and sudden death (Davies et al. 1986b) are described. These are rare but may be related to the high-energy delivery. Attemps have therefore been made to limit these effects by using lower energies in an attempt to modify rather than destroy AV nodal conduction (McComb et al. 1985) as well as to increase the destructive efficiency of energy delivery. There have been two main approaches to this latter aim which have either been to use an active fixation electrode in an attempt to direct and increase the intensity of delivered energy (Holt et al. 1985b) or to modify the method of energy delivery (Cunningham et al. 1986). The latter approach has essentially removed the inductor from a conventional "Lown" defibrillator, thereby converting the waveform of the energy delivered from a truncated exponential pulse into a pure exponential pulse and increasing the local current delivery during a discharge. Using this modification of the technique, the associated unwanted extreme rises in pressure are avoided and normal AV conduction has successfully been destroyed with as little as 3.6 J (total energy), approx. 100th of the energy used for conventional "fulguration" of AV nodal conduction.

The delivery, via a catheter electrode, of radiofrequency has been used to ablate intracardiac tissue in the treatment of a variety of cardiac arrhythmias (HUANG et al. 1987). The technique is relatively painless and so can be performed without a general anaesthetic. The preliminary results are encouraging but the technique needs further evaluation.

2. Direct Surgical Ablation

After precise mapping, there is a choice of complementary antiarrhythmic surgical techniques. Traditional dissection with a scalpel and cryothermy may, alone or in combination, be used to divide, resect or isolate the source of the arrhythmia. Laser therapy has been applied to human ventricular tachycardia at open heart surgery (SVENSON et al. 1987) and percutaneously in dogs, successfully modifying AV nodal conduction (NARULA et al. 1985). Experience with the technique is still limited and further data are required to establish its clinical value.

A variety of surgical techniques exist for treating supraventricular and junctional tachycardias. Whilst the most readily available method of therapy for refractory atrial arrhythmias is now catheter ablation of atrioventricular conduction to isolate the source from the ventricles, a surgical alternative has recently been described whereby a corridor of right atrium connecting the sinus and AV nodes is isolated from other atrial tissue, thereby allowing sinus depolarizations to be exclusively conducted to the ventricles via the AV node (GUIRAUDON et al. 1985). This restores sinus control of heart rate but without the benefit of mechanical atrioventricular synchrony. It has, as yet, rarely been performed and is unlikely to gain great popularity because of the success of alternative therapies in controlling a non-life-threatening situation.

Large series now exist describing the successful destruction of accessory AV pathways by a variety of surgical techniques. Most use dissection alone or in combination with cryothermy. After the site of the accessory pathway (usually an epicardial structure) has been identified as outlined above, then it may be destroyed by dissection from within (SEALY et al. 1974), cryothermy from within (GALLAGHER et al. 1977) or a combination of epicardial AV groove dissection and cryothermy (GUIRAUDON et al. 1984). This last technique is gaining widespread popularity because opening the heart is not necessary so that cardiopulmonary bypass may sometimes be avoided, the risks of air entering the systemic circulation are minimized and it has improved the results for destroying the previously difficult septal pathways. Finally, for atrioventricular re-entry tachycardia, electrical current has been applied in a similar way to catheter ablation but using a disc held in the AV groove at surgery with the shock delivered between this and a cutaneous plate electrode (BOKERIA et al. 1987). By virtue of the larger surface area of the disc, there is no explosion. This has been reported by only one centre and the numbers are as yet too small for the method to be widely adopted.

The management of AV nodal re-entry tachycardia has been revolutionised by the advent of precise para-AV nodal dissection to destroy the retrograde limb of the re-entry circuit (ROSS et al. 1985). The series of those who originated the method now exceeds 50 patients with no failures being encountered when the original dissection was adhered to. The technique has already been successfully modified with the application of para-AV nodal cryoablation (Klein et al. 1985). Surgical resection and pacing are alternatives rarely used for therapy of atrial tachycardias.

Ventricular subendocardial resection is the usual method for destroying the substrate for ventricular tachycardia, usually present in scarred endocardium after myocardial infarction (Josephson et al. 1979). It consists of slicing thin slivers of tissue from the area involved, preserving as much as possible of the ventricular wall. This has largely superseded other techniques such as a simple ventriculotomy through an identified circuit (Guiraudon et al. 1974; Spurrell et al. 1975; Fontaine et al. 1976, 1977) and an encircling ventriculotomy (Guiraudon et al. 1978) in which an endocardial incision was made around the source of tachycardia, thus isolating it from the remainder of the ventricles. Subendocardial resection has gained popularity because of its improved success whilst preserving ventricular function compared with the other methods. However, one example of surgical isolation of a source of tachycardia, that of right ventricular isolation (Guiraudon et al. 1983), is highly successful when performed, with the resultant impairment to right ventricular function seeming to matter little clinically, presumably because the interventricular septum is spared. However, long-term follow up is still awaited.

Cryothermy is performed by applying a probe to the tissue to be destroyed and cooling this to $-60°$ C for a least 2 min. Shorter periods result in reversible damage and may be used for "cryomapping". The tissue is destroyed by the rewarming injury to a localized area. Thus, larger probes produce larger lesions. The lesions resulting from cryothermy are uniform (so not themselves arrhythmogenic) and retain considerable tensile strength. Thus the technique may compliment dissection in areas where the scalpel is likely to cause unwanted damage such as at the base of a papillary muscle.

D. Conclusions

A variety of often complementary therapies thus exist as potential alternatives to antiarrhythmic drugs in the treatment of almost all cardiac arrhythmias. The improvements seen in such therapies continue at such a rate that, in the absence of similar advances toward the perfect antiarrhythmic drug, their use will only increase in the foreseeable future.

References

Bokeria LA, Mikhailin AS, Revishvili LJ et al. (1987) Epicardial electric shock ablation of accessory pathways in preexcitation syndrome. In: Fontaine G, Scheinman MM (eds) Ablation in cardiac arrhythmias. Futura, New York

Boyd EGCA, Holt PM (1985) An investigation into the electrical ablation technique and a method of electrode assessment. PACE 8:815

Burchell HB, Frye RL, Anderson MW, McGoon DC (1967) Atrioventricular and ven-

triculoatrial excitation in WPW syndrome type B: temporary ablation at surgery. Circulation 36:663
Camm J, Ward D (1983) Pacing for tachycardia control. Telectronics, London
Camm J, Ward D, Cory-Pearce R et al. (1979) The successful cryosurgical treatment of paroxysmal ventricular tachycardia. Chest 75:621
Camm J, Ward D, Gainsborough J, Spurrell RAJ (1980) A microcomputer-based tachycardia termination system—a preliminary report. J Med Eng Technol 4:80
Cassidy DM, Vassallo JA, Buxton JE et al. (1984) The value of catheter mapping during sinus rhythm to localize site of origin of ventricular tachycardia. Circulation 69:1103
Cobb FR, Blumenschein SD, Sealy WC et al. (1968) Successful surgical interruption of the bundle of Kent in a patient with Wolff-Parkinson-White syndrome. Circulation 38:1018
Coumel P (1970) Different modes of pacing in the long term management of paroxysmal tachycardia. In: Sandoe E, Olesen KH, Flensted-Jensen E (eds) Symposium on cardiac arrhythmias. Astra, Södertölje
Coumel P, Flammang D, Attuel P, Leclercq JF (1979) Sustained intraatrial reentrant tachycardia. Electrophysiologic study of 20 cases. Clin Cardiol 2:167
Cunningham D, Rowland E, Rickards AF (1986) A new low energy power source for catheter ablation. PACE 9(II):1384
Davies DW, Wainwright RJ, Tooley MA et al. (1986a) Detection of pathological tachycardia by analysis of electrogram morphology. PACE 9:200
Davies DW, Nathan AW, Camm AJ (1986b) Three deaths after attempted high energy ablation of ventricular tachycardia. Br Heart J 55:506
Davies DW, Ward DE, Nathan AW, Camm AJ (1987a) Fulgurative ablation of accessory pathways in humans. In: Fontaine G, Scheinman MM (eds) Ablation in cardiac arrhythmias. Futura, New York
Davies DW, Nathan AW, Caplin JL and Camm AJ (1987b) Provocation of ventricular tachycardia by high energy catheter ablation of normal atrioventricular conduction. Br Heart J 57:579
Denes P, Wu D, Dhingra R, Amat-Y-Leon F et al. (1975) Dual atrioventricular nodal pathways. Br Heart J 37:1069
Dreifus LS, Nichols H, Morse D et al. (1968) Control of recurrent tachycardia of WPW syndrome by surgical ligature of the A-V bundle. Circulation 38:1030
Durrer D, Schoo L, Schuilenberg R, Wellens HJI (1967) The role of premature beats in the initiation and termination of supraventricular tachycardia in the Wolff-Parkinson-White syndrome. Circulation 36:644
Echt DS, Armstrong K, Schmidt P et al. (1985) Clinical experience, complications, and survival in 70 patients with the automatic implantable cardioverter/defibrillator. Circulation 71:289
Fisher JD, Brodman R, Kim SG et al. (1984) Attempted non-surgical electrical ablation of accessory pathways via the coronary sinus in the Wolff-Parkinson-White syndrome. J Am Coll Cardiol 4:685
Fontaine G, Guiraudon G, Frank R et al. (1976) Epicardial mapping and surgical treatment in six cases of resistant ventricular tachycardia not related to coronary artery disease. In: Wellens HJJ, Lie KI, Janse MJ et al. (eds) The conduction system of the heart. Stenfert, Kroese, Leiden
Fontaine G, Guiraudon G, Frank R (1977) Stimulation studies and epicardial mapping in ventricular tachycardia: study of mechanism and selection for surgery. In Kulbertus HE (ed) Re-entrant Arrhythmias.
Fontaine G, Tonet JL, Frank R et al. (1987) Treatment of resistant ventricular tachy-

cardia by endocavitary fulguration associated with antiarrhythmic therapy. In: Fontaine G and Scheinman MM (eds) Ablation in cardiac arrhythmias. Futura, New York

Gallagher JJ Sealy WC, Anderson RW et al. (1977) Cryosurgical ablation of accessory atrioventricular connections: a method for correction of the preexcitation syndrome. Circulation 55:471

Guiraudon G, Frank R, Fontaine G (1974) Interêt des cartographies dans le traitement chirugical des tachycardies ventriculaires rebelles recidivants. Nouv Press Med 3:321

Guiraudon G, Fontaine G, Frank R et al. (1978) Encircling endocardial ventriculotomy: a new surgical treatment for life-threatening ventricular tachycardias resistant to medical treatment following myocardial infarction Ann Thorac Surg 26:438

Guiraudon GM, Klein GJ, Gulamhusein SS et al. (1983) Total disconnection of the right ventricular free wall: surgical treatment of right ventricular tachycardia associated with right ventricular dysplasia. Circulation 67:463

Guiraudon GM, Klein GJ, Gulamhusein S et al. (1984) Surgical repair of Wolff-Parkinson-White syndrome: a new closed-heart technique. Ann Thorac Surg 37:67

Guiraudon GM, Campbell CS, Jones DL et al. (1985) Combined sino-atrial node atrioventricular isolation: a surgical alternative to His bundle ablation in patients with atrial fibrillation. Circulation 72[III]:220

Harken AH, Horowitz LN, Josephson ME (1980) Comparison of standard aneurysmectomy and aneurysmectomy with directed endocardial resection for the treatment of recurrent sustained ventricular tachycardia. J Thorac Cardiovasc Surg 8:527

Hartzler GO (1983) Electrode catheter ablation of refractory focal ventricular tachycardia. J Am Coll Cardiol 2:1107

Holt PM, Smallpeice C, Deverall PB et al. (1985a) Ventricular arrhythmias. A guide to their localisation. Br Heart J 53:417

Holt PM, Boyd EGCA, Crick JCP, Sowton E (1985b) Low energies and helifix electrodes in the successful ablation of atrioventricular conduction. PACE 8:639

Huang SKS (1987) Use of radiofrequency energy for catheter ablation of the endomyocardium: a prospective energy source. J Electrophysiol 1:78

Jackman WM, Friday KJ, Scherlag BJ et al. (1983) Direct endocardial recording from an accessory atrioventricular pathway: localization of the site of block, effect of antiarrhythmic drugs, and attempt at surgical ablation. Circulation 68:906

Josephson ME, Horowitz LN, Farshidi A et al. (1978) Continuous local electrical activity: a mechanism of recurrent ventricular tachycardia. Circulation 57:659

Josephson ME, Harken AH, Horowitz LN et al. (1979) Endocardial excision. A new surgical technique for the treatment of recurrent ventricular tachycardia. Circulation 60:1430

Josephson ME, Horowitz LN, Spielman SR et al. (1982a) The role of catheter mapping in the preoperative evaluation of ventricular tachycardia. Am J Cardiol 49:207

Josephson ME, Waxman HL, Cain ME et al. (1982b) Ventricular activation during ventricular endocardial pacing. II. Role of pace-mapping to localize origin of ventricular tachycardia. Am J Cardiol 50:11

Kappenberger L, Sowton E (1981) Programmed stimulation for longterm treatment and non-invasive investigation of recurrent tachycardia. Lancet i:909

Klein GJ, Guiraudon GM, Perkins DG et al. (1985) Controlled cryothermal injury to the AV node: feasibility for AV nodal modification. PACE 8:630

Krikler DM, Curry PVL, Buffet J (1976) Dual-demand pacing for reciprocating atrioventricular tachycardia. Br Med J 1:1114

Malik M, Davies DW, Camm AJ (1986) Limiting factors in the use of DDD pacema-

kers to prevent junctional reentry: computer modelling experiments. Clin Prog Electrophysiol Pacing 4:137
Marchlinski FE, Josephson ME (1984) Appropriate diagnostic studies for arrhythmia surgery. PACE 7:902
Massumi RA, Kistin AD, Tawakkol AA (1967) Termination of reciprocating tachycardia by atrial stimulation. Circulation 36:637
McComb JM, McGovern BA, Garan H et al. (1985) Modification of atrioventricular conduction using low energy transcatheter shocks. J Am Coll Cardiol 5:454
Mirowski M, Reid PR, Mower MM et al. (1980) Termination of malignant ventricular arrhythmias with an implanted automatic defibrillator in human beings. New Engl J Med 303:322
Mirowski M, Reid PR, Mower MM et al. (1985) Clinical experience with the automatic implantable defibrillator. Arch Mal Coeur 78:39
Morady F, Scheinman MM, Winston SA et al. (1987) Transcatheter ablation of posteroseptal accessory pathways. In: Fontaine G, Scheinman MM (eds) Ablation in cardiac arrhythmias. Futura, New York
Moss AJ, McDonald J (1971) Unilateral cervicothoracic sympathetic ganglionectomy for the treatment of long Q-T interval syndrome. New Engl J Med 285:903
Narula OS, Boveja BK, Cohen DM et al. (1985) Laser catheter-induced atrioventricular nodal delays and atrioventricular block in dogs: acute and chronic observations. J Am Coll Cardiol 5:259
Nathan A, Hellestrand K, Bexton R et al. (1982) Clinical evaluation of an adaptive tachycardia intervention pacemaker with automatic cycle length adjustment. PACE 5:201
Nathan AW, Creamer JE, Davies DW, Camm AJ (1986) Clinical experience with a software based tachycardia reversion pacemaker. PACE 9[II]:1312
Olson WH, Bardy GH (1986) Cycle length and morphology patterns at the onset of spontaneous ventricular tachycardia and fibrillation. PACE 9:284
Ross DL, Johnson DC, Denniss AR et al. (1985) Curative surgery for atrioventricular junctional ("AV nodal") reentrant tachycardia. J Am Coll Cardiol 6:1383
Scheinman M, Morady F, Hess DS et al. (1982) Catheter induced ablation of the atrioventricular junction to control reciprocating supraventricular arrhythmias. JAMA 243:851
Schofield PM, Bowes RJ, Brooks N, Bennett DH (1986) Exercise capacity and spontaneous heart rhythm after transvenous fulguration of atrioventricular conduction. Br Heart J 56:358
Sealy WC, Wallace AJ, Ramming KP et al. (1974) An improved operation for the definitive treatment of the Wolff-Parkinson-White syndrome. Ann Thorac Surg 17:107
Silka MJ, Gillette PC, Garson A (1985) Transvenous catheter ablation of a right atrial automatic ectopic tachycardia. J Am Coll Cardiol 5:999
Sowton E, Leatham A, Carson P (1964) The suppression of arrhythmias by artifical pacemaking. Lancet ii:1098
Sowton E (1984) Clinical results with the Tachylog antitachycardia pacemaker. PACE 7:1313
Spurrell RAJ (1975) Artifical cardiac pacemakers. In: Krikler DM, Goodwin JF (eds) Cardiac arrhythmias. Saunders, London
Spurrell RAJ (1976) Future aspects of cardiac pacing. In: Luderitz B (ed) Cardiac Pacing. Springer, Berlin/Heidelberg/New York
Spurrell RAJ, Sowton E (1976) Pacing techniques in the management of supraventricular tachycardia. Part 2. J Electrocardiol 9:89
Spurrell RAJ, Sowton E, Deuchar D (1973) Ventricular tachycardia in 4 patients evalu-

ated by programmed electrical stimulation of the heart and treated in 2 patients by surgical division of anterior radiation of left-bundle branch. Br Heart J 35:1014
Spurrell RAJ, Yates AK, Thorburn CW et al. (1975) Surgical treatment of ventricular tachycardia after epicardial mapping studies. Br Heart J 37:115
Spurrell RAJ, Bexton R, Nathan A et al. (1982) Implantable automatic scanning pacemaker for termination of supraventricular tachycardia. Am J Cardiol 49:753
Svenson RH, Selle JG, Gallagher JJ et al. (1987) Neodymium: YAG laser potocoagulation: a potentially useful method for intraoperative ablation of arrhythmogenic foci. In: Fontaine G, Scheinman MM (eds) Ablation in Cardiac Arrhythmias. Futura, New York
Vedel J, Frank R, Fontaine G et al. (1979) Block auriculoventriculaire intra-hisian definitif induit au cours d'une exploration endoventriculaire droite. Arch Mal Coeur 72:107
Vergara G, Disertori M, Inama G et al. (1985) Sustained ventricular tachycardia: low-energy transcatheter intracardiac cardioversion. Efficacy, reliability and tolerance also in comparison with ventricular burst. G Ital Cardiol 15:862
Ward DE, Camm AJ, Spurrell RAJ (1979) The response of regular reentrant supraventricular tachycardia to right heart stimulation. PACE 2:586
Wellens HJJ (1975) Contribution of cardiac pacing to our understanding of the Wolff-Parkinson-White syndrome. Br Heart J 37:231
Wellens HJJ, Duren DR, Lie KI (1976) Observations on mechanisms of ventricular tachycardia in man. Circulation 54:237
Zipes DP, Jackman WM, Heger JJ et al. (1982) Clinical transvenous cardioversion of recurrent life-threatening ventricular tachycardias and termination of ventricular fibrillation in patients using a catheter electrode. Am Heart J 103:789
Zipes DP, Heger JJ, Miles WM (1984) Early experience with a implantable cardioverter. New Engl J Med 311:485
Zoll PM, Linenthal AJ, Zarsky LRN (1960) Ventricular fibrillation. Treatment and prevention by external electric currents. New Engl J Med 262:105

Factors Involved in Arrhythmogenesis

CHAPTER 22

Alpha-Adrenoceptors in Arrhythmogenesis

J. C. McGrath

A. Introduction

Since alpha-adrenoceptors perform many physiological roles in the nervous and cardiovascular systems it is inevitable that, theoretically, their activation or blockade will impinge on any dysrhythmogenic process. Empirical observation also clearly shows that activation of the sympathetic nervous system can be dysrhythmogenic (Chap. 12), either when it is excessive or when some defect of cardiac function already exists, and that drugs which are known to be alpha-adrenoceptor antagonists can alleviate this. At present the fundamental basis of this phenomenon is not entirely clear. Consequently an opportunity may exist, by clarifying this, to increase the rational basis of antiarrhythmic therapy. To this end, this chapter reviews the roles of alpha-adrenoceptors in the heart or with a bearing on cardiac function, with particular emphasis on possible influences on normal or abnormal cardiac rhythm.

Activation of alpha-adrenoceptors can influence normal or abnormal cardiac function via:

I. Effects at various sites within the heart
II. Alteration of haemodynamics via effects on blood vessels outside the heart
III. Effects on blood composition particularly hormones
IV. Effects on autonomic nerve activity directly within the CNS, within the reflex arc or indirectly through nervous reflexes consequent on factors I–III
V. Combinations of I–III such as might arise from the haemodynamic consequences of altered hormone status

Clearly, this wide range of effects will produce physiological end points with regard to cardiac rhythm which are diverse and in opposition. Any assessment of the role of alpha-adrenoceptors in the genesis, maintenance or correction of any particular form of arrhythmia will depend on a balance of factors rather than a straightforward influence in one direction.

This chapter will approach this issue by mapping the possible relevant influences of alpha-adrenoceptors and the theoretical and experimental approaches which are involved in this.

B. Multiple Sites of Action for Catecholamines

I. Separation of Alpha- and Beta-Adrenoceptors

Separation of alpha- and beta-adrenoceptors in functional tests can be achieved by using the relative potency of antagonists and agonists (Ahlquist 1948; Furchgott 1972; Gillespie 1980). This is most satisfactory when a relative potency series can be established both for antagonists and agonists. Where this is not possible, the potency of antagonists versus catecholamines or versus other more selective agonists provides a more reliable index than agonist potency alone. Similarly, corresponding ligands are available which differentiate between alpha- and beta-ligand-binding sites: again antagonists provide a clearer definition than do agonists (Hoffman and Lefkowitz 1980; Bylund and U'Prichard 1983).

However, there seem to be no major problems in differentiating between alpha- and beta-adrenoceptors with either functional or ligand-binding tests. If a functional response is produced by a catecholamine or sympathetic nerve stimulation, blockade by phentolamine or propranolol, but not by the other, is adequate proof of alpha or beta, respectively. Any difficulties which do arise are at the level of subdivision of the alpha-adrenoceptor into subtypes. The major problem which arises in the literature occurs when complete reliance is placed on agonists alone. For example phenylephrine is often used as a selective alpha-adrenoceptor agonist (alpha$_1$); yet this compound can act as an indirect sympathomimetic, releasing noradrenaline from adrenergic nerves (Luchelli-Fortis and Langer 1974). It can also act as an agonist at beta-adrenoceptors (Lefevre et al. (1977).

II. Alpha-Adrenoceptor Subtypes

1. Definition

When alpha-adrenoceptors were found to modulate the release of neurotransmitters from nerves, it quickly became apparent, in the first few preparations which were tested, that both agonist and antagonist potency series differed at prejunctional and postjunctional sites (Langer 1979; Starke 1977). At this early stage a few compounds became established as markers for the two types and the terms alpha$_1$- and alpha$_2$-adrenoceptors were coined to refer respectively to (1) the postjunctional receptors, e.g. on rabbit main pulmonary artery, at which phenylephrine was more potent than clonidine as an agonist and prazosin was more potent than yohimbine as an antagonist, and (2) the prejunctional receptor at which these potencies were reversed, clonidine being more potent than phenylephrine and yohimbine more potent than prazosin. Having established this principle on a small number of preparations, several further selective agonists and antagonists were synthesized, or discovered from among the available drugs, consolidating this classification. With these tools, on investigating a wider range of preparations, it was found that these two subtypes, as defined by the drug potencies, could be found at either pre-

Table 1. Drugs on which the subdivision of alpha-adrenoceptors is currently based. Approximate "windows of selectivity" are indicated for concentration in vitro and doses in vivo. Effects obtained above these ranges are unlikely to be selective. Evidence for each subtype is obtained by establishing the antagonist potency series preferably against a catecholamine or, in the case of a mixture of receptor types, against one of the "selective" agonists. Confirmation can be obtained from the agonist potency series, although this is often unreliable. Comments indicate problems associated with the use of each drug

Drug	In Vitro	In Vivo	Comment.
Selective alpha$_1$-antagonists			
Prazosin	0.1 n*M*-100 n*M*	0.01-1 mg/kg	
Corynanthine	100 n*M* - 3 μ*M*	0.3 -3 mg/kg	5-HT-antagonism
Selective alpha$_2$-antagonists			
Rauwolscine	100 n*M*-300 n*M*	0.05-0.5 mg/kg	5-HT-antagonist, narrow window of selectivity
Yohimbine	100 n*M*-300 n*M*	0.05-0.5 mg/kg	5-HT-antagonist, very narrow window of selectivity
Potency series for alpha$_1$			
Prazosin > corynanthine ≧ yohimbine ≧ rauwolscine			
Potency series for alpha$_2$			
Rauwolscine ≧ yohimbine ≫ corynanthine and prazosin			
Selective alpha$_1$-agonists			
Phenylephrine	1 n*M*- 30 n*M*	0.03- 10 μg/kg	At high concentrations: beta- and alpha$_2$-agonism, indirect sympathomimetic
Amidephrine	1 n*M*- 1 μ*M*	0.03-500 μg/kg	Relatively untried in heart, partial agonist in some preps
Methoxamine	10 n*M*- 300 n*M*	0.3 -100 μg/kg	low potency, partial agonist
Cirazoline	1 n*M*- 300 n*M*	0.03-100 μg/kg	Relatively untried in heart, high affinity for alpha$_2$
Selective alpha$_2$-agonists (all have alpha$_1$ potency at high concentrations)			
Xylazine	10 n*M*- 100 n*M*	1 -500 μg/kg	Low potency
UK14304	1 n*M*- 30 n*M*	0.03- 10 μg/kg	Relatively untried in heart
Clonidine	1 n*M*- 10 n*M*	0.03- 0.1 μg/kg	Partial agonism at alpha$_2$ and alpha$_1$, makes window of selectivity small or non-existent
B-HT 933	10 n*M*- 100 n*M*	1 -500 μg/kg	Low potency, relatively untried in heart
Potency series for alpha$_1$			
Noradrenaline ≧ phenylephrine ≧ cirazoline ≧ amidephrine > methoxamine ≧ clonidine ≧ xylazine and UK14304 and B-HT 933			
Potency series for alpha$_2$			
(Establishment of noradrenaline's prejunctional alpha$_2$ potency is controversial) Noradrenaline ≧ UK14304 ≧ clonidine > xylazine ≧ B-HT 933 ≧ phenylephrine > methoxamine and amidephrine and cirazoline			

junctional or postjunctional sites (Docherty et al. 1979; McGrath 1982; Starke 1981). Subsequently, the categorization relied solely on pharmacological criteria. Ligand-binding sites were found which corresponded to the two subtypes shown by these functional pharmacological tests (Bylund and U'Prichard 1983).

The drugs on which the subdivision of alpha-adrenoceptors is currently based are shown in Table 1. There are indications, for both $alpha_1$- and $alpha_2$-adrenoceptors and the equivalent binding sites, that each subtype might not be the same everywhere even though they are clearly defined as $alpha_1$ or $alpha_2$ by the above criteria. The absolute potencies of the marker drugs are found to vary in different tissues or different species, even using the nearest estimates which can be gained of their dissociation constants, and this extends to their order of potency, which is more difficult to account for on theoretical grounds (Ruffolo et al. 1980). This has not yet led to a general reclassification or further subclassification but it is considered at present that, within the fairly convincing groupings of $alpha_1$ and $alpha_2$, there may be differences with regard to affinity of some compounds between different locations (Nahorski et al. 1985; Drew 1985; Waterfall et al. 1985; Alabaster et al. 1986; Flavahan and Vanhoutte 1986).

2. Antagonists Used

The most commonly employed selective $alpha_1$-adrenoceptor antagonist is prazosin (Cambridge et al. 1977; Karliner et al. 1979). In both functional tests and ligand binding tests, it has high affinity for $alpha_1$ sites, with pA_2 values or dissociation constants (log values) ranging from 8 to 11 (Drew 1985; Flavahan and Vanhoutte 1986), but it has low affinity for other receptors and binding sites including the other adrenoceptors, $alpha_2$-, $beta_1$- and $beta_2$-, receptors for histamine, 5-HT or dopamine. Other compounds used as selective $alpha_1$-antagonists must be judged against this standard particularly with regard to potency and to lack of affinity for other receptors.

Other compounds which may be used as selective $alpha_1$-adrenoceptor antagonists and their main drawbacks are indicated in Table 2. There is no antagonist of $alpha_2$-adrenoceptors equivalent to prazosin as a potent and selective standard. Originally the drug of choice was yohimbine (Starke 1977) and subsequently this was replaced by rauwolscine, a stereoisomer of yohimbine,

Table 2. "Selective $alpha_1$-adrenoceptor antagonists" used to back-up standard compounds indicated in Table 1

	Advantages	Disadvantages
Azapetine	High selectivity	Old compound, scanty recent data
BE 2254	High potency	Relatively untried
Doxazosin	High potency	
Indoramin	Good selectivity	Relatively low potency
Labetalol	Good $alpha_1$/$alpha_2$ selectivity	Beta-antagonism

Table 3. Newer "selective $alpha_2$-antagonists"

	Advantages	Disadventages
Idazoxan	Most studied of the group Clinical data available	$Alpha_1$-agonism
Imiloxan	Good selectivity	Low potency
Piperoxan	Old compound; clinical data available	Relatively low selectivity
BDF-6143	High potency and selectivity	$Alpha_1$-agonism Inhibits noradrenaline release
WY 26703	High $alpha_1$/$alpha_2$ selectivity	5-HT antagonism[a]
DG5128	Reasonable selectivity	Low potency Relatively untested
SKF86466	Good potency, high selectivity	Relatively untested

[a] Data are not available on this for most of the other compounds
For a more detailed evaluation see CLARK et al. (1986)

which is slightly more selective than yohimbine between $alpha_2$- and $alpha_1$-adrenoceptors (WEITZELL et al. 1979) but neither drug has a high enough margin of selectivity for $alpha_1$ and $alpha_2$ to provide a definitive test for $alpha_2$-(c.f. $alpha_1$-)-adrenoceptors (McGRATH 1984). They share also the common disadvantage that they are antagonists of 5-HT receptors (LAMBERT et al. 1978).

Within the past 10 years, several new compounds have appeared which possess selectivity for $alpha_2$-adrenoceptors (Table 3). Those which have been in use longest and therefore have been tested in a wider variety of preparations tend to lack potency, e.g. imiloxan, or to possess partial agonism, e.g. idazoxan, which limits their usefulness (CLARK et al. 1986). In addition, data are not always available on their selectivity with regard to other receptors such as beta, histamine, dopamine and 5-HT.

One means of defining a response as $alpha_2$-adrenoceptor-mediated is to use one of the less-selective but well-tried compounds such as rauwolscine, or even phentolamine, but first to administer a high dose of a selective $alpha_1$-adrenoceptor antagonist such as prazosin to eliminate the $alpha_1$-adrenoceptor-mediated component (DOCHERTY and McGRATH 1980a; GRANT et al. 1985a). This avoids the problems of interpretation encountered when using non-selective $alpha_2$-antagonists on their own.

3. Agonists Used

The compound most commonly employed as a "selective" $alpha_1$-adrenoceptor agonist is phenylephrine. However, it does not have a high degree of selectivity between $alpha_1$- and $alpha_2$-, $beta_1$- and $beta_2$-adrenoceptors (FLAVAHAN and McGRATH 1981a, 1982). This limits its usefulness although the extensive background information on the compound may justify its use, provided that its effects are confirmed by the use of antagonists. Several other reasonably selective $alpha_1$-adrenoceptor agonists are available as shown in Table 1.

There is no equivalent standard $alpha_2$-adrenoceptor agonist. When the initial distinction between $alpha_1$- and $alpha_2$-adrenoceptors was made, clonidine was commonly employed as a "selective" $alpha_2$-adrenoceptor agonist (Starke 1977; Langer 1979). It is now known that clonidine has affinity for, and intrinsic activity at, $alpha_1$-adrenoceptors (Ruffolo et al. 1982). However since it is a partial agonist, the functional effects which it initiates through $alpha_1$-adrenoceptors appear only in preparations where there is a substantial receptor reserve. This factor may be of relevance to cardiac $alpha_1$-adrenoceptors, which can play a minor role compared with beta-adrenoceptors under normal conditions, but whose contribution may be modified by pathophysiological factors which increase their receptor reserve, thus allowing greater expression of an alpha-adrenoceptor-mediated response. As with the $alpha_1$-adrenoceptor-mediated responses to phenylephrine, it may in some circumstances be possible to use a relatively unselective agonist like clonidine provided that the receptor can be defined using antagonists. In this case, however, since there is a paucity of potent and selective $alpha_2$-adrenoceptor antagonists, proof that a drug such as clonidine is acting on $alpha_2$-adrenoceptors may rely as much on the lack of effect of prazosin as on blockade by $alpha_2$-adrenoceptor antagonists.

Several other compounds with selectivity for $alpha_2$-adrenoceptors are available as shown in Table 1.

Since many of the supposedly selective agonists are less potent than noradrenaline and adrenaline and since they are often partial agonists, there is often merit in using the natural catecholamines as the agonist when studying a particular phenomenon and to retain "selectivity" experimentally by blocking the receptors which are not under consideration. This may be pharmacologically more complex but has the merit that it may provide some physiologically relevant information.

4. Postulated Alternative Criteria for Distinguishing Between Subtypes

In the few years since it has been known that both subtypes of alpha-adrenoceptor may mediate postjunctional events and hence produce the same end response such as vasoconstriction, attempts have been made to find differences in the transduction pathways which lie between the different recognition sites (receptors) and the common response (usually elevation of intracellular free calcium ion concentration). This has taken two main forms.

First there is the possibility that each subtype may use the same second messenger wherever it is found but that this messenger will be different from that used by the other subtype: this has been tested in semipurified membrane preparations where the production of the second messengers in the presence of agonists can be detected and has thus been limited in the types of system tested. Early results suggested that $alpha_1$ might use, exclusively, phosphatidyl inositol and that $alpha_2$ might use, exclusively, inhibition of adenylate cyclase (Berthelsen and Pettinger 1977; Fain and Garcia-Sainz 1980).

However, when the hypothesis was tested in a wider variety of tissues, the ubiquitous nature of the phenomenon no longer held (NAHORSKI et al. 1985; MICHELL 1985). Nevertheless, there may be sufficient difference in any given effector system that modulation within the transduction process may be sufficient to cause the highlighting of one receptor subtype, particularly in pathological conditions.

Secondly, and less specifically, but with more immediate relevance to the present topic, it was found that several drugs which affect ion channels or other modulatory factors which influence the end response could differentially affect the response mediated by the two receptors. Such selective toxicity could reflect some fundamental difference in the transduction processes involved: this was tested in more complex intact cell or even in vivo preparations where the end-organ response was measured, usually contraction of isolated blood vessels or changes in blood pressure which were interpreted as reflecting vasoconstriction. In most cases a common factor was that alpha$_2$-adrenoceptor-mediated responses were more susceptible to blockade by "physiological antagonism" than were those to alpha$_1$-adrenoceptor activation. Examples of this were (1) removal of the facilitatory influence of angiotensin II by angiotensin-converting enzyme inhibitors or angiotensin-receptor antagonists (HATTON and CLOUGH 1982; TIMMERMANS et al. 1982), (2) calcium channel blockers (DE MEY and VANHOUTTE 1981; VAN MEEL et al. 1981; GODFRAIND et al. 1982) and hypoxia (McGRATH et al. 1982). However, in each case, after the initial demonstration with a limited number of agonists, when further agonists were tested, the general nature of the phenomenon was shown to be unsound with regard to a complete split between the receptor subtypes. Rather, there was differentiation between responses to different agonists, particularly with regard to time course of the response (O'BRIEN et al. 1985; GRANT et al. 1985b; O'BRIEN and McGRATH 1987). Although a debate continues on the subject, argument now centres on whether the explanation in some, if not all, cases is that receptor reserve is an important factor: where this is less for the alpha$_2$ system, the responses are in effect less robust and more susceptible to attenuation (RUFFOLO et al. 1984; LEW ANGUS 1985).

The general conclusion from this, which is relevant here, is that in any given situation responses to the two alpha-adrenoceptor subtypes may well be differentially modulated by physiological factors, or by drugs which modify the physiological environment in which the receptors operate (ion channel function, etc.). This may be highly pertinent to the role of the receptors physiologically or pathologically. It is still worthwhile to seek unique aspects of the transduction pathways for the individual subtypes particularly where both occur in the same organ or tissue and the information may have some pragmatic use. It would, however, on current evidence, be incorrect to attempt to classify the receptors on this basis or to assume that the differential effect will be the same as in other tissues. This applies particularly to the alpha-adrenoceptors in cardiac muscle whose physiological and pathological roles are relatively unknown.

5. Special Factors Relevant to the Heart

Much of the information available on the properties of alpha-adrenoceptors has been obtained from non-cardiac tissue. By definition, the recognition site itself should be similar but specialization of the transduction and peculiarities of the functional response of cardiac muscle may produce distinctive phenomena involving alpha-adrenoceptors. Some of these are considered below.

a. Electrophysiology

The central role of the spontaneous initiation and spread of electrical excitation in timing and coordinating the contractile function of the heart is a facet which demands a purely modulatory role of the adrenoceptors, in changing the characteristics of the action potential. Since the specialization of the action potential in different parts of the heart comes about largely through the relative predominance of the conductance to different ions (see Chaps. 1, 2, 12), then the effects produced by alpha-adrenoceptors (and their relationship to beta-adrenoceptor-mediated effects) will depend on the influence which alpha-adrenoceptors can exert on the various ion channels, in the particular cells. This is not an area which has been comprehensively explored in a wide variety of cardiac cell-types and species, largely due to the predominant interest in beta-adrenoceptors and, as mentioned above, the difficulties which this entails for attempts to study alpha-adrenoceptors.

As will be discussed below in Sect. C, this is of central importance to the present topic of arrhythmia. Presumably the main questions on the mechanism involved following alpha-adrenoceptor activation, particularly in the genesis of arrhythmia, should concern whether they can affect the following processes and whether there are circumstances in which they do so physiologically or pathophysiologically: (1) conductance of K^+ and possibly of other ions in pacemaker tissue (changes which underly the positive chronotropic effect); (2) Ca^{2+} conductance in the atrioventricular node and Na^+ conductance in the fast conducting bundles (affecting the coordinated activation of contractile muscle and any re-entry phenomena associated with tissue damage); and (3) Ca^{2+} conductance in relation to availability of intracellular free calcium ions to initiate contraction (in relation to the positive inotropic effects). Apart from the known delayed inactivation of the inward calcium current (Miura et al. 1978; Bruckner and Scholz 1984; Dukes and Vaughan-Williams 1984a, b), there is little positive information concerning the influence of alpha-adrenoceptors on these systems (see Chaps. 1–3; Vaughan-Williams 1985). Consequently it is not clear whether the latter is a critical element in the role of alpha-adrenoceptors in arrhythmogenesis or is simply the only currently available electrophysiological information of relevance. In particular there is virtually no information on possible electrophysiological influences of alpha-adrenoceptors in cardiac ischaemia and there is no electrophysiological evidence on the positive chronotropic effect which can be clearly demonstrated in rat heart in vivo (Flavahan and McGrath 1982) or in vitro (Flavahan and McGrath unpublished).

b. Transduction Processes, Second Messengers and Receptor Regulation: Comparison of Alpha and Beta

Although the net responses of cardiac muscle to $alpha_1$- and beta-adrenoceptor activation are superficially similar, i.e. species-dependent chronotropic and positive inotropic, there are obvious superficial differences in the responses which point to likely differences in the transduction process utilized by each of the two types of receptor ($alpha_2$ are not relevant here since there is no evidence for their influence on cardiac muscle). The differences here seem to be even greater than those between $alpha_1$ and $alpha_2$ (see Sect. B. II.4 above). The differences with regard to inotropism have recently been susscinctly reviewed (BRUCKNER et al. 1985). In contrast to beta-adrenoceptor-mediated effects, the inotropic response to $alpha_1$ is not accompanied by faster force generation or quicker relaxation. Instead contraction is prolonged (see BENFEY 1982), presumably reflecting the delayed inactivation of the inward calcium current.

There is also considerable negative evidence showing that the physiological events (increase in magnitude without change in time course of slow inward current), known to be activated by beta, are not activated by $alpha_1$. Since the beta-mediated effects are believed to involve stimulation of adenylate cyclase with resultant activation of protein kinases and phosphorylation of slow channel proteins (REUTER 1974, 1983; TSIEN 1977; JAKOBS 1979; BRUCKNER et al. 1985), by implication $alpha_1$-adrenoceptor activation does not. This is supported by further negative observations showing that $alpha_1$ activation does not alter levels of cAMP (see SCHOLZ 1980): corollaries of this include resistance to phosphodiesterase inhibitors (SCHUMANN et al. 1975) and to agents which themselves diminish cAMP-related positive effects such as muscarinic agonists or adenosine (HATTORI and LEVI 1984).

More positively, by analogy with other $alpha_1$ systems known to involve inositol lipid breakdown (MICHELL 1985; LEFKOWITZ and CARON 1986), increased phosphoinositide turnover attributable to $alpha_1$-adrenoceptor activation has been found with noradrenaline in cardiomyocytes purified from rat ventricle (BROWN et al. 1985) and with methoxamine in rat atria (SEKAR and ROUFOGALIS 1984). This suggests that the general (though not necessarily ubiquitous) utilization by $alpha_1$-adrenoceptors of stimulation of inositol lipid hydrolysis as the step following receptor activation extends to cardiac muscle. In this case, the second messengers would be inosotol 1,4,5-trisphosphate, which commonly leads to intracellular release of Ca^{2+} and 1,2-diacylglycerol, which contributes to control of cell function through activation of protein kinase C (MICHELL 1985). It is not known whether either of these messengers, or a further unknown consequence of receptor activation, leads to the alteration of slow channel function. Experiments on cultured neonatal rat myocytes indicate a link between receptors activated by phenylephrine (associated with a *negative* chronotropic effect) and a pertussis toxin-specific regulatory protein (STEINBERG et al. 1985): this may or may not pertain to a further second messenger.

Since $alpha_1$-adrenoceptor-mediated cardiac responses persist in ischaemic conditions, yet superficially similar beta-adrenoceptor-mediated responses do not (Corr and Witkowski 1983), it is possible that the second messenger/transduction processes involved with $alpha_1$ can serve as a reserve mechanism under conditions which are metabolically unfavourable for the formation of the regulatory GTP complex necessary for expression of the beta-adrenoceptor's influence on adenylate cyclase (Vaughan Williams 1985). These relatively favourable circumstances for expression of the $alpha_1$-adrenoceptor's response might be assisted also by an increase in receptor number in ischaemia (Witkowski and Corr 1984).

Thus, in both vascular smooth muscle ($alpha_1$ and $alpha_2$) and cardiac muscle cells ($alpha_1$ and beta), similar physiological end points can be obtained via different transduction processes activated by different adrenoceptors. Consequently, in each case the systems are subject to different modulatory influences, so that physiological or pathological conditions can differentially change their engagement.

Another mechanism for altering the response mediated through a receptor system is a change in receptor number, leading to a change in sensitivity to the agonist. There seem to be differential effects on $alpha_1$ and beta systems in the heart. In general, animal experiments with ligand-binding techniques show that it is more difficult to "downregulate" (reduce number of receptors) $alpha_1$ compared with beta. It is possible to "upregulate" (increase the number of receptors) $alpha_1$-binding sites by chronic treatment with beta-blockers (Mugge et al. 1985a, b), suggesting that $alpha_1$ can somehow compensate for the loss of beta. Conversely, prolonged treatment with a beta-adrenoceptor agonist which causes downregulation of beta-adrenoceptors leaves the $alpha_1$-adrenoceptor-mediated response intact (Hayes et al. 1984). There is an interesting clinical parallel to this since patients with congestive heart failure show downregulation of beta-binding sites but less or no change in $alpha_1$ (see Ruffolo and Kopia 1986). Thus it should be borne in mind in treating this condition that a possible increased involvement of $alpha_1$-adrenoceptors in the heart might alter the effects of therepeutic agents, e.g. of inotropes activating the various receptors or antagonists used to block them. Similarly, in treating failing arrhythmic hearts the possibility of heterogeneous up- or down-regulation of alpha-adrenoceptor subtypes or, more likely, an unusual and heterogeneous dominance of alpha over beta, should be considered.

6. Relevance of Alpha-Adrenoceptors Outside the Heart

a. Central Nervous System

Clearly, alpha-adrenoceptors in the CNS exert an influence on efferent sympathetic nerve activity, e.g. this is the basis of the action of antihypertensives of the clonidine type (Schmitt 1977; Schmitt and Laubie 1983), and it has been argued that the effects of alpha-adrenoceptor agonists in the CNS are generally beneficial in congestive heart failure and ischaemic heart disease (Giles et al. 1985). A corollary of this is that, if ongoing physiological stimulation of the receptors which clonidine activates occurs normally, then antagonists with affinity for this receptor(s) could increase sympathetic activity. This has been demonstrated in the anaesthetized dog with yohimbine (ANDREJAK et al. 1983).

b. Blood Vessels—Haemodynamics

In experiments which involve in vivo preparations, the characterization of alpha-adrenoceptors is complicated by the influence of alpha-adrenoceptors outside the heart. Pre- and postcapillary resistance and venous capacitance are all modulated via alpha-adrenoceptors activated by sympathetic nerves or by circulating catecholamines, which act either directly on smooth muscle or by modulating transmitter release from sympathetic nerves. Consequently, both preload and afterload are influenced by alpha-adrenoceptors. Variations in these factors are particularly important under conditions where the heart is failing. Consequently their contribution must be considered when assessing the role of cardiac alpha-adrenoceptors in a clinical or experimental situation.

c. Hormones and Metabolism

In addition to their direct haemodynamic influence, alpha-adrenoceptors are also involved in the release of several hormones, e.g. insulin (NAKADATE et al. 1980; NAKAKI et al. 1980), growth hormone and calcitonin (BROWN et al. 1985). Although alpha-adrenoceptor-mediated effects on cellular metabolism are less comprehensively documented than those mediated by beta-adrenoceptors, they are important in several organs including the heart itself (CLARK et al. 1983). Consequently the hormonal composition of blood reaching the heart together with direct effects on metabolism may both be altered by alpha-adrenoceptor influence with resulting consequences for cardiac performance, particularly under stress (see also Sect. B.III.1).

7. Consequences of Diversity of Alpha-Adrenoceptor Subtypes

If, within a species, there are only two subtypes of alpha-adrenoceptors, for each of which any given agonist or antagonist has a uniform affinity and any agonist a uniform efficacy, then "selectivity" between the two subtypes will be the major factor determining the net effect of a drug on the cardiac rhythm. For example, the overall effects of all highly selective alpha$_1$-adrenoceptor antagonists would be the same, subject to distribution and metabolism. However, if a drug can make further differentiation, such as (1) between pre- and postjunctional alpha$_2$-adrenoceptors, (2) between alpha$_1$-adrenoceptors mediating different effects on cardiac muscle or (3) between cardiac and vascular receptors, then the pharmacological consequences and hence the therapeutic potential of drugs which currently appear similar will radically alter. There is little evidence for this at present although it has recently been postulated that a new alpha-adrenoceptor antagonist SK&F 104078 can selectively block postjunctional alpha$_2$-adrenoceptors (HIEBLE et al. 1986).

The next section deals with the various sites at which alpha-adrenoceptor activation may occur within the heart as a basis for judging how a drug with a particular spectrum of affinity or efficacy at the subtypes may influence cardiac arrhythmia.

III. Separate Sites of Action for Alpha-Adrenoceptors Within the Heart

Within the heart, alpha-adrenoceptors mediate diverse effects since they are located on different cell types. Even within each cell type, they can medaite different responses either because of the presence of more than one subtype of

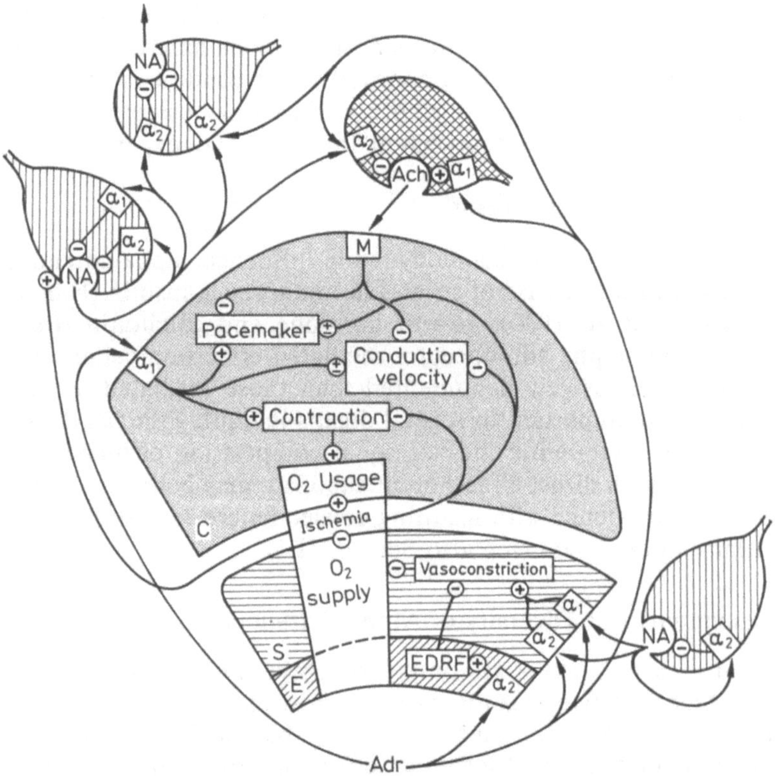

Fig. 1. Diagrammatic representation of the various sites at which subtypes of alpha-adrenoceptors can influence structures within the heart. These sites include cardiac muscle cells (C), coronary vascular smooth muscle cells (S), coronary vascular endothelium (*E*) and the autonomic nerve varicose terminals which serve cardiac and vascular muscle: adrenergic (shown releasing transmitter noradrenaline, NA) and cholinergic (shown releasing transmitter acetylcholine, Ach, to act on muscarinic receptors, M) The influence of the receptor activation is indicated on various aspects of cell function and on ischaemia, shown as a balance between O_2 supply from the blood vessels and usage by the cardiac muscle. Stimulation of a function or enhancement of release of a substance is indicated by ⊕. Inhibition of a function or attenuation of release of a substance is indicated by ⊖. The second adrenergic varicosity at the top, releasing transmitter away from the cardiac cell, indicates crosstalk between terminals serving different muscle cells

alpha-adrenoceptor or simply because a given subtype initiates a train of events which leads to more than one detectable consequence. As shown in Fig. 1, the main sites which must be considered are cardiac muscle, coronary vascular smooth muscle, the nerves which serve both of these and the endothelium of the vascular smooth muscle.

1. Cardiac Muscle

Since cardiac muscle is highly specialized in several directions, even if all alpha-adrenoceptors utilized the same second messenger and subsequent cascade of events, the modulation of cellular function would vary with the cell type. This could usefully be considered by subdividing cardiac muscle along functional lines using a division into pacemaker activity, speed of conduction, force of contraction and metabolism. Unfortunately, the available evidence concentrates mostly on inotropic effects so that other actions tend to be relatively ignored.

The significance of alpha-adrenoceptor-mediated changes in cardiac function has been subject to considerable debate. Under normal conditions the major adrenoceptor on heart rate is clearly beta so that the demonstration of an alpha-adrenoceptor-mediated component requires either blockade of the beta-adrenoceptor-mediated element or the use of a highly selective alpha-adrenoceptor agonist. It is more difficult to differentiate between alpha- and beta-adrenoceptor-mediated effects when these are functionally similar, as they are in the heart, each mediating an increase in contractility, in contrast to those other tissues in which alpha and beta act in opposition.

It has proved relatively easy to demonstrate convincingly alpha-adrenoceptor-mediated effects in abnormal conditions which disturb the balance between alpha and beta in alpha's favour. Two such chronic conditions are hypothyroidism (Nakashima et al. 1971; Ishac et al. 1983; Bilezikian and Loeb 1983; Fox et al. 1985) and diabetes (Jackson et al. 1986). Important acute examples, for the present topic, are ischaemia (Corr et al. 1981), low heart rate (Endoh and Shumann 1975), cholinergic nerve stimulation (Endoh and Motomura 1979) and adenosine (Endoh and Yamashita 1980; Hattori and Levi 1984). It is not yet entirely clear whether it is easier to demonstrate this response in hypothyroid or diabetic rats because of a loss of beta-adrenoceptor response or a gain in alpha-adrenoceptor response. Attempts to clarify this by assessing receptor numbers have, so far, failed to give a clear answer (Bilezikian and Loeb 1983; Williams et al. 1983; Latifpour and McNeill 1984; Fox et al. 1985).

A major stumbling block in studies of alpha-adrenoceptors in the heart has been the difficulty of finding suitable alpha-adrenoceptor agonists which are highly potent and are selective between alpha- and beta-adrenoceptors. Phenylephrine has generally been used since it is of the same order of potency as the catecholamines at alpha$_1$-adrenoceptors but is relatively weak on beta-adrenoceptors. However, in the heart where the beta-mediated adrenoceptor response is so dominant, phenylephrine has sufficient efficacy at beta-adrenoceptors to produce responses which are mediated by a mixture of alpha- and beta-adrenoceptors over the same dose or concentration range. Thus, a convincing demonstration of $alpha_1$-adrenoceptor-mediated cardiac effects in normal animals tends to require not only the use of phenylephrine as the agonist but also the presence of beta-blockade. For example, an $alpha_1$-adrenoceptor-mediated positive chronotropic action of phenylephrine or adrenaline can be shown in the pithed rat, but not in the rabbit, in the presence of propranolol (Flavahan and McGrath 1981b). For this reason, many workers use methoxamine although this has a higher degree of toxicity on other systems than has phenylephrine.

The positive inotropic action mediated by $alpha_1$-adrenoceptors is well established and has been reviewed extensively (Govier 1967; Benfey and Varma 1967; Scholz 1980). There is no doubt that appropriately selective agonists can increase force of contraction via such receptors, but the physiological relevance of this remains subject to debate (Schumann 1980; Benfey 1982).

Since these receptors have the potential to mediate positive inotropic effects of catecholamines, the possibility must be considered that this form of activation can increase the workload on the myocardium, particularly in the failing heart. This is pertinent to the hypothesis that $alpha_1$-adrenoceptor-mediated effects survive well in ischaemic conditions (Corr et al. 1981; discussed below in Sect. C.). It has been suggested also that $alpha_1$-adrenoceptors may stimulate a Ca^{2+}-sensitive glucose uptake in cardiac muscle cells (Clark and Patten 1984), as part of a general pattern of coordination of glucose metabolism by alpha-adrenergic mechanisms (Clark et al. 1983). If this survives in ischaemic conditions it could accelerate ischaemic damage.

There is now a need to define the ways in which the hormonal and metabolic environment can differentially affect these alpha- and beta-adrenoceptor-mediated responses. Since the available information suggests that the beta-adrenoceptor leads to activation of adenylate cyclase and $alpha_1$-adrenoceptors have been implicated mainly with the second messenger systems associated with phosphatidyl inositol, it seems quite possible that a variety of circumstances might exist which reduce the importance of beta-adrenoceptor-mediated modulatory factors but increase those through $alpha_1$-adrenoceptors. Thus the $alpha_1$-adrenoceptor-mediated system in cardiac muscle may have survival value in "last resort" circumstances where beta-adrenoceptors have become ineffective either through the effects of local conditions on their transduction mechanism or through downregulation of receptor number.

An interesting example is the way in which responses mediated through the alpha- and beta-adrenoceptors are differentially affected by activation of muscarinic receptors. This has been shown for both inotropic and chronotropic effects.

The $alpha_1$-adrenoceptor-mediated positive inotropic effect is more resistant than is the beta-adrenoceptor-mediated effect, to cancellation by the negative chronotropic effect of cholinergic stimulation (Endoh and Motomura 1979; Inui et al. 1982). This probably arises because the beta-adrenoceptor-mediated response involves activation, and the cholinergic response deactivation, of adenylate cyclase, whereas the $alpha_1$-adrenoceptor involved here does not influence this system, instead delaying the inactivation of the inward calcium current (MIURA et al. 1978; BRUCKNER and SCHOLZ 1984) by an alternative transduction mechanism: by analogy with other systems this may be phosphatidyl inositol. Hence the $alpha_1$ system is relatively independent of cholinergic influence. The same factor may apply to alpha-adrenoceptor-mediated responses being resistant to adenosine (BRUCKNER et al. 1985).

In the case of the inotropic effect, the physiological interaction between $alpha_1$ and cholinergic influence may be small since the direct inotropic influence of cholinergic stimulation (as distinct from inhibition of sympathetic nerves) is considered to be of relatively minor physiological importance. However, the observation of the phenomenon serves to underline the differences in the receptor-excitation coupling between $alpha_1$- and beta-adrenoceptors. Where circumstances do arise which heighten parasympathetic inotropic influence, then $alpha_1$-mediated positive inotropic effects might be expected to become important.

To date, there are no reports of positive or negative inotropic effects which can be clearly ascribed to $alpha_2$-adrenoceptor activation. Several $alpha_2$-adrenoceptor agonists which are imidazolines (or related heterocyclics) can decrease the heart rate in the rat but this action is not blocked by $alpha_2$-adrenoceptor antagonists (FLAVAHAN and MCGRATH 1982; OHGUCHI et al. 1984). Such negative chronotropic effects can actually be enhanced by $alpha_1$-adrenoceptor antagonists such as prazosin since the small positive chronotropic effect caused by $alpha_1$-adrenoceptor activation in rats is suppressed (FLAVAHAN and MCGRATH 1982). The negative chronotropic action is a high-dose effect and may be unrelated to activation of adrenoceptors but rather related to the intracellular modulatory effects of imidazolines on cell function (HARRISON et al. 1986).

In contrast to the inotropic actions, postjunctional interactions on pacemaker cells between parasympathetic, cholinergic, muscarinic effects and sympathetic, adrenergic $alpha_1$-adrenoceptor-mediated effects are likely to be of immediate physiological significance. In the face of muscarinic activation either by carbachol in vitro on isolated, spontaneously beating, rat atria or by vagal stimlation in vivo in pithed rats, the positive chronotropic effect of phenylephrine loses its beta component but not its $alpha_1$ component (FLAVAHAN and MCGRATH, unpublished observations). Possibly, during simultaneous activation of cholinergic and adrenergic nerves to the heart (or cholinergic nerves + circulating adrenaline), the only restraint on complete cholinergic dominance is the $alpha_1$-adrenoceptor-mediated positive chronotropic response. Furthermore, it is in conditions of low cardiac frequency, which cholinergic activity would induce, that the $alpha_1$-mediated inotropic effect is most effective (ENDOH and SHUMANN 1975). Thus, an $alpha_1$-adrenoceptor antagonist would allow parasympathetic dominance. In conditions of sympathetic activation in which reflex activation of the parasympathetic system exists (e.g. short-term elevation of blood pressure) $alpha_1$-adrenoceptor antagonists might reduce, and hence stabilize, heart rate.

There are, therefore, several circumstances in which the $alpha_1$-adrenoceptor-mediated excitatory effects on cardiac muscle cells (e.g. delayed repolarization) can become a relatively greater proportion of the adrenergic response, including ischaemia, cholinergic activation, increased adenosine levels, hypothyroidism and diabetes. The likeliest common link derives from the utilization by alpha-adrenoceptors of a different transduction process from that used by beta-adrenoceptors, allowing a backup mechanism when the beta-activated system fails (Vaughan Williams 1985).

A further complication, which can disguise the alpha$_1$-adrenoceptor-mediated positive chronotropic response, is that in some circumstances alpha$_1$-agonists such as phenylephrine can initiate a negative chronotropic response through the release of acetylcholine. This effect is blocked by selective alpha$_1$-adrenoceptor antagonists. An action of alpha$_1$-adrenoceptors which causes the release of acetylcholine is suggested since the response (1) occurs in pithed rats in the absence of autonomic tone, (2) is potentiated by eserine, (3) is effectively blocked by atropine and (4) is partly blocked by ganglion blockade or tetrodotoxin. This points to the release of acetylcholine from the postganglionic parasympathetic terminals but at least a part of this may be initiated by transmitter release from the preganglionic neurone endings in the ganglia (Flavahan and McGrath 1982).

It is not known what physiological role, if any, is played by this "indirect parasympathomimetic" action, but it demonstrates that activation of alpha$_1$-adrenoceptors has a potential both to mimic the sympathetic system by directly increasing heart rate and contraction and to activate the parasympathetic system thereby decreasing heart rate. This would seem to provide ideal circumstances for the genesis of arrhythmia by catecholamines particularly under pathological conditions where adjacent areas may be functioning to different extents and hence open to different influences, causing heterogeneity of polarization. A bradycardic action of alpha$_1$-adrenoceptor agonists has also been demonstrated in isolated rabbit atria, an action which microelectrode recording from sinoatrial cells has indicated as being due exclusively to delayed repolarization possibly as a consequence of delaying the decline of $[Ca]_i$, in turn resulting from a delayed inactivation of inward calcium current (Dukes and Vaughan Williams 1984a, b). It is clear that this bradycardic action is a direct one exerted on the myocardium, and not due to an indirect release of acetylcholine, because Ach shortens APD.

In contrast to the well-explored dromotropic influence of muscarinic receptors, there has been little study of possible influence of alpha-adrenoceptors on speed of conduction of the action potential in cardiac muscle.

2. Coronary Vascular Smooth Muscle

Oxygen supply to the myocardium depends entirely on the resistance to flow offered by its vasculature (relative to that offered by other beds). The factors affecting coronary vascular smooth muscle tone were recently reviewed before the appreciation that there were subtypes of postjunctional alpha-adrenoceptor (Feigl 1983). The tone of the coronary vascular smooth muscle is influenced by a dual adrenoceptor population, alpha-mediating constriction and beta-mediating relaxation. The relative importance of the different subtypes of alpha-adrenoceptor seems to vary with species and with the experimental conditions under which they are studied. Evidence has been presented for both subtypes in the coronary vessels of dog (HOLTZ et al. 1982; HEUSCH et al. 1984a), guinea pig (DECKER and SCHWARTZ 1985), monkey and human (TODA 1986).

In experimental studies of the role of sympathetic or adrenoceptor mechanisms in the compromised heart, e.g. stenosis or other flow reduction, much of the interpretation centres on the influence that alpha-adrenoceptors have on the various coronary vessels. This applies particularly to interpretation of the effects of alpha-adrenoceptor antagonists on cardiac function and distribution of blood flow. These analyses are difficult. They observe an overall effect on cardiac function or blood flow distribution in situations where the influences of autonomic nervous tone and blood pressure must be taken into account and where localized ischaemia is deliberately created. It is necessary to attempt such studies in order to predict the overall effects which alpha-adrenoceptor antagonists might have if used clinically to treat cardiac dysfunction including arrhythmia. Caution should, however, be exercised in accepting the general applicability of the postulated explanations, for several reasons.

First, the pharmacology is not straightforward because many of the drugs employed such as yohimbine and phentolamine can block receptors other than the types desired, and the desired recognition sites on other (unintended) cells, e.g. $alpha_2$ on nerves or endothelium. Secondly, it is becoming obvious that both $alpha_1$ and $alpha_2$ subtypes are present on, and functional in, coronary vessels but their distribution varies with species and within a given species according to the particular location in the bed. Thirdly, the particular structure of each vessel, its size, its innervation and its location with respect to metabolically active cardiac cells will dictate its response to activation of these receptors and whether it will fall more under the influence of circulating catecholamines or its own nerves. Fourthly, how the excitatory, constrictor effect of activation of alpha-adrenoceptors interacts with the inhibitory, dilatory influences caused by ischaemia is not well understood.

The basic problems are illustrated in Fig. 2. An important aspect that this does not show, both because it is not known and is in any case likely to vary with species, concerns the proportions of the alpha-adrenoceptor subtypes which are present (and are functional during local ischaemia) at the various vascular sites. Almost any consequence for distribution of blood flow caused by selective alpha-adrenoceptor antagonists can be proposed by varying the subtypes in such a model. Indeed this may explain both how local heterogeneities of blood flow can lead to arrhythmias and also why widely different interpretations of experimental results have been made by those studying the phenomenon.

For these reasons it is doubtful whether experiments on hearts in vivo and possibly even perfused hearts in vitro are capable of providing information on the functional capacity of the alpha-adrenoceptor subtypes in coronary vessels. However, the direct method of in vitro analysis of isolated vessels is extremely difficult with the small collaterals and resistance arterioles which are

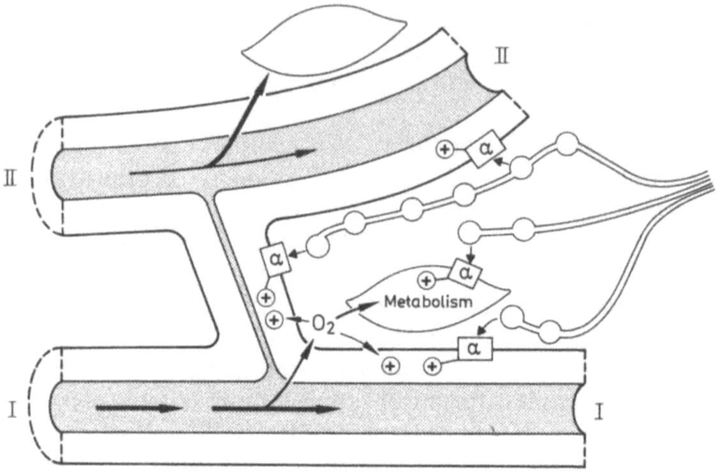
II
II
α
α
α
Metabolism
O2
α
I
I

a

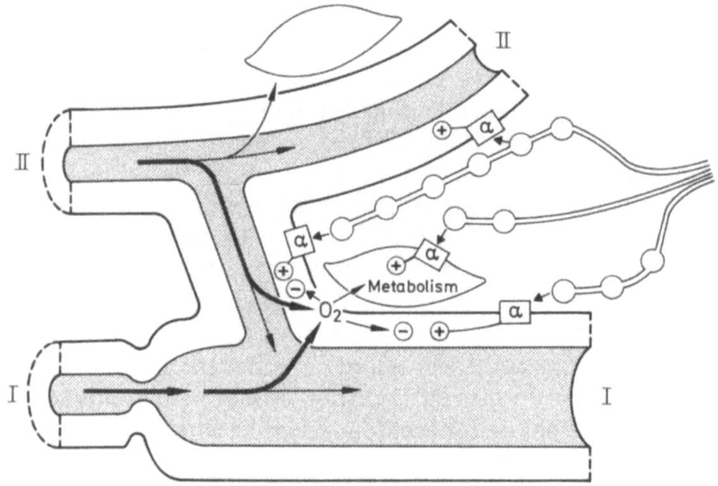
II
II
α
α
α
Metabolism
O2
α
I
I

b

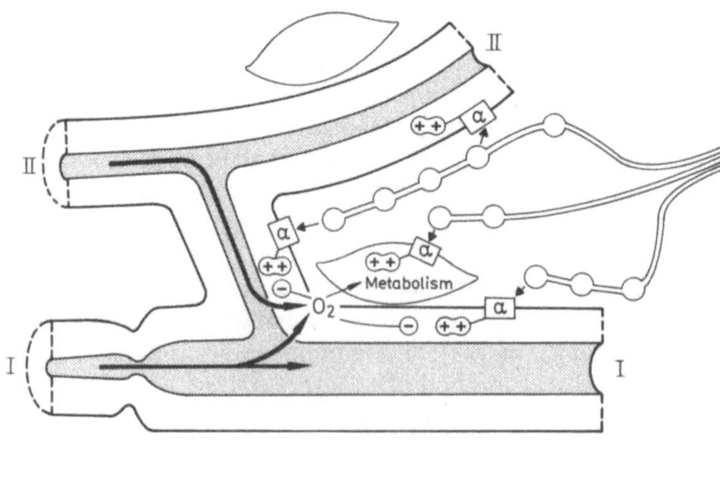
II
II
α
α
α
Metabolism
O2
α
I
I

c

of such importance in the physiological distribution and pathological redistribution of coronary blood flow: the required information is simply not available at present. Consequently, the best source available may be the macroscopic description of the effects of drugs or nerve stimulation on flow changes and it should not be surprising that different authors, using different experimental techniques, should observe different phenomena and hence reach different conclusions. It is very difficult to extrapolate accurately from such experimental results to provide general therapeutic guidance.

Where the blood supply has been compromised and the area is ischaemic, the remaining patent vessels will be subject to a variety of vasodilator influences. If a vasoconstrictor influence through alpha-adrenoceptors was then effective it would lead to further restriction of blood supply and consequent damage to the area (see Fig. 2). Under such conditions the potential exists for an improvement in blood supply brought about by the appropriately selective alpha-adrenoceptor antagonists. It is therefore of some importance to know which subtype is relevant in such pathophysiological conditions. On the other hand, alpha-adrenoceptor-mediated vasoconstriction in healthy regions of the heart or in the rest of the body may be, by increasing resistance elsewhere, en-

---

◀

Fig. 2. Diagrammatic representation of the factors to be considered when assessing the theoretical involvement of alpha-adrenoceptors in experimental reduction of part of the coronary blood flow. Two arterial vessels (I and II) serving adjacent areas are shown, connected by a collateral vessel. It is assumed that all of the blood vessels have alpha-adrenoceptors, which mediate vasoconstriction, when the sympathetic nervous system is activated, and that there is normally basal neurogenic sympathetic tone together with some basal myogenic tone resulting from the "metabolite" composition of the extracellular environment (indicated for simplicity as O_2 but in fact multifactorial). No assumptions are made about alpha-adrenoceptor subtypes. *Arrows* in the vessels indicate blood flow/oxygen supply. *Arrows* and *symbols* elsewhere indicate the influence of neurotransmitters or "metabolites" on vascular smooth muscle or cardiac muscle: excitatory ⊕; inhibitory ⊖. **a** normal resting conditions are mildly constrictory with collaterals closed. **b** Applying a partial stenosis to vessel I creates ischaemia in the area which it serves, causing metabolic vasodilation of vessel I and of the collateral, allowing some mild "steal" from adjacent normal areas. **c** If the stenosis is applied together with increased sympathetic tone, the result will depend on the balance between metabolic vasodilation and alpha-adrenergic vasoconstriction. This cannot be predicted. The consequence shown is that vasoconstriction in the normal area served by vessel II is greater than in the ischaemic area so that the pathway of least resistance to blood flow, via the collateral, "steals" a significant proportion of vessel II's blood. This will set up an unstable situation, causing ischaemia around vessel II and hence a countersteal, with the potential for oscillation of flow between the two routes. The influence upon this of alpha-adrenoceptor blockade by "selective" antagonists is even more difficult to predict since there is little information on the distribution of receptor subtypes and whether or not they would be differentially affected by the metabolic environment. If, for example, the collaterals had different receptors from the main resistance vessels, and these were differentially susceptible or resistant to ischaemia, almost any distribution of blood flow could be hypothesized based on this diagram, and selective alpha$_1$- or alpha$_2$-adrenoceptor antagonists can produce almost any redistribution

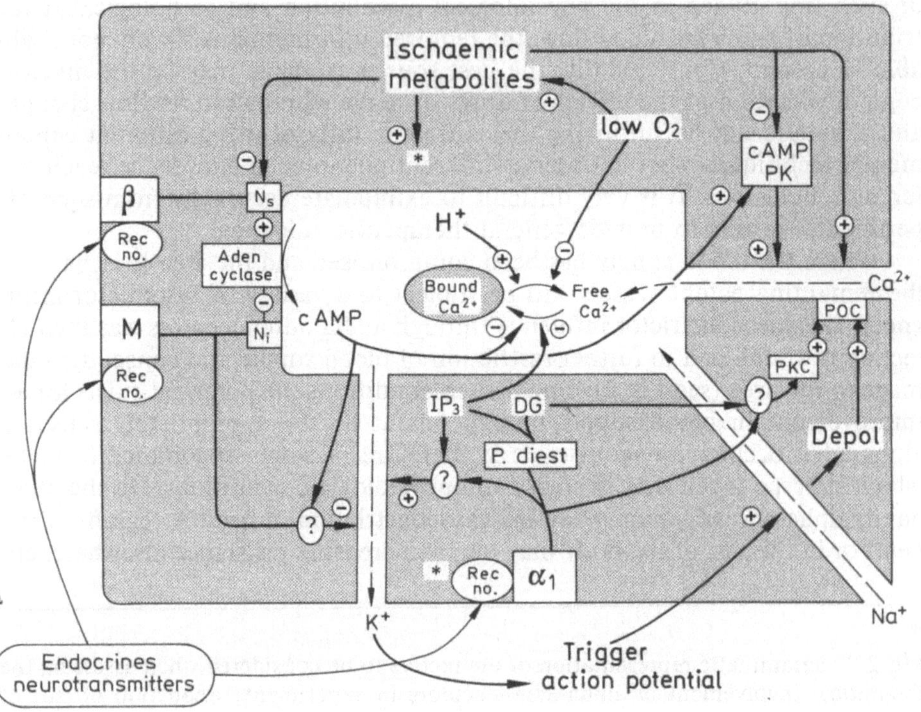
Ischaemic metabolites
low O2
cAMP PK
*
β
Rec no.
Ns
Aden cyclase
Ni
M
Rec no.
H+
Bound Ca2+
Free Ca2+
cAMP
Ca2+
POC
PKC
IP3
DG
P. diest
Depol
?
α1
Rec no.
*
K+
Na+
Trigger action potential
Endocrines neurotransmitters

a

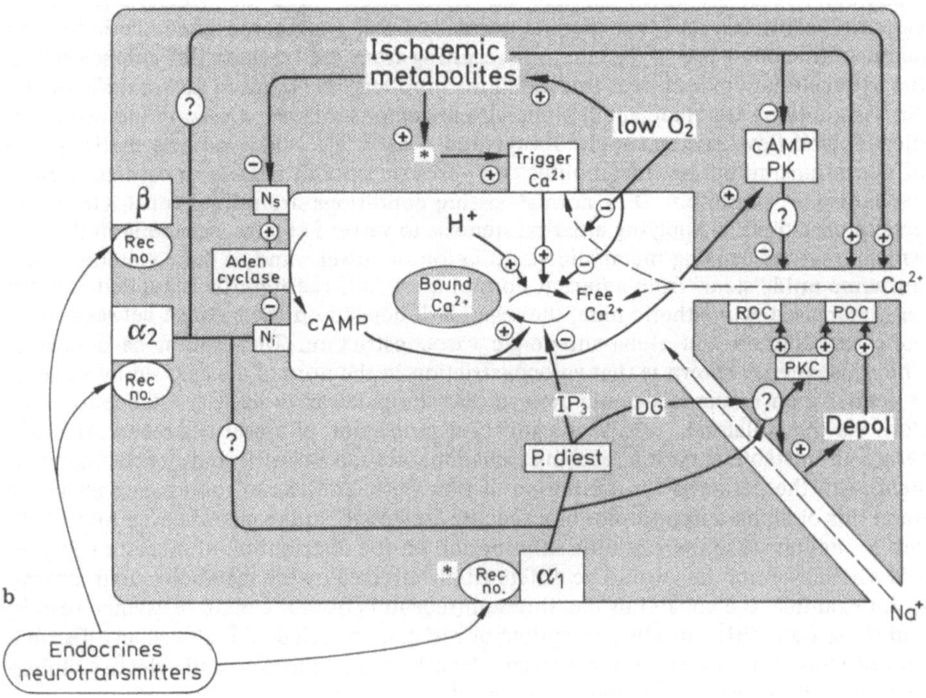
Ischaemic metabolites
low O2
cAMP PK
*
Trigger Ca2+
β
Rec no.
Ns
Aden cyclase
Ni
α2
Rec no.
H+
Bound Ca2+
Free Ca2+
cAMP
Ca2+
ROC
POC
PKC
IP3
DG
P. diest
Depol
?
α1
Rec no.
*
Na+
Endocrines neurotransmitters

b

suring a diversion of blood to the dilated ischaemic area (coronary "steal"). In such a case alpha-blockade may be counterproductive in the attempt to supply blood to the ischaemic area. In practice, it is a balance between these two factors which is important. It is also possible to demonstrate that sympathetic/alpha-adrenoceptor activation can prevent some types of "steal" by selective vasoconstriction which distributes blood preferentially to ischaemic areas (HIRSHFIELD et al. 1974; CHIARIELLO et al. 1977) or transmurally (BUFFINGTON and FEIGL 1983; NATHAN and FEIGL 1986).

A working hypothesis has been proposed (HEUSCH et al. 1986) synthesizing many of these points. This starts with the experimental demonstration that large coronary vessels have predominantly alpha$_1$-adrenoceptors and small vessels have predominantly alpha$_2$, based on information from blood flow studies in anaesthetized dogs (HEUSCH et al. 1984a; SAEED et al. 1985). It goes

◀

Fig. 3a, b. A theoretical framework for the differential influence of ischaemia, hormones and cholinergic activation on cellular responses to activation of different adrenoceptors, showing the main differences between **(a)** cardiac muscle cells and **(b)** vascular smooth muscle cells. Note that the various influences of ischaemia are often in physiological antagonism to each other as well as to the consequences of receptor activation. **a** Cardiac muscle: alpha$_1$- and beta-adrenoceptors both produce responses, either chronotropic or inotropic according to the cell's functional specialization, and this is opposed physiologically by muscarinic (M) cholinergic activation. Since the two adrenoceptors utilize different second messengers and further coupling steps, the responses which they modulate are open to separate physiological antagonism. In particular the beta-mediated response is susceptible to attenuation by factors which inhibit adenylate cyclase function, while alpha$_1$ may be particularly open to facilitation of its activation of Ca^{2+} channels in circumstances which attenuate adenylate cyclase. **b** Smooth muscle: alpha$_1$ and alpha$_2$ are both excitatory, while beta is inhibitory, to cellular contraction. Since the two alpha subtypes utilize different second messengers and further coupling steps, the responses which they modulate could be open to separate physiological antagonism. However, the outcome is much less predictable than in the case of the cardiac muscle cell since there is no obvious point at which one or the other is more vulnerable and since less confidence can be placed in the second messenger systems than in the case of cardiac beta. The differential effects of O_2 and H^+ on the early and late "phases" of alpha-adrenoceptor-mediated pressor responses suggest that ischaemic conditions can divert the response from one coupling process to another. In particular, facilitated activation of Ca^{2+} channels by metabolites might be accompanied by inhibition of intracellular Ca^{2+} release. Stippled areas indicate cell membrane and intracellular sites of Ca^{2+} binding. (P. diest), phosphodiesterase involved in inositol lipid hydrolysis to form inositol 1,4,5 triphosphate (IP_3) and 1,2 diacylglycerol (DG). (PKC), protein kinase, C; (cAMP PK), cAMP-dependent protein kinase: each of these may lead to the phosphorylation of Ca^{2+}-channel proteins. (N_s), (N_i) regulatory proteins, respectively stimulating and inhibiting adenylate cyclase function. (Rec No.), receptor number for each receptor: this is susceptible to extracellular factors and to metabolites (*); alpha$_1$ in smooth muscle can enhance intracellular Ca^{2+} release. (ROC), "receptor-operated Ca^{2+} channel"; (POC), "potential-operated Ca^{2+} channel"; (Depol), cellular depolarization; (?), evidence weak or hypothetical; (+), excitatory influence; (−) inhibitory influence on to demonstrate that either sympathetic nerve stimulation or alpha$_2$-adrenoceptor agonists can produce vasoconstriction distal to an experimental stenosis, which persists even when the coronary reserve has been exhausted, i.e. when the local dilator factors produced by ischaemia would produce complete relaxation of the vascular smooth muscle were it not for the excitatory alpha-mediated effect (Deussen et al. 1985; Heusch et al. 1985b). This can be opposed by alpha$_2$-antagonists including rauwolscine and phentolamine (Heusch et al. 1985b). In contrast the effects of the alpha$_1$-adrenoceptor agonist methoxamine and the antagonist prazosin were interpreted as indicating that alpha$_1$ influence is exerted mainly on large vessel resistance. Furthermore, resistance in ischaemic areas was not influenced by alpha$_1$-activation; rather the constriction which this produced elsewhere favoured redistribution to the ischaemic area (Heusch et al. 1983; Heusch et al. 1984a).

A possible explanation for the survival of alpha$_2$ but not alpha$_1$ in ischaemic conditions is that ischaemia has differential effects on alpha$_1$- and alpha$_2$-mediated vascular responses, hypoxia attenuating alpha$_1$ and acidosis favouring alpha$_2$ (McGrath et al. 1982; Grant et al. 1985b), although this has not been tested on coronary vessels. This is, of course, the opposite fate for alpha$_1$ with regard to ischaemia from that in cardiac muscle, where alpha$_1$ is a relatively good survivor (see previous Sect. B.III.1). Consequently, in an ischaemic area, the alpha$_1$-mediated vasoconstriction would be lost but the stimulation of cardiac muscle cells would remain, both effects having acute, local, survival value. A theoretical framework for these differential modulations of responses via different receptors is set out in Fig. 3.

It would follow from this hypothesis, if correct, that alpha$_2$-adrenoceptor antagonists should assist perfusion of compromised areas but that alpha$_1$-antagonists should not. So far this has not received support from other groups of workers. There has been a contradictory report, also in anaesthetized dog, which suggests that adverse coronary steal can occur in ischaemic myocardium with alpha$_2$-antagonism (rauwolscine) but not with alpha$_1$-antagonism (doxazosin) (Dart et al. 1987b). Extrapolation to humans clearly depends on species differences in the distribution of alpha-adrenoceptor subtypes. For example, in isolated large vessels of dog, only alpha$_1$ are found, whereas in man and monkey there is evidence for both subtypes (Toda 1986). As noted above, there may be no general basis for the therapeutic prescription of alpha$_1$- or alpha$_2$-antagonists with regard to beneficial redistribution of blood flow in the failing heart; rather the available information should be regarded as supplementary to any other rational basis for their use.

3. Coronary Vascular Endothelium

The tone of vascular smooth muscle is under the control not only of receptors on smooth muscle responding to blood-borne factors and neurotransmitters, but is also subject to modulation by inhibitory and excitatory factors released from the vascular endothelium. The greatest attention so far has been paid to endothelial-derived relaxant factors (EDRFs) (Furchgott 1984; Vanhoutte 1986).

The endothelial factors exert an influence over vascular smooth muscle contraction induced by activators, including alpha-adrenoceptor agonists (MARTIN et al. 1986) and sympathetic nerve stimulation (SPOKAS and FOLCO 1984). This is important in assessment of vascular sensitivity to agonists but was overlooked until Furchgott's dicovery of EDRF in the late 1970s. Earlier studies should thus be viewed in this light. In the absence of the endothelium most blood vessels are more sensitive to agonists because of the loss of spontaneous release of EDRF. This applies particularly to weak agents, such as partial agonists, whose effects can be very effectively attenuated by EDRF. Since the extent to which spontaneous release of EDRF occurs in vivo is presently unknown, this has two serious consequences for assessment of the functional integrity of vascular receptor systems in coronary vessels. First, responsiveness of vessels must be tested with and without endothelium: failure of a vessel to contract to an agonist with endothelium present does not mean that the smooth muscle is insensitive. Secondly, ways must be found to assess the role of EDRF in vivo so that factors which modulate its release can be determined: for example, how does EDRF release and its effectiveness against alpha-adrenoceptor-mediated excitation survive in ischaemic conditions?

A further element is the possible release of endothelial-derived factors by agonists themselves. In coronary vascular smooth muscle of dog and pig, activation of alpha$_2$-adrenoceptors can release EDRF with consequent relaxation of a precontracted vessel (COCKS and ANGUS 1983). This property is not shared by all other blood vessels, e.g. rat aorta (GODFRAIND et al. 1985) and so may be of particular relevance to the control of coronary blood flow. Thus circumstances may exist where alpha$_2$-adrenoceptors on the endothelium (relaxant) and the coronary vascular smooth muscle cells (contractile) can produce opposing influences on coronary vascular tone. It may be of relevance that the endothelial alpha$_2$-adrenoceptors are likely to be activated only by catecholamines in the plasma whereas the vascular receptors would be more likely to be subject to the additional influence of sympathetic nerves. Since the two sets of alpha$_2$-adrenoceptors may be subject to different modulatory factors, the potential seems to exist for conflicting alpha$_2$-adrenoceptor-mediated influences and thus for opposite effects of alpha$_2$-adrenoceptor antagonists on different coronary blood vessels (in addition to the further different influences of alpha$_2$-"autoreceptors" on nerve terminals: see next section).

4. Nerves

All of the known types of autonomic nerve ending within the heart are subject to modulation of their transmitter release by alpha-adrenoceptors. This includes the sympathetic nerves innervating the myocardium and the coronary blood vessels as well as the parasympathetic nerves innervating the myocardium. Thus, all forms of neurotransmission are subject to influence via alpha-adrenoceptors from circulating catecholamines or the noradrenaline released from the sympathetic nerves. This applies whether their own postjunctional action is mediated by noradrenaline, acetylcholine or their cotransmitters, or whether they utilize postjunctional receptors which are alpha, beta, muscarinic, purinergic or peptidergic. In this respect, crosstalk between the nerve terminals may be of considerable importance in establishing a uniform response throughout the region of the heart or in the precise localization of the pacemaker (see Fig. 1).

a. Sympathetic Nerves in the Myocardium

Release of noradrenaline from cardiac sympathetic nerves and the corresponding postjunctional response can be attenuated by activation of alpha$_2$-adrenoceptors by exogenous agents and this can be prevented by the administration of alpha$_2$-adrenoceptor antagonists (Armstrong and Boura 1973; Antonaccio et al. 1974; Drew 1976; Langer et al. 1977) confirming the receptor subtype as alpha$_2$. When the nerves are stimulated with a train of pulses at a sufficiently high frequency, alpha$_2$-adrenoceptor antagonists can increase transmitter output, suggesting that the neurotransmitter noradrenaline has been restricting its own release by action at these prejunctional alpha$_2$-adrenoceptors ("autoreceptors").

The currently predominant view is that the potential for negative feedback certainly exists (Starke 1977, 1981; Gillespie 1980; Rand et al. 1982), although the general applicability of the hypothesis has been challenged and it has been suggested for several tissues, including heart, that part of the increase in transmitter output produced by alpha-adrenoceptor antagonists might be unrelated to blockade of such a negative feedback system (Kalsner 1982; Fuder et al. 1983; Baker et al. 1984; Robie 1984). In practical terms the importance of this distinction for cardiac function would lie in the physiological conditions which are requirements for potentiation by alpha$_2$-adrenoceptor antagonists. For example, in the feedback hypothesis, there will be a minimum frequency below which the individual pulses are too far apart in time to interact, whereas this would not necessarily follow if antagonists simply had their own direct action on transmitter output. In either case, the net effect of prejunctional alpha-blockade would be an increase in sympathetic transmission with the consequent effects of increased activation of the tissue (via whatever postjunctional receptors are activated by noradrenaline and its cotransmitters).

The relatively high frequencies needed to activate alpha$_2$-mediated feedback might suggest that the mechanism operates as a rate-limiting safety valve in extremes of sympathetic activation. It is indeed difficult to demonstrate functional feedback in many preparations without resort to high stimulation frequencies or pharmacological tricks such as blockade of the neuronal uptake of noradrenaline (see below). However, one circumstance which might produce physiological conditions for activation of the system is exercise. It has been shown in conscious dogs that chronotropic and inotropic activation caused by exercise is potentiated by phentolamine and yohimbine, but less so by prazosin, which the authors have interpreted as indicating that prejunctional alpha$_2$-adrenoceptors are activated in exercise and consequently alpha$_2$-adrenoceptor antagonists will cause an increased and less-controlled release of noradrenaline which can then act through beta-adrenoceptors to stimulate the various cardiac cells (HEYNDRICKX et al. 1984). However, there is evidence from several laboratories that during exercise coronary vessels are vasoconstricted via both alpha$_1$- and alpha$_2$-adrenoceptors (VATNER 1984; SAEED et al. 1985; GWIRTZ et al. 1986; WOODMAN and VATNER 1986) and that part of this is due to circulating catecholamines originating in the adrenal medulla (CHILIAN et al. 1986). This illustrates the problems involved in unravelling the different alpha-sites in the functioning heart.

There is evidence from rat and rabbit heart that part of the negative feedback control of atrial noradrenaline release and hence of control of heart rate may be mediated through alpha$_1$-adrenoceptors (KOBINGER and PICHLER 1982; DOCHERTY 1983, 1984; SMITH et al. 1983). Although these preceding authors regarded this as the minor influence in autoregulation and its relevance to physiological conditions or in humans are unexplored, another group consider that, at least under their experimental conditions with isolated rat atria, alpha$_1$- and alpha$_2$-adrenoceptors contributed to similar degrees in autoinhibitor feedback regulation (STORY et al. 1985). If this did occur in humans then, in theory, selective alpha$_1$-antagonists such as prazosin could enhance sympathetic transmitter release and thus, where the postjunctional receptors were other than alpha$_1$ (e.g. alpha$_2$, beta or due to cotransmitter purines or peptides), could enhance the effectiveness of sympathetic nerve activity.

The physiological significance of the prejunctional alpha-adrenoceptors may not lie predominantly with local control at each varicosity but may be intimately concerned with the spread of activity throughout the branching varicose terminals of each sympathetic neurone (STJÄRNE 1985). Also if adjacent varicosities, arising from different neurones, can influence each other, this provides a means of crosstalk which could stabilize transmitter release over a region in the myocardium wider than that served by one neurone. This important function could be disrupted by antagonists of the prejunctional receptors leading, particularly in the pacemaker region, to a potential for arrhythmia (DAVEY 1986).

These receptors are also subject to activation by circulating catecholamines so that there is also a "long-loop" interaction between the hormonal and direct neural aspects of the sympathetic system.

Since the endogenous feedback control of noradrenaline release persists for only a short time after each nerve impulse, it is frequency related, requiring a sufficiently high frequency for the effect still to be present when the next action potential arrives at the nerve terminal. However, the presence of neurotransmitter noradrenaline at the terminal region can be prolonged if the neuronal reuptake process of noradrenaline is inhibited. This allows feedback to occur at lower frequencies. Thus, it can be argued that feedback occurs to a small extent under physiological conditions where the frequency is too low but that, in the presence of a blocker of neuronal uptake, feedback may then occur at physiological frequencies (DOCHERTY and MCGRATH 1979, 1980b).

Since a very wide variety of pharmacological agents block the neuronal uptake of noradrenaline either as a side effect or as their main mode of action, either the experimental conditions in animal studies or other drugs given to patients may induce a greater degree of feedback control. Examples in wide clinical use are the muscle relaxant pancuronium bromide, and the anaesthetic ketamine, both of which can be shown to potentiate cardiac sympathetic responses and enhance the degree of $alpha_2$-mediated autofeedback (Clanachan and McGrath 1976; Docherty and McGrath 1978). Consequently there is a greater potential for $alpha_2$-adrenoceptor blockade to increase sympathetic effectiveness when such drugs are present. The same may apply in general to ischaemic tissue where uptake is blocked metabolically. Certainly, yohimbine can increase nerve-induced noradrenaline output from ischaemic dog heart (Forfar et al. 1983) and in ischaemic rat heart the uptake process has been suggested to go into reverse as "carrier-mediated efflux", actually releasing noradrenaline (Schomig et al. 1985). Thus in the ischaemic heart autofeedback is likely to be of increased importance, provided that ischaemia is not sufficiently advanced that neural function fails altogether. This scenario could be used to argue against the use of $alpha_2$-adrenoceptor antagonists in the ischaemic heart since the increased sympathetic response could be detrimental.

It should be noted that the blockade of uptake will not necessarily radically alter sympathetic effectiveness on its own because the increased persistence of the neurotransmitter will be balanced by its reduced output due to the feedback process: interruption of feedback will of course change this. An additional consequence of this situation is that noradrenaline would become relatively more important compared with any other cotransmitters. Also the increased distance over which the transmitter exerts its influence may increase the degree of interneuronal crosstalk and also change the receptor population activated by the nerves.

Clearly this "cardiac autonomic plexus" is open to intervention and modulation by $alpha_2$-adrenoceptor antagonists, and considerable regional variation in the dominant neurotransmitter/receptor system, can be initiated by localised ischaemic damage or by drugs intervening in this situation. For example, it has been suggested that in ischaemic areas the neuronal uptake system for noradrenaline becomes ineffective due to intracellular hypoxia and acidity, even allowing the passage of noradrenaline out of the nerves through the "uptake" system (carrier-mediated efflux) (Schomig et al. 1985). In such areas the neurotransmission will be disrupted as described above and, in addition, localized hypersensitivity to catecholamines should occur.

b. Sympathetic Nerves in Coronary Blood Vessels

The same factors apply to sympathetic nerves serving the coronary vascular smooth mucle as those discussed above for the myocardium. However, even less is known about the role of the alpha-adrenoceptor-mediated feedback in these nerves. If feedback is effective in physiological or pathological conditions, then its effect will be to restrict sympathetic activity and thus to encourage vasodilation. Blockade of this system would result in increased vasoconstriction. Again this might be considered detrimental in the failing heart.

c. Parasympathetic Nerves in the Myocardium

Although the postganglionic parasympathetic terminals do not release catecholamines as transmitters, they possess prejunctional alpha$_2$-adrenoceptors which restrict release of their neurotransmitters (see GILLESPIE 1980). This, along with the corresponding muscarinic receptors on sympathetic nerves (MUSCHOLL 1973), allows crosstalk between sympathetic and parasympathetic terminals. Again, the precise physiological role of these receptors is not well established, but clearly where circumstances exist to allow their activation these receptors would serve to reduce the parasympathetic influence and allow sympathetic dominance. Their blockade by alpha$_2$-adrenoceptor antagonists would allow a greater parasympathetic influence. Clearly for these receptors to be effective the timing of the arrival of sympathetic and parasympathetic impulses in the interacting neurones will be important. However, the blockade of the neuronal uptake of noradrenaline will reduce this factor and increase the probability of the occurence of sympathetic modulation of the parasympathetic terminal.

There is, however, some debate concerning the importance of prejunctional alpha-adrenoceptors on cardiac parasympathetic nerves, most of the experimental evidence coming from parasympathetic nerves in the gut (see GILLESPIE 1980), e.g. in guinea pig right atria clonidine and noradrenaline fail to inhibit chronotropic responses to vagal stimulation (LEW and ANGUS 1983). Interestingly there is an observation on the effect of methoxamine on acetylcholine release in rat atria which suggests that alpha$_1$-adrenoceptor activation inhibits nerve-induced acetylcholine release (WETZEL et al. 1985).

In summary, the above discussion suggests that the influence of the prejunctional alpha$_2$-adrenoceptors on neurotransmission in the heart can favour either sympathetic or parasympathetic control of cardiac function and that this can be intensified by blockade of the neuronal reuptake of noradrenaline. The net effect in any given situation would have to be considered on its own merits but a major factor would be whether the parasympathetic or sympathetic system was dominant within the heart in conjunction with the levels of circulating catecholamines. From the teleological viewpoint, it is reasonable that a sympathetic transmitter should inhibit parasympathetic nerves, and that a parasympathetic transmitter should inhibit sympathetic nerves. In practice the process is difficult to analyse in detail, and the final resultant effect of blockade of individual receptors becomes impossible to predict.

5. Platelet Aggregation

It is noted above that several acute local conditions or chronic pathological states can alter the proportionate influence of alpha- and beta-adrenoceptors in the heart. Further extrinsic influences relevant to alpha-adrenoceptor function in the heart are the effects of platelet aggregation and the associated thrombus formation. Human (and other species) platelets have well-defined alpha$_2$-binding sites which seem to be involved in the platelet aggregation caused by adrenaline (Nahorski et al. 1985) and this property is now exploited as a model or marker for regulation of alpha$_2$-adrenoceptor number. Unfortunately a change in receptor number on circulating elements does not guarantee that a similar change occurs in other tissues.

When the platelets start to aggregate and form a thrombus a complex cyclical series of reactions takes place involving the release of several substances from platelets, including adenine nucleotides, 5-HT, platelet-activating factor, thrombin, thromboxane A_2 and vasopressin, each of which can release EDRF and hence produce vasodilation but which is also capable of causing vasoconstriction directly on vascular smooth muscle. In addition prostacyclin, derived from the endothelium, can inhibit aggregation. The ultimate response may depend on whether or not the endothelium is intact and functioning. If not, as may be the case in various forms of coronary artery disease, e.g. involving atheromatous plaque formation, then vasoconstriction or coronary artery spasm may occur, adding to arterial occlusion (Vanhoutte 1986). Thus alpha$_2$-adrenoceptor function, by facilitating aggregation, could contribute to coronary arterial spasm.

Several theoretical means are available by which spasm resulting from adrenaline-induced thrombus formation can be contained. Thus several pharmacological interventions are possible. This can be indirect, by inhibiting platelet function, e.g. by infusion of prostacyclin (Bertha and Folts 1984), or direct, by attempting to inhibit the stimulus to aggregation with alpha-adrenoceptor antagonists. This latter approach has, however, met with mixed success. In studies involving the observation of cyclical blood flow reductions caused by experimental coronary stenosis in anaesthetized dogs, the hypothesis was tested that occlusion was caused by platelet aggregation facilitated by alpha-adrenoceptor activation by circulating catecholamines. Although platelet aggregation could be confirmed as a contributory factor, prazosin was not effective, phentolamine was not very effective unless very high doses were given and yohimbine was effective in only some of the studies. Since ketanserin, which is an alpha$_1$-adrenoceptor antagonist and 5-HT-receptor antagonist, was effective, the simplest conclusion is that 5-HT is more likely to be involved than catecholamines and that this explains the effects of phentolamine and yohimbine (Bush et al. 1984). However, another group with similar alpha-blocker data, but no potent 5-HT antagonists, favour a "membrane-stabilizing" effect of the "antagonists" (Bralet et al. 1985). A third group, again with roughly similar data, attributed the effectiveness of a compound, nicergoline, to a combination of alpha-adrenoceptor blockade and inhibition of platelet phospholipase but do not consider the possibility of 5-HT involvement (Bolli et al. 1984; Bolli et al. 1985). Overall the evidence of alpha-adrenoceptor involvement in these experimental models is not strong, but alleviation of any aggravation of coronary vascular blockade remains a possible contributory factor on the coronary beneficial side of alpha-adrenoceptor blockade.

6. Differences Between Circulating and Local Neurotransmitter Catecholamines

It is worth drawing attention to some specific aspects of the differences between the alpha-adrenoceptor-mediated effects of circulating catecholamines in the bloodstream and locally released neurotransmitter noradrenaline. In general, circulating catecholamines will have access to all alpha-adrenoceptors irrespective of location, whereas the neurotransmitter will reach only a limited anatomical area. On the other hand, the concentration achieved in the extracellular space due to the circulating catecholamines, while it is uniform throughout the heart, will be relatively low.

The tissues most likely to experience the highest concentration from circulating catecholamines would be the endothelium and the first few adjacent smooth muscle cells. In contrast, the neurotransmitter achieves very high concentrations in restricted areas. For a given phenomenon mediated by alpha-adrenoceptors therefore it is necessary to consider the extent of the innervation and the sensitivity to catecholamines in order to interpret the possible role of neurotransmitters vis-à-vis circulating catecholamines. It is common to find non-innervated regions where the sensitivity to catecholamines is relatively higher. For example even within the wall of a large blood vessel the smooth muscle nearest to the nerves on the adventitial side may be less sensitive to catecholamines than the layers distant from the nerves but adjacent to the endothelium (GRAHAM and KEATINGE 1975).

C. Experimental Approaches to the Role of Alpha-Adrenoceptors in Arrhythmia

The above discussion illustrates the extensive but fragmented current knowledge of factors related to alpha-adrenoceptors which could affect the function of the heart. However, the experimental approaches to understanding how they do contribute to arrhythmia have not produced an agreed set of interacting causative factors which can form a fixed prescriptive basis for rational therapy.

Two points are clear. First—sympathetic activation can precipitate arrhythmia (MALING and MORAN 1957; HOFFMAN and SINGER 1967). Secondly, alpha-blockers reduce the incidence of some arrhythmias (LEIMDORFER 1953; SHERIDAN et al. 1980; MAZE and SMITH 1983; AUBRY et al. 1985).

Historically it had been assumed that the effects of sympathetic nerve stimulation to the myocardium were beta-adrenoceptor-mediated and intervention on the basis of reducing myocardial work had concentrated on beta-blockade. Similarly the early evidence pointed to alpha-adrenoceptor-mediated vasoconstriction and beta-adrenoceptor-mediated vasodilation in coronary blood vessels with only alpha activated by nerves, so that intervention aimed at increasing coronary blood flow would utilize alpha-blockade while beta would be considered neutral or counterproductive. It was not clear whether alpha-blockade would assist coronary blood flow since it would also vasodilate other beds competing for the blood supply and its reduction of arterial blood pressure would not always be welcome especially in ambulant patients.

However, the empirical clinical observations that alpha-blockade can reduce the incidence of some arrhythmias, particularly when these have a sympathetic component, require explanation. Consequently an experimental search has been underway for the contribution of alpha-adrenoceptor activation to arrhythmia and for the basis of the antiarrhythmic action of alpha-adrenoceptor antagonists. The two were not necessarily connected since the antagonists might have antiarrhythmic actions unconnected with their known, straightforward, alpha-adrenoceptor antagonism. Most of this work pertains to four main factors all based on the premise that arrythmia will arise from inhomogeneity of the properties of adjacent areas of the myocardium and alpha-adrenoceptor activation will initiate this or will exacerbate it when some defect already exists or is experimentally induced by reduction of blood/perfusate flow:

1. Direct myocardial alpha-adrenoceptor-mediated effects which could disrupt regular initiation and spread of cardiac electrical activity and which would be present to differing degrees according to the metabolic state of the area
2. Alpha-adrenoceptor-mediated vasoconstrictor effects directly on coronary vascular smooth muscle, which could restrict coronary blood flow to ischaemic areas, presumably because the receptor/transduction system is somehow hypereffective there
3. Alpha-adrenoceptor-facilitated thrombus formation, resulting in local restriction of coronary blood flow and possibly involving coronary vasospasm
4. A role for alpha-adrenoceptors in arrhythmias resulting from reperfusion: this is really a variant on (1) and (2), the distinction not always being made clear.

The distinctions between these are often difficult to unravel. Some lines of work led only to a demonstration that a particular agent precipitated or blocked experimental arrhythmias without necessarily locating the mode of action. Unfortunately often the more controlled were the experimental conditions and the more precise was the conclusion on mode of action, the more diverse were the conclusions reached by different groups of workers. There are, however, few real inconsistencies in the observations as shown below in considering some of the main approaches.

The group of Dart, Riemersma and coworkers has carried out a series of studies relating the loss of functional effectiveness of adrenergic nerves, which occurs during ischaemia, to the presence of arrhythmias. This clearly relates to any hypothesis concerning the involvement of either pre- or postjunctional alpha-adrenoceptors in arrhythmia. They demonstrate (using anaesthetized dogs and isolated perfused rat heart) that, during the development of ischaemia and on reperfusion, arrhythmias occur primarily during periods when at least some of the adrenergic nerves are functionally active or when ischaemia has induced spontaneous release of noradrenaline by carrier-mediated efflux (DART et al. 1984; FORFAR et al. 1985; DART und RIEMERSMA 1985; DART et al. 1987a). This implies that arrhythmia can be precipitated by local inhomogeneities in the loss of adrenergic function caused by adjacent areas becoming hypoxic at different rates, leading to inhomogeneity of function in cardiac myocytes.

Their main concern was not with postjunctional alpha-adrenoceptors but with effective transmission in general, which is likely to be a mixture of $alpha_1$ and beta. From the work of others (e.g. CORR and WITKOWSKI 1983), an enhancement of the role of $alpha_1$-adrenoceptors in the ischaemic areas is likely, and their demonstration of the temporal coincidence of adrenergic effectiveness and arrhythmia is consistent with $alpha_1$-adrenoceptor activation in ischaemic areas being a prime cause of adrenergic-induced arrhythmia. They deal with alpha-adrenoceptors only in terms of the likely role of prejunctional alpha-adrenoceptors in enhancing noradrenaline output (as shown by the effects of the mixed $alpha_1$/$alpha_2$-antagonist yohimbine), and how this is altered by hypoxia or drugs which block the neuronal uptake of noradrenaline. They note that yohimbine increases the incidence of arrhythmias in ischaemic hearts and attribute this to enhancement of sympathetic transmission by blockade of $alpha_2$-adrenoceptor-mediated negative feedback (FORFAR et al. 1985). This would indicate caution in the use of $alpha_2$-adrenoceptor antagonists to remove the aggravation of poststenotic myocardial ischaemia, which it is suggested can be caused by $alpha_2$-mediated poststenotic coronary vasoconstriction (HEUSCH et al. 1986; see also Sect. B.III.2); however, their data are not inconsistent with a role for postjunctional $alpha_2$-adrenoceptor-mediated coronary vasoconstriction during ischaemia.

CORR and WITKOWSKI have carried out a series of studies, mainly in anaesthetized animals, into the electrophysiological and other functional and biochemical effects in cardiac muscle which can be modulated by $alpha_1$-adrenoceptors (confirmed with the $alpha_1$-adrenoceptor-selective antagonist BE-2254) (reviewed in CORR and WITKOWSKI 1983; WITKOWSKI and CORR 1984). They concentrate on effects mediated directly on cardiac muscle, dismissing the possible role of $alpha_1$-adrenoceptor-mediated coronary vasoconstriction on the basis of an early study in the series (SHERIDAN et al. 1980), but it must be noted that global flow was measured so that vasoconstriction in small vessels causing heterogeneity of flow in the microcirculation was not excluded. They clearly show that cardiac muscle $alpha_1$-adrenoceptors are effective under ischaemic conditions and bring up several interesting possibilities concerning the nature of the biochemical events activated by the receptors, including the hypothesis that accumulation of metabolic products in ischaemic myocardium, such as lysophosphatides and long-chain acyl carnitines, might modulate receptor number or interfere with transduction of $alpha_1$-adrenoceptors. They do not, however, make a strong case to exclude the involvement of coronary vascular alpha-adrenoceptors in the genesis of arrhythmia. Thus their important contribution to the direct $alpha_1$-mediated stimulation of cardiac muscle should be viewed together with the evidence on coronary vascular alpha-adrenoceptors as suggested by DAVEY and co-workers (AUBRY et al. 1985; DAVEY 1986). In particular, the suggested role of acyl carnitines in modulating alpha-adrenoceptor function may relate critically to both alpha-mediated stimulation of cardiac muscle and to the control of the coronary vasculature since Spedding and co-workers have shown that acyl carnitines can modulate Ca^{2+} channels in a manner that both produces inotropic effects involving Ca^{2+} entry and modifies direct vasoconstrictor and endothelial-mediated vasodilator processes (Mir and Spedding 1986; Spedding et al. 1986; Bigaud and Spedding 1987; Duncan et al. 1987). This points to a general mechanism which may be involved in arrhythmias involving "Ca^{2+} overload" (use of "class IV" antagonists) (Spedding and Mir 1987). A consequence may be that the role of $alpha_1$-adrenoceptors in aggravation of arrhythmias may be more important than has been realized and that this can be tackled therapeutically either by the use of $alpha_1$-adrenoceptor antagonists or by finding ways of eliminating the link between the metabolic products (e.g. acyl carnitines) and facilitation of Ca^{2+} channel opening (see Fig. 3).

Sheridan et al. have studied the influence of $alpha_1$-adrenoceptors on cardiac function, including arrhythmia, in isolated, perfused guinea pig hearts, anaesthetized cats and patients with angina, using primarily indoramin (as selective $alpha_1$-antagonist) and methoxamine (as selective $alpha_1$-agonist) (Sheridan et al. 1980; Penny et al. 1985; Sheridan et al. 1986). They show that reperfusion arrhythmias can be alleviated by $alpha_1$-adrenoceptor antagonists and attribute this at least in part to improvement of myocardial blood flow; involving a general haemodynamic improvement from effects outside the heart and also improved perfusion due to coronary vasodilation through removal of catecholamine-induced effects on coronary vascular smooth muscle. This $alpha_1$-mediated effect occurs in the ischaemic myocardium even of isolated hearts where no nerve stimulation is present and presumably arises from an ischaemia-induced release of catecholamines from nerves such as has been demonstrated in isolated perfused rat heart (Schomig et al. 1985). These authors do not highlight possible $alpha_1$-adrenoceptor-mediated effects on cardiac muscle cells but a contribution from this source, particularly the removal by $alpha_1$-antagonists of inhomogeneity of function in neighbouring areas (Corr and Witkowski 1983), is not inconsistent with their data. Similarly there is no direct inconsistency with a role for postjunctional $alpha_2$-adrenoceptor-mediated coronary vasoconstriction during ischaemia.

Further support for the physiological involvement of vasoconstrictor alpha-adrenoceptors in coronary vessels comes from a series of studies by Vatner and co-workers, and several other groups, in instrumented conscious dogs (reviewed by Vatner 1983). This showed that, during near-maximal exercise, coronary metabolic vasodilation is restrained by a constrictor effect which is abolished by phentolamine (indicating non-specific alpha-adrenoceptor involvement) (Murray and Vatner 1979). In these dogs, large coronary arteries could be shown to be constricted by methoxamine, indicating more directly that $alpha_1$-adrenoceptor activation is vasoconstrictor in some coronary vessels (Vatner et al. 1980). The physiological involvement of coronary vasopressor $alpha_1$-adrenoceptors has been confirmed with prazosin in hypotensive anaesthetized dogs (Jones et al. 1986).

The experimental evidence for the physiological involvement of both myocardial and coronary vascular $alpha_1$-adrenoceptors in vivo is therefore strong and there is no case to highlight one or the other as the prime aggravating factor in arrhythmia.

D. Summary and Rationales for Utilizing Alpha-Adrenoceptor Blockade in Arrhythmia

In summary, it seems, at the time of writing, that $alpha_1$-adrenoceptor activation is functionally effective particularly in the ischaemic heart. It can and does contribute to some forms of arrhythmia. This is attributable to effects on *both* the myocardium and the coronary blood vessels, which can be initiated by either adrenergic nerves within each structure or by circulating agonists (natural catecholamines or synthetic drugs). Heterogeneity of cardiac cell function in ischaemic and adjacent healthy areas is exacerbated by the $alpha_1$-mediated stimulation of ischaemic cells and the attenuation of blood flow. There is therefore a dual basis for the use of selective $alpha_1$-adrenoceptor antagonists in alleviating such dysfunction. The only qualifications to this are, first, to consider the possibility that there may be circumstances in which $alpha_1$-blockade will divert blood away from ischaemic regions because they, specifically, are susceptible to $alpha_2$-adrenoceptor-mediated vasoconstriction, secondly, to consider the desirability of the overall haemodynamic effects of the antagonist and, thirdly, to remember that the hypothesis that there are prejunctional $alpha_1$-autoreceptors has not yet been thoroughly explored.

$Alpha_2$-adrenoceptors have no obvious function on cardiac muscle cells. They are present on sympathetic and parasympathetic nerve terminals, and can modulate their transmitter output. They are present also on coronary vascular smooth muscle, where they can initiate vasoconstriction. There are insufficient data from which to decide whether these receptors contribute to arrhythmias and hence whether their blockade would be beneficial. The effects of specifically blocking these receptors have not yet been fully elucidated largely because sufficiently selective compounds have only recently become available and have not yet been tested in models of arrhythmia. The drugs which have been employed, the diastereoisomers yohimbine and rauwolscine, are antagonists also at $alpha_1$ and 5-HT-receptors so that part of their action could be attributable to blockade of $alpha_1$-adrenoceptors on myocardium or coronary vessels, or 5-HT receptors which modulate thrombus formation. With regard to this latter topic of thrombus formation it seems that there is little evidence for an alpha-adrenoceptor involvement but this should be considered as a possible side effect (beneficial or otherwise) of drugs which are used to assess adrenoceptors but which are not greatly selective.

E. Future Directions

There is no doubt that alpha-adrenoceptors are functionally active in ischaemic and/or arrhythmic hearts (more so than under normal conditions) and that some forms of arrhythmia can be attenuated by drugs the primary known action of which is alpha-adrenoceptor blockade, particularly $alpha_1$-adrenoceptor blockade. This use of antagonists remains, however, semi-empirical since it is not absolutely clear whether one or both of the likely sites of action (cardiac muscle and coronary vascular smooth muscle) are involved. There is a clear case, on this evidence alone, to investigate further the involvement of alpha-adrenoceptor subtypes in the heart for two main reasons. First, in the short term, to clarify whether there is a case for the use of selective alpha-adrenoceptor antagonists in arrhythmia, whether this be selective $alpha_1$- or selective $alpha_2$-antagonists. Secondly, and more importantly in the long run, to establish the roles of alpha-adrenoceptor subtypes in the heart both in health and in disease. This is important because it contributes to the general fund of knowledge on cardiac function and can be applied to the treatment of arrhythmias and other problems where alpha-adrenoceptor blockade may not be relevant.

Other questions deserving attention are:

Why do $alpha_1$-adrenoceptors in the myocardium and $alpha_2$-adrenoceptors in vascular smooth muscle remain functionally active in ischaemia? Is there survival value in this? If so, will this be removed by therapy which blocks it? Is there an order of survival of receptor-linked mechanisms in ischaemia which has come about through evolution, i.e. beta most vulnerable, followed by the less vulnerable $alpha_1$, and the least vulnerable $alpha_2$? If so, have we yet to find the vital $alpha_2$-adrenoceptor-mediated process in the heart?

What are the transduction steps linked to the various adrenoceptors and what different modulating factors are they open to? In particular how do the second messengers lead to modification of ion-channel function? If we find this information will this lead to more specific biochemical targeting of drugs or will the mechanisms be too general, the recognition site of the receptor remaining the best hope for selective intervention? Can we stop the facilitation of alpha-adrenoceptor-mediated stimulation caused by metabolic products, thereby eliminating the need for alpha-blockade (and probably also for several other types of antiarrhythmic drug)?

Will alpha-adrenoceptors in the heart turn out to have a slightly different recognition site from those elsewhere, perhaps by virtue of being linked to specific second messenger systems, thereby allowing the development of more selective antagonists? Will the bio-phase for cardiac alpha-adrenoceptors have some unique properties which allow selective targeting of antagonists?

As methods develop to allow manipulation of receptor number, either through interference with the physiological control processes for this at the cell surface or through pharmacogenetics, will we be in a position of sufficient knowledge of the physiological and pathological roles of cardiac alpha-adrenoceptors to say whether there would be advantage in deliberate up- or downregulation of alpha- or beta-adrenoceptors?

Overall, it appears unlikely that we have heard the last of alpha-adrenoceptors in relation to arrhythmia.

References

Ahlquist RP (1948) A study of adrenotropic receptors. Am J Physiol 153:586-600

Alabaster VA, Peters CJ, Keir RF (1986) Comparison of potency of alpha 2 antagonists in vitro: evidence for heterogeneity of alpha-2 adrenoceptors. Br J Pharmacol 88:607-615

Andrejak M, Ward M, Schmitt H (1983) Cardiovascular effects of yohimbine in anaesthetized dogs. Eur J Pharmacol 94:219-228

Antonaccio MJ, Halle J, Kerwin L (1974) Functional significance of alpha-stimulation and alpha-blockade on responses to cardiac nerve stimulation in anesthetized dogs. Life Sci 15:765-777

Armstrong JM, Boura ALA (1973) Effects of clonidine and guanethidine on peripheral symphathetic nerve function in the pithed rat. Br J Pharmacol 47:850-852

Aubry ML, Davey MJ, Petch B (1985) Cardioprotective and antidysrhythmic effects of alpha-1-adrenoceptor blockade during myocardial ischaemia and reperfusion in the dog. J Cardiovasc Pharmacol 7:S93-S102

Baker DJ, Drew GM, Hilditch A (1984) Presynaptic alpha-adrenoceptors: do exogenous and neuronally released noradrenaline act at different sites? Br J Pharmacol 81:457-464

Benfey BG (1982) Function of myocardial alpha-adrenoceptors. Life Sci 31:101-112

Benfey BG, Varma DR (1967) Interactions of sympathomimetic drugs, propranolol and phentolamine, on atrial refractory period and contractility. Br J Pharmacol 30:603-611

Bertha BG, Folts JD (1984) Inhibition of epinephrine-exacerbated coronary thrombus formation by prostacyclin in the dog. J Lab Clin Med 103:204-214

Berthelsen S, Pettinger WA (1977) A functional basis for classification of alpha-adrenergic receptors. Life Sci 21:595-606

Bigaud M, Spedding M (1987) Inhibition of the effects of endothelial-derived relaxant factor (EDRF) in aorta by palmityl carnitine. Br J Pharmacol 89:540P.

Bilezikian JP, Loeb JN (1983) The influence of hyperthyroidism and hypothyroidism on alpha- and beta-adrenergic receptor systems and adrenergic responsiveness. Endocr Rev 4:378-388

Bolli R, Ware JA, Brandon TA, Weilbaecher AG, Mace ML Jr (1984) Platelet-mediated thrombosis in stenosed canine coronary arteries: inhibition by nicergoline, a platelet-active alpha-adrenergic antagonist. J Am Coll Cardiol 3:1417-1426

Bolli R, Brandon TA, Mace ML Jr, Weilbaecher DG (1985) Influence of alpha-adrenergic blockade on platelet-mediated thrombosis in stenosed canine coronary arteries. Cardiovasc Res 19:146-154

Bralet J, Didier J, Moreau D, Opie LH, Rochette L (1985) Effect of alpha-adrenoceptor antagonists (phentolamine, nicergoline and prazosin) on reperfusion arrhythmias and noradrenaline release in perfused rat heart. Br J Pharmacol 84:9-18

Brown JH, Buxton IL, Brunton LL (1985) Alpha-1-adrenergic and muscarinic cholinergic stimulation of phosphoinositide hydrolysis in adult rat cardiomyocytes. Circ Res 57:532-537

Brown MJ, Struthers AD, Di Silvio L, Yeo T, Ghatei M, Burrin JM (1985) Metabolic and haemodynamic effects of alpha-2-adrenoceptor stimulation and antagonism in man. Clin Sci 68:137s-139
Bruckner R, Scholz H (1984) Effects of alpha-adrenoceptor stimulation with phenylephrine in the presence of propranolol on force of contraction, slow inward current and cyclic AMP content in the bovine heart. Br J Pharmacol 82:223-232
Bruckner R, Mugge A, Scholz H (1985) Existence and functional role of alpha-one adrenoceptors in the mammalian heart. J Mol Cell Cardiol 17:639-645
Buffington CW, Feigl EO (1983) Effect of coronary artery pressure on transmural distribution of adrenergic coronary vasoconstriction in the dog. Circ Res 53:613-621
Bush LR, Campbell WB, Kern K, Tilton GD, Apprill P, Ashton J, Schmitz J, Buja LM, Willerson JT (1984) The effects of alpha 2-adrenergic and serotonergic receptor antagonists on cyclic blood flow alterations in stenosed canine coronary arteries. Circ Res 55:642-652
Bylund DB, U'Prichard DC (1983) Characterisation of alpha-1 and alpha-2 adrenergic receptors. Int Rev Neurobiol 24:343-431
Cambridge D, Davey MJ, Massingham R (1977) Prazosin, a selective antagonist of post-synaptic alpha-adrenoceptors. Br J Pharmacol 59:514-515P
Chiariello M, Ribeiro LGT, Davis MA, Maroko PR (1977) 'Reverse coronary steal' induced by coronary vasoconstriction following coronary artery occlusion in dogs. Circulation 56:809P
Chilian WM, Harrison D, Haws CW, Snyder WD, Marcus ML (1986) Adrenergic coronary tone during submaximal exercise in the dog is produced by circulating catecholamines. Evidence for adrenergic denervation supersensitivity in the myocardium but not in coronary vessels. Circ Res 58:68-82
Clanachan AS, McGrath JC (1976) Effects of ketamine on the peripheral autonomic nervous system of the rat. Br J Pharmacol 58:247-252
Clark MG, Patten GS (1984) Adrenergic regulation of glucose metabolism in rat heart. A calcium-dependent mechanism mediated by both alpha- and beta-adrenergic receptors. J Biol Chem 259:15204-15211
Clark MG, Patten GS, Filsell OH, Rattigan S (1983) Co-ordinated regulation of muscle glycolysis and hepatic glucose output in exercise by catecholamines acting via alpha-receptors. FEBS Lett 158:1-6
Clark RD, Michel AD, Whiting RL (1986) Pharmacology and structure-activity relationships of alpha-2-adrenoceptor antagonists. In: Ellis GP, West GB (eds) Progress in medicinal chemistry, vol 23. Elsevier, Amsterdam, pp 1-40
Cocks TM, Angus JA (1983) Endothelium-dependent relaxation of coronary arteries by noradrenaline and serotonin. Nature 305:627-630
Corr PB, Shayman JA, Kramer JB, Kipnis RJ (1981) Increased alpha-adrenergic receptors in ischemic cat myocardium. J Clin Invest 67:1232-1236
Corr PB, Witkowski FX (1983) Potential electrophysiologic mechanisms responsible for dysrhythmias associated with reperfusion of ischemic myocardium. Circulation 68: Suppl I:I16-I24
Dart AM, Riemersma RA (1985) Neurally mediated and spontaneous release of noradrenaline in the ischemic and reperfused rat heart. J Cardiovasc Pharmacol 7:45-49
Dart AM, Schomig A, Dietz R, Mayer E, Kubler W (1984) Release of endogenous catecholamines in the ischemic myocardium of the rat. Part B. Effect of sympathetic nerve stimulation. Circ Res 55:702-706
Dart AM, Riemersma RA, Schomig A, Ungar A (1987a) Metabolic requirements for release of endogenous noradrenaline during myocardial ischaemia and anoxia. Br J Pharmacol 90:43-50

Dart AM, Riemersma RA, Russell DC (1987b) Differential effects of alpha-1 and alpha-2 adrenoceptor blockade on regional myocardial blood flow in acutely ischaemic myocardium. Br J Pharmacol 90:28P

Davey MJ (1986) Alpha adrenoceptors—an overview. J Mol Cell Cardiol (Suppl V) 18:1-16

Decker N, Schwartz J (1985) Postjunctional alpha-1 and alpha-2 adrenoceptors in the coronaries of the perfused guinea-pig heart. J Pharmacol Exp Ther 232:251-257

De Mey J, Vanhoutte PM (1981) Uneven distribution of postjunctional alpha-1-and alpha-2-like adrenoceptors on canine arterial and venous smooth muscle. Circ Res 48:875-884

Deussen A, Heusch G, Thamer V (1985) Alpha-2-adrenoceptor-mediated coronary vasoconstriction persists after exhaustion of coronary dilator reserve. Eur J Pharmacol 115:147-153

Docherty JR (1983) An investigation of presynaptic alpha-adrenoceptor subtypes in the pithed rat heart. Br J Pharmacol 78:655-657

Docherty JR (1983) An investigation of presynaptic alpha-adrenoceptor subtypes in the pithed rat heart and in the rat isolated vas deferens. Br J Pharmacol 82:15-23

Docherty JR, McGrath JC (1978) Sympathomimetic effects of pancuronium bromide on the cardiovascular system of the pithed rat: a comparison with the effects of drugs blocking the neuronal uptake of noradrenaline. Br J Pharmacol 64:589-599

Docherty JR, McGrath JC (1979) An analysis of some factors influencing alpha-adrenoceptor feed-back at the sympathetic junction in the rat heart. Br J Pharmacol 66:55-63

Docherty JR, McGrath JC (1980a) A comparison of pre- and post-junctional potencies of several alpha-adrenoceptor agonists in the cardiovascular system and anococcygeus of the rat: evidence for 2 types of postjunctional alpha-adrenoceptor. Naunyn Schmiedebergs Arch Pharmacol 312:107-116

Docherty JR, McGrath JC (1980b) An examination of factors influencing adrenergic transmission in the pithed rat, with special reference to noradrenaline uptake mechanisms and post-junctional alpha-adrenoceptors. Naunyn Schmiedebergs Arch Pharmacol 313:101-111

Docherty JR, MacDonald A, McGrath JC (1979) Further sub-classification of alpha-adrenoceptors in the cardiovascular system, vas deferens and anococcygeus of the rat. Br J Pharmacol 67:421-422P

Drew GM (1976) Effects of alpha-adrenoceptor agonists and antagonists on pre- and postsynaptically located alpha-adrenoceptors. Eur J Pharmacol 36:313-320

Drew GM (1985) What do antagonists tell us about alpha-adrenoceptors? Clin Sci 68:15s-19s

Dukes ID, Vaughan Williams EM (1984 a) Effects of selective alpha-one-, alpha-two-, beta-one- and beta-two-adrenoceptor stimulation on potentials and contractions in the rabbit heart. J Physiol 355:523-546

Dukes ID, Vaughan Williams EM (1984 b) Electrophysiological effects of alpha-adrenoceptor antagonists in rabbit sino-atrial node, cardiac Purkinje cells and papillary muscles. Br J Pharmacol 83:419-426

Duncan GP, Patmore L, Spedding M (1987) Positive inotropic effects of palmitoyl carnitine in embryonic chick heart. Br J Pharmacol 89:757P

Endoh M, Motomura S (1979) Differentiation by cholinergic stimulation of positive inotropic actions mediated via alpha- and beta-adrenoceptors in the rabbit heart. Life Sci 25:759-768

Endoh M, Schumann HJ (1975) Frequency-dependence of the positive inotropic effect of methoxamine and naphazoline mediated by alpha-adrenoceptors in the isolated rabbit papillary muscle. Naunyn Schmiedebergs Arch Pharmacol 287:377-389

Endoh M, Yamashita S (1980) Adenosine antagonizes the positive inotropic action mediated via beta, but not alpha-adrenoceptors in the rabbit papillary muscle. Eur J Pharmacol 65:445-448

Fain JN, Garcia-Sainz A (1980) Role of phosphatidylinositol in alpha-1 and of adenylate cyclase in alpha-2 effects of catecholamines. Life Sci 26:1183-1195

Feigl EO (1983) Coronary physiology. Physiol Rev 63:1-205

Flavahan NA, McGrath JC (1981a) Demonstration of simultaneous alpha-1-, alpha-2-, beta-1- and beta-2-adrenoceptor-mediated effects of phenylephrine in the cardiovascular system of the pithed rat. Br J Pharmacol 72:585P

Flavahan NA, McGrath JC (1981b) Alpha-1-adrenoceptors can mediate chronotropic responses in the rat heart. Br J Pharmacol 73:586-588

Flavahan NA, McGrath JC (1982) Alpha-1 adrenoceptor activation can increase heart rate directly or decrease it indirectly through parasympathetic activation. Br J Pharmacol 77:319-328

Flavahan NA, Vanhoutte PM (1986) Alpha-1 adrenoceptor subclassification in vascular smooth muscle. Trends Pharmacol Sci 7:347-349

Forfar JC, Riemersma RA, Oliver MF (1983) Alpha-adrenoceptor control of norepinephrine release from acutely ischaemic myocardium: effects of blood flow, arrhythmias, and regional conduction delay. J Cardiovasc Pharmacol 5:752-759

Forfar JC, Russell DC, Riemersma RA (1985) Control of myocardial catecholamine release during acute ischemia. J Cardiovasc Pharmacol 5:S33-S39

Fox AW, Juberg EN, May JM, Johnson RD, Abel PW, Minneman KP (1985) Thyroid status and adrenergic receptor subtypes in the rat: comparison of receptor density and responsiveness. J Pharmacol Exp Ther 235, 715-723

Fuder H, Muscholl E, Spemann R (1983) The determination of presynaptic pA2 values of yohimbine and phentolamine on the perfused rat heart under conditions of negligible autoinhibition. Br J Pharmacol 79:109-119

Furchgott RF (1972) The classification of adrenoceptors (adrenergic receptors). In: Blaschko H, Muscholl E (eds) An evaluation from the standpoint of receptor theory. Springer, Berlin Heidelberg New York (Handbook of experimental pharmacology, vol 33)

Furchgott RF (1984) The role of endothelium in the responses of vascular smooth muscle to drugs. Annu Rev Pharmacol Toxicol 24:175-197

Giles TD, Thomas MG, Sander GE, Quiroz AC (1985) Central alpha-adrenergic agonists in chronic heart failure and ischemic heart disease. J Cardiovasc Pharmacol 7:S51-55

Gillespie JS (1980) Section II. Effects on the autonomic and on the central nervous system. A. Presynaptic receptors in the autonomic nervous system. Springer, Berlin Heidelberg New York, pp 169-205 (Handbook of experimental pharmacology, vol 54)

Godfraind T, Miller RC, Socrates Lima J (1982) Selective alpha-one- and alpha-two-adrenocepter agonist induced contractions and $^{45}Ca^{++}$ fluxes in the rat isolated aorta. Br J Pharmacol 77:597-604

Godfraind T, Egleme C, Osachie IA (1985) Role of endothelium in the contractile response of rat aorta to alpha-adrenoceptor agonists. Clin Sci 68:65s-71s

Govier WC (1967) A positive inotropic effect of phenylephrine mediated through alpha adrenergic receptors. Life Sci 6:1361-1365

Graham JM, Keating WR (1975) Responses of inner and outer muscle of the sheep carotid artery to injury. J Physiol (Lond) 247:437-482

Grant TL, Flavahan NA, Greig J, McGrath JC, McKean CE, Reid JL (1985a) Attempts to uncover subtypes of alpha-adrenoceptors and associated mechanisms by using sequential administration of blocking drugs. Clin Sci 68:25s-30s

Grant TL, McGrath JC, O'Brien (1985 b) The influence of blood gases on alpha-1- and alpha-2-adrenoceptor-mediated pressor responses in the pithed rat. Br J Pharmacol 86:69-77

Gwirtz PA, Overn SP, Mass HJ, Jones CE (1986) Alpha-1-adrenergic constriction limits coronary flow and cardiac function in running dogs. Am J Physiol 250:1117-1126

Harrison SM, Lamont Christine, Miller DJ (1986) Carnosine and other natural imidazoles enhance muscle Ca sensitivity and are mimicked by caffeine and AR-L 115s. J Physiol (Lond) 371:197P

Hatton R, Clough DP (1982) Captopril interferes with neurogenic vasoconstriction in the pithed rat by angiotensin dependent mechanisms. J Cardiovasc Pharmacol 4:116-123

Hattori Y, Levi R (1984) Adenosine selectively attenuates H2- and beta-mediated cardiac responses to histamine and norepinephrine: an unmasking of H1- and alpha-mediated responses. J Pharmacol Exp Ther 231:215-223

Hayes JS, Pollock GD, Fuller RW (1984) In vivo cardiovascular responses to isoproterenol, dopamine and tyramine after prolonged infusion of isoproterenol. J Pharmacol Exp Ther 231:633-639

Heusch G, Yoshimoto N, Heegemann H, Thamer V (1983) Interaction of methoxamine with compensatory vasodilation distal to coronary stenoses. Arzneimittelforschung 33:1647-1650

Heusch G, Deussen A, Schipke J, Thamer V (1984 a) Alpha-1- and alpha-2-adrenoceptor-mediated vasoconstriction of large and small canine coronary arteries in vivo. J Cardiovasc Pharmacol 6:961-968

Heusch G, Deussen A, Schipke J, Thamer V (1984 b) Adenosine, dipyridamole and isosorbide dinitrate are ineffective to prevent alpha-2-adrenergic vasoconstriction distal to severe coronary stenoses. Pflugers Arch 402:R27

Heusch G, Deussen A, Schipke J, Vogelsgang H, Hoffman V, Thamer V (1985 a) Role of cardiac sympathetic nerves in the genesis of myocardial ischemia distal to coronary stenoses. J Cardiovasc Pharmacol 7:S13-18

Heusch G, Deussen A, Thamer V (1985 b) Cardiac sympathetic nerve activity and progressive vasoconstriction distal to coronary stenoses: feed-back aggravation of myocardial ischemia. J Auton Nerv Syst 13:311-326

Heusch G, Schipke J, Thamer V (1986) Sympathetic mechanisms in poststenotic myocardial ischemia. J Cardiovasc Pharmacol 8:S33-S40

Heyndrickx GR, Vilaine JP, Moerman EJ, Leusen I (1984) Role of prejunctional alpha-2-adrenergic receptors in the regulation of myocardial performance during exercise in conscious dogs. Circ Res 54:683-693

Hieble JP, Sulpizio AC, Nichols AJ, DeMarinis RM, Pfeiffer FR, Lavanchy PG, Ruffolo RR (1986) Pharmacological differentiation of pre- and postjunctional $alpha_2$-adrenoceptors. J Hypertension 4(Suppl 6):S189-S192

Hirshfield JW, Borer JS, Goldstein RE, Barrett MJ, Epstein SE (1974) Reduction in severity and extent of myocardial infarction when nitroglycerin and methoxamine are administered during coronary occlusion. Circulation 49:291

Hoffman BB, Lefkowitz RJ (1980) Radioligand binding studies of adrenergic receptors: new insights into molecular and physiological regulation. Annu Rev Pharmacol Toxicol 20:581-608

Hoffman BF, Singer DH (1967) Appraisal of the effects of catecholamines on cardiac electrical activity. Ann NY Acad Sci 139:914-939

Holtz J, Saeed M, Sommer O, Bassenge E (1982) Norepinephrine constricts the canine coronary bed via postsynaptic alpha-2-adrenoceptors. Eur J Pharmacol 82:199-202

Inui J, Brodde OE, Schumann HJ (1982) Influence of acetylcholine on the positive inotropic effect evoked by alpha- or beta-adrenoceptor stimulation in the rabbit heart. Naunyn Schmiedebergs Arch Pharmacol 320:152-159

Ishac EJ, Pennefather JN, Handberg GM (1983) Effect of changes in thyroid state on atrial alpha- and beta-adrenoceptors, adenylate cyclase activity, and catecholamine levels in the rat. J Cardiovasc Pharmacol 5:396-405

Jackson CV, McGrath GM, McNeill JH (1986) Alterations in alpha-1-adrenoceptor stimulation of isolated atria from experimental diabetic rats. Can J Physiol Pharmacol 64:145-151

Jakobs KH (1979) Inhibition of adenylate cyclase by hormones and neurotransmitters. Mol Cell Endocrinol 16:147-156

Jones CE, Liang IY, Maulsby MR (1986) Cardiac and coronary effects of prazosin and phenoxybenzamine during coronary hypotension. J Pharmacol Exp Ther 236:204-211

Kalsner S (1982) Evidence against the unitary hypothesis of agonist and antagonist action at presynaptic adrenoceptors. Br J Pharmacol 77:375-380

Karliner JS, Barnes P, Hamilton CA, Dollery CT (1979) Alpha-1-adrenergic receptors in guinea pig myocardium: identification by binding of a new radioligand, (3H)-prazosin. Biochem Biophys Res Commun 90:142-149

Kobinger W, Pichler L (1982) Presynaptic activity of the imidazolidine derivate ST 587, a highly selective alpha-1 adrenoceptor agonist. Eur J Pharmacol 82:203-206

Lambert GA, Lang WJ, Friedman E, Meller E, Gershon S (1978) Pharmacological and biochemical properties of isomeric yohimbine alkaloids. Eur J Pharmacol:49:39-48

Langer SZ (1979) Presynaptic regulation of catecholamine release. Biochem Pharmacol 23:1793-1800

Langer SZ, Adler-Graschinsky E, Giorgio O (1977) Physiological significance of alpha-adrenoceptor mediated negative feedback mechanism regulating noradrenaline release during nerve stimulation. Nature 265:648-650

Latifpour J, McNeill JH (1984) Cardiac autonomic receptors: effect of long-term experimental diabetes. J Pharmacol Exp Ther 230:242-249

Lefevre F, Fenard S, Cavero I (1977) Vascular beta-adrenoceptor stimulating properties of phenylephrine. Eur J Pharmacol 43:85-88

Lefkowitz RJ, Caron MG (1986) Regulation of adrenergic receptor function by phosphorylation. J Mol Cell Cardiol 18:885-895

Leimdorfer A (1953) Prevention of cardiac arrhythmias by regitine. Arch Int Pharmacodyn Ther 94:119-126

Lew MJ, Angus JA (1983) Clonidine and noradrenaline fail to inhibit vagal induced bradycardia. Evidence against prejunctional alpha-adrenoceptors on vagal varicosities in guinea pig right atria. Naunyn Schmiedebergs Arch Pharmacol 323:228-232

Lew MJ, Angus JA (1985) Alpha-1- and alpha-2-adrenoceptor mediated pressor responses: are they differentiated by calcium antagonism or by functional antagonism? J Cardiovasc Pharmacol 7:401-408

Luchelli-Fortis MA, Langer SZ (1974) Reserpine-induced depletion of the norepinephrine stores: is it a reliable criterion for the classification of the mechanism of action of sympathomimetic amines? J Pharmacol Exp Ther 188:640-653

McGrath JC (1982) Commentary: evidence for more than one type of postjunctional alpha-adrenoceptor. Biochem Pharmacol 31:467-484

McGrath JC (1984) Alpha-adrenoceptor antagonism by apoyohimbine and some observations on the pharmacology of alpha-adrenoceptors in the rat anococcygeus and vas deferens. Br J Pharmacol 82:769-781

McGrath JC, Flavahan NA, McKean CE (1982) Alpha-1- and alpha-2-adrenoceptor-

mediated pressor and chronotropic effects in the rat and rabbit. J Cardiovasc Pharmacol 4 S101-S107
Maling HM, Moran NC (1957) Ventricular arrhythmias induced by sympathomimetic amines in unanesthetized dogs following coronary artery occlusion. Circ Res 5:409-413
Martin W, Furchgott RF, Villani GM, Jothianandan D (1986) Depression of contractile responses in rat aorta by spontaneously released endothelium-derived relaxing factor. J Pharmacol Exp Ther 237:529-538
Maze M, Smith CM (1983) Identification of receptor mechanism mediating epinephrine-induced arrhythmias during halothane anaesthesia in the dog. Anesthesiology 59:322-326
Michell RH (1985) Inositol lipid breakdown as a step in alpha-adrenergic stimulus-response coupling. Clin Sci 68:43s-46s
Mir AK, Spedding M (1986) Palmitoyl carnitine. A lipid metabolite produced in ischaemia, activates Ca++ channels in smooth muscle. Br J Pharmacol 88:381P
Miura Y, Inui J, Imamura H (1978) Alpha-adrenoceptor-mediated restoration of calcium-dependent potential in the partially depolarized rabbit papillary muscle. Naunyn Schmiedebergs Arch Pharmacol 301:201-205
Mugge A, Reupcke C, Scholz H (1985 a) Changes of myocardial alpha-1- and beta-adrenoceptor density in rats pretreated with propylthiouracil (PTU) or propranolol (PROP). Naunyn Schmiedebergs Arch Pharmacol 329:R52
Mugge A, Reupcke C, Scholz H (1985 b) Increased myocardial alpha-1-adrenoceptor density in rats chronically treated with propranolol. Eur J Pharmacol 112:249-252
Murray PA, Vatner SF (1979) Alpha adrenoceptor attenuation of coronary vascular response to severe exercise in the conscious dog. Circ Res 45:654-660
Muscholl E (1973) Regulation of catecholamine release. The muscarinic inhibitory mechanism. In: Usdin E, Snyder SH (eds) Frontiers in catecholamine research. Pergamon, Oxford, 537-542.
Nahorski SR, Barnett DB, Cheung YC (1985) Alpha-adrenoceptor-effector coupling: affinity states or heterogeneity of the alpha-2-adrenoceptor? Clin Sci 68:39s-42s
Nakadate T, Nakaki T, Muraki T, Kato R (1980) Adrenergic regulation of blood glucose levels—possible involvement of post-synaptic alpha-2-type adrenergic receptors regulating insulin release. J Pharmacol Exp Ther 215:226-230
Nakaki R, Nakadate T, Kato R (1980) Alpha-2-adrenoceptors modulating insulin release from isolated pancreatic islets. Naunyn Schmiedebergs Arch Pharmacol 313:151-153
Nakashima M, Maeda K, Sekiya A, Hagino Y (1971) Effect of hypothyroid status on myocardial responses to sympathomimetic drugs. Jpn J Pharmacol 21:819-825
Nathan HJ, Feigl EO (1986) Adrenergic vasoconstriction lessens transmural steal during coronary hypoperfusion. Am J Physiol 250:H645-653.
O'Brien JW, McGrath JC (1987) Blockade by nifedipine of responses to intravenous bolus injection or infusion of alpha-1- and alpha-2-adrenoceptor agonists in the pithed rat. Br J Pharmacol (in press)
O'Brien JW, Flavahan NA, Grant TL, McGrath JC, Marshall RJ (1985) Influence of blood gases, Ca^{2+}-entry blockade and angiotensin converting enzyme inhibition on pressor responses to alpha-adrenoceptor agonists: evidence in vivo for subtypes of response independent of receptor subtype? Clin Sci 68:99s-104s
Ohguchi S, Sotobata I, Oguro K, Nakashima M (1984) Changes in the effects of clonidine on left atrium and hindlimb vasculature of rats in various thyroid states. A study of the responsiveness of alpha 2-adrenoceptors in the cardiovascular system. Jpn Heart J 25:425-437
Penny WJ, Culling W, Lewis MJ, Sheridan DJ (1985) Antiarrhythmic and electrophysio-

logical effects of alpha adrenoceptor blockade during myocardial ischaemia and reperfusion in isolated guinea-pig heart. J Mol Cell Cardiol 17:399-409
Rand MJ, McCulloch MW, Story DF (1982) Feedback modulation of noradrenergic transmission. Trends Pharmacol Sci 3:8-11
Reuter H (1974) Localization of beta adrenergic receptors, and effects of noradrenaline and cyclic nucleotides on action potentials, ionic currents and tension in mammalian cardiac muscle. J Physiol [Lond] 242:429-451
Reuter H (1983) Calcium channel modulation by neurotransmitters, enzymes and drugs. Nature 301:569-574
Robie NW (1984) Controversial evidence regarding the functional importance of presynaptic alpha receptors. Fed Proc 43:1371-1374
Ruffolo RR Jr, Kopia GA (1986) Importance of receptor regulation in the pathophysiology and therapy of congestive heart failure. Am J Med 80:67-72
Ruffolo RR Jr, Yaden EL, Waddell JE (1980) Receptor interactions of imidazolines. V. Clonidine differentiates postsynaptic alpha adrenergic receptor subtypes in tissues from the rat. J Pharmacol Exp Ther 213:557-559
Ruffolo RR Jr, Yaden EL, Waddell JE, Ward JS (1982) Receptor interactions of imidazolines. XI. Alpha-adrenergic and antihypertensive effects of clonidine and its methylene-bridged analog, St 1913. Pharmacology 25:187-201
Ruffolo RR Jr, Morgan EL, Messick K (1984) Possible relationship between receptor reserve and the differential antagonism of alpha-1- and alpha-2-adrenoceptor-mediated pressor responses by calcium channel antagonists in the pithed rat. J Pharmacol Exp Ther 230:587-594
Saeed M, Holtz J, Elsner D, Bassenge E (1985) Sympathetic control of myocardial oxygen balance in dogs mediated by activation of coronary vascular alpha-2 adrenoceptors. J Cardiovasc Pharmacol 7:167-173
Schmitt H (1977) The pharmacology of clonidine and related products. In: Gross F (ed) Antihypertensive agents. Springer, Berlin Heidelberg New York, pp 299-396 (Handbook of experimental pharmacology, vol 39)
Schmitt H, Laubie M (1983) Adrenoceptors and central cardiovascular regulation. In: Kunos G (ed) Adrenoceptors and catecholamine action, part B. Wiley, New York, 219-264
Scholz H (1980) Effects of beta- and alpha-adrenoceptor activators and adrenergic transmitter releasing agents on the mechanical activity of the heart. In: Szekeres L (ed) Adrenergic activators and inhibitors. Springer, Berlin Heidelberg New York, pp 51-733 (Handbook of experimental pharmacology, vol 54/1)
Schomig A, Dart AM, Dietz R, Kubler W, Mayer EC (1985) Paradoxical role of neuronal uptake for the locally mediated release of endogenous noradrenaline in the ischemic myocardium. J Cardiovasc Pharmacol 7:540-544
Schumann HJ (1980) Are there alpha-adrenoceptors in the mammalian heart? Trends Pharmacol Sci 1:195-197
Schumann HJ, Endoh M, Brodde OE (1975) The time course of the effects of beta- and alpha-adrenoceptor stimulation by isoprenaline and methoxamine on the contractile force and cAMP level of the isolated rabbit papillary muscle. Naunyn Schmiedebergs Arch Pharmacol 289:291-302
Sekar MC, Roufogalis BD (1984) Comparison of muscarinic and alpha-adrenergic receptors in rat atria based on phosphoinositide turnover. Life Sci 35:1527-1533
Sheridan DM, Penkoske PA, Sobel BE, Corr PB (1980) Alpha adrenergic contributions of dysrhythmia during myocardial ischemia and reperfusion in cats. J Clin Invest 65:161-171

Sheridan DJ, Thomas P. Culling W, Collins P (1986) Antianginal and haemodynamic effects of alpha-1-adrenoceptor blockade. J Cardiovasc Pharmacol 8:S144-S150
Smith EF, Schaffran R, Kluth M (1983) Comparative effects of alpha-1- and alpha-2-adrenoceptor blockers on catecholamine overflow and cardiac responses in sympathetically-stimulated rabbit hearts. Naunyn Schmiedebergs Arch Pharmacol 322:294
Spedding M, Mir AK (1987) Direct activation of Ca^{2+} channels by palmitoyl carnitine, a putative endogenous ligand. Br J Pharmacol 92:457-468
Spedding M, Schine V, Schoeffter P, Miller RC (1987) Calcium channel activation does not increase the release of endothelial-derived relaxant factor (EDRF) although tonic release of EDRF may modulate calcium channel activity in smooth muscle. J Cardiovasc Pharmacol 8:1130-1137
Spokas EG, Folco GC (1984) Intima-related vasodilatation of the perfused rat caudal artery. Eur J Pharmacol 100:211-217
Starke K (1977) Regulation of noradrenaline release by presynaptic receptor systems. Rev Physiol Biochem Pharmacol 77:1-124
Starke K (1981) Alpha-adrenoceptor subclassification. Rev Physiol Biochem Pharmacol 88:199-236
Steinberg SF, Drugge ED, Bilezikian JP, Robinson RB (1985) Acquisition by innervated cardiac myocytes of a pertussis toxin-specific regulatory protein linked to the alpha-1 receptor. Science 230:186-188
Stjarne L (1985) Scope and mechanisms of control of stimulus-secretion coupling in single varicosities of sympathetic nerves. Clin Sci 68:77s-81s
Story DF, Standford-Starr CA, Rand MJ (1985) Evidence for the involvement of alpha-1 adrenoceptors in negative feedback regulation of noradrenergic transmitter release in rat atria. Clin Sci 68:111s-115s
Timmermans PBMWM, Wilffert B, Kalkman HO, Thoolen MJMC, Van Meel JCA, De Jonge A, Van Zwieten PA (1982) Selective inhibition of alpha-2-adrenoceptor-mediated vasoconstriction in vivo by captopril and MK-421. Br J Pharmacol 75:135P
Toda N (1986) Alpha-adrenoceptor subtypes and diltiazem actions in isolated human coronary arteries. Am J Physiol 250:H718-724
Tsien RW (1977) Cyclic AMP and contractile activity in heart. Adv Cyclic Nucleotide Res 8:363-420
Vanhoutte PM (1986) Could the absence or malfunction of vascular endothelium precipitate the occurence of vasospasm? J Mol Cell Cardiol 18:679-689
Van Meel JCA, De Jonge A, Kalkman HO, Wilffert B, Timmermans PBMWM, Van Zwieten PA (1981) Vascular smooth muscle contraction initiated by postsynaptic alpha-2-adrenoceptor activation induced by an influx of extracellular calcium. Eur J Pharmacol 69:205-208
Vatner SF (1983) Alpha-adrenergic regulation of the coronary circulation in the conscious dog. Am J Cardiol 52:15A-21A
Vatner SF (1984) Alpha-adrenergic tone in the coronary circulation of the conscious dog. Fed Proc 43:2867-2872
Vatner SF, Pagani M, Manders WT, Pasipoularides AD (1980) Alpha adrenergic vasoconstriction and nitroglycerin vasodilation of large coronary arteries in the conscious dog. J Clin Invest 65:5-14
Vaughan Williams EM (1985) Cardiac electrophysiological effects of selective adrenoceptor stimulation and their possible roles in arrhythmias. J Cardiovasc Pharmacol 7:S61-S64

Waterfall JF, Rhodes KF, Lattimer N (1985) Studies of alpha-2-adrenoceptor antagonist potency in vitro: comparisons in tissues from rats, rabbits, dogs and humans. Clin Sci 68:21s-24s
Weitzell R, Tanaka T, Starke K (1979) Pre- and postsynaptic effects of yohimbine stereoisomers on noradrenergic transmission in the pulmonary artery of the rabbit. Naunyn Schmiedebergs Arch Pharmacol 308:127-136
Wetzel GT, Goldstein D, Brown JH (1985) Acetylcholine release from rat atria can be regulated through an alpha-1-adrenergic receptor. Circ Res 56:763-766
Williams RS, Schaible TF, Scheuer J, Kennedy R (1983) Effects of experimental diabetes on adrenergic and cholinergic receptors of rat myocardium. Diabetes 32:881-886
Witkowski FX, Corr PB (1984) Mechanisms responsible for arrhythmias associated with reperfusion of ischemic myocardium. Ann NY Acad Sci 427:187-198
Woodman OL, Vatner SF (1986) Noradrenaline-induced coronary vasoconstriction is mediated by both alpha-1- and alpha-2-adrenoceptors in conscious dogs. J Mol Cell Cardiol 18:155P

CHAPTER 23

Adrenergic Arrhythmogenesis and the Long Q-T Syndrome

P. J. SCHWARTZ and S. G. PRIORI

A. Introduction

It has become progressively evident that increases in sympathetic activity favour the onset of malignant cardiac arrhythmias, particularly when acting in a diseased myocardium (SCHWARTZ and STONE 1982; CORR et al. 1986). In this chapter two conditions that involve in a causal relationship the sympathetic nervous system and life-threatening arrhythmias will be examined: acute myocardial ischaemia and the idiopathic long Q-T syndrome.

B. Acute Myocardial Ischaemia

I. Pathophysiology

Myocardial ischaemia excites both vagal (RECORDATI et al. 1971; THOREN 1978) and sympathetic afferent fibres of cardiac origin, eliciting a variety of reflex responses. The cardiac sympathetic sensory endings (MALLIANI et al. 1973; CASATI et al. 1979) are mechanoreceptors normally excited by mechanical events, but their activity can be further enhanced by chemical substances, like bradykinin (LOMBARDI et al. 1981), which are known to be released in the ischaemic heart. This activation elicits an excitatory cardiocardiac reflex (MALLIANI et al. 1969). This reflex, which takes place within a few seconds of ischaemia, plays an important role in the genesis of the early ventricular arrhythmias. This is shown by the fact that interruption of its afferent limb, by section of the dorsal roots from the eighth cervical segment to the fifth thoracic segment, is capable of greatly reducing the arrhythmias associated with short-lasting coronary occlusion (SCHWARTZ et al. 1976a).

The excitation of cardiac sympathetic afferents not only leads to an increase in efferent cardiac sympathetic activity, but can also reflexly and selectively inhibit the activity of efferent cardiac vagal fibres (SCHWARTZ et al. 1973). This sympathovagal reflex has the potential of impairing the vagally mediated maintenance of an optimal heart rate (CHADDA et al. 1974), thus facilitating the occurrence of a dangerous tachycardia.

It is noteworthy that the afferent limbs of most cardiocardiac sympathetic reflexes seem to be preferentially distributed through left-sided nerves (PAGANI et al. 1974), which makes these reflexes dependent to a major extent on an intact left stellate ganglion.

The cardiac vagal sensory endings, which seem to be preferentially distributed in the posteroinferior wall of the left ventricle (Thames et al. 1978), can reflexly lower both heart rate and blood pressure. When these vagally mediated depressor reflexes are overwhelming, they can further reduce coronary flow to the ischaemic areas, eventually resulting in either asystole or ventricular fibrillation. On the other hand, powerful, but not excessive, vagal reflexes during acute myocardial ischaemia can have a protective effect, largely through an antagonism of the arrhythmogenic influence of adrenergic activity (Verrier and Lown 1981). More recently, two sets of data have been reported that contribute significantly to the concept of a "protective" effect of vagal activity. Vagal stimulation in anaesthetized cats strikingly reduces the incidence of reperfusion arrhythmias, an effect largely dependent on the attendant reduction in heart rate (Zuanetti et al. 1987). Vagal stimulation in *conscious* dogs with a healed myocardial infarction (Schwartz et al. 1984) prevents ventricular fibrillation induced by acute myocardial ischaemia during an exercise stress test (De Ferrari et al. 1987). Conversely, a depressed baroreflex sensitivity, a marker of impaired vagal reflex activity, is associated with a greater susceptibility to ventricular fibrillation in postmyocardial infarction dogs (Billman et al. 1982; Schwartz et al. 1988) and probably also in patients with a myocardial infarction (La Rovere et al. 1988).

Thus, a critical factor for survival during acute myocardial ischaemia is represented by the balance and respective weight of the sympathetic and vagal tonic and reflex activity.

The effects of sympathetic discharges most relevant to life-threatening arrhythmias are those affecting the electrophysiological properties of the heart, the coronary circulation and the level of heart rate. Those aspects have already been discussed in detail elsewhere (Schwartz and Stone 1982; Schwartz and Zaza 1986) and they will only be briefly outlined here.

1. Sympathetic Activity and Cardiac Electrophysiology

Ventricular arrhythmias can be generated either by the direct action of catecholamines at the cellular level or by the consequences of these actions on the activation pattern of the myocardial syncytium. Electrical inhomogeneity of the myocardium, enhanced by adrenergic stimulation (Han and Moe 1964), favours the development of re-entrant circuits. Also the coexistence of zones at different potentials, during the activation cycle, can lead to ectopic activity triggered by an "injury current" (Janse et al. 1980). Adrenergic stimulation is associated with a beta-mediated shortening of the action potential duration (Matoba et al. 1979) and of the refractory period (Kralios et al. 1975). Of importance is the fact that activation of the cardiac sympathetic nerves increases the dispersion of the ventricular refractory periods, while systemic infusion of catecholamines has the opposite effect (Han and Moe 1964). To complicate the picture further, there is the concept, now generally accepted, that the left-sided sympathetic nerves are dominant in the control of the electrophysiological properties of the heart (Schwartz 1984). Particularly relevant were those studies that showed that unilateral right stellectomy lowers the threshold for ventricular fibrillation (Schwartz et al. 1976b) and shortens ventricular refractory periods (Schwartz et al. 1977) while unilateral left stellectomy has the opposite effect and is protective in terms of cardiac electrical stability.

In a recent study (Janse et al. 1985) DC extracellular electrograms were recorded from 60 left ventricular epicardial sites during 5-min periods of coronary arterial occlusions in dogs. In each animal the recordings made during control occlusions were compared with those made under various interventions on the sympathetic nervous system. Local TQ segment potential changes, markers of the degree of membrane depolarization of the myocardium under the electrode, and activation sequences were analysed. Left stellate stimulation, resulting in the appearance of ventricular tachyarrhythmias, increased the degree of TQ segment depression and enlarged the area on which TQ segment changes were observed, indicating a more severe and widely distributed membrane depolarization. Opposite changes were observed after left stellectomy. Interestingly, the results were not significantly different whether the anterior descending or the circumflex coronary artery was occluded. In this study, left stellate stimulation was associated with a marked reduction of conduction delay in the ischaemic zone, despite the greater TQ segment depression. This improvement of conduction, although rather unexpected on the basis of a more severe membrane depolarization during left stellate stimulation, can be explained by the previous observation that action potentials can be conducted faster in this situation because of the reduction of the gap between resting membrane potential and excitation threshold (Peon et al. 1978; Dominguez and Fozzard 1970). An improvement in impulse conduction should make the genesis of subepicardial re-entrant circuits less likely. This poses again the puzzling question of whether re-entry or the enhancement of focal acitivity is responsible for the occurrence of ventricular tachycardia and fibrillation during early myocardial ischaemia and sympathetic activation.

There is also evidence for an involvement of the adrenergic nervous system in a more recently described mechanism for cardiac arrhythmias, the so-called triggered activity (Wit and Rosen 1986). Early studies performed in vitro showed that superfusion of Purkinje fibers (Cranefield and Aronson 1974; Kimura et al. 1987) and atrial fibers (Johnson et al. 1986) with adrenergic agonists elicited delayed afterdepolarizations and ventricular ectopy. More recently, we (Priori et al. 1988) recorded delayed afterdepolarizations in vivo from the endocardial surface of anaesthetized cats during electrical stimulation of the left stellate ganglion, thus demonstrating a direct link between activation of cardiac nerves and afterdepolarizations.

The causal relationship between sympathetic hyperactivity and ventricular arrhythmias is definitely established, and the main underlying mechanisms have been fairly well delineated; nonetheless, the precise conditions under which each of them may become operant remain to be defined.

2. Sympathetic Activity and Coronary Circulation

Besides the wide acceptance of the importance of local metabolic regulation, an increasing number of reports have indicated the ability of sympathetic activation to increase coronary resistance, independently from its effects on contractility and oxygen requirements. In 1967 FEIGL reported that left stellate stimulation could induce alpha-mediated coronary vasoconstriction (FEIGL 1967). In 1978 FEIGL again specifically analysed the relationship between metabolic and neurogenic components and showed that sympathetic vasoconstriction could be observed even when the metabolic demand of the myocardium was enhanced (MOHRMAN and FEIGL 1978). A resting alpha- and beta-adrenergic tone on coronary circulation has been demonstrated also during adenosine-induced vasodilation (VLAHAKES et al. 1982). Recently HEUSCH et al. have reported that in the vascular bed distal to various degrees of coronary stenosis the predominant response to sympathetic stimulation is continuously shifted from metabolic dilation to alpha-2-mediated adrenergic vasoconstriction (HEUSCH et al. 1985). As a consequence, sympathetic activation would enhance the difference in coronary resistance between the normal and ischaemic zone, resulting in a steal of blood with further impairment of myocardial perfusion.

The significance of these studies is limited by the use of anaesthesia. The physiological relevance of these data became apparent when it was shown that in the conscious dog the reactive hyperaemic response to a 10 s circumflex ar-

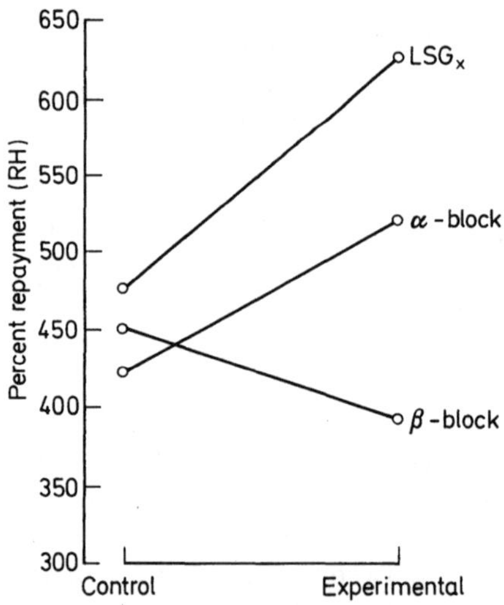

Fig. 1. Reactive hyperaemia in conscious dogs before and after left stellectomy (*LSGx*), phentolamine (*α-block*) and propranolol (*β*-block). The data points represent the means of 130 trials in 16 dogs. They indicate that left stellectomy and alpha-adrenergic blockade increase reactive hyperaemia, a variable that relates to the capability of the coronary bed to dilate, whereas beta-adrenergic blockade has an opposite effect. The same kind of responses were found when heart rate was kept constant by pacing. (SCHWARTZ and STONE 1982)

tery occlusion was greatly enhanced by left stellectomy or alpha-blockade, was not modified by right stellectomy and was reduced by beta-blockade (SCHWARTZ and STONE 1977) (Fig. 1). This study, besides demonstrating a tonic adrenergic restraint on the ability of coronary arteries to dilate in response to metabolic stimuli, suggested a dominance of the left-sided sympathetic nerves in the control of coronary flow. This could be explained simply by the preferential distribution of left-sided nerves to the posterior wall of the left ventricle (KRALIOS et al. 1975); however, a later study strongly supported the concept of left dominance by showing that the effects on coronary flow of

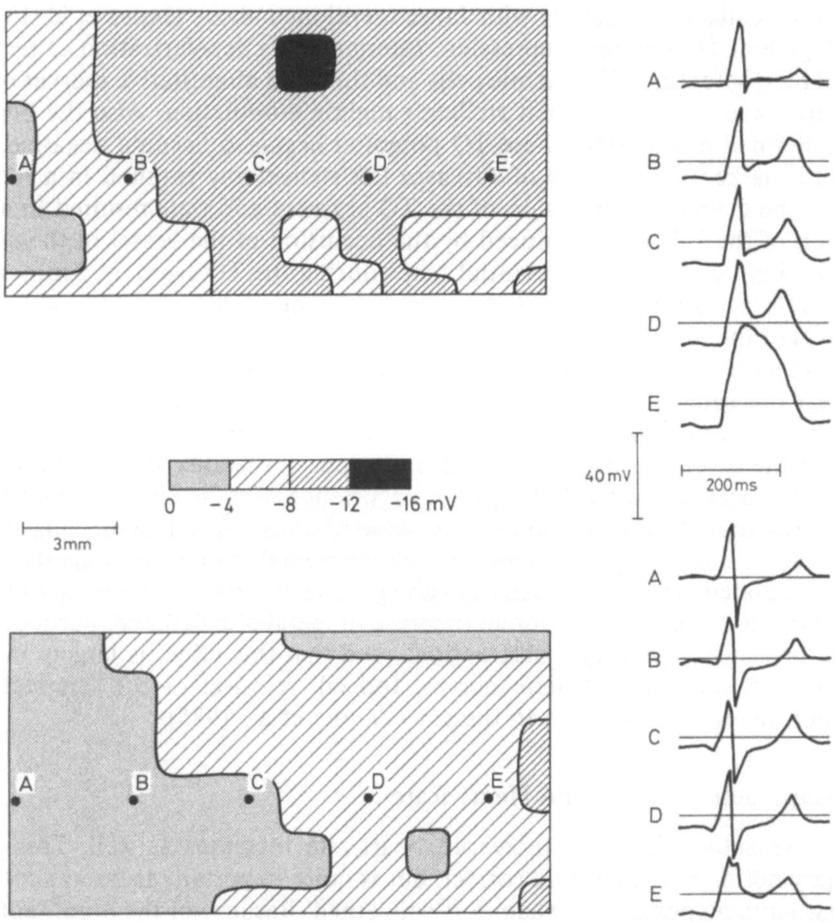

Fig. 2. Selected direct-current electrograms recorded 2 min after coronary occlusion from sites *A* to *E*. On the *left*, the area from which 60 electrograms were simultaneously recorded is indicated, and iso potentials during the TQ segment are shown. The *upper panel* depicts the situation during the control occlusion (second occlusion), and the *lower panel* shows the situation 2 min after a subsequent occlusion was made when the left stellate ganglion had been removed. Note that the degree of ischaemic changes is less marked after stellectomy. (JANSE et al. 1985)

left stellate stimulation prevail on those of right stellate stimulation on the entire heart (RINKEMA et al. 1982). Even if an increase in myocardial metabolic requirements is induced in normal hearts by exercise, left stellectomy or alpha-blockade is followed by an increase in coronary flow (SCHWARTZ and STONE 1979), demonstrating tonic adrenergic vasoconstrictor tone even in physiological conditions.

It has also been shown that cyclical changes in coronary blood flow, resulting from spontaneous disaggregation of platelet aggregates, at the site of coronary arterial stenosis, were abolished by left stellectomy and were enhanced by left stellate stimulation (RAEDER et al. 1981). These results probably depend on a catecholamine-mediated fluctuation in platelet aggregability.

In the collaborative study with Janse mentioned above (JANSE et al. 1985), the degree of TQ segment depression, interpreted as depolarization of resting membrane potential of ischaemic cells and thus as a quantitative marker of ischaemia, was increased by left stellate ganglion stimulation. Also the border zone, defined as the area where TQ potential began to become negative, increased markedly. In 10 out of 13 cases left stellectomy resulted in a reduction of the degree of depression of the TQ segment and thus resulted in a reduction of the ischaemia induced by the occlusion of the vessel in the same animal. Figure 2 shows isopotential maps during the TQ segment 2 min after a control occlusion and 2 min after a subsequent occlusion performed after removal of the left stellate ganglion. During the control occlusion, a large part of the myocardium covered by the electrode has TQ potential of −8 to 16 mV, whereas this same area has TQ potential of −4 to −8 mV during the occlusion performed after left stellectomy. Thus, the absence of the left stellate ganglion reduces the severity of ischaemia and the extent of the area affected in the ischaemic process or, at least, delays the occurrence of these changes. This may be of particular clinical relevance in the case of transient ischaemic episodes.

Thus, a considerable amount of experimental evidence suggests that alpha-mediated sympathetic activity can significantly reduce the ability of the coronary bed to compensate for an increase in metabolic demand, particularly in the presence of coronary obstructions, and that this effect is largely mediated by left-sided sympathetic nerves. Indeed, removal of the left stellate ganglion reduces infarct size in the cat (VANOLI et al. 1985).

3. Sympathetic Activity and Heart Rate

When sympathetic activity increases, heart rate increases as well. This elementary notion is not entirely correct, but is quite important as far as survival during acute myocardial ischaemia is concerned. Because of the dominant effect of the parasympathetic nervous system at the sinus node level, when both sympathetic and vagal activity are increased heart rate decreases or does not change. Classic examples are those of the diving reflex or asphyxia where there are simultaneously bradycardia and hypertension because of the concomitant activation of vagal and sympathetic nerves. Thus, when sympathetic activity increases, the effect on heart rate depends on the level of vagal activity. However, when sympathetic hyperactivity results in a heart rate increase, by far the most common event, this has deleterious effects on cardiac rhythm. Several observations consistently indicate that during myocardial ischaemia excessively high levels of heart rate greatly facilitate the occurrence of ventricular fibrillation (CHADDA et al. 1974; SCHWARTZ et al. 1984; VERRIER et al. 1974). For instance, as discussed elsewhere (SCHWARTZ et al. 1984) the level of heart rate at the time of a coronary artery occlusion can often overcome any protective mechanism.

Accordingly, interventions that reduce heart rate during myocardial ischaemia, either indirectly by antagonizing the beta-adrenergic effects or directly by a negative chronotropic effect (e.g. acetylcholine), can be expected to lower the incidence of ventricular fibrillation.

II. From Pathophysiology to Antiarrhythmic Interventions

The considerations made above, although necessary for understanding the background for adrenergic arrhythmogenesis, need to be translated into animal models specifically designed for the reproducible induction of cardiac arrhythmias in order to assess the efficacy of a variety of potentially protective interventions. Some animal models in which sympathetically mediated life-threatening arrhythmias can be consistently induced and in which there is already information on the results with an adequate number of antiarrhythmic drugs or manoeuvres are now available and will be described here in some detail together with their clinical implications.

III. Animal Models for Adrenergic Arrhythmias

1. Myocardial Ischaemia and Left Stellate Ganglion Stimulation

In alpha-chloralose-anaesthetized cats a 2-min occlusion of the anterior descending branch of the left coronary artery is performed. The left stellate ganglion is stimulated for 30 s, starting at the end of the 1st min of ischaemia (Fig. 3). Neither the 2-min coronary artery occlusion nor the left stellate ganglion stimulation usually induce arrhythmias if performed separately. However, when they are performed together life-threatening arrhythmias occur in a high percentage of animals. A characteristic feature of this animal model is the consistency and reproducibility of the arrhythmias in most experiments (SCHWARTZ and VANOLI 1981). Indeed, when a specific type of ventricular arrhythmia occurs in three consecutive trials the same arrhythmia can be reproducibly induced for seven to eight consecutive trials (Fig. 4). This is critical because the large interindividual variability makes it almost essential for studies on the efficacy of antiarrhythmic drugs to be performed with internal control analysis avoiding group comparisons.

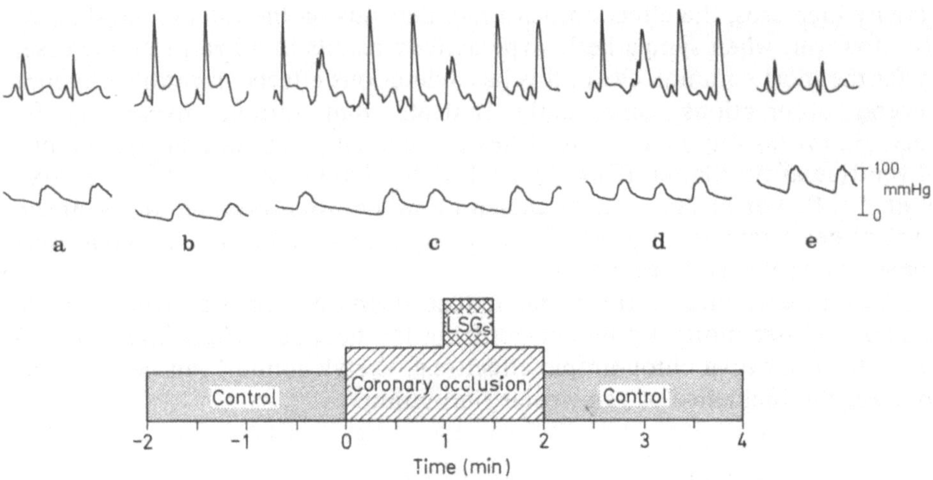

Fig. 3. Diagram of experimental model with example of ECG changes. Letters indicate different phases of occlusion stimulation protocol during which ECG was recorded. Ventricular arrhythmias appear during left stellate ganglion stimulation, they persist during the last part of occlusion, and ECG tracing returns to control baseline a few seconds after coronary artery occlusion release. LSG_s, left stellate ganglion stimulation. (SCHWARTZ and VANOLI 1981)

In this preparation we evaluated different drugs from each of the four classes of antiarrhythmic agents (SCHWARTZ et al. 1985a) as described by Vaughan Williams (VAUGHAN WILLIAMS 1970). The drugs used were: lidocaine (2 mg/kg + 4.3 mg/kg/per hour), mexiletine (3.6 mg/kg) and propafenone (2 mg/kg + 2 mg/kg/per hour), the beta-blocker propranolol (0.2 mg/kg); the class III agent amiodarone (1.5 or 3 mg/kg) and the calcium channel blockers nifedipine (0.02 mg/kg), verapamil (0.2 mg/kg) and diltiazem (0.1 mg/kg + 0.2 mg/per hour/kg). In addition two alpha-blockers were used: prazosin (0.1 mg/kg) and phenoxybenzamine (1 mg/kg). The overall results showed that lidocaine (Fig. 5), mexiletine (Fig. 6) and propafenone (Fig. 7) did not affect the incidence of ventricular fibrillation (VF) nor that of ventricular tachycardia (VT). Propranolol (79%; Fig. 8), prazosin (60%; Fig. 9) and phenoxybenzamine (55%; Fig. 10) significantly reduced the occurrence of both VF and VT; however, this protection, although good, was partial. Amiodarone (Fig. 11), as well as verapamil (Fig. 12) and diltiazem (Fig. 13), totally prevented the induction of VT and VF while nifedipine (Fig. 14) afforded protection in 10/13 cases (77%), thus presenting an antiarrhythmic effect comparable to that of antiadrenergic agents.

An interesting result was the complete failure of class I antiarrhythmic drugs. This finding might appear surprising since at least lidocaine is widely used to control arrhythmias in the coronary care units. However, lidocaine is generally administered hours after the beginning of the ischaemic episode, on

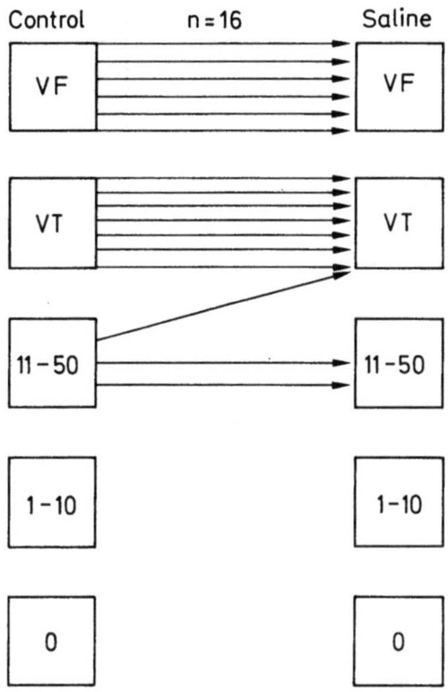

Fig. 4. Reproducibility of arrhythmias in the experimental model. *Each square* represents the average result of at least three trials in each cat. *Each arrow* represents one experiment. *Left*, squares show results observed during three coronary occlusions plus sympathetic stimulation performed in control condition, that is, before drug administration. *Right*, squares show results observed when three additional trials are performed after drug administration. It is evident that the injection of saline solution does not modify the response to coronary occlusion plus sympathetic stimulation. *VF*, ventricular fibrillation; *VT*, ventricular tachycardia. *Numbers in the squares* indicate number of premature ventricular contractions). (Schwartz et al. 1986)

a substrate that is different from that present during the very early phase of ischaemia. Indeed, in the clinical setting the administration of lidocaine very shortly after onset of symptoms has been unsuccessful (Adgey and Webb 1979). Mexiletine worsened the type of arrhythmias in some of our experiments; this finding fits with two clinical trials in which mexiletine given to post-MI patients did not decrease the incidence of sudden death in the treated compared with the placebo group; actually the mortality was slightly but not significantly higher in the treated groups (Chamberlain et al. 1980; Impact Research Group 1984). Our data suggest that class I antiarrhythmic agents, even if effective in treating chronic ventricular arrhythmias, may not be a rational choice for the prevention of sudden death in patients with ischaemic heart disease.

The antiadrenergic compounds showed a fairly good protective effect. The antiarrhythmic action of propranolol was not entirely dependent on the reduction of heart rate, because it was still present in most experiments when heart rate was kept at the pre-drug levels by cardiac pacing. When alpha-blockers did not afford protection, the consequent administration of propranolol totally

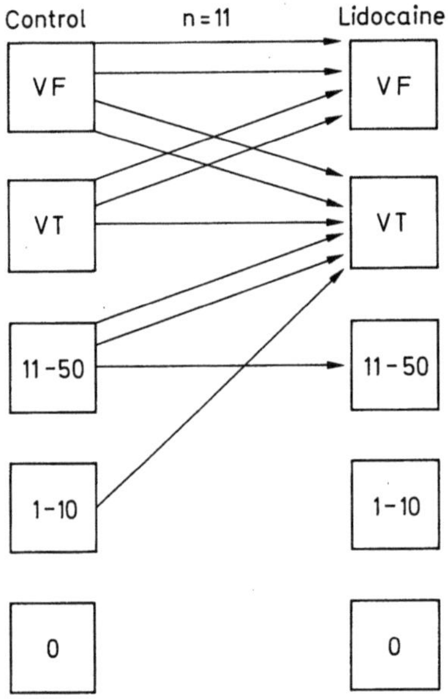

Fig. 5

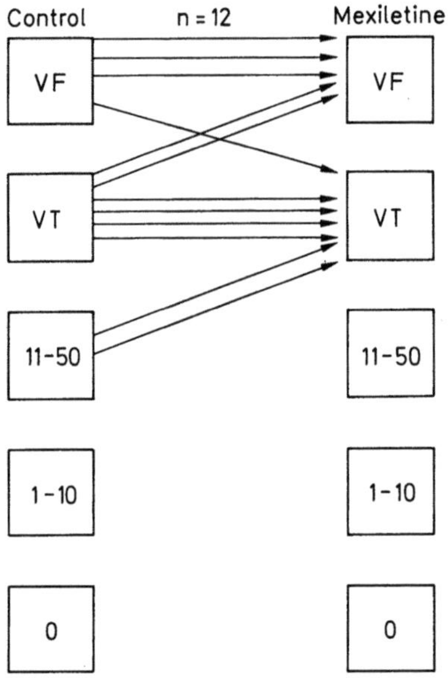

Fig. 6

Fig. 5. Effect of lidocaine in 11 animals. (SCHWARTZ et al. 1985a)
Fig. 6. Effect of mexiletine in 12 animals. (SCHWARTZ et al. 1985a)

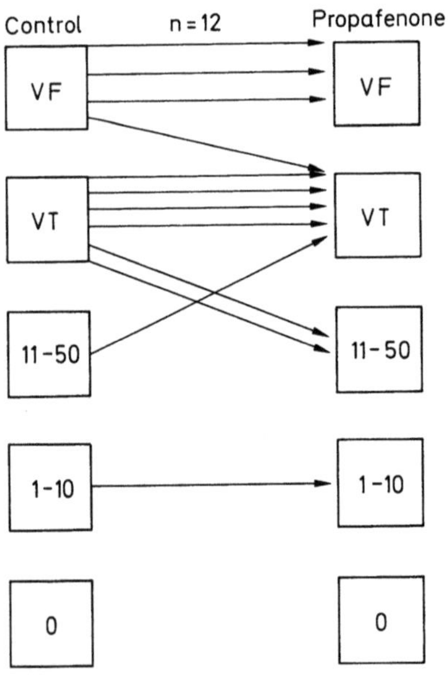

Fig. 7

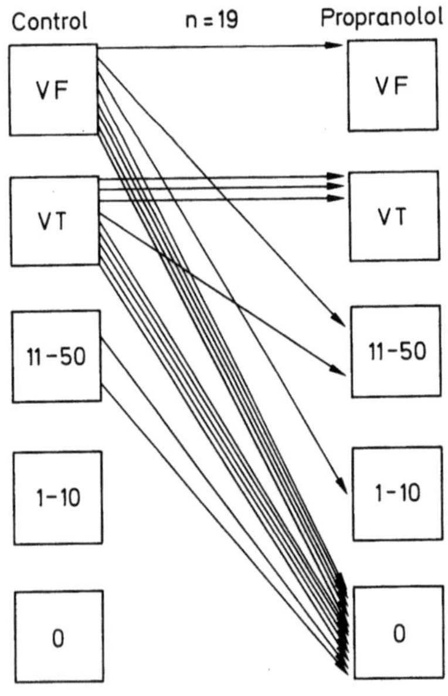

Fig. 8

Fig. 7. Effect of propafenone in 12 animals. (SCHWARTZ et al. 1985a)
Fig. 8. Effect of propranolol in 19 animals. (SCHWARTZ et al. 1985a)

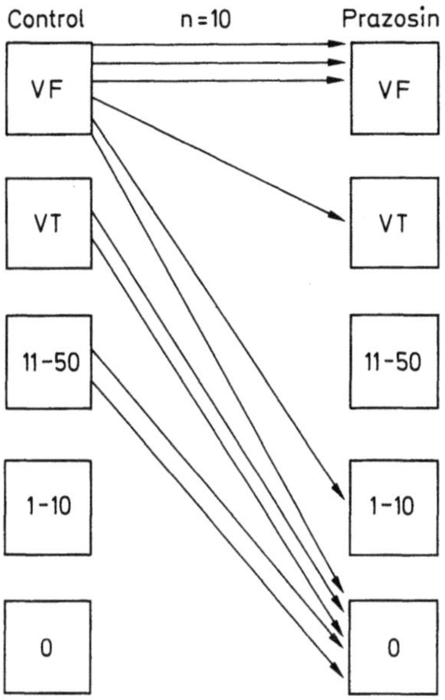

Fig. 9

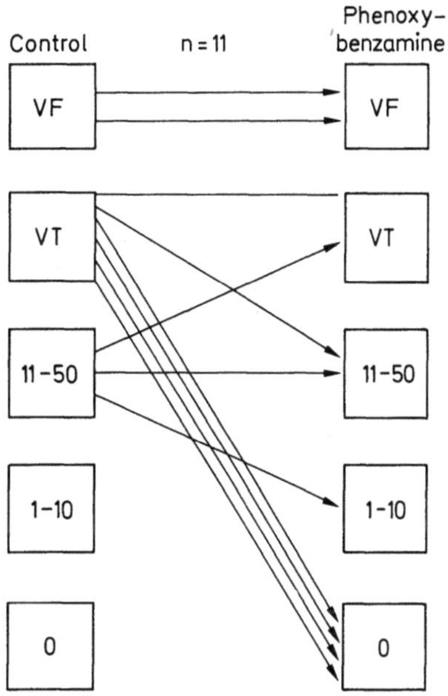

Fig. 10

Fig. 9. Effect of prazosin in ten animals. (SCHWARTZ et al. 1985a)
Fig. 10. Effect of phenoxybenzamine in 11 animals. (SCHWARTZ et al. 1985a)

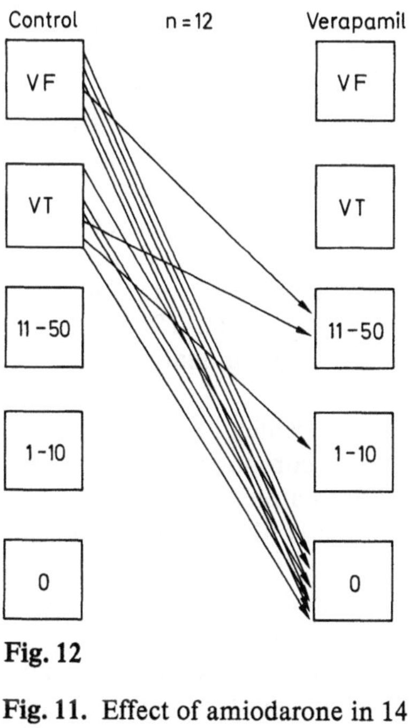

Fig. 12

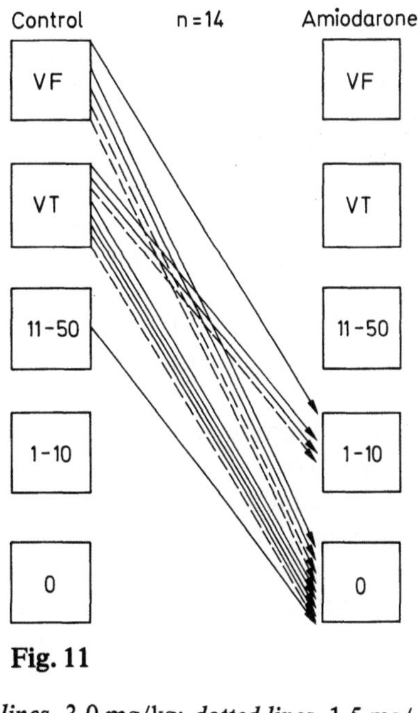

Fig. 11

Fig. 11. Effect of amiodarone in 14 animals. *Solid lines*, 3.0 mg/kg; *dotted lines*, 1.5 mg/kg. (SCHWARTZ et al. 1985a)
Fig. 12. Effect of verapamil in 12 animals. (SCHWARTZ et al. 1985a)

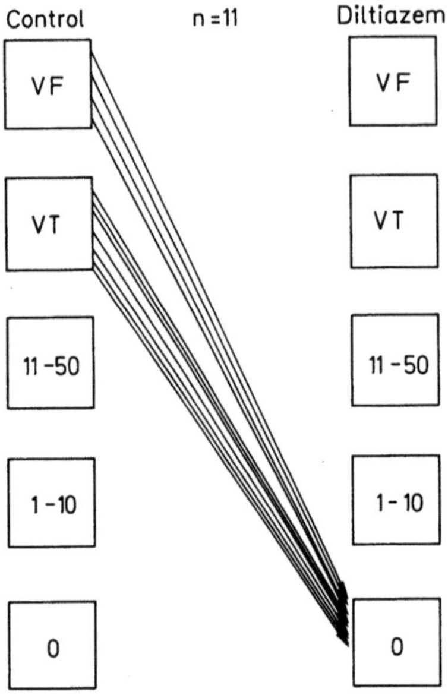

Fig. 13

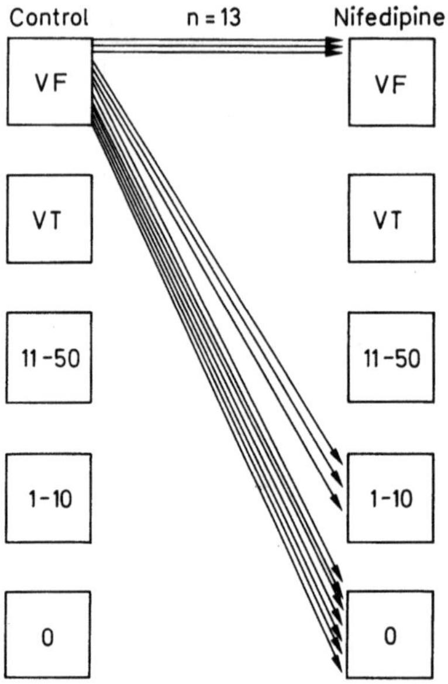

Fig. 14

Fig. 13. Effect of diltiazem in 11 animals. (Schwartz et al. 1986)
Fig. 14. Effect of nifedipine in 13 animals. (Priori et al. 1988b)

prevented arrhythmias. This finding lends further support to the concept previously demonstrated (Elharrar et al. 1979; Sheridan et al. 1980) that both alpha- and beta-adrenergic mechanisms contribute significantly to the arrhythmias associated with high sympathetic activity. The protective effect of beta-blockers in the prevention of sudden death has already been confirmed by most clinical studies in postinfarction patients (Hjalmarson 1984). Surgical denervation, of course, produces alpha- and beta-blocking effects; a selective cardiac surgical denervation such as that obtained by high thoracic left sympathectomy has been recently evaluated clinically on the basis of multiple and solid experimental evidence. In a group of postmyocardial infarction patients at high risk for sudden cardiac death (Schwartz et al. 1985b), high thoracic left sympathectomy reduced the incidence of sudden death from 22% to 3.5%.

Amiodarone was extremely effective. This drug, besides having the electrophysiological properties of class III antiarrhythmic compounds, is a coronary vasodilator and also counteracts the effects of both alpha- and beta-adrenergic stimulation: the combination of these properties is likely to have played a major role in its efficacy in this study.

Three calcium entry blockers have been evaluated in this preparation because they have rather different properties. Verapamil has been extensively investigated in models of ischaemia-induced arrhythmias, and it has been found to prevent the profibrillatory effect of both left stellate ganglion stimulation and catecholamine infusion (Verrier et al. 1983). Moreover, a direct antiadrenergic effect, possibly exerted by an action on alpha-receptors, has been postulated (KARLINERS et al. 1982). Less obvious is the mechanism underlying the protective effect of diltiazem (SCHWARTZ et al. 1986); its potential antiadrenergic activity has not been investigated, but its action on conduction delay (PETER et al. 1983) or the coronary vasodilating properties might explain the striking protective effect. The fact that nifedipine, a drug thought to act only on the ischaemic substrate, was also rather effective was somewhat surprising. It is important to note that nifedipine was able to prevent the further decrease in ventricular fibrillation threshold that occurs when the left stellate ganglion is stimulated during coronary artery occlusion: such an effect was not present during ischaemia alone or during stellate stimulation alone, suggesting that nifedipine acts largely by preventing the extension of the ischaemic area produced by the increase in coronary resistance induced by sympathetic stimulation (PRIORI et al. 1988b).

In summary, in this animal model the most effective drugs share a significant non-competitive antiadrenergic effect and a salutary action on coronary flow. Those drugs which specifically interact with either alpha- or beta-recep-

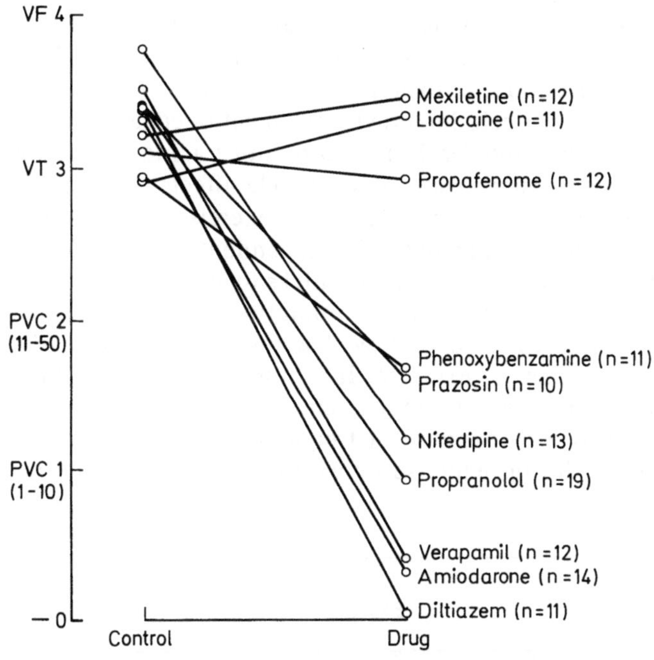

Fig. 15. Effect of antiarrhythmic drugs tested in the model of myocardial ischaemia and sympathetic stimulation ($n = 125$). Arbitrary grading assigned the following values for each type of arrhythmias: ventricular fibrillation, 4; ventricular tachycardia, 3; frequent (11-50) premature ventricular contractions, 2; sporadic (1-10) premature ventricular contractions, 1; no arrhythmias, 0. For each group of experiments the control score is obtained by averaging the results observed in all the animals during the three trials before drug administration. The score of the post drug trials is obtained by averaging the results observed in all the animals during the last three trials

tors as prazosin and propranolol afforded good but incomplete protection, while those which are devoid of both of the previously mentioned actions were unable to prevent the ventricular arrhythmias generated in our model (Fig. 15).

These data concur with the reduction in mortality rate in postmyocardial infarction patients obtained with beta-adrenergic-blocking agents; they may contribute to explain the failure of the clinical trials with class I antiarrhythmic drugs; they also suggest that other groups of drugs, such as calcium antagonists and amiodarone, may be considered for the prevention of sudden death in coronary artery disease patients, particularly in those with normal left ventricular function.

2. Myocardial Ischaemia, Exercise and a Healed Myocardial Infarction

A transient episode of acute myocardial ischaemia occurring in a heart with a healed myocardial infarction is an important cause of sudden death in man. This consideration, the important triggering role of sympathetic hyperactivity, and the need of overcoming limitations of anaesthetized open-chest preparations, have guided the development of the following experimental protocol.

One month after the production of an anterior myocardial infarction, dogs are engaged in an exercise stress test on a motor-driven treadmill. At the beginning of the last minute of exercise whenever heart rate is close to 210–220 beats/min, a balloon occluder previously positioned around the circumflex coronary artery is inflated for 2 min to produce acute myocardial ischaemia (Fig. 16). This short-lasting ischaemic episode affects the last minute of exercise and the first minute after cessation of exercise. This sequence of events allows separation between arrhythmias dependent on cessation of exercise or on release of occlusion. This model mimicks to some extent what may happen to a patient with a prior myocardial infarction who engages in exercise and has a brief reduction in coronary flow, leading to acute myocardial ischaemia, cardiac pain and arrest of exercise. It is important to note that such a brief coronary occlusion does not induce ventricular arrhythmias at rest whereas, when it is coupled with exercise, it frequently elicits malignant arrhythmias that occur either during exercise or within seconds from its cessation. The

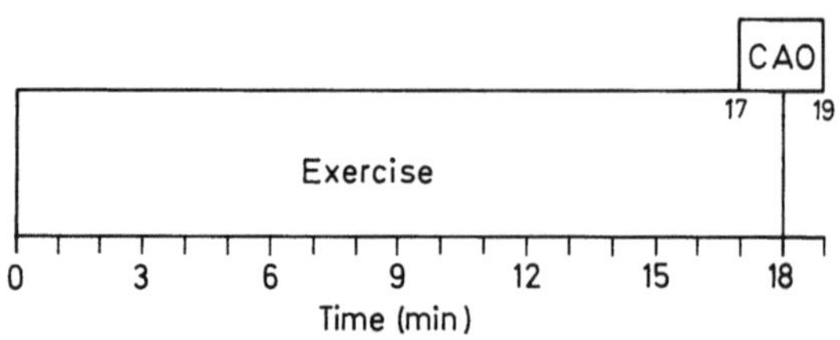

Fig. 16. Outline of the experimental protocol. At the beginning of the last minute of an exercise stress test, with progressively increasing workload on a treadmill, the circumflex coronary artery was occluded for 2 min. After 1 min of exercise plus ischaemia, the treadmill was stopped and myocardial ischaemia continued for the 2nd min after cessation of exercise. *CAO*, coronary arterial occlusion. (SCHWARTZ and STONE 1982)

dogs run on a treadmill with large steel plates placed across their chests so that electrical defibrillation can be performed within 10–20 s from the onset of ventricular fibrillation. When the lethal arrhythmia occurs during the exercise plus ischaemia test (susceptible animals), it will be reproducibly induced for up to 3–4 months; similarly if an animal does not develop ventricular fibrillation during this test (resistant), it will continue to survive whenever the test is repeated. The high reproducibility of either sudden death or survival in this model allows the evaluation of interventions with an internal control analysis.

The incidence of sudden death during the exercise and ischaemia test is high, as 90/161 dogs (60 %) developed VF under these conditions. Because of the complexity of the model and the time necessary for each study, only a limited number of potentially protective interventions have been evaluated so far. Left stellectomy provided excellent protection, as all 14 animals studied survived the test. This observation fits with the considerable protection afforded by high thoracic left sympathectomy in a group of post-MI patients as mentioned above.

The beta-blocker propranolol (1 mg/kg i.v.) was also effective in reducing the occurrence of ventricular fibrillation; in fact only 2/18 dogs, 11 %, fibrillated during the test but 2 additional animals died of pump failure during the exercise plus ischaemia protocol; therefore the overall mortality was 20 %. These data suggest that also in this animal model an interference with sympathetic activity, particularly when selective, as with left-sided denervation, is protective. However, they also suggest that when the heart is entirely deprived of its adrenergic support, the decreased risk of malignant arrhythmias is associated with increased risk of major left ventricular dysfunction. This becomes evident in the present animal model in which acute myocardial ischaemia is superimposed on a previous myocardial infarction and during an increased performance as required during exercise. It has already been reported that (Billman et al. 1985) the animals susceptible to ventricular fibrillation have a greater degree of left ventricular dysfunction compared with the resistant animals; and that this is not evident at rest but is unmasked by exercise and is exaggerated by the administration of beta-blockers.

Amiodarone was the other drug tested in this model. It was orally administered in 11 dogs: 50 mg/kg for 3 days and 35 mg/kg for 4 additional days. All the animals survived the exercise plus ischaemia test, thus confirming the protective effect showed in the anaesthetized preparation (Gagnol et al. 1983).

In summary, these data indicate that protection from sudden death can be reduced by combining an antiischaemic with an antiadrenergic effect. As to the latter it seems important that it does not significantly interfere with left ventricular function. Indeed, the extent of the underlying coronary disease is likely to affect the results and the survival provided by general or selective antiadrenergic therapy.

3. Behavioural Stress and Life-Threatening Arrhythmias

Behavioural stress increases the risk for ventricular fibrillation particularly in an ischaemic heart, and primarily through an activation of the sympathetic nervous system. This complex and important relationship, which also involves significant changes in coronary circulation (Billman and Randall 1981; Verrier et al. 1987), has been extensively studied and reviewed by Verrier (Verrier and Lown 1984; Verrier 1986) and colleagues.

In their initial studies, a simple classical aversive conditioning protocol was employed (Rabinowitz and Lown 1978). Dogs were exposed to two different environments: a cage in which the animal was left largely undisturbed, and a Pavlovian sling in which the animal received a single 5-J transthoracic shock at the end of each experimental period for three successive days. The two environments were compared on days 4 and 5. At these times, the dogs in the sling were restless, salivated excessively and exhibited somatic tremor, sinus tachycardia and increased mean arterial blood pressure. In the cage, as evidenced by behavioural signs and haemodynamic variables, the animals appeared relaxed. Transferring the animals from the non-aversive to the aversive environment resulted in a substantial reduction in vulnerable period threshold. These findings indicate that psychological stress profoundly lowers the cardiac threshold for ventricular fibrillation (Verrier and Lown 1984). Under these circumstances, three pharmacological approaches have been found protective. Beta-adrenergic blockade with propranolol or with the cardioselective agents tolamolol and metoprolol completely prevented the effects of aversive conditioning on the vulnerable period threshold (Verrier and Lown 1984). A reduction of cardiac sympathetic activity through a tryptophan-induced modulation of serotoninergic neurotransmission in the brain is also capable of increasing the threshold for repetitive extrasystole, a measure of vulnerability to ventricular fibrillation. Also an alternate approach, namely the enhancement of vagal tone by the administration of morphine, was protective (DeSilva et al. 1978). This effect, partly due to vagal antagonism of the profibrillatory influence of augmented adrenergic input to the heart and partly to the sedative action, is evident only in the stressful environment. Indeed, administration of morphine, in the non-stressful environment where the sympathetic activity is reduced, has no effect (Verrier 1986).

These observations were made in normal hearts. Studies exist on the efficacy of some pharmacological interventions when behavioural provocation of ventricular arrhythmias occurs during myocardial ischaemia. It has been reported (Skinner et al. 1975) that, whereas farm pigs adapted to a laboratory environment are well protected from ventricular fibrillation during acute myocardial ischaemia, beta-adrenergic blockade with propranolol provided no protection in unadapted animals. This surprising observation is at odds with most experimental and clinical observations. It could perhaps be explained by assuming an inadequate blockade of the adrenergic effects on the heart or by the unproven involvement of extra-adrenergic factors in the antifibrillatory effects of psychological adaptation (Verrier and Lown 1984).

By contrast, a protective effect of beta-adrenergic blockade against life-threatening ventricular arrhythmias associated with acute coronary artery occlusion in dogs exposed to a variety of behavioural stresses has been reported (ROSENFELD et al. 1978). Beta-blockade with the cardioselective agent tolamolol afforded a substantial protection. Importantly, the quaternary analogue of propranolol, UM 272, which is devoid of beta-adrenergic blocking properties, but exerts direct local anaesthetic effects on the heart, did not provide an antiarrhythmic protection. Thus, the beneficial effect of tolamolol appears to result from its antiadrenergic action rather than from a non-specific effect on myocardial tissue. In the same study an antiarrhythmic effect of the antianxiety drug diazepam was also demonstrated.

Thus, the majority of studies on stress-induced malignant arrhythmias indicate that considerable protection can be conferred by directly or indirectly interfering with the cardiac sympathetic activity.

C. Idiopathic Long Q-T Syndrome

The idiopathic long Q-T syndrome (LQTS) is a well-defined clinical entity (SCHWARTZ et al. 1975; MOSS and SCHWARTZ 1982; SCHWARTZ 1985). In its most typical form it is characterized by a prolongation of the Q-T interval ($Q\text{-}T_c > 440$ ms) and by the occurrence of syncopal episodes. These episodes, usually triggered by stress conditions associated with sympathetic hyperactivity such as physical exercise or fear or sudden noises, are due to "torsade de pointes", which often degenerates into ventricular fibrillation. Typically, the affected patients have several of these episodes and, eventually, if left without proper treatment, die suddenly. Indeed, for patients either untreated or treated without antiadrenergic intervention the mortality in the 1st year after the first syncopal episode is 22% (SCHWARTZ and LOCATI 1985), a dramatically high figure especially if one considers that the average age at first syncope is 16 ± 14 years. Other characteristic features of these patients, which involve a reduced heart rate and abnormalities of ventricular repolarization, have been described in detail elsewhere (SCHWARTZ 1985; SCHWARTZ and MALLIANI 1975; DEAMBROGGI et al. 1986) as well as the diagnostic criteria for doubtful cases (SCHWARTZ 1985). The most likely pathogenetic mechanism for LQTS is the one proposed in 1975 (SCHWARTZ et al. 1975) and referred to as "sympathetic imbalance" hypothesis. A congenital imbalance in the cardiac sympathetic innervation due to a primary lower-than-normal activity through the right-sided nerves would reflexly (SCHWARTZ 1984) result in a higher-than-normal activity of the very arrhythmogenic left-sided nerves. This type of imbalance lenghtens the Q-T interval and favours the onset of malignant arrhythmias (SCHWARTZ 1984). This mechanism is generally (BIGGER 1980; VAUGHAN WILLIAMS 1982) thought to be the one operating in LQTS; it seems fair to note that this hypothesis, although supported by many experimental and clinical observations, remains unproven. Indeed, the possibility of an intracardiac abnormality (SCHWARTZ 1986) (defective regulation of transmembrane potassium currents) which would decrease cardiac electrical stability and sensitize the heart to catecholamines has never been excluded. In this case, sympathetic activation would still act as a trigger for life-threatening arrhythmias. As to the underlying cellular electrophysiological basis for the arrhythmias of LQTS an intriguing possibility, supported by clinical (SCHECHTER et al. 1984) and experimental data (BRACHMANN et al. 1983; LEVINE et al. 1985), has been recently proposed by Lazzara and associates (JACKMAN et al. 1984). Torsade de pointes would result from afterdepolarizations that reaching the threshold produce repetitive excitation (triggered activity). Both early and delayed afterdepolarizations have been implicated. Early afterdepolarizations have been related to prolongation of Q-T interval induced by agents that lengthen repolarization and they are enhancend by long pauses; delayed afterdepolarizations are enhanced by increases in heart rate and correlate with the clinical characteristics of the idiopathic LQTS being elicited by adrenergic activation (JOHNSON et al. 1986; PRIORI and SCHWARTZ 1987). These delayed afterpolarizations appear to depend importantly on the calcium current. If this hypothesis is correct, beta-adrenergic blockade would be effective by reducing the slow inward calcium current as a consequence of blocking the calcium channel through a reduction in cyclic AMP (SCHECHTER et al. 1984). All of these hypotheses are practically important because they provide a rational basis for the therapy of LQTS, as described below. The efficacy of any given treatment requires an adequate knowledge of the natural history of a disease and of the long-term effects of the therapies employed. When dealing with an uncommon, if not rare, disease this becomes difficult (SCHWARTZ 1985). Based on the initial survey of over 200 LQTS patients (SCHWARTZ et al. 1975) and on the institution of the worldwide prospective registry (SCHWARTZ 1983; MOSS et al. 1985), data are now available on approximately 800 LQTS patients. The mortality in the symptomatic patients is extremely high (71%); it remains elevated when non-adrenergic therapies are used (35%) but it decreases dramatically when beta-blockers (6%) or left stellectomy (7%) are employed. It has to be noted that the surgical denervation has been reserved for those patients resistant to beta-blockers who clearly constitute a subgroup at even higher risk for sudden death. These figures suffer from the lack of a precise time relationship, even if it is fair to consider them valid for a period of between 5 and 10 years. How-

Table 1. Treatment and mortality in long Q-T syndrome ($n = 762$)

	Patients (n)	Mortality (n_0)
Asymptomatic	196	–
Unknown therapy	93	–
No therapy	157	71%
Miscellaneous therapy	50	35%
Beta-blockers	214	6%
Left Stellectomy	51	7%

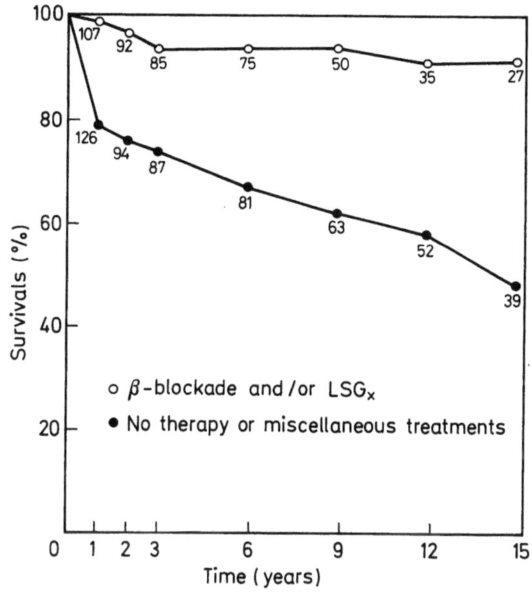

Fig. 17. Effect of therapy on the survival, after the first syncopal episode, of 233 patients affected by the idiopathic LQTS. The protective effect of beta-adrenergic blockade and of left stellectomy (*LSGx*) is dramatically evident. For example the mortality 3 years after the first syncope is 6% in the group treated with antiadrenergic interventions and is 26% in the group treated differently or not treated. Fifteen years after the first syncope the respective mortality is 9% and 53%. o, beta-blockade and/or LSGx; •, no therapy or miscellaneous treatments. (SCHWARTZ and LOCATI, 1985).

ever, precise figures exist for 233 LQTS patients for whom the data for the first syncopal episode, for the therapy and for follow-up information are available (SCHWARTZ and LOCATI 1985). It is evident that more than 20% of the patients untreated or treated without antiadrenergic intervention die within 1 year (Fig. 17). When this is compared with the survival obtained with either beta-blockade or high thoracic left sympathectomy (the human equivalent of left stellectomy), it is evident that these therapies have radically modified the prognosis for the symptomatic LQTS patients. There is no evidence that one beta-blocker agent is superior to others; however, the largest experience is with propranolol and this remains the drug of choice even if drugs with a longer half-life such as nadolol seem equally effective. For propranolol the recommended dosage is 3 mg/kg; higher doses have also been used. If syncope occurs again despite full-dose beta-blockade the therapy of choice becomes left stellectomy. Details for the technical aspects and of the surgical intervention have been described elsewhere (SCHWARTZ 1986). The use of other therapies in these high-risk patients has to be viewed as experimental and kept only for those patients (approximately 5% of the entire LQTS population with syncope) who continue to have syncopal episodes despite combined beta-blockade and left sympathectomy. Some success has been reported with propafenone; however, this drug has been mostly employed in so-called benign families and great caution is necessary before translating these observations to the typical LQTS cases with frequent episodes of malignant arrhythmias. Amiodarone has been considered of potential value with the idea of uniformly prolonged action potential duration, but the few LQTS patients who received this drug continued to have syncopal episodes and had to be treated more conventionally. On the other hand, there are no data so far on the potential value of Penticainide, an antiarrhythmic drug that shortens the Q-T interval particularly when pro longed (PRIORI et al. 1987; PALA et al. 1987). The case with calcium entry blockers seems to be different. If afterdepolarizations were indeed involved in the genesis of the ventricular tachyarrhythmias of LQTS, then drugs such as verapamil could be effective. Indeed, in a few patients resistant to beta-blockade verapamil, given prior to proceeding with left sympathectomy, was able to interrupt the frequent occurrence of malignent arrhythmia (SCHWARTZ and LOCATI unpublished). If confirmed, these preliminary observations would be highly relevant to the pathophysiology and the management of the long Q-T syndrome.

References

Adgey AAJ, Webb SW (1979) The treatment of ventricular arrhythmias in acute myocardial infarction. Br J. Hosp Med 21:356

Bigger JT Jr (1980) Mechanisms and diagnosis of arrhythmias. In: Braunwald E (ed) Heart disease. Saunders, Philadelphia, pp 691-743

Billman GE, Randall DC (1981) Mechanisms mediating the coronary vascular response to behavioral stress in the dog. Circ Res 48:214-223

Billman GE, Schwartz PJ, Stone HL (1982) Baroreceptor reflex control of heart rate. A predictor of sudden cardiac death. Circulation 66:874-880

Billman GE, Schwartz PJ, Gagnol JP, Stone HL (1985) The cardiac response to submaximal exercise in dogs susceptible to sudden cardiac death. J Appl Physiol 59:890-897

Brachmann J, Scherlag BJ, Rosenstrauk LV, Lazzara R (1983) Bradycardia-dependent triggered activity: relevance to drug induced multiform ventricular tachycardia. Circulation 68:846-856

Casati R, Lombardi F, Malliani A (1979) Afferent sympathetic unmyelinated fibers with left ventricular ending in cats. J Physiol (Lond) 292:135-148

Chadda KD, Bamka VS, Helfant RH (1974) Rate dependent ventricular ectopia following acute coronary occlusion. Circulation 49:654-658

Chamberlain DA, Jewitt DE, Julian DG, Campbell RWF, Boyle DMC, Shanks RG (1980) Oral mexiletine in high risk patients after myocardial infarction. Lancet 2:1324

Cranefield PF, Aronson RS (1974) Initiation of sustained rhythmic activity by single propagated action potentials in canine Purkinje fibers. Circ Res 34:477-481

Corr PB, Yamada KA, Witkowski FX (1986) Mechanisms controlling cardiac autonomic function and their relation to arrhythmogenesis. In: Fozzard HA, Haber E, Jennings RB, Katz AM, Morgan HE (eds) The heart and the cardiovascular system. Raven, New York, pp 1343-1404

De Ambroggi L, Bertoni T, Locati E, Stramba-Badiale M, Schwartz PJ (1986) Mapping of the body surface potential in patients with the idiopathic long Q-T syndrome. Circulation 74:1334-1345

De Ferrari AM, Vanoli E, Stramba-Badiale M, Foreman RD, Schwartz PJ (1987) Vagal stimulation and sudden death in conscious dogs with a healed myocardial infarction. Circulation 76 (IV):426

De Silva RA, Verrier RL, Lown B (1978) The effects of psychological stress and vagal stimulation with morphine on vulnerability to ventricular fibrillation (VF) in the conscious dog. Am Heart J 95:197-203

Dominguez G, Fozzard HA (1970) Influence of extracellular K^+ concentration on cable properties and excitability of sheep cardiac Purkinje fibers. Circ Res 26:565-574

Elharrar V, Watanabe AM, Molello J, Besch HR Jr, Zipes D (1979) Adrenergically mediated ventricular fibrillation in Probucol-treated dogs: roles of alpha and beta adrenergic receptors. Pace 2:435-443

Feigl EO (1967) Sympathetic control of coronary circulation. Circ Res 20:262-271

Gagnol JP, Schwartz PJ, Billman GE, Stone HL (1983) Hemodynamic response to exercise following oral administration of amiodarone. Fed Proc 42:1289

Han J, Moe JK (1964) Nonuniform recovery of excitability in ventricular muscle. Circ Res 14:44-60

Heusch G, Deussen A, Schipke J, Vogelsang H, Hoffmann V, Thamer V (1985) Role of the cardiac sympathetic nerves in the genesis of myocardial ischemia distal to coronary stenosis. J Cardiovasc Pharmacol 7:S13-18

Hjalmarson I (1984) Beta-blockers effectiveness in post infarction: an antiarrhythmic or antiischemic effect. Ann NY Acad Sci vol 427:101-111

Impact Research Group International Mexiletine and Placebo Antiarrhythmic Coronary Trial (1984) Report on arrhythmia, and other findings. J Am Coll Cardiol 4:1148-1163

Janse MJ, Van Capelle FG, Morsink H, Kleber AG, Wilms-Skopman F, Cardinal R, Naumann D'Alnoncourt C, Durrer D (1980) Flow of "injury" current and pattern of excitation during early ventricular arrhythmias in acute regional myocardial ischemia in isolated porcine and canine hearts. Circ Res 47:151-165

Janse MJ, Schwartz PJ, Wilms-Skopman F, Peters RJG, Durrer D (1985) Effects of unilateral stellate ganglion stimulation and ablation on electrophysiologic changes induced by acute myocardial ischemia in dogs. Circulation 72:585-595

Jackman WM, Clark M, Friday KJ, Aliot EM, Anderson J, Lazzara R (1984) Ventricular tachyarrhythmias in the long Q-T syndrome. Med Clin North Am 68:1079-1109

Johnson N, Danilo P, Wit AL, Rosen MR (1986) Characteristics of initiation and termination of catecholamine-induced triggered activity in atrial fibers of the coronary sinus. Circulation 74:1168-1179

Karliners JS, Motulsky HJ, Dunlyp J, Heller Brown J, Insel PA (1982) Verapamil competitively inhibits alpha adrenergic and muscarinic but not beta adrenergic receptors in rat myocardium. J Cardiovasc Pharmacol 4:515-520

Kimura S, Bassett AL, Kohyat-Kozlouskis PL, Myerburg RJ (1987) Automaticity, triggered activity and responses to adrenergic stimulation in cat subendocardial Purkinje fibers after healing of myocardial infarction. Circulation 75:651-660

Kralios FA, Martin L, Burgess MJ, Millar K (1975) Local ventricular repolarization changes due to sympathetic nerve-branch stimulation. Am J Physiol 228:1621-1626

La Rovere MT, Specchia G, Mortara A, Schwartz PJ (1988) Baroreflex sensitivity, clinical correlates and cardiovascular mortality among patients with a first myocardial infarction. Circulation 78: October

Levine, JH, Spear JF, Guarnieri T, Weisfeldt ML, De Langen CDJ, Becker LC, Moore EN (1985) Cesium chloride induced long Q-T syndrome: demonstration of afterdepolarization and triggered activity in vivo. Circulation 72:1092-1103

Lombardi F, Della Bella P, Casati R, Malliani A (1981) Effects of intracoronary administration of bradykinin on the impulse activity of afferent sympathetic unmyelinated fibers with left ventricular endings in the cat. Circ Res 48:69-75

Malliani A, Schwartz PJ, Zanchetti A (1969) A sympathetic reflex elicited by experimental coronary occlusion. Am J Physiol 217:703-709

Malliani A, Recordati G, Schwartz PJ (1973) Nervous activity of afferent cardiac sympathetic fibers with atrial and ventricular endings. J Physiol (Lond) 229:457-469

Matoba T, Toshima H, Nagae K, Yamazaki S (1979) Changes of ventricular monophasic action potential duration by stellate ganglion stimulation in dogs. Jpn Heart J 20:477-484

Mohrman DE, Feigl EO (1978) Competition between sympathetic vasoconstriction and metabolic vasodilation in the canine coronary circulation. Circ Res 42:79-86

Moss AJ, Schwartz PJ (1982) Delayed repolarization (Q-T or QTU prolongation) and malignant ventricular arrhythmias. Mod Concepts Cardiovasc Dis 51:85-90

Moss AJ, Schwartz PJ, Crampton RS, Locati E, Carleen E (1985) The long Q-T syndrome: a prospective international study. Circulation 71:17-21

Pagani M, Schwartz PJ, Banks R, Lombardi F, Malliani A (1974) Reflex responses of sympathetic preganglionic neurones initiated by different cardiovascular receptors in spinal animals. Brain Res 68:215-225

Pala M, Locati E, Priori SG, Munoz A, Schwartz PJ (1987) Q-T interval shortening by Penticainide in patients with ventricular arrhythmias with and without the long Q-T syndrome. Circulation 76 (IV):1650

Peon J. Ferrier GR, Moe JK (1978) The relationship of excitability to conduction velocity in canine Purkinje tissue. Circ Res 43:125-135

Peter T, Fujimoto T, Hamamoto H, Mandel WJ (1983) Comparative study of the effect of slow channel inhibiting agents on ischemia-induced conduction delay as relevant to the genesis of ventricular fibrillation. Am Heart J 106:1023-1028

Priori SG, Zuanetti G and Schwartz PJ (1988) Ventricular fibrillation induced by the interaction between acute myocardial ischemia and sympathetic hyperactivity: effect of Nifedipine. Am Heart J 116:37-43

Priori SG, Bonazzi O, Facchini M, Varisco T, Schwartz PJ (1987) Antiarrhythmic efficacy of Penticainide and comparison with Disopyramide, Flecainide, Propafenone and Mexiletine by acute oral drug testing. Am J Cardiol 60:1068-1072

Priori SG, Mantica M, Schwartz PJ (1988) Delayed afterdepolarizations elicited in vivo by left stellate ganglion stimulation. Circulation 78:178-185.

Priori SG, Schwartz PJ (1987) Pharmacologic and neurogenic induction of delayed afterdepolarizations in vivo, Pace 10:1033

Rabinowitz SH, Lown B (1978) Central neurochemical factors related to serotonic metabolism and cardiac ventricular vulnerability for repetitive electrical activity. Am J Cardiol 41:516-522

Raeder EA, Verrier RL, Lown B (1981) Influence of the autonomic nervous system on coronary blood flow during partial stenosis. Am Heart J 104:249-253

Recordati G, Schwartz PJ, Pagani M, Malliani A, Brown AM (1971) Activation of cardiac vagal receptors during myocardial ischemia. Experientia 27:1423-1424

Rinkema LE, Thomas JX Jr, Randall WC (1982) Regional coronary vasoconstriction in response to stimulation of stellate ganglia. Am J Physiol 243:410-415

Rosen MR, Moak JP, Damiano B (1984) The clinical relevance of afterdepolarizations. In: Greenberg HM, Kulbertus HE, Moss AJ, Schwartz PJ (eds) Clinical aspects of life-threatening arrhythmias. Ann N Y Acad Sci 427:84-93

Rosenfeld J, Rosen MR, Hoffman BF (1978) Pharmacologic and behavioral effects on arrhythmias that immediately follow abrupt coronary occlusion: a canine model of sudden coronary death. Am J Cardiol 41:1075-1082

Schechter E, Freeman CC, Lazzara R (1984) Afterdepolarizations as a mechanism for the long Q-T syndrome: electrophysiologic studies of a case. JACC 3:1556-1561

Schwartz PJ (1983) The idiopathic long Q-T syndrome: the need for a prospective registry. Eur Heart J 4:529-551

Schwartz PJ (1984) Sympathic imbalance and cardiac arrhythmias. In: Randall WC (ed) Nervous control of cardiovascular function. Oxford University Press, New York, pp 225-252

Schwartz PJ (1984) The rationale and the role of left stellectomy for the prevention of malignant arrhythmias. In: Greenberg HA, Kulbertus HE, Moss AJ, Schwartz PJ (eds) Clinical aspects of life-threatening arrhythmias. Ann NY Acad Sci 427:199-221

Schwartz PJ (1985) Idiopathic long Q-T syndrome: progress and questions. Am Heart J 109:399-411

Schwartz PJ (1986) Prevention of arrhythmias in the long Q-T syndrome. In: Kulbertus HE (ed) Medical management of cardiac arrhythmias, Churchill, Edinburgh, pp 153-162

Schwartz PJ, Locati E (1985) The idiopathic long Q-T syndrome: pathogenetic mechanisms and therapy. Eur Heart J 6 [Suppl D]:103-114

Schwartz PJ, Malliani A (1975) Electrical alternation of the T-wave clinical and experimental evidence of its relationship with the sympathetic nervous system and with the long Q-T syndrome. Am Heart J 89:45-50

Schwartz PJ, Stone HL (1977) Tonic influence of the sympathetic nervous system on myocardial reactive hyperemia and on coronary blood flow distribution. Circ Res 41:51-58

Schwartz PJ, Stone HL (1979) Effects of unilateral stellectomy upon cardiac performance during exercise in dogs. Circ Res 44:637-645

Schwartz PJ, Stone HL (1982) The role of autonomic nervous system in sudden coronary death. Ann NY Acad Sci 382:162-180

Schwartz PJ, Stone HL (1985) The analysis and modulation of autonomic reflexes in the prediction and prevention of sudden death. In: Zipes DP, Jalife J (eds) Cardiac electrophysiology and arrhythmias. Grune and Stratton, New York, pp 165-176

Schwartz PJ, Vanoli E (1981) Cardiac arrhythmias elicited by interaction between acute myocardial ischemia and sympathetic hyperactivity: a new experimental model for the study of antiarrhythmic drugs. J Cardiovasc Pharmacol 3:1251-1259

Schwartz PJ, Zaza A (1986) The rational basis and the clinical value of selective cardiac sympathic denervation in the prevention of malignant arrhythmias. Eur Heart J 7 (Suppl. A):107-118

Schwartz PJ, Pagani M, Lombardi F, Malliani A, Brown AM (1973) A cardiocardiac sympatho-vagal reflex in the cat. Circ Res 32:215-220

Schwartz PJ, Periti M, Malliani A (1975) The long Q-T syndrome. Am Heart J 89:378-390

Schwartz PJ, Foreman RD, Stone HL, Brown AM (1976a) Effect of dorsal root section on the arrhythmias associated with coronary occlusion. Am J Physiol 231:923-928

Schwartz PJ, Snebold NG, Brown AM (1976b) Effects of unilateral cardiac sympathetic denervation on the ventricular fibrillation threshold. Am J Cardiol 37:1034-1040

Schwartz PJ, Verrier RL, Lown B (1977) Effect of stellectomy and vagotomy on ventricular refractoriness in dogs. Circ Res 40:536-540
Schwartz PJ, Billman GE, Stone HL (1984) Autonomic mechanisms in ventricular fibrillation induced by myocardial ischemia during exercise in dogs with healed myocardial infarction. Circulation 69:790-800
Schwartz PJ, Vanoli E, Zaza A, Zuanetti G (1985a) The effect of antiarrhythmic drugs on life-threatening arrhythmias induced by the interaction between acute myocardial ischemia and sympathetic hyperactivity. Am Heart J 109:937-948
Schwartz PJ, Motolese M, Pollavini G, Malliani A, Bartorelli C, Zanchetti A and the Sudden Death Italian Prevention Group (1985b) Surgical and pharmacological antiadrenergic interventions in the prevention of sudden death after a first myocardial infarction. Circulation 72:III-358
Schwartz PJ, Priori SG, Vanoli E, Zaza A, Zuanetti G (1986) Efficacy of Diltazem in two feline experimental models of sudden cardiac death. J Am Coll Cardiol 8:661-668
Schwartz PJ, Vanoli E, Stramba-Badiale M, DeFerrari GM, Billman Ge, Foreman RD (1988) Autonomic mechanisms and sudden death. New insights from analysis of baroreceptor reflexes in conscious dogs with and without a myocardial infarction. Circulation 78: October
Sheridan DG, Penkoske PA, Sobel BE, Corr PB (1980) Alpha adrenergic contributions to dysrhythmia during myocardial ischemia and reperfusion in cats. J Clin Invest 65:161-171
Skinner JL, Lie MI (1975) Modification of ventricular fibrillation latency following coronary artery occlusion in the conscious pig. The effects of psychological stress and beta-adrenergic blockade. Circulation 51:656-667
Thames MD, Klopfenstein HS, Abboud FM, Mark AL, Walker JL (1978) Preferential distribution of inhibitory cardiac receptors with vagal afferents to the inferoposterior wall of the left ventricle activated during coronary occlusion in the dog. Circ Res 43:512-519
Thoren P (1978) Vagal reflexes elicited by left ventricular C-fibers during myocardial ischemia in cats. In: Schwartz PJ, Brown AM, Malliani A, Zanchetti A (eds) Neural mechanisms in cardiac arrhythmias. Raven, New York, p 179-186
Vanoli E, Zaza A, Zuanetti G, Stramba-Badiale M, Cerati D, Schwartz PJ (1985) Unilateral stellectomy and infarct size. Eur Heart J 6:118
Vaughan Williams EM (1970) Classification of antiarrhythmic drugs. In: Sandoe E, Flensted-Jensen E, Olesen KH (eds) Symposium on cardiac arrhythmias. Astra Sodertalje, p 449
Vaughan Williams EM (1982) Q-T and action potential duration. Br Heart J 47:513-514
Verrier RL, Thompson PL, Lown B (1974) Ventricular vulnerability during sympathetic stimulation: role of heart rate and blood pressure. Cardiovasc Res 8:602-610
Verrier RL (1986) Neurochemical approaches to the prevention of ventricular fibrillation. Fed Proc 45:2191-2196
Verrier RL, Lown B (1981) Autonomic nervous system and malignant cardiac arrhythmias. In: Wiener H, Hofer MA, Stunkard AJ (eds) Brain behaviour and bodily disease. Raven, New York, p 273
Verrier RL, Lown B (1984) Behavioral stress and cardiac arrhythmias. Annu Rev Physiol 46:155-176
Verrier RL, Raeder EA, Lown B (1983) Use of calcium channel blockers after myocardial infarction. Potential cardioprotective mechanisms. In: Kulbertus HE, Wellens HJJ (eds) The first year after a myocardial infarction. Futura, New York, p 341

Verrier RL, Hagestad EL, Lown B (1987) Delayed myocardial ischemia induced by anger. Circ 75:249-254

Vlahakes GJ, Baer RW, Uhlig PN, Verrier ED, Bristow JD, Hoffman JIE (1982) Adrenergic influence in the coronary circulation of conscious dogs during maximal vasodilation with adenosine. Circ Res 51:371-384

Wit AL, Rosen MR (1986) Afterdepolarizations and triggered activity. In Fozzard HA, Haben E, Jennings RB, Morgan HE, Katz AM (eds) The heart and the cardiovascular system. Scientific foundations. Raven, New York, pp 1449–1490

Zuanetti G, De Ferrari GM, Priori SG, Schwartz PJ (1987) Protective effect of vagal stimulation on reperfusion arrhythmias in cats. Circ Res 61:429-435

CHAPTER 24a

Effects of Cardiac Glycosides at the Cellular Level

S. HERZIG and H. LÜLLMANN

A. Introduction

The therapeutic use of cardiac glycosides is based on their beneficial effects on two kinds of cardiac disease, namely congestive heart failure and certain cardiac rhythm disturbances. On the other hand, cardiac arrhythmias are frequent, and occasionally become serious side effects of cardiac glycoside administration. Thus, they contribute to the narrow therapeutic range, which necessitates cautious handling and close monitoring of this drug therapy. In spite of an extremely large body of knowledge about the effects of cardiac glycosides on biochemical preparations, isolated organs, intact animals and healthy as well as diseased humans, the mechanisms responsible for the observed effects are still incompletely understood.

Although there is now little, if any, doubt that the target of cardiac glycoside action is the membrane-bound enzyme (Na^++K^+)-activated, Mg^{2+}-dependent adenosine triphosphatase ((Na^++K^+)-ATPase) (SCHATZMANN 1953; SKOU 1957), the complex biochemical regulation and the multiplicity of physiological functions of this enzyme make it very difficult to interpret experiments with its specific inhibitors, the cardiac glycosides.

Within the space of this section, only a few aspects of cardiac glycoside action can be discussed, which are considered to be important for the arrhythmogenic or for the antiarrhythmic properties of this class of drugs. For more thorough information on cardiac glycosides and (Na^++K^+)-ATPase function, the reader is referred to another volume of this handbook series (vol. 56), and to a number of review articles (LENDLE 1935; LEE and KLAUS 1971; THOMAS 1972; GLYNN and KARLISH 1975; SCHWARTZ et al. 1975; AKERA and BRODY 1978; LÜLLMANN and PETERS 1979; GILLIS and QUEST 1980; AKERA and BRODY 1982; GLITSCH 1982; POWIS 1983; HANSEN 1984; KAPLAN 1985).

B. Effects of Cardiac Glycosides on Cellular Electrolytes

Since the maintenance of transmembrane ionic gradients is an important function of (Na^++K^+)-ATPase, and since these gradients represent the driving force of electrophysiological events, changes in cellular electrolyte levels are likely to represent the causal link between the primary site of action of cardiac glycosides and their effects on the electrical properties of excitable tissues. Hence, findings about the influence of cardiac glycosides on the most important ions Na^+, K^+ and Ca^{2+} are discussed in this section. Attention will be paid to the concentration dependency of these effects.

I. Effects of Cardiac Glycosides on Cytosolic Sodium

In many recent studies performed with ion-sensitive microelectrodes on isolated cardiac preparations, it has been demonstrated that cardiac glycosides, in positive inotropic concentrations, elevate the cytosolic sodium ion activity (a^i_{Na}) (VASSALLE and LEE 1984; ABETE and VASSALLE 1985; LEE et al. 1985). The observed increase is concentration dependent (ELLIS 1977; DEITMER and ELLIS 1978; LEE et al. 1980; LEE and DAGOSTINO 1982), and a new stable value is obtained, which lies only a few millimoles above the control level. The time course has been reported to correspond to (LEE and DAGOSTINO 1982; WASSERSTROM et al. 1983), or to dissociate from (BOYETT and LEVY 1985), the time course of development of positive inotropy. The elevation of a^i_{Na} seems to be too small for a significant alteration of the sodium electrochemical gradient, but due to the very steep interrelationship between a^i_{Na} and the activity of the sodium pump (EISNER et al. 1981) the increase in a^i_{Na} is sufficient to stimulate the uninhibited fraction of sodium pumps to an extent which is necessary for maintaining the sodium homeostasis (AKERA and BRODY 1982; EISNER et al. 1983; KIM et al. 1984; HERZIG et al. 1985b). From a study on rat ventricular myocardium, it becomes particularly clear that the increment in a^i_{Na} is not a sign of toxicity: due to the presence of two populations of binding sites with markedly different affinities towards ouabain in this organ (ADAMS et al. 1982; FINET et al. 1982; HERZIG and MOHR 1984), the threshold ouabain concentration for positive inotropy is about 1000-fold lower than the lowest toxic concentration. Even in the non-toxic concentration range, an elevation of a^i_{Na} has been observed in rat ventricle (GRUPP et al. 1985). In conclusion, a slight increase of the free cytosolic sodium ion concentration is well documented with non-toxic concentrations of cardiac glycosides. It has been speculated to be of significance for the positive inotropic effect (GLITSCH et al. 1970; DAUT 1982; EISNER et al. 1984; LEE et al. 1985), but it is less likely to be a primary reason for changes in electrophysiological parameters.

II. Effects of Cardiac Glycosides on Cellular Potassium

Due to the coupled transport function of $(Na^+ + K^+)$-ATPase, a decrease in the cellular potassium ion concentration should result from a partial inhibition of sodium pumps by cardiac glycosides. This decrease would be negligible compared with the high intracellular concentration of this ion, and it would be difficult to be observed, if it amounted to the same order of magnitude (a few millimoles) as does the increment in a^i_{Na}. Indeed, many studies on cellular potassium content (KLAUS et al. 1962; see LEE and KLAUS 1971) failed to show a significant alteration caused by non-toxic concentrations of cardiac glycosides. Ion-sensitive microelectrode measurements gave positive results only when toxic concentrations were employed (MIURA and ROSEN 1978). Radioisotope investigations using ^{42}K or the bioequivalent ^{86}Rb revealed that the uphill ion transport is concentration dependently impaired (TUTTLE et al. 1962; HOUGEN and SMITH 1978). This impairment has been found to be related to higher glycoside concentrations (KLAUS et al. 1962; TUTTLE et al. 1962; BROWN et al. 1983; LECHAT et al. 1983), which likewise cause toxicity. Thus, under non-toxic conditions, cardiac glycosides affect neither the cellular potassium concentrations nor its overall transmembrane movements to a measurable degree. On the other hand, changes of action potential duration (KLINE and KUPERSMITH 1982) or resting membrane potential (MIURA and ROSEN 1978) have been attributed to an increase in the potassium concentration in a diffusion-restricted area of the extracellular space $[K^+]_{cleft}$ (BAUMGARTEN and ISENBERG 1977). This area is close to the outer surface of the cell membrane, and it could be represented by the T-tubular system. Experimental evidence for such an increment induced by cardiac glycosides has been obtained using K^+ -sensitive electrodes (KUNZE 1977) placed at the intimate surface of the myocardial cells. A sustained, evenly distributed elevation of $[K^+]_{cleft}$, however, is only possible if a continuous loss of potassium takes place from the intracellular compartment to the extracellular space. A significant potassium loss, however, can be demonstrated only with toxic cardiac glycoside concentrations (see above). Yet, it cannot be excluded that in spite of a lack of clear effects on cellular potassium some non-toxic actions of cardiac glycosides on electrical properties of the cells might emerge from slight changes in the transmembrane potassium gradients.

III. Effects of Cardiac Glycosides on Cellular Calcium Transients

The inotropic effect of cardiac glycosides is not associated with changes of the total cellular calcium levels (KLAUS et al. 1962; CARRIER et al. 1974; KASPAREK et al. 1981). Studies with ^{45}Ca have shown that the rate of exchange of calcium is enhanced by non-toxic concentrations without alterations of the total cellular calcium content, suggesting an increased releasability of calcium from cellular stores involved in excitation-contraction coupling (CARRIER et al. 1974; KASPAREK et al. 1981). By means of calcium-sensitive microelectrodes (SHEU and FOZZARD 1982), or of calcium-sensitive dyes like the photoprotein aequorin (ALLEN and BLINKS 1978; WIER and HESS 1984; MORGAN 1985), a marked increase in the mean cytosolic free $[Ca^{2+}]$, or in the cellular calcium-transients, respectively, has been reported to occur after addition of non-toxic cardiac glycoside concentrations. The influence of an elevated cytosolic Ca^{2+}-concentration on the electrical properties of cardiac preparations may be quite complex under non-toxic and toxic conditions and will be discussed in Sect. C.

IV. Toxic Effects of Cardiac Glycosides on Cellular Electrolyte Contents

In contrast to the small alterations of ionic balance caused by moderate concentrations of cardiac glycosides (see above), the toxic effects of these drugs are clearly apparent as an impairment of the transmembrane gradients of Na^+

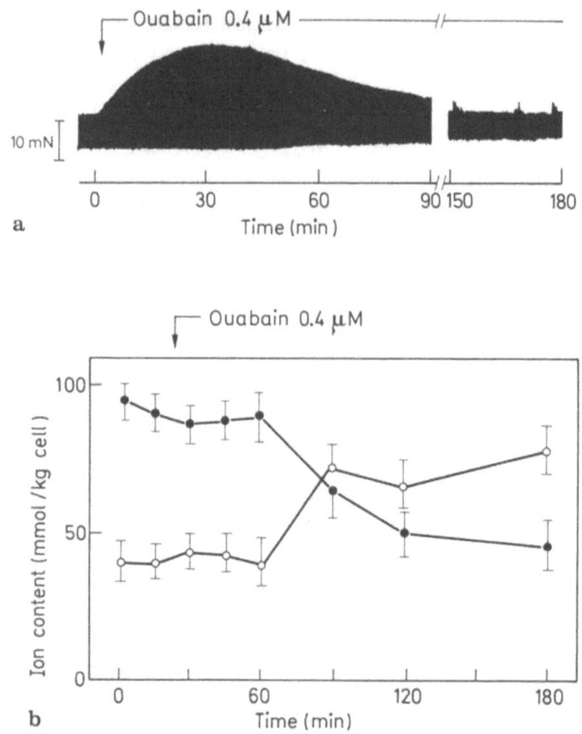

Fig. 1. Effect of a toxic concentration of ouabain (0.4 μ*M*) on isolated, electrically stimulated (2-Hz) guinea pig auricles. **a** A typical mechanogram: after a transient maximum of contractile force, negative inotropy, an increase in diastolic tension and extrasystoles take place; **b** The time course of the ion contents under the above-mentioned conditions. Cellular K^+ (*filled circles*) and Na^+ (*open circles*) were determined by means of flame photometry ($n \geqq 8$; redrawn from BENTFELD et al. 1977)

and K^+ (Fig. 1). This abolition of ion homeostasis, occurring when the positive inotropy has vanished, is the consequence of a continuous imbalance between passive ion fluxes and active transport carried by $(Na^+ + K^+)$-ATPase. It is widely accepted that the major alterations of ion concentrations are the molecular basis of digitalis toxicity (VICK and KAHN 1957; TUTTLE et al. 1961; KLAUS et al. 1962; see also ARONSON and GELLES 1977; NAYLER and NOACK 1981; WEINGART 1981; Sect. C). On the other hand, the small changes found in the positive inotropic concentration range have led to many suggestions concerning a mode of action unrelated to $(Na^+ + K^+)$-ATPase inhibition (HUANG et al. 1979; ERDMANN et al. 1980; NOBLE 1980). More likely, it is the ability of $(Na^+ + K^+)$-ATPases to compensate for a certain degree of inhibition, because of the large reserve capacity of this enzyme, which leads to the observation that ionic gradients are essentially unaltered under non-toxic conditions (BENTFELD et al. 1977; LÜLLMANN and PETERS 1979; AKERA and BRODY 1982).

C. Direct Effects of Cardiac Glycosides on Cardiac Electrical Properties

I. Contribution of the Electrogenic Sodium Pump to Cardiac Electrophysiological Properties

The main function of the sarcolemmal ($Na^+ + K^+$)-ATPase is to compensate for the passive transmembrane movements of Na^+ and K^+, thereby providing the basis for the excitability of cells. Besides this, the observation of a transport ratio near to 3:2 for Na^+ versus K^+ of ($Na^+ + K^+$)-ATPase of erythrocytes (SEN and POST 1964; GARRAHAN and GLYNN 1967) suggests that a net charge movement is also carried by the pump. It is by now widely accepted that the Na^+, K^+ pump is electrogenic also in cardiac tissue (DELEZE 1960; ISENBERG and TRAUTWEIN 1974), but since the pump stoichiometry is not exactly known (THOMAS 1972; GLITSCH 1982), and might be variable in this tissue (AKERA et al. 1981), the relationship between ion transport and outward current generated by the Na^+, K^+ pump remains to be elucidated. Nevertheless, this outward current is believed to contribute to the level of resting membrane potential by a few millivolts (HORRES et al. 1979; GLITSCH 1982); it should tend to hyperpolarize the cells, thereby delaying slow diastolic depolarization (VASSALLE 1970; NOMA and IRISAWA 1974), and shortening the action potential duration (PAGE and STORM 1965; GADSBY and CRANEFIELD 1979). The degree of pump current activation during the course of an action potential might differ in two ways from the degree present during the interval. On the one hand, an activation due to a transiently elevated Na^+ concentration close to the inner leaflet of the sarcolemma (AKERA et al. 1976) or due to the potential dependence of the pump (GADSBY et al. 1985; HASUO and KOKETSU 1985) might occur. On the other hand, the elevated $[Ca^{2+}]_i$ during systole could inhibit the pump (SCHWARTZ et al. 1975; LÜLLMANN and PETERS 1979; AKERA and BRODY 1982), presumably by interference with binding of Mg^{2+} and Na^+ to the enzyme (TOBIN et al. 1973). A further, indirect influence of Na^+, K^+-pump activity upon cardiac electrical properties might reside in its ability to maintain the transmembrane Na^+ gradient. This gradient could contribute to the maintenance of the Ca^{2+} gradient, which has been speculated to be controlled in part by means of sarcolemmal sodium/calcium exchange (see MULLINS 1981). Thus, impairment of Na^+, K^+-pump function might lead to alterations of electrophysiological parameters, which are mediated by an elevated systolic $[Ca^{2+}]_i$. Indeed, some effects of cardiac glycosides on cardiac electrical events resemble those of an elevated $[Ca^{2+}]_0$ (LEE and KLAUS 1971). In conclusion, the long-term effect of Na^+, K^+-pump activity on cellular electrolyte gradients provides the chemical and electrical basis for cardiac cells to maintain their resting membrane potential and to depolarize quickly upon an appropriate electrical stimulus. During a cardiac cycle, short-term variations of Na^+, K^+-pump activity may alter the time course of repolarization and, to a small extent of a few millivolts, the level of resting membrane potential. Additionally, by the linkage of cellular calcium metabolism to the function of the Na^+, K^+

pump, alterations of pump activity might indirectly cause changes of electrophysiological parameters.

II. Effects of Cardiac Glycosides on Resting Membrane Potential, Diastolic Depolarization and Action Potential Configuration

The effect of a non-toxic concentration of ouabain on contractile force, resting membrane potential (RMP) and action potential duration (ADP) of guinea pig papillary muscles is depicted in Fig. 2. After addition of the drug, there is a short-lasting small prolongation of APD, until a slowly developing decrease of APD occurs at all levels of repolarization. Resting membrane potential and action potential amplitude (not depicted) remain unchanged. A si-

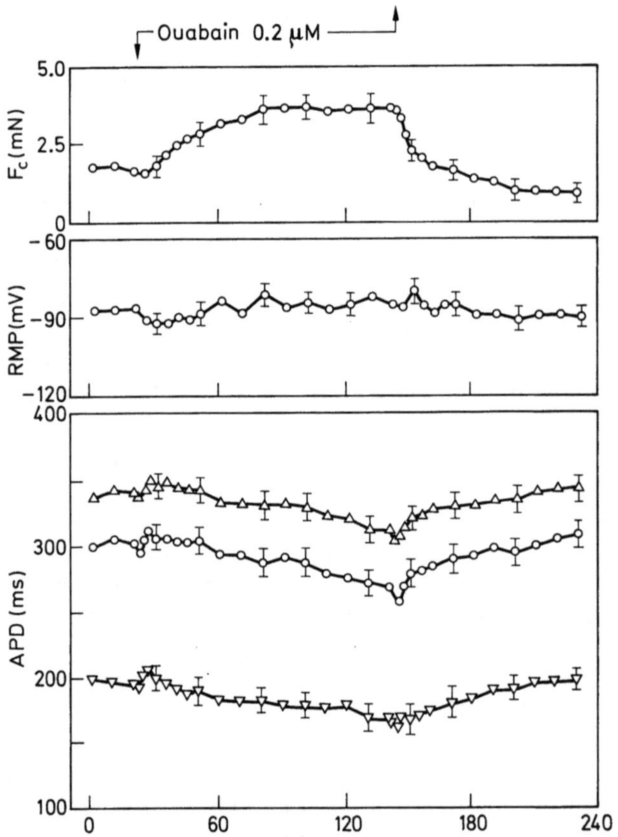

Fig. 2. Effect of a non-toxic ouabain concentration (0.2 μ*M*) on contractile force (*upper panel, circles*), resting membrane potential (*RMP, middle panel, circles*) and action potential duration (*ADP, lower panel*) at 20 % (*inverted triangles*), 50 % (*circles*) and 90 % of repolarization (*upright triangles*) of isolated, electrically stimulated (0.5-Hz) guinea pig papillary muscles ($\bar{x} \pm s_{\bar{x}}$, $n = 6 - 9$). The effects were reversible upon washout. (Data kindly supplied by H.-M. BRINKMANN and U. RAVENS 1976)

milar biphasic effect on the duration of action potentials in the presence of cardiac glycosides has often been reported in working myocardium (DUDEL and TRAUTWEIN 1958; KASSEBAUM 1963) as well as in Purkinje fibres (KASSEBAUM 1963; ISENBERG und TRAUTWEIN 1974). The transient initial prolongation has been interpreted as a sign of decreased repolarizing Na^+, K^+-pump current during phase 2 and 3 of the action potential (ISENBERG and TRAUTWEIN 1974). A temporary net inhibition of ($Na^+ + K^+$)-ATPase is what has to be expected even in the presence of non-toxic cardiac glycoside concentrations, because the temporary imbalance between sodium influx and active outward transport should lead to an increase of $[Na^+]_i$, which increases the activity of uninhibited enzyme molecules, until the overall ($Na^+ + K^+$)-ATPase activity again counterbalances the passive movements. The slowly developing, permanent shortening has been speculated to result from a $[Ca^{2+}]_i$-induced increase of the K^+-conductance (ISENBERG 1975; BASSINGTHWAIGHTE et al. 1976), from a direct effect of cardiac glycosides on K^+-permeability (LÜLLMANN and RAVENS 1973), from an increase in K^+-permeability due to an elevated $[K^+]_{cleft}$ (HOFFMAN and BIGGER 1985) or from an activation of a $[Na^+]_i$-activated K^+-channel (KAMEYAMA et al. 1984).

Toxic concentrations of cardiac glycosides cause a diversity of effects on electrophysiological parameters in isolated cardiac preparations. Besides the marked reduction of action potential duration, a decrease in resting membrane potential (DUDEL and TRAUTWEIN 1958; ROSEN et al. 1973) and in upstroke velocity V_{max} (KASSEBAUM 1963; ROSEN et al. 1973) have been consistently observed. These effects can be attributed to the impairment of transmembrane Na^+ and K^+ gradients and, probably less importantly, to the inhibition of the outward current carried by the Na^+, K^+ pump. In preparations of the impulse-generating and impulse-conducting system, which reveal a slow diastolic depolarization during phase 4, an increase in depolarization rate and a reduction of the maximum diastolic potential have been found (VASSALLE et al. 1962; HOGAN et al. 1973). An additional phenomenon, which can be elicited by toxic concentrations of cardiac glycosides, is the occurrence of delayed afterdepolarization during phase 4 (FERRIER et al. 1973). These afterdepolarizations, also called oscillatory afterpotentials, are generated by a transient inward current (LEDERER and TSIEN 1976) and can initiate propagated action potentials. They are accompanied by oscillatory changes in tension (CANNELL and LEDERER 1986) and $[Ca^{2+}]_i$ (EISNER and VALDEOLMILLOS 1986), can be augmented by an elevated $[Ca^{2+}]_0$ (FERRIER and MOE 1973; KASS et al. 1978) and diminished by agents reducing calcium-overload of the cells (FERRIER and MOE 1973; ROSEN et al. 1974; KOJIMA and SPERELAKIS 1986). Calcium liberation from membranous stores may be of significance in the genesis of these afterdepolarizations and concomitant aftercontractions (KASS et al. 1978), which are considered to be a sign of intracellular calcium-overload (REITER 1962). Thus, the toxic manifestations of cardiac glycosides on cardiac electrophysiological properties are largely compatible with their deleterious effects on Na, K homeostasis. Cellular calcium-overload may play a significant additional role. In contrast, the minor changes observed with non-toxic concentrations do not present conclusive evidence for a net inhibition of

Na, K transport. Compartmentalized or transient changes of ion concentrations, particularly the well-known increase of intracellular systolic Ca^{2+} transients, may account for the non-toxic effects on action potential configuration.

III. Possible Mechanisms of Ectopic Activity Elicited by Cardiac Glycosides

The arrhythmogenic action of toxic concentrations of cardiac glycosides can be seen in isolated cardiac preparations from various origins as well as in intact animals or humans. The underlying ectopic activity may have several reasons (VASSALLE et al. 1963; ROSEN et al. 1975; WEINGART 1981). The prerequisites for intracardial circus movement of excitation, slowed conductivity and shortened refractoriness can both be present in working myocardium as well as in the His-Purkinje system. In addition to the decrease of the resting membrane potential and of the Na^+ gradient, a decrease in nexal conductance due to increased $[Ca^{2+}]_i$ has been proposed to contribute to the impaired conductivity (WEINGART 1977; TAKAYANAGI and JALIFE 1986). The shortened refractory period results from a decrease in action potential duration. A second possibility resides in the effect on slow diastolic depolarization of specialized fibres, which shortens the cycle length of automaticity, thereby precipitating ectopic activity originating from these tertiary centres (VASSALLE et al. 1963; WITTENBERG et al. 1972; ROSEN et al. 1973). Thirdly, propagated impulses originating from delayed afterdepolarizations are very likely to be a primary event of ventricular arrhythmias (FERRIER et al. 1973; HOGAN et al. 1973; LEDERER and TSIEN 1976, CRANEFIELD 1977). Purkinje fibres seem to be more susceptible than working myocardium to some of these mechanisms involved in cardiac toxicity of digitalis (VASSALLE et al. 1962), but the cellular basis of this phenomenon still remains unknown (PALFI et al. 1978; BROWN et al. 1985).

IV. Possible Mechanisms of Direct Antiarrhythmic Effects of Cardiac Glycosides

Most of the antiarrhythmic effects of cardiac glycosides are considered to be caused by the influence of these drugs on extracardiac tissue, especially on the autonomic nervous system (see GILLIS and QUEST 1980; Sect. D). Evidence in favour of a direct antiarrhythmic effect on the myocardium, e.g. in the case of atrial flutter or atrial fibrillation, comes from studies on the denervated (FARAH and LOOMIS 1950) or transplanted (LEACHMAN et al. 1971) heart. The electrophysiological effects of therapeutic digitalis concentrations on normal atrial preparations, however, do not provide a clue to understanding a possible beneficial effect in these types of tachyarrhythmic activity in the atria. These two statements are not necessarily controversial: the beat frequency of myocardial cells can determine whether a certain cardiac glycoside concentration is still non-toxic or already toxic. The relationship between frequency of electrical stimulation and threshold ouabain concentration of intoxication

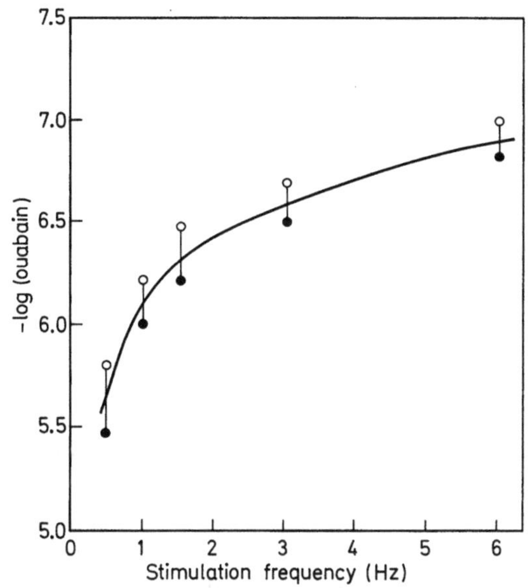

Fig. 3. Dependency of the threshold ouabain concentration to induce toxicity in electrically stimulated guinea pig left atria within 3 h of incubation upon the frequency of stimulation. The *upper points* (*open circles*) represent the respective highest non-toxic concentration, whereas the *lower points* (*filled circles*) represent the respective lowest toxic concentration ($n = 5 - 6$; S. HERZIG, unpublished)

in isolated guinea pig left atria is demonstrated in Fig. 3. The susceptibility to ouabain toxicity is markedly increased with increasing beat frequency, due to the higher degree of specific binding and to the lower reserve capacity of the sodium pump (BENTFELD et al. 1977; HERZIG et al. 1985a). From this, it is conceivable that in vivo a certain cardiac glycoside concentration is in the non-toxic range for the ventricles, whereas atrial cells participating in flutter or fibrillation become intoxicated, because they are excited at a frequency of several hundred per minute. These cells may be temporarily rendered inexcitable, and circus movement of excitation may be interrupted (MENDEZ and MENDEZ 1957). Although this explanation is rather speculative at present and has not yet been further substantiated by experimental evidence, it might be worthwhile to consider the proposed inhomogeneity of cardiac glycoside actions in the heart under conditions where determinants of susceptibility are unevenly distributed, as it is the case for the beat frequency in atrial flutter and fibrillation, which coexist with a comparatively slow ventricular rate.

D. Neurally Mediated Effects of Cardiac Glycosides

Due to the ubiquitous physiological significance of ($Na^{+} + K^{+}$)-ATPase, a variety of extracardiac effects can be elicited by cardiac glycosides in isolated preparations and intact organisms. Alterations of the function of the autonomic nervous system seem to be of relevance for the antiarrhythmic properties of these drugs. The so-called vagal effects of cardiac glycosides on the atrium and AV node of the heart in situ are well known (see Lendle 1935) and of major importance for the therapeutic use of digitalis. They can be abolished by surgical or pharmacological elimination of the autonomic influence on the heart (Chai et al. 1967; Ten Eick and Hoffman 1969a; Goodman et al. 1975; Hariman and Hoffman 1982). The mechanisms involved in the actions of cardiac glycosides on the nervous system, however, are complex and still a matter of debate. For instance, an increased sensitivity of the carotid sinus and the aortic baroreceptors (Quest and Gillis 1971; Quest and Gillis 1974; Gillis et al. 1975; Saum et al. 1976), activation of vagal centres, inhibition of peripheral sympathetic influences (Mendez et al. 1961) and increased susceptibility of cardiac cells to acetylcholine have been claimed to be of significance for the decrease in sinus rate, atrial refractory period and atrioventricular conductivity (see Gillis and Quest 1980). Even in isolated atrial tissue, cardiac glycosides are able to induce acetylcholine-like alterations, which can be blocked by atropine (Hordof et al. 1978). The molecular basis of the neurally mediated effects of cardiac glycoside has been investigated in a number of studies, which shall be briefly reviewed in this paragraph.

I. Cardiac Glycoside Effects on Neurotransmitter Release

Cardiac glycosides have been frequently shown to elicit release of different neurotransmitters from preparations of a variety of organs, including motor nerve terminals, adrenal glands, brain and diverse organs containing peripheral autonomic nerve endings. The concentrations necessary to affect transmitter release were consistently quite high, being sufficient to impair transmembrane transport of Na^+ and K^+ (Pocock 1983b; Wada et al. 1986) and to inhibit an isolated ($Na^+ + K^+$)-ATPase preparation almost completely (Sweadner 1985). Cardiac glycoside-induced neurotransmitter release has frequently been reported to be reduced with a lowered concentration of the extracellular sodium ion concentration (Birks and Cohen 1968; Nakazato et al. 1978; Pocock 1983a; Sweadner 1985; Nakazato et al. 1986). Concerning the requirement of external calcium for cardiac glycoside-induced neurotransmitter release, the experimental results are divergent: abolition, mitigation and sustainment of effects were reported in Ca^{2+}-free media (see Tan and Powis 1985). Many proposals have been made about the molecular mechanism (see Powis 1983). Norepinephrine release might take place via the plasmalemmal sodium-catecholamine cotransport system due to a reduction of the transmembrane sodium gradient (Paton 1973; Rutledge 1978; Lorenz et al. 1980; Sweadner 1985; Tan and Powis 1985). On the other hand, exocytotic release of acetylcholine as well as of catecholamines has been supposed to be mediated by an elevated $[Ca^{2+}]_i$ (Banks 1967; Baker and Crawford 1975; Garcia et al. 1980), which has been speculated to result from alterations of ion movements through plasmalemmal sodium/calcium exchange (Birks and Cohen 1968; Baker and Crawford 1975; Nakazato et al. 1978; Esquerro et a. 1980; Wada et al. 1986), from labilization of binding of calcium to intracellular stores (ELMQUIST and FELDMAN 1965), from depolarization of the cell membrane (WADA et al. 1985; TAN and POWIS 1985), or from reduction of active calcium outward transport (POCOCK 1983b). A direct link between $(Na^+ + K^+)$-ATPase activity and neurotransmitter release has also been suggested (VIZI 1972; VIZI and VYSKOCIL 1979; VIZI et al. 1982). More than one of these mechanisms may be present in one tissue (LORENZ et al. 1980), and the prevailing mechanism may depend on the concentration of cardiac glycoside (TAN and POWIS 1985) or on the type of tissue investigated (SWEADNER 1985). In short, a facilitating action of high cardiac glycoside concentrations on the release of neurotransmitter of various origins can be unequivocally demonstrated in vitro. The molecular mechanisms responsible are still a matter of debate, as is the question of the significance of these effects for the action of therapeutic concentrations of cardiac glycosides on the nervous system in vivo.

II. Cardiac Glycoside Effects on the Sensitivity to Neurotransmitters

A second possible explanation for the "neuroexcitatory effects" of cardiac glycosides would be an increased susceptibility of postsynaptic structures towards the neurotransmitter. This idea is supported by some studies which demonstrate an increased end-organ response to a given amount of added acetylcholine in the presence of a cardiac glycoside. In sympathetic ganglia (KONZETT and ROTHLIN 1952; PERRY and REINERT 1954), in isolated heart preparations (PERRY and REINERT 1954; GAFFNEY et al. 1958; TODA and WEST 1966) and in the heart in situ (LANG et al. 1985), the effects of acetylcholine are amplified in the presence of apparently non-toxic cardiac glycoside concentrations. An increased sensitivity of postsynaptic membranes to acetylcholine due to cardiac glycosides has also been shown at the neuromuscular junction (WESTERMANN 1954; GAGE 1965). Within the heart, intracardiac parasympathetic ganglion cells are probably more affected than the myocardial cells (PERRY and REINERT 1954; FEINAUER et al. 1986). The observations that nerve cells are more easily excited by a near-threshold chemical or electrical stimulus in the presence of a cardiac glycoside have been interpreted in terms of the inhibitory action on $(Na^+ + K^+)$-ATPase: a reduction of the Na^+ and K^+ gradients across the nerve cell membrane (TEN EICK and HOFFMAN 1969b) or a reduced electrogenic sodium pump current (SAUM et al. 1976) could lead to a diminution of the membrane potential of nerve cells, rendering previously subthreshold impulses sufficient to cause an action potential. It should be emphasized, however, that in many studies the possibility of involvement of cardiac glycoside-induced neurotransmitter release could not be excluded, since ganglionic transmission was interposed between the site of stimulation and the site of response measurement. Neurotransmitter release evoked by physiological stimuli might be enhanced by lower cardiac glycoside concentrations than those necessary to elicit an increase of spontaneous release (POWIS 1983; c.f. Sect. D.I). In short, apparently postsynaptically located neuroexcitatory effects of cardiac glycosides are detectable at various sites of the nervous system. The concentrations eliciting these effects are rather low, suggesting that they take place also at therapeutic concentrations. The underlying molecular events are probably related to the interaction between cardiac glycosides and $(Na^{+}+K^{+})$-ATPase, but further experimental evidence is required for a sufficient understanding of the mode of the excitatory cardiac glycoside action in nerves.

E. Concluding Remarks

The therapeutic value and the undesired side effects of cardiac glycoside administration to patients have been well known for a long time. A vast amount of experimental studies helped to characterize the changes induced by cardiac glycosides in isolated organs. The molecular events determining the interaction between cardiac glycosides and $(Na^{+}+K^{+})$-ATPase have been fairly well elucidated by means of biochemical techniques. At the cellular level, however, many questions about cardiac glycoside action await clarification, mainly because more knowledge is needed concerning the mechanisms of cell functions, like excitation-contraction coupling and excitation-secretion coupling, and concerning the role played by $(Na^{+}+K^{+})$-ATPase within these processes.

References

Abete P, Vassalle M (1985) The role of intracellular sodium activity in the inotropy potentiation among high $[Ca]_0$, norepinephrine, and strophanthidin. Arch Int Pharmacodyn Ther 278:87–96

Adams RJ, Schwartz A, Grupp G, Grupp I, Lee SW, Wallick ET, Powell T, Twist VW, Gathiram P (1982) High-affinity ouabain binding site and low-dose positive inotropic effect in rat myocardium. Nature 296:167–169

Akera T, Brody TM (1978) The role of Na^{+}, K^{+}-ATPase in the inotropic action of digitalis. Pharmacol Rev 29:187–220

Akera T, Brody TM (1982) Myocardial membranes: regulation and function of the sodium pump. Annu Rev Physiol 344:375–388

Akera T, Bennett RT, Olgaard MK, Brody TM (1976) Cardiac Na^{+}, K^{+}-adenosine triphosphatase inhibition by ouabain and myocardial sodium: a computer simulation. J Pharmacol Exp Ther 199:287–297

Akera T, Yamamoto S, Temma K, Kim DH, Brody TM (1981) Is ouabain-sensitive rubidium or potassium uptake a measure of sodium pump activity in isolated cardiac muscle? Biochim Biophys Acta 640:779–790

Allen DG, Blinks JR (1978) Calcium transients in aequorin-injected frog cardiac muscle. Nature 273:509–513

Aronson RS, Gelles JM (1977) The effect of ouabain, dinitrophenol, and lithium on the pacemaker current in sheep cardiac Purkinje fibers. Circ Res 40:517–524

Baker BF, Crawford AC (1975) A note of the mechanism by which inhibitors of the sodium pump accelerate spontaneous release of transmitter from motor nerve terminals. J Physiol (Lond) 247:209–226

Banks P (1967) The effect of ouabain on the secretion of catecholamines and on the intracellular concentration of potassium. J Physiol (Lond) 193:631-637

Bassingthwaighte JB, Fry CH, McGuigan JAS (1976) Relationship between internal calcium and outward current in mammalian ventricular muscle; a mechanism for the control of the action potential duration. J Physiol (Lond) 262:15-37

Baumgarten CM, Isenberg G (1977) Depletion and accumulation of potassium in the extracellular clefts of cardiac Purkinje fibers during voltage clamp hyperpolarization and depolarization. Pflügers Arch 368:19-31

Bentfeld M, Lüllmann H, Peters T, Proppe D (1977) Interdependence of ion transport and the action of ouabain in heart muscle. Br J Pharmacol 61:19-27

Birks R, Cohen MW (1968) The influence of internal sodium on the behaviour of motor nerve endings. Proc R Soc Lond [Biol] 170:401-421

Boyett MR, Levy AJ (1985) Cardiac glycoside inotropy dissociated from intracellular Na^+activity (a^i_{Na}) in isolated cardiac Purkinje fibres of the sheep. J Physiol (Lond) 365:66P

Brinkmann HM, Ravens U (1976) Dependence of the ouabain-induced changes in action potentials and force of contraction on stimulation frequency. Naunyn Schmiedebergs Arch Phamacol 293:R20

Brown L, Werdan K, Erdmann E (1983) Consequences of specific [^{3}H]ouabain binding to guinea pig left atria and cardiac cell membranes. Biochem Pharmacol 32:423-435

Brown L, Hug E, Wagner G, Erdmann E (1985) Comparison of ouabain receptors in sheep myocardium and Purkinje fibres. Biochem Pharmacol 34:3701-3710

Cannell MB, Lederer WJ (1986) The arrhythmogenic current i_{TI} in the absence of electrogenic sodium-calcium exchange in sheep cardiac Purkinje fibres. J Physiol (Lond) 374:201-219

Carrier GO, Lüllmann H, Neubauer L, Peters T (1974) Significance of a fast exchanging superficial calcium fraction for the regulation of contractile force in heart muscle. J Mol Cell Cardiol 6:333-347

Chai CY, Wang HH, Hoffman BF, Wang SC (1967) Mechanisms of bradycardia induced by digitalis substances. Am J Physiol 212:26-34

Cranefield PF (1977) Action potentials, afterpotentials, and arrhythmias. Circ Res 41:415-423

Daut J (1982) The role of intracellular sodium ions in the regulation of cardiac contractility. J Mol Cell Cardiol 14:189-192

Deitmer JW, Ellis D (1978) The intracellular sodium activity of cardiac Purkinje fibres during inhibition and reactivation of the Na-K pump. J Physiol (Lond) 284:241-259

Deleze J (1960) Possible reasons for drop of resting potential of mammalian heart preparations during hypothermia. Circ Res 8:553-557

Dudel J, Trautwein W (1958) Elektrophysiologische Messungen zur Strophanthinwirkung am Herzmuskel. Arch Exp Pathol Pharmakol 232:393-407

Eisner DA, Valdeolmillos M (1986) A study of intracellular calcium oscillations in sheep cardiac Purkinje fibres measured at the single cell level. J Physiol (Lond) 372:539-556

Eisner DA, Lederer WJ, Vaughan-Jones RD (1981) The dependence of sodium pumping and tension on intracellular sodium activity in voltage-clamped sheep Purkinje fibres. J Physiol (Lond) 317:163-187

Eisner DA, Lederer WJ, Vaughan-Jones RD (1983) Comments on "Active transport and inotropic state in guinea pig left atrium". Circ Res 53:834-835

Eisner DA, Lederer WJ, Vaughan-Jones RD (1984) The quantitative relationship be-

tween twitch tension and intracellular sodium activity in sheep cardiac Purkinje fibres. J Physiol (Lond) 355:251-266
Elmquist D, Feldman DS (1965) Effects of sodium pump inhibitors on spontaneous acetylcholine release at the neuromuscular junction. J Physiol (Lond) 181:498-505
Ellis D (1977) The effects of external cations and ouabain on the intracellular sodium activity of sheep heart Purkinje fibres. J Physiol (Lond) 273:211-240
Erdmann E, Philipp G, Scholz H (1980) Cardiac glycoside receptor, $(Na^+ + K^+)$-ATPase activity and force of contraction in rat heart. Biochem Pharmacol 29:3219-3229
Esquerro E, Garcia AG, Hernandez M, Kirpekar SM, Prat JC (1980) Catecholamine secretory response to calcium reintroduction in the perfused cat adrenal gland treated with ouabain. Biochem Pharmacol 29:2669-2673
Farah A, Loomis TA (1950) The action of cardiac glycosides on experimental auricular flutter. Circulation 2:742-748
Feinauer M, Lindmar R, Löffelholz K, Ullrich B (1986) Ouabain enhances release of acetylcholine in the heart evoked by unilateral vagal stimulation. Naunyn Schmiedebergs Arch Pharmacol 333:7-12
Ferrier GR, Moe GK (1973) Effect of calcium on acetylstrophanthidin-induced transient depolarizations in canine Purkinje tissue. Circ Res 33:509-515
Ferrier GR, Saunders JH, Mendez C (1973) A cellular mechanism for the generation of ventricular arrhythmias by acetylstrophanthidin. Circ Res 32:600-609
Finet M, Noel F, Godfraind T (1982) Inotropic effect and binding sites of ouabain to rat heart. Arch Int Pharmacodyn Ther 256:168-170
Gadsby DC, Cranefield PF (1979) Electrogenic sodium extrusion in cardiac Purkinje fibers. J Gen Physiol 73:819-837
Gadsby DC, Kimura J, Noma A (1985) Voltage dependence of Na/K pump current in isolated heart cells. Nature 315:63-65
Gaffney TE, Kahn JB, van Mannen EF, Acheson GH (1958) A mechanism of the vagal effect of cardiac glycosides. J Pharmacol Exp Ther 122:423-429
Gage PW (1965) Effect of cardiac glycosides on neuromuscular transmission. Nature 205:84-85
Garcia AG, Hernandez M, Horga JF, Sanchez-Garcia P (1980) On the release of catecholamines and dopamine-β-hydroxylase evoked by ouabain in the perfused cat adrenal gland. Br J Pharmacol 68:571-583
Garrahan PJ, Glynn IM (1967) The stoichiometry of the sodium pump. J Physiol (Lond) 192:217-235
Gillis RA, Quest JA (1980) The role of the nervous system in the cardiovascular effects of digitalis. Pharmacol Rev 31:19-97
Gillis RA, Quest JA, Thibodeaux H, Clancy MM, Evans DE (1975) Neural mechanisms involved in acetylstrophanthidin-induced bradycardia. J Pharmacol Exp Ther 193:336-345
Glitsch HG (1982) Electrogenic Na pumping in the heart. Annu Rev Physiol 44:389-400
Glitsch HG, Reuter H, Scholz H (1970) The effect of internal sodium concentration on calcium fluxes in isolated guinea-pig auricles. J Physiol (Lond) 209:25-43
Glynn IM, Karlish SDJ (1975) The sodium pump. Annu Rev Physiol 37:13-55
Goodman DJ, Rossen RM, Cannon DS, Rider DK, Harrison DC (1975) Effect of digoxin on atrioventricular conduction. Studies in patients with and without cardiac autonomic innervation. Circulation 51:251-256
Grupp I, Im WB, Lee CO, Lee SW, Pecker MS, Schwartz A (1985) Relation of sodium

pump inhibition to positive inotropy at low concentrations of ouabain in rat heart muscle. J Physiol (Lond) 360:149-160

Hansen O (1984) Interaction of cardiac glycosides with ($Na^+ + K^+$)-activated ATPase. A biochemical link to digitalis-induced inotropy. Pharmacol Rev 36:143-162

Hariman RJ, Hoffman BF (1982) Effects of ouabain and vagal stimulation on sinus nodal function in conscious dogs. Circ Res 51:760-768

Hasuo H, Koketsu K (1985) Potential dependence of the electrogenic Na^+-pump current in bullfrog atrial muscle. Jpn J Physiol 35:89-100

Herzig S, Mohr K (1984) Action of ouabain on rat heart: comparison with its effect on guinea-pig heart. Br J Pharmacol 82:135-142

Herzig S, Krey U, Lüllmann H, Mohr K (1985a) Is digitalis binding to intact myocardium cooperative? Trends in Pharmacological Sciences 7:432-433

Herzig S, Lüllmann H, Mohr K (1985b) On the cooperativity of ouabain-binding to intact myocardium. J Mol Cell Cardiol 17:1095-1104

Hoffman BF, Bigger JT (1985) Digitalis and allied cardiac glycosides. In: Goodman Gilman A, Goodman LS, Rall TW, Murad F (eds) Goodman and Gilman's The pharmacological basis of therapeutics, 7th edn. Macmillan, New York, p 716-747

Hogan PM, Wittenberg SM, Klocke FJ (1973) Relationship of stimulation frequency to automaticity in the canine Purkinje fiber during ouabain administration. Circ Res 32:377-383

Hordof AJ, Spotnitz A, Mary-Rabine L, Edie RN, Rosen MR (1978) The cellular electrophysiologic effects of digitalis on human atrial fibers. Circulation 57:223-229

Horres CR, Aiton JF, Lieberman M, Johnson EA (1979) Electrogenic transport in tissue cultured heart cells. J Mol Cell Cardiol 11:1201-1205

Hougen TJ, Smith TW (1978) Inhibition of myocardial monovalent cation active transport by subtoxic doses of ouabain in the dog. Circ Res 42:856-863

Huang W, Rhee HM, Chiu TH, Askari A (1979) Re-evaluation of the relationship between the positive inotropic effect of ouabain and its inhibitory effect on ($Na^+ + K^+$)-dependent adenosine triphosphatase in rabbit and dog hearts. J Pharmacol Exp Ther 211:571-582

Isenberg G (1975) Is potassium conductance of cardiac Purkinje fibres controlled by $[Ca^{2+}]_i$? Nature 253:273-274

Isenberg G, Trautwein W (1974) The effect of dihydro-ouabain and lithium ions on the outward current in cardiac Purkinje fibres. Pflügers Arch 250:41-54

Kameyama M, Kakei M, Sato R, Shibasaki T, Matsuda H, Irasawa H (1984) Intracellular Na^+ activates a K^+ channel in mammalian cardiac cells. Nature 309:354-356

Kaplan JH (1985) Ion movements through the sodium pump. Annu Rev Physiol 47:535-544

Kasparek R, Lüllmann H, Peters T (1981) Influence of ouabain on Ca exchangeability in resting atria of guinea-pig. Eur J Pharmacol 73:371-374

Kass RS, Lederer WJ, Tsien RW, Weingart R (1978) Role of calcium ions in transient inward current and aftercontractions induced by strophanthidin in cardiac Purkinje fibres. J Physiol (Lond) 281:187-208

Kassebaum DG (1963) Electrophysiological effects of strophanthin in the heart. J Pharmacol Exp Ther 140:329-338

Kim D, Barry WH, Smith TW (1984) Kinetics of ouabain binding and changes in cellular sodium content, ^{42}K transport and contractile state during ouabain exposure in cultured chick heart cells. J Pharmacol Exp Ther 231:326-333

Klaus W, Kuschinsky G, Lüllmann H (1962) Über den Zusammenhang zwischen positiv inotroper Wirkung von Digitoxigenin, Kaliumflux und intrazellulären Ionenkonzentrationen im Herzmuskel. Arch Exp Pathol Pharmakol 242:480- 496

Kline RP, Kupersmith J (1982) Effects of extracellular potassium accumulation and sodium pump activation on autonomic canine Purkinje fibres. J Physiol (Lond) 324:507-533
Kojima M, Sperelakis N (1986) Effects of calcium channel blockers on ouabain-induced oscillatory afterpotentials in organ-cultured young embryonic chick hearts. Eur J Pharmacol 182:65-73
Konzett H, Rothlin (1952) Effect of cardioactive glycosides on a symphathetic ganglion. Arch Int Pharmacodyn Ther 89:343-352
Kunze DL (1977) Rate-dependent changes in extracellular potassium in the rabbit atrium. Circ Res 41:122-127
Lang J, Lakhai M, Timour Cha Q, Faucon G (1985) Vagal role in the potentiation by Ca^{2+} ions of the action of cardiac glycosides on the atrial specialized tissue. J Pharmacol 16:125-137
Leachman RD, Cokkinos DVP, Cabrera R, Leatherman LL, Rochelle DG (1971) Response of the transplanted, denervated human heart to cardiovascular drugs. Am J Cardiol 27:272-276
Lechat P, Malloy CR, Smith TW (1983) Active transport and inotropic state in guinea pig left atrium. Circ Res 52:411-422
Lederer WJ, Tsien RW (1976) Transient inward current underlying arrhythmogenic effects of cardiotonic steroids in Purkinje fibres. J Physiol (Lond) 263:73-100
Lee CO, Dagostino M (1982) Effect of strophanthidin on intracellular Na ion activity and twitch tension of constantly driven canine Purkinje fibres. Biophys J 40:185-198
Lee KS, Klaus W (1971) The subcellular basis for the mechanism of inotropic action of cardiac glycosides. Pharmacol Rev 23:193-261
Lee CO, Kang SH, Sokol JH, Lee KS (1980) Relation between intracellular Na ion activity and tension of sheep cardiac Purkinje fibres exposed to dihydro-ouabain. Biophys J 29:315-330
Lee CO, Abete P, Pecker M, Sonn JK, Vassalle M (1985) Strophanthidin inotropy: role of intracellular sodium ion activity and sodium-calcium exchange. J Mol Cell Cardiol 17:1043-1053
Lendle L (1935) Digitaliskörper und verwandte Substanzen (Digitaloide). In: Häubner W, Schüller J (eds) Handbook of experimental pharmacology [Suppl. vol 1] pp 11-241 Springer, Berlin
Lorenz RR, Powis DA, Vanhoutte PM, Shepard JT (1980) The effects of acetylstrophanthidin and ouabain on the synaptic adrenergic neuroeffector junction in canine vascular smooth muscle. Circ Res 487:845-854
Lüllmann H, Peters T (1979) Action of cardiac glycosides on the excitation-contraction coupling in heart muscle. Prog Pharmacol 2:1-58
Lüllmann H, Ravens U (1973) The time courses of the changes in contractile force and in transmembrane potentials induced by cardiac glycosides in guinea-pig papillary muscle. Br J Pharmacol 49:377-390
Mendez C, Mendez R (1957) The action of cardiac glycosides on the excitability and conduction velocity of the mammalian atrium. J Pharmacol Exp Ther 121:402-413
Mendez C, Aceves J, Mendez R (1961) Inhibition of cardiac acceleration by cardiac glycosides. J Pharmacol Exp Ther 131:191-198
Miura DS, Rosen MR (1978) The effects of ouabain on the transmembrane potentials and intracellular potassium activity of canine cardiac Purkinje fibres. Circ Res 42:333-338
Morgan JP (1985) The effects of digitalis on intracellular calcium transients in mam-

malian working myocardium as detected with aequorin. J Mol Cell Cardiol 17:1065-1075
Mullins LJ (1981) Ion transport in heart. Raven, New York
Nakazato Y, Ohga A, Onoda Y (1978) The effect of ouabain on noradrenaline output from peripheral adrenergic neurones of isolated guinea-pig vas deferens. J Physiol (Lond) 278:45-54
Nakazato Y, Ohga A, Yamada Y (1986) Facilitation of transmitter action on catecholamine output by cardiac glycoside in perfused adrenal gland of guinea-pig. J Physiol (Lond) 374:475-491
Nayler W, Noack EA (1981) Influence of cardiac glycosides on electrolyte exchange and content in cardiac muscle cells. In: Greeff K (ed) Cardiac glycosides. Springer, Berlin Heidelberg New York pp 407-436 (Handbook of experimental pharmacology, vol 56/I)
Noble D (1980) Review: mechanism of action of therapeutic levels of cardiac glycosides. Cardiovasc Res 14:495-514
Noma A, Irisawa H (1974) Electrogenic sodium pump in rabbit sinoatrial node cell. Pflügers Arch 35:177-182
Page E, Storm SR (1965) Cat heart muscle in vitro. VIII: active transport of sodium in papillary muscles. J Gen Physiol 48:957-972
Palfi FJ, Besch HR, Watanabe AM (1978) Ouabain sensitivity of the Na^+, K^+-ATPase activity from single bovine cardiac Purkinje fiber and adjacent papillary muscle. J Mol Cell Cardiol 10:1149-1155
Paton DM (1973) Mechanism of efflux of noradrenalin from adrenergic nerves in rabbit atria. Br J Pharmacol 49:614-627
Perry WLM, Reinert H (1954) The action of cardiac glycosides on autonomic ganglia. Br J Pharmacol 9:324-328
Pocock G (1983a) Ionic and metabolic requirements for stimulation of secretion by ouabain in bovine adrenal medullary cells. Mol Pharmacol 23:671-680
Pocock G (1983b) Ion movements in isolated bovine medullary cells treated with ouabain. Mol Pharmacol 23:681-697
Powis DA (1983) Cardiac glycosides and autonomic neurotransmission. J Auton Pharmacol 3:127-154
Quest JA, Gillis RA (1971) Carotid sinus reflex changes produced by digitalis. J Pharmacol Exp Ther 177:650-661
Quest JA, Gillis RA (1974) Effect of digitalis on carotid sinus baroreceptor activity. Circ Res 35:247-254
Reiter M (1962) Die Entstehung von "Nachkontraktionen" im Herzmuskel unter Einwirkung von Calcium und von Digitalisglycosiden in Abhängigkeit von der Reizfrequenz. Arch Exp Pathol Pharmakol 242:497-507
Rosen MR, Gelband H, Hoffman BF (1973) Correlations between effects of ouabain on the canine electrocardiogram and transmembrane potentials of isolated Purkinje fibers. Circulation 47:65-72
Rosen MR, Ilvento JP, Gelband H, Merker C (1974) Effects of verapamil on electrophysiologic properties of canine cardiac Purkinje fibers. J Pharmacol Exp Ther 189:414-422
Rosen MR, Wit AL, Hoffman BF (1975) Electrophysiology and pharmacology of cardiac arrhythmias. IV. Cardiac antiarrhythmic and toxic effects of digitalis. Am Heart J 89:391-399
Rutledge CO (1978) Effect of metabolic inhibitors and ouabain on amphetamine- and potassium-induced release of biogenic amines from isolated brain tissue. Biochem Pharmacol 27:511-516

Saum WR, Brown AM, Tuley FH (1976) An electrogenic sodium pump and baroreceptor function in normotensive and spontaneously hypertensive rats. Circ Res 39:497-505

Schatzmann HJ (1953) Herzglykoside als Hemmstoffe für den aktiven Kalium- und Natriumtransport durch die Erythrocytenmembran. Helv Physiol Acta 11:346-354

Schwartz A, Lindenmayer GE, Allen JC (1975) The sodium-potassium adenosine triphosphatase: pharmacological, physiological and biochemical aspects. Pharmacol Rev 27:3-134

Sen AK, Post RL (1964) Stoichiometry and localization of adenosine triphosphate-dependent sodium and potassium transport in the erythrocyte. J Biol Chem 239:345-352

Sheu SS, Fozzard HA (1982) Transmembrane Na^+ and Ca^{2+} electrochemical gradients in cardiac muscle and their relationship to force development. J Gen Physiol 80:325-351

Skou JC (1957) The influence of some cations on an adenosine triphosphatase from peripheral nerves. Biochim Biophys Acta 23:349-401

Sweadner KJ (1985) Ouabain-evoked norepinephrine release from intact rat sympathetic neurons: evidence for carrier-mediated release. J Neurosci 5:2397-2406

Takayanagi K, Jalife J (1986) Effects of digitalis intoxication on pacemaker rhythm and synchronization in rabbit sinus node. Am J Physiol 250:H567-H578

Tan CM, Powis DA (1985) Cardiac glycosides, calcium and the release of neurotransmitter from peripheral noradrenergic nerves. Naunyn Schmiedebergs Arch Pharmacol 329:1-9

Ten Eick RE, Hoffman BF (1969a) Chronotropic effect of cardiac glycosides in cats, dogs, and rabbits. Circ Res 25:365-378

Ten Eick RE, Hoffman BF (1969b) The effect of digitalis on the excitability of autonomic nerves. J Pharmacol Exp Ther 169:95-108

Thomas RC (1972) Electrogenic sodium pump in nerve and muscle cells. Physiol Rev 52:563-594

Tobin T, Akera T, Baskin SI, Brody TM (1983) Calcium ion and sodium- and potassium-dependent adenosine triphosphatase: its mechanism of inhibition and identification of the E_1-P intermediate. Mol Pharmacol 9:336-349

Toda N, West TC (1966) The influence of ouabain on cholinergic responses in the sinoatrial node. J Pharmacol Exp Ther 153:104-113

Tuttle RS, Witt PN, Farah A (1961) The influence of ouabain on intracellular sodium and potassium concentrations in the rabbit myocardium. J Pharmacol Exp Ther 133:281-287

Tuttle RS, Witt PN, Farah A (1962) Therapeutic and toxic effects of ouabain on K^+ fluxes in rabbit atria. J Pharmacol Exp Ther 137:24-30

Vassalle M (1970) Electrogenic suppression of automaticity in sheep and dog Purkinje fibers. Circ Res 27:361-377

Vassalle M, Lee CO (1984) The relationship among intracellular sodium activity, calcium, and strophanthidin inotropy in canine cardiac Purkinje fibers. J Gen Physiol 83:287-307

Vassalle M, Karis J, Hoffman BF (1962) Toxic effects of ouabain on Purkinje fibers and ventricular muscle fibers. Am J Physiol 203:433-439

Vassalle M, Greenspan K, Hoffman BF (1963) An analysis of arrhythmias induced by ouabain in intact dogs. Circ Res 13:132-148

Vick RL, Kahn JB (1957) The effects of ouabain and veratridine on potassium movement in the isolated guinea pig heart. J Pharmacol Exp Ther 121:389-401

Vizi ES (1972) Stimulation by inhibition of (Na^+-K^+-Mg^{2+})-activated ATPase of acetylcholine release in cortical slices from rat brain. J Physiol (Lond) 226:95–117
Vizi ES, Vyskocil F (1979) Changes in total and quantal release of acetylcholine in the mouse diaphragm during activation and inhibition of membrane ATPase. J Physiol (Lond) 286:1-14
Vizi ES, Török T, Seregi A, Serföző P, Adam-Vizi V (1982) Na-K-activated ATPase and the release of acetylcholine and noradrenaline. J Physiol (Paris) 78:399-406
Wada A, Izumi F, Yanagihara N, Kobayashi H (1985) Modulation by ouabain and diphenylhydantoin of veratridine-induced ^{22}Na influx and its relation to ^{45}Ca influx and the secretion of catecholamines in cultured bovine adrenal medullary cells. Naunyn Schmiedebergs Arch Pharmacol 328:273-278
Wada A, Takara H, Yanagihara N, Kobayashi H, Izumi F (1986) Inhibition of Na^+-pump enhances carbachol-induced influx of $^{45}Ca^{2+}$ and secretion of catecholamines by elevation of cellular accumulation of $^{22}Na^+$ in cultured bovine adrenal medullary cells. Naunyn Schmiedebergs Arch Pharmacol 332:351-356
Wasserstrom JA, Schwartz DJ, Fozzard HA (1983) Relation between intracellular sodium and twitch tension in sheep cardiac Purkinje strands exposed to cardiac glycosides. Circ Res 52:697-705
Weingart R (1977) The actions of ouabain on intercellular coupling and conduction velocity in mammalian ventricular muscle. J Physiol (Lond) 264:341-365
Weingart R (1981) Influence of cardiac glycosides on electrophysiologic processes. In: Greeff K (ed) Cardiac glycosides. Springer, Berlin Heidelberg New York pp 221-254 (Handbook of experimental pharmacology, vol 56/I)
Westermann E (1954) Zur Digitaliswirkung auf die neuromuskuläre Übertragung. Arch Exp Pathol Pharmakol 222:398-407
Wier WG, Hess P (1984) Excitation-contraction coupling in cardiac Purkinje fibres: effects of cardiotonic steroids on the intracellular [Ca^{2+}] transient, membrane potential, and contraction. J Gen Physiol 83:395-414
Wittenberg SM, Gandel P, Hogan PM, Kreuzer W, Klocke FJ (1972) Relationship of heart rate to ventricular automaticity in dogs during ouabain administration. Circ Res 30:167-176

CHAPTER 24b

Clinical Efficacy of Cardiac Glycosides for Arrhythmias

T. J. CAMPBELL

A. Introduction

Plant extracts containing cardiac glycosides have been used as medicines and poisons for thousands of years (HOFFMAN and BIGGER 1980). WILLIAM WITHERING, who published his famous treatise on digitalis in 1785, believed it to be primarily a diuretic although he did comment specifically on its remarkable ability to reduce abnormally fast heart rates, and we can be reasonably confident that some of his patients had atrial fibrillation (KRIKLER 1985). In the early part of this century, the work of Lewis, Mackenzie, Cushny and others led to digitalis becoming the standard treatment for atrial fibrillation (HOFFMAN and BIGGER 1980), and it remains the most popular remedy for this arrhythmia.

In the remainder of this section we will briefly and systematically review the use of cardiac glycosides as therapeutic agents for a range of arrhythmias. The emphasis will be on digoxin because it has been most intensively studied in clinical trials. The very important subject of the use of cardiac glycosides as inotropic agents will not be dealt with.

B. Specific Arrhythmias

I. Atrial Ectopic Beats

These do not generally require any treatment. Their onset sometimes indicates atrial dilatation and incipient cardiac failure. If this is the case, then diuretic or vasodilator therapy may be appropriate. On occasions (for example in the first few days after cardiac surgery), atrial ectopic activity may precede the onset of atrial fibrillation or flutter. Digitalis is often used in such patients although the author is unaware of any hard evidence for its efficacy. Similarly, digitalis is frequently recommended in standard textbooks for symptomatic atrial ectopic activity (for example ZIPES 1984).

II. Atrial Fibrillation and Flutter

These arrhythmias remain by far the most frequent arrhythmic indications for the use of cardiac glycosides. They are widely regarded as the agents of first choice, with other modalities such as cardioversion, class Ia agents or beta-adrenergic blocking drugs being reserved for refractory cases (WEINER et al. 1983; see also Chaps. 9, 11, 12).

Given that the major electrophysiological effects of digitalis on atrial muscle involve the shortening of action potential duration and hence refractory period (see Chap. 2), it is difficult to conceive of a direct antifibrillatory action for this group of drugs. PICK (1957) claimed a direct converting effect but most other workers have suggested that any tendency to promote reversion to sinus rhythm must be on the basis of a slowed ventricular rate (vagal effect on atrioventricular node) and hence haemodynamic improvement (WEINER et al. 1983).

Certainly in the specific case of atrial flutter it is a common clinical observation that treatment with digitalis may be followed by conversion to atrial fibrillation. This is not surprising in view of the expected reduction in refractory periods, and indeed may be of therapeutic value at times. The ventricular rate is often easier to control in atrial fibrillation than atrial flutter.

Convincing evidence that digitalis actually promotes reversion to sinus rhythm is hard to find. Certainly if patients with recent onset atrial fibrillation are given large doses of digoxin, many of them revert to sinus rhythm over the subsequent 12–16 h. Actual figures vary widely. CAMPBELL and MORGAN (1980) studied patients with acute atrial fibrillation or flutter after cardiac surgery. Five out of 15 reverted to sinus rhythm after 12 h of treatment with digoxin alone. In a larger study with similar patients, 45 out of 201 returned to sinus rhythm within 2 h of an intravenous bolus of digoxin. Those remaining in their arrhythmia at 2 h were then given disopyramide, which proved significantly better than digoxin alone (GAVAGHAN et al. 1985). WEINER et al. (1983) reported successful reversion with digoxin within 16 h in 40/47 episodes of atrial fibrillation of less than 1 week's duration.

It is, however, a common observation that many patients with recent-onset atrial flutter or fibrillation revert spontaneously to sinus rhythm. Despite this, there is a remarkable lack of placebo-controlled studies of the efficacy of digoxin in these arrhythmias. This is due at least in part to the fact that digitalis became recognized as standard therapy for atrial fibrillation before the concept of placebo-controlled clinical trials was established.

Three very recent studies have addressed this problem. KNOWLTON et al. (1986) reported the first randomized study of digoxin against placebo in 40 patients with recent onset (less than 7 days) atrial fibrillation. Over the 16-h study period 52 % of 21 patients randomized to oral digoxin and 42 % of 19 patients on placebo reverted to sinus rhythm. The mean times to reversion were 5.7 h in the digoxin group and 4.0 h in placebo. Neither of these differences was significant.

The other two studies do not directly compare digitalis with placebo but they do also call into question its usefulness in managing atrial fibrillation. GRANDE et al. (1986) showed that quinidine alone was significantly better than any of digoxin alone, digoxin plus quinidine and no therapy, in maintaining sinus rhythm in patients electrically cardioverted from atrial fibrillation. Similarly, STEINBECK et al. (1986) found digoxin alone to be less effective than digoxin plus quinidine or digoxin plus flecainide in preventing episodes of paroxysmal atrial fibrillation.

In summary then, while most would agree that cardiac glycosides may be of benefit in slowing the ventricular response to atrial fibrillation/flutter, there are very few data supporting its continued use as the sole or major therapy for these arrhythmias.

III. Supraventricular Tachycardia

The majority of cases of supraventricular tachycardia (SVT) are now known to be due either to re-entry within the atrioventricular node or to macrore-entry involving an atrioventricular nodal bypass tract. In both situations intravenous therapy with either verapamil or a beta-adrenergic blocking agent will be successful in reverting over 90 % of cases to sinus rhythm. Digitalis is still recommended by some authorities for acute therapy of the intranodal variety but not as a first-line agent in adults. It may, however, be useful for prophylactic therapy (ZIPES 1984) and is still widely used as a drug of first choice in infants and children with SVT (FRIEDMAN 1984).

The use of digitalis in patients with SVT or atrial fibrillation involving antegrade conduction down an atrioventricular nodal bypass tract is potentially dangerous and should be avoided if possible (SELLERS et al. 1977). This is because of the potential ability of digitalis to shorten the refractory period of the bypass tract and permit extremely rapid ventricular response rates which may degenerate into ventricular fibrillation.

IV. Ventricular Arrhythmias

Cardiac glycosides have not generally been regarded as useful agents for ventricular arrhythmias and indeed are well known to be capable of causing them. There is, however, some evidence that these agents may be of therapeutic benefit in certain patients with ventricular ectopic beats. LOWN et al. (1977) reported that acetylstrophanthidin was effective in suppressing ventricular ectopic activity in approximately one-half of patients studied over several hours. GRADMAN et al. (1983) later reported on 13 patients with chronic ventricular ectopy. The seven patients with normal left ventricular function had a reduction in mean ectopic frequency from $69 \pm 58/1000$ beats to 20 ± 18. This was reported as significant ($P < 0.05$) although, in view of the excellent prognosis of such patients untreated, the small numbers, and the difficulty inherent in demonstrating a true drug-induced reduction in ectopic frequency (see Chap. 9; 18), the clinical significance of this finding is doubtful. The six patients with depressed left ventricles showed no antiarrhythmic benefit.

C. Conclusions

Apart from certain specific uses such as controlling the ventricular rate in chronic atrial fibrillation and perhaps in the therapy of supraventricular tachycardia in infants and children, there is very little role for cardiac glycosides today as antiarrhythmic agents. Their major place in cardiology is as inotropic agents in supporting patients with chronic cardiac failure.

References

Campbell TJ, Morgan JJ (1980) Treatment of atrial arrhythmias after cardiac surgery with intravenous disopyramide. Aust NZ J Med 10:644-649

Friedman WF (1984) Congenital heart disease in infancy and childhood. In: Braunwald E (ed) Heart Disease, 2nd edn. Saunders, Philadelphia pp 941-1023

Gavaghan TP, Feneley MP, Campbell TJ, Morgan JJ (1985) Atrial tachyarrhythmias after cardiac surgery: results of disopyramide therapy. Aust NZ J Med 15:27-32

Gradman AH, Cunningham M, Harbison BS, Berger HJ, Zaret BL (1983) Effects of oral digoxin on ventricular ectopy and its relation to left ventricular function. Am J Cardiol 51:765-769

Grande P, Sonne B, Pedersen A (1986) A controlled study of digoxin and quinidine in patients DC reverted from atrial fibrillation to sinus rhythm. Circulation 74 (Suppl II):101

Hoffman BF, Bigger JT (1980) Digitalis and allied cardiac glycosides. In: Gilman AG, Goodman LS, Gilman A (eds) The pharmacologic basis of therapeutics, 6th edn. Macmillan, New York pp 729-760

Knowlton AA, Falk RH, Bernard S, O'Brien JL, Gotlieb NE, Battinelli NJ, Apstein CS (1986) The efficacy of digoxin in the conversion of atrial fibrillation to normal sinus rhythm. J Am Coll Cardiol 7:159a

Krikler DM, (1985) The foxglove, "the old woman from Shropshire", and William Withering. J Am Coll Cardiol 5:3A-9A

Lown B, Graboys TB, Podrid PJ, Cohen BH, Stockman MB, Gaughan CE (1977) Effect of a digitalis drug on ventricular premature beats. N Engl J Med 296:301-306

Pick A (1957) Digitalis and the electrocardiogram. Circulation 15:603-608

Sellers TD, Bashore TM, Gallagher JJ (1975) Digitalis in the pre-excitation syndrome. Analysis during atrial fibrillation. Circulation 56:260-267

Steinbeck KG, Doliwa R, Bach P (1986) Cardiac glycosides for paroxysmal atrial fibrillation? Circulation 74 (Suppl II):100

Weiner P, Bassan MM, Jarchovsky J, Iusim S, Plavnick L (1983) Clinical course of acute atrial fibrillation treated with rapid digitalization. Am Heart J 105:223-227

Zipes DP (1984) Specific arrhythmias: diagnosis and treatment. In: Braunwald E (ed) Heart disease, 2nd edn. Saunders, Philadelphia, pp 683-743

CHAPTER 25

Eicosanoids and Arrhythmogenesis

J. R. PARRATT

A. Introduction

Most standard approaches to antiarrhythmic therapy, for example in myocardial ischaemia, involve the administration of synthetic drugs that directly modify ionic exchange across myocardial cell membranes. However, there may be less direct ways of reducing the severity of such arrhythmias. These depend on modulating the release, or cellular effects, of biochemical and metabolic factors that themselves influence sarcolemmal ionic flux. Thus, such ischaemia-induced arrhythmias may result in part from the release of endogenous arrhythmogenic substances. A good example of an approach which attempts to modulate the effects of such an arrhythmogenic biochemical factor is the use of beta-adrenoceptor-blocking drugs to antagonize the effects of noradrenaline and adrenaline released during myocardial ischaemia. There are other less well appreciated possibilities. Examples include the ability of certain antagonists of both 5-HT and opioid peptides to reduce the severity of ischaemic arrhythmias in experimental animals (COKER et al. 1986; FAGBEMI et al. 1982; PARRATT and SITSAPESAN 1986).

Another possible approach to antiarrhythmic therapy in this wider sense arises from the hypothesis that, besides potentially arrhythmogenic substances, protective "endogenous antiarrhythmic substances" are also released under conditions of myocardial ischaemia and reperfusion. This concept of the myocardial release of endogenous protective agents was first enunciated by FÖRSTER (1976). He suggested, partly on the basis of the efficacy of various prostaglandins (and prostacyclin) to protect against chemically induced arrhythmias in experimental animals, that endogenous prostaglandins might have a functional role in protecting the myocardium. The purpose of this present chapter is to review the evidence that arachidonic acid products of the cyclooxygenase pathway might be involved in the generation of those arrhythmias that result from ischaemia and reperfusion which, in the clinical situation, are responsible for sudden cardiac death. In brief, the evidence suggests that some of these products are likely to be arrhythmogenic; it might therefore be a helpful antiarrhythmic strategy to inhibit their synthesis or to prevent their effects at receptor level. In one sense this approach is no different from that which led to the use of beta-adrenoceptor-blocking drugs to reduce the incidence and severity of arrhythmias after acute myocardial infarction and to reduce mortality in patients recovering from infarction (secondary prevention); it simply extends this principle to possible endogenous arrhythmogenic

Table 1. Biochemical factors possibly involved in early ischaemia and reperfusion-induced arrhythmias

Catecholamines (difference between local and systemic release?)
Cyclic adenosine monophosphate (cAMP)
Adenosine
Free radicals
Arachidonic acid derivatives:
a) Cyclooxygenase products (prostaglandins, prostacyclin, thromboxane)
b) Lipoxygenase products
Vasoactive peptides (e.g. opioid peptides)
PAF-acether
Plasma kinins
Histamine
5-Hydroxytryptamine
Angiotensin
Some of the above (e.g. adenosine, prostacyclin) may be protective

agents other than noradrenaline. On the other hand, some arachidonic acid derived products might act as "endogenous antiarrhythmic agents", in which case it might be helpful to attempt to increase the availability of such substances generated during ischaemia, for example by increasing their synthesis or inhibiting their breakdown. This concept would thus involve the therapeutic support of already existing defence mechanisms for the myocardium.

This chapter concentrates on arachidonic acid derived products but it should not be forgotten that a wide variety of other vasoactive agents with possible effects on arrhythmogenesis may be released under conditions of myocardial ischaemia, or be washed out from the myocardium during any subsequent spontaneous or artificially induced reperfusion. Some of these substances are listed in Table 1. Another aspect which must be emphasized concerns the close interrelationships between many of these biochemical factors. For example, adenosine and prostacyclin (and related stable synthetic prostacyclin analogues) inhibit myocardial noradrenaline release from sympathetic neurones (ZYLKA et al. 1981; HIRCHE et al. 1982) and prostacyclin also inhibits thromboxane-induced platelet aggregation. Further, adrenaline potentiates thromboxane release from platelets (PURCHASE et al. 1986) and the thromboxane agonist U46619 potentiates noradrenaline release from sympathetic nerves (TRACHTE 1986).

B. Activation and Modulation of the Arachidonic Acid Cascade in Myocardial Ischaemia and in Reperfusion and Their Relation to Arrhythmias

A simplified, schematic account of the metabolism of arachidonic acid is given in Fig. 1. During myocardial ischaemia an early event is degradation of membranal phospholipids by various phospholipase enzymes. This leads not

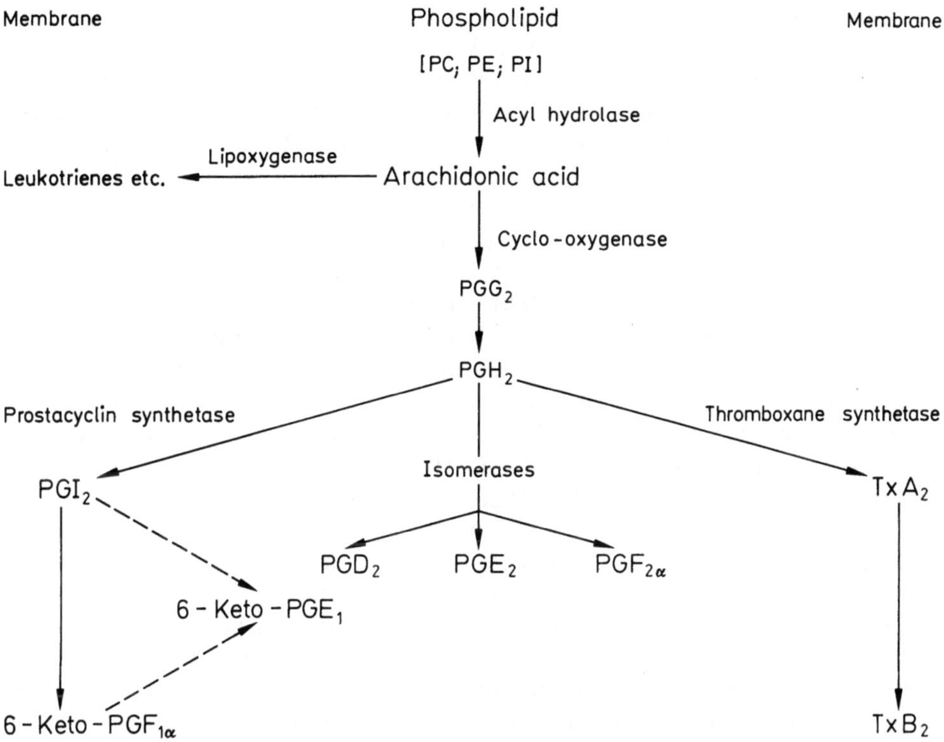

Fig. 1. Simplified summary of the arachidonic acid cascade outlining those products of the cyclooxygenase pathway (thromboxane, TxA_2 and prostacyclin, PGI_2) that may be involved in determining the severity of ischaemia and reperfusion-induced arrhythmias

only to the arachidonic cascade reaction through both the cyclooxygenase and lipoxygenase pathways, but also to other events which may have a bearing on arrhythmogenesis. These include the accumulation of lysophospholipids in the ischaemic myocardial zone and the initiation of the phosphatidylinositol response. It is clear that some amphiphiles, such as lysophosphoglycerides and long chain acyl carnitines, are factors which may well be involved in arrhythmogenesis (Corr and Sobek 1982; McGrath, this volume, Chap. 22). Phospholipase A_2 is itself arrhythmogenic in cultured heart cells (Wenzel and Innis 1983) and antiphospholipase procedures such as the administration of coenzyme Q_{10} (Ajioka et al. 1986), mepacrine (Otani et al. 1986) and dexamethasone all either reduce ischaemia or reperfusion-induced arrhythmias and/or reperfusion-induced myocardial injury.

There is good experimental and clinical evidence for the rapid activation of the arachidonic acid cascade in both experimental and clinical myocardial ischaemia. Some of the early evidence has been reviewed (Coker 1982). In brief, it is that PGE_2 and PGF_α appear in increasing amounts in coronary sinus blood following occlusion of a coronary artery. This release commences during the first 20–30 min of ischaemia but does not appear to be related to the generation of the early phase of arrhythmias that result from coronary artery obstruction (Parratt and Coker 1980; Coker et al. 1981a). Of more interest is the finding that a more rapid and substantial release of the breakdown products of thromboxane and prostacyclin occurs very soon (minutes) after the onset of myocardial ischaemia (Coker et al. 1981b) and that this release appears to be related to the occurrence and severity of early arrhythmias (Coker et al. 1981c). The myocardial release of thromboxane and prostacyclin also occurs in patients with both stable and variant angina pectoris (Tada et al. 1981) and acute myocardial infarction (MacDonald et al. 1983; Tada et al. 1985; Henriksson et al. 1986). Thromboxane production under these conditions probably reflects a generalized increase in platelet aggregability, especially since catecholamine levels are also increased at this time and adrenaline potentiates thromboxane release from platelets (Purchase et al. 1986). Thromboxane generation by myocardial tissue has also been reported (Brandt et al. 1984; Mehta and Mehta 1985). Evidence is accumulating that thromboxane may be one of the endogenous arrhythmogenic substances released during myocardial ischaemia and that prostacyclin may be one of the endogenous antiarrhythmic factors generated by the ischaemic myocardium. The evidence comes from studies relating the amounts of these eicosanoids generated during ischaemia to the severity of ischaemic and reperfusion arrhythmias (Coker et al. 1981c; Parratt et al. 1987), from studies involving dietary modifications of membranal phospholipids with linoleic acid (Leprán et al. 1981a) and from experiments designed to modify the balance between the relative amounts of thromboxane and prostacyclin generated.

The evidence regarding the relationship between relative amounts of thromboxane and prostacyclin released and the severity of ischaemia-induced arrhythmias applies only to those arrhythmias that arise within 2 min of coronary artery occlusion. Although there is no evidence as yet that total amounts of eicosanoids generated during the entire occlusion period relate to arrhythmia severity either during ischaemia or in any subsequent reperfusion period, there are a number of experimental studies indicating that feeding animals (rats) with a diet rich in linoleic acid (usually as sunflower seed oil) reduces the severity of arrhythmias resulting from coronary artery occlusion (Leprán et al. 1981a; Szekeres et al. 1982, in conscious rats; McLennan et al. 1985 in anaesthetized rats). For example, in the Lepran et al. 1981a study, survival was 81% in those rats treated with 12% sunflower seed oil for 3 months (LAR diet) compared with only 19% in the controls (assessed at 20 min of occlusion). This treatment altered the lipid content of myocardial cell membranes, leading to an increase in 20:4 and 22:4 fatty acids in membrane phospholipids and a reduction in 22:5 and 22:6 fatty acids; the ratio of n-6/n-3 fatty acids was thus increased by up to seven times. Since this protection is prevented if rats are pretreated just before coronary artery occlusion with the cyclooxygenase inhibitor indomethacin, it is suggested that this beneficial effect of the LAR diet is due to alterations in prostaglandin synthesis. Although such a linoleic acid rich diet reduces the amount of thromboxane released by human platelets (Needleman et al. 1982) there is no evidence as yet that it beneficially modifies the ratio of myocardial prostacyclin (or

PGE_2) to thromboxane release in favour of prostacyclin. Indeed, there is some evidence suggesting that prostacyclin release from cultured endothelial cells is reduced following enrichment with linoleic acid (SPECTOR et al. 1980).

The most satisfactory evidence for the importance of the prostacyclin/thromboxane balance as one determinant of the severity of ischaemia- and reperfusion-induced arrhythmias comes from drug studies involving selective antagonism or synthesis inhibition of thromboxane and the selective "boosting" of myocardial prostacyclin production. This evidence will now be reviewed:

C. Evidence for the Arrhythmogenic Effect of Thromboxane A_2

I. The Thromboxanemimetic U46619 Induces Arrhythmias When Given During Myocardial Ischaemia

There are two studies that indicate that stimulation of TxA_2 receptors induces ventricular ectopic activity in dogs. One of these studies relates to a time when spontaneous, early ischaemia-induced ectopic activity had subsided (PARRATT and WAINWRIGHT 1986a) and both studies used a thromboxanemimetic (U46619) rather than the in vivo generation of TxA_2. In greyhound dogs the direct local injection of U46619 into the ischaemic myocardium (through a catheter inserted into the occluded left anterior descending coronary artery) resulted in increases in peripheral coronary pressure and also in ventricular ectopic activity (single or multiple ectopic beats and, occasionally, ventricular tachycardia; PARRATT and WAINWRIGHT 1986a). In this respect U46619 was considerably (10–50 times) less active than noradrenaline. In anaesthetized rats subjected to coronary artery occlusion (method described by CLARK et al. 1980) an intravenous infusion of U46619 (which had no effect on either systemic arterial blood pressure or heart rate) increased the severity of early ischaemia-induced ventricular arrhythmias; both the number of ventricular ectopic beats and the duration of ventricular tachycardia (VT) were significantly increased in animals receiving U46619 compared with control, coronary artery ligated rats; the incidence and duration of ventricular fibrillation VF were also somewhat increased (PARRATT et al. 1987).

There is one study, the results of which suggest that thromboxane A_2 may not be involved in ischaemic and reperfusion arrhythmias. BURKE et al. (1982) used very high doses of a rather unspecific drug, imidazole, to inhibit thromboxane synthesis. This treatment failed to modify either ischaemic or reperfusion arrhythmias in mongrel dogs subjected to coronary artery occlusion (for 1 h) followed by reperfusion. However, this proved not to be a severe model for arrhythmias (none of the dogs fibrillated) and, more importantly, prostacyclin breakdown products were not measured. It is of course possible that thromboxane might not be involved in the relatively benign arrhythmias which arise during reperfusion following prolonged periods of ischaemia.

II. Effects of Selective Inhibition of Thromboxane Synthesis

We have used three highly selective inhibitors of thromboxane synthetase (COKER et al. 1982; COKER and PARRATT 1983a; COKER 1984) and the results obtained with two of these (dazoxiben and dazmegrel) are illustrated in Table 2. That there was indeed selective inhibition of thromboxane production is demonstrated by the fact that the elevated coronary venous levels of thromboxane that normally occur following coronary artery occlusion were not seen in dogs treated with dazoxiben (or dazmegrel). In contrast, the amounts of 6-keto $PGF_{1\alpha}$ (an index of prostacyclin production) found in local coronary venous blood (draining predominantly from the ischaemic region) were unchanged by these drugs.

This selective inhibition of thromboxane production had a marked effect on the incidence of reperfusion-induced VF and thus of survival from an ischaemia-reperfusion insult. Table 2 shows that the incidence of reperfusion-induced VF in control animals was between 78% and 88%; this was reduced to 12.5% (with dazoxiben) and 29% (with dazmegrel). Total survival was thus increased from 12.5% to 20% (in the controls) to between 71% and 88% in those animals in which thromboxane generation had been suppressed. However, this marked protective effect is only seen when the drugs are given before the ischaemic insult. There is no reduction in reperfusion-induced VF (and no increase in total survival) when, for example, dazmegrel is given *after* the onset of ischaemia but (15 min) *before* reperfusion (COKER and PARRATT 1985a), even though there is a significant reduction in local coronary venous thromboxane concentrations within 5 min of dazmegrel administration. This result shows that (1) dazmegrel has no "direct" antiarrhythmic action (con-

Table 2. Effects of selective inhibition of thromboxane synthesis (dazoxiben, dazmegrel) and of selective thromboxane antagonists (AH 23848, BM 13.177) on the severity of arrhythmias resulting from coronary artery occlusion (ischaemia) and release (reperfusion) in anaesthetized greyhound dogs. (Data from COKER und PARRATT 1983a, 1985b; COKER 1984; PARRATT and WAINWRIGHT 1986b)

	Ventricular ectopic beats	VF (on reperfusion)	Survival (%)
Controls	875 ± 264	7/8	13
Dazoxiben	511 ± 141	1/8[a]	88[a]
Controls	832 ± 158	7/9	20
Dazmegrel	193 ± 126[a]	2/7[a]	71[a]
Controls	736 ± 155	7/8	11
AH 23848	339 ± 111	2/8	67
Controls	1084 ± 159	6/7	10
BM 13.177	544 ± 179[b]	4/9[c]	50[b]

[a] $P<0.01$; [b] $P<0.05$; [c] $P<0.07$

firmed by an absence of effect on action potentials recorded in vitro from ventricular muscle cells) and (2) that it is the generation of thromboxane early in ischaemia and its effects at that time that determines the severity of reperfusion-induced arrhythmias. This is confirmed by studies (see below) with a thromboxane antagonist, AH 23848. Although this afforded marked protection if given prior to the onset of ischaemia it also, like inhibition of thromboxane synthesis, failed to protect when given just prior to reperfusion.

Two recent studies have confirmed an antiarrhythmic effect following suppression of thromboxane generation. In anaesthetized cats the thromboxane synthetase inhibitor CGS-13080 attenuated, in a dose-dependent manner, the decrease in VF threshold that occurs following coronary artery occlusion and reduced the incidence of ischaemia-induced ventricular ectopic activity (O'Connor et al. 1986). In sedated, conscious dogs subjected to pneumatic occlusion of the left circumflex coronary artery, the thromboxane synthetase inhibitors RO22-4679 and U-63557A strikingly reduced the incidence of ventricular fibrillation from 33%–38% (in the vehicle-treated controls) to 6% and 0% respectively (Hammon and Oates 1986). Since this protection was abolished by indomethacin the conclusion was drawn that this increased survival was due to shunting of endoperoxides to other prostaglandins, including prostacyclin. Prostanoids were not measured but this is an important study since it shows, for the first time, a protective effect of this class of drug in a relevant conscious animal model of sudden cardiac death.

III. Effects of Thromboxane Receptor Blocking Drugs

An alternative approach to preventing the formation of thromboxane during ischaemia would be to antagonize the effects of formed and released thromboxane at receptor level. Two recently introduced thromboxane antagonists have been used, AH23848, ([1_α(I),2_β,5_α]-(+)-7-[5-[(1,1′-biphenyl)-4-yl]methoxy]-2-(4-morpholinyl)-3-oxocyclopentyl]-4-heptenoic acid and BM 13,177 (4-[2-(benzenesulphonamido)-ethyl]-phenoxyacetic acid; Coker and Parratt 1985b, Parratt and Wainwright 1986b). These compounds were administered intravenously to greyhound dogs (in doses that prevented the haemodynamic effects of U46619) prior to the onset of ischaemia or reperfusion. The results are also summarized in Table 2. As with the specific thromboxane synthetase inhibitors (Table 2) the results show that blockade of thromboxane receptors slightly decreases the severity of ischaemic arrhythmias and markedly reduces the incidence of reperfusion-induced VF; total survival is increased from 10%–11% in the controls to between 50% and 67% ($P<0.05$). In this model AH23848 is about 12 times more active than BM 13,177. AH23848 did not significantly reduce the incidence of reperfusion-induced VF when given 25 min *after* the onset of ischaemia (i.e. 15 min prior to reperfusion; Coker and Parratt 1985b). This result is thus similar to the effect of giving dazmegrel at this time. Taken together they strongly suggest that reperfusion-induced VF is dependent on the release and actions of thromboxane A_2 (or some consequence of these actions) *during* ischaemia. Since prevention of further release late in ischaemia or blockade of the thromboxane receptors immediately prior to reperfusion is ineffective, it suggests that it is the relatively early release of thromboxane that triggers further events eventually leading (on reperfusion) to VF. The most likely such event would appear to be platelet aggregation leading to a more pronounced degree of ischaemia; BM 13,177 has indeed been shown to reduce certain enzymatic and electrocardiographic consequences of coronary artery occlusion in anaesthetized cats (SCHRÖR and THIEMERMANN 1986). Coronary vasoconstriction cannot as yet be ruled out as a partial explanation for an arrhythmogenic effect of thromboxane. Not surprisingly, in view of its short half-life, there is no evidence as yet that thromboxane has any direct effects on cardiac muscle action potentials. A study (KRAMER et al. 1985) of a stable thromboxane analogue revealed no effect on any measured action potential parameter in slices of ventricular muscle removed from infarcted and normal regions of the canine myocardium.

D. Evidence for an Antiarrhythmic Effect of Prostacyclin

I. Myocardial Prostacyclin Generation and Early Ischaemia and Reperfusion-Induced Ventricular Arrhythmias

If it is the balance between the amounts of thromboxane and prostacyclin generated that is an important determining factor governing the severity of early ischaemia and reperfusion-induced arrhythmias, then an alternative approach to thromboxane suppression as an antiarrhythmic procedure in these circumstances would be enhancement of prostacyclin generation.

1. Antiarrhythmic Effects of Prostacyclin (and Related Stable Derivatives) During Myocardial Ischaemia and Reperfusion

The apparent controversy in the literature regarding the effect of prostacyclin on ventricular arrhythmias is almost certainly due to the potential antiarrhythmic property of this prostanoid being masked by a reflex increase in cardiac sympathetic drive resulting from its powerful vasodilator properties. Large doses given intravenously are usually arrhythmogenic (AU et al. 1980; DIX et al. 1979; KOWEY et al. 1982) but, especially if the fall in blood pressure is not pronounced, prostacyclin reduces the severity of ischaemia-induced arrhythmias in rats (AU et al. 1979; COKER and PARRATT 1981; JOHNSTON et al. 1983), pigs (HIRCHE et al. 1982; ZYLKA et al. 1981) and dogs (AU et al. 1979; RIBEIRO et al. 1981; COKER and PARRATT 1983b; KOWEY et al. 1982; JOUVE et al. 1985 in anaesthetized dogs; STARNES et al. 1982; FIEDLER and MARDIN 1985a, 1986 in conscious dogs). The incidence of ventricular fibrillation during ischaemia was particularly reduced (e.g. from 60% to 25% in the JOUVE et al. (1985) study and from 100% to 18%-25% in the FIEDLER and MARDIN 1986 study). Further, this protection was observed whether ischaemia was induced by mechanical or thrombotic coronary artery occlusion.

In contrast to this protective effect of exogenous prostacyclin against early ischaemia and reperfusion-induced arrhythmias, it has minimal effects against those arrhythmias that arise once the infarct has fully developed (HASHIMOTO et al. 1983 in conscious dogs) or against ventricular arrhythmias induced by programmed electrical stimulation in patients with apparently healthy hearts (BREMBILLA-PERROT et al. 1985).

The potential antiarrhythmic effects of more stable prostacyclin analogues have not been so fully examined. The intracoronary administration of iloprost (ZK 36374) reduced the severity of ischaemia-induced arrhythmias and markedly suppressed reperfusion-induced VF in anaesthetized greyhound dogs (COKER and PARRATT 1983b) and the systemic administration of iloprost reduced the incidence of VF induced by intracoronary microspheres in anaesthetized pigs (de LANGEN et al. 1985). Reperfusion-induced VF is also reduced in rat isolated Langendorff-perfused hearts pretreated with iloprost (VAN GILST et al. 1985). Iloprost also reduces arrhythmias induced by aconitine and G-strophanthin in isolated cardiac preparations from rabbits and frogs (AKSULU et al. 1985). Another stable prostacyclin analogue, CG 4203, suppresses ischaemia-induced VF in anaesthetized rats subjected to coronary artery ligation (MÜLLER et al. 1984).

The exact mechanism of the antiarrhythmic effects of prostacyclin and related stable analogues is uncertain. There are several possibilities. A direct membranal cytoprotective mechanism has been invoked (NAYLER et al. 1984) as have inhibition of platelet aggregation (and reduced thromboxane release; KOWEY et al. 1983; MICHAEL et al. 1986), reduced release of noradrenaline from cardiac sympathetic nerves (ZYLKA et al. 1981; SCHRÖR et al. 1981a), a reduction in myocardial oxygen demand, in the degree and extent of ischaemia and in resultant cellular damage (SCHRÖR et al. 1982), inhibition of free radicle production (THIEMERMANN et al. 1984), inhibition of adrenaline-stimulated platelet aggregation (BERTHA and FOLTS 1984) and inhibition of adrenaline-stimulated myocardial lipid peroxidation (HERBACZYNSKA-CEDRO and GORDON-MAJSZAK 1986) and coronary vasodilatation. Most studies have failed to demonstrate any significant direct effect of prostacyclin on cardiac muscle action potentials. Some of these possible mechanisms for the protective effect of prostacyclin and iloprost in myocardial ischaemia have been discussed in depth by SCHRÖR et al. (1981b) and by SMITH et al. (1984). In isolated cardiac muscle (atria, papillary muscles) low concentrations of PGA_1, PGE_1, PGE_2 and, to a lesser extent, $PGF_{2\alpha}$ and prostacyclin increase the resting membrane potential, the rate of depolarization (V_{max}) and the overshoot of transmembrane action potentials (APs) (KECSKEMETI et al. 1973, 1978; KECSKEMETI 1980). This ability modestly to improve depressed conduction velocity might explain why several prostaglandins (PGE_1, PGE_2, $PGF_{2\alpha}$, PGA_2) have been shown to be antiarrhythmic in several different species against arrhythmias induced by a variety of chemical agents including calcium, aconitine, noradrenaline and ouabain (ZIJLSTRA et al. 1972; FÖRSTER et al. 1973; FÖRSTER 1976; MEST and FÖRSTER 1978; KELLIHER and GLENN 1973; MADAN et al. 1974; SOMBERG et al. 1977). This antiarrhythmic effect is rather modest and is probably due to the ability of these prostaglandins to improve depressed conduction since higher concentrations of all those prostaglandins so far studied reduce V_{max} and decrease APs in isolated cardiac muscle preparations (KECS-

KEMETI 1980). One interesting recent finding (SZEKERES 1986, personal communication) is that as long as 2 h after the i.m. administration of the stable 7-oxo-prostacyclin in dogs, ventricular refractory period was markedly prolonged.

2. Promotion of Prostacyclin Generation as an Antiarrhythmic Procedure in Acute Myocardial Ischaemia and in Reperfusion

This was first demonstrated for the Bayer compound BAY g 6575 (nafazatrom) in 1984 (COKER and PARRATT 1984a). Nafazatrom has both antithrombotic (SEUTER et al. 1979) and antimetastatic activity (HONN and DUNN 1982). In thrombosis it acts by stimulating prostacyclin release from blood vessel walls (VERMYLEN et al. 1979; CHAMONE et al. 1981), including the coronary vessels (COKER and PARRATT 1984a), and by inhibiting its breakdown by 15-hydroxy-prostaglandin dehydrogenase (WONG et al. 1982). It also inhibits 5-lipoxygenase (HONN and DUNN 1982).

In dogs, the most pronounced effect of nafazatrom administration is a dramatic reduction in the incidence of VF (from 86% in the controls to only 14%) induced by reperfusion following a 40-min coronary artery occlusion (COKER and PARRATT 1984a) and a marked reduction in ventricular ectopic activity following reperfusion after a 6-h coronary artery occlusion (FIEDLER et al. 1985b). The incidence and severity of both early and late ischaemia-induced arrhythmias are only slightly modified (COKER and PARRATT 1984a; FIEDLER 1983b). These effects are thus rather similar to those seen following the local intracoronary administration of prostacyclin (see above) and may be due to a marked increase in the amount of prostacyclin generated by the ischaemic myocardium. Nafazatrom has also been shown to exhibit antiarrhythmic activity in conscious rats (FIEDLER 1983a) and to reduce myocardial ischaemic damage in rats (FIEDLER 1983a), rabbits (FIEDLER 1984), dogs (SHEA et al. 1984; FIEDLER et al. 1985; FIEDLER and MARDIN 1985b) but not pigs (FIEDLER 1985). The most likely explanations for this reduced myocardial damage are (1) a shift in the prostacyclin/thromboxane balance and (2) 5-lipoxygenase inhibition leading to inhibition of inflammatory mediator release from activated neutrophils.

There are other antithrombotic agents which may have a mechanism of action similar to that of nafazatrom in that they "promote" prostacyclin production. These include trapidil (OHNISHI et al. 1981; SAKANASHI et al. 1983), difibrotide, which reduces the severity of ventricular tachyarrhythmias in cats subjected to coronary artery occlusion (LÖBEL and SCHRÖR 1985; THIEMERMANN et al. 1985) and the angiotensin-converting enzyme inhibitors captopril (DÜSING et al. 1983) and ramipril. The stimulation of prostacyclin production by ACE inhibitors may be secondary to kinin production. Captopril has been shown to reduce ventricular arrhythmias induced by coronary artery occlusion and one mechanism could be the "promotion" of prostacyclin.

Prostacyclin "promotion" may also contribute to the antiarrhythmic actions of some calcium antagonists (e.g. nifedipine and diltiazem, COKER and PARRATT 1983c; MEHTA et al. 1986), some beta-adrenoceptor-blocking drugs

(KAHLEN and SCHRÖR 1982) and the antianginal drugs carbochromen and molsidomine, which also reduce both ischaemia and reperfusion-induced arrhythmias in anaesthetized dogs (FIEDLER et al. 1985a).

E. Antiarrhythmic Effects of Cyclooxygenase Inhibitors

I. Studies with Aspirin

Most emphasis has been on aspirin; there have been relatively few studies, except in anaesthetized rats, with other inhibitors of cyclooxygenase. Indeed, there is some evidence that aspirin differs from other cyclooxygenase inhibitors and the protection afforded by this drug may be related only in part to a transient, dose-dependent inhibition of platelet cyclooxygenase.

The first indication that aspirin reduces the severity of arrhythmias was the study of MOSCHOS et al. (1972) in anaesthetized dogs. They induced a thrombus in a major branch of the left coronary artery with an electrode catheter. This resulted in ventricular ectopic activity and a high incidence of VF (five of eight control animals). The incidence of arrhythmias, especially VF, was significantly reduced in the presence of aspirin (30 or 60 mg/kg either orally or intravenously).

This beneficial effect was not due to a reduction in the size of the thrombus formed; indeed thrombus weight was slightly increased by aspirin. In a follow-up study (MOSCHOS et al. 1973) it was demonstrated that this reduction in the incidence of arrhythmias (83% down to 30%) and especially of VF (67% down to 12%) was associated with a reduction in the extent of microcirculatory platelet thrombosis rather than of arterial thrombus formation. The suggestion was made that arrhythmias were induced by multiple platelet emboli, aggregating in intramyocardial vessels, mechanically obstructing collateral blood flow, particularly to the conducting system. The higher incidence of ^{51}Cr-labelled platelets within the ischaemic region (average radioactivity ratio of ischaemic:normal areas of 3.3) was much reduced (to 1.35) following the oral administration of aspirin (approximately 30 mg/kg per day for 7 days prior to the induction of thrombosis). That this cannot be the only explanation for an antiarrhythmic effect of aspirin was demonstrated by the fact that the ventricular arrhythmias resulting from non-thrombotic coronary occlusions (induced with a balloon catheter) were also reduced by the oral administration of aspirin (30 mg/kg daily for 7 days). Thus the incidence of VF in these animals was reduced from 39% to only 5% (MOSCHOS et al. 1978, 1980a). An antiarrhythmic action of aspirin in various dog models has been confirmed by JOHNSON et al. (1981), in a coronary artery embolization model and by COKER et al. (1981d) in greyhound dogs following mechanical obstruction of a major branch of the left coronary artery. In the latter study the reduced incidence of ventricular ectopic beats during ischaemia was observed with much lower doses of aspirin than those used previously (1–4 mg/kg) and was linked with reduced myocardial thromboxane (but not prostacyclin) production. Aspirin was also shown to reduce the severity of ventricular arrhythmias when given intravenously *after* the onset of ischaemia (Coker and Parratt 1984b) and markedly to reduce the incidence of reperfusion-induced VF (Coker and Parratt 1984b, 1985c). However, this protection was lost if coronary artery occlusion was induced 24 h after the intravenous administration of aspirin, i.e. at a time when the myocardial generation of *both* thromboxane and prostacyclin was inhibited (Coker and Parratt 1985c).

In anaesthetized rats subjected to coronary artery ligation aspirin (given intravenously) in doses of 1-100 mg/kg reduced the number of ventricular ectopic beats, the incidence and duration of VT, the incidence and duration of VF (often a non-terminal event in this species) and reduced mortality (Coker and Parratt 1981; Fagbemi 1984). These effects were especially marked if aspirin was combined with a beta-adrenoceptor-blocking drug such as propranolol, practolol or oxprenolol (Fagbemi 1985), the beta-antagonist preferentially suppressing the immediate (phase la) early ischaemic arrhythmias (within 0-10 min in this model of occlusion) whilst aspirin was more effective on the later, phase 1b, arrhythmias (those occurring from 15-30 min of occlusion). In conscious rats aspirin, in doses from 4-200 mg/kg orally or 100 mg/kg intravenously, reduced the incidence and severity of early postocclusion arrhythmias and increased survival (Leprán et al. 1981b; Johnston et al. 1983). For example, with the higher dose used (200 mg/kg) survival was increased from only 33% in the controls to 67% (Leprán et al. 1981b). Later arrhythmias (those occurring up to 4 h of occlusion) were not significantly reduced (Johnston et al. 1983).

As indicated above, there is good evidence that cyclooxygenase inhibition is not the only, or even the predominant, cause of this protection afforded by aspirin in the early stages of myocardial ischaemia. This evidence is:

1. The doses required for protection in both dogs and rat models are often extremely high (100-200 mg/kg) whereas selective inhibition of thromboxane synthesis can be achieved, at least in the dog, by doses as low as 1 mg/kg (Coker and Parratt 1984b).
2. There is a clear difference between indomethacin and aspirin in the conscious rat dietary studies of Lepran et al. (1981a). In these the protection afforded by supplementing the diet with linoleic acid increased survival following coronary artery occlusion from 19% to 81%. The oral administration of indomethacin 1 h prior to occlusion abolished this protective effect (survival rate 12%) whereas oral aspirin administration (200 mg/kg 1 h prior to occlusion) did not influence survival rate (83%). One possible explanation for this result is that one mechanism of aspirin protection was some effect other than on cyclooxygenase.
3. The salicylate ion is itself antiarrhythmic in the conscious rat model (in doses lower than those of acetylsalicylic acid; Lepran et al. 1981b) and is more active than aspirin at inhibiting arrhythmias produced by reoxygenation of hypoxic rat heart cells in tissue culture (Wenzel and Innis 1983). In sheep ventricular fibres sodium salicylate prolongs the action potential and reverses the shortening of action potential duration induced by ouabain (Cohen et al. 1979). It would be of interest to determine whether either aspirin or sodium salicylate is able to prevent the shortening of the

action potential duration that occurs in isolated ventricular cells subjected to conditions (increased K^+, decreased PO_2 and pH) that mimic those occurring during ischaemia (BOACHIE-ANSAH et al. 1986).

Perhaps the most likely explanation for the beneficial effect of aspirin in the animal arrhythmia models outlined above is a combination of thromboxane synthetase inhibition, inhibition of platelet aggregation and perhaps, with higher doses, a direct effect of the salicylate ion on ischaemic myocardial cells. Which of these predominates would have a bearing on the dose of aspirin used in, for example, secondary prevention trials in patients with myocardial infarction and in the treatment of unstable angina. In both of these situations aspirin administration (but in a variety of doses) reduces mortality (e.g. LEWIS et al. 1983; CAIRNS et al. 1985; MAY et al. 1982). However, there are other possible mechanisms for an antiarrhythmic action of aspirin. These include prevention of the cyclical reduction in coronary blood flow in the presence of a fixed stenosis (FOLTS et al. 1976, 1982), an increase in epicardial coronary collateral blood flow (DAVENPORT et al. 1981), inhibition of platelet trapping in ischaemic myocardial areas (RUF et al. 1980), inhibition of FFA release (MOSCHOS et al. 1978) and decreased myocardial loss of K^+ (MOSCHOS et al. 1978).

II. Studies with Other Cyclooxygenase Inhibitors

Evidence for an antiarrhythmic effect of other cyclooxygenase inhibitors is much less convincing. In anaesthetized rats subjected to coronary artery occlusion, sodium meclofenamate and indomethacin (FAGBEMI 1984) especially when combined with beta-adrenoceptor-blocking drugs (FAGBEMI 1985) reduced the incidence and severity of early ischaemic arrhythmias. Flurbiprofen is inactive in this model (COKER and PARRATT 1981). Indomethacin and sulphinpyrazone also increase survival in conscious rats subjected to coronary artery occlusion (BRUNNER et al. 1980; LEPRÁN et al. 1981b) although, somewhat surprisingly, the protective effect of indomethacin is not observed in rats simultaneously fed on a diet rich in linoleic acid, despite the fact that this diet is itself protective (LEPRÁN et al. 1981a). Arrhythmias produced by reoxygenation of isolated neonatal rat heart myocytes are also prevented by these two drugs (WENZEL and INNIS 1983). Sulphinpyrazone also prevents adrenochrome-induced ventricular arrhythmias and "sudden death" in anaesthetized rats (BEAMISH et al. 1981) and adriamycin-induced arrhythmias in rabbits (RABKIN and OHMAE 1980).

In contrast to these generally beneficial effects in rats there is little evidence for an antiarrhythmic effect of cyclooxygenase inhibitors (apart from aspirin and sulphinpyrazone) in other species. Ibuprofen and indomethacin increase the sensitivity of dogs to the arrhythmogenic effect of ouabain whilst indomethacin (MOSCHOS et al. 1980a), sodium meclofenamate and flurbiprofen (PARRATT and WAINWRIGHT, unpublished data) fail to modify the severity of either ischaemia or reperfusion-induced ventricular fibrillation in dogs subjected to non-thrombotic coronary artery occlusion.

Antiarrhythmic effects of sulphinpyrazone have been demonstrated in anaesthetized cats (KELLIHER et al. 1980; DIX et al. 1982) and dogs subjected to coronary artery occlusion (MOSCHOS et al. 1980b; POVALSKI et al. 1980) but not in pigs subjected to the same procedure (STÄUBLI et al. 1984). This protective action does not appear to be due to a direct action on cardiac cell membranes (BENDITT et al. 1980). It may be due to preferential maintenance of prostacyclin production in blood vessels despite platelet cyclooxygenase inhibition (LIVIO et al. 1980).

References

Ajioka M, Nagai S, Ogawa K, Satake T, Sugiyama S, Ozawa T (1986) The role of phospholipase in the genesis of reperfusion arrhythmia. J Electrocardiol 19:165–172

Aksulu HE, Ercan ZS, Turker RK (1985) Further studies on the antiarrhythmic effect of iloprost. Arch Int Pharmacodyn Ther 277:223–234

Au TLS, Collins EA, Harris CJ, Walker MJA (1979) The action of prostaglandin I_2 and prostaglandin E_2 on arrhythmias produced by coronary occlusion in the rat and dog. Prostaglandins 18:707–720

Au TLS, Harvie CJ, Johnston K, MacLeod BA, Walker MJA (1980) The effect of prostaglandin infusions on arrhythmic and other responses to coronary artery ligation. In: Förster, W (ed) Prostaglandins and thromboxanes. Fischer, Jena, pp 37–53

Beamish RE, Dhillon KS, Singal PK, Dhalla NS (1981) Protective effect of sulfinpyrazone against catecholamine metabolite adrenochrome-induced arrhythmias. Am Heart J 102:149–152

Benditt DG, Grant AO, Hutchison ABS, Strauss HC (1980) Electrophysiological effects of sulfinpyrazone on canine cardiac Purkinje fibers. Can J Physiol Pharmacol 58:738–742

Bertha BG, Folts JD (1984) Inhibition of epinephrine-exacerbated coronary thrombus formation by prostacyclin in the dog. J Lab Clin Med 103:204–214

Boachie-Ansah G, Kane KA, Parratt JR (1986) Effects of adenosine, ATP and noradrenaline on normal and mildly "ischaemic" sheep cardiac Purkinje action potentials. Br J Pharmacol 88:309P

Brandt R, Nowak J, Sonnenfeld T (1984) Prostaglandin formation from exogenous precursor in homogenates of human cardiac tissue. Basic Res Cardiol 79:135–141

Brembilla-Perrot B, Terrier De La Chaise A, Clozel JP, Cherrier F, Faivre G (1985) Proarrhythmic and antiarrhythmic effects of intravenous prostacyclin in man. Eur Heart J 6:609–614

Brunner L, Stepanek J, Brunner H (1980) Reduction of mortality by sulfinpyrazone after experimental myocardial infarction in the rat. J Pharm Pharmacol 32:714–715

Burke SE, Antonaccio MJ, Lefer AM (1982) Lack of thromboxane A_2 involvement in the arrhythmias occurring during acute myocardial ischaemia in dogs. Basic Res Cardiol 77:411–422

Cairns JA, Gent M, Singer J, Finnie K, Froggatt GM, Holder DA, Jablonsky G, Kostuk WJ, Melendez LJ, Myres MG, Sackett DL, Sealey BJ, Tanser PH (1985) Aspirin, sulfinpyrazone, or both in unstable angina. N Engl J Med 313:1369–1375

Chamone DAF, van Damme B, Carreras LO, Vermylen J (1981) Increased release of vascular prostacyclin-like activity after long-term treatment of diabetic rats with Bay g6575. Haemostasis 10:297–303

Clark C, Foreman MI, Kane KA, McDonald FM, Parratt JR (1980) Coronary artery ligation in anaesthetised rats as a method of the production of experimental dys-

rhythmias and for the determination of infarct size. J Pharmacol Methods 76:504-506
Cohen I, Noble D, Ohba M, Ojeda C (1979) The interaction of ouabain and salicylate on sheep cardiac muscle. J Physiol (Lond) 297:187-205
Coker SJ (1982) Early arrhythmias arising from acute myocardial ischaemia; possible involvement of prostaglandins and thromboxanes. In: Parratt JR (ed) Early arrhythmias resulting from myocardial ischaemia. Oxford University Press, New York, pp 219-237
Coker SJ (1984) Further evidence that thromboxane exacerbates arrhythmias: effects of UK38485 during coronary artery occlusion and reperfusion in anaesthetized greyhounds. J Mol Cell Cardiol 16:633-641
Coker SJ, Parratt JR (1981) The effects of prostaglandins E_2, $F_{2\alpha}$, prostacyclin, flurbiprofen and aspirin on arrhythmias resulting from coronary artery ligation in anaesthetized rats. Br J Pharmacol 74:155-159
Coker SJ, Parratt JR (1983a) Effects of dazoxiben on arrhythmias and ventricular fibrillation induced by coronary artery occlusion and reperfusion in anaesthetised greyhounds. Br J Clin Pharmacol 15:87S-95S
Coker SJ, Parratt JR (1983b) Prostacyclin-antiarrhythmic or arrhythmogenic? Comparison of the effects of intravenous and intracoronary prostacyclin and ZK36374 during coronary artery occlusion and reperfusion in anaesthetised greyhounds. J Cardiovasc Pharmacol 5:557-567
Coker SJ, Parratt JR (1983c) Nifedipine reduces arrhythmias but does not alter prostanoid release during coronary artery occlusion and reperfusion in anaesthetised greyhounds. J Cardiovasc Pharmacol 5:406-417
Coker SJ, Parratt JR (1984a) The effects of nafazatrom on arrhythmias and prostanoid release during coronary artery occlusion and reperfusion in anaesthetized greyhounds. J Mol Cell Cardiol 16:43-52
Coker SJ, Parratt JR (1984b) Aspirin in the early stages of myocardial infarction. In: Royal Society of Medicine International Congress and Symposium Series 71:71-75
Coker SJ, Parratt JR (1985a) Relationships between the severity of myocardial ischaemia, reperfusion-induced ventricular fibrillation, and the late administration of dazmegrel or nifedipine. J Cardiovasc Pharmacol 7:327-334
Coker SJ, Parratt JR (1985b) AH23848, a thromboxane antagonist, suppresses ischaemia and reperfusion-induced arrhythmias in anaesthetized greyhounds. Br J Pharmacol 86:259-264
Coker SJ, Parratt JR (1985c) Does inhibition of thromboxane synthesis account for the antiarrhythmic activity of aspirin in canine myocardial ischaemia and reperfusion? J Mol Cell Cardiol 17 [Suppl 3] Abstract 109
Coker SJ, Marshall RJ, Parratt JR, Zeitlin IJ (1981a) Does the local myocardial release of prostaglandin E_2 or $F_{2\alpha}$ contribute to the early consequences of acute myocardial ischaemia? J Mol Cell Cardiol 13:425-434
Coker SJ, Ledingham IMcA, Parratt JR, Zeitlin IJ (1981b) Thromboxane B_2 and 6-keto $PGF_{1\alpha}$ release into coronary venous blood: effect of coronary artery occlusion. J Physiol (Lond) 316:12P
Coker SJ, Parratt JR, Ledingham IMcA, Zeitlin IJ (1981c) Thromboxane and prostacyclin release from ischaemic myocardium in relation to arrhythmias. Nature 291:323-324
Coker SJ, Ledingham IMcA, Parratt JR, Zeitlin IJ (1981d) Aspirin inhibits the early myocardial release of thromboxane B_2 and ventricular ectopic activity following coronary artery occlusion in dogs. Br J Pharmacol 72:593-595

Coker SJ, Parratt JR, Ledingham IMcA, Zeitlin IJ (1982) Evidence that thromboxane contributes to ventricular fibrillation induced by reperfusion of the ischaemic myocardium. J Mol Cell Cardiol 14:438-485
Coker SJ, Dean HG, Kane KA, Parratt JR (1986) The effects of 205-930, a 5-HT antagonist, on arrhythmias and catecholamine release during canine myocardial ischaemia and reperfusion. Eur J Pharmacol 127:211-218
Corr PB, Sobel BE (1982) Amphiphilic lipid metabolism and ventricular arrhythmias. In: Parratt JR (ed) Early arrhythmias resulting from myocardial ischaemia. Oxford University Press, New York, pp 199-218
Davenport N, Goldstein RE, Capurro N, Lipson LC, Bonow RO, Shulman NR, Epstein SE (1981) Sulfinpyrazone and aspirin increase epicardial coronary collateral flow in dogs. Am J Cardiol 47:848-854
De Langen CDJ, van Gilst WH, Wesseling H (1985) Sustained protection by iloprost of the porcine heart in the acute and chronic phases of myocardial infarction. J Cardiovasc Pharmacol 7:924-928
Dix RK, Kelliher GJ, Jurkiewicz N, Lawrence T (1979) The influence of prostacyclin in coronary occlusion induced arrhythmia in cats. Prostaglandins Med 3:173-184
Dix RK, Kelliher GJ, Jurkiewicz N, Smith JB (1982) Effect of sulfinpyrazone on ventricular arrhythmia, prostaglandin synthesis, and catecholamine release following coronary artery occlusion in the cat. J Cardiovasc Pharmacol 4:1068-1076
Düsing R, Scherag R, Landsberg G, Glanzer K, Kramer HJ (1983) The converting enzyme inhibitor captopril stimulates prostacyclin synthesis by isolated rat aorta. Eur J Pharmacol 91:501-504
Fagbemi SO (1984) The effect of aspirin, indomethacin and sodium meclofenamate on coronary artery ligation arrhythmias in anaesthetized rats. Eur J Pharmacol 97:283-287
Fagbemi O (1985) The effects of the combined administration of β-adrenoceptor antagonists and non-steroidal anti-inflammatory drugs on ligation-induced arrhythmias in rats. Br J Pharmacol 85:361-365
Fagbemi O, Leprán I, Parratt JR, Szekeres L (1982) Naloxone inhibits early arrhythmias resulting from acute coronary ligation. Br J Pharmacol 76:504-506
Fiedler VB (1983a) Reduction of myocardial infarction and dysrhythmic activity by nafazatrom in the conscious rat. Eur J Pharmacol 88:263-267
Fiedler VB (1983b) The effects of oral nafazatrom (= BAY g6575) on canine coronary artery thrombosis and myocardial ischemia. Basic Res Cardiol 78:266-280
Fiedler VB (1984) Reduction of acute myocardial ischemia in rabbit hearts by nafazatrom. J Cardiovasc Pharmacol 6:318-324
Fiedler VB (1985) Failure of nafazatrom to reduce infarct size and arrhythmias in a porcine model of acute coronary occlusion. Eur J Pharmacol 114:189-196
Fiedler VB, Mardin M (1985a) Influence of prostacyclin on coronary thrombosis and myocardial ischemia in conscious canine experiments. Arch Int Pharmacodyn Ther 278:114-127
Fiedler VB, Mardin M (1985b) Effects of nafazatrom and indomethacin on experimental myocardial ischemia in the anesthetized dog. J Cardiovasc Pharmacol 7:983-989
Fiedler VB, Mardin M (1986) Prostacyclin prevents ventricular fibrillation in a canine model of sudden cardiac death. Basic Res Cardiol 81:40-53
Fiedler VB, Kettenbach B, Gobel H, Nitz R-E (1985a) Protection by carbochromen and molsidomine against arrhythmias occurring during coronary artery occlusion and reperfusion in dogs. J Cardiovasc Pharmacol 7:964-970
Fiedler VB, Mardin M, Perzborn E, Grützmann R (1985b) The effects of nafazatrom

in an acute occlusion-reperfusion model of canine myocardial injury. Naunyn-Schmiedebergs Arch Pharmacol 331:267-274
Folts JD, Crowell EB, Rowe GG (1976) Platelet aggregation in partially obstructed vessels and its elimination with aspirin. Circulation 54:365-370
Folts JD, Gallagher K, Rowe GG (1982) Blood flow reductions in stenosed canine coronary arteries: vasospasm or platelet aggregation? Circulation 65:248-255
Förster W (1976) Prostaglandins and prostaglandin precursors as endogenous antiarrhythmic principles of the heart. Acta Biol Med Germ 35:1101-1112
Förster W, Mest H-J, Mentz P (1973) The influence of $PGF_{2\alpha}$ on experimental arrhythmias. Prostaglandins 3:895-904
Hammon JW, Oates JA (1986) Interaction of platelets with the vessel wall in the pathophysiology of sudden cardiac death. Circulation 73:224-226
Hashimoto K, Shibuya T, Imai S (1983) Effects of prostaglandins on late coronary ligation arrhythmias in the dog. Jpn J Pharmacol 33:1035-1039
Henriksson P, Wennmalm A, Edhag O, Vesterqvist O, Green K (1986) In vivo production of prostacyclin and thromboxane in patients with acute myocardial infarction. Br Heart J 55:543-548
Herbaczynska-Cedro K, Gordon-Majszak W (1986) Attenuation by prostacyclin of adrenaline-stimulated lipid peroxidation in the myocardium. Pharmacol Res Commun 81:321-332
Hirche HJ, Friedrich R, Kebbel V, McDonald F, Zylka V (1982) Early arrhythmias, myocardial extracellular potassium and pH. In: Parratt JR (ed), Early arrhythmias resulting from myocardial ischaemia. Oxford University Press, New York, pp 113-124
Honn KV, Dunn JR (1982) Nafazatrom (Bag g6575) inhibition of tumor cell lipoxygenase activity and cellular proliferation. FEBS Lett 139:65-68
Johnson GJ, Heckel R, Leis LA, Franciosa J (1981) Effect of inhibition of platelet function with carbenicillin or aspirin on experimental canine sudden death. J Lab Clin Med 98:660-672
Johnston KM, MacLeod BA, Walker MJA (1983) Effects of aspirin and prostacyclin on arrhythmias resulting from coronary artery ligation and on infarct size. Br J Pharmacol 78:29-37
Jouve R, Puddu PE, Langlet F, Guillen J-C, Gautier T, Cano J-P, Serradimigni A (1985) Epoprostenol (PGI_2) prevents postischemic ventricular fibrillation and improves outcome in a canine model of sudden death. J Pharmacol 16:139-157
Kahlen T, Schrör K (1982) Mepindolol protection of prostacyclin formation. Subsequent increase in arachidonic acid-induced prostacyclin release in isolated guinea pig heart. Eur J Pharmacol 82:81-84
Kecskemeti V (1980) Effects of prostacyclin-sodium and prostacyclin-ethyl ester on cardiac transmembrane potentials. In: Fischer, Förster W (ed) Prostaglandins and thromboxanes. Jena, pp 107-109
Kecskemeti V, Kelemen K, Knoll J (1973) Effect of prostaglandin E_1 on the cardiac transmembrane potentials. Eur J Pharmacol 24:289-295
Kecskemeti V, Kelemen K, Knoll J (1978) Comparative effects of prostaglandin F_2 and A_2 on the cardiac transmembrane potentials. Acta Biol Med Germ 37:821-824
Kelliher GJ, Glenn TM (1973) Effect of PGE_1 on ouabain induced arrhythmias. Eur J Pharmacol 24:410-416
Kelliher GJ, Dix RK, Jurkiewicz N, Lawrence TL (1980) Effect of sulfinpyrazone on arrhythmia and death following coronary occlusion in cats. In: McGregor M, Mustard JF, Oliver MF, Sherry S (eds) Cardiovascular actions of sulfinpyrazone: basic and clinical research. Symposium Specialists, Miami, pp 193-209

Kowey PR, Verrier RL, Lown B (1982) Effects of prostacyclin (PGI_2) on vulnerability to ventricular fibrillation in the normal and ischemic canine heart. Eur J Pharmacol 80:83-91

Kowey PR, Verrier RL, Lown B, Handin RI (1983) Influence of intracoronary platelet aggregation on ventricular electrical properties during partial coronary artery stenosis. Am J Cardiol 51:596-602

Kramer JB, Davis AG, Dean R, McCluskey ER, Needleman P, Corr PB (1985) Thromboxane A_2 does not contribute to arrhythmogenesis during evolving canine myocardial infarction. J Cardiovasc Pharmacol 7:1069-1076

Leprán I, Nemecz GY, Koltai M, Szekeres L (1981a) Effect of a linoleic acid-rich diet on the acute phase of coronary occlusion in conscious rats: influence of indomethacin and aspirin. J Cardiovasc Pharmacol 3:847-853

Leprán I, Koltai M, Szekeres L (1981b) Effect of non-steroidal anti-inflammatory drugs in experimental myocardial infarction in rats. Eur J Pharmacol 69:235-238

Lewis HD, Davis JW, Archibald DG (1983) Protective effects of aspirin against acute myocardial infarction and death in men with unstable angina: results of a Veterans Administration co-operations study. N Engl J Med 309:396-403

Livio M, Villa S, De Gaetano G (1980) Long-lasting inhibition of platelet prostaglandin but normal vascular prostacyclin generation following sulphinpyrazone administration to rats. J Pharm Pharmacol 32:718-719

Löbel P, Schrör K (1985) Selective stimulation of coronary vascular PGI_2 but not of platelet thromboxane formation by defibrotide in the platelet perfused heart. Naunyn-Schmiedebergs Arch Pharmacol 331:125-130

Madan BR, Gupta RS, Madan V (1974) Actions of prostaglandins E_1, E_2, $F_{2\alpha}$ in ouabain-induced arrhythmia and maximal electroshock seizures. Indian J Med Res 62:1647-1651

May GS, Eberlein KA, Furberg CD, Passamani ER, De Mays DL (1982) Secondary prevention after myocardial infarction: a review of long-term trials. Prog Cardiovasc Dis 24:331-352

McDonald BR, Jones PBB, Russell RGG, Radford J, Martin JF (1983) Urinary 6-oxo prostaglandin $F_{1\alpha}$ in myocardial infarction. Br Med J 287:727

McLennan PL, Abeywardena MY, Charnock JS (1985) Influence of dietary lipids on arrhythmias and infarction after coronary artery ligation in rats. Can J Physiol Pharmacol 63:1411-1417

Mehta J, Mehta P (1985) Prostacyclin and thromboxane A_2 production by human cardiac atrial tissues. Am Heart J 109:1-3

Mehta J, Mehta P, Ostrowski N (1986) Calcium blocker diltiazem inhibits platelet activation and stimulates vascular prostacyclin synthesis. Am J Med Sci 291:20-24

Mest H-J, Förster W (1978) The antiarrhythmic action of prostacyclin (PGI_2) on aconitine induced arrhythmia in rats. Acta Biol Med Germ 37:827-828

Michael LH, Hunt JR, Lewis RM, Entman ML (1986) Myocardial ischemia:platelet and thromboxane concentrations in cardiac lymph and the effects of ibuprofen and prostacyclin. Circ Res 59:49-55

Moschos CB, Lahiri K, Peter A, Jesrani MU, Regan TJ (1972) Effect of aspirin upon experimental coronary and non-coronary thrombosis and arrhythmia. Am Heart J 84:525-530

Moschos MD, Lahiri K, Lyons M, Weisse AB, Oldewurtel HA, Regan TJ (1973) Relation of microcirculatory thrombosis to thrombus in the proximal coronary artery: effect of aspirin, dipyridamole, and thrombolysis. Am Heart J 86:61-68

Moschos CB, Haider B, DeLa Cruz C, Lyons MM, Regan TJ (1978) Antiarrhythmic effects of aspirin during nonthrombotic coronary occlusion. Circulation 57:681-684

Moschos CB, Haider B, Escobinas AJ, Gandhi A, Regan TJ (1980aa) Chronic use of aspirin versus indomethacin during non-thrombotic myocardial ischemia: effects on survival. Am Heart J 100:647-652

Moschos CB, Escobinas AJ, Jorgensen OB (1980b) Effects of sulfinpyrazone in ischemic myocardium. In: McGregor M, Mustard JF, Oliver MF, Sherry S (eds) Symposia Specialists, Miami, pp 175-187

Müller B, Schneider J, Hennies HH, Flohe L (1984) Cardioprotective action of the new stable epoprostenol analogue CG 4203 in rat models of cardiac hypoxia and ischemia. Arzneimittelforschung 34:1506-1509

Nayler WG, Purchase M, Dusting GJ (1984) Effect of prostacyclin infusion during low-flow ischaemia in the isolated perfused rat heart. Basic Res Cardiol 79:125-134

Needleman SW, Spector AA, Hoak JC (1982) Enrichment of human platelet phospholipids with linoleic acid diminishes thromboxane release. Prostaglandins 24:607-622

O'Connor KM, Friehling TD, Kelliher GJ, MacNab MW, Wetstein L, Kowey PR (1986) Effect of thromboxane synthetase inhibition on vulnerability to ventricular arrhythmia following coronary occlusion. Am Heart J 111:683-688

Ohnishi H, Kosuzume H, Hayashi Y, Yamaguchi K, Suzuki Y, Itoh R (1981) Effects of trapidil on thromboxane A_2-induced aggregation of platelets, ischemic changes in heart and biosynthesis of thromboxane A_2. Prostaglandins and Medicine 6:269-281

Otani H, Engelman RM, Breyer RH, Rousou JA, Lemeshow S, Das DK (1986) Mepacrine, a phospholipase inhibitor. J Thorac Cardiovasc Surg 92:247-254

Parratt JR, Coker SJ (1980) The significance of prostaglandin and thromboxane release in acute myocardial ischaemia. In: Förster W (ed) Prostaglandins and Thromboxanes Fischer, Jena, pp 21-25

Parratt JR, Sitsapesan R (1986) Stereospecific antiarrhythmic effect of opioid receptor antagonists in myocardial ischaemia. Br J Pharmacol 87:621-622

Parratt JR, Wainwright CL (1986a) Ventricular arrhythmias induced by local injections of vasoconstrictors following coronary occlusion. Br J Pharmacol 88:397P

Parratt JR, Wainwright CL (1986b) The effects of the thromboxane antagonist BM13.177 on ischaemia and reperfusion induced arrhythmias in dogs. Br J Pharmacol 88:291P

Parratt JR, Coker SJ, Wainwright CL (1987) Eicosanoids and susceptibility to ventricular arrhythmias during myocardial ischaemia and reperfusion. J Mol Cell Cardiol 19, Suppl V, 55-66

Povalski HJ, Olson R, Kopia S, Furness P (1980) Comparative effects of sulfinpyrazone and aspirin in the coronary occlusion-reperfusion dog model. In: McGregor M, Mustard JF, Oliver MF, Sherry S (eds) Cardiovascular actions of sulfinpyrazone. Symposium Specialists, Miami, pp 153-171

Purchase M, Dusting GJ, Li DMF, Read MA (1986) Physiological concentrations of epinephrine potentiate thromboxane A_2 release from platelets in the rat isolated heart. Circ Res 58:172-176

Rabkin SW, Ohmae M (1980) Effect of sulfinpyrazone on adriamycin induced acute cardiotoxic arrhythmias in rabbits. Pharmacol Res Commun 12:196-204

Ribeiro LGT, Brandon TA, Hopkins DG, Reduto LA, Taylor AA, Miller RR (1981) Prostacyclin in experimental myocardial ischaemia: effects on hemodynamics, regional myocardial blood flow, infarct size and mortality. Am J Cardiol 47:835-840

Ruf W, McNamara JJ, Suehiro A, Suehiro G, Wickline S (1980) Platelet trapping in myocardial infarct in baboons: therapeutic effect of aspirin. Am J Cardiol 46:405-412

Sakanashi M, Yoshikawa Y, Akiyoshi R, Itoh C, Kitamura Y, Niho T, Ohnishi H (1983) Possible antiarrhythmic activities of trapidil. Arzneimittelforschung 33:215-217

Schrör K, Thiemermann C (1986) Treatment of acute myocardial ischaemia with a selective antagonist of thromboxane receptors (BM 13.177). Br J Pharmacol 87:631-637

Schrör K, Addicks K, Darius H, Ohlendorf R, Rosen P (1981a) PGI_2 inhibits ischemia-induced platelet activation and prevents myocardial damage by inhibition of catecholamine release from adrenergic nerve terminals. Evidence for cAMP as common denominator. Thromb Res 21:175-180

Schrör K, Ohlendorf R, Darius H (1981b) Beneficial effects of a new carbacyclin derivative, ZK 36374, in acute myocardial ischemia. J Pharmacol Exp Ther 219:243-249

Schrör K, Darius H, Ohlendorf R, Matzky R, Klaus W (1982) Dissociation of antiplatelet effects from myocardial cytoprotective activity during acute myocardial ischemia in cats by a new carbacyclin derivative (ZK 36375). J Cardiovasc Pharmacol 4:554-561

Seuter F, Busse WD, Meng K, Hoffmeister F, Möller E, Horstmann H (1979) The antithrombotic activity of BAY g6575. Arzneimittelforschung 29:54-59

Shea MJ, Murtagh JJ, Jolly SR, Abrams GD, Pitt B, Lucchesi BR (1984) Beneficial effects of nafazatrom on ischemic reperfused myocardium. Eur J Pharmacol 102:63-70

Smith EF, Gallenkämper W, Beckmann R, Thomsen T, Mannesmann G, Schrör K (1984) Early and late administration of a PGI_2-analogue, ZK 36374 (iloprost): effects on myocardial preservation, collateral blood flow and infarct size. Cardiovasc Res 18:163-173

Somberg JC, Bounous H, Cagin N, Anagnostopoulos L, Levitt B (1977) The influence of prostaglandins E_1 and E_2 on ouabain cardiotoxicity in the cat. J Pharmacol Exp Ther 203:480-484

Spector AA, Hoak C-L, Fry GL, Denning EM, Stoll LL, Smith JB (1980) Effect of fatty acid modification on prostacyclin production by cultured human endothelium cells. J Clin Invest 65:1003-1012

Starnes VA, Primm RK, Woosley RL, Oates JA, Hammon JW (1982) Administration of prostacyclin prevents ventricular fibrillation following coronary occlusion in conscious dogs. J Cardiovasc Pharmacol 4:765-769

Stäubli RC, Baur HR, Althaus U, Pop HP, Wehrli HP, Gurtner HP (1984) Influence of sulfinpyrazone on infarct size, hemodynamics, and arrhythmia following coronary occlusion in the pig. J Cardiovasc Pharmacol 6:829-832

Szekeres L, Lepran I, Boros E, Takats I, Koltai M (1982) The effect of non-steroid anti-inflammatory drugs and of linoleic acid-rich diet on early arrhythmias resulting from myocardial ischaemia. In: Parratt JR (ed) Early arrhythmias resulting from myocardial ischaemia. Oxford University Press, New York, pp 239-249

Tada M, Kuzuya T, Michitoshi I, Kodama K, Mishima M, Yamada M, Inui M, Abe H (1981) Elevation of thromboxane B_2 levels in patients with classic and variant angina pectoris. Circulation 64:1107-1115

Tada M, Hoshida S, Kuzuya T, Michitoshi I, Minamino T, Abe H (1985) Augmented thromboxane A_2 generation and efficacy of its blockade in acute myocardial infarction. Int J Cardiol 8:301-312

Thiemermann C, Steinhagen-Thiessen E, Schrör K (1984) Inhibition of oxygen-centered free radical formation by the stable prostacyclin-mimetic iloprost (ZK 36374) in acute myocardial ischaemia. J Cardiovasc Pharmacol 6:365-366

Thiemermann C, Löbel P, Schrör K (1985) Usefulness of defibrotide in protecting ischemic myocardium from early reperfusion damage. Am J Cardiol 56:978-982
Trachte GJ (1986) Thromboxane agonist (U46619) potentiates norepinephrine efflux from adrenergic nerves. J Pharmacol Exp Ther 237:473-477
van Gilst WH, Terpstra JA, de Langen CDJ (1985) Ventricular arrhythmias and purine loss upon reperfusion of ischemic myocardium: comparison of ZK 36374 and diltiazem. In: Schrör K, (ed) Prostaglandins and other eicosanoids in the cardiovascular system. Karger, Basel, pp 207-212
Vermylen J, Chamone DAF, Verstraete M (1979) Stimulation of prostacyclin release from vessel wall by Bay g6575, an antithrombotic compound. Lancet I:518-520
Wenzel DG, Innis DJ (1983) Protection from arrhythmias of cultured heart cells by nonsteroid anti-inflammatory drugs. Pharmacol Res Commun 15:167-172
Wong PY-K, Chao PH-W, McGiff JC (1982) Nafazatrom (Bay g-6575), an antithrombotic and antimetastatic agent, inhibits 15-hydroxyprostaglandin dehydrogenase. J Pharmacol Exp Ther 223:757-780
Zijlstra WG, Brunsting JR, Ten Hoor P, Vergrosen AJ (1972) Prostaglandin E_1 and cardiac arrhythmia. Eur J Pharmacol 18:392-395
Zylka V, Addicks K, Deutsch HJ, Friedrick R, Griebenow R, Hirche HJ (1981) The antiarrhythmic effect of prostacyclin (PGI_2) in severe myocardial ischemia in the pig heart. Pflugers Arch 389:[Suppl Rl]

CHAPTER 26

Possible Role of Lipids and of Free Radicals in Arrhythmogenesis

O. D. Mjøs

A. Free Fatty Acids and Arrhythmias

During the initial phase of an acute myocardial infarction in man, plasma concentrations of free fatty acids (FFAs) are often raised in excess of 1000 μ*M*/litre (Oliver et al. 1968). The rise in plasma FFAs is probably due to increased sympathoadrenal activity with consequent enhanced lipolysis/FFA mobilization from adipose tissue. Raised plasma FFAs have been associated with the development of serious ventricular arrhythmias and sudden death following acute myocardial infarction (Oliver et al. 1968). Kurien and Oliver (1970) postulated that perhaps the raised plasma concentrations of FFAs were directly responsible for the development of arrhythmias rather than an indirect effect through increased sympathoadrenal activity.

Free fatty acids are bound to two high-affinity sites on plasma albumin. It is suggested that at higher plasma concentration of FFAs than 1200 μ*M*/litre, representing approximately an FFA/albumin molar ratio of 2:1, the loosely bound fatty acids may accumulate in the ischaemic myocardium and provoke arrhythmias. Several groups have confirmed the association between raised plasma FFAs and arrhythmias in acute myocardial infarction (Gupta et al. 1969; Reimann and Schwandt 1971; Takano 1976), but the association has not been found by others (Ravens and Jipp 1972; Hagenfeldt and Wester 1973). This question has been a matter of controversy over the past 15-20 years.

Indirect support for the FFA hypothesis came from a study by Rowe et al. (1975), using a nicotinic acid analogue, 5-fluoro-3-hydroxy-methylpyridine, to inhibit lipolysis, thereby lowering raised plasma FFA concentrations in patients with acute myocardial infarction. Administration of the nicotinic acid analogue was carried out within 5 h after onset of symptoms. This did *not* reduce overall incidence of ventricular tachycardia or R-on-T or premature ventricular beats. But when the arrhythmias were related to the degree of fall in plasma FFAs, important differences appeared:

In patients where plasma FFAs fell quickly and remained below the upper normal level of 800 μ*M*/litre *none* had ventricular tachycardia, while the incidence of ventricular tachycardia was 80% in those where the nicotinic acid analogue did not reduce plasma FFAs. This difference could not be attributed to catecholamines per se since total plasma concentration of catecholamines was not significantly different in groups with effective and ineffective antilipolytic therapy. Rowe et al. (1975) concluded that "high concentrations of plasma FFAs can themselves be arrhythmogenic, independent of changes in plasma catecholamines". However, it must be borne in mind that the groups/subgroups were small; a much larger study would have been required to demonstrate an effect on ventricular fibrillation and death.

Over the past 2 decades many *experimental* studies have been carried out to test the FFA hypothesis. In some studies ventricular arrhythmias have been induced in dogs after coronary artery occlusion, by elevating plasma FFAs using Intralipid and heparin (Kurien et al. 1969, 1971), although this was not confirmed by Opie et al. (1971). Attempts have been made to explain these differences through different experimental models. This controversy has, however, not been definitely settled. The most serious criticism against all these studies was unphysiologically high plasma FFA levels. It is now known that this was due to a laboratory artefact, caused by a high rate of continuing in vitro lipolysis after the collection of blood (Riemersma 1982). Therefore the true in vivo levels of plasma FFA were moderate in these studies. In more recent studies Riemersma (1987) increased plasma FFAs using the technique developed by Greenough et al. (1969) to raise plasma FFA. This technique allowed the infusion of fatty acid sodium salts by binding them "physiologically" to circulating albumin. However, no arrhythmogenic effect could be demonstrated by raising plasma FFAs in dogs with acute coronary artery occlusion (Riemersma 1987).

In other experimental studies antilipolytic treatment reduced the incidence of arrhythmias and ventricular fibrillation during acute myocardial ischaemia (Smith and Duce 1974; Brownsey and Brundt 1977).

In conclusion, antilipolytic therapy has a mild antiarrhythmic effect in acute myocardial infarction in both clinical and experimental settings. However, experimental studies using intravascular lipolysis or infusion of FFAs to raise plasma FFAs have not convincingly demonstrated arrhythmogenic effects of FFAs. The reasons for this apparent discrepancy remain to be found.

Electrophysiological effects of high FFA concentrations have been extensively studied (Platou et al. 1981) during normoxic and hypoxic conditions, but less so during ischaemia. Earlier studies indicated enhanced automaticity by arachidonic acid, but not by octanoic, oleic or linoleic acid (Borbola et al. 1976, 1977). The effects of arachidonic acid could be prevented by albumin and were attributed to the formation of peroxides (Borbola et al. 1977). Studies with peroxidized fatty acids have supported this (Kim et al. 1985). In other studies the authors stress the importance of underlying hypoxia and claim reduction in ventricular fibrillation threshold (Murnaughan 1981), action potential duration and effective refractory period (Hough and Gevers 1975; Cowan and Vaughan Williams 1977; Murnaughan 1981) effected by FFAs. Most authors, however, used very high FFA/albumin molar ratios, with the exception of the study by Cowan and Vaughan Williams (1977), who found that FFAs did not induce *directly* any electrophysiological effect which could be arrhythmogenic.

Raised plasma FFAs also reduce myocardial contractility during ischaemia (Henderson et al. 1970; Kjekshus and Mjøs 1972; Liedke et al. 1978) and increase myocardial oxygen consumption (Mjøs 1971; Simonsen and Kjekshus

1978). The cellular mechanisms underlying these and the possible arrhythmogenic effects of FFAs are not definitely established. They have been linked to intracellular accumulation of FFAs, long-chain acyl CoA or carnitine esters, causing cellular derangements and inhibition of enzyme systems. An alternative hypothesis has recently been proposed by RIEMERSMA (1987). He suggested that a fatty acid-triglyceride energy-wasting cycle operates at different rates within the ischaemic heart, causing differential rates of K^+ loss, leading to ventricular fibrillation.

B. Lysophosphoglycerides and Arrhythmias

Lysophosphoglycerides are catabolites of sarcolemmal phospholipids. They are amphiphiles and have been implicated in the mediation of arrhythmias induced by ischaemia. SOBEL et al. (1978) presented evidence that lysophosphoglycerides (LPGs), including both lysophosphatidyl choline (LPC) and lysophosphatidyl ethanolamine (LPE), accumulated in the ischaemic myocardium. They found that LPG increased by 53% in ischaemic compared with control myocardium within 10 min after the onset of coronary occlusion in the cat. LPC bound to albumin induced marked electrophysiological derangements (see below) in isolated tissue in vitro (CORR et al. 1979), analogous to alterations characteristic of ischaemic myocardium in vivo. Neither glycerophosphoryl choline nor free fatty acids, both catabolites of LPC, induced alterations of the transmembrane action potential resembling those elicited by LPC. Thus, it appeared that the accumulation of LPGs within ischaemic myocardium may contribute to the electrical instability of the ischaemic heart.

In the presence of albumin (0.4 mM) a concentration of 1.5 mM LPG induced marked electrophysiological derangements in isolated tissue, resembling closely changes in ischaemic tissue in vivo including:

1. Reduction of maximum diastolic potential, V_{max} of phase 0, amplitude overshoot and action potential duration
2. Fractionation of the upstroke of phase 0 of the action potential and unresponsiveness to external stimulation
3. Enhanced automaticity at normal and reduced membrane potentials
4. A rightward and downward shift in the membrane response curve with a prolongation of conduction time
5. An increase in the ratio of effective refractory period to action potential duration (APD) such that the effective refractory period persists beyond the APD, resulting in postrepolarization refractoriness (CORR et al. 1979)

These effects would cause heterogeneity of recovery in the ischaemic regions, and current flow into partially depolarized cells from fully repolarized neighbours could initiate premature excitation in the latter.

The mechanisms by which LPGs exert their effect on membranes have not been elucidated completely. Since LPGs are assymmetrical compounds with hydrophobic and hydrophilic constituents, they can profoundly influence membrane functional integrity, which is exclusively dependent on the physical characteristics of phospholipid constituents. Potentially detrimental effects of LPG during ischaemia may include loss of intracellular K^+, contributing to the observed increase in extracellular K^+ (Corr and Sobel 1983).

Sarcolemmal phospholipids undergo continuous turnover. But accumulation of their lysophospholipid catabolites is prevented under physiological conditions by either reacylation to form diacyl phospholipids or hydrolysis of the lysophospholipids to form glycerophosphoryl choline (Corr and Sobel 1983). Several mechanisms may be responsible for the accumulation of lysophospholipids during ischaemia. These include, e.g. enhanced phospholipase A_2 activity, inhibition of microsomal lysophospholipase activity and de novo synthesis of LPC.

In conclusion, these observations support the hypothesis that even small concentrations of lysophosphoglycerides may be important contributors to electrophysiological deterioration of the ischaemic heart and to the evolution of malignant ventricular arrhythmia. The arrhythmogenic effects of LPGs are significantly enhanced by concomitant acidosis, and additive to those induced by long-chain acyl carnitine and FFAs. This suggests that modest changes in tissue or plasma concentrations of several metabolites may elicit substantial electrophysiological consequences in combination, particularly in a milieau with reduced pH (Corr and Sobel 1983).

C. Free Radicals and Arrhythmias

Free radicals have recently received attention in medicine because of new knowledge about their potential toxic effects. Free radicals are by definition molecular species which contain unpaired electrons. This gives them properties of high chemical reactivity and of chain reaction.

There are many biological oxidations, both enzymatical and spontaneous, which generate oxygen free radicals (Fridovich 1978; Freeman and Crapo 1982). The best-known oxygen radicals are *superoxide anion, hydrogen peroxide* and *hydroxyl radical*, and they are all products of the univalent reduction of oxygen to water. The toxicity of oxygen radicals is probably due to their potent effects on cell membrane function. In particular they cause peroxidation of lipids (Meerson et al. 1982; Rao et al. 1983) and oxidation of sulphhydryl groups (Freeman and Crapo 1982) in the cell membrane which will alter ionic permeability and enzyme function. Oxygen radicals also affect intracellular calcium sequestration (Hess et al. 1981), and these changes might be expected to lead to the development of arrhythmias.

Against the toxicity of oxygen radicals there exists in all aerobic cells a defence system consisting of specific enzymes and antioxidants usually called *scavengers*. The most important of these scavengers is probably *superoxide dismutase*, which catalyzes the dismutation of superoxide to hydrogen peroxide (Brawn and Fridovich 1980). *Catalase*, a peroxisomal enzyme, catalyzes the conversion of hydrogen peroxide into water and oxygen. The cytoplasmic selenium-containing enzyme *glutathione peroxidase* reduces both lipid peroxides and hydrogen peroxide.

Considerable evidence has accumulated over recent years suggesting that oxygen radicals may be involved in the pathophysiology of myocardial tissue damage induced by ischaemia and reperfusion. It is well established that during ischaemia the concentration of *hypoxanthine* increases substantially due to the degradation of adenine nucleotides. There is also an ischaemia-induced conversion of *xanthine dehydrogenase* to *xanthine oxidase* (CHAMBERS et al. 1985). Upon reperfusion, increased production of oxygen radicals by xanthine oxidase may then overload the natural scavenging mechanisms of the cells (McCORD 1984), causing myocardial damage. Recently YTREHUS et al. (1986) reported harmful myocardial changes effected by oxygen radicals generated by hypoxanthine and xanthine oxidase in the isolated perfused rat heart. Conversely protective effects of oxygen radical scavengers and specific enzymes have been demonstrated by several investigators (SHLAFER et al. 1982; STEWART et al. 1983; JOLLY et al. 1984; YTREHUS et al. 1985).

A variety of molecular mechanisms have been suggested to induce arrhythmias during both ischaemia and reperfusion (OPIE et al. 1978; CORR and WITKOWSKI 1983; MANNING and HEARSE 1984). To this list of potential arrhythmogenic factors, MANNING et al. (1984) recently added oxygen radicals formed by the action of the enzyme xanthine oxidase. Free oxygen radicals may compromise membrane ion pump activity and lead to local electrophysiological derangements that trigger ventricular arrhythmias.

MANNING et al. (1984) investigated the possibility that xanthine oxidase-linked free radical production has a role in the genesis of arrhythmias during ischaemia and reperfusion using an anaesthetized open-chest rat preparation. *Allopurinol* pretreatment reduced the incidence of ventricular tachycardia during ischemia from 88% to 50% ($P < 0.05$) and the number of premature ventricular complexes from 471 ± 120 to 116 ± 46 ($P < 0.02$). But the treatment had no effect upon the incidence or duration of ventricular fibrillation or upon mortality. In contrast, far more dramatic protection was observed during *reperfusion* after 5 min ischaemia. Allopurinol treatment reduced the incidence of ventricular fibrillation from 67% to 11% ($P < 0.01$), reduced the mean duration of fibrillation from 230 ± 70 to 14 ± 1 sec ($P < 0.05$) and reduced mortality by half although this did not reach statistical significance. Thus allopurinol pretreatment afforded some protection against ischaemia-induced arrhythmias, but a higher degree of protection against reperfusion-induced arrhythmias. The authors proposed that free radical formation by xanthine oxidase, or its consequences, is a new factor to consider in relation to the vulnerability of tissue to ischaemia and particularly to reperfusion-induced arrhythmias.

WOODWARD and ZAKARIA (1985) studied the possible role of oxygen-free radicals in the development of reperfusion arrhythmias using a 10-min period of coronary ligation followed by reperfusion in the isolated rat heart. Superoxide dismutase, glutathione or ascorbic acid, when given before coronary ligation, attenuated the development of reperfusion arrhythmias. Mannitol and catalase did not have any significant effect on reperfusion arrhythmias when given alone, but they did potentiate the antiarrhythmic effect of superoxide dismutase. Glutathione, and a combination of superoxide dismutase, catalase and mannitol, also reduced the incidence of reperfusion-induced ventricular fibrillation when given just before reperfusion. Their results provided evidence that free radicals are produced and may be important in the genesis of reperfusion-induced arrhythmias in the isolated rat heart.

In a recent study Bernier et al. (1986) have further investigated the possible involvement of free radicals in the production of reperfusion-induced arrhythmias in the isolated, perfused rat heart with transient coronary artery occlusion. They showed that when added to the perfusion fluid, each of six agents (superoxide dismutase, catalase, methionine, glutathione, mannitol and desferrioxamine) known to be free radical scavengers or to inhibit free radical production, significantly reduced the incidence of reperfusion-induced ventricular fibrillation and, in many cases, the incidence of reperfusion-induced ventricular tachycardia. The mean duration of sinus rhythm during reperfusion was also increased significantly. Conversely, under conditions where, in the control group, the incidence of reperfusion arrhythmias was lowered by increasing perfusate potassium to 6.5 mM, the addition of the free radical-generating system *$FeCl_3$· adenosine diphosphate* to the perfusion fluid increased dramatically the incidence of reperfusion-induced ventricular fibrillation and tachycardia. Simultaneous perfusion with $FeCl_3$· adenosine diphosphate and superoxide dismutase, catalase, mannitol, methionine, or desferrioxamine again reduced the incidence of reperfusion-induced arrhythmias and increased the duration of sinus rhythm during the reperfusion phase. Thus Bernier et al. (1986) provided more circumstantial evidence for an involvement of free radicals in the genesis of reperfusion-induced arrhythmias.

It is, however, important to question where the radicals are formed and where they exert their cytotoxic action. The protective interventions used in the study of Bernier et al. (1986) vary enormously with respect to their molecular weight and lipophilicity, and while some, e.g. methionine, might be expected to gain rapid access to the cytoplasm of the myocyte, others, e.g. superoxide dismutase and catalase, would be expected to be restricted to the vascular space. Considering the high molecular weight of superoxide dismutase and catalase, it might be questioned how events occurring in the myocyte could be influenced by scavengers in the vascular space. A possible explanation may be derived from a study of Jarasch et al. (1981), who demonstrated that the bulk of the myocardial xanthine oxidase is located in the vascular endothelium. Thus, superoxide produced in endothelial cells during reperfusion and hydroxyl radicals derived from it might well cause injury to the endothelium or adjacent contractile or conducting cells. This could be prevented by scavengers or inhibitors of free radical production which are present in the vascular and extravascular space.

Bernier et al. (1986) concluded that, although free radicals may play a role in the genesis of reperfusion arrhythmias, the evidence is circumstantial and must remain so until techniques such as electron spin resonance can be exploited to identify and measure free radical production in the beating heart.

In a recent study Riva et al. (1987) provided evidence that exogenous superoxide dismutase can act as an antiarrhythmic agent during reperfusion in vivo using anaesthetized rats. Hearts were subjected to 7 min of regional ischaemia followed by 10 min of reperfusion. In the control group (saline) 73 % of the hearts fibrillated during reperfusion, 20 % had atrioventricular block and 47 % died as a result of ventricular arrhythmias. Superoxide dismutase, administered as an intravenous bolus 2 min prior to reperfusion, exerted a marked protective effect. At its most effective dose (10 mg/kg body wt., i.e. 27 000 IU/body wt.) reperfusion-induced ventricular fibrillation was reduced to 33 %, reperfusion-induced atrioventricular block was eliminated and mortality was reduced to 7 % ($P < 0.05$). The protective effects were, however, very dose dependent and at higher doses superoxide dismutase exhibited no antiarrhythmic actions during reperfusion. The reason for this is not known. Taken together with previous in vitro findings, this study lends further support to the proposition that oxygen-derived free radicals may play a role in the induction of serious cardiac arrhythmias and that anti-free radical interventions, even given *after* the onset of ischaemia, can be highly protective.

The precise mechanism for the antiarrhythmic effects of superoxide dismutase remains to be resolved. Its biological role is to catalyze the dismutation of superoxide:

$$O_2^{-\bullet} + O_2^{-\bullet} + 2\,H^+ \rightarrow O_2 + H_2O_2$$

Superoxide initiates a variety of injuries, including lipid peroxidation, which might lead to changes in membrane permeability and loss of potassium ions. This might in turn increase electrical instability and account for the initiation of ventricular fibrillation (RIVA et al. 1987). Hydrogen peroxide, the product of the superoxide dismutase reaction, and the hydroxyl radical are also candidates for cell injury (FREEMAN and CRAPO 1982). The hydroxyl radical, which is very reactive, arises from superoxide and hydrogen peroxide:

$$O_2^{-\bullet} + H_2O_2 \rightarrow OH^- + OH^\bullet + O_2$$

Thus the elimination of hydrogen peroxide and hydroxyl radicals might therefore be expected to result in more protection than the elimination of only the superoxide radical. This notion is supported by BERNIER et al. (1986) and WOODWARD and ZAKARIA (1985), showing that the enzyme catalase and the hydroxyl radical scavenger mannitol were effective in reducing reperfusion-induced ventricular fibrillation in vitro. However, the precise molecular mechanisms by which superoxide dismutase or catalase prevent reperfusion-induced arrhythmias are not known.

The mechanisms responsible for reperfusion-induced malignant arrhythmias are distinctly different from those underlying arrhythmia due to ischaemia alone (PENKOSKE et al. 1978). In contrast to arrhythmia induced by sustained coronary occlusion, arrhythmia induced by reperfusion is associated with significant improvement in midmyocardial conduction, synchronous depolarization in the ischaemic regions and an elevated idioventricular rate. Since different arrhythmogenic mechanisms may underlie arrhythmia due to ischaemia alone in contrast to arrhythmia due to reperfusion, the prophylactic or therapeutic interventions required for each may be different.

Reperfusion-induced arrhythmias are a routinely experienced clinical phenomenon. Thus, regional reperfusion after only minutes of ischaemia occurs during the relief of coronary spasm, and this may trigger serious ventricular arrhythmias and sudden cardiac death. Reperfusion of the whole heart, after minutes or a few hours, is used during cardiac surgery with cardiopulmonary bypass and ischaemic cardiac arrest, often resulting in serious rhythmic disturbances. Reperfusion-induced arrhythmias may also arise during angioplastic or thrombolytic procedures.

The clinical occurrence and associated hazard of reperfusion arrhythmias, together with the realization of the differences between reperfusion- and ischaemia-induced arrhythmias, provide a considerable stimulus for the investigation of the mechanisms involved. Further studies are needed to establish a definite role for free radicals in this context. It is possible that superoxide dismutase, possibly in combination with other antifree radical interventions, may be of considerable therapeutic value.

References

Bernier M, Hearse DJ, Manning AS (1986) Reperfusion-induced arrhythmias and oxygen-derived free radicals: studies with anti-free radical interventions and a free radical generating system in the isolated perfused rat heart. Circ Res 58:331-340

Borbola J Jr, Papp GJ, Szekeres L (1976) Effect of free fatty acids on the automaticity of the sinus node and Purkinje fibres. In: Szekeres L, Papp GJ (eds) Symposium on pharmacology of the heart. Akademiai Kiado, Budapest, pp 75-80

Borbola J Jr, Süsskand K, Siess M, Szekeres L (1977) The effect of arachidonic acid in isolated atria of guinea pigs. Eur J Pharmacol 41:27-36

Brawn K, Fridovich I (1980) Superoxide dismutase: threat and defence. Acta Physiol Scand 492 [Suppl]:9-18

Brownsey RW, Brundt RV (1977) The effet of adrenaline-induced endogenous lipolysis upon the mechanical and metabolic performance of ischaemically perfused rat hearts. Clin Sci Mol Med 53 [Suppl]:513-521

Chambers DE, Parks DA, Patterson G, Roy R, McCord JM, Yoshida S, Parmley LF, Downey JM (1985) Xanthine oxidase as a source of free radical damage in myocardial ischemia. J Mol Cell Cardiol 17:145-152

Corr PB, Sobel BE (1983) The concentration dependence of the electrophysiological effects of amphiphiles. In: Refsum H, Jynge P, Mjøs OD (eds) Myocardial ischaemia and protection. Churchill Livingstone, Edinburgh, pp 90-100

Corr PB, Witkowski FX (1983) Potential electrophysiologic mechanisms responsible for dysrhythmias associated with reperfusion of ischemic myocardium. Circulation 68 [Suppl 1]:16-24

Corr PB, Cain ME, Witkowski FW, Price DA, Sobel BE (1979) Potential arrhythmogenic electrophysiological derangements in canine Purkinje fibers induced by lysophosphoglycerides. Circ Res 44:822-832

Cowan JW, Vaughan Williams EM (1977) The effects of palmitate on intracellular potentials recorded from Langendorff-perfused guinea-pig hearts in normoxia and hypoxia, and during perfusion at reduced rate of flow. J Mol Cell Cardiol 9:327-342

Freeman BA, Crapo JD (1982) Free radicals and tissue injury. Lab Invest 47:412-426

Fridovich I (1978) The biology of oxygen radicals. Science 201:875-880

Greenough WB III, Crespin SR, Steinberg D (1969) Infusion of long-chain fatty acid anions by continuous-flow centrifugation. J Clin Invest 48:1923-1933

Gupta DK, Young R, Jewett DE, Hartog M, Opie LH (1969) Increased plasma free-fatty-acid concentrations and their significance in patients which acute myocardial infarction. Lancet 2:1209-1213

Hagenfeldt L, Wester PO (1973) Plasma levels of individual free fatty acids in acute myocardial infarction. Acta Med Scand 194:357-362

Henderson AH, Most AS, Parmley WW, Gorlin R, Sonnenblick EH (1970) Depression of myocardial contractility in rats by free fatty acids during hypoxia. Circ Res 26:439-449

Hess ML, Okabe E, Kontos HA (1981) Proton and free oxygen radical interaction with the calcium transport system of cardiac sarcoplasmic reticulum. J Mol Cell Cardiol 13:767-772

Hough FS, Gevers W (1975) Catecholamine release as mediator of intracellular enzyme activation in ischaemic perfused rat hearts. S Afr Med J 49:538-543

Jarasch ED, Grund C, Bruder G, Heid HW, Keenan TW, Francke WW (1981) Localization of xanthine oxidase in mammary gland epithelium and capillary endothelium. Cell 25:67-82

Jolly SR, Kane WJ, Bailie MB, Abrams GD, Lucchesi BR (1984) Canine myocardial reperfusion injury. Its reduction by the combined administration of superoxide dismutase and catalase. Circ Res 54:277-285

Kim RS, Bihler I, Labella FS (1985) Calcium-translocating and cardiotonic properties of oxidation-products of linoleic-acid. Can J Physiol Pharmacol 63:1392-1397

Kjekshus JK, Mjøs OD (1972) Effect of free fatty acids on myocardial function and metabolism in the ischemic dog heart. J Clin Invest 51:1767-1776

Kurien VA, Oliver MF (1970) A metabolic cause of arrhythmias during acute myocardial ischaemia. Lancet 1:813-815

Kurien VA, Yates PA, Oliver MF (1969) Free fatty acids, heparin, and arrhythmias during experimental myocardial infarction. Lancet 2:185-187

Kurien VA, Yates PA, Oliver MF (1971) The role of free fatty acids in the production of ventricular arrhythmias after acute coronary artery occlusion. Eur J Clin Invest 1:225-241

Liedke AJ, Nellis S. Neely JR (1978) Effects of excess free fatty acids on mechanical and metabolic function in normal and ischemic myocardium in swine. Circ Res 43:652-661

Manning AS, Hearse DJ (1984) Reperfusion-induced arrhythmias: mechanisms and prevention. J Mol Cell Cardiol 16:459-470

Manning AS, Coltart DJ, Hearse DJ (1984) Ischemia- and reperfusion-induced arrhythmias in the rat: effects of xanthine oxidase inhibition with allopurinol. Circ Res 55:545-548

McCord JM (1984) Are free radicals a major culprit? In: Hearse DJ, Yellon DM (eds) Therapeutic approaches to infarct size limitation. Raven, New York, pp 209-218

Meerson FZ, Kagan VE, Kozlov YP, Belinka LM, Arkhipenko YV (1982) The role of lipid peroxidation in pathogenesis of ischaemic damage and the antioxidant protection of the heart. Basic Res Cardiol 77:465-485

Mjøs OD (1971) Effect of free fatty acids on myocardial function and oxygen consumption in intact dogs. J Clin Invest 50:1386-1389

Murnaghan MF (1981) Effect of free fatty acids on the ventricular arrhythmia threshold in the isolated heart of the rabbit. Br J Pharmacol 73:909-915

Oliver MF, Kurien VA, Greenwood TW (1968) Relation between serum free-fatty-acids and arrhythmias and death after acute myocardial infarction. Lancet I:710-714

Opie LH, Norris RN, Thomas M, Holland AJ, Owen P, van Noorden S (1971) Failure of high concentrations of circulating free fatty acids to provoke arrhythmias in experimental myocardial infarction. Lancet I:818-822

Opie LH, Nathan D, Lubbe WF (1978) Biochemical aspects of arrhythmogenesis and ventricular fibrillation. Am J Cardiol 43:131-148

Penkoske PA, Sobel BE, Corr PB (1978) Disparate electrophysiological alterations accompanying dysrhythmias due to coronary occlusion and reperfusion in the cat. Circulation 58:1023-1035

Platou ES, Myhre ESP, Refsum H, Mjøs OD (1981) Free fatty acids and the electrophysiology of the dog heart in situ: effects of isoprenaline, nicotinic acid and lipid emulsion with heparin. Clin Physiol 1:553-563

Rao PS, Cohen MV, Mueller HS (1983) Production of free radicals and lipid peroxides in early experimental myocardial ischaemia. J Mol Cell Cardiol 15:713-716

Ravens KG, Jipp P (1972) Die freien Plasmafettsäuren in der Frühphase eines Myocardinfarkts. Arzneimittelforschung 22:1831-1835

Reimann R, Schwandt P (1971) Frischer Herzinfarkt und freie Fettsäuren. Dtsch Med Wochenschr 96:93-96

Riemersma RA (1986) Raised plasma non-esterified fatty acids (NEFA) during ischaemia: implications for arrhythmias. In: Stam H, van der Vusse GJ (eds) Lipid metabolism in the normoxic and ischaemic heart. Steinkopff, Darmstadt, pp 177-186

Riemersma RA, Logan RL, Russel DC, Smith HJ, Simpson J, Oliver MF (1982) Effect of heparin on plasma free fatty acid concentrations after acute myocardial infarction. Br Heart J 48:134-139

Riva E, Manning AS, Hearse DJ (1987) Superoxide dismutase and the reduction of reperfusion-induced arrhythmias: in vivo dose-response studies in the rat. Cardiovascular Drugs and Therapy. 1:131-139.

Rowe MJ, Neilson JMM, Oliver MF (1975) Control of ventricular arrhythmias during myocardial infarction by antilipolytic treatment using a nicotinic-acid analogue. Lancet 1:295-308

Shlafer M, Kane PF, Wiggins VY, Kirsh MM (1982) Possible role for cytotoxic oxygen metabolites in the pathogenesis of cardiac ischemic injury. Circulation 66 [Suppl 1]:185-192

Simonsen S, Kjekshus JK (1978) The effect of free fatty acids on myocardial oxygen consumption during atrial pacing and catecholamine infusion in man. Circulation 58:484-491

Smith ER, Duce BR (1974) Anti-arrhythmic and serum free fatty acid lowering effects of 5-fluoro-nicotinyl alcohol following experimentally induced myocardial infarction in the dog. Cardiovasc Res 8:550-561

Sobel BE, Corr PB, Robison AK, Goldstein RA, Witkowski FX, Klein MS (1978) Accumulation of lysophosphoglycerides with arrhythmogenic properties in ischemic myocardium. J Clin Invest 62:546-553

Stewart JR, Blackwell WH, Crute SL, Loughlin V, Greenfield LJ, Hess ML (1983) Inhibition of surgically induced ischemia/reperfusion injury. J Thorac Cardiovasc Surg 86:262-272

Takano S (1976) Genetic studies on the arrhythmia in acute myocardial infarction with special reference to serum free fatty acid level. Jpn Circ J 40:287-297

Woodward B, Zakaria MNM (1985) Effect of some free radical scavengers on reperfusion induced arrhythmias in the isolated rat heart. J Mol Cell Cardiol 17:485-493

Ytrehus K, Gunnes S, Myklebust R, Mjøs OD (1985) Protection by superoxide dismutase (SOD) and catalase of the isolated perfused rat heart during long time cardioplegia. J Mol Cell Cardiol 17 [Suppl 3]:183 (Abstract)

Ytrehus K, Myklebust R, Mjøs OD (1986) Influence of oxygen radicals generated by xanthine oxidase in the isolated perfused rat heart. Cardiovasc Res 20:597-603

CHAPTER 27

Clinical and Pharmacologic Characterization and Treatment of Potentially Malignant Arrythmias of Chronic Chagasic Cardiomyopathy

P. A. CHIALE and M. B. ROSENBAUM

A. Introduction

Chagas' disease is a widespread parasitosis affecting most Latin American countries (CARCAVALLO 1975; COURA 1966; MARINKELLE 1975; OSIMANI 1972; PONCE and ZELEDON 1973). The etiologic agent is a protozoan, *Trypanosoma cruzi*.[1] This parasite is commonly transmitted to humans by hematophagous insects generically known as *"triatomes,"* although transmission by blood transfusion is not infrequent, and congenital infection can also occur.

The acute period of infection (which is often acquired during the first 2 decades of life) lasts 1 or 2 months and is usually symptomless. It can be roughly estimated that more than 90 % of infected individuals enter the chronic period of infection (and can develop the illness) without warning. This is an unfortunate fact, because parasitologic cure by specific drugs (which would prevent late visceral damage caused by chronic infection) can, at present, only be achieved during the acute period.

Most of the chronically infected individuals never show discernible evidence of the disease, in spite of the fact that the infection lasts all life long. A variable percentage (up to 30 %, with strong regional differences) develop, after 10, 20, or more years, symptoms and/or signs of visceral damage. This is what, in a strict sense, should be called chronic Chagas' disease.

The digestive forms of the disease (mainly megaesophagous and megacolon), thought to be related to destruction of parasympathetic nerve fibers, are not uncommon. Yet, the most frequent and important clinical manifestation is a severe chronic *myocarditis*.

Several hypotheses have been proposed for the pathogenesis of chagasic myocardial damage. The most recent studies support an outstanding role of *autoimmune* mechanisms (PERALTA 1981; RIBEIRO DOS SANTOS and HUDSON 1980; SACHS and LANFRANCHI 1978).

In 1964, ROSENBAUM coined the term *"panmyocarditis"* to describe the fact that the microscopic foci of chagasic myocardial damage were scattered

[1] *Trypanosoma cruzi* is an hematic parasite during its period of dissemination, but essentially it is a histic parasite which invades preferentially the myocardial cells and the nerve tissue. In these tissues the trypanosoma is transformed into a leishmania, the only form in which it persists and reproduces itself.

throughout the heart. He also called attention to the coexistence of areas of myocytic degeneration, inflammatory infiltration, and fibrosis as suggesting a permanently evolving process. The intracellular forms of the parasite were found to be extremely scarce, further favoring the immunologic hypothesis.

Depending on the magnitude, localization, and perhaps also the time course of development of these widespread foci of myocardial damage, quite different clinical, electrocardiographic, and radiologic manifestations can be observed. Usually, chronic chagasic myocarditis behaves as a dilated cardiomyopathy, with marked cardiac enlargement and characteristic electrocardiographic abnormalities, and can evolve into *congestive heart failure*. However, other clinical forms are not infrequent; for instance, severe electrocardiographic alterations with a normal or nearly normal cardiac size. The "panmyocarditis" concept explains the myriad of electrocardiographic abnormalities often found in the advanced stages of the disease. Sinus node and atrial involvement can induce different manifestations of so-called *sick sinus syndrome*: severe depression of sinus node activity, *sinoatrial block, tachycardia-bradycardia* syndrome, and *atrial fibrillation* or flutter. Atrial damage can also account for intraatrial conduction disturbances and grossly abnormal P waves. The ventricular conducting system is often involved. *Right bundle branch block* and *left anterior hemiblock* (singly as well as together) are the commonest conduction disturbances. A further impairment of the conducting system can lead to different degrees of *atrioventricular block*, invariably showing *wide QRS* complexes. The damage of working ventricular myocardium can result in abnormal *Q waves* and primary abnormalities of ventricular repolarization (*ST segment elevation*, "ischemic" T waves). Ventricular premature beats are also a conspicuous expression of the disseminated foci of myocardial damage.

In its most advanced stages, chronic chagasic myocarditis leads to disability and death, mostly as a consequence of heart failure and sudden death. In a follow-up study carried out in San Felipe, Brazil, 58% of deaths in the chagasic population were due to refractory heart failure, and 37% to *sudden death* (PRATA 1975); and 17.6% of chagasic individuals autopsied by REIS LOPES et al. (1976) in Uberaba, Brazil, had experienced sudden death apparently without any previous symptoms of cardiac disease. Although conclusive proofs are still lacking, it is generally admitted that ventricular fibrillation must be the commonest precipitating event of sudden death (REIS LOPES et al. 1974; ROSENBAUM 1964). As in other cardiac disorders, this event can be heralded by premonitory ventricular arrhythmias. In fact, a number of chagasic patients present with ventricular arrhythmias resembling the "potentially malignant ventricular arrhythmias" (i.e., implying an enhanced risk of cardiac sudden death) which have been described for ischemic heart disease and other cardiomyopathies (BIGGER 1983). Thus, chagasic patients with such ventricular arrhythmias have been shown to develop symptomatic, sustained, and recurrent ventricular tachycardia and/or ventricular fibrillation, and sudden death (CHIALE 1982 et al.; KASKI et al. 1981).

In this chapter we will address different aspects of the potentially malignant ventricular arrhythmias of chronic chagasic myocarditis. We will describe consecutively the clinical characterization and context; the response to antiarrhythmic drugs with dissimilar electrophysiologic and pharmacologic properties as an approach to the underlying arrhythmogenic mechanisms; and the long-term control obtained with amiodarone. Finally, we will discuss briefly the indications and potential value of combined antiarrhythmic treatments and surgical therapy.

B. Clinical Context and Characterization

I. Electrocardiographic Features

Figure 1b illustrates the electrocardiographic appearance of what we call potentially malignant ventricular arrhythmias, in a patient with severe chagasic myocardiopathy. Ventricular premature complexes are extremely frequent, multiform, with copious repetitive forms (couplets and runs of ventricular tachycardia) and R on T phenomena.

In a previous report (Chiale et al. 1982), we analyzed the electrocardiographic characteristics of such arrhythmias in 28 chagasic patients with severe chronic myocarditis and ventricular premature beats showing at least two of the following criteria: (1) multiform; (2) repetitive; and (3) closely coupled, with R on T phenomenon. This is obviously a highly selected group showing the most bizarre ventricular arrhythmias. Hence, the findings that will be described below should not be extrapolated to less serious ventricular arrhythmias, which can also be found in chagasic patients. In 203 conventional electrocardiograms with rhythm strips lasting 1-5 min, obtained at variable intervals (3-30 days) during a 3-month period, the frequency of ventricular premature beats ranged between 0.2 and 6 every 10 beats. Figure 2 depicts the percentage of patients and electrocardiograms showing multiform ventricular premature beats, repetitive forms, and R on T phenomenon. Multiform ventricular extrasystoles were found to be virtually ubiquitous (100 % of the patients and 97.04 % of the electrocardiograms). The electrocardiographic configuration of the ventricular extrasystoles was extremely varied, suggesting that they arose from different sites of both right and left ventricles. This is another obvious expression of the widespread foci of ventricular damage. Couplets were also found in all patients and in a lesser percentage of electrocardiograms (79.31 %). The percentage of patients and electrocardiograms with runs of ventricular tachycardia (64.28 % and 42.85 %, respectively) and R on T phenomenon (46.43 % and 21.67 %, respectively) were considerably less. From the preceding analysis it is clear that multiformity of ventricular extrasystoles is the most constant electrocardiographic feature. As will be described below, this is still true even in the presence of an extremely reduced number of ventricular premature beats caused by spontaneous variability or antiarrhythmic therapy.

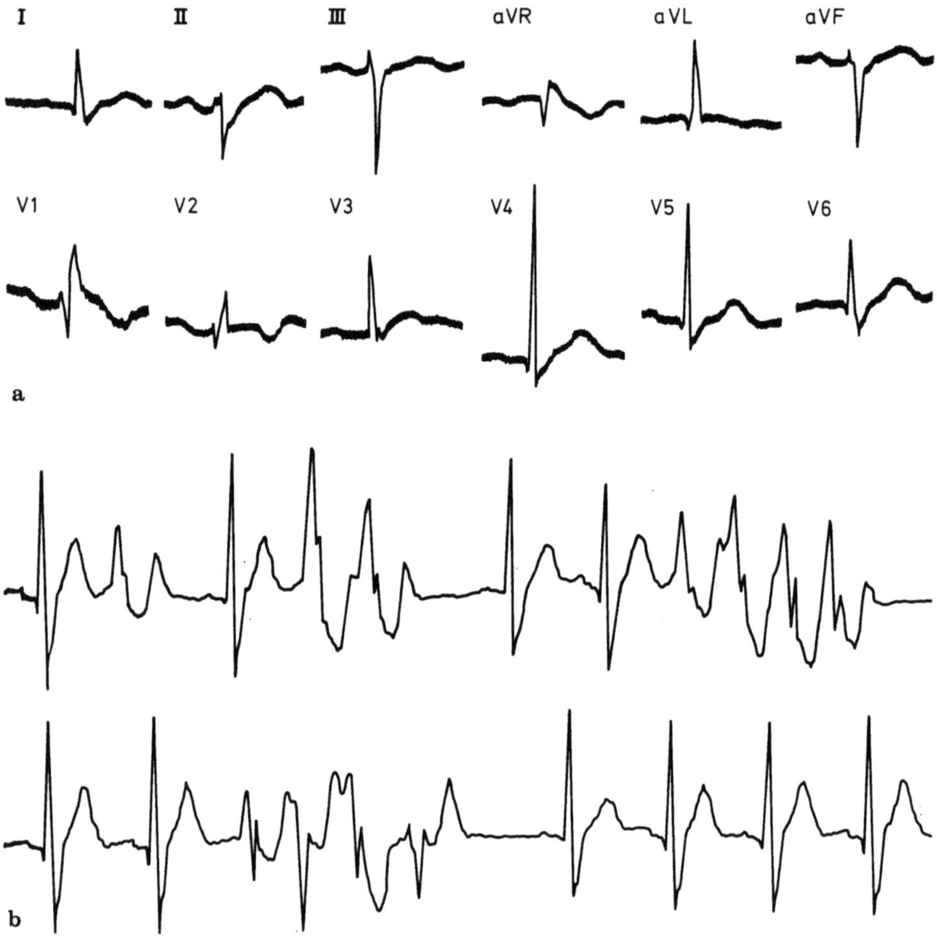

Fig. 1 a. Electrocardiograms obtained during sinus rhythm in a chagasic patient with potentially malignant ventricular arrhythmias. Note the presence of right bundle branch block and left anterior hemiblock. **b** (same patient) A rhythm strip, selected from a 24-h ambulatory Holter recording, shows the main electrocardiographic features of ventricular arrhythmias. Ventricular premature beats are frequent, multiform, repetitive and early, with R on T phenomenon

II. Clinical Context

Table 1 summarizes the symptoms and radiologic and electrocardiographic findings in 41 cases of chronic chagasic myocarditis with potentially malignant ventricular arrhythmias. Several facts should be emphasized. Firstly, despite the alarming electrocardiographic appearance of the arrhythmias, a relatively small number of patients presented with severe symptoms which could be attributed to sustained ventricular arrhythmias: only eight of them had a history of recurrent syncopal attacks and six others complained of dizziness. Secondly, there was evidence of congestive heart failure in 11 patients (26.82 %). Chest X-ray examination however, showed moderate or severe car-

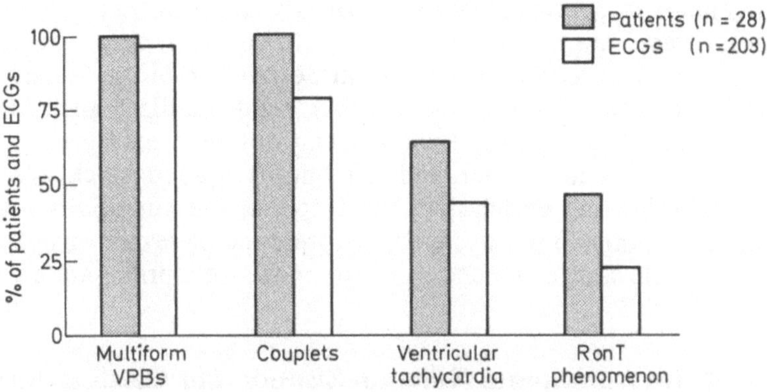

Fig. 2. Percentage of patient and electrocardiograms showing complex ventricular arrhythmias during a 3-month follow-up period. See text for details

Table 1. Forty-one chagasic patients with potentially malignant ventricular arrhythmias: clinical, radiologic, and electrocardiographic findings

Symptoms	Patients (n)
Recurrent syncope	8
Dizziness	6
Palpitations	22
CHF	11
Total	38
Cardiac enlargement	**Patients (n)**
Moderate	17
Severe	16
Total	33
Conduction disturbances	**Patients (n)**
RBBB and LAH	18
LAH	6
RBBB	4
RBBB and LPH	1
LBBB	1
First degree AV block	1
Total	31

CHF, congestive heart failure; LAH, left anterior hemiblock; LBBB, left bundle branch block; LPH, left posterior hemiblock; RBBB, right bundle branch block diac enlargement in 33 cases (80.48%). Thirdly, intraventricular conduction disturbances were the most frequent electrocardiographic abnormalities (75.6% of patients). In accordance with our description of the "chagasic electrocardiographic pattern" (ROSENBAUM 1964), right bundle branch block and left anterior hemiblock (mainly their association) were extremely common, whereas left posterior hemiblock and left bundle branch block were exceedingly rare. This picture emphasizes the fact that the potentially malignant ventricular arrhythmias often occur in the presence of severe myocardial damage, which is relevant to the choice of the most appropriate antiarrhythmic therapy.

III. Role of the Autonomic Nervous System and Cardiac Rate

In our preselected group of 28 chagasic patients (see Sect. B.I) we assessed the relative influence of daily activities on the main electrocardiographic features of the ventricular arrhythmias by 24-h ambulatory *Holter* recordings (CHIALE et al. 1982).

In 16 cases (57.14%), the frequency of ventricular premature beats and complex forms (couplets, runs of ventricular tachycardia, multiform and early ventricular extrasystoles) showed no discernible modifications related to slight or moderate efforts, emotions, rest or sleep, or to the correlative changes in heart rate (Fig. 3). This behavior suggests that the underlying arrhythmogenic mechanisms remained similarly operative irrespective of the prevailing autonomic tone and cardiac rate. However, since many of these patients had congestive heart failure, the possible role of a persistently increased adrenergic tone in stabilizing the arrhythmias cannot be totally ruled out.

Conversely, in a slightly smaller group (39.28% of cases), ventricular arrhythmias were considerably aggravated (increasing the frequency of both ventricular premature beats and complex forms) by physical and emotional stress, leading to faster cardiac rates, and were markedly decreased at rest and during sleep, when cardiac rate was substantially slower (Fig. 3b). Only multiformity of ventricular extrasystoles remained practically unmodified. In some of these patients undergoing an exercise stress test, slight exercising without load or with loads up to 300 kgm precipitated frequent runs of ventricular tachycardia or "chaotic" ventricular activity. It appears that in this group of patients the arrhythmogenic mechanisms were favored by sympathetic discharges and restrained by an enhanced vagal tone. Nevertheless, obvious methodologic limitations prevented us from determining whether the changes in ventricular ectopic activity were due to the direct electrophysiologic effects of the autonomic mediators on ventricular tissues, or resulted simply from changes in heart rate. This is an issue that deserves further investigation.

Only one patient showed a significant increment of ventricular premature beats, couplets, and runs of ventricular tachycardia during night sleep, at a time when cardiac rate was slow. This exceptional behavior would suggest the participation of arrhythmogenic mechanisms associated with a critical slowing of cardiac rate (dispersion of ventricular refractoriness or bradycardia-dependent conduction disturbances).

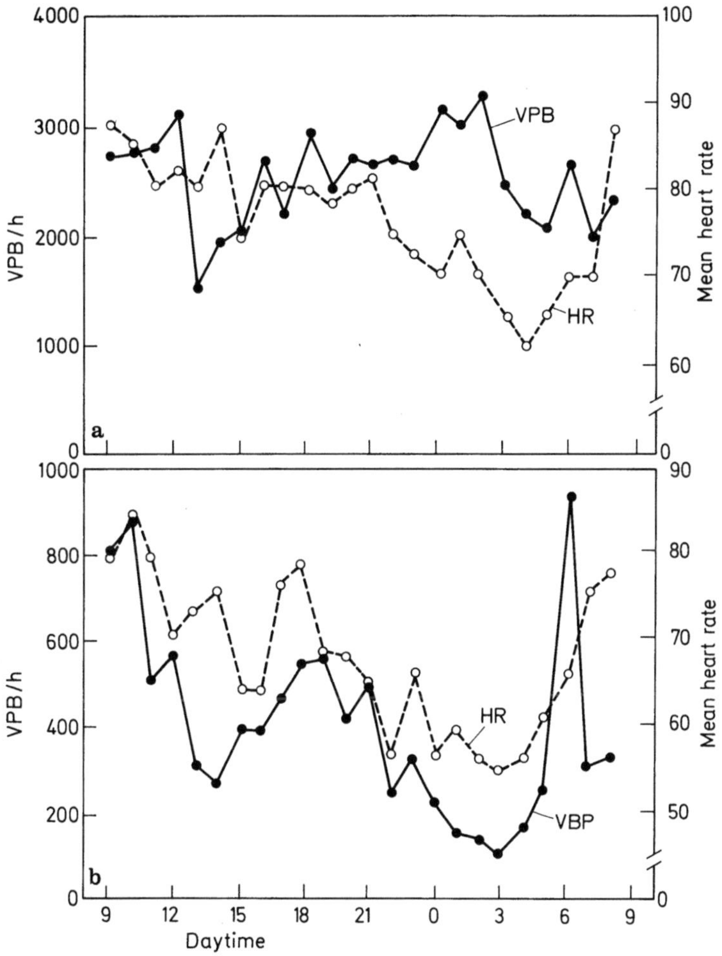

Fig. 3a, b. Relationships between ventricular arrhythmias and heart rate. **a** The number of ventricular premature beats was persistently high, irrespective of changes in cardiac rate; **b** the number of ventricular ectopic beats was high when cardiac rate was fast, and vice versa (except for brief periods). *VPB*, ventricular premature beats; *HR*, heart rate

IV. Spontaneous Variability. "The Chagasic Model"

In the above-mentioned study, the potentially malignant ventricular arrhythmias were found to be remarkably persistent and steadfast. Thus, regarding short-term variability (during a 24-h period), the maximal hourly reduction in the number of ventricular premature beats (related to the 1st h of recording) was less than 50 % in half of the 16 cases without significant circadian variations, and greater than 90 % in only one of these patients. On the other hand, a maximal hourly reduction in the frequency of the ventricular premature beats greater than 90 % occurred during sleep in 8 out of the 12 cases showing marked circadian variations. However, when sleeping hours were excluded,

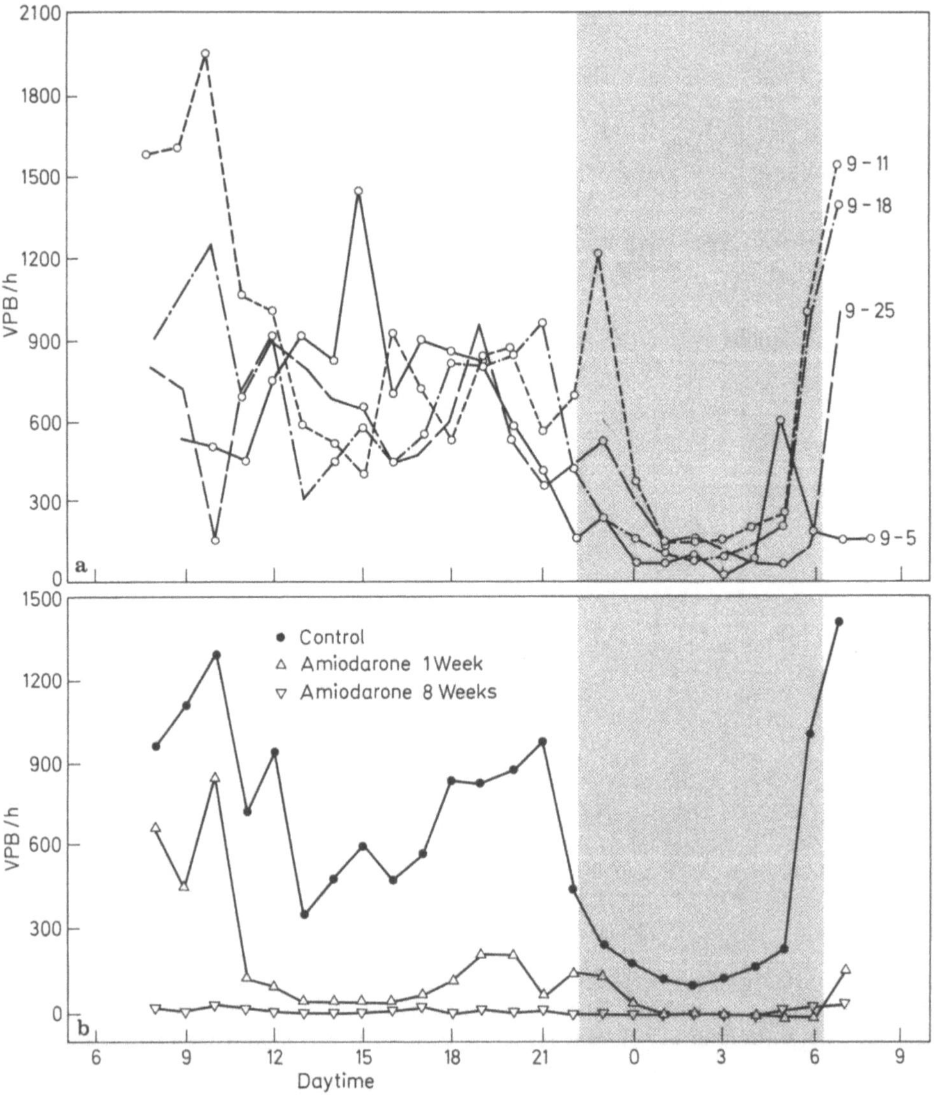

Fig. 4a, b. Reproducibility of the hourly distribution of ventricular premature beats showing marked circadian variations. **a** The hourly distribution of ventricular extrasystoles in four 24-h ambulatory Holter recordings obtained at weekly intervals; **b** (same patient) it illustrates the persistence of circadian variations in spite of a great reduction in the frequency of ventricular premature beats caused by amiodarone. **Shaded areas** indicate sleeping hours the degree of variation became similar to that observed in the former group of patients. Spontaneous changes in the frequency of couplets and runs of ventricular tachycardia as well as in the occurrence of R on T phenomenon were found to parallel those described for the total number of ventricular premature beats. On the other hand, multiformity of ventricular extrasystoles remained virtually unmodified in all cases, even when their number was markedly reduced.

An analogous behavior of the arrhythmias could be demonstrated when intermediate and long-term spontaneous variability were evaluated, respectively, by 24-h ambulatory Holter recordings obtained at weekly intervals during 4 weeks, and 10–24 months after the first study. The maximum changes in the frequency of ventricular premature beats ranged between +39% and −39% in intermediate-term studies, and between +57.6% and −59.1% in long-term evaluations. The spontaneous variability of couplets, runs of ventricular tachycardia, R on T phenomenon and multiform ventricular extrasystoles was found to be similarly unimportant. Furthermore, as depicted in Fig. 4, the hourly distribution of ventricular premature beats and complex forms also showed an impressive reproducibility.

When the figures described above were compared with the data from several studies in patients with "frequent" ventricular premature beats caused by ischemic heart disease and other cardiomyopathies (WINKLE 1978; WINKLE et al. 1978a; WINKLE et al. 1978b), it was manifest that the spontaneous variability of the arrhythmias was much less in chronic chagasic myocarditis than in other cardiac diseases. The outstanding persistence and stability of "chagasic" ventricular arrhythmias virtually excludes a major problem, often hindering an accurate evaluation of antiarrhythmic treatments: the therapeutic or proarrhythmic effects mimicked by spontaneous variability (WINKLE 1978). Therefore, these arrhythmias provide a useful and reliable clinical model (which we call "the chagasic model") to test the efficacy of antiarrhythmic drugs in the treatment of potentially malignant ventricular arrhythmias and the prevention of sudden cardiac death. Nonetheless, it should be emphasized that this is indeed a highly exacting model. In the presence of the serious underlying myocardial damage caused by Chagas' disease, it could be anticipated that many antiarrhythmic agents should not only be ineffective, but also prone to induce dangerous and even life-threatening cardiac side effects. Antiarrhythmic drugs passing this arduous test can be confidently used in most of the other cardiac conditions leading to severe ventricular arrhythmias.

C. Pharmacologic Responses. An Approach to the Arrhythmogenic Mechanisms Underlying Chagasic Ventricular Arrhythmias

In most chagasic patients with potentially malignant ventricular arrhythmias the assessment of the most probable arrhythmogenic mechanisms is utterly impossible. An abnormally increased ventricular automaticity should partici-

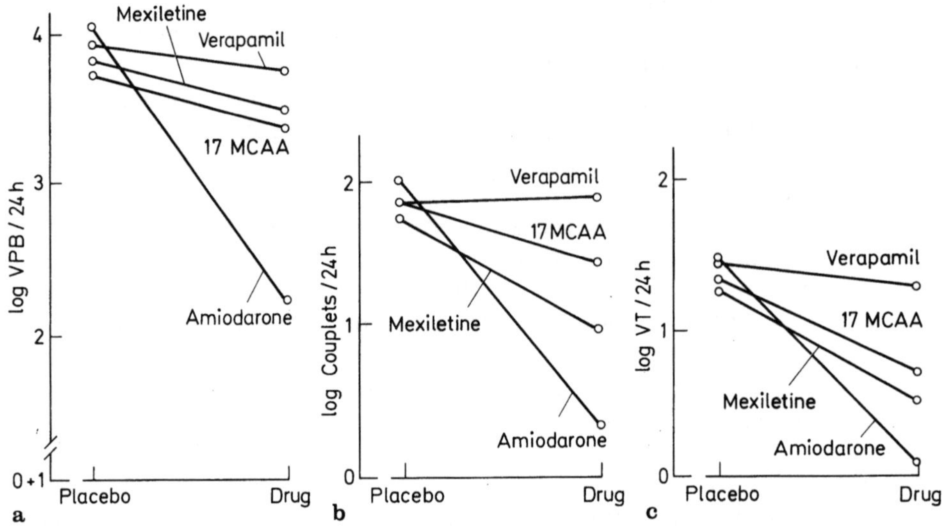

Fig. 5a–c. Comparative antiarrhythmic effects of verapamil, 17-monochloracetilajmaline (*17 MCAA*), mexiletine, and amiodarone on the number of ventricular premature beats **(a)**, couplets **(b)**, and bursts of ventricular tachycardia **(c)**, expressed in the vertical scale as the decimal logarithm, in 14 chagasic patients. Amiodarone was much more effective than the other drugs. *1*, 10; *2*, 100; *3*, 1000; *4* 10 000. [Partially modified from Haedo et al. (1986) with permission]

pate to some extent because relatively early ventricular escapes and idioventricular tachycardia favored by relatively slow sinus rates are not uncommon during night sleep (Haedo and Rosenbaum 1985, personal communication). In addition, ventricular *parasystole* has also been shown to occur (Faria et al. 1983; Nau et al. 1985). In a few cases the electrophysiologic behavior of some of the multiple ectopic foci would suggest the probable participation of reentrant reflecting mechanisms (Nau 1985, personal communication). However, in most cases, the apparently random coexistence of multiple ectopic foci (as re-

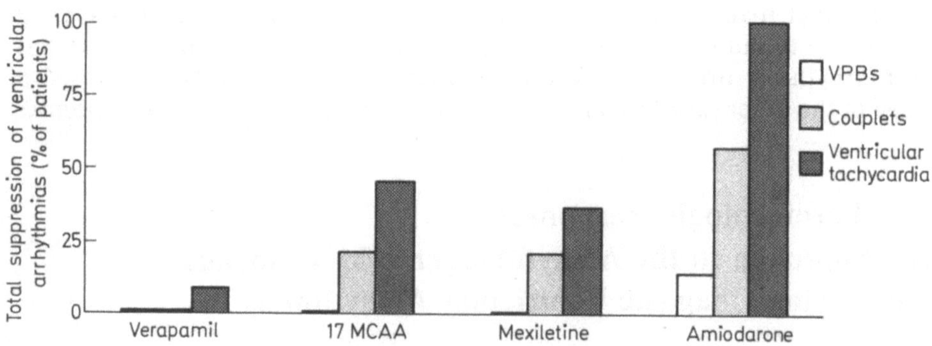

Fig. 6. Percentage of patients with total suppression of ventricular arrhythmias caused by verapamil, 17-monochloracetilajmaline, mexiletine, and amiodarone treatment. See text for details

vealed by the extremely varied electrocardiographic configurations) precludes a systematic electrophysiologic appraisal. As a further approach to the mechanisms of these arrhythmias, we evaluated comparatively the effects of 4 antiarrhythmic drugs with different electrophysiologic properties (mexiletine, 17 monochloracetilajmaline, amiodarone, and verapamil) in 14 chagasic patients with severe ventricular arrhythmias, by serial 24-h ambulatory Holter recordings (HAEDO et al. 1986). A broad enough spectrum of pharmacologic responses could be demonstrated. The following description is based mostly on the results of that study, which are summarized in Figs. 5, 6.

I. Unresponsiveness to Calcium Blocking Agents

Verapamil, administered at daily doses of 320 mg during a week, failed to control the arrhythmias. The mean reduction in the total number of ventricular premature beats (30.8 %), couplets (18.6 %), and runs of ventricular tachycardia (30.9 %) was nonsignificant. Multiform ventricular extrasystoles as well as R on T phenomenon remained also unmodified. In addition, total suppression of ventricular tachycardia was achieved in only 1 of 11 patients (9.09 %), whereas couplets and, obviously, ventricular premature beats persisted in all the 14 patients.

Verapamil induced marked *sinus bradycardia* or *sinoatrial block* at rest or during sleep in 11 of the 14 cases. Conversely, in none of the patients were the intraventricular conduction disturbances aggravated by the drug. The failure of verapamil to develop any antiarrhythmic action or to induce (or aggravate) conduction disturbances would suggest that the calcium channel, and so-called slow responses, do not play a significant role in the mechanisms of the ventricular arrhythmias and slow conduction that are conspicuous features of chronic chagasic myocardiopathy.

It should be mentioned that three patients developed overt congestive heart failure during verapamil administration. This serious complication reveals that in patients with severe myocardial involvement this drug can have significant detrimental effects on cardiac contractility.

II. Partial Response to Sodium Channel Blocking Agents

Two sodium channel blocking drugs, *mexiletine* and *17-monochloracetilajmaline* (which were administered at daily doses of 800 mg and 1200 mg, respectively, 1 week each), showed a similarly partial antiarrhythmic activity. Thus, both drugs caused a significant reduction of runs of ventricular tachycardia (mean, 86.4 % with mexiletine; 78.3 % with 17-monochloracetilajmaline) and couplets (mean, 87.7 % during mexiletine treatment; 73.6 % during 17-monochloracetilajmaline administration). However, total suppression of repetitive forms was achieved in only a small percentage of cases (Fig. 6). R on T phenomenon was suppressed in 42.8 % and 28.5 % of the patients, respectively, during mexiletine and 17-monochloracetilajmaline therapy. On the other hand, multiform ventricular extrasystoles remained unmodified, whereas the mean reduction in the total number of ventricular premature beats was irrevelant.

As with verapamil administration, marked sinus bradycardia or transient sinoatrial block were also induced by mexiletine (seven patients) and 17-monochloracetilajmaline (ten patients). Furthermore, transitory right bundle branch block developed in four patients with mexiletine and, in three of them, also with 17-monochloracetilajmaline, an effect that was never observed during verapamil administration.

The incomplete antiarrhythmic action developed by these two drugs, in conjunction with the induction of right bundle branch block, would suggest that chagasic ventricular arrhythmias and conduction disturbances may be related, at least to some extent, to depressed fast responses, probably caused by scattered zones of partial depolarization of the myocardium and the conducting tissue of the heart. It could be thought that the incomplete antiarrhythmic effect may be due to a relatively weak blockade of the sodium channel and, consequently, that a stronger depressing effect on sodium channel conductance should be accompanied by a much more effective antiarrhythmic action. However, in other groups of chagasic patients with similarly severe ventricular arrhythmias, the antiarrhythmic effects of two powerful blockers of the sodium channel, *flecainide* and *propafenone*, were found to be similar to those of mexiletine and 17-monochloracetilajmaline (Pastori and Rosenbaum 1985, personal communication). Moreover, the former two drugs showed a greater tendency to develop proarrhythmic effects. In fact, *quinidine* (another similarly powerful blocker of the sodium channel), induced sudden death, probably due to ventricular fibrillation, in a few chagasic patients with severe ventricular arrhythmias treated by us several years ago.

III. Beta-Blockers. Useful Agents for Very Selected Cases

As mentioned above, chagasic ventricular arrhythmias may be considerably worsened by sympathetic discharges. Under such circumstances, long-term blockade of beta-adrenergic receptors would appear as a logical therapeutic alternative. However, beta-blockers may have deleterious effects on cardiac inotropism, particularly in the presence of heart failure or marked cardiac enlargement commonly present in the most advanced stages of chronic chagasic myocardiopathy. Therefore, the use of beta-blockers should be restricted to patients without heart failure or severe cardiomegaly. In fact, in a recent study we assessed the antiarrhythmic efficacy of *nadolol* (a beta-blocking agent endowed with a relatively long half-life) in five chagasic patients fulfilling those conditions (Schmidberg et al. 1985). Nadololol was administered orally at doses ranging between 40 and 120 mg/day during 2–6 months. In three out of the five patients, ventricular arrhythmias where abolished and could no longer be demonstrated during repeated exercise stress tests. One patient developed severe sinus bradycardia in spite of being treated with a relatively low dose of the drug. Further studies are necessary to define the usefulness and safety of beta-blocking agents in chagasic ventricular arrhythmias. The possible benefits of low doses of beta-blockers in combined antiarrhythmic schedules also deserves to be assessed.

IV. Singular Efficacy of Amiodarone

In the study of HAEDO et al. (1986), amiodarone developed much greater antiarrhythmic effects than those of the sodium channel modifiers. A meaningful reduction of ventricular premature beats and repetitive forms could be documented after 4 weeks of amiodarone administration at a daily dose of 800 mg (Fig. 5). Thus, ventricular premature beats were decreased by 97.8 %, with two patients (14.28 %) showing total suppression of ventricular ectopy (Fig. 6). Runs of ventricular tachycardia were abolished in 100 % of cases, whereas couplets were reduced by 98.1 %, with complete suppression in 57.14 % of patients. Suppression of R on T phenomenon was obtained in 78.5 % of cases. On the other hand, no discernible effects on multiformity of ventricular extrasystoles could be demonstrated.

As described for the other drugs tested, amiodarone induced periods of marked sinus bradycardia or sinoatrial block in eight cases, and, like the two sodium channel blockers, it caused transient right bundle branch block in three patients.

The mechanisms accounting for the greater antiarrhythmic potency of amiodarone remain still undefined. This is certainly not due to a stronger blocking action on the sodium channel. Therefore, the remarkable effectiveness of amiodarone should be attributed to some other properties. In our comparative study, amiodarone was the only drug causing a consistent slowing of mean cardiac rate. It could be thought that this effect may contribute to the more powerful antiarrhythmic activity of this drug, at least in patients showing ventricular arrhythmias with marked circadian variations. Nevertheless, even in those cases, strong reduction or suppression of the arrhythmias also occurred in the presence of relatively fast heart rates. It is not unlikely that certain distinctive electrophysiologic properties of amiodarone could play a role in abolishing the arrhythmias, particularly the repetitive forms. For example, it has been shown that amiodarone causes a significant lengthening of the absolute refractory period at the expense of the relative refractory period of the normal Purkinje fibers (ELIZARI et al. 1980). Thus, premature impulses would show either total block or rapid propagation and this, in turn, would preclude slow conduction, which is an essential condition for reentrant or reflecting mechanisms. Finally, a noncompetitive antiadrenergic action could also provide an additional antiarrhythmic mechanism.

D. Long-term Control of Chagasic Ventricular Arrhythmias with Amiodarone

Since the treatment of chagasic ventricular arrhythmias must be life-lasting, long-term evaluation is required to confirm the apparent efficacy and safety of a drug, encountered during short- or intermediate-term administration, as well as to determine the influences of controlling the arrhythmias on symptoms and sudden mortality.

Recently, we evaluated the results of long-term administration of amiodarone in 24 chagasic patients with severe ventricular arrhythmias (CHIALE et al. 1984). Eight of them had had recurrent syncope and two others complained of frequent dizzy spells. Seven patients had overt congestive heart failure and ten cases showed severe cardiac enlargement.

Evaluation of antiarrhythmic effects of amiodarone was carried out by serial 24 h ambulatory Holter recordings. After a control recording, amiodarone was given orally ot doses of 600 or 800 mg/day. The continuous recordings were initially reiterated at weekly intervals, and the dose adjusted (up to 800–1 000 mg/day) after 4–6 weeks if repetitive forms had not been abolished. Once the manifest maximal antiarrhythmic effect[2] was achieved, continuous recordings were obtained at 15- to 90-day intervals until 6 months of treatment, and every 3–6 months thereafter. The mean follow-up was 26.6 ± 4 months (range, 2–55 months).

I. Long-term Antiarrhythmic Effects of Amiodarone

In the pretreatment studies, the frequency of ventricular premature beats varied between 3780 and 61 733/24 h (mean, 17 137 ± 3055). All patients had multiform ventricular extrasystoles and countless couplets, whereas salvos of

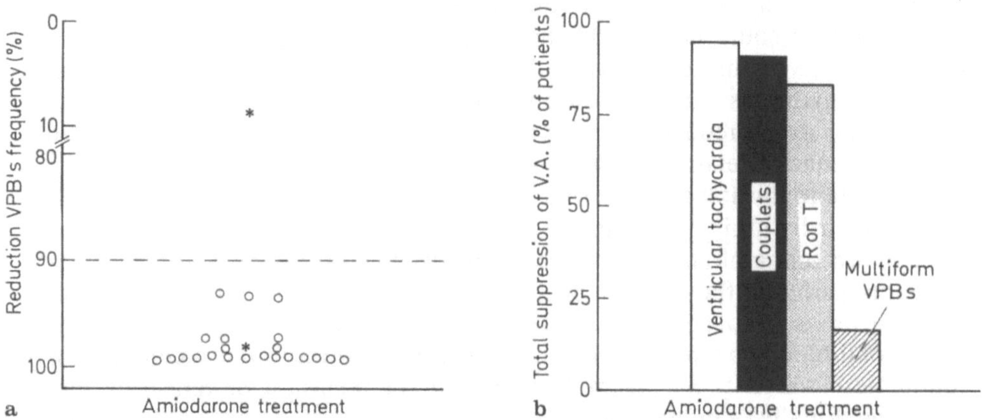

Fig. 7a, b. Long-term antiarrhythmic effects of amiodarone in 24 patients with severe chagasic myocardiopathy. **a** A more than 90% reduction in the number of ventricular premature beats was obtained in 23 patients. Repetitive forms (*stars*) persisted in only two cases [Reproduced from CHIALE et al. (1984) with authorization]. **b** Percentage of patients showing total suppression of complex ventricular extrasystoles during long-term amiodarone treatment. *V.A.*, ventricular arrhythmias. See text for details

[2] We arbitrarily defined the manifest maximal antiarrhythmic effect as the one occurring whenever two or more successive continuous recordings showed a mean intervariation in the frequency of ventricular premature beats of less than 2.5% in relation to the control study, and similar presence or absence of repetitive forms.

ventricular tachycardia were observed in 31 and R on T phenomenon was documented in 17 patients.

Long-term results of amiodarone treatment were similar to those reported by HAEDO et al. in an intermediate-term evaluation (HAEDO et al. 1986). A substantial and sustained reduction (93.2%-99.9%; mean, 94.8%) in the frequency of ventricular premature beats could be documented (Fig. 7a). Couplets were abolished in 22 patients (91%) and bursts of ventricular tachycardia in 20 of 21 cases (95%). Suppression of R on T phenomenon was achieved in 15 of 17 patients (88.23%). On the other hand, multiformity of ventricular extrasystoles remained virtually unmodified (Fig. 7b).

There were only two nonresponders (represented by stars in Fig. 7a). In one, no discernible antiarrhythmic effects could be demonstrated; in the other, couplets and runs of ventricular tachycardia persisted in spite of a 98.2% reduction in the frequency of ventricular premature beats. This latter patient died suddenly during amiodarone therapy.

II. Dose-Response Relations, "Abeyance Period" and Persistence of Antiarrhythmic Protection

The initial dose (600 or 800 mg/day) was enough to achieve the control of ventricular arrhythmias in 9 patients, and had to be increased up to 800-1000 mg/day in 13 patients. The mean dose required to attain the manifest maximal antiarrhythmic effect was 782 mg. In four responders, lowering the dose to 600 or 400 mg once the desired antiarrhythmic action was reached led to a recrudescence of the arrhythmias.

The extremely long half-life of amiodarone (HOLT et al. 1983) accounts for the considerable delay (3-26 weeks; mean 7.4 ± 1 weeks) in achieving the "steady state" for the maximal antiarrhythmic action, in spite of using relatively high doses.

Discontinuance of amiodarone after 3-12 months of uninterrupted and successful administration allowed us to determine that its suppressive action on repetitive forms persisted up to 28-45 days, at a time when the frequency of ventricular premature beats was still manifestly lower than before treatment.

When the response of chagasic ventricular arrhythmias to amiodarone was matched with the response of other clinical arrhythmias, it became ostensible that the former were the least sensitive to this drug. The mean minimal effective dose was higher, the abeyance period lasted longer, and persistence of antiarrhythmic protection was briefer in chagasic patients than in patients with recurrent paroxysmal atrial tachyarrhythmias, potentially dangerous ventricular arrhythmias related to ischemic heart disease, and sustained, recurrent, and symptomatic ventricular tachycardia (ROSENBAUM et al. 1983). Nevertheless, it should be pointed out that the therapeutic goals and the criteria for evaluation of treatment were not the same in the different group of patients. In any event, this relative insensitivity of chagasic ventricular arrhythmias to amiodarone emphasizes the severe nature of the chagasic model.

III. Does Control of Potentially Malignant Ventricular Arrhythmias with Amiodarone Prevent Sudden Death in Chagasic Patients?

This fundamental question cannot be answered yet. However, some indirect data would suggest that amiodarone improves the prognosis of chagasic patients with advanced cardiomyopathy and severe ventricular arrhythmias. In fact, recurrent syncopal episodes were impeded in seven of eight patients, and symptoms were also suppressed in two others with dizziness caused by brief bursts of ventricular tachycardia or ventricular flutter. Moreover, the only patient suffering from syncope in whom amiodarone was unable to eradicate the repetitive forms experienced sudden death. The day before, a 24-h ambulatory Holter recording showed only 125 ventricular extrasystoles and a symptomatic episode of multiform ventricular tachycardia, which lasted 12 s and reverted spontaneously.

It has been estimated that around 20% of chagasic patients die within 2 years after the diagnosis has been made (Rosenbaum 1964). The rate of mortality attributable to chagasic cardiomyopathy in our 22 responders to amiodarone was clearly lower (9%) during the 26.6 month follow-up.

Our findings are in accordance with the results reported in other patients with life-threatening ventricular tachyarrhythmias (Nademanee et al. 1981; Heger et al. 1981) as well as in survivors of prehospital ventricular fibrillation (Nademanee et al. 1983). Nevertheless, further properly programmed prospective studies in larger groups of patients are still necessary to ascertain the role of the pharmacologic control of potentially malignant ventricular arrhythmias in the prevention of sudden cardiac death in chronic Chagas' disease.

IV. Side Effects

A wide spectrum of side effects was observed. Five patients presented cardiac side effects. Two cases showed marked sinus bradycardia during sleep, one had occasional episodes of sinoatrial block, and two developed right bundle branch block. Symptomatic second-degree AV block occurred in one patient after 3 years of treatment, probably as a consequence of the natural course of the disease, and a permanent pacemaker had to be implanted. None of the patients, even those with heart failure, showed detrimental hemodynamic effects; on the contrary, the control of ventricular arrhythmias was accompanied by a significant hemodynamic improvement in three patients with heart failure.

One patient developed *thyrotoxicosis* after 10 months of treatment, which was accompanied by recrudescence of the arrhythmias. He was successfully managed with antithyroid drugs, without reducing the dose of amiodarone. Two patients developed violaceous *facial discoloration*, two others had mild gastric discomfort, and all patients presented *corneal microdeposits*, but without worsening of visual acuity. In spite of this variety of side effects, in none of these patients did the treatment have to be discontinued. However, it should be stressed that serious and limiting adverse reactions, such as interstitial *pneumonitis*, can occasionally be seen (Heger et al. 1981; Sobol and Rakita 1982).

E. Potential Usefulness of Combined Antiarrhythmic Therapy

In some chagasic patients, the pharmacologic control of ventricular arrhythmias with a single antiarrhythmic agent cannot be achieved. This can be related to a "true" refractoriness of the arrhythmias, or to unbearable side effects precluding the administration of relatively high doses of the drugs. In such circumstances, combination of antiarrhythmic drugs may be warranted to obtain a greater antiarrhythmic activity and/or to reduce or suppress limiting side effects.

The usefulness of combined antiarrhythmic schedules to treat "refractory" ventricular arrhythmias has been demonstrated in several studies (DUFF et al. 1983; LEAHEY et al. 1980; WALEFFE et al. 1980). However, in chagasic patients, most of the possible combinations of antiarrhythmic drugs can easily induce serious cardiac side effects, including impairment of the hemodynamic conditions, severe depression of sinus node activity, and sometimes worsening of ventricular arrhythmias. Therefore, in these cases, combined antiarrhythmic treatment must be carefully selected for achieving the two aims mentioned above.

Recently, we have treated two chagasic patients with severe ventricular arrhythmias by combining low doses of amiodarone and mexiletine (400 mg/day each). The two patients had been previously treated with relatively high doses of each drug, but the antiarrhythmic response was modest and/or the treatment had to be discontinued due to intolerable side effects (tremor, nausea, and vomiting). The response to combined treatment was really impressive. Total suppression of ventricular ectopy was obtained and side effects were no longer observed. In an ongoing study, we intend to corroborate these promising observations.

F. Surgical Treatment

It could be thought that surgical procedures have a rather limited usefulness in the treatment of chagasic ventricular arrhythmias. In fact, multiple sources of ventricular ectopy often coexist in chagasic patients, and surgical treatment of all these arrhythmogenic foci is virtually unimaginable. Thus, persistence of severe ventricular arrhythmias has been reported after apical aneurysmectomy (JORGE et al. 1978). However, some chagasic patients develop symptomatic, recurrent, and sustained *monomorphic* ventricular tachycardia unresponsive to different pharmacologic treatments. In such cases, surgical treatment is totally justified. In fact, apical aneurysmectomy has been demonstrated to be life-saving in a couple of patients treated by LAMOURIER et al. 1975. Nevertheless, this can only be considered as a temporary or partial solution, and long-term antiarrhythmic therapy may be advisable for controlling additional ectopic activity.

G. Final Remarks

Chagas' disease is one of the most critical public health problems in Latin American countries, where an estimated 10–12 million people are infected, and approximately 35 million are at a risk of acquiring the infection (Brener 1982). However, this may become an important matter of concern in nonendemic areas, due to the strong migrational currents of Latin Americans. In fact, some recent reports have called attention to the development of symptomatic Chagas' disease in Latin American immigrants residing in the United States, where autochthonous cases are a rarity (Kirchhoff and Neva 1985; Pearlman 1983). It is unknown to what extent *blood transfusion* may also contribute to uncontrolled transmission of chagasic infection beyond its "natural" geographic limits.

Chagasic cardiomyopathy, the most frequent and serious consequence of chronic chagasic infection, leads to disability and death as a result of refractory heart failure or sudden death. A significant number of chagasic individuals, even asymptomatic, have persistent and stable severe ventricular arrhythmias, which can herald the occurrence of ventricular fibrillation and sudden death. Control of these premonitory arrhythmias is of the utmost importance, since this may improve significantly the long-term prognosis of chagasic patients with advanced cardiomyopathy. Although this tends to resolve a very partial aspect of the entire problem, the results obtained in chagasic patients may have much more extended repercussions if applied to other chronic cardiac diseases underlying potentially malignant ventricular arrhythmias and sudden death.

Eradication of Chagas' disease is a complex and multifarious problem, with strong cultural, social, and economic implications. It will be resolved only through appropiate sanitary programs embracing all these aspects.

Acknowledgments. W thank Prof. Dr. José Jalife, MD, and Dr. Jorge Leal, MD, for reading the manuscript. The secretarial skill of Joanne Scheible is also appreciated. This work was supported in part by the Fundación de Investigaciones Cardiológicas Einthoven.

References

Bigger JT Jr (1983) Definition of benign versus malignant ventricular arrhythmias: targets for treatment. Am J Cardiol 52:47C–54C

Brener Z (1982) Recent developments in the field of Chagas' disease. Bull WHO 60:463–473

Carcavallo RV (1975) Aspects of the epidemiology of Chagas' disease in Venezuela and Argentina. In: Pan American Health Organization (ed) American trypanosomiasis research. Pan American Health Organization, Washington DC, pp 374–383

Chiale PA, Halpern MS, Nau GJ, Przybylski J, Tambussi AM, Lázzari JO, Elizari MV, Rosenbaum MB (1982) Malignant ventricular arrhythmias in chronic chagasic myocarditis. PACE 5:162–172

Chiale PA, Halpern MS, Nau GJ, Tambussi AM, Przybylski J, Lázzari JO, Elizari MV, Rosenbaum MB (1984) Efficacy of amiodarone during long-term treatment of malignant ventricular arrhythmias in patients with chronic chagasic myocarditis. Am Heart H 107:656-665

Coura JR (1966) Contribuçao ao estudo da doença de Chagas no estado de Guanabara. Rev Bras Malariol Doenças Trop 18:9-15

Duff HJ, Roden D, Primum K, Oates JA, Woosley RL (1983) Mexiletine in the treatment of resistant ventricular arrhythmias: enhancement of efficacy and reduction of dose related side effects by combination with quinidine. Circulation 67:1124-1128

Elizari MV, Levi RJ, Novakoski A, Lázzari JO, Vetulli HM (1980) Cellular effects of antiarrhythmic drugs. Remarks on methodology. VIII European Congress of Cardiology, Paris, p 11

Faria CAF, Veloso C, Carvallo CA (1983) Parassístole ventricular dupla espontánea em paciente portador de cardiopatía chagasica. Relato de um caso. Arq Bras Cardiol 41:385-388

Haedo AH, Chiale PA, Bandieri JD, Lázzari JO, Elizari MV, Rosenbaum MB (1986) Comparative antiarrhythmic efficacy of verapamil, 17-monochloracetilajmaline, mexiletine and amiodarone in patients with severe chagasic myocarditis: relation with the underlying arrhythmogenic mechanisms. J Am Coll Cardiol. 7:1114-1120

Heger JJ, Prystowsky EN, Jackman WM, Naccarelli GV, Warfel KA, Rinkenberger RL, Zipes DP (1981) Amiodarone. Clinical efficacy and electrophysiology during long-term therapy for recurrent ventricular tachycardia or ventricular fibrillation. N Engl J Med 305:539-547

Holt DW, Tucker GT, Jackson PR, Storey GCA (1983) Amiodarone pharmacokinetics. Am Heart J 106:788-797

Jorge PAR, Nogueira EA, Bittencourt LAK, Fortuna ABP, Terzi RGG, Vieira RW (1978) Arritmia cardiaca a lesao apical na cardiopatia chagasica cronica. Consideraçoes a respeito de un caso. Arq Bras Cardiol 31:135-137

Kaski JC, Girotti LA, Messuti H, Rutitzky B, Rosenbaum MB (1981) Long-term management of sustained, recurrent, symptomatic ventricular tachycardia with amiodarone. Circulation 64:273-279

Kirchhoff LV, Neva FA (1985) Chagas' disease in Latin American immigrants. JAMA 254:3058-3060

Lamourier EN, Herrmann JLV, Martinez Filho EE, Buffolo E, De Andrade JCS, Korkes H, Schubsky V, Ferreira C, Barcellini A, Portugal OP (1975) Aneurismectomia como tratamento de taquiarritmias refratarias en pacientes portadores de aneurisma ventricular de etiologia chagasica. Arq Bras Cardiol 28:549-555

Leahey EB Jr, Heissenbuttel RH, Giardina EGV, Bigger JT Jr (1980) Combined mexiletine and propranolol treatment of refractory ventricular tachycardia. Br Med J [Clin Res] 281:357-358

Marinkelle CJ (1975) Epidemiology of Chagas' disease in Colombia. In: Pan American Health Organization (ed) American trypanosomiasis research. Pan American Health Organization, Washington DC, pp 340-345

Nademanee K, Hendrickson BS, Cannon DS, Goldreyer BN, Singh BN (1981) Control of refractory life-threatening ventricular arrhythmias by amiodarone. Am Heart J 101:759-768

Nademanee K, Singh BN, Cannon DS, Weiss S, Feld G, Stevenson WG (1983) Control of sudden recurrent arrhythmic deaths: role of amiodarone. Am Heart J 106:895-901

Nau GJ, Acunzo RS, Saleme JP, Aldariz A, Elizari MV, Rosenbaum MB (1985) Carac-

terísticas clínicas, evaluación y pronóstico de las parasistolias ventriculares. Rev Arg Cardiol 53:553 (Abstract)
Osmani JJ (1972) Epidemiología de la enfermedad de Chagas en Uruguay. In: Secretaría de Salud Pública de la República Argentina (ed) Simposio Internacional sobre Enfermedad de Chagas. Buenos Aires, pp 209-214
Pearlman JD (1983) Chagas' disease in Northern California. No longer an endemic diagnosis. Am J Cardiol 75:1057-1060
Peralta JM (1981) Auto-antibodies and chronic Chagas' heart disease. Trans R Soc Trop Med Hyg 75:568-569
Ponce C, Zeledon R (1973) La enfermedad de Chagas en Honduras. Bol Of Sanit Panam 75:239-247
Prata A (1975) Natural history of chagasic cardiomyopathy. In: Pan American Health Organization (ed) American trypanosomiasis research. Washington DC, p 191
Reis Lopes E, Chapadeiro E, Almeida HC, Rocha A, Prata A (1974) Doença de Chagas e morte subita. X Congreso Brasileiro de patologia e i encontro Luso-Brasileiro de anatomia patologica. Curitiba, Paraná, Brazil (Abstract)
Reis Lopes E, Chapadeiro E, Almeida HC, Rocha A (1976) Contribuçao ao estudo de anatomia patologica dos coraçoes de chagasicos falecidos subitamente. Rev Soc Bras Med Trop 9:269-279
Ribeiro dos Santos R, Hudson L (1980) *Trypanosoma cruzi*: immunological consequences of parasite modification of host cells. Clin Exp Immunol 40:36-41
Rosenbaum MB (1964) Chagasic myocardiopathy. Prog Cardiovasc Dis 7:199-225
Rosenbaum MB, Chiale PA, Haedo AH, Elizari MV (1983) Ten years of experience with amiodarone. Am Heart J 101:957-964
Sachs RN, Lanfranchi J (1978) Cardiomyopathies primitives et anomalies immunitaires. Coeur Med Int 17:193-198
Schmidberg JM, Acunzo RS, Nau GJ, Elizari MV, Rosenbaum MB (1985) Eficacia del nadolol en el tratamiento de las arritmias ventriculares inducidas o agravadas por la ergometría. Rev Arg Cardiol 53:S130 (Abstract)
Sobol SM, Rakita L (1982) Pneumonitis and pulmonary fibrosis associated with amiodarone treatment: a possible complication of a new antiarrhythmic drug. Circulation 65:819-822
Waleffe A, Mary-Rabine L, Legrand V, Demoulin JCL, Kulbertus HE (1980) Combined mexiletine and amiodarone treatment of refractory recurrent ventricular tachycardia. Am Heart J 100:788-793
Winkle RA (1978) Antiarrhythmic drug effect mimicked by spontaneous variability of ventricular ectopy. Circulation 57:1116-1120
Winkle RA, Meffin PJ, Harrison DC (1978a) Long-term tocainide therapy for ventricular arrhythmias. Circulation 57:1008-1011
Winkle RA, Gradman AH, Fitzgerald JW (1978b) Antiarrhythmic drug effect assessed from ventricular arrhythmias reduction in the ambulatory electrocardiogram and treadmill test: comparison of propranolol, procainamide, and quinidine. Am J Cardiol 42:473-484

CHAPTER 28

Autonomic Mechanisms in Cardiac Rhythm and Arrhythmias

W. SPINELLI and M. R. ROSEN

A. Introduction

Our purpose in this chapter is to review the cellular electrophysiologic effects of autonomic stimulation, and to demonstrate its complexity in two ways: by describing the changes in cardiac autonomic interactions that occur with growth and development, and by describing how autonomic mediators modulate cardiac arrhythmias.

B. Autonomic Effects at the Cellular Level

I. Sympathetic Effects

1. Beta-Adrenergic Stimulation

The beta-adrenergic effects of catecholamines are attributed to interaction with $beta_1$-adrenergic receptors (WATANABE et al. 1982). Radioligand studies have also shown the presence of $beta_2$-adrenergic receptors in significant number on cardiac cells (MINNEMAN et al. 1979; MANALAN et al. 1981). The function of these receptors is not yet clear. Although some studies have indicated that $beta_2$-receptors may be involved in the chronotropic action of catecholamines (TUTTLE et al. 1976; CARLSSON et al. 1977), other studies have shown that $beta_2$-receptors are mostly confined to fibroblasts (LAU et al. 1980). Pharmacological studies using terbutaline, a nonspecific beta-agonist with predominant $beta_2$ characteristics, have also indicated that the observed chronotropic and inotropic actions of beta-agonists are due to their interaction with the $beta_1$-receptor (DANILO and ROSEN 1982). The electrophysiologic effects of stimulation of individual types of adrenoceptor were described recently by DUKES and VAUGHAN WILLIAMS (1984).

In the SA node, beta-adrenergic stimulation increases the slope of phase 4 depolarization, resulting in a faster rate of pacemaker firing. The effects are accompanied by an increase in maximum diastolic potential, and in the rate of rise and amplitude of the upstroke of the transmembrane action potential (HOFFMAN and CRANEFIELD 1960). In the AV node, beta-adrenergic stimulation enhances the velocity of impulse propagation and shortens the effective refractory period. These effects are accompanied by an increase in the amplitude and rate of rise of the action potential, without a significant alteration in the maximum diastolic potential (WIT et al. 1975; CRANEFIELD 1975). Reflecting these effects at the cellular level, beta-adrenergic stimulation reduces the electrocardiographic P-R interval, which is an indicator of the AV conduction time (LEVY and ZIESKE 1969) and the A-H interval (SPEAR and MOORE 1973). Stimulation of beta-adrenergic receptors in the ventricular conducting system increases the slope of phase 4 depolarization of spontaneous pacemakers in the His bundle and in Purkinje fibers (VASSALLE 1979; ROSEN et al. 1977). Recent data indicate that the depolarization during phase 4 may be related to a time-dependent activation of an inward current (i_f or i_h) carried mainly by Na^+ and activated by hyperpolarization (DIFRANCESCO 1981 a, b). Beta-adrenergic stimulation shifts the activation range to a more positive voltage; thus i_f activates more completely and more rapidly following the final phase of repolarization. This, in turn, results in more rapid depolarization during phase 4. A similar ionic current has been proposed to contribute to phase 4 depolarization in SA nodal cells (BROWN and DIFRANCESCO 1980). The increase in the slope of phase 4 in Purkinje fibers is thought to be mediated by an intracellular increase in the level of cyclic AMP, as shown by iontophoresis of cyclic AMP into Purkinje cells (TSIEN 1974). However, other experiments in the same tissue indicate that catecholamines can increase automaticity without altering cyclic AMP levels (DANILO et al. 1978). Ordinarily, the spontaneous automaticity of ventricular specialized conducting cells is suppressed by the faster rate of firing of the SA node (VASSALLE 1979). In Purkinje fibers, Beta-adrenergic agonists increase plateau height and duration (ROSEN et al. 1977) by augmenting the slow inward current (i_{si}) (REUTER 1979), and increase the slope of repolarization. The latter action is due to the increase in activation of an outward K^+ current (i_{xi}) (TSIEN et al. 1972). Catecholamines have similar actions on the plateau and repolarization phases of atrial and ventricular myocytes, although the changes in normally polarized fibers are less marked. The observed increase in i_{si} is probably the major mechanism for the positive inotropic effect of beta-adrenergic stimulation (NATHAM and BEELER 1975). In depolarized Purkinje fibers, as well as in atrial and ventricular fibers, beta-adrenergic stimulation may induce hyperpolarization (WIT et al. 1975), which is thought to be due to a (beta$_2$-mediated) stimulation of electrogenic pumping (VASSALLE and BARNABEI 1971). All the effects of beta-adrenergic stimulation thus far described can be antagonized by non-selective beta-adrenergic blocking agents, such as propranolol.

2. Alpha-Adrenergic Stimulation

The presence and function of alpha-adrenergic receptors in the myocardium has aroused controversy (OSNES et al. 1978). Physiologic and pharmacologic studies have indicated a role of alpha-adrenergic stimulation in increasing contractility in human atrium (SCHÜMANN et al. 1978). Radioligand binding studies have shown the presence of alpha-adrenergic receptors in the myocardium (YAMADA et al. 1980; KUPFER et al. 1982). The electrophysiologic effects of alpha-adrenergic stimulation on cardiac cells have been described more recently. Unlike beta-agonists, alpha-adrenergic agonists have been shown to prolong action potential duration in Purkinje fibers without significantly altering the maximum diastolic potential (GIOTTI et al. 1973). Furthermore, unlike beta-agonists, alpha-adrenergic agonists in low concentrations ($<5 \times 10^{-8}$ *M*) depress the slope of phase 4 depolarization, and thus the rate of firing in normally polarized Purkinje (ROSEN et al. 1977; POSNER et al. 1976) and atrial specialized fibers (MARY-RABINE et al. 1978). This action on automaticity is biphasic in that higher concentrations of alpha-adrenergic agonists tend to have beta-adrenergic effects as well, and increase automaticity via beta-receptor stimulation. The negative chronotropic effects observed at low concentrations of agonists can be antagonized by the alpha-adrenergic blocker phentolamine, but not by the beta-blocker propranolol (ROSEN et al. 1977). More specifically, the effect of alpha-adrenergic stimulation on automaticity seems to be mediated by alpha$_1$-receptors (ROSEN et al. 1984). In these studies, the action of phenylephrine, a nonspecific alpha-agonist, was blocked by the alpha$_1$-blocker prazosin but not by the predominant alpha$_2$-blocker yohimbine. As the concentrations of prazosin and yohimbine used did not have any direct action on the transmembrane potential, the observed effects were attributed to their interaction with alpha$_1$- and alpha$_2$-adrenergic receptors (see Chap. 22). The mechanism of the chronotropic action has not been elucidated yet. Studies of K^+ fluxes have shown that low concentrations of epinephrine and norepinephrine that can depress automatic rate also depress $^{42}K^+$ uptake without affecting $^{42}K^+$ efflux in canine cardiac Purkinje fibers (VASSALLE and BARNABEI 1971; POSNER et al. 1976). Recent studies conducted in primary cultures of rat myocytes and in intact rat ventricles have shown that the negative chronotropic effect of phenylephrine on the cultures is not associated with any significant change in the levels of cyclic AMP (STEINBERG et al. 1985). These studies have also provided evidence that the alpha$_1$-receptor is probably coupled to the effector mechanism through an inhibitory guanine nucleotide protein (GILMAN 1987). This protein appears similar to the GTP regulatory protein G in that it is ADP-ribosylated and has a molecular weight of about 41000. Study of neonatal rat cardiac myocytes in primary culture, as well as in culture with sympathetic nerves, has indicated both that sympathetic innervation induces the alpha-adrenergic chronotropic action (i.e. in the absence of innervation alpha-agonists *increase* automaticity) and the innervation also increases the levels of the G_i-like regulatory protein (STEINBERG et al. 1985). More recent studies of adult. Purkinje fibers have shown that if the G_1-like protein is ADP ribosylated by pertussis toxin, then the alpha-adrenergic effect reverts from negative to positive chronotropy. Hence, it appears that the alpha-adrenergic effect on automaticity not only is dependent on a specific regulatory protein, but also that the levels of the protein, in turn, are modulated by the sympathetic innervation (ROSEN et al. 1987).

Several observations have indicated a role of alpha-receptor stimulation in pathologic conditions such as hyperthyroidism and hypothyroidism (CIARALDI and MARINETTI 1977), and ischemia (CORR et al. 1981). In the ischemic cat heart, the increase in alpha-receptor density was accompanied by premature ventricular depolarizations and fibrillation, which could be prevented by treatment with the alpha-blocker prazosin, but not by the beta-blocker propranolol (SHERIDAN et al. 1980). In contradistinction to the effects observed in atrial myocardium and in the ventricular specialized conducting system, no alpha$_1$-adrenergic effect was identifiable on the automaticity of SA cells in studies by HEWETT and ROSEN (1985). Studies in depolarized canine Purkinje fibers (HEWETT and ROSEN 1985) and in depolarized guinea pig ventricular cells (HUME and KATZUNG 1978) did not show any effect of alpha stimulation. It has been speculated that the alpha-adrenergic action on automaticity may be voltage-dependent. If so, alpha stimulation would be expected to depress automaticity in cells with negative (−50 mV) maximum diastolic potential, possibly acting on the i_f current, and have no effect, or even increase automaticity (AMERINI et al. 1984) in cells with maximum diastolic potential more positive than −50 mV.

II. Parasympathetic Effects

All the effects of acetylcholine that will be described are mediated through stimulation of muscarinic receptors and are blocked by atropine. Parasympathetic innervation in the supraventricular tissue is much richer than in the ventricles and the vagal effects are qualitatively and quantitatively different. In SA node, muscarinic stimulation increases the maximum diastolic potential and depresses the slope of phase 4 depolarization, resulting in a negative chronotropic effect (HOFFMAN and CRANEFIELD 1960). The cellular effects result from an increase in outward K^+ conductance (TRAUTWEIN et al. 1956) and a decrease of inward flow of Ca^{2+} and Na^+ through slow channels (GARNIER et al. 1978; GILES and NOBLE 1976). In the cells of the AV junction, acetylcholine, by depressing the slow inward current and increasing the outward K^+ current, depresses the rate of rise and the amplitude of the action potential, thus slowing the propagation of the impulse through the node. The effect is also accompanied by an increase of the refractory period of the nodal cells (HOFFMAN and CRANEFIELD 1975). Characteristically, acetylcholine mostly affects cells of the upper (AN) and midnodal (N) zone; cells of the lower part of the node (NH) are little affected. In atrial myocardial fibers, muscarinic stimulation shortens the action potential duration, tends to hyperpolarize the membrane potential, and shortens the refractory period. As for the nodal cells, both effects are thought to result from the action of acetylcholine on i_{si} and the outward K^+ conductance. The reduction of the slow inward current, together with the shortening of the action potential duration, produce a significant negative inotropic effect (TEN EICK et al. 1976). In fibers that undergo a significant hyperpolarization, acetylcholine may increase conduction velocity significantly. This effect is variable and very limited in normally polarized fibers (STRAUSS et al. 1977).

Although early anatomic studies failed to substantiate the parasympathetic innervation of the ventricles, it is now firmly established that the ventricles receive a substantial vagal innervation. Such evidence is based on both anatomic (STRAUSS et al. 1977; KENT et al. 1974) and biochemical data. Radioligand binding studies in several species have shown that, although the density of muscarinic receptor in the atria is six to nine times higher, the ventricles, due to their larger mass, contain 60% of the total number of cardiac muscarinic receptors (FIELDS et al. 1978). Despite these observations, the direct action of acetylcholine on the ventricular conducting system is rather modest. In the dog, for example, very high concentrations of acetylcholine (10^{-5}–10^{-4} *M*) produce only a minor shortening of the action potential duration (GADSBY et al. 1978). The effects on phase 4 depolarization and automaticity are more marked. In canine Purkinje fibers, acetylcholine induces a concentration-dependent depression of automaticity with an EC_{50} of about 10^{-7} *M* (BAILEY et al. 1972; DANILO P JR et al. 1978). Although acetylcholine can reduce automaticity in both supraventricular and ventricular tissue, the effect is especially relevant in the SA and AV nodes. This may be explained by the fact that whereas the vagal innervation of the conducting system is relatively dense at the level of the proximal bundle branches, it becomes progressively more sparse in the distal Purkinje fiber system (KENT et al. 1974).

In the ventricular myocardium, the direct effect of acetylcholine is negligible (HOFFMAN and SUCKLING 1953). However, experimental evidence shows that strong vagal stimulation can regulate the inotropic response of the ventricle following sympathetic stimulation (LEVY 1971). Considerable evidence indicates that parasympathetic modulation of sympathetic stimulation can occur at several different levels (WATANABE and LINDEMANN 1984). Sympathetic and parasympathetic fibers lie very close to each other (JACOBOWITZ et al. 1967) and there is evidence that prejunctional muscarinic receptors are present on sympathetic terminals (SHARMA and BAREIJEE 1978). Furthermore, both sympathetic and muscarinic receptors are present on the sarcolemma of myocardial cells (JONES et al. 1979). Thus, acetylcholine may regulate the release of epinephrine from sympathetic terminals by interacting with prejunctional muscarinic receptors present on these terminals. Acetylcholine may also interact with postjunctional muscarinic receptors on myocardial cells to modulate the biochemical effects of released or circulating catecholamines (BAILEY et al. 1972). Clear electrophysiologic effects following muscarinic stimulation can be demonstrated in the presence of beta-adrenergic stimulation. Thus, muscarinic agonists can antagonize the electrophysiologic effects of beta stimulation both in Purkinje fibers and in ventricular muscle in vitro (BAILEY et al. 1979; INUI and IMAMURA 1977) and in vivo (VERRIER and LOWN 1978).

Despite this experimental evidence, it is not yet clear to what extent the vagally induced release of acetylcholine in ventricular myocardium affects the electrophysiologic properties of human ventricles. The data available at present, obtained in several animal models, only allow the speculation that vagal effects may be profound, especially in conditions of high sympathetic tone such as, for example, after myocardial infarction. Vagal stimulation alone, in the absence of an increase in sympathetic tone, was found to be without effect on ventricular vulnerability to fibrillation (VERRIER and LOWN 1978). However, more recent clinical studies have shown that in the presence of beta-adrenergic receptor blockade muscarinic receptor blockade by atropine can significantly shorten effective and functional refractory periods in human ventricles (PRYSTOWSKY et al. 1981). In clinical studies of sustained ventricular tachycardia, a reflex increase in vagal tone, produced by increasing blood pressure with phenylephrine, resulted in termination of the tachycardia. Pretreatment with atropine blocked the conversion of the ventricular tachycardia by phenylephrine, suggesting an important vagal role in the antiarrhythmic effect (WAXMAN and WALD 1977). In studies of ventricular automaticity conducted in dogs, vagal stimulation caused AV nodal block and a prolongation of ventricular asystole. However, the suppression of the ventricular pacemakers during vagal stimulation was a function of the rate and duration of stimulation of the ventricles prior to nodal block, and not the result of a direct vagal effect on the ventricles (VASSALLE et al. 1967 a, b). In conclusion, most of the available evidence, although supportive of the hypothesis of a significant vagal action, also stresses the importance of changes in heart rate in modifying ventricular arrhythmias.

C. Developmental Changes in Cardiac-Autonomic Interactions

In the past 10 years, it has become increasingly clear that the action of autonomic transmitters on cardiac electrophysiologic function undergoes significant age-related changes. The similar time course for the development of functional innervation and for the changes in responses to autonomic transmitters suggests that the onset of innervation might mediate these changes (PAPPANO 1977; SPERALAKIS and PAPPANO 1983). Cardiac sympathetic innervation begins at an early stage of development and, in several mammalian species, is still incomplete at birth, and reaches the final stage only during neonatal life (FRIEDMAN et al. 1968; PAPPANO 1977; LIPP and RUDOLPH 1972; STANDEN 1978). On the contrary, parasympathetic innervation of the heart precedes sympathetic innervation, but it is still controversial whether the parasympathetic nervous system is functionally mature at birth in dogs, chicks, and humans (WALKER 1969; URTHALER et al. 1980). The development of postsynaptic autonomic receptors, which has been shown to precede innervation (PAPPANO 1977), has been reviewed recently by PAPPANO (1984).

The effects of muscarinic and alpha- and beta-adrenergic stimulation on the transmembrane action potential undergo important developmental changes. We will concentrate on the effect on phase 4 depolarization, although age-related differences in the action of autonomic mediators have also been shown for other phases of the transmembrane action potential (ROSEN et al. 1977; DANILO P JR et al. 1978; MOAK et al. 1986). In Purkinje fibers obtained from adult dogs, acetylcholine induces the previously described depression of automaticity at a threshold concentration of 1×10^{-7} *M*. However, in Purkinje fibers from neonatal dogs, acetylcholine shows a biphasic action. At low concentrations (10^{-9} *M*), it increases the slope of phase 4 depolarization and, consequently, the rate of automatic firing, while, at higher concentrations, the usual depression of automaticity is seen (DANILO P JR et al. 1978). The effect observed in neonatal fibers is not due to liberation of catecholamines, as blocking concentrations of propranolol do not prevent the increase in automaticity. Similarly, by using cimetidine, an H_2-receptor blocker, it was shown that the increase in rate does not result from acetylcholine-induced release of histamine. Finally, the increased action of acetylcholine is mediated by interaction with muscarinic receptors.

The beta-adrenergic effects on automaticity also change during development. The increase in rate produced by identical concentrations of isoproterenol or epinephrine is smaller in adult than in neonatal Purkinje fibers (ROSEN et al. 1977). In fact, with isoproterenol, the EC_{50} for the increase in automaticity is fivefold lower in neonates (SPINELLI et al. 1986). Other studies conducted in neonatal rat hearts and in dog Purkinje fibers have also indicated a higher sensitivity to the beta-blocking effects of propranolol in younger animals (SPINELLI et al. 1986; ROSEN et al. 1979). The basis for these age-related differences in agonist and antagonist sensitivity is not known, although it is possible that in the neonate, due to the incomplete development of the sympathetic innervation at birth, there is the equivalent of denervation-induced supersensitivity to the agonist effects of catecholamines.

As previously mentioned, muscarinic stimulation can counteract the electrophysiologic effects of beta-adrenergic stimulation in canine Purkinje fibers (BAILEY et al. 1979). Developmental differences in this sympathetic-parasympathetic interaction have recently been described (MOAK et al. 1986). Beta-adrenergic stimulation with isoproterenol, alone, shortens the action potential duration in adult fibers, whereas, in those of neonates, isoproterenol prolongs the action potential duration. Muscarinic stimulation with acetylcholine significantly inhibits the effect of isoproterenol in adult fibers; however in neonatal fibers, even high concentrations of acetylcholine (10^{-5} *M*) do not show any significant antagonism. The differences in the time course of development and maturity of the sympathetic and parasympathetic nervous systems may provide an explanation for these results.

The effects of alpha-adrenergic agonists on automaticity also undergo age-related changes which seem to depend on the state of the autonomic innervation. It has been described previously that the alpha-adrenergic agonist phenylephrine has a biphasic effect on automaticity. In adult and neonatal Purkinje fibers, low concentrations of phenylephrine depress the slope of phase 4 depolarization, while higher concentrations have the opposite effect (ROSEN et al. 1977). It must be noted that phenylephrine is not a very selective α-agonist, since β-receptors are activated by the higher concentrations (DUKES and VAUGHAN WILLIAMS, 1984; Chap. 22). Although the concentrations of agonist that cause slowing of frequency and the magnitude of the slowing are the same at both ages, in adults, 75 % of the tested fibers showed the decrease in rate, while in neonates (up to 10-day-old dogs) only 50 % of the fibers had a decrease in rate (ROSEN et al. 1977). The difference in percentage between neonates and adults suggests that the responsiveness to alpha-agonists may depend on the state of the autonomic innervation. In fact, other studies showed that in very immature newborn dogs, when the sympathetic innervation is still developing, alpha-adrenergic stimulation *increases* rather than decreases Purkinje fiber automaticity (REDER et al. 1984). Similar age-dependent results following alpha stimulation have also been obtained in studies of automaticity in intact rat ventricles. Again, such chronotropic responses are blocked by phentolamine but not by propranolol (DRUGGE et al. 1985). The hypothesis that the state of sympathetic innervation may change the response to alpha-adrenergic stimulation is further supported by studies in cultured, noninnervated rat myocytes, where phenylephrine only caused a positive chronotropic response. However, when the myocytes were cocultured with sympathetic neurons until sympathetic innervation of the myocytes could be demonstrated, the response to alpha stimulation changed to a depression of automaticity (DRUGGE et al. 1985). Recent studies have also shown that the negative chronotropic response is not explained by variations in the levels of acetylcholine, norepinephrine, epinephrine, or dopamine in the culture medium. The results of these studies also indicated that the trophic effects of the nerve cells cannot be explained by the release of a still unidentified substance in the culture medium, but require close nerve-muscle association. In fact, muscle cells grown in the same petri dish with sympathetically innervated muscle cells, to allow conditioning of the muscle cells by the neurons, do not acquire a negative chronotropic response to alpha stimulation (DRUGGE and ROBINSON 1987).

Although these age-related alterations have been clearly documented only in dog and rat hearts so far, they might conceivably reflect and explain the different patterns of arrhythmias clinically reported in younger individuals (FERRER 1977).

D. Relationship of Autonomic Stimulation to the Cellular Mechanisms of Arrhythmias

The cellular mechanism for arrhythmias includes abnormal impulse initiation and abnormal propagation. The former group includes arrhythmias generated by automaticity and afterdepolarizations; the latter includes arrhythmias induced by reentry and reflection. In this discussion, we will concentrate on the effects of autonomic interactions. A more detailed description of the various mechanisms and their interrelationships is included in Chap. 12, 22.

I. Abnormal Impulse Initiation

1. Automaticity

Abnormal impulse initiation may result from the expression of a normal automatic mechanism or from the occurrence of abnormal automaticity. Normal automaticity is the result of pacemaker currents operating in the sinus node, in specialized atrial cells, and in the ventricular conducting system. In contrast, abnormal automaticity is the result of phase 4 depolarization at levels of transmemembrane potential that are less negative than the normal maximum diastolic potential of any tissue.

Clearly, sinus bradycardia and tachycardia can be considered due to normal automatic mechanisms. In cases of profound bradycardia, the ectopic rhythms that emerge are probably generated by the normal automaticity of specialized cells in the atria, the AV junction, or the ventricular conducting system. There is some controversy as to whether normal automaticity can produce a rhythm sufficiently fast to be life threatening. Sympathetic stimulation can increase the automatic rate of the ventricles up to 120 beats/min (HOFFMAN and CRANEFIELD 1960). Experiments conducted in dogs show that during right stellate ganglion stimulation the heart may reach rates in excess of 200 beats/min (RANDALL 1977). The effects of vagal and adrenergic interventions have been discussed in more detail in the preceding paragraphs and can be summarized by saying that acetylcholine, by binding to muscarinic receptors, depresses the slope of phase 4 depolarization, and thus decreases automaticity. The situation with adrenergic agonists is more complicated: beta-receptor stimulation, by enhancing the slope of phase 4 and shortening APD, increases automaticity, while alpha-adrenergic stimulation has the opposite (DUKES and VAUGHAN WILLIAMS 1984) effect by delaying repolarization.

Impulses can also be generated spontaneously by fibers whose maximum diastolic potential is reduced to about -50 to -60 mV. There is clear evidence that most fibers, including atrial and ventricular muscle fibers, can generate automatic rhythms by developing spontaneous phase 4 depolarization under appropriate experimental interventions (SURAWICZ and IMANISHI 1976; KATZUNG and MORGENSTERN 1977). Of clinical importance is the observation that under conditions of ischemia and following myocardial infarction, maximum diastolic potential decreases in some myocardial fibers which then may show spontaneous phase 4 depolarization. Such conditions are also encountered in tissue obtained from diseased human atria and ventricles where fibers depolarized to maximum diastolic potentials of -50 to -60 mV show phase 4 depolarization and automaticity (HORDORF et al. 1976; MARY-RABINE et al. 1980; DANGMAN and HOFFMAN 1983). The effects of autonomic regulation on these pacemakers is similar to what has already been observed in the case of normal automaticity. Acetylcholine will depress phase 4 depolarization while catecholamines will increase it, and consequently will increase automaticity. The action of both acetylcholine and catecholamines was accompanied by a significant hyperpolarization of the membrane (MARY-RABINE et al. 1978). In contrast to what has been observed in normal Purkinje fibers, alpha-agonists do not depress automaticity in depolarized canine Purkinje fibers (HEWETT and ROSEN 1985) or in depolarized guinea pig ventricular myocardium (HUME and KATZUNG 1978), and may actually increase automaticity (AMERINI et al. 1984).

As with normal automaticity, there are doubts whether partially depolarized cells can generate very rapid rhythms. Most examples of abnormal automaticity studied in vitro have shown rates only slightly faster than those observed with normal automaticity. Rhythms generated by abnormal automaticity, however, are not readily suppressed by overdrive stimulation (DANGMAN and HOFMAN 1983). As membrane potential attains more positive values, there is less entry of Na^+ ion via the fast inward current during the action potential uptroke (i. e., the action potential becomes more dependent on slow inward Ca^{2+} current). Since Na_i^+ is a major stimulus for electrogenic Na^+/K^+ pumping, the reduced Na^+ entry in depolarized tissues results in a lesser stimulus for overdrive suppression. Microelectrode studies of the myocardium from the infarcted zone have shown depolarized membrane potentials and automaticity which is very sensitive to the chronotropic effects of catecholamines and relatively resistant to overdrive pacing (LE MAREC et al. 1985).

2. Afterdepolarizations

Afterdepolarizations are depolarizing potentials that occur during or following repolarization.

Afterdepolarizations have been described as "early" if they occur during phase 2 and/or 3 of repolarization (CRANEFIELD 1975). There are many possible causes for early afterdepolarizations. They may be due to a decrease in outward, repolarizing current carried by K^+ and/or an increase in inward current carried by Na^+ or Ca^{2+}. Since the total membrane conductance during phase 2 and the beginning of phase 3 is no greater than during diastole, a minor perturbation of the balance between outward and inward currents could reverse or accelerate the normal course of repolarization. Early afterdepolarizations are seen most frequently at low rates of stimulation. When the rate becomes faster, repolarization accelerates and early afterdepolarizations decrease in frequency and amplitude. Clearly, a vagally induced decrease in heart rate might be expected to favor the development of early afterdepolarizations (countered in part, by the acetylcholine-induced enhancement of repolarization secondary to increased K conductance). A sympathetic-induced increase in the rate of impulse initiation might be expected to suppress early afterdepolarizations, but this action might be counteracted by the increase in slow inward current induced by beta-adrenergic stimulation. For example, high concentrations of catecholamines have been shown to induce early depolarizations in Purkinje fibers (BROOKS et al. 1955). The final effect of vagal and/or sympathetic actions then depends on the net balance between outward and inward currents.

An early afterdepolarization can attain sufficient magnitude to initiate a premature response or a train of action potentials. The resulting rhythms can last for a variable amount of time and respond in a variable manner to overdrive stimulation. Early afterdepolarization-induced rhythms that occur at high levels of membrane potentials are more easily suppressed by overdrive than rhythms that arise from a low (less negative) level of membrane potential (DAMIANO and ROSEN 1984). In general, the characteristics of these rhythms triggered by pacing resemble automatic rhythms generated by maintained depolarization. Consequently, the effects of vagal and sympathetic stimulation are similar to those already described for abnormal automaticity.

It has been speculated that early afterdepolarizations may initiate the arrhythmia called torsade de pointes (SMITH and GALLAGHER 1980). This arrhythmia, most frequently seen in the setting of an abnormally long electrocardiographic Q-T interval (possibly, but not necessarily, due to delayed repolarization), also occurs more readily following a long diastolic interval. Studies using agents such as CsCl which induce early afterdepolarizations in vitro (DAMIANO and ROSEN 1984), have precipitated arrhythmias resembling torsade de pointes in the dog (BRACHMAN et al. 1983) and, during this arrhythmia, catheter techniques have been used to record early afterdepolarizations (LEVINE et al. 1985).

Afterdepolarizations have been described as delayed if they appear when the membrane potential has been restored to a maximum diastolic potential that is no more than 10–15 mV less negative than normal. Delayed afterdepolarizations can be induced by digitalis toxicity (FERRIER et al. 1973; ROSEN et al. 1973b), by high levels of catecholamines, or by myocardial infarction (DANGMAN et al. 1982). They have been recorded in normal canine coronary sinus (WIT and CRANEFIELD 1977), in fibers of the AV valves (WIT and CRANEFIELD 1976), as well as in diseased human atrium (MARY-RABINE et al. 1980) and ventricle (DANGMAN et al. 1982). Such afterdepolarizations can reach threshold and initiate a premature response and, under certain conditions, may even start a train of action potentials. More specifically, as the pacing frequency increases the amplitude of delayed afterdepolarizations also increases and the coupling intervals to the action potentials which induce them shorten (ROSEN et al. 1973a). Therefore, if one afterdepolarization reaches threshold and causes a premature response, it is also likely that the next will reach threshold and a train of self-sustained action potentials will result. Thus, while early afterdepolarizations might induce bradycardia-dependent tachyarrhythmias, delayed afterdepolarizations will tend to induce tachycardia-dependent tachyarrhythmias. Acting through this rate-dependent mechanism, the autonomic nervous system can play an important role. For example, with a decrease in heart rate, delayed afterdepolarizations might decrease in amplitude and not reach threshold, and the resulting abnormal rhythm might be prevented. Delayed afterdepolarizations are increased in amplitude by agents that increase $[Ca^{2+}]_i$ and/or $[Na^+]_i$. Thus, beta-agonists, by increasing i_{si}, augment the amplitude of the depolarization (HEWETT and ROSEN 1984). Alpha-agonists have no effect at normal levels of $[Ca^{2+}]_0$ (HEWETT and ROSEN 1984); but in the presence of elevated levels of $[Ca^{2+}]_0$ they will also enhance the amplitude of the afterdepolarizations (KIMURA et al. 1984). Acetylcholine, presumably by depressing i_{si} or by increasing outward K^+ current, has the opposite effect (HASHIMOTO and MOE 1973). The increase in K^+ permeability tends to hyperpolarize the membrane (provided that it is positive to E_K) and, because the amplitude of the delayed afterdepolarizations decreases with hyperpolarization (FERRIER 1980), this effect of acetylcholine might contribute to the suppression of delayed afterdepolarizations.

II. Abnormal Impulse Propagation

Abnormal conduction can result in the *slowing* of *impulse conduction* and, in some cases, the *blocking* of impulse propagation in parts of the conducting system. A typical case, illustrating the influence of the autonomic nervous system, is the conduction block in the AV node following vagal stimulation, which might result in the escape of a supraventricular or ventricular focus and the appearance of abnormal rhythms. Acetylcholine can slow or completely block conduction through the node. This effect is counteracted by the beta-adrenergic actions of catecholamines.

Tachycardia may involve reentry (MOE 1975; WIT and CRANEFIELD 1978; HOFFMAN and ROSEN 1981). The prerequisites for inducing reentry in one experimental model are: first, a site of unidirectional conduction block; second, presence of a pathway in which the retrograde impulse propagates at a velocity sufficiently slow to allow the termination of refractoriness in the tissue that is being approached by the retrograde impulse.

Basically, there are two classes of reentry (HOFFMAN and ROSEN 1981). In one type, which may be called ordered or stable reentry, there is a defined anatomic pathway (cf. in the Wolff-Parkinson-White arrhythmia). In the other type, random reentry, the pathway or pathways change from cycle to cycle. Fibrillation involves random reentry and provides a good example of the importance of heart rate in arrhythmogenesis. Fibrillation may be induced by any accelerating tachycardia, regardless of type or cause (WIGGERS 1940; SCHERF 1947). With the acceleration of rate, not all fibers undergo a similar shortening of action potential duration and effective refractory period, and this inhomogeneity of refractoriness sets favorable conditions for reentry and, eventually, fibrillation.

By controlling the heart rate, the autonomic nervous system can influence stable reentry. One-way block responsible for the initiation of reentry may respond variably to changes of heart rate. For example, due to the long duration of refractoriness of diseased or partially depolarized fibers (GETTES and REUTER 1974), the impulse may propagate at a long cycle length, but show unidirectional block at a short cycle length. Alternatively, at a still shorter cycle length, the block may become complete, terminating the reentry. Thus, it is possible to hypothesize a range of cycle lengths that could cause one-way block allowing reentry. At longer or shorter cycle lengths, the critical requirements of conduction velocity and refractoriness would not be met and the reentry would terminate (FRAME and HOFFMAN 1984).

Besides these effects related to the heart rate, the autonomic nervous system has direct effects which could influence the reentrant pathway at various sites which could either favor reentry or terminate it. Beta-receptor stimulation, by increasing i_{si}, may cause slow responses to propagate beyond an area of total block and reestablish reentry. Acetylcholine, by depressing i_{si} and by antagonizing the sympathetic input, may have the opposite effect. Alternatively, beta stimulation can hyperpolarize the membrane, by enhancing electrogenic pumping, and thereby improve conduction and eliminate unidirectional block. Acetylcholine, in atrial fibers, may conceivably have the same effect by increasing K^+ conductance which will tend to hyperpolarize the membrane.

Another condition in which autonomic transmitters can alter anterograde conduction is when phase 4 depolarization causes phase 0 to take off from a sufficiently positive potential to reduce the rate of rise of the action potential and the velocity of propagation. In this condition, acetylcholine can improve conduction by depressing the slope of phase 4 depolarization and thus by increasing the membrane potential at which phase 0 is initiated. In general, therefore, the effects of autonomic stimulation on reentry are difficult to predict.

Finally, one must consider the effect of autonomic neurotransmitters on the action potential duration and refractoriness. Catecholamines, by stimulation of alpha- and beta-receptors, can either prolong or shorten action potential duration, although the latter action is predominant. Such effects are particularly prominent in the Purkinje system, where ordered reentry has been induced experimentally (WIT and CRANEFIELD 1978). On the contrary, in atrial tissue, acetylcholine markedly shortens the action potential duration and may have profound effects on the timing characteristics of the reentrant circuit. Such effects of vagal stimulation have been clearly documented in the "leading circle" model of atrial reentry (ALLESSIE et al. 1977). All these examples are meant to stress the complexity of the relationship between abnormal conduction and the action of the autonomic nervous system.

E. Conclusions

We have discussed the cellular mechanisms responsible for the normal cardiac impulse and arrhythmias, as well as the effects of autonomic stimulation on these. The effects of autonomic agonists are complex in that via actions on their receptors throughout the atrium and ventricles, as well as their effects on heart rate, they can induce or depress both normal rhythms and arrhythmias. The complexity of autonomic actions is further attested to by the fact that not only receptors, but a sequence of second messengers, regulatory proteins, and enzymes, all play a role in the physiologic expression of autonomic actions. It is obvious that whereas many questions have been answered concerning autonomic effects on the heart, many more remain that will require a great deal more investigation for their solution.

Acknowledgment. Certain of the studies referred to were supported by USPHS-NHLBI grants HL-28958, HL-28223, and HL-33727.

References

Allessie MA, Bonke FIM, Schopman FJG (1977) Circus movement in the rabbit atrial muscle as a mechanism of tachycardia. III. The "leading circle" concept: a new model of circus movement in cardiac tissue without the involvement of an anatomical obstacle. Circ Res 41:9–18

Amerini S, Piazzesi G, Mugelli A (1984) Alpha-adrenoceptor stimulation enhances automaticity in barium-treated cardiac Purkinje fibers. Arch Int Pharmacodyn Ther 270:97–105

Bailey JC, Greenspan K, Elizari MV, Anderson GJ, Fisch C (1972) Effects of acetylcholine on automaticity and conduction in the proximal portion of the His-Purkinje specialized conduction system of the dog. Circ Res 30:210–216

Bailey JC, Watanabe AM, Besch HR, Lathrop DA (1979) Acetylcholine antagonism of the electrophysiological effects of isoproterenol on canine cardiac Purkinje fibers. Circ Res 44:378-383

Brachman J, Scherlag B, Rosenshtraukh LV, Lazzara R (1983) Bradycardia-dependent triggered activity: relevance to drug-induced multiform ventricular tachycardia. Circulation 68:846-856

Brooks CMC, Hoffman BF, Suckling EE, Orias O (1955) Excitability of the heart. Grune & Stratton, New York

Brown H, DiFrancesco D (1980) Voltage-clamp investigations of membrane currents underlying pace-maker activity in rabbit sinoatrial node. J Physiol (Lond) 308:331-351

Carlsson E, Dahlof CG, Hedberg A, Persson H, Tangstrand B (1977) Differentiation of cardiac chronotropic and inotropic effects of beta-adrenoceptors agonists. Naunyn Schmiedebergs Arch Pharmacol 300:101-105

Ciaraldi T, Marinetti GV (1977) Thyroxine and propylthiouracil effects in vivo on alpha and beta adrenergic receptors in rat heart. Biochem Biophys Res Commun 74:984-991

Corr PB, Shayman JA, Kramer JB, Kipnis RJ (1981) Increased alpha-adrenergic receptors in ischemic cat myocardium: a potential mediator of electrophysiological derangements. J Clin Invest 67:1232-1236

Cranefield PF (1975) The conduction of the cardiac impulse. Futura, Mt. Kisco, pp 135-137, 199-231, 243-263

Damiano BP, Rosen MR (1984) Effects of pacing on triggered activity induced by early afterdepolarizations. Circulation 69:1013-1025

Dangman KH, Hoffman BF (1983) Studies on overdrive stimulation of canine cardiac Purkinje fibers: maximal diastolic potential as a determinant of the response. J Am Coll Cardiol 2:1183-1190

Dangman KH, Danilo P Jr, Hordof AJ, Mary-Rabine L, Reder RF, Rosen MR (1982) Electrophysiologic characteristics of human ventricular and Purkinje fibers. Circulation 65:362-368

Danilo P Jr, Vulliemoz Y, Verosky M, Rosen MR (1978) Epinephrine-induced automaticity of canine Purkinje fibers and its relationship to the adenylate cyclase adenosine 3′,5′-monophosphate system. J Pharmacol Exp Ther 205:175-182

Danilo P Jr, Rosen TS (1982) Effects of terbutaline on cardiac automaticity and contractility. J Clin Pharmacol 22:223-230

Danilo P Jr, Rosen MR, Hordof AJ (1978) Effects of acetylcholine on the ventricular specialized conducting system of neonatal and adult dogs. Circ Res 43:777-784

DiFrancesco D (1981a) A new interpretation of the pace-maker current in calf Purkinje fibers. J Physiol (Lond) 314:359-376

DiFrancesco D (1981b) A study of the ionic nature of the pace-maker current in calf Purkinje fibers. J Physiol (Lond) 314:377-393

Drugge ED, Robinson RB (1987) The trophic influence of sympathetic neurons on the cardiac alpha adrenergic response requires close nerve-muscle association. Dev Pharmacol Ther 10:47-59

Drugge ED, Rosen MR, Robinson RB (1985) Neuronal regulation of the development of the alpha-adrenergic chronotropic response in the rat heart. Circ Res 57:415-423

Dukes ID, Vaughan Williams EMI (1984) Effects of selective α_1-, α_2-, β_1- and β_2-adrenoceptor stimulation on potentials and contractions in the rabbit heart. J Physiol (Lond) 355:523-546

Ferrer PL (1977) Arrhythmias in the neonate. In: Roberts NK, Gelband H (eds) Cardiac arrhythmias in the neonate, infant and child. Appleton, New York, pp 265-316

Ferrier GR (1980) Effects of transmembrane potential on oscillatory afterpotentials induced by acetylstrophanthidin in canine ventricular tissue. J Pharmacol Exp Ther 215:332-341

Ferrier GR, Saunders J, Mendez C (1973) A cellular mechanism for the generation of ventricular arrhythmias by acetylstrophanthidin. Circ Res 32:600-609

Fields JZ, Roeske WR, Morkin E, Yamamura HI (1978) Cardiac muscarinic cholinergic receptors. Biochemical identification and characterization. J Biol Chem 253:3251-3258

Frame LH, Hoffman BF (1984) Mechanisms of tachycardia. In: Surawicz B, Pratap Reddy C, Prystowsky EN (eds) Tachycardias. Martinus Nijhoff, Boston, pp 7-36

Friedman WF, Pool PE, Jacobowitz D, Seagren SC, Braunwald E (1968) Sympathetic innervation of the developing rabbit heart—biochemical and histochemical comparison of fetal, neonatal, and adult myocardium. Circ Res 23:25-32

Gadsby DC, Wit AL, Cranefield PF (1978) The effect of acetylcholine on the electrical activity of canine cardiac Purkinje fibers. Circ Res 1978:29-35

Garnier D, Nargeot J, Ojeda C, Rougier O (1978) The action of acetylcholine on background conductance in frog atrial trabeculae. J Physiol (Lond) 274:381-396

Gettes LS, Reuter H (1974) Slow recovery from inactivation of inward currents in mammalian myocardial fibers. J Physiol (Lond) 240:703-704

Giles WR, Noble SJ (1976) Changes in membrane currents in bullfrog atrium produced by acetylcholine. J Physiol (Lond) 261:103-123

Gilman AG (1987) G Proteins: transducers of receptor-generated signals. Ann Rev Biochem 56:615-649

Giotti A, Ledda F, Mannaioni PF (1973) Effects of noradrenaline and isoprenaline, in combination with alpha- and beta-receptor blocking substances, on the action potential of cardiac Purkinje fibers. J Physiol (Lond) 229:99-113

Hashimoto K, Moe GK (1973) Transient depolarizations induced by acetylstrophanthidin in specialized tissue of dog atrium and ventricle. Circ Res 32:618-624

Hewett KW, Rosen MR (1984) Alpha and beta adrenergic interactions with ouabain-induced delayed afterdepolarizations. J Pharmacol Exp Ther 229:188-192

Hewett KW, Rosen MR (1985) Developmental changes in the rabbit sinus node action potential and its response to adrenergic agonists. J Pharmacol Exp Ther 235:308-312

Hoffman BF, Cranefield PF (1960) Electrophysiology of the heart. McGraw-Hill, New York, pp 104-174

Hoffman BF, Rosen MR (1981) Cellular mechanisms for cardiac arrhythmias. Circ Res 49:1-15

Hoffman BF, Suckling EE (1953) Cardiac cellular potentials: effect of vagal stimulation and acetylcholine. Am J Physiol 173:312-320

Hordof AJ, Edie R, Malm JR, Hoffman BF, Rosen MR (1976) Electrophysiologic properties and response to pharmacologic agents of fibers from diseased human atria. Circulation 54:774-779

Hume JR, Katzung BG (1978) The effects of alpha and beta adrenergic agonists upon depolarization-induced ventricular automaticity. Proc West Pharmacol Soc 21:77-81

Inui J, Imamura H (1977) Effects of acetylcholine on calcium-dependent electrical and mechanical responses in the guinea-pig papillary muscle partially depolarized by potassium. Naunyn Schmiedebergs Arch Pharmacol 229:1-7

Jacobowitz DM, Cooper T, Barner HB (1967) Histochemical and chemical studies of the localization of adrenergic and cholinergic nerves in normal and denervated cat hearts. Circ Res 20:289-298

Jones LR, Besch HR Jr, Fleming JW, McConnaughey MM, Watanabe AM (1979) Separation of vesicles of cardiac sarcolemma from vesicles of cardiac sarcoplasmic reticulum. J Biol Chem 254:530-539

Katzung BG, Morgenstern JA (1977) Effects of extracellular potassium on ventricular automaticity and evidence for a pacemaker current in mammalian ventricular myocardium. Circ Res 40:105-111

Kent KM, Epstein SE, Cooper T, Barner HB (1974) Cholinergic innervation of the canine and human ventricular conducting system: anatomic and electrophysiologic correlations. Circulation 50:948-955

Kimura S, Cameron JS, Kozlovskis PL, Bassett AL, Myerburg RJ (1984) Delayed afterdepolarizations and triggered activity induced in feline Purkinje fibers by alpha-adrenergic stimulation in the presence of elevated calcium levels. Circulation 70:1074-1082

Kupfer LE, Robinson RB, Bilezikian JP (1982) Identification of $alpha_1$-adrenergic receptors in cultured rat myocardial cells with a new iodinated $alpha_1$-adrenergic antagonist, [^{125}I]IBE 2254. Circ Res 51:250-254

Lau YH, Robinson RB, Rosen MR, Bilezikian JP (1980) Subclassification of beta-adrenergic receptors in cultured rat cardiac myoblasts and fibroblasts. Circ Res 47:41-48

Le Marec H, Dangman KH, Danilo P Jr, Rosen MR (1985) An evaluation of automaticity and triggered activity in the canine heart one to four days after myocardial infarction. Circulation 71:1224-1236

Levine JH, Spear JF, Guarnieri T, Weisfeldt ML, de Langen CDJ, Becker LC, Moore EN (1985) Cesium chloride-induced long QT syndrome: demonstration of afterdepolarizations and triggered activity in vivo. Circulation 72:1092-1103

Levy MN (1971) Sympathetic-parasympathetic interactions in the heart. Circ Res 29:437-445

Levy MN, Zieske H (1969) Autonomic control of cardiac pacemaker activity and atrioventricular transmission. J Appl Physiol 27:465-470

Lipp JAM, Rudolph AM (1972) Sympathetic nerve development in the rat and guinea-pig heart. Biol Neonate 21:76-82

Manalan AS, Besch HR, Watanabe AM (1981) Characterization of [^{3}H]($\pm$) carazol binding to beta-adrenergic receptor subtypes in canine ventricular myocardium and lung. Circ Res 49:326-336

Mary-Rabine L, Hordof AJ, Bowman FO, Malm JR, Rosen MR (1978) Alpha and beta adrenergic effects on human atrial specialized conducting fibers. Circulation 57:84-90

Mary-Rabine L, Hordof AJ, Danilo P Jr., Malm JR, Rosen MR (1980) Mechanisms for impulse initiation in isolated human atrial fibers. Circ Res 47:267-277

Minneman KP, Hegstrand LR, Molinoff PB (1979) The pharmacological specificity of $beta_1$ and $beta_2$ adrenergic receptors in rat heart and lung in vitro. Mol Pharmacol 16:21-33

Moak JP, Reder RF, Danilo P Jr, Rosen MR (1986) Developmental changes in the interaction of cholinergic and beta-adrenergic agonists on the electrophysiological properties of canine Purkinje fibers. Pediatr Res 20:613-618

Moe GK (1975) Evidence for reentry as a mechanism for cardiac arrhythmias. Rev Physiol Biochem Pharmacol 72:56-66

Natham D, Beeler GW (1975) Electrophysiologic correlates of the inotropic effects of isoproterenol in canine myocardium. J Mol Cell Cardiol 7:1-15

Osnes JB, Refsum H, Skomedal T, Oye I (1978) Qualitative differences between beta-adrenergic and alpha-adrenergic inotropic effects in rat heart muscle. Acta Pharmacol Toxicol 42:235-247

Pappano AJ (1977) Ontogenic development of autonomic neuroeffector transmission and transmitter reactivity in embryonic and fetal hearts. Pharmacol Rev 29:3-33

Pappano AJ (1984) The development of postsynaptic cardiac autonomic receptors and their regulation of cardiac function during embryonic, fetal, and neonatal life. In: Speralakis N (ed) Physiology and pathophysiology of the heart. Martinus Nijhoff, Boston, pp 355-375

Posner P, Farrar E, Lambert CR (1976) Inhibitory effects of catecholamines in canine cardiac Purkinje fibers. Am J Physiol 231:1415-1420

Prystowsky EN, Jackman WM, Rinkenberger RL, Heger JJ, Zipes DP (1981) Effect of autonomic blockade on ventricular refractoriness and atrioventricular nodal conduction in humans. Evidence supporting a direct cholinergic action on ventricular muscle refractoriness. Circ Res 49:511-518

Randall WC (1977) Sympathetic control of the heart. In: Randall WC (ed) Neuronal regulation of the heart. Oxford University, New York, pp 45-94

Reder RF, Danilo P Jr, Rosen MR (1984) Developmental changes in alpha adrenergic effects on canine Purkinje fiber automaticity. Dev Pharmacol Ther 7:94-108

Reuter H (1979) Properties of two inward membrane currents in the heart. Ann Rev Physiol 41:413-424

Rosen MR, Gelband H, Hoffman BF (1973a) Correlation between effects of ouabain in the canine electrocardiogram and transmembrane potentials of isolated Purkinje fibers. Circulation 47:65-72

Rosen MR, Gelband H, Merker C, Hoffman BF (1973b) Mechanisms of digitalis toxicity: effects of ouabain on phase 4 of Purkinje fiber transmembrane potential. Circulation 47:681-689

Rosen MR, Hordof AJ, Ilvento JP, Danilo P Jr (1977) Effect of adrenergic amines on electrophysiological propeties and automaticity of neonatal and adult canine Purkinje fibers: Evidence for alpha and beta-adrenergic actions. Circ Res 40:390-400

Rosen T, Lin M, Spector S, Rosen MR (1979) Maternal, fetal and neonatal effects of chronic propranolol administration in the rat. J Pharmacol Exp Ther 208:118-122

Rosen MR, Weiss RM, Danilo P Jr (1984) Effects of alpha adrenergic agonists and blockers on Purkinje fiber transmembrane potentials and automaticity in the dog. J Pharmacol Exp Ther 231:566-571

Rosen MR, Steinberg SF, Chow YK, Bilezikian JP, Danilo P Jr (1987) The role of a pertussis-toxin sensitive protein in the modulation of canine Purkinje fiber automaticity. Circ Res 62:315-323

Scherf D (1947) Studies on auricular tachycardia caused by aconitine administration. Proc Soc Exp Biol Med 64:233-239

Schümann HJ, Wagner J, Knorr A, Reidemeister JC, Sadony V, Schramm G (1978) Demonstration in human atrial preparations of alpha-adrenoceptors mediating positive inotropic effects. Naunyn Schmiedebergs Arch Pharmacol 302:333-336

Sharma VK, Banerjee SP (1978) Presynaptic muscarinic cholinergic receptors. Nature 272:276-278

Sheridan DJ, Penkoske PA, Sobel BE, Corr PB (1980) Alpha adrenergic contributions to dysrhythmia during myocardial ischemia and reperfusion in cats. J Clin Invest 65:161-171

Smith WM, Gallagher JJ (1980) "Les torsades de pointes": an unusual ventricular arrhythmia. Ann Intern Med 93:578-584

Spear JF, Moore EN (1973) Influence of brief vagal and stellate nerve stimulation on pacemaker activity and conduction within the atrioventricular conduction system of the dog. Circ Res 32:27-41

Speralakis N, Pappano AJ (1983) Physiology and pharmacology of developing heart cells. Pharmacol Ther 22:1-39
Spinelli W, Danilo P Jr, Buchthal SD, Rosen MR (1986) Developmental changes in the effects of beta-adrenergic blocking concentrations of propranolol on canine Purkinje fibers. Dev Pharmacol Ther 9:412-426
Standen NB (1978) The postnatal development of adrenoreceptor responses to agonists and electrical stimulation in rat isolated atria. Br J Pharmacol 64:83-89
Steinberg SF, Drugge ED, Bilezikian JP, Robinson RB (1985) Innervated cardiac myocytes acquire a pertussis toxin-specific regulatory protein functionally linked to the alpha$_1$-receptor. Science 230:186-188
Strauss HC, Prystowsky EN, Scheinman MM (1977) Sino-atrial and atrial electrogenesis. Prog Cardiovasc Dis 19:385-404
Surawicz B, Imanishi S (1976) Automatic activity in depolarized guinea pig ventricular myocardium: characteristics and mechanisms. Circ Res 39:751-759
Ten Eick R, Nawrath H, McDonald TF, Trautwein W (1976) On the mechanisms of the negative inotropic effect of acetylcholine. Pflugers Arch 361:207-213
Trautwein W, Kuffler SW, Edwards C (1956) Changes in membrane characteristics of heart muscle during inhibition. J Gen Physiol 40:135-145
Tsien RW (1974) Adrenaline-like effects produced by intracellular iontophoresis of cyclic AMP in cardiac Purkinje fibers. Nature New Biol 245:120-122
Tsien RW, Giles WR, Greengard P (1972) Cyclic AMP mediated the action of adrenaline on the action potential plateau of cardiac Purkinje fibers. Nature New Biol 240:181-183
Tuttle RR, Hillman CC, Toomey RE (1976) Differential beta adrenergic sensitivity of atrial and ventricular tissue assessed by chronotropic, inotropic and cyclic AMP responses to isoprenaline and dobutamine. Cardiovasc Res 10:452-458
Urthaler F, Walker AA, James TN (1980) Changing negative inotropic effects of acetylcholine in maturing canine cardiac muscle. Am J Physiol 238:H1-7
Vassalle M (1979) Electrogenesis of the plateau and pacemaker potential. Annu Rev Physiol 41:425-440
Vassalle M, Barnabei O (1971) Norepinephrine and potassium fluxes in cardiac Purkinje fibers. Pflugers Arch 322:287-303
Vassalle M, Caress DL, Slovin AJ, Stuckey JH (1967 a) On the cause of ventricular asystole during vagal stimulation. Circ Res 20:228-241
Vassalle M, Vagnini FJ, Gourin A, Stuckey JH (1967 b) Suppression and initiation of idoventricular automaticity during vagal stimulation. Am J Physiol 212:1-7
Verrier RL, Lown B (1978) Sympathetic-parasympathetic interactions and ventricular electrical stability. In: Swartz PJ, Brown AM, Malliani A, Zanchetti A (eds) Neural mechanisms in cardiac arryhthmias. Raven, New York, pp 75-85
Walker D (1969) Functional development of the autonomic innervation of the human heart. Biol Neonate 25:31-42
Watanabe AM, Lindemann JP (1984) Mechanisms of adrenergic and cholinergic regulation of myocardial contractility. In: Speralakis N (ed) Physiology and pathophysiology of the heart. Martinus Nijhoff, Boston, pp 377-404
Watanabe AM, Jones LR, Manalan AS, Besch HR Jr (1982) Cardiac autonomic receptors: recent concepts from radiolabeled ligand-binding studies. Circ Res 50:161-174
Waxman MB, Wald RW (1977) Termination of ventricular tachycardia by an increase in cardiac vagal drive. Circulation 56:385-391
Wiggers CS (1940) The mechanism and nature of ventricular fibrillation. Am Heart J 20:399-412

Wit AL, Cranefield PF (1976) Triggered activity in cardiac muscle fibers of the simian mitral valve. Circ Res 38:85-98

Wit AL, Cranefield PF (1977) Triggered and automatic activity in the canine coronary sinus. Circ Res 41:435-445

Wit AL, Cranefield PF (1978) Reentrant excitation as a cause of cardiac arrhythmias. Am J Physiol 235:H1-H17

Wit AL, Hoffman BF, Rosen MR (1975) Electrophysiology and pharmacology of cardiac arrhythmias. IX. Cardiac electrophysiological effects of beta adrenergic receptor stimulation and blockade. Part A. Am Heart J 90:521-533

Yamada S, Yamamura HI, Roeske WR (1980) Characterizations of $alpha_1$ adrenergic receptors in the heart using [^{3}H]WB4101: effect of 6-hydroxydopamine treatment. J Pharmacol Exp Ther 215:176-185

Epilogue

In planning this volume the main objective was to provide aspiring cardiologists with basic information on the mode of action of antiarrhythmic drugs, to facilitate a rational approach to therapy. Since no single remedy can be both safe and efficacious in all circumstances, an understanding of what individual drugs actually do is essential if combinations of different types of agent are to be exhibited. Indeed, combinations of similarly acting drugs may increase efficacy without a corresponding multiplication of toxicity. In addition, the clinical pharmacology of all types of antiarrhythmic compound has been described by experts, so that the book can serve as a practical manual for physicians responsible for the clinical management of cardiac arrhythmias. Basic scientists involved in the development and testing of new drugs may also find this information useful in providing an insight into the utility of the compounds they may create or study.

An emergent theme is the mulitfactorial nature of arrhythmogenicity. Even unstressed normal individuals have extrasystoles or brief arrhythmic episodes with surprising frequency, and under emotional or physical stress the propensity of high sympathetic drive to induce cardiac arrhythmias is striking. Nevertheless such episodes do not lead in normal people to disability or malfunction, and they can be abolished by quite small doses of a beta-blocker. It is reasonable to assume, therefore, that in patients with symptomatic or life-threatening arrhythmias additional factors are operative, providing, as a consequence of congenital abnormality or as a sequel to disease, a substrate of anatomical or functional disorder which increases the probability of re-entry or ectopic activity. The multiplicity of factors involved argues against the suggestion that human arrhythmias are initiated by "the" arrhythmogenic current. The experimental conditions under which triggered activity or late after-depolarizations can be observed are extreme, and since the heart must beat continuously for survival, the trigger is always present. In many patients arrhythmias resembling those that occur spontaneously can be initiated by appropriate stimuli without any alteration of the normal humoral environment. Thus, while a transient inward current during diastole may be involved in the coupled extrasystoles associated with digitalis intoxication, there is no reason to invoke such currents as initiators of other arrhythmias not linked with drug abuse.

Another dominant theme is the propensity of antiarrhythmic drugs to in stigate or exacerbate arrhythmias in a small proportion of patients. Even high concentrations do not induce arrhythmias in animals or normal humans, so that here again additional factors existing in a few patients may increase the risk of toxicity. An abnormal pathway for re-entry may exist, but the circuit may only become conducting if the transit time is lengthened by a drug which slows conduction velocity. The most serious arrhythmia associated with antiarrhythmic therapy is torsade de pointes. There is no universally accepted explanation of the nature and origin of torsade, but it is frequently associated with prolongation of the Q-T interval. Drugs which widen the QRS will prolong Q-T, without any effect on the J-T interval or on APD (notably class Ic compounds). Conversely drugs which have little effect on conduction velocity or QRS can prolong Q-T by delaying repolarization. APD is shorter in the epicardium than in the endocardium (hence the concordant T wave, Chap. 1), and in the ventricular septum it is shorter at the apex than at the base. The cells which depolarize last repolarize first, but how they "know" that they are at the end of the line is a mystery. It may be speculated that this disparity evolved to prevent re-excitation of proximal cells by current flow into still depolarized cells at the end of the conduction pathway. If repolarization of the distal cells is delayed, either by slowed conduction or by prolonged APD, this disparity may be lost, permitting excitation of a proximal Purkinje cell by still depolarized myocardium. The left bundle has three major divisions, and if parts of these were depolarized in sequence, the point of entry into the system would precess around the base of the left ventricle, explaining the alternation of R and S dominance characteristic of torsade de pointes. A more stable circuit would constitute a ventricular flutter, but some investigators describe "sine-wave" type ECGs as torsade even though there is no twisting—a semantic point of no great importance if the fundamental nature of the arrhythmia is the same. Whatever explanation is accepted, the fact has emerged that torsade is sometimes associated with antiarrhythmic therapy, especially with the use of drugs which delay conduction or prolong Q-T, or which do both (quinidine) at high concentration.

The prevention and control of arrhythmias by drugs is only palliative, and, with the exception of long-term beta-blockade after myocardial infarction, has not often been shown to prolong life. There is, consequently, an increasing trend to employ ablative or surgical techniques instead of, or as an adjunct to, drug therapy, in the hope of removing the source of the arrhythmia. For this reason a chapter on this topic has been added. Although not concerned, strictly speaking, with antiarrhythmic drugs, its inclusion may be justified on the ground that the advisability and efficacy of drug therapy must be evaluated against alternative forms of treatment. With all their limitations, antiarrhythmic drugs, by controlling symptoms and improving function, can increase the quality, if not the duration, of life for many people, and it is only by understanding fully the manner in which they act that the use of the currently available remedies can be optimized and new and better compounds developed.

E.M. Vaughan Williams Oxford, May 1987

Subject Index

MIX
Papier aus verantwortungsvollen Quellen
Paper from responsible sources
FSC® C105338

If you have any concerns about our products,
you can contact us on
ProductSafety@springernature.com

In case Publisher is established outside the EU,
the EU authorized representative is:
Springer Nature Customer Service Center GmbH
Europaplatz 3, 69115 Heidelberg, Germany

Printed by Libri Plureos GmbH
in Hamburg, Germany